A SUMMARY OF PHASOR RELATIONS[...]

MW00389855

Element	Phasor relationships	Impedance diagram	Admittance diagram	Phasor diagram	Time functions	Waveforms

$\dfrac{\mathbf{V}}{\mathbf{I}} = Z = R$

$(X = 0)$

$Y = \dfrac{1}{R}$

$(B = 0)$

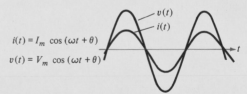

$i(t) = I_m \cos(\omega t + \theta)$

$v(t) = V_m \cos(\omega t + \theta)$

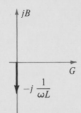

$\dfrac{\mathbf{V}}{\mathbf{I}} = Z = j\omega L$

$(R = 0)$

$(X = \omega L)$

$Y = -j\dfrac{1}{\omega L}$

$(G = 0)$

$\left(B = -\dfrac{1}{\omega L}\right)$

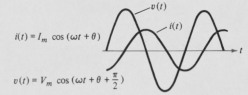

$i(t) = I_m \cos(\omega t + \theta)$

$v(t) = V_m \cos\left(\omega t + \theta + \dfrac{\pi}{2}\right)$

$\dfrac{\mathbf{V}}{\mathbf{I}} = Z = -j\dfrac{1}{\omega C}$

$(R = 0)$

$\left(X = -\dfrac{1}{\omega C}\right)$

$Y = j\omega C$

$(G = 0)$

$(B = \omega C)$

$i(t) = I_m \cos(\omega t + \theta)$

$v(t) = V_m \cos\left(\omega t + \theta - \dfrac{\pi}{2}\right)$

	Inductor	Capacitor	Resistor
Impedance	$Z_L(j\omega) = j\omega L$	$Z_C(j\omega) = -j\dfrac{1}{\omega C}$	$Z_R = R$
Reactance	$X_L = \omega L$	$X_c = \dfrac{-1}{\omega C}$	0
Admittance	$Y_L(j\omega) = \dfrac{1}{Z_L(j\omega)} = -j\dfrac{1}{\omega L}$	$Y_C(j\omega) = \dfrac{1}{Z_C(j\omega)} = j\omega C$	$Y_R(j\omega) = \dfrac{1}{Z_R} = \dfrac{1}{R}$
Conductance G	0	0	$\dfrac{1}{R}$
Susceptance B	$-\dfrac{1}{\omega L}$	ωC	0

An Introduction to
CIRCUIT ANALYSIS

McGraw-Hill Series in Electrical Engineering

Consulting Editor

Stephen W. Director, Carnegie-Mellon University

Circuits and Systems

Communications and Signal Processing

Control Theory

Electronics and Electronic Circuits

Power and Energy

Electromagnetics

Computer Engineering

Introductory

Radar and Antennas

VLSI

Circuits and systems

Consulting Editor
Stephen W. Director, Carnegie-Mellon University

An Introduction to CIRCUIT ANALYSIS
A Systems Approach

Donald E. Scott

Associate Professor of
Electrical and Computer Engineering
University of Massachusetts

McGraw-Hill Book Company
New York   St. Louis
San Francisco   Auckland   Bogotá
Hamburg   Johannesburg   London   Madrid   Mexico   Milan
Montreal   New Delhi   Panama   Paris
São Paulo   Singapore   Sydney   Tokyo   Toronto

ABOUT THE AUTHOR   The author earned his B.S.E.E. and M.S. in E.E. degrees from the University of Connecticut and the Ph.D. degree from Worcester Polytechnic Institute. He has worked for the General Electric Company and consults in the areas of applications of control systems and mathematical modeling. He recently spent five months at the University of Puerto Rico's School of Engineering in Mayaguez where a large portion of this work was written.

An Introduction to CIRCUIT ANALYSIS:
A Systems Approach

1234567890 KPHKPH 89432109876

ISBN 0-07-056127-3

This book was set in Times Roman by *Santype International Limited*. The editors were *Sanjeev Rao* and *Steven Tenney*; the designer was *Ben Kann*; the production supervisor was *Diane Renda*. The drawings were done by *J & R Services, Inc.*
Arcada Graphics/Kingsport was printer and binder.

Library of Congress Cataloging-in-Publication Data

Scott, Donald E.
    An introduction to circuit analysis.

    1. Electric circuit analysis. I. Title.
TK454.S37 1987   621.3815′3   86-15360
ISBN 0-07-056127-3

Dedicated to:

DR. CECIL EDWARD SCOTT
Engineer, Teacher, Gentleman

CONTENTS

PREFACE

This book is designed for use as a teaching text in a two-semester introductory course for electrical and/or computer systems engineering students. The reader is not presumed to have had any previous experience with electrical circuits.

Normally this course is taught in the second or third year of a four-year undergraduate curriculum. Therefore, a knowledge of the calculus through integration of trigonometric functions is presumed. Also, some knowledge of linear algebra (development of determinants; inverse matrices) would be helpful. Since students may not yet have seen the latter material, a brief treatment of this topic is included in an appendix.

The book contains an unusually large number of worked out numerical examples. Therefore the instructor who uses it can devote less class time to the drudgery of numerical manipulation and more to the basic derivations, conceptual underpinnings, applicability, and limitations of the techniques being discussed.

Most students will not have taken a course in differential equations at the time they enter this course, therefore no knowledge of that topic is presumed. A basic physics course that introduces electromagnetism would be useful as background, but is not at all prerequisite to understanding the material presented here.

In order to make students aware of the wide-ranging, general applicability of the electrical engineering principles that they are learning, mechanical, thermal, and other analogies are presented fairly early and reappear randomly in examples and homework problems throughout the book.

The first EE course is almost universally considered a prerequisite to electronics and control theory courses. This text demonstrates the applicability of basic circuit analysis in both those areas. For example, the student quickly sees diodes, load lines, linear-equivalent transistor models and nonideal op amps, as well as unit-impulse responses, block diagrams, mechanical systems and state-variable analysis as natural applications and extensions of the fundamental material. Thus, the student is helped to understand that his/her electrical engineering education is a continuum of interdependent ideas, not simply a collection of unconnected courses.

The relative amount of "design" work in their various courses is always a concern of electrical engineering faculty. The inclusion in this text of active and passive filter design and an extensive treatment of nonideal op amps as well as other design-related material is in direct response to that concern.

The first two chapters present all the basics of circuit analysis using resistive circuits containing dc sources. The third chapter describes the set of time functions most often encountered in circuit analysis. Particular attention is paid to the unit-impulse and unit-step functions and the exponential function. The effective value of a periodic function is derived in this chapter and, as an application, the rms value of a sinusoid is determined.

Energy storage elements and several of their analogs are presented in Chapter 4. The physical basis of inductance and capacitance is presented only to an extent great enough to (1) give the reader a feel for why these elements might, in reality, be nonlinear and (2) to give insight into what constitutes mutual inductance.

Chapter 5 is the start of a pedagogical development that begins with the p operator, proceeds through transfer functions, block diagrams (in Chapter 8), state variables (in Chapter 9), and concludes in the final section of the book with the Laplace transform solution of linear state equations. This logical trail has been intentionally interwoven into the structure of this work.

Chapters 6 and 7 treat first- and second-order circuits, placing particular emphasis on the distinction between zero-input and zero-state responses as well as between natural, particular (forced), and complete responses.

Chapter 8, on operational amplifiers, stresses the utility of this device as a circuit element within a larger network. The analog computer is mentioned in only one section of that chapter and, even then, only briefly. In addition the block-diagram description of systems is discussed in detail. This material is put to immediate use in the following Chapter (9) on state-variable analysis.

Most engineering faculties believe that students should make use of computers in this first EE course. However, many well-meaning attempts by authors, and professors alike, to inject computer applications into this fundamental subject matter often seem to be little more than awkward and unnecessary additions, without which the course could be more efficiently taught.

A *natural and necessary use of the computer* occurs in Chapter 9 in the solution of nonlinear state equations. Here it is clearly demonstrated that nonlinear state equations yield to *exactly* the same numerical integration methods as do linear ones. The study of linear state equations is presented in Chapter 9 as a natural outgrowth of the study of linear first- and second-order systems in Chapters 6 and 7 and at *the same pedagogical level* as that material. The usefulness of state-space techniques in affording *insight* into the behavior of (both linear and nonlinear) systems without having to obtain closed form solutions is stressed.

In addition, an appendix on SPICE is included. This easy-to-learn and powerful circuit analysis computer package (or its equivalent) is becoming almost universally available in universities and industry. Examples from the body of the text are solved via SPICE in that appendix. Students should come to perceive the computer as being a friendly helpmate rather than just one more hurdle for them to overcome. SPICE's ability to instantly perform many of the same time domain analyses, frequency domain analyses (Bode plots), and Fourier series calculations as the students are being asked to do makes it an ideal adjunct to this course.

Chapter 10 introduces the frequency domain. Topics such as the typical phasor and sinusoidal immittance techniques are presented as well as immittance loci, resonance, and a section on three-phase systems.

Chapter 11 further develops the concept of the s plane (which was introduced in Chapter 7) via a discussion of complex exponential input functions. Topics include transfer functions, poles and zeros, Bode plots, and quality factor.

Chapter 12, on two-port networks, covers more applications than is usual in an introductory text. In some presentations students are only very briefly exposed to the existence of z parameters, y parameters, and possibly the hybrid parameters, leaving them with the vague promise that they will see this material really applied later in electronics. However, the intent in this book is to show the practical utility of this material by immediately designing passive and active filters.

The order of presentation of the material in the final three chapters is designed intentionally to make for easier comprehension of these concepts by the student. The progression from Fourier series (which is in turn a logical extension of the sinusoidal-phasor analysis of Chapter 10) through Fourier transforms and then, via the reasoned introduction of the damping exponential factor, to the Laplace transform, makes logical sense to the student.

Although not specifically labeled as such, the first section of the book (Chapters 1–9 inclusive) presents analysis techniques of the time domain. Chapters 10–15 concern the frequency domain. This natural division is convenient and appropriate for a two-semester treatment of this material.

The book is written in a conversational, informal style, and with many numerical examples. An effort has been made to keep to an absolute minimum phrases such as "it may be shown that" or "this is left to the student as an exercise."

ACKNOWLEDGMENTS   The author owes a tremendous debt of gratitude to many individuals without whom this work would never have come to fruition. At McGraw-Hill, Sanjeev Rao, Steven Tenney, B. J. Clark, Bob Schuyler, and Mike Slaughter all have given encouragement and very helpful advice.

Dr. Lewis E. Franks and Dean Charles E. Hutchinson both have been most helpful in reviewing sections of this work, and making suggestions of topics to be included. But, more important, both freely volunteered their strong personal support at an absolutely critical juncture. A very special note of thanks to both of them.

Professor G. Dale Sheckels graciously made available many of the problems and examples that have been included. Professor Dan Schaubert reviewed the computer programs for the IBM PC and compatibles that are included. Dr. Keith Carver volunteered some much needed graphics and has been very understanding of my need to spend time on this project. Many unsung graduate and undergraduate students have checked and rechecked for errors. (The responsibility for any remaining errors is solely mine.)

My good friend, Dr. Leonard Bobrow, has offered moral support and advice, has reviewed portions of the material, and has generally been of great comfort to me throughout this effort.

To Professors K. S. P. Kumar, Carl H. Osterbrock, Syed Nasar, Don G. Daugherty, and Thomas W. Moore, who spent many hours reviewing this material, my most heartfelt thanks. Virginia McWhorter's diligent, detailed, and constructive efforts to help me in this work are most especially appreciated.

My most sincere thanks also go to Mrs. Pat Moriarty and to Ms. Jacqueline Orsini-Bonini who typed the major part of the manuscript.

"They also serve who only stand and wait."—To my wife, Annis, and my family, who have long awaited the end of my labors as a time to become reacquainted with husband and father—my love and thanks.

Donald E. Scott

chapter 1
RESISTIVE CIRCUITS

A system is a collection of interconnected devices or elements that, acting together, perform some specified objective. Every electric circuit is a system, but not every system is a circuit. The goal of this text is to familiarize you with electric circuits as a type of system. Many of the techniques that we will discuss are applicable in the analysis of other types of systems as well. Although our main concern is the electric circuit—its design and analysis—we will see that the mathematical and engineering techniques that we will use are equally of use in analyzing mechanical systems, fluid systems, and many other types as well. One of the things that make some electrical engineers as sought after as they are is their ability to apply these techniques in a wide variety of different problem areas. For example, mechanical vibration problems and electric oscillations are really different manifestations of the same basic mathematical principles. So one of our main goals will be to recognize the broad usefulness of the methods we will be discussing.

But we must learn to walk before we can have the fun of running and leaping. So, in this first chapter, we discuss the simplest type of electrical system: the resistive circuit containing only constant sources. If you have had previous experience with electric circuits, so much the better. But if you have not, just read along carefully. We will assume that you have never seen or heard of anything electrical before.

1.1   introduction

CHARGE   The study of resistive circuits is concerned with the movement of a quantity called electric charge through a closed *system* of elements and ideal conductors called a *circuit*. Charge is measured in coulombs (C) and is given the symbol q. It can be positive or negative.

EXAMPLE 1.1.1

a. What is a "unit charge"?
b. What charge is on an electron?
ANS.:   (a) $q = +1$ C;   (b) $q = -1.602 \times 10^{-19}$ C.

CURRENT   The rate at which positive charge moves past a given observation point is called the electric current. It is measured in amperes (A) and is given the symbol i.

$$i = \frac{dq}{dt}$$

Note that if, in an electric conductor, 1 C moves past an observation point each second, we say that a current of 1 A exists (Figure 1.1.1).

EXAMPLE 1.1.2

In a conductor a charge of 3 C moves toward the right every $\frac{1}{2}$ s. At the same time a charge of -5 C moves toward the left every $\frac{1}{3}$ s. The total current I toward the right is given by

$$I = \frac{3}{\frac{1}{2}} - \frac{-5}{\frac{1}{3}} = 21 \text{ A}$$

Note it is the net flow of charge that is important. Negative charge flowing toward the left contributes positively to the total current I which flows toward the right.

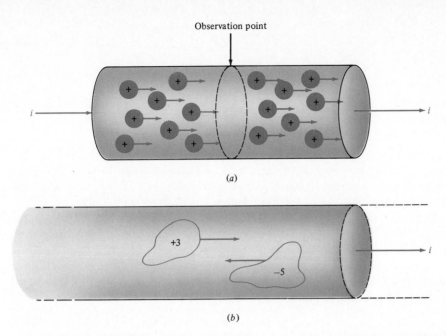

Observation point

(a)

(b)

figure 1.1.1
(a) Positive charges moving past an observation point constitute an electric current. (b) Negative charge flowing toward the left contributes positively to the net current directed toward the right.

If, in a conductor, the current is defined to be in one direction but the net positive charge actually flows in the other direction, we say the current is negative.

VOLTAGE

As charge flows through a network it moves through points having different electric potential. Consider a given charge to be at a point in a circuit where a high electric potential exists. This is analogous to a mass located at a high altitude: it has the potential to do work. The work that must be done on a unit charge (1 C) in order to move it from a point of low potential to a point of high potential is called the *voltage rise* between those two locations. This voltage rise is given the symbol V and is measured in volts.

$$\text{Voltage} = \frac{\text{work}}{\text{unit charge}} = \frac{\text{joules}}{\text{coulomb}}$$

SOURCES

Electric networks, in their simplest form, are called circuits and consist of interconnected electrical elements. We now consider two of the various possible types of such elements: the voltage source and the current source (see Fig. 1.1.2).

INDEPENDENT VOLTAGE SOURCE

An *ideal independent voltage source* is an electrical element that maintains a specified voltage rise across its terminals regardless of the current it carries. The voltage rise is from the terminal having the minus sign up to the terminal having the positive sign. This is true regardless of the magnitude and direction of the current passing through it.

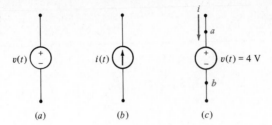

(*a*)   (*b*)   (*c*)

Such a source is called *ideal* because in the real world even the best obtainable voltage sources have output voltages that do indeed change to some degree as the magnitude of current drawn from the source increases. The typical 1.5-V dry cell battery, for example, is far from being a perfect voltage source. Its output voltage drops quickly from 1.5 V toward zero if it is forced to carry a large current. Even so, the ideal independent voltage source is a very useful *first approximation* of the way most real voltage sources behave. If a more accurate model is needed in any given application, we shall learn how to add additional elements to the ideal voltage source to make it approximate more closely any real (nonideal) voltage source that we have to deal with.

EXAMPLE 1.1.3

1. The voltage rise from *b* to *a* is $+4$ V.

$i(t)$

a

$v(t) = 4$ V

b

2. The voltage rise from *a* to *b* is -4 V.
3. The voltage at *a* with respect to the voltage at *b* is 4 V. We denote this as $V_{ab} = v(t) = 4$ V.

The three statements in Example 1.1.3 are true regardless of whether $i = +3$ A, or $i = +3 \times 10^6$ A, or $i = -17 \times 10^{10}$ A. The voltage $v(t)$ is specified that way to remind you that it is, in general, a quantity that may vary as a function of time.

At any instant of time the voltage across an ideal voltage source plotted versus the current through that source is a straight line parallel to the i_{ab} axis. It intersects the v axis at the value V. Such a plot of v versus i for any element is called its *volt-ampere characteristic*. Any point on this locus gives a pair of values (I, V) that can coexist through and across, respectively, that element (Figure 1.1.3).

If a voltage source is connected into a circuit, the magnitude and direction of the current that will, as a result, go through it are determined by the electrical properties of the rest of the circuit, not solely by the voltage source itself.

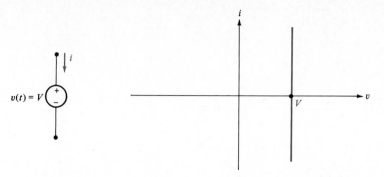

figure 1.1.3
The volt-ampere characteristic of an ideal
voltage source.

figure 1.1.4
An alternative way to show a time-invariant
(constant) (dc) voltage source.

A voltage source of 0 V has a volt-ampere characteristic that lies along the current axis. Such a source maintains both its terminals at the same voltage regardless of the current through it. It is, therefore, acting like a piece of wire, a perfect conductor.

If a voltage source is not a function of time, it is called a *direct current* (dc)† voltage source. Such constant voltage sources are sometimes drawn as in Figure 1.1.4.

Note that a voltage source with a perfect conductor connected across its terminals (a, b) is an impossibility. It is a contradiction in terms. On the one hand, the voltage source guarantees that there is a voltage rise of the specified value $v(t)$ from one terminal up to the other. On the other hand, connecting the perfect conductor‡ across those terminals guarantees they are at the same voltage. Obviously both these conditions cannot coexist simultaneously. In the real world such a circuit cannot be constructed. See Figure 1.1.5.

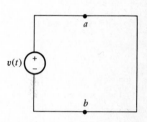

figure 1.1.5
An impossible situation (a contradiction in terms).

An ideal current source is an element that will maintain a specific charge flow regardless of the resulting voltage that appears across its terminals. Thus for a current source $i(t)$ we see that the volt-ampere characteristic looks like Figure 1.1.6. If a current source is connected into a circuit, the magnitude and polarity of

INDEPENDENT CURRENT SOURCE

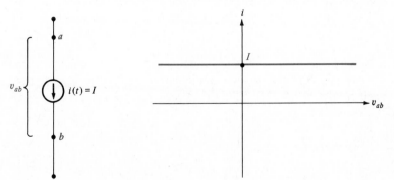

figure 1.1.6
The volt-ampere characteristic of an ideal current source.

---

† This is poor nomenclature because there is nothing *direct* about this source. It is simply constant. It does not vary with time.

‡ When, in a normally functioning circuit, a wire is connected between two different points (a and b), then electric charge has a new *short* path to get from a to b. Such a conductor is termed a *short circuit*.

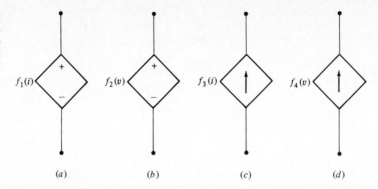

figure 1.1.7
(a) A current-dependent voltage source
(CDVS), (b) a voltage-dependent voltage
source (VDVS), (c) a current-dependent
current source (CDCS), and (d) a
voltage-dependent current source (VDCS).

figure 1.1.7
(a) A current-dependent voltage source
(CDVS), (b) a voltage-dependent voltage
source (VDVS), (c) a current-dependent
current source (CDCS), and (d) a
voltage-dependent current source (VDCS).

the resulting voltage that appears across its terminals are a function of the electrical properties of the rest of the circuit.

A current source of 0 A has a volt-ampere characteristic that lies along the voltage axis. Such a source maintains this current of 0 A (prohibits any charge flow) regardless of the voltage that may be found across its terminals. It is, therefore, acting like and is equivalent to an *open circuit*.

A current source guarantees the existence of $i(t)$ amperes and therefore cannot ever be open-circuited. There has to be somewhere for $i(t)$ to go. A current source has to have a circuit of some sort connected to it that will accept its specified $i(t)$.

DEPENDENT SOURCES

*must also have
an independent
source in circuit
to drive.*

A type of voltage source exists whose value $v(t)$ depends on the value of some other variable (either voltage or current) somewhere else in the circuit. Similarly, there exists a type of current source whose value $i(t)$ is a function of some voltage or current elsewhere in the circuit. The symbols that are used in drawing these *dependent sources* are shown in Figure 1.1.7. Each of these sources has all the properties of a regular independent source except that its instantaneous numerical value is not prespecified, but rather depends on the value of some voltage or current somewhere else. In Figure 1.1.8 the value of the voltage-dependent voltage source (VDVS) is proportional to the value of V_C, and the value of the current-dependent current source (CDCS) is proportional to the value of I_1.

UNITS

The system of units we will use in this text is called the Standard International (SI) system. This is a type of metric system that is universally used by electrical engineers. It used to be called the rationalized MKSC system. The major units of this system are

figure 1.1.8
A circuit containing a voltage-dependent
voltage source and a current-dependent
current source.

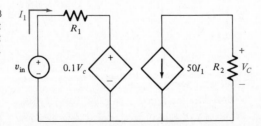

$$\text{Length} = \text{meter}$$

$$\text{Mass} = \text{kilogram}$$

$$\text{Time} = \text{second}$$

$$\text{Charge} = \text{coulomb}$$

Associated with this system is the following set of prefixes each of which has the listed multiplying effect and abbreviation:

Prefix	Multiplying effect	Abbreviation	Example
pico	10^{-12}	p	2 pA
nano	10^{-9}	n	2 nA
micro	10^{-6}	μ	2 μV
milli	10^{-3}	m	2 mV
centi	10^{-2}	c	2 cV
kilo	10^{3}	k	2 kV
mega	10^{6}	M	2 MV
giga	10^{9}	G	3 GHz

For example, 1 km is equal to 1000 m.

In general, we will use lowercase letters, v and i, for example, to denote quantities that may vary with time. Capital letters such as I_0, V_{cd}, or $(V_3)_{rms}$ are used to denote constants. This might be a constant such as the numerical value of a time-invariant source. It might also be a constant which describes in some way a time-varying source, for example, the peak value of $v(t) = 10 \sin 3t$, which is $V_{pk} = 10$.

NOTATION

DRILL PROBLEM

In the circuit of Figure 1.1.8, $V_C = -10$ V and $I_1 = 3$ A.
a. What is the value of the current flowing upward in resistor R_2?
b. What is the resulting value of the VDVS?
ANS.:   (a) 150 A; (b) -1 V.

Ohm's law **1.2**

Charge moves easily in good conductors such as most metal wires but with difficulty through poorer conductors called resistors. That is to say, it requires no energy to move a charge from one place to another in a perfect conductor. Thus, the voltage rise (work per unit charge) from any one location to another in or along a perfect conductor is zero. On the other hand, passage of electric charge through a resistor causes the molecules of the resistive material to vibrate. It takes energy to cause this vibration. Therefore, each unit of charge that passes through a resistor will give up some energy to the resistor. So if a current exists in

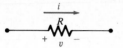

figure 1.2.1
Voltage and current polarities.

a resistor, an electric potential difference v will exist across that resistor. In fact we define a resistor as being any electrical element wherein the value of the terminal voltage is proportional to the value of the terminal current in it.

Resistors are usually made of carbon or silicon. The resistance R of any linear resistor is defined as being the proportionality constant between the two quantities v and i shown in Figure 1.2.1. Thus:

$$\boxed{v = i \times R} \tag{1.2.1}$$

(Note this mathematical relationship is *only valid for the polarities shown in Figure 1.2.1*.) R is measured in ohms (Ω).

EXAMPLE 1.2.1

In Figure 1.2.1 find v if $R = 2\ \Omega$ and $i = 3$ A.
ANS.: $v = 3(2) = 6$ V.

EXAMPLE 1.2.2

In Figure 1.2.1 find v if $R = 2\ \Omega$ and $i = -3$ A.
ANS.: $v = -3(2) = -6$ V.

figure 1.2.2
If v is a positive quantity, plus charges at point a have a higher potential energy than they (or similar charges) have when they are at point b.

As in the two examples above, voltage polarity can be defined with plus and minus signs. The point with the plus is assumed to have the higher potential. Thus in Figure 1.2.2 if point a is at a higher potential than point b, we say v is a positive quantity. On the other hand, if we are told v is -5 V, then we know that point b is 5 V higher than point a. An alternative way of denoting the voltage rise from b to a is v_{ab}, which reads: "the voltage at point a with respect to the voltage at point b."

Ohm's law by itself allows us to solve many circuits.

EXAMPLE 1.2.3

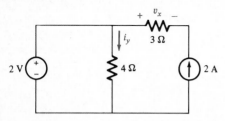

figure 1.2.3
The circuit of Example 1.2.3.

The 2-A current source forces a current through the 3-Ω resistor from right to left. Therefore,

$$v_x = -iR = -2(3)$$
$$= -6 \text{ V}$$

Also the voltage source impresses 2 V across the 4-Ω resistor. Thus, $i_y = 2 \text{ V}/4\ \Omega = 0.5$ A.

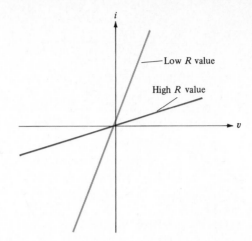

figure 1.2.4
Volt-ampere characteristic of a linear resistor
of *R* ohms.

Every time we use Ohm's law we must make *absolutely certain* that v and i are as shown in Figure 1.2.1. Otherwise an error in algebraic signs will occur.

If a resistor can be described by a single fixed value of resistance R, independent of the value of the current passing through it and the voltage across it, then its *volt-ampere characteristic* (plot of v versus i) is a straight line which passes through the origin (see Figure 1.2.4).

Such a resistor is called *linear*. If current is plotted on the ordinate (vertically) and voltage on the abscissa (horizontally), then the reciprocal of the slope of the characteristic is the resistance value of the element. If either (1) the characteristic is not a straight line and/or (2) it does not pass through the origin, then the resistor is *nonlinear*. The resistance of a nonlinear resistor is a function of the current through it.

As a consequence of the two necessary requirements for a resistor to be linear, all such linear resistors exhibit two other extremely important properties:

* Additivity   If, in a linear resistor, a current i_1 produces a voltage v_1 and a current i_2 produces a voltage v_2, then a current $i_1 + i_2$ will produce the voltage $v_1 + v_2$.

* Homogeneity   If, in a linear resistor, a current i_1 produces a voltage v_1, then a current ki_1 will produce the voltage kv_1.

These properties are demonstrated graphically in Figure 1.2.5.

Consider the circuit element shown in Figure 1.2.6. This element might be a   **POWER**
voltage source, current source, or resistor (or a number of other things). Assuming the values of both the current i and the voltage v to be positive quantities, it is clear that the electric charge (which is flowing from left to right) is losing energy as it passes through the element. When a unit charge flows into the element, it possesses v joules more energy than each unit charge flowing out. That is, in fact, how we defined what voltage is:

$$v = \frac{\text{work (J)}}{\text{charge (C)}} \qquad (1.2.2)$$

figure 1.2.5
Volt-ampere characteristics demonstrating
(*a*) additivity and (*b*) homogeneity.

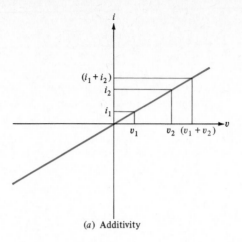

(*a*) Additivity

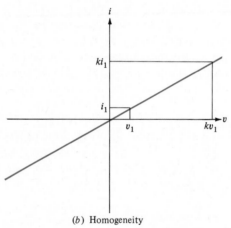

(*b*) Homogeneity

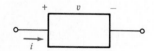

figure 1.2.6
A general circuit element with terminal
voltage *v* and terminal current *i*.

$$P_{V_t} = P_{R_1} + P_{R_2} + P_{R_3}$$

Recall that electric current is given by

$$i = \frac{\text{charge (C)}}{\text{time (s)}} \tag{1.2.3}$$

The product of voltage and current thus has the dimensions of power *p*, that is,

$$p = v \times i = \text{volts} \times \text{amperes}$$

$$= \frac{\text{work}}{\text{charge}} \times \frac{\text{charge}}{\text{time}} = \frac{\text{work (J)}}{\text{time (s)}}$$

$$= \text{power (watts)} \tag{1.2.4}$$

The product $p = v \times i$ is therefore the power (energy per unit time) delivered by the electric charge to the element into which it is flowing. The unit of power in the SI system is the watt (W).

What an element does with energy flow depends on what kind of element it is. Resistors immediately turn this power into heat and liberate it into the air. Certain types of sources (wet cell batteries) convert most of it into chemical energy and store it. Other types of elements store the energy in various other ways. But one thing is clear. If v and i are defined as in Figure 1.2.6, then their product,

$$p = v \times i$$

is the power delivered *to* that element.

If the value of p associated with some element turns out to be a negative quantity, it indicates that the element is actually delivering power back to the circuit. For example, when a car battery is being used to crank over the engine, power is being delivered by the battery to the circuit. The value of p associated with the battery is negative in this case. The value of p associated with the starter motor is positive.

Substituting Ohm's law, equation (1.2.1), into (1.2.4) yields either

$$p = vi = (iR)i$$
$$= i^2 R \qquad \text{❋ BEST Eq. to use} \qquad (1.2.5)$$

$$p = vi \text{ or} = v\left(\frac{v}{R}\right)$$

$$= \frac{v^2}{R} \qquad (1.2.6)$$

DRILL PROBLEM

In the circuit of Figure 1.2.3 find the power dissipated by
a. The 4-Ω resistor
b. The 3-Ω resistor
ANS.: (a) 1 W; (b) 12 W.

Kirchhoff's voltage law (KVL) **1.3**

Consider the circuit shown in Figure 1.3.1. In this circuit an ideal conductor connects the positive terminal of the voltage source to the upper terminal of the resistor. An ideal conductor is a path of zero resistance so points a and b are at the same electric potential (same voltage) for any value of current i. Similarly, points c and d are at the same potential. Therefore, since point a is $v(t)$ volts above point c because of the voltage source, point b is $v(t)$ volts higher than point d. Applying Ohm's law at the resistor we see that

$$i(t) = \frac{v(t)}{R}$$

Clearly then, if we go around the circuit in a clockwise direction starting at, say, point c, we first experience a voltage rise from c to a of $v(t)$ volts. Then we go through a drop (a negative rise) from b to d. Following is an example of a general law of circuit theory called Kirchhoff's voltage law.

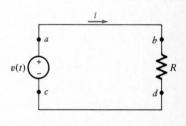

A simple circuit.

> KVL: At any instant of time, the algebraic sum of the voltage rises around any closed path is zero.

The word *drops* may be substituted for *rises* in the statement of the law with no loss of validity. A drop is a negative rise. Thus, the algebraic sum equation that would result is simply the same one that is obtained by summing rises except that it is multiplied by -1. This does not change the equation because the sum is equal to zero. We may proceed either clockwise or counterclockwise and we may sum either rises or drops. But once we have decided which quantity (rises or drops) we will consider to be positive, we must stick with that convention until the complete equation (complete path) has been written.

EXAMPLE 1.3.1

See Figure 1.3.2.

Direction of path	Quantity summed	
Clockwise	Rises	$+10 - v_1 - v_2 = 0$
Clockwise	Drops	$-10 + v_1 + v_2 = 0$
Counterclockwise	Rises	$v_2 + v_1 - 10 = 0$
Counterclockwise	Drops	$-v_2 - v_1 + 10 = 0$

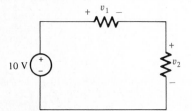

figure 1.3.2
The circuit of Example 1.3.1.

EXAMPLE 1.3.2

See Figure 1.3.3.

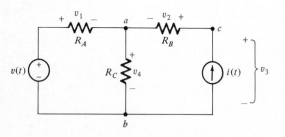

figure 1.3.3
Example 1.3.2.

All the following statements are true:

$$v(t) - v_1 + v_2 - v_3 = 0$$

$$v_4 + v_1 - v(t) = 0$$
$$v_3 - v_2 - v_4 = 0$$

Kirchhoff's voltage law is one of the fundamental mathematical tools that will help us to solve for the voltages and currents that exist in any given circuit.

EXAMPLE 1.3.3

Solve for i (Figure 1.3.4).

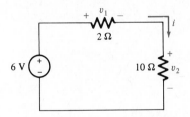

figure 1.3.4
The circuit of Example 1.3.3.

ANS.:

$$v_1 + v_2 - 6 = 0$$
$$2i + 10i = 6$$
$$12i = 6$$
$$i = 0.5 \text{ A}$$

Kirchhoff's voltage law demonstrates that two voltage sources v_1 and v_2 in series are equivalent to a single voltage source whose value v_{eq} is equal to the algebraic sum of the two individual sources as shown in Figure 1.3.5. Write a KVL summation around the circuit shown in part (*a*) of the figure.

$$v_1 + v_2 - v_a = 0 \qquad\qquad (1.3.1)$$

or

$$v_a = v_1 + v_2 \qquad\qquad (1.3.2)$$

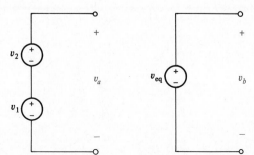

figure 1.3.5
Equivalent voltage source for two voltage sources in series.

(*a*)   (*b*)

Also in Figure 1.3.5b:

$$v_b = v_{eq} \qquad (1.3.3)$$

So v_a will equal v_b if and only if

$$v_{eq} = v_1 + v_2 \qquad (1.3.4)$$

Notice that, when writing KVL summations, there does not have to be a physically closed connection of elements, just a closed *path*. In equation (1.3.1) above, note that there is no element on the right side of the path.

DRILL PROBLEM

In Figure 1.3.3, if $v_4 = 4$ V, $R_B = 2\,\Omega$, and the value of the current source $i(t) = 1$ A, what is the value of voltage v_3?

ANS.:   6 V.

1.4 Kirchhoff's current law (KCL)

We define a *node* to be any point in a circuit where the terminals of two or more elements are connected together. Charge cannot be stored at a node; therefore at any instant, the total current directed *into* any given node must be exactly balanced out by an equal total current directed *out of* that node. This fact is stated formally by the following.

> KCL: At any instant of time, the algebraic sum of all currents directed into any node is zero.

EXAMPLE 1.4.1

See Figure 1.4.1.

$$-i_1 + i_2 + i_3 - i_4 = -4 \qquad i_a + i_b + i_c + i_d = 0$$

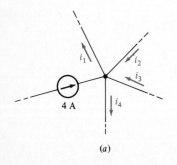

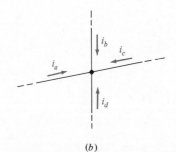

figure 1.4.1
Two examples of nodes. Kirchoff's current law applies at all such locations.

(a)

(b)

Of course, we could say that the sum of all currents *leaving* a node must be zero. This would result in a zero sum equation that is multiplied through by -1.

EXAMPLE 1.4.2

Solve for the voltage across R_2 (Figure 1.4.2).

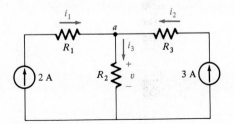

figure 1.4.2
The circuit of Example 1.4.2.

ANS.:   Write KCL at node a:

$$i_1 + i_2 - i_3 = 0$$

Since the 2-A source flows in R_1,

$$i_1 = 2$$

Similarly,

$$i_2 = 3$$

Substituting for i_1 and i_2 into the KCL equation:

$$2 + 3 - i_3 = 0$$
$$i_3 = 5$$

Applying Ohm's law at R_2,

$$v = 5R_2$$

EXAMPLE 1.4.3

Find the voltage across and the current through each element in the circuit (Figure 1.4.3).

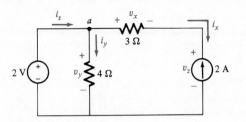

figure 1.4.3
The circuit of Example 1.4.3.

ANS.:   By inspection,

$$v_y = 2$$

therefore

$$i_y = \tfrac{2}{4} = 0.5 \text{ A}$$

and

$$i_x = -2$$

therefore

$$v_x = -6 \text{ V}$$

Then, by KCL at node a:

$$i_s = i_y + i_x = \tfrac{1}{2} - 2$$
$$= -\tfrac{3}{2} \text{ A}$$

Via KVL around the right-hand path:

$$v_y - v_x - v_z = 0.$$

So

$$v_z = 8 \text{ V}.$$

DRILL PROBLEM

Write a KCL summation at node a in the circuit of Figure 1.3.3.

ANS.: $\dfrac{v_1}{R_A} + \dfrac{v_2}{R_B} = \dfrac{v_4}{R_C}$

1.5   resistors in series

Any two elements are called *in series* if and only if:

1. One terminal of each element is connected to a common node.
2. No other element is connected to that node.

EXAMPLE 1.5.1

The two resistors shown in Figure 1.5.1 are in series because both requirements 1 and 2 are met by this connection of elements.

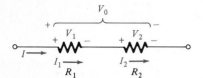

figure 1.5.1
Two resistors connected in series.

When any two elements are connected in series, it follows that the current in one is always identical to the current in the other. Also, the total voltage across the two elements is equal to the sum of their individual voltages.

Given the series connection of two resistors shown in Fig. 1.5.2a, can we find a single equivalent resistance value that would produce the same V, I relationship? For example, if V is produced by a voltage source, we seek a single resistance R_{eq} that would draw the same value current I as does the series combination. Or similarly, if I is produced by a current source, what single R_{eq} would produce the

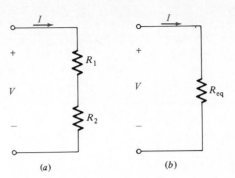

figure 1.5.2
(*a*) Resistors in series and (*b*) their equivalent resistance.

(*a*)   (*b*)

same V as the series combination? In any event, by KVL,

$$IR_1 + IR_2 = V$$

$$I(R_1 + R_2) = V$$

$$\frac{V}{I} = R_1 + R_2$$

In order for a single equivalent resistor to produce the same ratio of voltage and current,

$$V = IR_{eq}$$

$$\frac{V}{I} = R_{eq}$$

Thus, $$R_{eq} = R_1 + R_2$$

Such an equivalence is called a *terminal equivalent circuit*, because if the two different circuits of Figure 1.5.2 were put into black boxes so that we could not see which was which, there would be no way to tell them apart by making voltage and/or current measurements *at their terminals*.

A series equivalent resistance is always larger than its largest component resistance.

DRILL PROBLEM

An R-ohm resistor is connected across an ideal voltage source. In order to reduce the value of the resulting current by 10 percent, what size resistor should be placed in series with the existing resistor?
ANS.: $R/9\ \Omega$

---

resistors in parallel **1.6**

Any two elements are called *in parallel* if and only if:

1. One terminal of each element is connected to a common node.
2. The other terminal of each element is connected to another common node.

EXAMPLE 1.6.1

The two resistors shown in Figure 1.6.1 are in parallel because both requirements 1 and 2 are met.

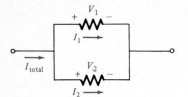

figure 1.6.1
Two resistors connected in parallel.

When any two elements are connected in parallel, it follows that the voltage across one is identical to the voltage across the other. Also, the total current into the combination is equal to the sum of the two individual currents.

Given two resistors in parallel we find their terminal equivalent resistance as follows (see Figure 1.6.2). For the parallel combination

$$I_1 + I_2 = I$$

$$\frac{V}{R_1} + \frac{V}{R_2} = I$$

$$V\left(\frac{1}{R_1} + \frac{1}{R_2}\right) = I$$

$$\frac{I}{V} = \frac{1}{R_1} + \frac{1}{R_2}$$

For the single equivalent resistor, by Ohm's law,

$$\frac{I}{V} = \frac{1}{R_{eq}}$$

Therefore,

$$\frac{1}{R_{eq}} = \frac{1}{R_1} + \frac{1}{R_2}$$

$$R_{eq} = \frac{R_1 R_2}{R_1 + R_2}$$

A parallel equivalent resistance is always smaller than its smallest component resistance.

figure 1.6.2
(a) Resistors in parallel and (b) their equivalent resistance.

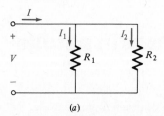

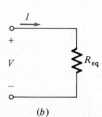

(a)   (b)

The reciprocal of resistance is called *conductance* and is given the symbol G.

$$G = \frac{1}{R}$$

The official SI unit for conductance is the siemen but *mho* (ohm spelled backward) is much more popular. Since we have just shown that, for two resistors in parallel,

$$\frac{1}{R_{eq}} = \frac{1}{R_1} + \frac{1}{R_2}$$

it follows that

$$G_{eq} = G_1 + G_2$$

Also note that, since for two resistors in series,

$$R_{eq} = R_1 + R_2$$

it follows that

$$\frac{1}{R_{eq}} = \frac{1}{R_1 + R_2}$$

$$G_{eq} = \frac{1}{\dfrac{1}{G_1} + \dfrac{1}{G_2}}$$

$$G_{eq} = \frac{G_1 G_2}{G_1 + G_2}$$

The inverted Greek omega, $\mho$, is used to symbolize conductance in mhos.

EXAMPLE 1.6.2

Find a single equivalent resistance R for the resistive network shown in Figure 1.6.3.

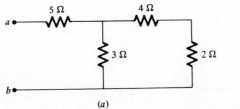
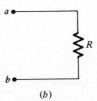

(a)   (b)

figure 1.6.3
(*a*) A resistive network and (*b*) its terminal equivalent circuit.

ANS.:   The 4- and 2-Ω resistors are in series. Their equivalent 6-Ω resistor is in parallel with the 3-Ω resistor. This equivalent 2-Ω [= 6(3)/(6 + 3)] resistor is in series with the 5-Ω resistor. Thus, $R_{eq} = 7$ Ω.

Note: In Figure 1.6.3*a* the 5- and 3-Ω resistors are not in series. They do not have identical currents. The 3- and 4-Ω resistors are *not* in parallel. They do not have identical voltages.

EXAMPLE 1.6.3

Find the equivalent terminal resistance R_{eq} of the circuit in Figure 1.6.4a.

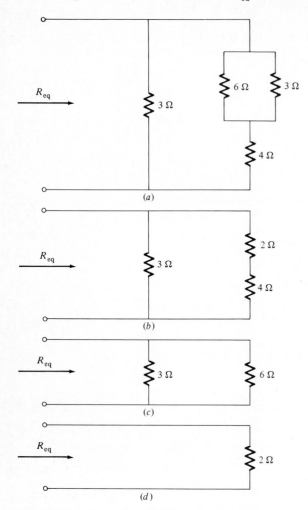

figure 1.6.4
(a) A resistive circuit. (b) to (d) Development of the terminal equivalent circuit.

ANS.:   The 6- and 3-Ω resistors are in parallel. Therefore we have $R_{parallel} = 6(3)/(6 + 3) = 2\ \Omega$ and the circuit reduces to that of Figure 1.6.4b.

The 2- and 4-Ω resistors are in series, so $R_{series} = 2 + 4 = 6\ \Omega$ and the circuit is as shown in Figure 1.6.4c.

Again there are 3- and 6-Ω resistors in parallel, so that finally $R_{eq} = 3(6)/(3 + 6) = 2\ \Omega$. See Figure 1.6.4d.

DRILL PROBLEM

When two resistors, one of which is double the resistance of the other, are connected in parallel, the total equivalent resistance is 1 kΩ. What are the resistance values of the two resistors?
ANS.:   1500 and 3000 Ω.

voltage division **1.7**

It often happens in analyzing a circuit that we know the total voltage across two series-connected resistors, but we then need to find the voltage across one of them. For example, in Figure 1.7.1 suppose we want to find v_2 given that we know v_0.

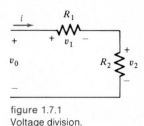

figure 1.7.1
Voltage division.

Since the equivalent resistance of two series-connected resistors is the sum of the two individual resistances, we can solve for i.

$$i = \frac{v_0}{R_{eq}} = \frac{v_0}{R_1 + R_2}$$

Thus, by applying Ohm's law at R_2,

$$v_2 = iR_2$$

$$= v_0 \frac{R_2}{R_1 + R_2}$$

Note that the quantity $R_2/(R_1 + R_2)$ must be less than unity. This quantity tells us what fraction of the total voltage v_0 falls across R_2. It is the value of R_2 itself divided by the sum of the two resistances. Similarly,

$$v_1 = \frac{R_1}{R_1 + R_2} v_0$$

A typical student error in the application of this technique is to forget that the two resistors must be in series. For example, in Figure 1.7.2, it is correct to write

$$v_4 = \frac{R_4}{R_3 + R_4} v_2$$

But it is *incorrect* to write

$$v_2 = \frac{R_2}{R_1 + R_2} v_0$$

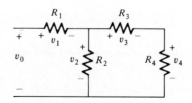

figure 1.7.2
A ladder network.

DRILL PROBLEM

In Figure 1.7.2 all resistances $= 1\ \Omega$ and $v_0 = 10$ V. Find the value of v_2 by voltage division.
ANS.:   4 V.

1.8 current division

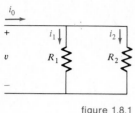

figure 1.8.1
Current division

Given the *total* current i_0 directed into two parallel connected resistors, we may solve for the current in either one of those resistors (Figure 1.8.1). The current i_0 obviously divides. What fraction of it goes into i_1? What fraction into i_2?

Utilizing the equivalent resistance of two resistors in parallel, we first solve for the voltage v.

$$v = i_0 R_{eq}$$

$$= i_0 \left(\frac{R_1 R_2}{R_1 + R_2} \right)$$

Then

$$i_1 = \frac{v}{R_1}$$

$$= i_0 \left(\frac{R_2}{R_1 + R_2} \right)$$

Similarly,

$$i_2 = i_0 \left(\frac{R_1}{R_1 + R_2} \right)$$

Thus, the quantities in parentheses in the last two equations represent the fractions of i_0 which become i_1 and i_2, respectively. Note that the fraction of i_0 that becomes i_1 is the ratio of the *other* resistance R_2 divided by the sum $R_1 + R_2$. This is a mathematical statement of the fact that if one of the two alternative paths that i_0 may take is made more difficult (higher resistance), then a larger fraction of i_0 will take the other path—"the path of least resistance."

DRILL PROBLEM

In the circuit of Figure 1.8.1, $i_1 = 0.3i_0$ and $v/i_0 = 2.1\ \Omega$. Find the values of R_1 and R_2.
ANS.: $R_1 = 7\ \Omega$ and $R_2 = 3\ \Omega$

1.9 the resistive ladder network

A circuit that often is seen in electrical engineering is the so-called resistive ladder network (Figure 1.9.1). Such circuits can have any finite number of resistors. If v_A (or i_A) is given, together with all resistance values, we usually have to solve for i_B and/or v_B. One method for doing this is to start at the right-hand side and replace the resistors such as R_{12} and R_{11} that are in series with their equivalent

figure 1.9.1
A resistive ladder network.

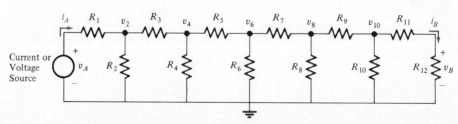

R_{eq}. Then this equivalent is seen to be in parallel with R_{10}, and so on. Continually combining resistances in series and then in parallel we work toward the left until the equivalent resistance across the source is found. Then, voltage division between R_1 and the remaining equivalent resistance yields the value of v_2. Thus we can find the current in R_2. We can do this over and over working toward the right until we find i_B and v_B.

Another method for obtaining a solution which saves about half this amount of work is described in the example that follows. This shortcut is predicated on the fact that if, in a circuit containing only resistors that obey Ohm's law and a single source, we multiply any current or voltage by some fixed number, then every current and voltage must be multiplied by that same factor.

EXAMPLE 1.9.1

Find i_5 given that $v_s = 6$ V (see Figure 1.9.2).

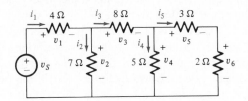

figure 1.9.2
The resistive ladder network of Example 1.9.1

ANS.: **1.** Assume a value for either i_5 or v_6. Let us assume $i_5 = 1$ A.
2. Find the value of v_s that would be necessary to produce the value assumed in part 1. If $i_5 = 1$ A, then

$$v_6 = (1\ \text{A})(2\ \Omega) = 2\ \text{V}$$

$$v_5 = 3(1) = 3\ \text{V}$$

$$v_4 = v_6 + v_5 = 5\ \text{V}$$

$$i_4 = \tfrac{5}{5} = 1\ \text{A}$$

$$i_3 = i_4 + i_5 = 2\ \text{A}$$

$$v_3 = 2(8) = 16\ \text{V}$$

$$v_2 = v_3 + v_4 = 16 + 5 = 21\ \text{V}$$

$$i_2 = \tfrac{21}{7} = 3\ \text{A}$$

$$i_1 = i_2 + i_3 = 2 + 3 = 5\ \text{A}$$

$$v_1 = 4i_1 = 20\ \text{V}$$

$$v_s = v_1 + v_2 = 41\ \text{V}$$

But, since v_s is given as being only 6 V, we see that our original i_5 estimate was too large. It would take 41 V at v_s to produce $i_5 = 1$ A. We actually have only $\tfrac{6}{41}$ of that value of v_s. Therefore, we will actually get only $\tfrac{6}{41}$ of our assumed value of i_5. Or

$$i_5 = 1 \times \tfrac{6}{41} = \tfrac{6}{41}\ \text{A}$$

a. If all the resistance values in Figure 1.9.2 were doubled, and at the same time v_s was doubled, what would happen to the value of voltage v_6?

b. What will be the value of the actual current i_1 directed out of the source in Figure 1.9.2?
ANS.: (a) It would double; (b) $\frac{30}{41}$ A.

1.10 node equations

Kirchhoff's current law (together with Ohm's law) provides a general systematic method for the solution of any resistive circuit. Consider the circuit shown in Figure 1.10.1. We select one node as our *reference* or *datum* node. Every other node voltage is then measured with respect to that reference node. The voltage of this node is then, by definition, zero. We think, alternatively, of connecting this node to the potential of our planet Earth.† Then if a voltage at some node is at $+3$ V, that means it is 3 V higher in electric potential than is the Earth. The symbol used to denote such a reference node is shown in Figure 1.10.2. Suppose we arbitrarily choose ground node d and define a consistent set of element voltages and currents (see Figure 1.10.3). By inspection we note that

$$v_a = 6 \text{ V} \qquad \text{and} \qquad v_c = 22 \text{ V}$$

At the node where the voltage remains unknown we write a KCL equation:

$$i_1 + i_2 + i_3 = 0$$

$$\frac{v_1}{6} + \frac{v_2}{2} + \frac{v_3}{3} = 0$$

where via KVL

$$v_1 = v_a - v_b = 6 - v_b$$

$$v_2 = v_c - v_b = 22 - v_b$$

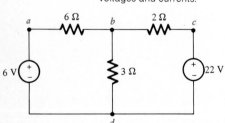

figure 1.10.1
A circuit to be solved for all unknown voltages and currents.

figure 1.10.2
Ground-node symbol.

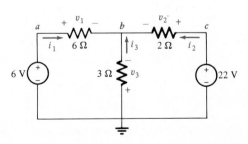

figure 1.10.3
The circuit of Figure 1.10.1 with element voltages and currents defined.

† In the United Kingdom the phrase *main earth* is used sometimes to describe the datum node. American engineers use the word *ground* to describe this node.

Thus,

$$\frac{6 - v_b}{6} + \frac{22 - v_b}{2} - \frac{v_b}{3} = 0$$

$$v_b = 12$$

Then we can find the currents:

$$i_1 = \frac{6 - v_b}{6} = -1 \text{ A}$$

$$i_2 = \frac{22 - v_b}{2} = +5 \text{ A}$$

$$i_3 = -\frac{v_b}{3} = -4 \text{ A}$$

In general, we see that the method involves writing a Kirchhoff current summation at any node where the voltage is unknown. Specifically:

1. Select a convenient datum node—ground it. Choose (all else being equal) the node having the largest number of voltage sources connected to it.
2. Write down one equation for each voltage source (*a*) by inspection (the value of the voltage at any node connected to ground by a voltage source equals the value of that source). (*b*) For any voltage source, neither of whose terminals are connected to ground, write a simple equation relating the node voltage at one terminal to the node voltage at the other terminal.
3. At each remaining node write a KCL summation. By Ohm's law convert the variables all to node voltages.
4. Solve the set of simultaneous equations obtained in steps 2 and 3.

EXAMPLE 1.10.1

Solve for all currents and voltages (Figure 1.10.4).

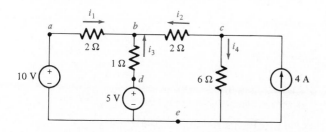

figure 1.10.4
The circuit of Example 1.10.1

ANS.:
Step 1
Node *e* has two voltage sources connected to it. All others have one or none. Ground node *e*.
Step 2a

$$v_a = 10 \quad \text{and} \quad v_d = 5$$

Step 3

At node *b*:
$$i_1 + i_2 + i_3 = 0$$

$$\frac{v_a - v_b}{2} + \frac{v_c - v_b}{2} + \frac{v_d - v_b}{1} = 0$$

At node *c*:
$$i_2 + i_4 = 4$$

$$\frac{v_c - v_b}{2} + \frac{v_c}{6} = 4$$

Step 4

$$\frac{10 - v_b}{2} + \frac{v_c - v_b}{2} + \frac{5 - v_b}{1} = 0$$

$$\frac{v_c - v_b}{2} + \frac{v_c}{6} = 4$$

or, multiplying the first equation by 2 and the second by 6:

$$10 - v_b + v_c - v_b + 10 - 2v_b = 0$$

$$3v_c - 3v_b + v_c = 24$$

Collecting terms,

$$4v_b - v_c = 20$$

$$-3v_b + 4v_c = 24$$

Solving the first equation for v_b,

$$v_b = \frac{v_c + 20}{4}$$

and substituting into the second,

$$-3\,\frac{v_c + 20}{4} + 4v_c = 24$$

Thus
$$v_c = 12 \text{ V}$$

and
$$v_b = \frac{12 + 20}{4} = 8 \text{ V}$$

Knowing the voltage at every node, we can solve for all element currents by Ohm's law.

$$i_1 = \frac{v_a - v_b}{2} = \frac{10 - 8}{2} = 1 \text{ A}$$

$$i_2 = \frac{v_c - v_b}{2} = \frac{12 - 8}{2} = 2 \text{ A}$$

$$i_3 = \frac{v_d - v_b}{1} = \frac{5 - 8}{1} = -3 \text{ A}$$

$$i_4 = \frac{v_c}{6} = \frac{12}{6} = 2 \text{ A}$$

We have seen that each voltage source in any given circuit reduces by one the number of independent unknown node voltages. This in turn reduces the number of independent simultaneous node equations that we must solve. Each voltage source that has one of its terminals connected to the datum node generates a single simple equation called a *degenerate* node equation (for example, $v_a = 10$ and $v_d = 5$ in Example 1.10.1).

If a voltage source is connected to a node at which we are writing a node equation, and the other terminal of that source is not connected to the reference (datum or ground) node,† then we cannot directly describe the current that source contributes to this node. (Recall that the value of the current in any voltage source is a function of the rest of the circuit into which the source is connected.) In such cases we must write an expression for the current in the voltage source by applying KCL at the *other terminal* of the source.

EXAMPLE 1.10.2

Solve the circuit in Figure 1.10.5 for all node voltages and all element currents. Use node equations.

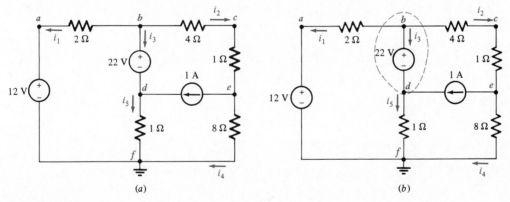

(a)   (b)

figure 1.10.5
(a) The circuit of Example 1.10.2. (b) The same circuit showing a closed surface (supernode) surrounding the ungrounded (floating) voltage source.

ANS.:   Note immediately that v_b and v_e are the only two independent unknown node voltages because one node voltage is already known due to the voltage source

$$v_a = 12 \tag{1.10.1}$$

and two others simply depend on v_b and v_e:

† Such an element is said to be *floating* above ground.

$$v_d = v_b - 22 \tag{1.10.2}$$

and
$$v_c = v_e + \frac{1}{4 + 1}(v_b - v_e) \tag{1.10.3}$$

These, then, are the degenerate node equations. If we can now simply write KCL summations at nodes b and e, we will generate the two independent simultaneous equations that enable us to solve the problem.

At node b:
$$i_1 + i_2 + i_3 = 0 \tag{1.10.4}$$

$$\frac{v_b - v_a}{2} + \frac{v_b - v_e}{4 + 1} + i_3 = 0 \tag{1.10.5}$$

We cannot write i_3 in terms of v_b and v_d using Ohm's law in the same manner in which we wrote i_1 and i_2. We must go to node d and use KCL there.

At node d:
$$i_3 = i_5 - 1$$

$$i_3 = \frac{v_d}{1} - 1 \tag{1.10.6}$$

Substituting equations (1.10.1) and (1.10.6) into (1.10.5) yields

$$\frac{v_b - 12}{2} + \frac{v_b - v_e}{5} + v_d - 1 = 0$$

or, using equation (1.10.2)
$$\tfrac{17}{10}v_b - \tfrac{1}{5}v_e = 29 \tag{1.10.7}$$

At node e:
$$i_2 = 1 + i_4$$

$$\frac{v_b - v_e}{4 + 1} = 1 + \frac{v_e}{8} \tag{1.10.8}$$

and, after collecting terms in v_e,

$$\tfrac{1}{5}v_b - \tfrac{13}{40}v_e = 1 \tag{1.10.9}$$

Equations (1.10.7) and (1.10.9) are two simultaneous algebraic equations. We can solve one of them for one variable and insert that into the other equation. Or we could use determinants (Cramer's rule).

$$\begin{bmatrix} \tfrac{17}{10} & -\tfrac{1}{5} \\ \tfrac{1}{5} & -\tfrac{13}{40} \end{bmatrix} \begin{bmatrix} v_b \\ v_e \end{bmatrix} = \begin{bmatrix} 29 \\ 1 \end{bmatrix}$$

where
$$\Delta = (-\tfrac{17}{10})(\tfrac{13}{40}) - (-\tfrac{1}{5})(\tfrac{1}{5}) = -\tfrac{41}{80}$$

Similarly,
$$\Delta_1 = 29(-\tfrac{13}{40}) + \tfrac{1}{5} = -\tfrac{369}{40}$$

$$\Delta_2 = \tfrac{17}{10} - \tfrac{29}{5} = -\tfrac{41}{10}$$

so that
$$v_b = \frac{\Delta_1}{\Delta} = 18 \text{ V}$$

$$v_e = \frac{\Delta_2}{\Delta} = 8 \text{ V}$$

and thus

$$v_d = v_b - 22 = -4 \text{ V}$$

$$v_c = \tfrac{1}{5}(v_b - v_e) + v_e = 10 \text{ V}$$

and

$$i_1 = \frac{v_b - v_a}{2} = 3 \text{ A}$$

$$i_2 = \frac{v_b - v_c}{4} = 2 \text{ A}$$

$$i_3 = -(i_1 + i_2) = -5 \text{ A}$$

$$i_4 = \frac{v_e}{8} = 1 \text{ A}$$

$$i_5 = \frac{v_d}{1} = -4 \text{ A}$$

The student should note that at every node KCL obtains numerically. Check it out.

Although we have defined Kirchhoff's current law as stating that the sum of the currents directed into any *node* is zero, we could have said the sum of the currents directed into any *closed region* of the circuit is zero. For example, in Figure 1.10.5*b*, we could sum the currents leaving the region that includes the 22-V source and nodes *b* and *d*:

$$i_1 + i_2 + i_5 - 1 = 0$$

$$\frac{v_b - 12}{2} + \frac{v_b - v_e}{5} + v_d - 1 = 0$$

$$\tfrac{17}{10}v_b - \tfrac{1}{5}v_e = 29$$

Note this result is identical to equation (1.10.7). So one way to treat any floating voltage source is to surround it with a closed surface† that includes the source and both nodes to which it is attached. Then a single independent node equation is obtained by writing a KCL summation of the currents directed out of (into) this surface.

EXAMPLE 1.10.3

Solve for all node voltages, element currents, and the power delivered to each element (Figure 1.10.6).
ANS.: First we draw a surface around the − 30-V source that includes nodes *c* and *d*.
Degenerate Equations:

Node *a*: $\qquad\qquad\qquad\qquad\qquad v_a = 60 \qquad\qquad\qquad\qquad\qquad$ (1.10.10)

Node *c*: $\qquad\qquad\qquad\qquad\qquad v_c = -30 + v_d \qquad\qquad\qquad\qquad$ (1.10.11)

---

† Such a closed surface is sometimes called a *supernode*.

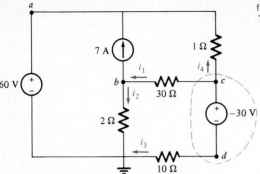

figure 1.10.6
The circuit of Example 1.10.3.

Independent Equations:

Node b:

$$i_1 = 7 + i_2$$

$$\frac{v_c - v_b}{30} = 7 + \frac{v_b}{2} \qquad (1.10.12)$$

Insert equation (1.10.11) into (1.10.12):

$$\frac{-30 + v_d - v_b}{30} = 7 + \frac{v_b}{2}$$

$$-30 + v_d - v_b = 210 + 15v_b$$

$$16v_b - v_d = -240 \qquad (1.10.13)$$

Supernode:

$$i_1 + i_3 + i_4 = 0$$

$$\frac{v_c - v_b}{30} + \frac{v_d}{10} + \frac{v_c - v_a}{1} = 0 \qquad (1.10.14)$$

Insert equations (1.10.10) and (1.10.11) into (1.10.14):

$$\frac{-30 + v_d - v_b}{30} + \frac{v_d}{10} - 30 + v_d - 60 = 0$$

$$-30 + v_d - v_b + 3v_d - 900 + 30v_d = 1800$$

$$-v_b + 34v_d = 2730 \qquad (1.10.15)$$

Thus (1.10.13) and (1.10.15) can be written:

$$\begin{bmatrix} 16 & -1 \\ -1 & 34 \end{bmatrix} \begin{bmatrix} v_b \\ v_d \end{bmatrix} = \begin{bmatrix} -240 \\ 2730 \end{bmatrix}$$

$$\Delta = 543$$

$$\Delta_1 = \begin{vmatrix} -240 & -1 \\ 2730 & 34 \end{vmatrix} = -8160 + 2730 = -5430$$

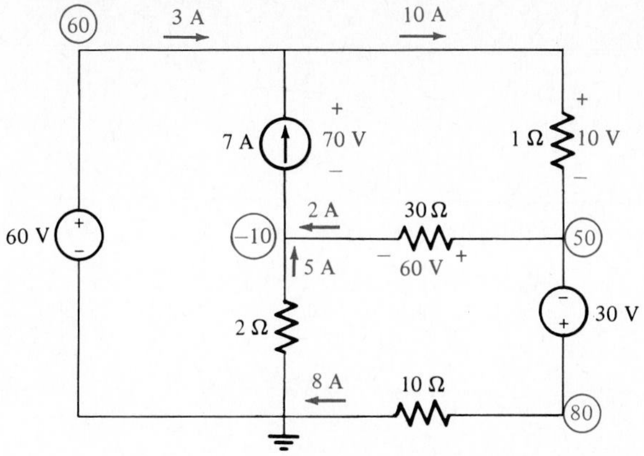

figure 1.10.7
The solution to Example 1.10.3.

$$\Delta_2 = \begin{vmatrix} 16 & -240 \\ -1 & 2730 \end{vmatrix} = 43{,}680 - 240 = 43{,}440$$

$$v_b = \frac{\Delta_1}{\Delta} = -10$$

$$v_d = \frac{43{,}440}{543} = 80$$

See Figure 1.10.7. The power delivered *to* each element is found as follows:

	$v \times i = p$	(W)
60-V source	$60 \times -3 =$	-180
7-A source	$70 \times -7 =$	-490
30-V source	$30 \times -8 =$	-240
1-Ω resistor	$10 \times 10 =$	100
2-Ω resistor	$-10 \times -5 =$	50
10-Ω resistor	$80 \times 8 =$	640
30-Ω resistor	$60 \times 2 =$	120
	Total	0

In general, node equations can be manipulated into a standard form wherein they exhibit certain symmetry properties. Consider the following example.

EXAMPLE 1.10.4

Write node equations for v_a and v_b in terms of current source values and conductances (Figure 1.10.8).
ANS.:

At node a: $$i_1 + i_2 = I_x$$

At node b: $$i_2 = i_3 + I_y$$

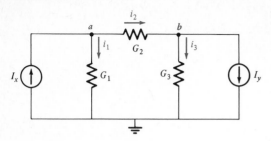

figure 1.10.8
The circuit of Example 1.10.4.

Simplifying,

$$(a) \quad v_a G_1 + (v_a - v_b)G_2 = I_x$$

$$(b) \quad (v_a - v_b)G_2 = v_b G_3 + I_y$$

and finally

$$(a) \quad v_a(G_1 + G_2) - v_b G_2 = I_x$$

$$(b) \quad -v_a G_2 + v_b(G_2 + G_3) = -I_y$$

Several properties of these equations may be noted:

1. Node equations consist of sums of products of node voltages times conductances. These terms are on one side of the equal sign; current sources are on the other side.
2. In the equation written for the ith node, the coefficient of v_i is the sum of all conductances connected to that node. The coefficients of all other voltages in this equation are negative. The magnitude of the coefficient of v_j is the conductance connecting node i to node j.
3. Any current sources connected to the ith node are included on the other side of the equal sign in the ith equation. If a current source forces charge into the ith node, it carries a positive algebraic sign. If a current source carries charge away from node i, it is negative.
4. Any voltage source having one terminal connected to ground appears in the equations simply by replacing the node voltage which it determines.

Voltage sources which have neither terminal connected to ground affect the symmetry properties of node equations in ways which are quite complicated.

The general symmetrical properties of node equations written in this standard form are useful more as a check on your work than as a shortcut method for writing them.

EXAMPLE 1.10.5

Write the node equations in standard form (see Figure 1.10.9).

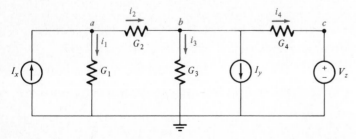

figure 1.10.9
The circuit of Example 1.10.5.

ANS.:

At node a:
$$i_1 + i_2 = I_x$$

At node b:
$$i_2 - i_3 - i_4 - I_y = 0$$

At node c:
$$v_c = v_z$$

(a) $v_a G_1 + (v_a - v_b)G_2 = I_x$

$v_a(G_1 + G_2) - v_b(G_2) = +I_x$ (1.10.16)

(b) $(v_a - v_b)G_2 - v_b G_3 - (v_b - v_c)G_4 - I_y = 0$

$-v_a G_2 + v_b(G_2 + G_3 + G_4) - v_c G_4 = -I_y$ (1.10.17)

$-v_a G_2 + v_b(G_2 + G_3 + G_4) - v_z G_4 = -I_y$

EXAMPLE 1.10.6

Solve for voltages v_a, v_b, and v_c in Figure 1.10.10.

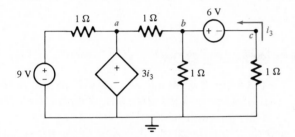

figure 1.10.10
The circuit of Example 1.10.6 containing a current-dependent voltage source.

ANS.: Node equation at a is

$$v_a = 3i_3 \qquad (1.10.18)$$

Node equation at b is

$$\frac{v_a - v_b}{1} + i_3 = \frac{v_b}{1} \qquad (1.10.19)$$

where

$$i_3 = \frac{-(v_b - 6)}{1} \qquad (1.10.20)$$

so $v_a = -3v_b + 18$; or substituting (1.10.20) into (1.10.19):

$$v_a - v_b - v_b + 6 = v_b$$

$$v_a + 6 = 3v_b$$

$$-3v_b + 18 + 6 = 3v_b$$

$$24 = 6v_b$$

$$4 = v_b$$

$$i_3 = -(4 - 6) = 2$$

so
$$v_a = 3i_3 = 6 \quad \text{and} \quad v_c = -2 \text{ V}$$

DRILL PROBLEMS

a. In the circuit of Figure 1.10.9 all conductances equal 1 ℧; $I_x = 12$ A, $I_y = 10$ A, and $V_z = 4$ V. Find the values of v_a, v_b, and v_c.

b. In the circuit shown in Figure 1.10.11, use node equations to find v_a, v_b, and v_c.

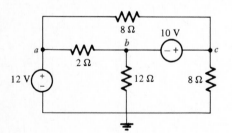

figure 1.10.11
The circuit of Drill Problem **b**

ANS.:   (a) 6 V, 0 V, 4 V; (b) 12 V, 6 V, 16 V.

1.11   loop and mesh equations

The systematic application of Kirchhoff's voltage law (KVL) to any electric circuit enables us to write a set of equations, the solutions of which can be used to calculate the current in each element in the circuit. In general, KVL states that the sum of the voltage rises around any closed path† is zero. If the path consists of a closed loop of elements, however long or circuitous, the resulting zero-valued voltage summation is called a *loop equation*.

Circuits that can be drawn on a piece of paper or constructed on a surface of silicon without any crossovers of conductors are called *planar circuits*. The smallest loops in a planar circuit are called *meshes*. Thus, all meshes are loops, but not all loops are meshes.

Setting the sum of the voltage rises (or drops) around any mesh equal to zero yields a *mesh equation*. We will soon be writing such equations. Although we do not have a really precise mathematical definition of what constitutes a mesh, we should be very clear as to which paths we follow when we write such equations.

EXAMPLE 1.11.1

See Figure 1.11.1.

1. Path *abehgda* is a loop but not a mesh.

---

† Such a path does not have to be made up of a closed sequence of connected elements. The path can jump from any node to any other, whether or not these two nodes are directly connected by an element.

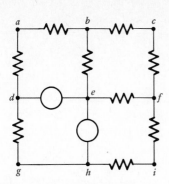

figure 1.11.1
The circuit of Example 1.11.1.

2. Path *befcb* is a mesh.
3. Path *ghedg* is a mesh.

For our purpose now, it is sufficient to recognize that meshes are analogous to the individual small panes of glass in a large window.

It is convenient to think of one continuous current flowing in each mesh of a circuit as in Figure 1.11.2. These currents are called *mesh currents*. The current in each element is the algebraic sum of all mesh currents flowing through that element. Thus, in Figure 1.11.2,

$$i_1 = i_a + i_b$$

$$i_2 = -i_b - i_c$$

$$i_3 = i_a - i_c$$

$$i_4 = -i_b$$

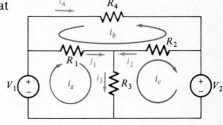

figure 1.11.2
Mesh currents and element currents.

So, mesh currents are really component currents which constitute the individual element currents.

EXAMPLE 1.11.2

In the circuit shown in Figure 1.11.3 find the mesh currents i_a, i_b, and i_c given that $i_1 = 4$ A, $i_2 = 2$ A and $i_3 = 2$ A.
ANS.:   By definition,

$$i_1 = i_a - i_b$$

$$i_2 = -i_b$$

$$i_3 = i_b - i_c$$

So

$$4 = i_a - i_b$$

$$2 = -i_b$$

$$2 = i_b - i_c$$

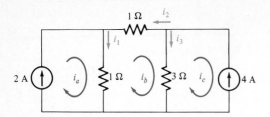

figure 1.11.3
The circuit in Example 1.11.2.

Substituting the fifth equation into the fourth and sixth equation:

$$4 = i_a + 2$$

$$2 = -2 - i_c$$

Thus $$i_a = 2 \quad \text{and} \quad i_c = -4$$

Note, for example, that the total element current i_1 consists of two components:

$$i_a = 2 \text{ A} \quad \text{and} \quad i_b = -2 \text{ A}$$

so that $$i_1 = i_a - i_b$$

$$= 2 - (-2) = 4 \text{ A}$$

We see also that $$i_2 = -i_b = 2$$

and $$i_3 = i_b - i_c$$

$$= -2 + 4 = 2 \text{ A}$$

as was given originally.

To solve an electric circuit by means of mesh currents, we assign one mesh current to each mesh. The direction chosen for each such current is arbitrary (either clockwise or counterclockwise). But, once we have selected that direction, we must stay with that definition until the end of the problem.

Next, we write the voltage across each resistor in terms of the mesh currents flowing through it. Using these voltages, we write a KVL summation around each mesh. We may sum either voltage rises or voltage drops. But, again, we must be consistent. If we start out summing rises, we must continue to do so until the end of that mesh's summation. There is no reason why we cannot sum rises in one mesh equation and then sum the drops while writing the next equation. But within each equation we cannot change from summing drops to summing rises or vice versa. We then solve the resulting set of simultaneous algebraic equations.

If we select the *same direction* for all the mesh currents (all of them clockwise or all of them counterclockwise), then the resulting equations will have certain symmetry properties:

1. In the equation written for the jth mesh, the coefficient of i_j is the sum of all the resistances around the jth loop. This coefficient is positive and all other terms are negative. The coefficient of mesh current i_k is the resistance that is common between mesh j and mesh k.

2. Voltage sources appear on the right side of the equation for any mesh in which they are located in the circuit. If the source aids the mesh current (of the mesh for which this equation is being written), then the source carries a plus sign. If the source bucks the mesh current, it is negative.

EXAMPLE 1.11.3

In Figure 1.11.4 solve for all unknowns by means of mesh equations.

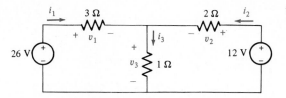

figure 1.11.4
The circuit of Example 1.11.3.

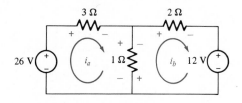

figure 1.11.5
Mesh currents i_a and i_b, assigned arbitrarily, determine the polarities of the component element voltages.

ANS.:

1. Assign a mesh current to each mesh as shown in Figure 1.11.5.
2. Write a KVL equation for each mesh. Note that each component mesh current in an element produces a component of the total voltage across that element:

$$26 - 3i_a - 1i_a + 1i_b = 0$$
$$1i_a - 1i_b - 2i_b - 12 = 0$$

or

$$4i_a - 1i_b = 26$$
$$-1i_a + 3i_b = -12$$

(Note symmetry properties)

Whence $i_a = 6$ A

and $i_b = -2$ A

Therefore, the element variables are quickly found to be:

$$i_1 = i_a = 6 \text{ A} \qquad v_1 = i_1 R_1 = 18 \text{ V}$$
$$i_2 = -i_b = 2 \text{ A} \qquad v_2 = i_2 R_2 = 4 \text{ V}$$
$$i_3 = i_a - i_b = 8 \text{ A} \qquad v_3 = i_3 R_3 = 8 \text{ V}$$

The presence of current sources usually reduces our work.

EXAMPLE 1.11.4

Solve for i_a, i_b, and i_c in Figure 1.11.6.

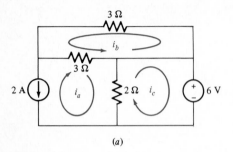

(a)

figure 1.11.6(a)
The circuits of Example 1.11.4

ANS.:
$$i_a = -2$$
$$3i_b + 3i_b + 3i_a = 0$$
$$2i_a - 2i_c - 6 = 0$$

Substituting the first equation into the other two:
$$6i_b + 3(-2) = 0$$
$$2(-2) - 2i_c - 6 = 0$$

Whence
$$i_b = 1 \text{ A}$$
and
$$i_c = -5 \text{ A}$$

Dependent sources are handled in the same way.

EXAMPLE 1.11.5

Solve for the mesh currents i_1, i_2, and i_3 in Figure 1.11.6b.

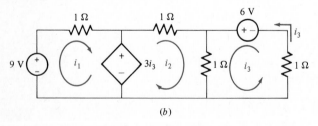

(b)

figure 1.11.6(b)
Example 1.11.5.

ANS.:

Mesh 1: $9 - 1i_1 - 3i_3 = 0$ or $-1i_1 + 0i_2 - 3i_3 = -9$

Mesh 2: $\qquad 3i_3 - 2i_2 - 1i_3 = 0 \qquad$ or $\qquad 0i_1 - 2i_2 + 2i_3 = 0$

Mesh 3: $\qquad 1i_2 + 1i_3 - 6 + 1i_3 = 0 \qquad$ or $\qquad 0i_1 + 1i_2 + 2i_3 = 6$

Solving for i_1, i_2, and i_3 by Cramer's rule:

$$\Delta = \begin{vmatrix} -1 & 0 & -3 \\ 0 & -2 & 2 \\ 0 & 1 & 2 \end{vmatrix} = -1 \begin{vmatrix} -2 & 2 \\ 1 & 2 \end{vmatrix} = -1(-4-2) = +6$$

$$\Delta_1 = \begin{vmatrix} -9 & 0 & -3 \\ 0 & -2 & 2 \\ 6 & 1 & 2 \end{vmatrix} = -9 \begin{vmatrix} -2 & 2 \\ 1 & 2 \end{vmatrix} + 6 \begin{vmatrix} 0 & -3 \\ -2 & 2 \end{vmatrix}$$

$$= -9(-4-2) + 6(-6) = +54 - 36 = 18$$

Therefore,
$$i_1 = \frac{\Delta_1}{\Delta} = 3 \text{ A.}$$

$$\Delta_2 = \begin{vmatrix} -1 & -9 & -3 \\ 0 & 0 & 2 \\ 0 & 6 & 2 \end{vmatrix} = -1 \begin{vmatrix} 0 & 2 \\ 6 & 2 \end{vmatrix} = 12$$

So $i_2 = \Delta_2/\Delta = 2$ A.

$$\Delta_3 = \begin{vmatrix} -1 & 0 & -9 \\ 0 & -2 & 0 \\ 0 & 1 & 6 \end{vmatrix} = -1 \begin{vmatrix} -2 & 0 \\ 1 & 6 \end{vmatrix} = 12$$

So $i_3 = \Delta_3/\Delta = 2$ A.

LOOP EQUATIONS

Although the presence of current sources often makes the solution computationally simpler, sometimes it is difficult to decide how to handle this situation. Also, nonplanar networks are almost impossible to solve because we have no good definition of a mesh for nonplanar circuits.

For example, how many equations are needed to solve the nonplanar network of Figure 1.11.7 completely? And how do we go about writing them? We can answer this question by studying the structure (topology) of the circuit. First we need some definitions.

∗ Graph   The schematic diagram showing all elements simply as lines.

∗ Node   A point at which two or more elements are connected.

∗ Branch   A single path, containing one simple element that connects two nodes. (Sometimes these are called *edges*.)

∗ Loop   A set of branches forming a closed path that passes through no node more than once.

figure 1.11.7
A circuit in the form of
a cube with elements on two body diagonals.

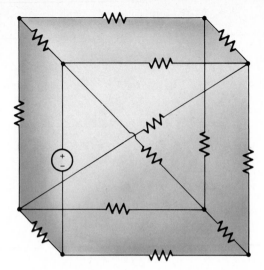

∗ Planar circuit   A circuit or network that may be drawn on a plane surface in such a way that no branch passes over or under another.

∗ Tree   A set of branches that connects to every node in the graph and contains no loops.

∗ Link (or chord)   A branch of a graph that is not in a specified tree.

Any given branch in a graph may or may not be a link—it depends on which tree is chosen for the graph.

EXAMPLE 1.11.6

Draw two possible trees for the circuit in Fig. 1.11.8a.
ANS.:   See Figure 1.11.8c and d.

Obviously if a graph has N nodes, then any tree will have $N - 1$ branches (since the first tree branch connects two nodes, and every additional tree branch connects to a single additional node).

Therefore if a graph has a total of B branches, the number of links L is:

$L =$ total number of branches − number of branches in a tree

$$L = B - (N - 1)$$

or $$L = B - N + 1$$

Consider any graph. Pick a tree. Define one and only one loop current in each link (it will pass through one or more tree branches; each tree branch may have several different loop currents passing through it). Since we have $B - N + 1$ links, we will have $B - N + 1$ different, independent paths. We write $\sum v = 0$ equations around each such path. This gives $B - N + 1$ independent equations.

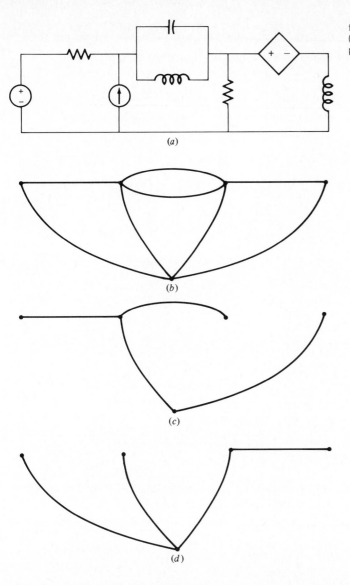

(a)

(b)

(c)

(d)

figure 1.11.8
(a) A circuit, (b) its graph, and (c) and (d) two possible trees.

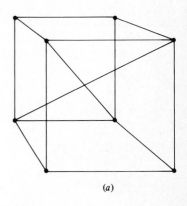

(a)

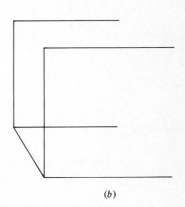

(b)

figure 1.11.9
(a) The graph and (b) a tree for the circuit of Figure 1.11.7.

EXAMPLE 1.11.7

How many loop equations must we write to solve the circuit in Figure 1.11.9?
ANS.: Since $B = 14$ and $N = 8$, the $B - N + 1$ number of links = 7. Therefore, we need $B - N + 1$ loop equations = 7.

When current sources appear in a circuit, always pick a tree so that the current source is a link. That way you immediately know the value of one of the loop currents.

EXAMPLE 1.11.8

Write a set of independent loop equations and solve for all element voltages and currents (Figure 1.11.10).

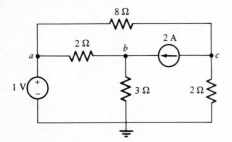

figure 1.11.10
The circuit of Example 1.11.8.

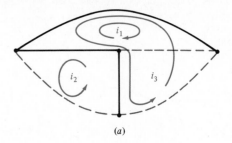

(a)

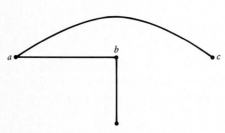

figure 1.11.11
A tree for the circuit in Figure 1.11.10.

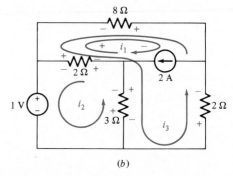

(b)

figure 1.11.12
(a) Choosing a loop current for each link. (b)
The loop currents written into the circuit
diagram.

ANS.:   Choose any tree such that the current source is a link (Figure 1.11.11). Define one and only one loop current in each link and write a KVL summation for each loop thus formed (Figure 1.11.12).

Loop 1: $$i_1 = 2 \tag{1.11.1}$$

Loop 2: $$+1 + 2(i_2 + i_1 - i_3) + 3(i_2 - i_3) = 0$$
$$-5i_2 + 5i_3 = 5$$
$$-i_2 + i_3 = 1 \tag{1.11.2}$$

Loop 3: $$-8i_3 + 8i_1 + 2i_1 + 2i_2 - 2i_3 + 3i_2 - 3i_3 - 2i_3 = 0$$
$$10i_1 + 5i_2 - 15i_3 = 0$$
$$2i_1 + i_2 - 3i_3 = 0$$
$$i_2 - 3i_3 = -4 \tag{1.11.3}$$

Solving equations (1.11.1) to (1.11.3) yields $i_1 = 2$ A, $i_2 = \frac{1}{2}$ A, $i_3 = \frac{3}{2}$ A (Figure 1.11.13).

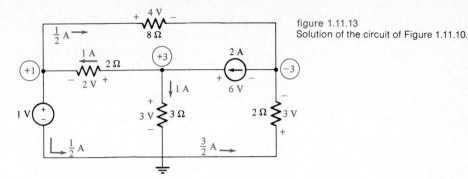

figure 1.11.13
Solution of the circuit of Figure 1.11.10.

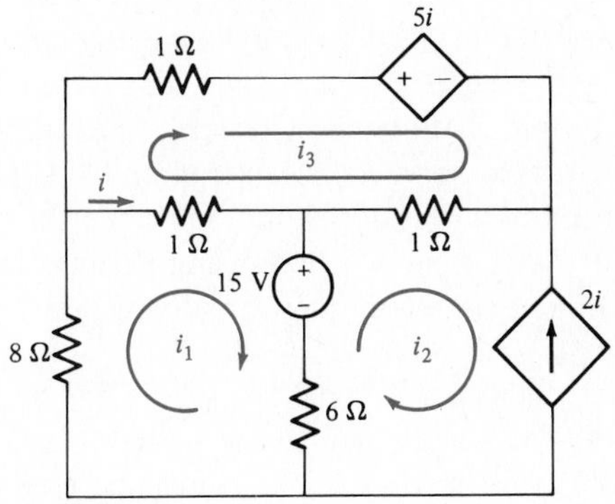

figure 1.11.14
The circuit of the drill problem.

The question of whether it is easier to use node equations or mesh (loop) equations to solve any given circuit cannot be answered in general. For some circuits, node equations are better—for others, mesh equations give a simpler solution.

Recall that to solve a circuit having N nodes by node equations we first choose a datum (ground) node and then write one node equation at each of the $N-1$ remaining nodes. If we use mesh equations, $B-N+1$ equations are needed.

In general, if students are at liberty to select their own methods of solution in any given problem, they are well advised to pause to consider which method, node equations or mesh equations, requires the solution of the fewest independent simultaneous equations.

DRILL PROBLEM

In the circuit shown in Figure 1.11.14, find the value of currents i_1, i_2, and i_3 by writing and solving the necessary mesh equations.
ANS.: -2, -2, -3.

1.12 **summary**

In this first chapter we have examined some of the basic concepts on which we will begin to build our study of the way electric circuits work. We have discussed charge, current, and voltage; also voltage and current sources, both dependent and independent. We have examined the volt-ampere characteristics of sources and resistors, because the graphical description of these (and most other) electrical elements is extremely helpful to us, as engineers, in understanding their properties. We shall make constant use of the graphical descriptions of elements throughout this book.

We have seen how Kirchhoff's voltage law gives rise to the analytical tools we call mesh equations and loop equations. The systematic application of Kirchhoff's current law enables us to write node equations. Both of these techniques enable us to completely solve any circuit made up of voltage and current sources and resistors. Finally we discovered, through a brief examination of network topology, how to determine the number of necessary loop equations we must write to solve nonplanar circuits. This same discussion also showed us the best way to write loop equations when the circuit contains current sources.

We are now ready, using this material as a base, to expand our knowledge of analytical techniques for solving simple circuits.

problems

1. The net charge that has passed by a given point in a wire is given by the expression $q(t) = 3 \sin 6t$ for $t > 0$. (No charge moved in the wire prior to $t = 0$.) What current $i(t)$ resulted in this $q(t)$?

2. At a given point in a wire the current varies with time as shown in Figure P1.2. How much net charge has moved past this point during the time interval $1 < t < 3$?

3. In the circuit in Figure P1.3, $v_{bd} = 4$ V. Find the values of i_1, i_2, i_3, v_{cd}, and V_s.

4. In problem 3, if $V_s = 20$ V, find v_{bd} and v_{cd}.

5. In Figure 1.1.8, $R_1 = 2 \ \Omega$, $V_C = 10$ V, and $I_1 = 2$ A. What is the value of v_{in}?

6. A positive electric charge, $q = 2$ C, moving from one point to another, goes through a voltage rise of 10 V. If this movement takes 3 ms, how much power is required to accomplish it?

7. A current of 4.5 A flows in a linear resistor R when the voltage across it is 6.9 V. When the current is increased to a new value, the voltage rises to a value of 55.2 V. (*a*) What is the new current value? (*b*) What value of conductance (in mhos), when placed in parallel

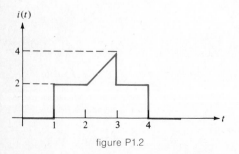

figure P1.2

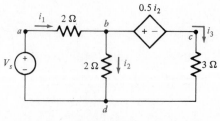

figure P1.3

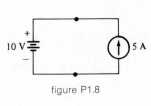

figure P1.8

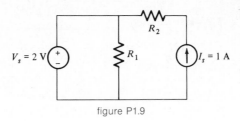

figure P1.9

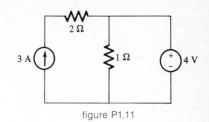

figure P1.11

with R, would produce this same new current value but at the original 6.9-V terminal voltage?

8. A 10-V battery is connected to a 5-A current source as shown in Figure P1.8. The current enters the positive terminal of the battery. Find the power *absorbed* by the: (a) voltage source and (b) current source.

9. Assuming that $R_1 = R_2 = 3\ \Omega$ in the circuit shown in Figure P1.9, find the following quantities: (a) the power dissipated in R_1, (b) the power dissipated in R_2, (c) the power supplied by V_s, and (d) the power supplied by I_s.

10. A light bulb rated at 1 W is connected to an ideal 3-V source. Assuming that the light bulb approximates a resistor, find: (a) the current drawn by the bulb and (b) the resistance of the bulb.

11. Find the power p delivered *to* each of the four elements in the circuit shown in Figure P1.11.

12. A 12-V battery is charged by supplying it with a constant current of 3 A for 2 h, and then by a current that drops linearly from 3 A to zero during the next hour. Assuming the battery voltage is constant at 12 V, find (a) The total charge in ampere-hours supplied to the battery. What is this in coulombs? (b) The average power delivered to the battery during the 3-h interval. (c) The total energy delivered to the battery. (d) At what time instant will the power being delivered to the battery be exactly 24 W?

13. A charging circuit for an automobile battery is shown in Figure P1.13. Find the (a) electric power generated inside the alternator, (b) power output delivered by the alternator at its terminals, (c) power absorbed by the battery cable (0.5 Ω), (d) power input to the battery, (e) power converted into chemical energy inside the battery, and (f) power dissipated as heat by the battery. (g) Of the power generated inside the alternator, what percentage actually charges the battery?

14. In the circuit shown in Figure P1.14, $v_1 = 20$ V, $v_2 = 10$ V, $R_1 = 10\ \Omega\ R_2 = 5\ \Omega$, and $R_3 = 15\ \Omega$. (a) Find the power delivered to each resistor. (b) Find the power delivered to the circuit by the sources.

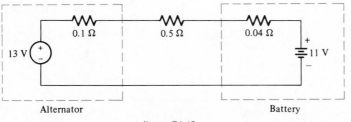

Alternator   Battery

figure P1.13

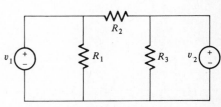

figure P1.14

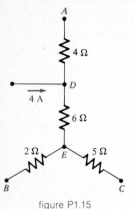

figure P1.15

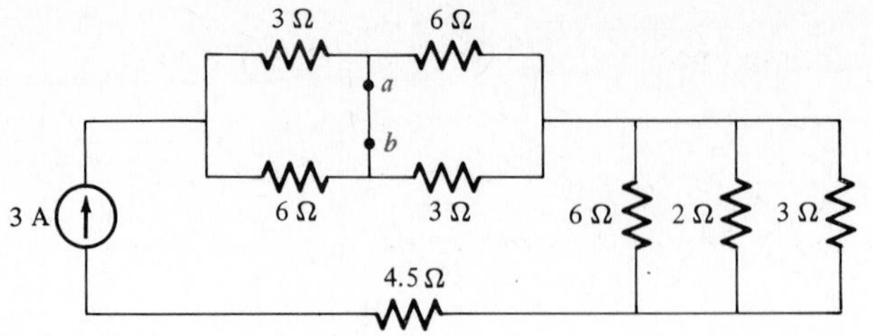

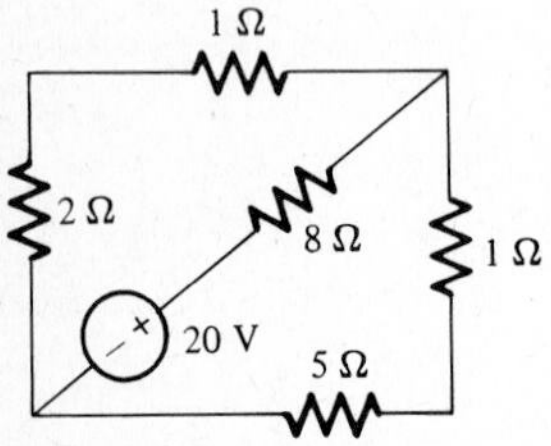

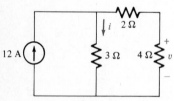

figure P1.20

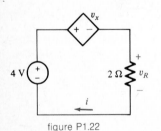

figure P1.21

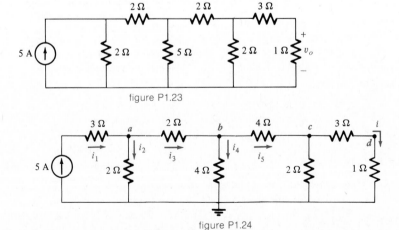

figure P1.22

figure P1.19

15. In the circuit shown in Figure P1.15 nodes A, B, and C are directly connected to the reference (ground) node. Find the voltages at nodes D and E. How much current leaves each of the nodes A, B, and C?

16. A 220-V dc source is connected by a two-wire transmission line to a 3-Ω load resistor. Each wire of the transmission line has a resistance of 0.08 Ω. Find (a) the power transmitted by the source, and (b) the cost of the power lost in the line in 24 h. (Assume the price of electricity is \$0.12/kWh.) (c) If we boost the source up to 2200 V and also raise the load resistance to maintain the same received power while using the same transmission line as before, then what is the cost of the daily loss of power in the line?

17. Find the equivalent resistance R_{eq} of three resistors R_1, R_2, and R_3 connected in series.

18. Find the equivalent resistance of three resistors connected in parallel.

19. Find the magnitude and direction of the current in the connection labeled a-b in Figure P1.19.

20. Find the power dissipated in the 8-Ω resistor in the circuit shown in Figure P1.20.

21. Using the principles of current and voltage division, find the current i and the voltage v in the circuit in Figure P1.21.

22. In the circuit in Figure P1.22 find v_R and i if (a) $v_x = 8i$ and (b) $v_x = 3v_R$.

23. Find the numerical value of the output voltage v_o in the circuit in Figure P1.23.

24. Find the numerical value of i in the circuit in Figure P1.24.

figure P1.23

figure P1.24

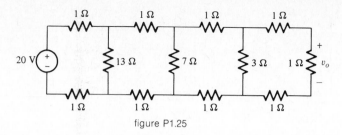

figure P1.25

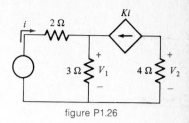

figure P1.26

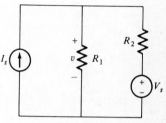

figure P1.27

25. Find the numerical value of the output voltage v_o in the circuit in Figure P1.25.

26. Given the circuit in Figure P1.26 (assume that $K = 8$ and $V_1 = 6$ V), (a) find V_2. (b) Assuming the ideal source on the left is a current source, what is its value? (c) If the ideal source is a voltage source, what is its value?

27. In the circuit shown in Figure P1.27 find the value of the voltage v by writing one equation. $R_1 = R_2 = 2$ Ω, $I_s = 1$ A, and $V_s = 4$ V.

28. Determine the value of v_x in the circuit shown in Figure P1.28.

29. Determine the value of v_2 in the circuit shown in Figure P1.29.

30. In the circuit shown in Figure P1.30, $R_1 = 4$ Ω, $R_2 = 2$ Ω, $R_3 = 1$ Ω, $V_s = 15$ V, and $I = 5$ A. Find v_B, the voltage across the current source.

31. In the circuit shown in Figure P1.31, all R's are 1-Ω resistors. If $V_A = 11$ V, and $V_B = 7$ V, find v_1 and v_2.

32. Write the node equations in variables v_a and v_b in Figure P1.32. Solve these equations for the values of v_a and v_b.

33. Solve the circuit in Figure 1.11.10 for voltages v_a, v_b, and v_c by means of node equations. Compare your answers with those shown in Figure 1.11.13.

34. Find the value of v_1, v_2, and the power delivered by the 9-V source in the circuit in Figure P1.34.

35. In the circuit shown in Figure P1.35 find the voltages v_a and v_b.

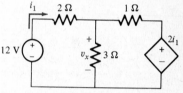

figure P1.28

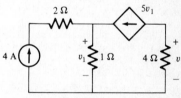

figure P1.29

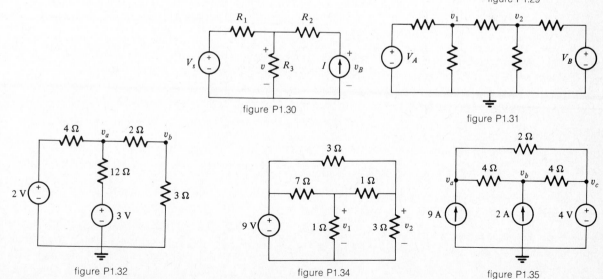

figure P1.30

figure P1.31

figure P1.32

figure P1.34

figure P1.35

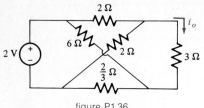

figure P1.36

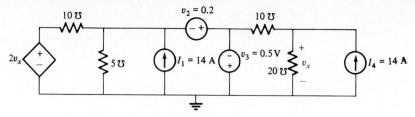

figure P1.37

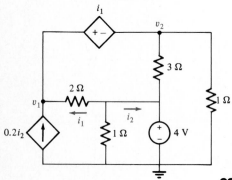

figure P1.38

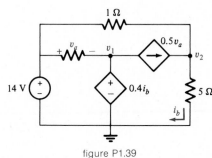

figure P1.39

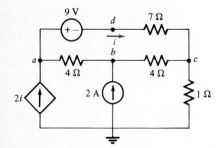

figure P1.40

36. Solve for the output current i_o in the circuit of Figure P1.36.

37. In the circuit shown in Figure P1.37 find (a) the power produced by each independent source and (b) the power produced by the dependent source.

38. Find the value of voltages v_1 and v_2 in the circuit in Figure P1.38.

39. Find the value of voltages v_1 and v_2 in the circuit in Figure P1.39.

40. Solve the circuit shown in Figure P1.40 for (a) the values of v_a, v_b, and v_c via node equations. (b) Repeat the solution to part (a) using mesh equations. Use clockwise mesh currents: i_1 in the lower left, i_2 in the lower right, and i_3 in the upper mesh.

41. Find the value of the voltage at each node of the circuit shown in Figure P1.41.

42. In the circuit shown in Figure P1.42 (a) solve for v_b in terms of the sources and the conductances. (b) In part (a), let $G_1 = G_2 = G_4 = 1$ ℧, $G_3 = 3$ ℧, $V_1 = 2$ V, and $V_2 = 4$ V. $H = \frac{2}{3}$. Find the power delivered to the circuit by the voltage source V_1.

43. The circuit shown in Figure P1.43 is in the shape of a three-dimensional cube wherein all resistance values are 1 Ω. The values of certain voltages and currents are given. Find the values of all other element currents and node voltages.

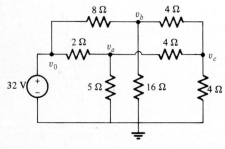

figure P1.41

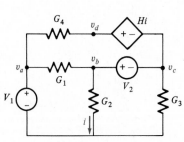

figure P1.42

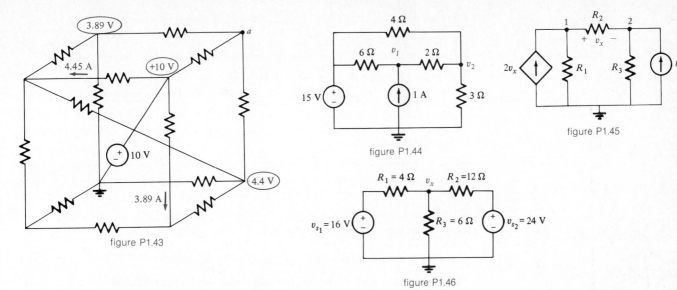

figure P1.43

figure P1.44

figure P1.45

figure P1.46

44. In the circuit shown in Figure P1.44 (*a*) solve for v_1 and v_2 by means of node equations. (*b*) Using clockwise loop currents, solve an appropriate set of loop equations for this circuit. Use the results to check your answers to part (*a*).

45. Write the node equations at nodes 1 and 2 in the circuit in Figure P1.45 and present the results in matrix form.

46. Using node equations in the circuit in Figure P1.46, find the voltage v_x with respect to ground.

47. Repeat the previous problem using mesh equations instead of node equations.

48. In the circuit of Figure P1.48 (*a*) find i_1 and i_2 via mesh equations and (*b*) solve the node equations necessary to check your answers to part (*a*).

49. In the circuit shown in Figure P1.49, find the values of i_1 and i_2 by first writing an equation for i_2 in terms of i_1 and then writing a KVL summation around the outer loop.

50. Using mesh equations in the circuit of Figure P1.50, find the voltage at node *A* with respect to ground. Check your answer by writing a node equation.

51. Solve for the mesh currents in the circuit in Figure P1.51. Your answer will be in terms of the voltage source *V*. All resistance values are 1 Ω.

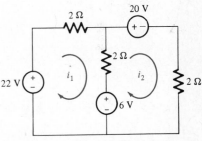

figure P1.48

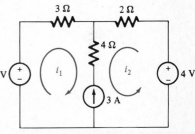

figure P1.49

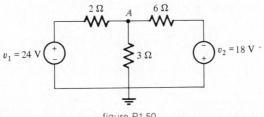

figure P1.50

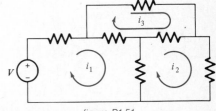

figure P1.51

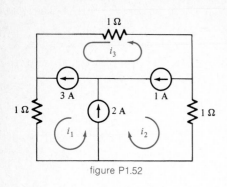

figure P1.52

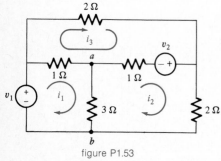

figure P1.53

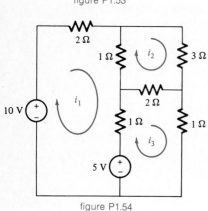

figure P1.54

52. Find the values of the three mesh currents i_1, i_2, and i_3 in the circuit of Figure P1.52.

53. In the circuit shown in Figure P1.53, $v_1 = 4$ V and $v_2 = 10$ V. (*a*) Determine the values of the mesh currents. (*b*) How much total power is delivered to the resistors?

54. Given the circuit shown in Figure P1.54, (*a*) write the mesh equations in matrix form (i.e., for $[R][i] = [v]$ find $[R]$, $[i]$, and $[v]$). (*b*) Solve for the numerical values of the elements of $[i]$.

55. Write the mesh equations for the circuit in Figure P1.55. (*a*) Write them in matrix form and (*b*) solve them.

56. Solve the circuit in Figure 1.10.6 in the text by means of loop equations as follows. Let one loop current equal 7 A. Then draw the two remaining loop currents in the clockwise direction.

57. For the network shown in Figure P1.57, draw a tree such that the current sources are links. Define loop currents such that one and only one loop current passes through each link. (*a*) Solve for the value of the loop currents. (*b*) What is the value of the power dissipated in the 2-Ω resistor?

58. The graph of a schematic has five nodes with single branches connecting every node to every other node. How many node equations and how many loop equations are required to solve this circuit?

59. Given the circuit shown in Figure P1.59, (*a*) draw a tree such that one independent unknown loop current is the only current (directed upward) in the 100-V source. (*b*) Write the loop equations necessary to solve this circuit. How many independent unknown loop currents are there? (*c*) Solve the equations in part (*b*). Find the value of the current in each branch and the voltage at each node.

60. (*a*) Solve for i_1, i_2, and i_3 in the following set of mesh equations:

$$4i_1 - i_2 - i_3 = 5$$

$$-i_1 + 6i_2 - 2i_3 = 0$$

$$-i_1 - 2i_2 + 4i_3 = 0$$

(*b*) Sketch a circuit from which these equations might have come.

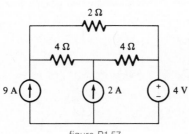

figure P1.57

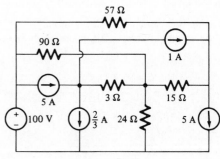

figure P1.59

figure P1.55

chapter 2
THE BASIC TOOLS OF CIRCUIT ANALYSIS

introduction 2.1

Learning to analyze circuits really consists of accumulating several basic analytical tools in our mental tool box and then pulling them out and using them singly or in conjunction with one another as the need arises. The most successful electrical engineer is the person who knows which tool is best to use in each case and is adept at using it. One of the most appealing things about electrical engineering is that there is not very much rote memorization of dry facts that has to be done by the student. An electrical engineer has only a few basic analytical tools at his or her disposal. But the best engineer is the one who can apply the correct analysis technique at the right time (and do so in a wide variety of problems).

In this chapter we examine a few of the basic tricks of the trade, analysis techniques that are used over and over, day in and day out, to solve circuit problems. We will make use of these analytical techniques, together with those we studied in Chapter 1, throughout this text, and you will use them throughout your career as an engineer. Their names are superposition, Thevenin and Norton's theorems, and the delta-wye transformation.

2.2 superposition

In the section on Ohm's law it was pointed out that linear resistors possess the additivity property. Consider the circuits shown in Figure 2.2.1. It is clear that equal voltages are impressed on R_1 in Figure 2.2.1*a* and *b*. Therefore, the same value of current i_a will exist in both these circuits. (Also recall that a 0-V voltage source is the electrical equivalent of an ideally conducting piece of wire; so circuits *a* and *b* are identical.)

Similarly, the circuits in Figure 2.2.1*c* and *d* are identical. The *additivity property* of resistor R_1 is shown in Figure 2.2.1*e*. If v_a produces i_a, and v_b produces i_b, then $v_a + v_b$ will produce $i_a + i_b$.

If we work backward through Figure 2.2.1, we may say the following: Assume that, in (*e*), v_a, v_b, and R are given and we wish to find the total current i_{total} in R_1. We realize from (*b*), (*d*), and (*e*) that this total current is the sum of two component currents: i_a is due to v_a acting alone, and i_b is due to v_b acting alone.

figure 2.2.1
The additivity property of linear resistors, the basis of superposition.

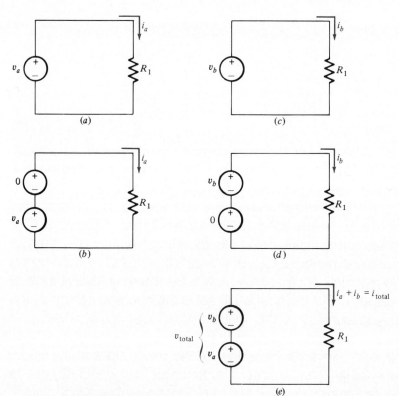

The principle of superposition thus consists of the following. Shut off (set to zero magnitude) *all independent sources* except one. Solve for the component of the sought-after current or voltage due to this source acting alone. Repeat this step for each independent source. Finally, add all components together to get the total value.

EXAMPLE 2.2.1

Find v_a and i in Figure 2.2.2a.

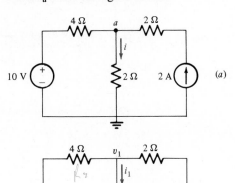

figure 2.2.2
An example of the application of the principle of superposition.

$$i_1 = 2\left(\frac{4}{\div 6}\right) = \frac{9}{3}$$

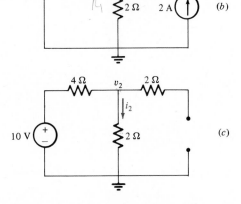

ANS.:   First set the voltage source to zero (by replacing it with its electrical equivalent—a short circuit) as in Figure 2.2.2b, and solve for the component current i_1 and voltage v_1. By current division we find

$$i_1 = 2\left(\frac{4}{4+2}\right) = \tfrac{4}{3} \text{ A}$$

and

$$v_1 = i_1 R = (\tfrac{4}{3})2 = \tfrac{8}{3} \text{ V}$$

Then set the current source to zero (by replacing it with its electrical equivalent—an open circuit) as in Figure 2.2.2c, and solve for components i_2 and v_2. By voltage division:

$$v_2 = 10\left(\frac{2}{2+4}\right) = \tfrac{10}{3} \text{ V}$$

and

$$i_2 = \frac{v_2}{R} = \tfrac{10}{3}(\tfrac{1}{2}) = \tfrac{5}{3} \text{ A}$$

So, finally,

$$v_a = v_1 + v_2$$
$$= \tfrac{8}{3} + \tfrac{10}{3} = 6 \text{ V}$$

and

$$i = i_1 + i_2$$
$$= \tfrac{4}{3} + \tfrac{5}{3} = 3 \text{ A}$$

EXAMPLE 2.2.2

This is a problem that can only be solved via superposition. Find the equivalent resistance R_{ab} at two adjacent nodes in an infinite network of 1-Ω resistors. See Figure 2.2.3.

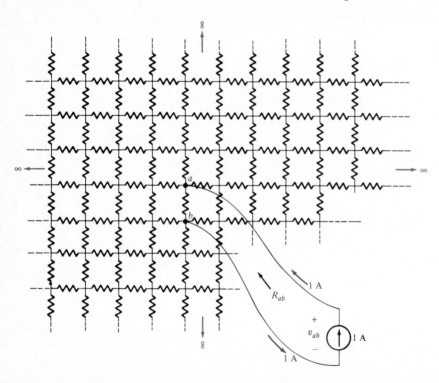

figure 2.2.3
Finding the equivalent resistance R_{ab} between two nodes a and b of a network of 1-Ω resistors which is of infinite extent; $R_{ab} = v_{ab}$.

ANS.:   We realize that if a 1-A current flows into terminal a and out of terminal b, then R_{ab} is given by Ohm's law: $R_{eq} = v_{ab}/1\text{A}$. Thus, we solve for R_{eq} by choosing $I = 1$ and solving for v_{ab}. First, ground the network at an infinite distance away from terminals a and b. Then connect a current source to force 1 A into node a and remove it at infinity. See Figure 2.2.4. Since the network is of infinite extent, this current will divide equally between the four 1-Ω resistors connected to node a. Note that this produces a $\tfrac{1}{4}$-V rise from b to a. Disconnect the source.

Now, connect a current source which will draw 1 A out from terminal b and deliver it to ground at infinity as in Figure 2.2.5. This current will produce equal $\tfrac{1}{4}$-A currents in the resistors connected to node b. This results in a voltage rise of $\tfrac{1}{4}$ V from b to a. By superposition we note that, since the total current in Figure

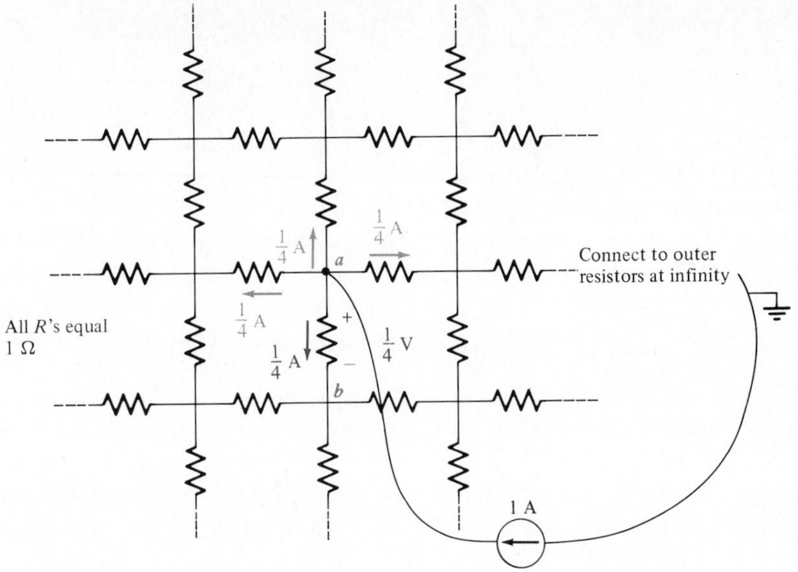

figure 2.2.4
A current of 1 A forced into terminal a.

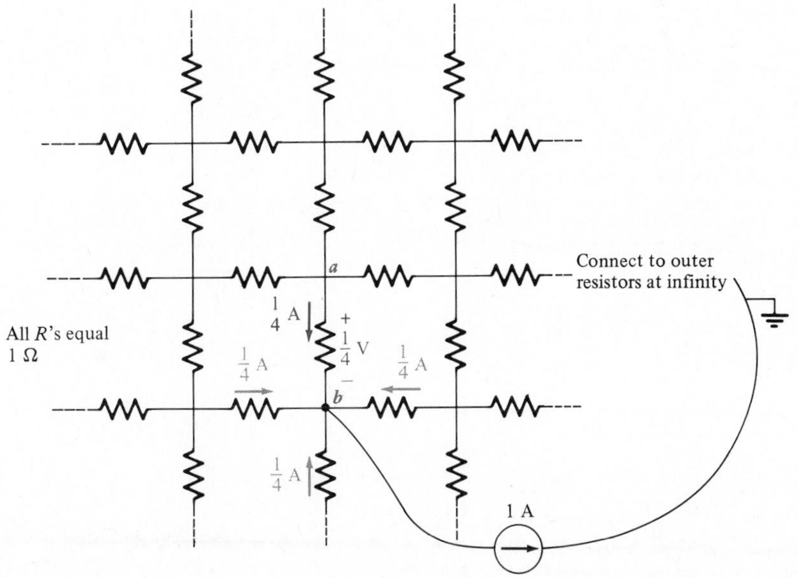

figure 2.2.5
A current of 1 A forced out of terminal b.

2.2.3 is the sum of the component sources in Figures 2.2.4 and 2.2.5, the total voltage rise from b to a in Figure 2.2.3 is the sum of the two component voltages. Therefore $v_{ab} = \frac{1}{4} + \frac{1}{4} = \frac{1}{2}$ V and $R_{eq} = \frac{1}{2}$ Ω.

DRILL PROBLEM

Reverse the direction of the 2-A source in Figure 2.2.2. What is the value of the voltage at node a?
ANS.: $\frac{2}{3}$ V

2.3 Thevenin's and Norton's theorems

Two-terminal networks are called *terminally equivalent* if the same current flows into both networks when their terminal voltages are equal, and/or if the same voltage appears across both pairs of terminals when identical currents are forced into both networks. Equivalent resistances of resistors in series or parallel are simple examples of such terminal equivalent circuits.

Thevenin and Norton expanded the concept of terminal equivalency to include circuits that also contain sources. They showed that any two-terminal network, such as the one in Figure 2.3.1, that contains linear resistors and sources (current, voltage, independent or dependent) has a terminal equivalent circuit of the form of either Figure 2.3.2*a* or *b*.

The network in Figure 2.3.1 can have any number of resistors and sources. If the same external electric circuit is connected to the terminals of all three networks in Figures 2.3.1 and 2.3.2, then the terminal currents i will all be equal.

Given an arbitrary two-terminal network, how is the value of R_{Th} and V_{Th} determined? The answer to this question lies in first recognizing what happens if a resistor of value $R = \infty$ ohms (an open circuit) is connected to all three circuits in Figures 2.3.1 and 2.3.2. (See Figure 2.3.3.) In all three of these circuits $i = 0$. Also, the terminal voltages must be identical if the circuits are to be terminal

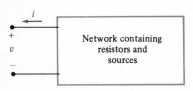

figure 2.3.1
A general linear two-terminal network.

figure 2.3.2
Terminal equivalent circuits: (*a*) Thevenin equivalent and (*b*) Norton equivalent.

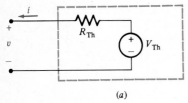

(*a*)

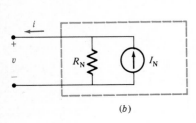

(*b*)

figure 2.3.3
Open circuits (infinite resistance) connected to the circuits of Figures 2.3.1 and 2.3.2. (*a*) The actual circuit. (*b*) The Thevenin equivalent circuit. (*c*) The Norton equivalent circuit.

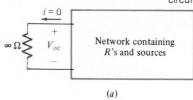

(*a*)

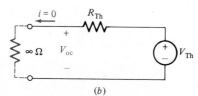

(*b*)

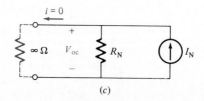

(*c*)

figure 2.3.4
Short circuits connected to the circuits of Figures 2.3.1 and 2.3.2.

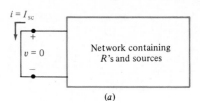

(*a*)

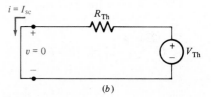

(*b*)

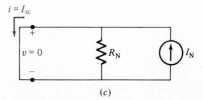

(*c*)

equivalents. In circuit *a* of Figure 2.3.3 we measure the voltage at the output terminals. Call this V_{oc}, the *open-circuit voltage*.

In circuit *b*, $v = V_{oc} = V_{Th}$ because, since $i = 0$, the voltage drop across resistor R_{Th} is zero.

In circuit *c*, $v = V_{oc} = I_N R_N$. Thus,

$$V_{oc} = V_{Th} = I_N R_N \qquad (2.3.1)$$

Suppose the value of the external resistor is changed to $R = 0$, a short circuit. See Figure 2.3.4. Call the current that flows out of the arbitrary network the short-circuit current I_{sc}. In circuit *b*,

$$i = I_{sc} = \frac{V_{Th}}{R_{Th}}$$

In circuit *c*
$$i = I_{sc} = I_N$$

Thus,
$$I_{sc} = I_N = \frac{V_{Th}}{R_{Th}}$$

or
$$V_{Th} = I_N R_{Th} \qquad (2.3.2)$$

Comparing equations (2.3.1) and (2.3.2), we note that

$$R_{Th} = R_N$$

The important things to remember about Thevenin and Norton equivalent circuits that have been demonstrated above are:

1. The voltage source in the Thevenin equivalent circuit is the open-circuit voltage.
2. The current source in the Norton equivalent circuit is the short-circuit current.
3. The series resistor in the Thevenin circuit is identical to the parallel resistor in the Norton circuit. Thus the name *output resistance* is equivalent to either R_{Th} and/or R_N.
4. The open-circuit voltage, the short-circuit current, and the resistance $R_{Th} = R_N$ are interrelated by Ohm's law: $V_{Th} = I_{sc} R_{Th}$.

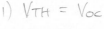

1) $V_{TH} = V_{OC}$
2) $I_N = I_{SC}$
3) $R_{TH} = R_N$
4) $V_{TH} = I_{SC} R_{TH}$

figure 2.3.5
Thevenin equivalent circuit with V_{Th} set equal to zero.

In addition, it should be noted that in the Thevenin equivalent circuit, Figure 2.3.2*a*, the value of R_{Th} can be observed from the terminals if and only if the internal voltage source is set equal to zero. Recall that a 0-V voltage source acts like a piece of wire (short circuit).

See Figure 2.3.5. The voltage source is set to 0 and replaced by an ideally conducting piece of wire.

Similarly, in the Norton equivalent circuit, if the current source is set to zero (by replacing it with an open circuit), R_N is seen at the terminals (see Figure 2.3.6).

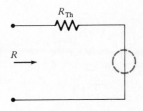

figure 2.3.6
Norton equivalent circuit with I_N set equal to zero.

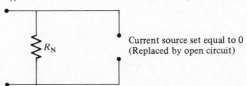

Current source set equal to 0
(Replaced by open circuit)

The equivalent output resistance of any arbitrary network can be found similarly as follows:

1. Set all independent sources inside the network to zero by replacing voltage sources with short circuits and current sources with open circuits. (Do *nothing* to dependent sources.)

2. Then determine the effective resistance seen at the terminals. The most general method for doing this is to drive 1 A into the network by connecting a 1-A current source at the terminals. Determine the resulting voltage at the terminals. This voltage is numerically equal to the resistance.

In other words, R_{Th} is found as follows (see Figure 2.3.7):

$$R_{Th} = \frac{V_o}{1} = V_o$$

Therefore one way to find R_{Th}, after setting all independent sources to 0, is to insert a 1-A current back into the network's + output terminal and then solve for V_o numerically. This numerical answer equals R_{Th}.

We can use this method as a check on our calculations of $V_{Th} = V_{oc}$ and $I_N = I_{sc}$ since $R_{Th} = V_{oc}/I_{sc}$. In general, then, in order to find either the Thevenin or the Norton terminal equivalent circuit (of any actual circuit that contains only linear resistors) one needs any two of the following (see Figure 2.3.8): the open-circuit voltage, the short-circuit current, the output resistance.

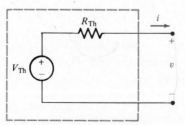

figure 2.3.7
Solving for $R_{Th} = R_N$ directly. Using this method $R_{Th} = R_N = V_o$.

Thevenin	Norton
$V_{oc} = v$ when $i = 0$	$I_{sc} = i$ when $v = 0$
$V_{oc} = V_{Th}$	$I_{sc} = I_N$
$I_{sc} = i$ when $v = 0$	$V_{oc} = v$ when $i = 0$
$I_{sc} = \dfrac{V_{Th}}{R_{Th}} = \dfrac{V_{oc}}{R_{Th}}$	$V_{oc} = I_N R_N = I_{sc} R_N$
$\dfrac{V_{oc}}{I_{sc}} = R_{Th}$	$\dfrac{V_{oc}}{I_{sc}} = R_N$

Therefore, $R_{Th} = R_N$.

WARNING   It should be carefully noted that the Thevenin and Norton circuits are *terminal equivalents* of some other (usually larger and more complicated) network. We

figure 2.3.8
Summary of interrelationships between Thevenin and Norton variables.

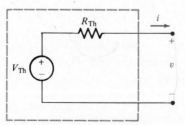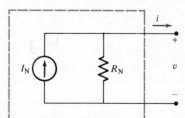

$$R_{Th} = R_N$$
$$V_{Th} = I_N R_N$$

cannot use these terminal equivalent circuits to answer questions about what is occurring *inside* the larger network. For example, if we are interested in knowing how much power is dissipated in the resistors inside the large network, we should not be tempted into thinking that this is equal to the power dissipated in R_{Th}. It is not. The Thevenin and Norton circuits are useful in determining the output voltage and current that appear at the *terminals* of the large network. They tell us *nothing* about the *internal* workings of the large network.

EXAMPLE 2.3.1

Find a Thevenin and a Norton terminal equivalent circuit for the circuit shown in Figure 2.3.9a.

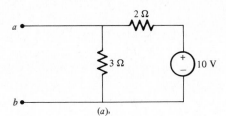

figure 2.3.9
(a) An arbitrary network. (b) Solving for I_{sc}. (c) Solving for $R_{Th} = R_N$.

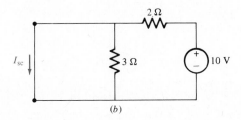

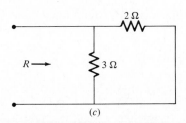

ANS.: Solve for the open-circuit voltage by voltage division in Figure 2.3.9a.

$$v_{ab} = V_{oc} = \left(\frac{3}{3+2}\right)10$$

$$V_{oc} = 6 = V_{Th}$$

Solve for the short-circuit current. In Figure 2.3.9b no current will flow in the 3-Ω resistor. Thus,

$$i = I_{sc} = \tfrac{10}{2} = 5 \text{ A} = I_N$$

The output resistance ($R_{Th} = R_N$) is found by replacing the 10-V source by a piece of wire, as in Figure 2.3.9c. Whereupon we see that

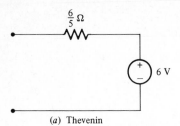

(a) Thevenin

figure 2.3.10
Terminal equivalents of the circuits in Figure 2.3.9a.

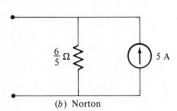

(b) Norton

$$R = R_{\text{Th}} = R_{\text{N}} = \frac{3(2)}{3 + 2} = \tfrac{6}{5}\ \Omega$$

Thus we have equivalent circuits as shown in Figure 2.3.10. As a check on our work we note that:

$$V_{\text{Th}} = I_{\text{N}} R_{\text{Th}}$$

$$6 = 5(\tfrac{6}{5}) \qquad \text{(checks)}$$

EXAMPLE 2.3.2

Solve for the current i in the 5-Ω resistor between nodes a and b in Figure 2.3.11a by means of Thevenin's theorem.

ANS.:   Remove the 5-Ω resistor as in Figure 2.3.11b and find the Thevenin equivalent circuit for the remainder of the network. Since i_1 is

$$i_1 = \frac{50 + 10 - 100}{4 + 2.2 + 1.5 + 2.3} = -\tfrac{40}{10} = -4\ \text{A}$$

V_{oc} is the sum of the 100-V source plus the voltage across the 4-Ω resistor:

$$V_{\text{oc}} = 100 + 4(-4)$$

$$= 84\ \text{V}$$

Solve for I_{sc} as in Figure 2.3.11c. By superposition

$$I_{\text{sc}} = I_{\text{sc}_1} + I_{\text{sc}_2}$$

where

$$I_{\text{sc}_1} = \frac{50 + 10}{2.2 + 1.5 + 2.3} = \tfrac{60}{6} = 10\ \text{A}$$

and

$$I_{\text{sc}_2} = \tfrac{100}{4} = 25\ \text{A}$$

so that

$$I_{\text{sc}} = 10 + 25 = 35\ \text{A}$$

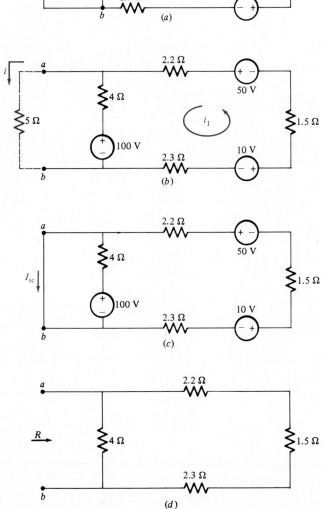

figure 2.3.11
The circuit of Example 2.3.2.

Solve for the output resistance by setting all independent sources in the network equal to zero and looking back into terminals a and b as in Figure 2.3.11d.

$$R = R_{\text{Th}} = R_{\text{N}} = \frac{4(2.2 + 1.5 + 2.3)}{4 + 2.2 + 1.5 + 2.3} = 2.4 \ \Omega$$

Note: Check that

$$V_{\text{oc}} = I_{\text{sc}} R_{\text{eq}}$$

$$84 = (35)(2.4) \qquad \text{(checks)}$$

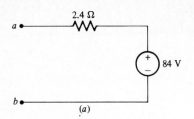

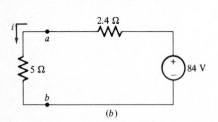

figure 2.3.12
(a) The Thevenin equivalent of the circuit in Figure 2.3.11b. (b) With the 5 = Ω load connected.

So we have a Thevenin equivalent as shown in Figure 2.3.12a. Connecting the 5-Ω resistor, we solve for i:

$$i = \frac{84}{2.4 + 5} = 11.35 \text{ A}$$

In some problems it is much easier to find one or two of the three variables V_{oc}, I_{sc}, R_0 than it is to find the other(s). Unfortunately there is no way to know ahead of time which variables are easy to find and which are not.

EXAMPLE 2.3.3

Find the Thevenin and the Norton equivalent circuits for Figure 2.3.13a.

ANS.: R_{Th}: we set the 5-A independent source equal to 0 (see Figure 2.3.13b), force 1 A in at the output terminals, and solve for v. Summing currents at the node between the R's:

$$-i + 2i + 1 = 0$$

$$i = -1 \text{ A}$$

Summing voltage rises around the outer loop, we see $v = R_{Th} = 0$ (see Figure 2.3.13c). Find V_{oc} (see Figure 2.3.13d). The sum of i's equals zero, so

$$5 - i + 2i + 0 = 0$$

$$i = -5$$

Therefore we have (see Figure 2.3.13e)

$$V_{oc} = -5R$$

Now, try to find I_{sc} (see Figure 2.3.13f). Summing currents into the upper node:

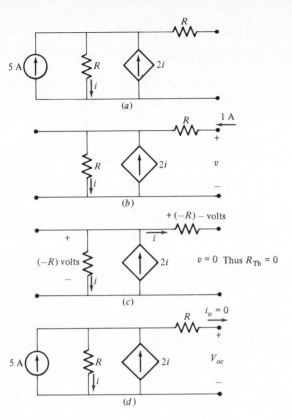

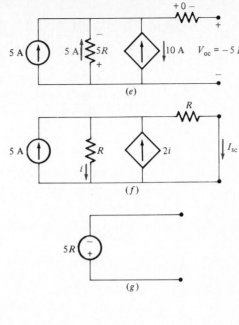

figure 2.3.13
The circuit of Example 2.3.3.

$$5 - i + 2i = I_{sc}$$

or

$$5 + i = I_{sc} \qquad (2.3.3)$$

Now sum the voltages around the loop in the network that contains both resistors:

$$iR - I_{sc}R = 0$$

$$i = I_{sc} \qquad (2.3.4)$$

There is only one possibility for the solution to (2.3.3) and (2.3.4):

$$i = I_{sc} = \infty$$

So the Thevenin equivalent circuit is shown in Figure 2.3.13g, and there really is no satisfactory Norton equivalent.

DRILL PROBLEM

In Figure 2.2.2 remove the 2-Ω resistor connected between node a and the reference node. Thevenize the rest of the circuit as seen from this new pair of terminals.
ANS.: $V_{oc} = 18$ V, $R_{Th} = 4$ Ω

2.4 applications of Thevenin and Norton equivalent circuits

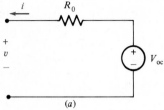

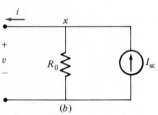

figure 2.4.1
(a) The Thevenin and (b) Norton equivalent circuits.

Consider the volt-ampere characteristic of the Thevenin and/or Norton equivalent circuits. Since these are terminal equivalents, their volt-ampere characteristics must, by definition, be identical.

Solving for the terminal voltage in Figure 2.4.1a via KVL:

$$v = V_{oc} - iR_0 \tag{2.4.1}$$

By algebra we then have

$$i = -\frac{1}{R_0}v + \frac{V_{oc}}{R_0} \tag{2.4.2}$$

and thus

$$i = -\frac{1}{R_0}v + I_{sc} \tag{2.4.3}$$

Equation (2.4.3) may be written directly by writing a KCL summation at node x in Figure 2.4.1b. This relationship between the variables i and v plots as a straight line with intercept $i = I_{sc}$ and slope $-1/R_0$. Graphically, this relationship is shown in Figure 2.4.2.

The straight line plotted in Figure 2.4.2 constitutes an uncountable and infinite set of points. Each point defines a pair of values (i, v) such that if a value of v is given, only one value of i is possible. Conversely, if i is specified, only one value of v can exist.

Suppose a resistor R_L is connected to terminals a-b of a Thevenin equivalent circuit (see Figure 2.4.3). The volt-ampere characteristic of the circuit to the right of a-b may be plotted as before. Then the volt-ampere characteristic of the circuit to the left of a-b may be plotted on the same axes provided we use the same definition of i and v for both plots: $i = v/R_L$. See Figure 2.4.4.

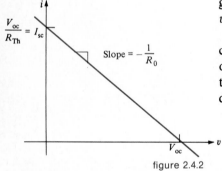

figure 2.4.2
Terminal volt–ampere characteristic of the Thevenin and Norton equivalent circuits.

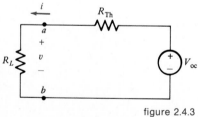

figure 2.4.3
A resistor connected to the output of a Thevenin equivalent circuit.

figure 2.4.4
Volt–ampere characteristic of the circuit (a) to the left and (b) to the right of terminals a–b in Figure 2.4.3.

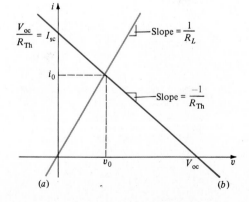

The intersection of the two straight-line plots defines the only point (i_0, v_0) which satisfies the requirements of both the circuit to the right of terminals a-b and the circuit to the left of a-b simultaneously.

EXAMPLE 2.4.1

Solve graphically for the values of i and v that will result in Figure 2.4.5a.

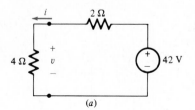

(a)

figure 2.4.5
The circuit and volt–ampere characteristic of Example 2.4.1.

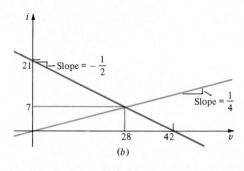

(b)

ANS.: Plot i versus v for the Thevenin circuit. The v intercept is $V_{oc} = 42$ V. The i intercept is $I_{sc} = V_{Th}/R_{Th} = \frac{42}{2} = 21$ A. Since v is also the voltage across the 4-Ω resistor and i the current through it, the Ohm's law relationship $i = v/4$ may be plotted on those same axes. See Figure 2.4.5b. The intersection $i = 7$, $v = 28$ satisfies both halves of the circuit.

EXAMPLE 2.4.2

Solve graphically for the voltage v and the current i in Figure 2.4.6a.

figure 2.4.6
The circuit and volt–ampere characteristics of Example 2.4.2.

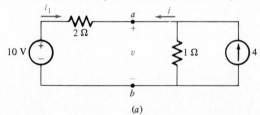

(a)

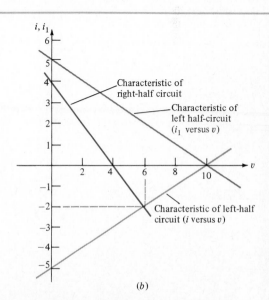

(b)

ANS.: For the circuit to the right of terminals *a-b* plot *i* versus *v*: $V_{oc} = (4\ A)(1\ \Omega) = 4$ V and $I_{sc} = 4$ A. See Figure 2.4.6*b*.

Plot on the same axes i_1 versus *v* for the circuit to the left of terminals *a-b*. $V_{oc} = 10$ V and $I_{sc} = 5$ A. (Note that *v* is defined the same for both halves of the circuit but that $i = -i_1$.) Therefore, we must invert the second volt-ampere plot as shown in Figure 2.4.6*b* so that both characteristics are in terms of *i* and *v*. The intersection $i = -2$ A and $v = 6$ V gives the solution.

This technique of plotting both the terminal volt-ampere characteristic of a Thevenin and/or Norton equivalent circuit and the circuit to which it is connected is much used in electronics.[†] Electronic devices are often nonlinear (their volt-ampere characteristics are not straight lines). See Figure 2.4.7.

EXAMPLE 2.4.3

An electronic device has a volt-ampere characteristic (*i* versus *v*) as shown in Figure 2.4.7*b*. It is connected to a circuit whose Thevenin terminal equivalent is a 5-V source in series with a 100-Ω resistor. Find the current *i* that will flow in the device.

ANS.: Plot the *i* versus *v* characteristic of the Thevenin circuit ($V_{oc} = 5$ V, $I_{sc} = \frac{5}{100} = 50$ mA). Graphically the value of *i* that will flow is $i_1 = 30$ mA.

[†] In electronics the emphasis is usually on the device itself and the Thevenin (or Norton) circuit is described as *loading* the device. Therefore the volt-ampere characteristic of the equivalent circuit is called the *load line*.

figure 2.4.7
(*a*) An electronic device with nonlinear volt–ampere characteristic connected to a Thevenin circuit. (*b*) The volt–ampere plots of the Thevenin circuit and the electronic device.

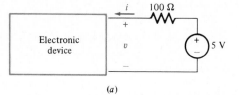

(*a*)

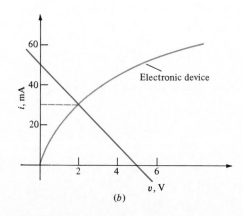

(*b*)

A diode is a nonlinear two-terminal element that is used very often in electric circuits. As a rough approximation, we usually visualize the diode as being an element that permits current flow in one direction and prohibits it in the other. It, therefore, acts like a switch—closed if current wants to go one way (the forward direction), open if current wants to go the other way. The symbol for a diode is shown in Figure 2.4.8.

In reality, the typical *pn*-junction diode (the most popular type) is not quite as perfect as this. It does allow a very small but nonzero current to flow in the *reverse direction* when it ought to be acting like a completely open circuit. When current is flowing in the *forward direction*, the diode ought to be acting like a perfect short circuit, but there is always a small nonzero voltage drop across the element in that case.

The actual terminal volt-ampere characteristic of a silicon *pn*-junction diode at room temperature is given, to a very high degree of accuracy, by the expression

$$i = I_0(e^{19.3v} - 1) \qquad (2.4.4)$$

where I_0 is called the *reverse saturation current*. The value of I_0 is different from diode to diode but is usually in the order of magnitude of 1 μA.

Equation (2.4.4) is used when a highly accurate numerical value must be calculated. In many applications the ideal (open or closed switch) model of a diode yields sufficient accuracy. For some other calculations where the ideal approximation is not accurate enough, but equation (2.4.4) is too tedious to use, we can approximate the silicon *pn*-junction diode as follows:

DIODES

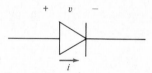

figure 2.4.8
A diode. If i is positive, v/i will be small and the device is said to be *forward-biased*. If v is negative, i will be almost zero and the device is said to be *back-biased*.

Polarity of diode voltage	Diode equivalent circuit
Forward bias	0.7-V source (independent of current)
Reverse (back) bias	Open circuit (or current source I_0)

A *Zener diode* is a *pn*-junction diode that has a volt-ampere characteristic similar to a regular diode except that it *breaks down* if the terminal voltage reaches a certain level in the reverse direction. This breakdown voltage is called the *Zener* voltage of the diode. Zener diodes are available in a wide range of different breakdown voltages.

If the breakdown voltage is exceeded, the diode acts like a short circuit. The Zener voltage for any given diode is a constant and, thus, is very useful in the design of regulated power supplies and other circuits wherein a reference voltage is needed.

DRILL PROBLEM

Find the numerical values of i and v in the circuit shown in Figure 2.4.9 to the nearest 0.1 mA and 0.1 V.

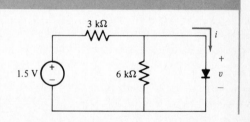

figure 2.4.9
A circuit containing a diode for the drill problem.

ANS.: (*Hint*: Plot, on the same set of axes, the diode's volt-ampere characteristic and the terminal characteristic of the Thevenin equivalent of the rest of the circuit.) $v = 0.3$ V and $i = 0.3$ mA.

$$Z_{IN} = Z_L^* \qquad R_{in} + jX_{in} = R_L - jX_L$$

2.5 maximum power transfer

Very often we have a network from which we want to receive as much power as possible. A typical example is the case of the hi-fi amplifier wherein we want to select the proper loudspeaker for this amplifier such that we get out as much audio power as we can.

In essence, the problem that we have is as follows (see Figure 2.5.1): What value of R_L do we select in order to maximize the power dissipated by R_L?

1. Get the Thevenin equivalent circuit of the network.
2. Find an expression for $p_{R_L}(t)$, the power dissipated in R_L as a function of R_L.
3. Set $\partial p_{R,L}/\partial R_L = 0$ and solve for R_L.

Thus, in Figure 2.5.1*b*,

$$V_o(t) = V_{\text{Th}}(t) \frac{R_L}{R_{\text{Th}} + R_L} \tag{2.5.1}$$

where V_o is the output voltage, and

$$p_{R_L}(t) = \frac{V_{\text{Th}}^2(t)}{R_L} \frac{R_L^2}{(R_{\text{Th}} + R_L)^2} \tag{2.5.2}$$

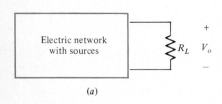

(a)

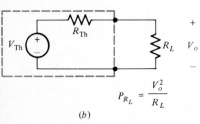

$$P_{R_L} = \frac{V_o^2}{R_L}$$

(b)

figure 2.5.1
Power transferred out of an electric network to a load resistor R_L.

so maximizing p_{R_L} by setting its derivative with respect to the quantity to be varied, R_L, equal to zero,

$$\frac{\partial p_{R_L}}{\partial R_L} = V_{\text{Th}}^2 \frac{(R_{\text{Th}} + R_L)^2 1 - R_L 2(R_{\text{Th}} + R_L)}{(R_{\text{Th}} + R_L)^4} = 0$$

$$R_{\text{Th}}^2 + 2R_{\text{Th}} R_L + R_L^2 - 2R_L R_{\text{Th}} - 2R_L^2 = 0$$

$$R_{\text{Th}}^2 - R_L^2 = 0$$

$$R_L = R_{\text{Th}} \tag{2.5.3}$$

A plot of p_{R_L} versus R_L has the form shown in Figure 2.5.2.

The conclusion we draw from equation (2.5.3) is the following: If we wish to obtain as much power as we possibly can from R_L, the best ohmic value for R_L is to set it equal to whatever the value of R_{Th} is. Under these conditions the power delivered to R_L will be

figure 2.5.2
Power delivered to R_L as a function of the value of R_L.

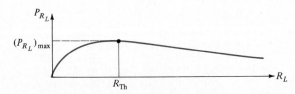

$$(p_{R_L})_{max} = \frac{V_o^2}{R_L}$$

$$= \frac{(V_{Th}/2)^2}{R_{Th}} = \frac{V_{Th}^2}{4R_{Th}}$$

The maximum value, $(p_{R_L})_{max}$, is called the *available power* since, if R_L is correctly chosen ($R_L = R_{Th}$), that amount of output power can be obtained. Under this maximum output power condition we say that the electric network and the load resistor are *matched*. Any other choice of value for R_L other than the matched condition will result in less than the maximum available power being delivered to R_L.

EXAMPLE 2.5.1

What value of R_L will maximize the power dissipated in R_L in Figure 2.5.3?

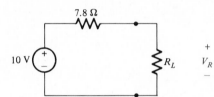

figure 2.5.3
The circuit of Example 2.5.1.

ANS.: $R_L = 7.8 \ \Omega$, and the power dissipated is

$$p_{R_L} = \frac{V_R^2}{R_L} = \frac{10^2\left(\dfrac{7.8}{7.8 + 7.8}\right)^2}{7.8} = 3.21 \ \text{W}$$

EXAMPLE 2.5.2

What value of R_{Th} will maximize the power dissipation in R_L in Figure 2.5.4?

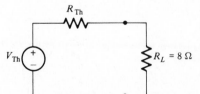

figure 2.5.4
The circuit of Example 2.5.2.

ANS.: $R_{Th} = 0$ (we do not seek R_L here but R_{Th}!!).

EXAMPLE 2.5.3

When it is being used as an amplifier in a *common-emitter* connection, a transistor has the equivalent circuit shown in Figure 2.5.5a.

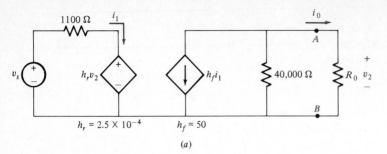

figure 2.5.5
(a) The transistor circuit of Example 2.5.3.(b)
The desired equivalent circuit. (c) Solving for
the open-circuit voltage.

(a)

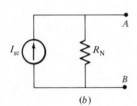

(b)

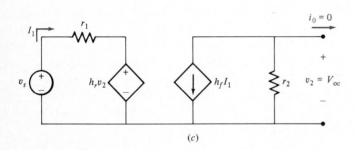

(c)

a. Make a Norton equivalent circuit of everything to the left of terminals A-B.

b. If v_s has the form $0.5 \sin t$, and $R_0 = 500\ \Omega$, what is $v_2(t)$?

ANS.: (a) First find the open-circuit voltage V_{oc}. Then find the short-circuit current I_{sc}. Get the R_N resistance via $R_N = V_{oc}/I_{sc}$.

V_{oc}:

Sum currents into the upper right-hand node in Figure 2.5.5c.

$$-h_f I_1 - \frac{v_2}{r_2} = 0 \qquad (2.5.4)$$

Sum voltages around the left-hand loop

$$v_s - I_1 r_1 - h_r v_2 = 0 \qquad (2.5.5)$$

Solve for I_1 in equation (2.5.4):

$$I_1 = \frac{-v_2}{r_2 h_f} \qquad (2.5.6)$$

Substitute (2.5.6) into (2.5.5):

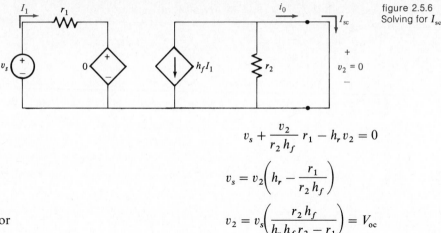

figure 2.5.6
Solving for I_{sc}.

$$v_s + \frac{v_2}{r_2 h_f} r_1 - h_r v_2 = 0$$

$$v_s = v_2\left(h_r - \frac{r_1}{r_2 h_f}\right)$$

or

$$v_2 = v_s\left(\frac{r_2 h_f}{h_r h_f r_2 - r_1}\right) = V_{oc}$$

Now find I_{sc} in Figure 2.5.6. Obviously $v_s = I_1 r_1$ and $-h_f I_1 - i_0 = 0$. Combining these two expressions (eliminating I_1):

$$i_0 = \frac{-h_f v_s}{r_1} = I_{sc} = I_N$$

R_N:

$$R_N = \frac{V_{oc}}{I_{sc}}$$

$$= \frac{v_s r_2 h_f}{h_r h_f r_2 - r_1}\left(-\frac{r_1}{h_f v_s}\right)$$

$$R_N = \frac{r_1 r_2}{r_1 - h_r h_f r_2}$$

As a check, find R_N directly as in Figure 2.5.7. Set independent sources (only) to zero. Thus,

$$I_1 r_1 + h_r v_2 = 0$$

and

$$h_f I_1 + \frac{v_2}{r_2} = 1$$

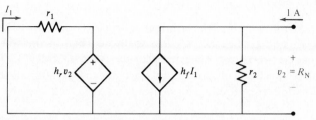

figure 2.5.7
Finding R_N directly.

Solve for I_1 in the first expression and insert into the second:

$$-h_f h_r \frac{v_2}{r_1} + \frac{v_2}{r_2} = 1$$

or

$$v_2 = \frac{r_1 r_2}{r_1 - h_f h_r r_2} = R_N = R_{Th}$$

Thus we see (using $I_{sc} R_N = V_{oc}$) that our answers for V_{oc}, I_{sc}, and R_N are mutually consistent.
(b) Insert numerical values (see Figure 2.5.8):

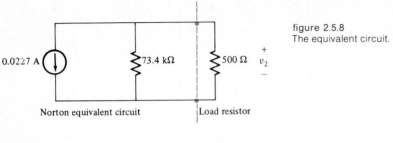

figure 2.5.8
The equivalent circuit.

Norton equivalent circuit   Load resistor

$$I_N = \frac{-50(0.5)}{1100} = -0.0227 \text{ A}$$

$$R_N = \frac{1100(4 \times 10^4)}{1100 - 2.5 \times 10^{-4} \times 50 \times 4 \times 10^4} = 73.4 \text{ k}\Omega$$

$$v_2 = -0.0227\left(\frac{73,400}{73,400 + 500}\right)500 = -11.3 \text{ V}$$

so that
$$v_2(t) = -11.3 \sin t$$

Note that the transistor, as used above, yields a voltage gain of $-11.3/0.5 = -22.6$. The minus sign indicates a reversal in voltage polarity.

DRILL PROBLEM

How much power is delivered to the 5-Ω load resistor in Figure 2.3.12b? If this load resistance could be varied, what would be the maximum obtainable output power?
ANS.: 644 W, 735 W

2.6 wye-delta transformation

We have discussed earlier how to find the terminal equivalent resistance of series-connected and parallel-connected resistors. Now we would like to find the terminal equivalent circuit for resistors connected as a *three*-terminal device. There are only two possible connections of resistors that are not trivial (after all series-parallel combinations have been found): these are called the delta and the wye. See Figure 2.6.1.

Let us assume that the delta resistances R_1, R_2, and R_3 are known. What values of R_a, R_b, and R_c will produce the same values of terminal equivalent resistance in the wye? To answer this question, simply set the three terminal equivalent resistances of both networks equal to each other.

$$R_{ab} = R_a + R_b = \frac{R_3(R_1 + R_2)}{R_1 + R_2 + R_3} \tag{2.6.1}$$

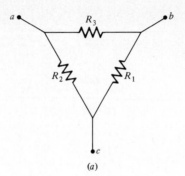

figure 2.6.1
Three-terminal resistive networks: (a) the
delta and (b) the wye connection.

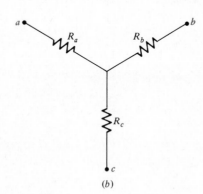

$$R_{bc} = R_b + R_c = \frac{R_1(R_2 + R_3)}{R_1 + R_2 + R_3} \qquad (2.6.2)$$

$$R_{ac} = R_a + R_c = \frac{R_2(R_1 + R_3)}{R_1 + R_2 + R_3} \qquad (2.6.3)$$

Multiplying the second equation by -1 and then adding all three yields:

$$R_a = \frac{R_2 R_3}{R_1 + R_2 + R_3} \qquad (2.6.4)$$

Similarly,

$$R_b = \frac{R_1 R_3}{R_1 + R_2 + R_3} \qquad (2.6.5)$$

and

$$R_c = \frac{R_1 R_2}{R_1 + R_2 + R_3} \qquad (2.6.6)$$

These are the wye resistances in terms of the delta resistances. These equations may be solved for the delta resistances in terms of the wye resistances:

$$R_1 = \frac{R_a R_b + R_b R_c + R_c R_a}{R_a} \qquad (2.6.7)$$

$$R_2 = \frac{R_a R_b + R_b R_c + R_c R_a}{R_b} \qquad (2.6.8)$$

$$R_3 = \frac{R_a R_b + R_b R_c + R_c R_a}{R_c} \qquad (2.6.9)$$

EXAMPLE 2.6.1

Find the equivalent resistance of the network shown in Figure 2.6.2a. All resistances are 1 Ω.

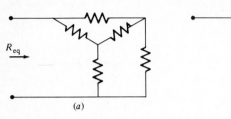
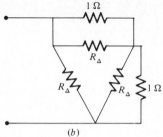
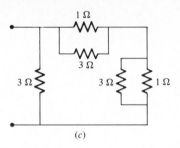

figure 2.6.2
(a) The circuit of Example 2.6.1. (b) Using the wye-delta transformation. (c) The simplified circuit.

ANS.: Perform a wye-delta transformation. See Figure 2.6.2b. From equations (2.6.7), (2.6.8), or (2.6.9):

$$R_\Delta = \frac{1 + 1 + 1}{1} = 3 \ \Omega$$

Thus we have the equivalent circuit shown in Figure 2.6.2c. 3 Ω in parallel with 1 Ω is

$$\frac{3(1)}{3 + 1} = \tfrac{3}{4} \ \Omega$$

So

$$R_{eq} = \frac{3(\tfrac{6}{4})}{3 + \tfrac{6}{4}} = 1 \ \Omega$$

DRILL PROBLEM

Suppose we have a delta circuit (Figure 2.6.1a) wherein all resistors are 30 Ω. We wish to build an equivalent wye. However, we are allowed to purchase only two new resistors. How can we utilize the original resistors and two new ones to design the wye?

ANS.: $R_Y = 10 \ \Omega$. Put the three 30-Ω resistors in parallel to make one 10-Ω resistance. Then buy two new 10-Ω resistors.

2.7 summary

In this chapter we have discussed many of the analytical tools that we will be using throughout our study of circuits and systems. Ideas such as linearity and superposition and the Thevenin and Norton theorems will be used again and again throughout this text. You will use them in your career in many ways other than to analyze simple resistive circuits with constant sources as we have done here.

At this point you should feel confident that given a circuit made up of resistors, current, and voltage sources (both independent and dependent), you can find the voltage at each node and the current in each element. You have seen several methods of attacking such circuits, e.g., mesh equations, loop equations, node equations, Thevenin and Norton theorems, superposition, and network simplification. Now we are ready to discuss the various kinds of time-varying sources that we usually encounter in circuits and systems.

problems

1. A voltage source V_s, a resistor R, and a current source I_s are all connected in parallel with each other. What fraction of the current from I_s goes through R?

2. In the circuit of Figure 2.3.11 find the component of the current that flows in the 5-Ω resistor due to each of the voltage sources.

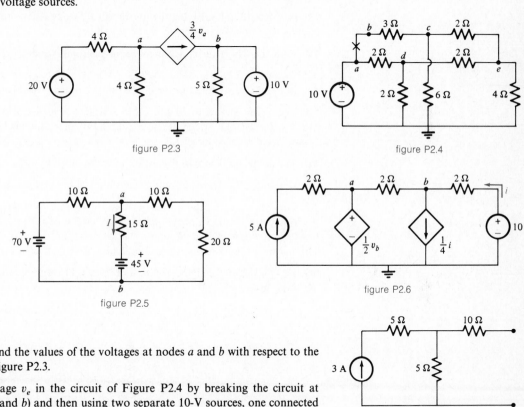

figure P2.3

figure P2.4

figure P2.5

figure P2.6

3. Use superposition to find the values of the voltages at nodes a and b with respect to the datum (ground) node in Figure P2.3.

4. Find the value of voltage v_e in the circuit of Figure P2.4 by breaking the circuit at point x (between points a and b) and then using two separate 10-V sources, one connected to a and one connected to b. Superposition can then be used to determine the value of v_e.

5. In the circuit of Figure P2.5 (a) find the component of the voltage v_{ab} that is due to the 70-V source. (b) Repeat part (a) for the 45-V source. (c) Find v_{ab}.

6. Use superposition to find the values of the voltages at nodes a and b with respect to ground in Figure P2.6.

7. Find the Thevenin and Norton equivalent circuits for the network shown in Figure P2.7.

8. Find I_{sc} for the circuit in Figure P2.8.

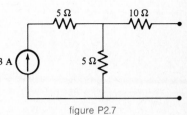

figure P2.7

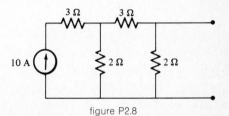

figure P2.8

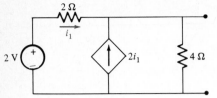

figure P2.9

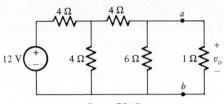

figure P2.10

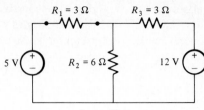

figure P2.11

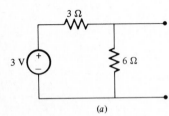

(a)

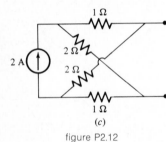

(b)

(c)

figure P2.12

9. Find R_{Th} for the circuit in Figure P2.9.

10. In the circuit of Figure P2.10 solve for the voltage v_o by first finding the Thevenin equivalent circuit to the left of terminals a and b. Then place the 1-Ω resistor across the equivalent circuit's terminals and solve for v_o.

11. In Figure P2.11 solve for the current in R_1 using Thevenin's theorem. Use any other method (e.g., node equations) to check your work.

12. Find the Thevenin and Norton equivalents for each of the three circuits in Figure P2.12.

13. In Figure P2.13, find the current in the 5-Ω resistor via Thevenin's theorem.

14. The circuit shown in Figure P2.14 is sometimes used to measure resistances and is called a Wheatstone bridge. Remove the 2-Ω resistor and Thevenize the remainder of the circuit. What is the value of I_g?

15. The nonlinear device in Figure P2.15 is described by $i_{NL} = 0.02v^2$. Find i and v.

16. Determine the Thevenin output resistance of the bridge shown in Figure P2.16.

17. For the circuit in Figure P2.17 (a) find the current I via KCL. (b) Find I via KVL. Use the loop currents shown in the figure. (c) What happens to the value of I if the *upper* 2-Ω resistor becomes 7 Ω? (d) Remove the 2-Ω resistor through which the current I flows. Thevenize the rest of the network. Recalculate I using this equivalent circuit.

18. For the circuit in Figure P2.18, find (a) the Thevenin equivalent for the circuit to the left of terminals a-b and (b) the Norton equivalent. (c) What is the maximum power that

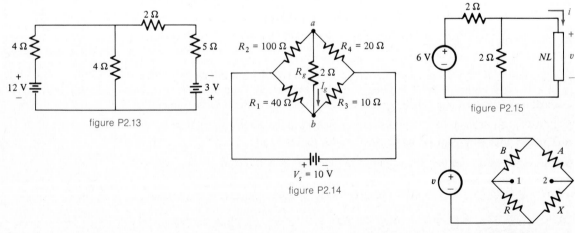

figure P2.13

figure P2.14

figure P2.15

figure P2.16

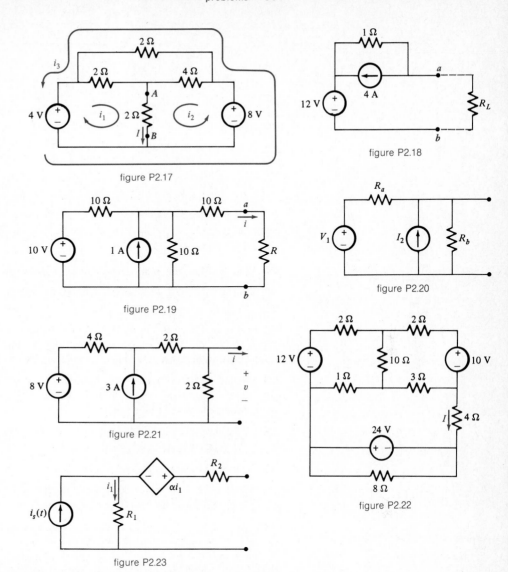

figure P2.17

figure P2.18

figure P2.19

figure P2.20

figure P2.21

figure P2.22

figure P2.23

can be supplied to a load resistor R_L by this circuit? (*d*) How much power is dissipated internally (to the left of terminals *a-b*) while maximum power is being drawn by R_L?

19. For the circuit in Figure P2.19 find the value of the current *i* if (*a*) $R = 15\ \Omega$, (*b*) $R = 10\ \Omega$, (*c*) $R = 5\ \Omega$, (*d*) $R = 0\ \Omega$.

20. Find the Thevenin and Norton equivalents of the circuit in Figure P2.20.

21. For the circuit in Figure P2.21 plot the volt-ampere characteristic (*i* versus *v*).

22. In Figure P2.22, use Thevenin's theorem to find the value of *I*.

23 Thevenize the circuit in Figure P2.23. If R_1 and R_2 are replaced with short circuits, what will the value of I_{sc} be?

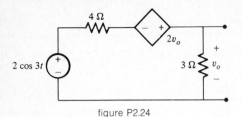

figure P2.24

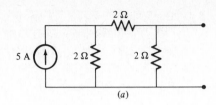

(a)

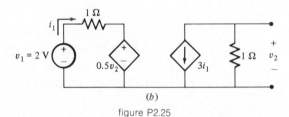

(b)

figure P2.25

24. Find the Thevenin equivalent of the circuit in Figure P2.24.

25. Find the Thevenin and Norton equivalents of the circuits shown in Figure P2.25.

26. In the circuit shown in Figure P2.25b (a) what is the value of the output voltage v_2? (b) what is the value of the output resistance of the rest of the circuit *as seen by the 1-Ω resistor*? (c) If the value of the 1-Ω resistor is changed to any value R, then what is the output voltage in terms of the input voltage v_1 and R?

27. For the circuit in Figure P2.27 (a) find the Norton and Thevenin equivalents. (b) Plot the volt-ampere characteristic. (c) What value of V_s will produce a 1-A current in a 1-Ω load resistor connected to the output terminals?

28. In the circuit in Figure P2.28 find the Thevenin equivalent circuit for everything to the left of terminals a-b. What can we say about maximum power transfer to R_L in this case?

29. Find the Thevenin and Norton equivalent of the circuit in Figure P2.29.

30. The reverse saturation current for a certain silicon *pn*-junction diode is 5×10^{-9} A. Find the forward voltage drop when the forward current is 10 mA.

31. Two silicon *pn*-junction diodes are connected in series across a constant voltage source so that they are both forward-biased. The reverse saturation current I_o of one of the diodes is 10 times that of the other. Assuming the current that flows is much greater than I_o, by how much does the voltage across one diode differ from that across the other?

32. The same two diodes as in problem 31 are connected in parallel (both forward-biased) across the same constant voltage source as before. Again, assume the current that flows in

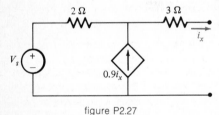

figure P2.27

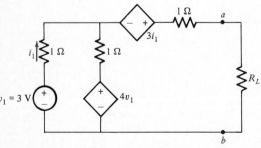

figure P2.28

figure P2.29

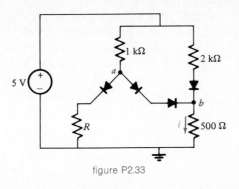

figure P2.33

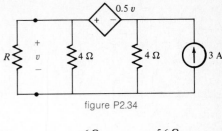

figure P2.34

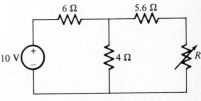

figure P2.35

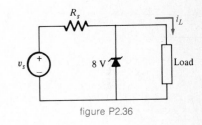

figure P2.36

each is much greater than the reverse saturation current. What is the ratio of the currents in the diodes?

33. One model of a *pn*-junction diode assumes the forward voltage drop in a conducting diode is always approximately 0.7 V and if this forward drop is less than 0.7 V, then no current will flow. (*a*) Sketch the volt-ampere characteristic of this model. (*b*) Using this model for the diodes in the circuit of Figure P2.33, find the value of the voltage at point *b* if $R = 1$ kΩ. (*c*) Repeat (*b*) for $R = 0$ Ω.

34. In the circuit in Figure P2.34 for what value of R will that resistor get maximum power delivered to it? What is the value of that power?

35. In the circuit in Figure P2.35 what value of R will draw the most power from the rest of the circuit? How much power is dissipated in the remaining resistors in the circuit when using that value of R?

36. The circuit diagram in Figure P2.36 contains an ideal† Zener diode whose breakdown voltage is 8 V. The supply voltage v_s can vary from 9 to 12 V and the load current can vary independently from 0 to 100 mA. The circuit designer placed the Zener there in an attempt to hold the load voltage at exactly 8 V for all combinations of supply voltage and load current. Find the optimum value of R_s that will produce the smallest current through the diode and still maintain regulation of the output voltage. What will be the maximum power dissipation in the diode using this value of R_s?

37. In the Wheatstone bridge circuit of Figure P2.14, let $R_1 = R_2 = R_3 = 1$ kΩ, $R_4 = (1 + x)$ kΩ, and $V_s = 12$ V. When $x = 0$, then $I_g = 0$ and the bridge is said to be *balanced*. (*a*) Find the Thevenin or Norton equivalent circuit seen by the resistor R_g (which is the input resistance of an ammeter placed there to measure I_g). (*b*) Find an expression for I_g in terms of R_g and x. (*c*) Show that, for small values of x, I_g is proportional to x and thus to any incremental resistance of the lower right-hand element.‡

† Zero resistance in its breakdown range.

‡ This circuit is convenient for measuring either absolute or differential deflections via substitution of strain gauges for R_3 and R_4.

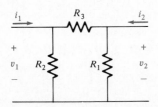
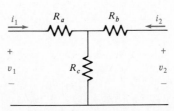

figure P2.39

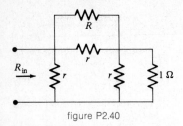

figure P2.40

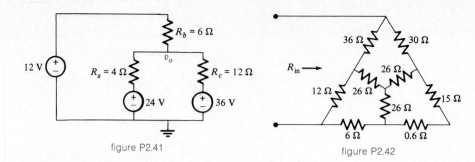

figure P2.41

figure P2.42

38. Suppose, in Figure P2.14, that $R_1 = R_3 = R_g = 1\ \Omega$, $R_2 = 6\ \Omega$, $R_4 = 3\ \Omega$ and $V_s = 10$ V. Use a delta-wye transformation on the 1-Ω resistors to solve for the value of I_g. Check your work by any other method.

39. Given a T network and a π network as shown in Figure P2.39, wherein $R_1 = 3\ \Omega$, $R_2 = 6\ \Omega$, and $R_a = 4\ \Omega$. Find the values of the other resistors that will make the networks terminally equivalent.

40. In the circuit shown in Figure P2.40 what is the value of R (in terms of r) such that $R_{in} = 1\ \Omega$? What is the minimum value of r in this case?

41. In the circuit in Figure P2.41 (*a*) convert the wye to a delta, solve for the currents in each source and then find v_o. (*b*) In which equivalent (wye or delta) is the resistor located that dissipates the most power? (*c*) What is the total dissipated power? (*d*) Check your answer to (*a*) via a node equation.

42. Find the value of the input resistance R_{in} of the circuit in Figure P2.42.

43. If $i_a = 2$ A and $i_b = 3$ A in Figure P2.43, what are the values of voltages v_{ab} and v_{cb}?

44. In Figure P2.5 assume that the 15-Ω resistor is variable from zero to 1 kΩ. What value(s) of this resistor results in current I being equal to or greater than 1 A?

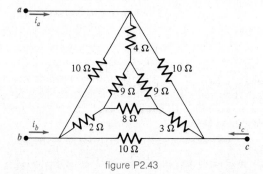

figure P2.43

chapter 3
SIGNALS

introduction 3.1

The voltages and currents in the circuits we have studied in this text so far have not varied with time. These so-called dc circuits are easy to analyze. But, if we do allow the magnitudes of the source(s) in circuits to change with time, this opens up vast new possible uses for these circuits. Transmission of information requires the use of quantities that vary in time. The study of control mechanisms is mainly concerned with the *dynamic* behavior of systems—that is, how the important physical quantities in those systems *vary with time*. The simple act of turning on a light switch *changes*, from one time to another, the value of the voltage applied to the lamp. There are many examples.

Although not every time-varying voltage or current is used for signaling the transmission of information from one point to another, it has become generally acceptable to use the word *signal* to describe any voltage or current, or any other quantity for that matter, that varies

figure 3.2.1
The unit-step function.

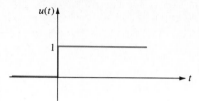

with time. Indeed, we even hear the words "dc signal" used occasionally to describe a constant voltage or current.

To describe a signal, then, we must specify how the quantity in question changes in time. We can do this by drawing a plot of its magnitude versus time t. Or we can specify its closed-form mathematical expression. Sometimes we do not know enough information about the signal to do either of these things. But then we might know that it never exceeds a certain value, for example, or maybe we might know some other statistical information about it.

Clearly there are an infinite number of possible shapes the plot of a voltage or current versus time may take on. However, there are a few waveshapes that do occur quite often. We will now take a look at some of these frequently used signal waveforms, so that when we encounter them later in the text we will be fully familiar with them. Also, we will often have to find the time integral and/or time derivative of these and other types of signals and so we shall do some examples of those kinds of operations on those signals as well.

3.2   the unit-step function

The unit-step function is a mathematical function that is equal to $+1$ (unity) for positive values of its argument. It is equal to zero for negative values of its argument. We will use the symbol $u(t)$ to describe the unit step that switches from 0 to $+1$ at $t = 0$. The switching process occurs instantaneously at $t = 0$ because the argument of $u(t)$ is time t, which changes from being a negative quantity to a positive quantity at $t = 0$. See Figure 3.2.1.

Any voltage or current that is switched on or off at some instant of time t_0 is easily described mathematically by using the unit-step function.

EXAMPLE 3.2.1

Write a mathematical expression for a voltage that switches from zero to 1 V at $t = 0$.
ANS.: $v(t) = u(t)$.

EXAMPLE 3.2.2

Write a mathematical expression for a current that changes instantaneously from 0 to 10 A at $t = 0$.
ANS.: $i(t) = 10u(t)$.

EXAMPLE 3.2.3

Find an expression for quantity q that goes from 0 to 5 at $t = 3$ s.
ANS.: $q(t) = 5u(t - 3)$. Remember the definition of $u(.)$: When the argument is positive, the $u(.)$ function is equal to $+1$. When the argument is negative, then $u(.)$ is equal to zero. See Figure 3.2.2a.

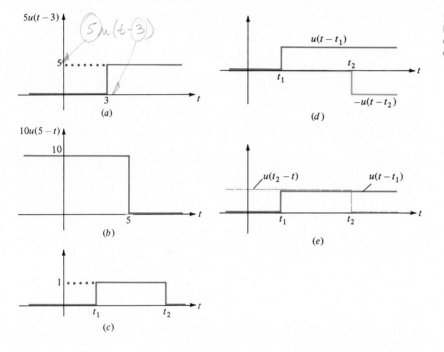

figure 3.2.2
How unit-step functions can be used to describe signals containing discontinuities.

So all we have to do when dealing with the unit-step function is to be very careful to determine for what values of t the argument is greater than zero. For only those values of t the unit-step function equals $+1$.

We can also use the step function to describe variables that turn *off* at some instant of time.

EXAMPLE 3.2.4

Write an expression for a current that switches from $+10$ to 0 at $t = 5$ s.
ANS.: $i(t) = 10u(5 - t)$. (See Figure 3.2.2b.)

The sum of two step functions and the product of two step functions are both very useful for describing quantities that switch on and off or that switch between two different nonzero levels. For example, a *gate* function (see Figure 3.2.2c), one that turns on at one time and off at another, can be written as the sum:

$$g(t) = u(t - t_1) - u(t - t_2) \qquad t_2 > t_1$$

(see Figure 3.2.2*d*), or as the product:

$$g(t) = u(t - t_1)u(t_2 - t)$$

(see Figure 3.2.2*e*).

EXAMPLE 3.2.5

Find $v_0(t)$ in Figure 3.2.3*a*. The switch closes at $t = 0$.

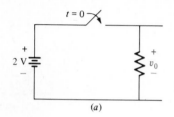

(a)

figure 3.2.3
(*a*) The circuit of Example 3.2.5; (*b*) $v_0(t)$.

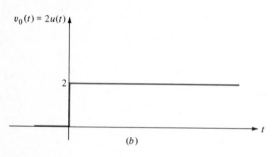

(b)

ANS.: $v_0 = 2u(t)$.

EXAMPLE 3.2.6

Given the function $f(t) = 1 - t$, sketch to scale in the interval $-2 \le t \le 2$:

a. $f(t)$
b. $f(t)u(t)$
c. $f(t)u(-t)$
d. $f(t)u(t - 1)$
e. $f(t) + u(t - 1)$
f. $f(t - 1)u(t)$
g. $f(t - 1)u(t + 1)$
h. $f(t)u(t + 1)u(1 - t)$

ANS.: See Figure 3.2.4.

Unit-step functions often appear in integrals. Their effect there is simply to modify the limits of integration.

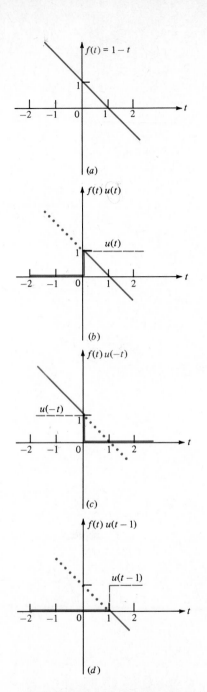

(a) $f(t) = 1 - t$

(b) $f(t) \, u(t)$, $u(t)$

(c) $f(t) \, u(-t)$, $u(-t)$

(d) $f(t) \, u(t-1)$, $u(t-1)$

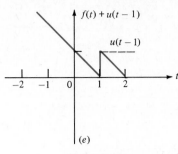

(e) $f(t) + u(t-1)$, $u(t-1)$

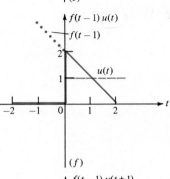

(f) $f(t-1) \, u(t)$, $f(t-1)$, $u(t)$

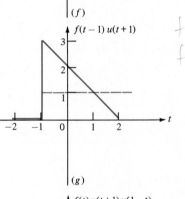

(g) $f(t-1) \, u(t+1)$

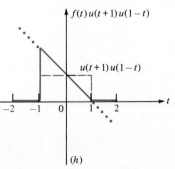

(h) $f(t) \, u(t+1) \, u(1-t)$, $u(t+1) \, u(1-t)$

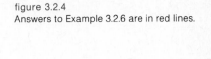

figure 3.2.4
Answers to Example 3.2.6 are in red lines.

$f(t) = 1 - t$
$f(t-1) = 1 - (t-1) = (2-t)$

EXAMPLE 3.2.7

Find $f(t) = \int_{-\infty}^{t} u(t-3) \, dt$ for all t and plot $f(t)$ versus t.
ANS.:   Before we do this problem let us remind ourselves that the t in the upper limit of integration is the *only* t to survive the integration process. (We will see many integrals of this sort and the reader should be very aware about what variables end up in the answer afterward.) To be explicit, let us write the integral as

$$f(t) = \int_{\tau=-\infty}^{\tau=t} u(\tau - 3) \, d\tau \qquad \text{for all } t$$

The t in the upper limit is the one that makes $f(.)$ into $f(t)$. The τ is the dummy variable of integration that gets substituted for (when we insert the limits). Now let's do the example:

First, suppose $t < 3$. Then, since $u(t - 3) = 0$ for $t < 3$,

$$f(t) = \int_{-\infty}^{t} u(\tau - 3) \, d\tau$$

$$= \int_{-\infty}^{t} 0 \, d\tau$$

$$= 0 \qquad \text{for } t < 3$$

Second, suppose $t > 3$. Then $u(t - 3)$ contains two segments so we do the integral in two parts:

$$f(t) = \int_{-\infty}^{3} u(\tau - 3) \, d\tau + \int_{3}^{t} u(\tau - 3) \, d\tau$$

$$= \int_{-\infty}^{3} 0 \, d\tau + \int_{3}^{t} 1 \, d\tau$$

$$= 0 + \tau \Big|_{3}^{t} = t - 3$$

So we say

$$f(t) = \begin{cases} 0 & \text{for } t \le 3 \\ t - 3 & \text{for } t \ge 3 \end{cases}$$

or

$$f(t) = (t - 3)u(t - 3)$$

See Figure 3.2.5.

It is very convenient when doing time integrals like the last example, where the lower limit is $-\infty$ and the upper limit is some arbitrary time t, to first sketch the integrand versus t and then to think about summing the area under this integrand curve. Start an infinite distance to the left and come along toward the right summing areas as you go. At any position (time t) the area already summed (to the left) is the value of the integral at time t.

DRILL PROBLEM

Evaluate the time integral of $tu(t - 3)u(5 - t)$ over the limits -5 to t_0, where the upper limit t_0 lies in the interval $4 < t_0 < 5$.
ANS.: $0.5(t_0 + 3)(t_0 - 3)$.

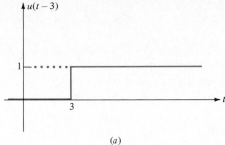

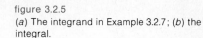

figure 3.2.5
(a) The integrand in Example 3.2.7; (b) the integral.

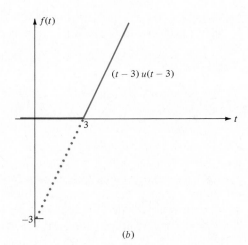

(b)

the unit-impulse function **3.3**

Consider a unit-area square-shaped pulse $f_p(t)$ whose time duration is Δ seconds and whose magnitude is $1/\Delta$. (See Figure 3.3.1.) If we allow Δ to approach zero, then the pulse gets taller and taller as it gets thinner and thinner. In the limit we have an infinitely tall, zero-width pulse, whose area is still unity. We call this the *unit impulse*. Thus, the unit impulse is defined as being

$$\delta(t) = \lim_{\Delta \to 0} f_p(t) \qquad (3.3.1)$$

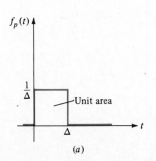

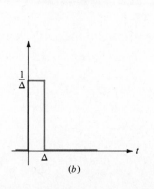

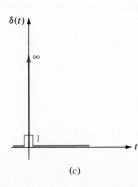

(a) (b) (c)

figure 3.3.1
(a) The unit pulse $f_p(t)$. (b) Allowing Δ to become smaller. (c) The limit as Δ approaches zero (which yields the unit impulse).

where
$$f_p(t) = \begin{cases} 1/\Delta & 0 < t < \Delta \\ 0 & \text{elsewhere} \end{cases} \tag{3.3.2}$$

The symbol δ is the small greek letter delta in honor of the physicist, Dirac. $\delta(t)$ is sometimes called the Dirac function. Note that the value of $\delta(t)$ is everywhere equal to zero *except* where its argument equals zero. Thus $\delta(t + 2)$ is a unit impulse located at $t = -2$ s.

Suppose we started out, in Figure 3.3.1a, with a pulse that was k/Δ units tall, but still Δ seconds wide. Then we would have a pulse with area $(k/\Delta)(\Delta) = k$. We could write this as

$$f(t) = kf_p(t)$$

Then taking the limit as Δ approaches zero:

$$\lim_{\Delta \to 0} f(t) = \lim_{\Delta \to 0} kf_p(t)$$

$$= k \lim_{\Delta \to 0} f_p(t)$$

$$= k\delta(t) \tag{3.3.3}$$

So we see that multiplying an impulse by a constant indicates that the *area* of the impulse is equal to that constant—not the height. (The height is already infinite—we cannot make that any bigger.) Note that when we draw an impulse we indicate that its height is infinite by writing the ∞ symbol next to an arrow pointing upward. A small box at its base reminds us that this pulse does have a finite area; the numerical value of that area is written at the base of the impulse.

Now, let us ask ourselves what the time derivative of the unit-step function is. Since a function's derivative is equivalent to the slope of the function, we see that the slope of $u(t)$ is everywhere equal to zero except it is infinite at $t = 0$. This sounds like a description of the unit impulse. But is the area equal to unity? To find out, let us approximate the unit-step function $u(t)$ by another function $\tilde{u}(t)$, such that

$$\lim_{\Delta \to 0} \tilde{u}(t) = u(t) \tag{3.3.4}$$

See Figure 3.3.2.

The derivative of $\tilde{u}(t)$ is the unit pulse $f_p(t)$:

$$\frac{d\tilde{u}}{dt} = f_p(t) \tag{3.3.5}$$

If we allow Δ to approach zero in Figure 3.3.2b, we see that, in the limit,

$$\lim_{\Delta \to 0} \tilde{u}(t) = u(t) \tag{3.3.6}$$

and

$$\lim_{\Delta \to 0} \frac{d\tilde{u}}{dt} = \frac{du}{dt} \tag{3.3.7}$$

Substituting (3.3.5) into (3.3.7),

$$\lim_{\Delta \to 0} f_p(t) = \frac{du}{dt}$$

figure 3.3.2
(*a*) The unit-step function. (*b*) An approximate unit step. (*c*) The time derivative of the waveform in (*b*). (*d*) The limiting shape of the waveform in (*c*) as Δ approaches zero.

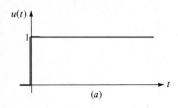

(*a*)

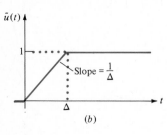

Slope $= \dfrac{1}{\Delta}$

(*b*)

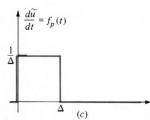

(*c*)

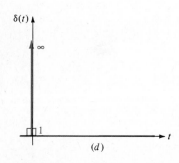

(*d*)

or
$$\delta(t) = \frac{du}{dt} \tag{3.3.8}$$

The impulse is obviously a rather strange function. We may write
$$v(t) = 10\delta(t) \tag{3.3.9}$$

on paper, but can we ever, in real life, get a voltage source to go from zero volts to *infinity volts* and back again to zero and do it all in *no time at all*? Of course not. But nor, for example, can we really ever build an ideal voltage source (one that will maintain its specified voltage absolutely regardless of the magnitude of the current it must deliver to whatever value of resistance we choose to put across it). These ideal cases are just that—ideals. But we can approximate them *sufficiently well* in many instances, for all practical purposes. They also turn out to be easier to work with mathematically than more accurate models of actual, buildable, voltage sources or other electrical elements. We *can* build voltage sources that go from 0 volts to 1 million volts and back again to 0 in about a millionth of a second—and that ought to be a good enough approximation of $\delta(t)$ for just about any purpose.

Let us investigate what happens if we include an impulse function in the integrand of an integral.

EXAMPLE 3.3.1

Find the value of
$$I = \int_{-\infty}^{\infty} 4\delta(t)\ dt \tag{3.3.10}$$

ANS.:   Since $\delta(t)$ is almost everywhere equal to zero, we may collapse the limits of integration to $0-$ (just prior to $t = 0$) and $0+$ (just after $t = 0$).

$$I = \int_{0-}^{0+} 4\delta(t)\ dt$$

$$= 4\int_{0-}^{0+} \delta(t)\ dt \tag{3.3.11}$$

The integral is simply the area of the unit impulse $\delta(t)$, and so
$$I = 4(1) = 4 \tag{3.3.12}$$

EXAMPLE 3.3.2

Find
$$I = \int_{-\infty}^{\infty} f(t)\ \delta(t - t_0)\ dt \tag{3.3.13}$$

ANS.:   This is the integral of the product of an arbitrarily shaped time function $f(t)$ multiplied by an impulse that occurs at $t = t_0$. See Figure 3.3.3.

The value of this product is zero everywhere except at $t = t_0$ [because $\delta(t)$ is zero everywhere except at

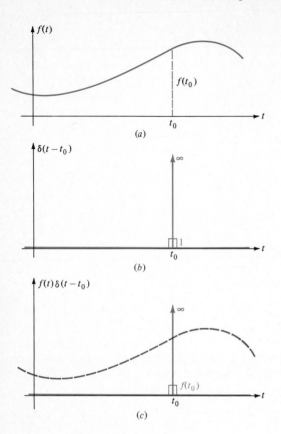

figure 3.3.3
(a) An arbitrary function $f(t)$. (b) The unit impulse $\delta(t - t_0)$. (c) The product of (a) and (b), which is an impulse of area $f(t_0)$.

$t = t_0$]. Thus we can write

$$I = \int_{t_0 -}^{t_0 +} f(t)\delta(t - t_0)\, dt \qquad (3.3.14)$$

Assuming $f(t)$ does not change over the infinitesimally small interval $(t_0 -) < t < (t_0 +)$, we can treat it like a constant and take it outside the integral.

$$I = f(t_0) \int_{t_0 -}^{t_0 +} \delta(t - t_0)\, dt \qquad (3.3.15)$$

Whence, since the area of the unit impulse is unity.

$$I = f(t_0) \qquad (3.3.16)$$

So we see that *the integral of the product of any function $f(t)$ with a unit impulse (over limits containing the impulse) is equal to the function evaluated at the time of occurrence of the impulse.* This, of course, presumes that $f(t)$ is single-valued at that time. Equating (3.3.13) and (3.3.16):

$$\int_{-\infty}^{\infty} f(t)\delta(t - t_0)\, dt = f(t_0) \qquad (3.3.17)$$

This is called by many authors the *sifting* property of the unit-impulse function.

EXAMPLE 3.3.3

Evaluate

$$f(t) = \int_{-\infty}^{t} 2tu(t-1)\delta(t-4)\,dt$$

ANS.: The integral is a function of t because of the upper limit being t. The presence of the impulse forces the integrand to be zero-valued at all t except $t = 4$. Thus, for $t < 4$,

$$f(t) = \int_{-\infty}^{t} 0\,dt$$

$$= 0 \qquad t < 4$$

And for $t > 4$,

$$f(t) = \int_{-\infty}^{t} 2tu(t-1)\delta(t-4)\,dt$$

$$= \int_{4-}^{4+} 2(4)(1)\delta(t-4)\,dt$$

$$= 8\int_{4-}^{4+} \delta(t-4)\,dt$$

$$= 8(1) = 8 \qquad t > 4$$

so

$$f(t) = 8u(t-4) \qquad \text{all } t$$

Clearly, if the unit impulse is the derivative of the unit-step function, then the integral of the unit impulse must be the unit step. Let us check: If $t > 0$,

$$\int_{-\infty}^{t} \delta(t)\,dt = \int_{0-}^{0+} \delta(t)\,dt = 1 \qquad t > 0$$

But if $t < 0$,

$$\int_{-\infty}^{t} \delta(t)\,dt = \int_{-\infty}^{t} 0\,dt = 0 \qquad t < 0$$

These can be combined in a single expression:

$$\int_{-\infty}^{t} \delta(t)\,dt = u(t) \qquad \text{for all } t \qquad\qquad \text{QED}$$

This is also true no matter when the impulse occurs, i.e.,

$$\int_{-\infty}^{t} \delta(t-t_0)\,dt = u(t-t_0) \qquad \text{for all } t$$

**THE UNIT-RAMP
FUNCTION** $r(t)$

Let us now consider the integral of the unit-step function:

$$\int_{-\infty}^{t} u(t - t_0) \, dt = \begin{cases} 0 & \text{for } t < t_0 \\ t - t_0 & \text{for } t > t_0 \end{cases}$$

$$= (t - t_0)u(t - t_0) \qquad \text{all } t$$

It is thus convenient to define the *unit-ramp* function

$$r(t) = tu(t)$$

or

$$r(t - t_0) = (t - t_0)u(t - t_0)$$

so that we can write

$$\int_{-\infty}^{t} u(t - t_0) \, dt = r(t - t_0)$$

So the unit-ramp function has slope equal to unity to the right of t_0, is zero-valued to the left of t_0, and is equal to zero at $t = t_0$. Thus,

$$\frac{d}{dt} r(t - t_0) = u(t - t_0)$$

There really is no reason why we could not continue on, defining a *unit-parabola* function and a *unit third-order* function, etc., and indeed their definitions are almost self-evident.

THE DOUBLET

The derivative of the unit impulse may be obtained via a limiting process analogous to that used for obtaining the derivative of the unit step (see Figure 3.3.2). Approximate the unit impulse by a trapezoidal pulse; note that the derivative of the approximating trapezoid consists of two rectangular pulses (one positive and one negative). In the limit the derivative is seen to be two simultaneous impulse-like waveforms—each with infinite area! Such a pulse is called a *doublet* and is of no practical use since no actual source can remotely approximate it in practice.

There exists then a set of functions, based on the impulse, that has an infinite number of members. Each member is the time integral of the previous function and the derivative of the succeeding one in the set. This set contains the doublet, the unit impulse, unit step, unit ramp, unit parabola, and so forth. Any integral of a sum of such functions may be written directly as a term-by-term replacement of each member function with its succeeding (integral) member function.

DRILL PROBLEM

Evaluate $v(4)$, where $v(t) = [10u(t - 1) - 5(t - 3) + \int_{-\infty}^{t} 5(t - 3)u(t - 1) \, \delta(t - 2) \, dt]u(t + 1)$.
ANS.: 0.

EXAMPLE 3.3.4

Find a closed-form expression for $f(t)$ below that is valid for all t:

$$f(t) = \int_{-\infty}^{t} \delta(t) - u(t) + 2u(t-1) - u(t-2) - \delta(t-2) \, dt$$

ANS.: $f(t) = u(t) - r(t) + 2r(t-1) - r(t-2) - u(t-2)$.

It is of crucial importance that the reader be able to sketch the integrand function and the resulting $f(t)$. Note that one is the integral and one the derivative of the other. Note that the *areas* of the impulses in the derivative (integrand) function are equal to (and of similar mathematical sign to) the *discontinuities* in the integral function. See Figure 3.3.4.

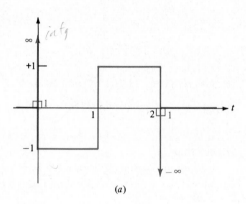

(a)

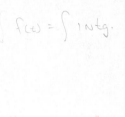

$f(t)$

(b)

figure 3.3.4
(a) The integrand function of Example 3.3.4.
(b) The integral $f(t)$.

the exponential 3.4

The base of natural logarithms is the irrational number called $e = 2.71 \cdots$. A signal that occurs again and again in electrical engineering, physics, and other sciences is that obtained by raising e to a power proportional to time, i.e.,

$$f(t) = Ae^{-at} \qquad\qquad (3.4.1)$$

Raising a real number to the zero power results in a value of unity. Thus the

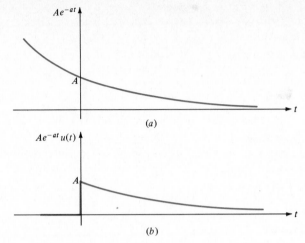

(a)

(b)

value of equation (3.4.1), at time equal to zero, is A.

$$f(0) = A \tag{3.4.2}$$

Note that for time less than zero, the exponential function given by (3.4.1) is greater than A.

If we multiply the right-hand side of equation (3.4.1) by the unit-step function $u(t)$, it is unchanged for t greater than zero but set equal to zero for t less than zero. (See Figure 3.4.1b.)

$$f(t) = Ae^{-at}u(t) \tag{3.4.3}$$

This is most often the form of the exponential function when we see it. The value of time t that makes the value of the exponent equal to -1 is called the *time constant* τ. So the value of t for which

$$-at = -1 \tag{3.4.4}$$

is the time constant $t = \tau$. So

$$\tau = \frac{1}{a} \tag{3.4.5}$$

The exponential function [equation (3.4.3)] evaluated at $t = \tau$ is therefore one eth of its value at $t = 0$.

$$f(\tau) = Ae^{-a\tau}u(t)$$

$$= Ae^{-a/a}(1)$$

$$f(\tau) = A\left(\frac{1}{e}\right) \tag{3.4.6}$$

We find the time derivatives and integrals of exponential functions as follows.

EXAMPLE 3.4.1

Find df/dt where $f(t) = Ae^{-at}$.

ANS.:
$$\frac{df}{dt} = -aAe^{-at} \tag{3.4.7}$$

See Figure 3.4.2.

EXAMPLE 3.4.2

Find df/dt where $f(t) = Ae^{-at}u(t)$.

ANS.: Use the derivative of the product rule

$$\frac{df}{dt} = A[e^{-at}\delta(t) - ae^{-at}u(t)]$$

$$= A[\delta(t) - ae^{-at}u(t)] \tag{3.4.8}$$

See Figure 3.4.3.

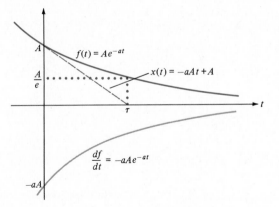

figure 3.4.2
$f(t) = Ae^{-at}$ and its time derivative.

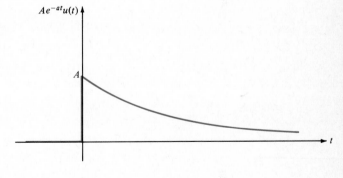

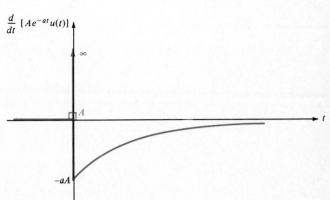

figure 3.4.3
$f(t) = Ae^{-at}u(t)$ and its time derivative. Note that the *area* of the impulse is equal to the *height* of the discontinuity in $f(t)$.

Note from Example 3.4.1 that a straight line having both the same slope and magnitude that

$$f(t) = Ae^{-at}u(t)$$

does at $t = 0+$ (just after $t = 0$) will reach zero at time equal to one time constant, $t = \tau$, i.e.,

$$\left.\frac{df}{dt}\right|_{t=0+} = -aAe^{-at}\Big|_{t=0+} = -aA$$

so

$$x(t) = -aAt + A$$

evaluated at $t = \tau = 1/a$ is

$$x(\tau) = \frac{-aA}{a} + A$$

$$= 0$$

(See Figure 3.4.2.)

EXAMPLE 3.4.3

Find the function $g(t)$ given by

$$g(t) = \int_{-\infty}^{t} f(t) \, dt$$

where

$$f(t) = Ae^{-at}u(t)$$

ANS.:

$$g(t) = \int_{-\infty}^{t} Ae^{-at}u(t) \, dt \tag{3.4.9}$$

If $t > 0$, then

$$g(t) = A \int_{0}^{t} e^{-at} \, dt$$

$$= A \frac{e^{-at}}{-a}\Big|_{t=0}^{t} = \frac{-A}{a}(e^{-at} - e^{(0)})$$

$$= \frac{A}{a}(1 - e^{-at}) \qquad \text{for } t > 0 \tag{3.4.10}$$

but if $t < 0$, then the step function in (3.4.9) has zero value and

$$g(t) = \int_{-\infty}^{t} 0 \, dt = 0 \qquad \text{for } t < 0 \tag{3.4.11}$$

Combining (3.4.10) and (3.4.11),

$$g(t) = \frac{A}{a}(1 - e^{-at})u(t) \qquad \text{for all } t \tag{3.4.12}$$

See Figure 3.4.4.

DRILL PROBLEM

Find the value of the time integral from $t = -\infty$ of $4e^{-t/2}[u(t) + \delta(t-3)]$ evaluated at $t = 2$ and at $t = 4$ s.
ANS.:   5.06, 7.81.

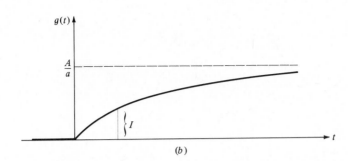

figure 3.4.4
The waveforms of example 3.4.3. (a)
$f(t) = Ae^{-at}u(t)$. (b) The integral of $f(t)$. The
area I in (a) equals the *height* I in (b).

sinusoid 3.5

One of the most often encountered waveforms is the sinusoid. For example, household electric power is sinusoidal and so, basically, are all radio signals. It is so common in fact that we shall devote an entire chapter later in this text to this function of time and the special methods that electrical engineers have developed for solving circuits that have sinusoidal sources.

The two trigonometric functions sin x and cos x are both equally useful in describing sinusoidal waveforms. Consider, for example, the sinusoidal function

$$f(t) = A \cos (\omega t + \theta) \tag{3.5.1}$$

where A, ω, and θ are constants (see Figures 3.5.1 and 3.5.2). A is called the *amplitude* and θ is the *phase angle*. The argument of any trigonometric function must be in radians. Thus $\omega t + \theta$ has the dimension radians, and so, therefore, must ωt and θ each by themselves be measured in radians.† Since time t is measured in seconds, it follows that ω has the dimensions radians per second, and is therefore called the *radian frequency* of the waveform. The length of time it takes for the argument of the function to go from 0 to 2π is called the *period T* of the function, i.e., if at time t_1,

$$\omega t_1 + \theta = 0 \tag{3.5.2}$$

---

† For convenience electrical engineers often write the phase angle in degrees rather than radians even though, strictly speaking, it is technically not correct, e.g., 4 cos $(377 + \pi/6)$ is often written 4 cos $(377t + 30°)$.

figure 3.5.1
The sinusoid $f(t) = A \cos (\omega t + \theta) = A \sin$
$(\omega t + \phi)$ plotted versus ωt. Note that
$\phi = \theta + \pi/2$.

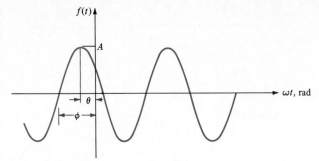

figure 3.5.1
The sinusoid $f(t) = A \cos (\omega t + \theta) = A \sin$
$(\omega t + \phi)$ plotted versus ωt. Note that
$\phi = \theta + \pi/2$.

figure 3.5.2
The sinusoid $f(t) = A \cos (\omega t + \theta) = A \sin$
$(\omega t + \phi)$ plotted versus t.

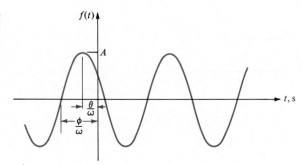

$$t_1 = \frac{-\theta}{\omega} \tag{3.5.3}$$

and at t_2,

$$\omega t_2 + \theta = 2\pi \tag{3.5.4}$$

$$t_2 = \frac{2\pi - \theta}{\omega} \tag{3.5.5}$$

then we define the period as being

$$T = t_2 - t_1 \tag{3.5.6}$$

$$= \frac{2\pi - \theta}{\omega} + \frac{\theta}{\omega} \tag{3.5.7}$$

$$= \frac{2\pi}{\omega} \tag{3.5.8}$$

The function repeats itself every 2π radians and therefore every T seconds. So we ought to be able to add an integer number of periods to t and have the value of the function be the same, i.e.,

$$f(t) = f(t + nT) \tag{3.5.9}$$

We can show this to be true by substituting $t + nT$ for t in equation (3.5.1) and thereby get

$$f(t + nT) = A \cos [\omega(t + nT) + \theta] \tag{3.5.10}$$

Using (3.5.8) we have

$$f(t + nT) = A \cos \left(\omega t + \omega n \frac{2\pi}{\omega} + \theta \right) \qquad (3.5.11)$$

After cancellation of the ω in (3.5.11) we note that the argument of (3.5.11) is equal to the argument of (3.5.1) except for the addition of an integer number of 2π which, of course, does not change the value of the function. Thus equation (3.5.9) is seen to be correct.

The interval T determines one cycle of the waveform and thus actually has the unit seconds per cycle. The inverse of T therefore has the unit cycles per second, or *Hertz*, and is called the *frequency f*. That is to say,

$$f = \frac{1}{T} \qquad (3.5.12)$$

and using (3.5.8),

$$f = \frac{\omega}{2\pi} \qquad (3.5.13)$$

or

$$\omega = 2\pi f \qquad (3.5.14)$$

EXAMPLE 3.5.1

What is the radian frequency of the sinusoidal voltage that constitutes normal household electricity in the United States?

ANS.:   In the United States, $\qquad\qquad f = 60$ Hz

Therefore, $\qquad\qquad\qquad\qquad\qquad \omega = 2\pi f$

$$= 377 \text{ rad/s}$$

We say that the sine function *lags* the cosine function by $\pi/2$ radians (or 90°) because $\cos \omega t$ is already up to its peak value of $+1$ at $\omega t = 0$ ($t = 0$ s), whereas it will take $\sin \omega t$ until $\omega t = \pi/2$ ($t = \pi/2\omega$ s) to get up to that value.

Similarly, we say that $\cos \omega t$ *leads* $\sin \omega t$ by $\pi/2$ radians (90°). The interrelationship is given by

$$\cos \omega t = \sin \left(\omega t + \frac{\pi}{2} \right)$$

EXAMPLE 3.5.2

Write the following time functions in terms of the other trigonometric function:

a. $10 \cos (377t + \pi/3)$ **b.** $156 \sin (10^6 t + \pi/3)$

ANS.:

(a) $10 \sin (377t + \pi/3 + \pi/2) = 10 \sin (377t + 5\pi/6)$

(b) $156 \cos (10^6 t + \pi/3 - \pi/2) = 156 \cos (10^6 t - \pi/6)$.

The sum of two sinusoidal waveforms that both have the same frequency is another sinusoidal waveform at that same frequency.

EXAMPLE 3.5.3

Two voltage sources v_1 and v_2 are connected in series. Find the total voltage v_0 across the two sources if $v_1(t) = 9 \cos (30t + \pi/4)$ and $v_2(t) = 12 \sin (30t - 3\pi/4)$.

ANS.: Use the trigonometric identities

$$\sin (a \pm b) = \sin a \cos b \pm \cos a \sin b \tag{3.5.15}$$

$$\cos (a \pm b) = \cos a \cos b \mp \sin a \sin b \tag{3.5.16}$$

to obtain

$$v_1(t) = 9[\cos (30t) \cos (\pi/4) - \sin (30t) \sin (\pi/4)]$$

$$v_2(t) = 12[\sin (30t) \cos (3\pi/4) - \cos (30t) \sin (3\pi/4)] \tag{3.5.17}$$

or

$$v_1(t) = 9[\cos 30t(0.707) - \sin 30t(0.707)]$$

$$= 6.36(\cos 30t - \sin 30t) \tag{3.5.18}$$

$$v_2(t) = 12[\sin 30t(-0.707) - \cos 30t(0.707)]$$

$$= -8.49(\sin 30t + \cos 30t) \tag{3.5.19}$$

Adding, we get

$$v_1(t) + v_2(t) = (6.36 - 8.49) \cos 30t + (-6.36 - 8.49) \sin 30t$$

$$= -2.13 \cos 30t - 14.85 \sin 30t \tag{3.5.20}$$

which is of the form

$$A(\sin \theta \cos 30t + \cos \theta \sin 30t) = A \sin (30t + \theta) \tag{3.5.21}$$

thus,

$$A \sin \theta = -2.13 \quad \text{and} \quad A \cos \theta = -14.85 \tag{3.5.22}$$

Dividing one equation by the other, we solve for $\tan \theta$ (and thus θ).

$$\frac{A \sin \theta}{A \cos \theta} = \tan \theta = \frac{-2.13}{-14.85}$$

$$\theta = 0.14 \tag{3.5.23}$$

Then using (3.5.23) in either expression in (3.5.22), we can solve for A:

$$A \sin \theta = -2.13$$

$$A = \frac{-2.13}{\sin (0.14)} = -15 \tag{3.5.24}$$

Using (3.5.23) and (3.5.24) in (3.5.21), we have

$$v_1(t) + v_2(t) = -15 \sin (30t + 0.14) \tag{3.5.25}$$

If the method of Example 3.5.3 seems rather complicated, it is! Electrical engineers have developed much quicker and easier ways to handle such problems.

But even at this point we can make certain observations about sinusoidal waveforms.

- Every sinusoidal time function can differ only in (at most) three ways from every other sinusoidal time function: amplitude, frequency, and phase angle. The general shapes are all the same.
- The pure sinusoidal waveform exists for a doubly infinite range of time, i.e., $-\infty < t < +\infty$. The waveform has been oscillating between $\pm A$ forever and will continue to oscillate forever.
- Multiplying a pure sinusoid by the unit-step function $u(t)$ creates a waveform that is zero-valued for negative values of t. Do not confuse $A \cos tu(t)$ with $A \cos t$. They are *very* different waveforms.

EXAMPLE 3.5.4

Find

$$f(t) = \frac{d}{dt} g(t)$$

where

$$g(t) = A \cos (\omega t + \theta)$$

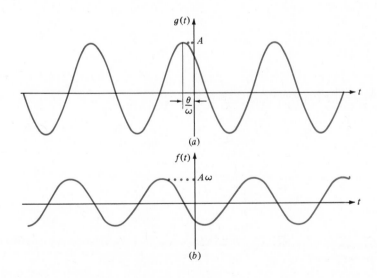

figure 3.5.3
(a) The waveform $g(t) = A \cos (\omega t + \theta)$.
(b) Its time derivative $f(t) = dg/dt$.

ANS.: $f(t) = -A\omega \sin (\omega t + \theta)$.

EXAMPLE 3.5.5

Find

$$f(t) = \frac{d}{dt} g(t)$$

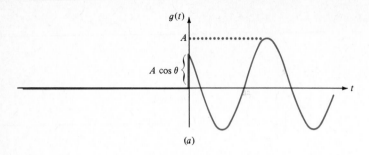

figure 3.5.4
(a) The waveform $g(t) = A \cos (\omega t + \theta)u(t)$.
(b) Its time derivative. Note that the area of the impulse is equal to the height of the discontinuity in the original waveform.

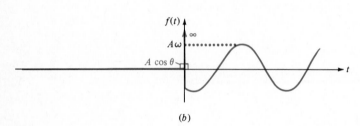

where

$$g(t) = A \cos (\omega t + \theta)u(t)$$

ANS.: Take the derivative of the product of two functions:

$$f(t) = A \cos (\omega t + \theta)\, \delta(t) - A\omega \sin (\omega t + \theta)u(t)$$

$$= A \cos \theta\, \delta(t) - A\omega \sin (\omega t + \theta)u(t)$$

See Figure 3.5.4.

EXAMPLE 3.5.6

Find

$$g(t) = \int_{-\infty}^{t} f(t)\, dt$$

where

$$f(t) = A \cos (\omega t + \theta)u(t)$$

ANS.:

$$g(t) = A \int_{0}^{t} \cos (\omega t + \theta)\, dt$$

$$= \frac{A}{\omega} \sin (\omega t + \theta)\Big|_{0}^{t} = \frac{A}{\omega} \left[\sin (\omega t + \theta) - \sin \theta \right]$$

See Figure 3.5.5.

EXAMPLE 3.5.7

Find

$$f(t) = \int A \sin (\omega t + \theta)\, dt$$

an indefinite integral.

ANS.:

$$f(t) = \frac{-A}{\omega} \cos (\omega t + \theta) + k$$

where k is the constant of integration.

The reader should notice very carefully in Figures 3.5.3 through 3.5.5 that each derivative waveform is equal to the slope, at every instant of time t of the corresponding integral function, and similarly, every integral waveform is equivalent to the total area accumulated (from the left) under its corresponding derivative waveform. Notice for example in Figure 3.5.5 the following observations:

- The slope of $g(t)$ is discontinuous at $t = 0$. It is zero for $t < 0$ and has a positive value immediately after $t = 0$ (at $t = 0+$). Thus $f(t)$ is discontinuous at $t = 0$.
- At every zero crossing of $f(t)$ the integral $g(t)$ is at an extremum (maximum or minimum) since its slope there is zero.
- The first zero crossing of the integral $g(t)$ occurs when equal amounts of positive and negative area under $f(t)$ have been accumulated.
- Since $f(t)$ contains no impulses, $g(t)$ has no discontinuities. Or one could say, since $g(t)$ is everywhere continuous, $f(t)$ contains no impulses.

DRILL PROBLEM

If $v_1(t) = 2 \sin (3t + 20°)u(t)$
a. Find an expression for $v_2(t) = dv_1/dt - 3v_1$.
b. Consider the time integral of v_2. About what dc value will this integral function oscillate?

ANS.: (a) $v_2 = 6\sqrt{2} \cos (3t + 65°)u(t) + 0.684\delta(t)$; (b) 0.684

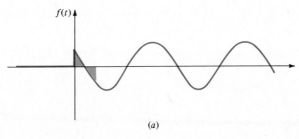

(a)

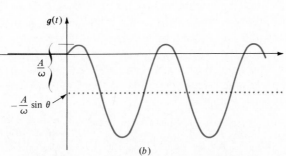

(b)

figure 3.5.5
(a) The waveform $f(t) = A \cos (\omega t + \theta)u(t)$.
(b) Its time integral $g(t) = \int_{-\infty}^{t} f(t)\, dt$.

3.6 periodic waveforms

Some signals repeat themselves regularly every T seconds. Each complete repetition is called one *cycle* of the waveform, and the time duration T of each cycle is called the *period* of the waveform. Such signals are, therefore, called periodic. Any sinusoid is, naturally, periodic, but there are a wide variety of nonsinusoidal periodic signals that we will be seeing as well.

For a signal to be truly periodic, it must exist for all time $(-\infty < t < +\infty)$.

EFFECTIVE VALUE OF PERIODIC WAVEFORMS Suppose a periodic current passes through a resistor R. The resistor will get hot—it will dissipate, each period of the waveform, a certain amount of energy. Let us ask the question: What value of *constant* current I_{dc} will dissipate energy at the same rate (will make the resistor equally as hot as the periodic current does)? The value of this constant current is called the *effective value* of the periodic current. It is the magnitude of dc current that has the *same power-producing capability* as does the periodic current.

We will now derive a method by which we can determine the effective value of any periodic waveform.

In one period T the energy delivered to R by any arbitrarily shaped (but periodic) current is $W_{per} = \int_{t_1}^{t_2} i_{per}^2 R \, dt$, where $t_2 - t_1 = T$.

The constant current I_{dc} delivers (during that same time T) an amount of energy:

$$W_{dc} = \int_{t_1}^{t_2} I_{dc}^2 R \, dt$$

Now, if $W_{per} = W_{dc}$, then

$$\int_{t_1}^{t_2} I_{dc}^2 R \, dt = \int_{t_1}^{t_2} i_{per}^2 R \, dt$$

Solving for I_{dc} by integrating the left-hand side and dividing through by R and T:

$$I_{dc} = \sqrt{\frac{1}{T} \int_{t_1}^{t_2} i_{per}^2 \, dt} = I_{rms}$$

Thus the constant current that is equivalent (insofar as its power-making capability is concerned) to the arbitrarily shaped periodic current is obtained by (1) squaring $i_{per}(t)$, (2) integrating over T seconds, (3) dividing by T, and (4) taking the square root.

This equivalent value is therefore called the *rms* or *root mean squared value* of the periodic current. A similar derivation can show that the effective value of any periodic voltage $v(t)$ is

$$V_{rms} = \sqrt{\frac{1}{T} \int_{t_1}^{t_2} v^2(t) \, dt}$$

EXAMPLE 3.6.1

Find the rms value of a square wave of ± 10-V magnitude. See Figure 3.6.1.

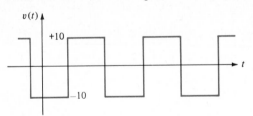

figure 3.6.1
The balanced square wave of Example 3.6.1.

ANS.:

$$V_{rms} = \sqrt{\frac{1}{T} \int_{t_1}^{t_1+T} 100 \ dt}$$

$$= \sqrt{\frac{100}{T} (t_1 + T - t_1)} = 10$$

This result does not surprise us very much when we realize that power delivered does not depend on the polarity of the voltage. This square wave is simply a constant 10 V which is periodically reversed in polarity. Of course its effective value is 10 V.

EXAMPLE 3.6.2

Show via this procedure (taking the root mean square) that the effective (rms) value of a sinusoid is the maximum value divided by $\sqrt{2}$.

ANS.:

$$x(t) = X_m \cos (\omega t + \theta)$$

$$x^2(t) = X_m^2 \cos^2 (\omega t + \theta) = \frac{X_m^2}{2} [1 + \cos (2\omega t + 2\theta)]$$

Integrate over one period of the x^2 waveform, i.e., from $2\omega t + 2\theta = 0 \rightarrow t_1 = -\theta/\omega$ to $2\omega t + 2\theta = 2\pi \rightarrow t_2 = (\pi - \theta)/\omega$ and divide by $t_2 - t_1 = \pi/\omega$:

$$\frac{1}{\pi/\omega} \frac{X_m^2}{2} \int_{-\theta/\omega}^{(\pi - \theta)/\omega} 1 + \cos (2\omega t + 2\theta) \ dt$$

$$\frac{\omega}{\pi} \left[\frac{X_m^2}{2} \left(\frac{\pi}{\omega} \right) + \frac{X_m^2}{2} \frac{\sin (2\omega t + 2\theta)}{2\omega} \Big|_{-\theta/\omega}^{(\pi - \theta)/\omega} \right]$$

$$\frac{\omega}{\pi} \left[\frac{X_m^2}{2} \left(\frac{\pi}{\omega} \right) + \frac{X_m^2}{4\omega} \underbrace{\sin (2\pi - 2\theta + 2\theta)}_{0} - \underbrace{\sin (-2\theta + 2\theta)}_{0} \right]$$

thus leaving $X_m^2/2 =$ mean squared value. Then, taking the square root, we are left with $X_m/\sqrt{2}$, as was to be shown.

EXAMPLE 3.6.3

Find the value of the power dissipated in a 1000-Ω resistor when a voltage $v(t) = 150 \cos (377t + \pi/3)$ is impressed on its terminals.

ANS.: If the voltage had been a constant V_{dc}, then the power would have been

$$P_{av} = \frac{V_{dc}^2}{R}$$

Since V_{rms} is the equivalent effective value of the sinusoidal voltage, this enables us to use this same expression:

$$P_{av} = \frac{V_{rms}^2}{R}$$

$$= \frac{(150/\sqrt{2})^2}{1000} = 11.25 \text{ W}$$

EXAMPLE 3.6.4

Find the rms value of the half-wave-rectified voltage sine wave shown in Figure 3.6.2.

ANS.:
$$v(t) = \begin{cases} V_m \sin \omega t & 0 < t < \pi/\omega \\ 0 & \pi/\omega < t < 2\pi/\omega \end{cases}$$

$$V_{rms} = \sqrt{\frac{1}{T} \int_0^T v^2(t) \, dt} \qquad \text{where } \omega = 2\pi f = \frac{2\pi}{T}$$

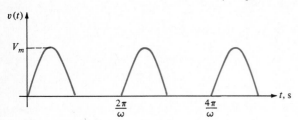

figure 3.6.2
The half-wave-rectified sinusoid of Example 3.6.4.

so
$$T = \frac{2\pi}{\omega}$$

$$= \sqrt{\frac{\omega}{2\pi} \int_0^{\pi/\omega} V_m^2 \sin^2 \omega t \, dt} = \sqrt{\frac{V_m^2 \omega}{4\pi} \int_0^{\pi/\omega} 1 - \cos 2\omega t \, dt}$$

$$= \sqrt{\frac{V_m^2 \omega}{4\pi} \left[\frac{\pi}{\omega} - \frac{\sin}{2\omega} \left(2\omega \, \frac{\pi}{\omega} \right) \right]} = \frac{V_m}{2}$$

Similarly, we can find the rms value of a full-wave-rectified sinusoid (see Figure 3.6.3):

$$v(t) = V_m \sin \omega t \qquad \text{for } 0 < t < \pi/\omega$$

and
$$v(t + n\pi/\omega) = v(t)$$

so
$$V_{rms} = \sqrt{\frac{\omega}{\pi} \int_0^{\pi/\omega} V_m^2 \sin^2 \omega t \, dt} = \frac{V_m}{\sqrt{2}}$$

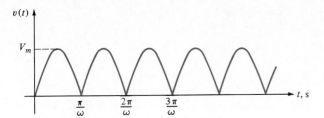

figure 3.6.3
A full-wave-rectified sinusoid.

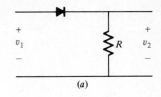

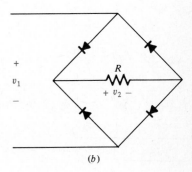

figure 3.6.4
Diode rectifier circuits: (*a*) half-wave;
(*b*) full-wave.

Half-wave and full-wave rectification of waveforms can be obtained from a variety of circuits. Two examples are shown in Figure 3.6.4.

FORM FACTOR

The ratio of the *effective* value of any periodic waveform to its *average* value is called its *form factor*. Some ammeters and voltmeters that supposedly yield the rms value of periodic signals are constructed such that they actually respond to the average value of the rectified† waveform. This value is then internally multiplied by the form factor of the rectified wave before being displayed. The form factor differs from one periodic waveform to another. Such an instrument, once calibrated for use on one waveshape, say sinusoids, will yield erroneous readings if used to measure a waveform that has a different form factor.

CREST FACTOR

Although a peaked periodic waveform will usually have a higher form factor than a flat-topped one, it is *not* necessarily true that, because periodic waveform $x(t)$ has a higher form factor than waveform $y(t)$, it will be more peaked than $y(t)$. A measure of the *peakedness* of a periodic signal is the *crest* (or *peak*) factor. The crest factor is the ratio of the *maximum* value of a periodic waveform to its *effective (rms)* value.

DRILL PROBLEM

A periodic function is defined as $v(t) = e^{-t}$ for $0 < t < 1$ and $v(t + n) = v(t)$ for all integer n. Find the rms value and the average value of $v(t)$.

ANS.: $V_{av} = 0.632$ V and $V_{rms} = 0.658$ V.

---

† Either full- or half-wave, depending on the instrument.

3.7 other types of signals

RANDOM SIGNALS

In order to contain information a signal must, in some way, change in a nondeterministic way. Voice, music, and computer output signals are neither pure sinusoids nor pure periodic waveforms. If they were, all we would need to do is to see one cycle of the signal and thereby know what the signal would look like for all future time. It would be nice to be able to know tomorrow's news (or stock prices!) today. But this is impossible. Any signal, then, that is capable of carrying meaningful information is in some way random. When this or that change in the signal occurs, it is "news," i.e., *information*. When an AM radio station is broadcasting only a pure sinusoid (called its *carrier*), it is on the air and we can locate it on the dial, but no announcer is speaking, no music is being played—no information is being transmitted. The study of random signals and information theory is well beyond the scope of this text, but some of the ideas and techniques that we will be discussing are basic to those more advanced topics. Signals that are not random are called *deterministic*—when we use that word we usually mean that we can write a closed-form mathematical expression for that quantity. For random signals, all we can do is to talk about their statistics, because random signals do not have the kind of totally predictable behavior that deterministic signals do.

ANALOG AND DIGITAL SIGNALS

The sinusoid, exponential, step, and impulse functions are all examples of what we call *analog* signals. In science and engineering the word analog means to act similarly, but in a different domain. For instance, the electric voltage at the output terminals of a hi-fi amplifier varies in exactly the same way (if it is a good system) as does the sound that activated the microphone that is feeding the amplifier. On the one hand we have air pressure that is rapidly varying with time (sound), and on the other we have an electric voltage $v(t)$ whose value at each and every instant of time is proportional to the corresponding instantaneous value of that sound pressure signal. In other words, an analog signal is a continuous† description or model of some other quantity.

A *digital* signal is one which is designed to take on only a certain number (usually two—in which case it is called a *binary* signal) of discrete values. Digital signals generally consist of pulses that occur at regular intervals. This periodicity determines what is called the *baud rate* of the signal. A certain specific sequence of pulses may stand for a certain letter in the alphabet or a certain number or a certain punctuation mark, etc. A collection of such sequences that includes the complete alphabet, and the numbers 0 to 9, is called a *code*. There are many such codes: the Morse, Baudot, and ASCII codes are three examples.

CONTINUOUS- AND DISCRETE-TIME SIGNALS

Signals can also be described as being either *continuous-time* signals or *discrete-time* signals. In our daily lives we observe the continuous passage of time at an

---

† This is not to say that an analog signal has to be continuous in the mathematical sense (i.e., having no abrupt jumps, or discontinuities), but simply that it is a *continuous description* of some physical quantity.

apparently smooth and unvarying rate. All the signals we have discussed in Sections 3.1 through 3.6 have been described as being functions of this continuous variable *t*. Such functions are called continuous-time signals.

On the other hand if, at discrete intervals of time, we record the instantaneous value of a continuous-time signal, we will produce a sequence of numbers. Such a sequence is called a discrete-time signal. (Clearly, such a string of numbers cannot describe any variation of the original continuous-time signal in between sampling instants.) Discrete-time signals are often encoded into digital signals (see above) which can then be processed by digital computers. Such computer processing often is designed to mimic the effects of more conventional electronic circuits on the original continuous-time signal. However, *digital-signal processing* can do things to discrete-time signals that standard electronic circuits cannot do to continuous-time signals.

Exciting new techniques for enhancement of video signals, high-fidelity audio signals, and radar images are being made possible by digital-signal processing techniques. In this text, however, we shall concentrate on continuous-time signals and systems. Once you have mastered the concepts involved in handling and treating continuous-time signals, it will be easier for you to understand discrete-time signals and systems.

ADDITIONAL SIGNALS

There are many other signals that electrical engineers work with very often: TV signals, radar signals, and even biological signals such as electrocardiogram and electroencephalogram waveforms. One very nice thing about being an electrical engineer, familiar with the techniques of signal analysis, is that it makes little difference where a signal comes from or how complicated it is. The techniques electrical engineers have developed to study and analyze signals are *widely applicable*. Thus we often find electrical engineers working usefully in areas quite diverse from where one might expect to find them (analyzing biological or economic signals, for example).

summary 3.8

In this chapter we have discussed several of the more popular types of time-dependent shapes that voltages and currents (and other variables) may have. This listing is by no means complete—there is really an infinite variety to the waveforms that engineers deal with. What we have presented here is simply a brief introduction to the ones that we see most often. Concepts such as the infinitely high but zero width pulse (the impulse) and the *effective value* of periodic functions are extremely important. The ideas that we discussed about the shape of the exponential function (its initial slope, its time constant) are crucial to further understanding how systems respond to the excitations (sources) that we will apply to them. Most of all, the idea that a *signal* is simply a quantity that may vary with time is fundamental to our understanding of communications—the transferral of information from one place to another.

problems

1. (a) Given $i_c(t) = u(t) - u(t - 1)$, find the time integral of i_c over the limits minus infinity to t. Sketch this integral function versus t for $-1 < t < 2$. (b) $u(1 - t) - u(t) = 1 - u(t) - u(t - 1)$ True or false? (c) Evaluate the time integral of the product $(t - 2)\delta(t - 1)$ over doubly infinite limits ($-\infty$ to $+\infty$).

2. Evaluate $\int_{-\infty}^{\infty} \delta(2 - t)(4t^2 + 1) \, dt$

3. Is it true that the following two functions are absolutely identical? $u(t) - u(t - 3)$ and $u(t)u(3 - t)$.

4. Sketch $i(t) = u(1 - t)$.

5. Sketch $i(t) = \delta(1 - t)$.

6. Evaluate $\int_{-1}^{+1} \delta(t) \sin(3t + 30°) \, dt$.

7. (a) Sketch $f(t) = tu(t) - (t - 2)u(t - 1) - 2u(t - 2) - \delta(t - 2)$ and evaluate the time integral of $f(t)$ from minus infinity to $t = 4$ s. (b) Evaluate $\int_{-1}^{+1} (6t^2 + t) \, \delta(t - 0.5) \, dt$

8. Sketch the following waveforms. Label important numerical values on the axes. (a) $i_1(t) = u(3 - t)$, (b) $i_2(t) = \delta(t - 3) + (t - 3)u(t - 3)$, and (c) $i_3(t) = u(1 - t) - u(2 - t)$. (d) Find an expression for the time integral of each of the above currents from $t = -\infty$ to time t. How much charge has been delivered by each at time $t = 4$ s?

9. (a) Express the functions shown in Figure P3.9 mathematically in terms of ramps, steps, and impulses. (b) Differentiate the functions of part (a). Sketch the resulting functions and write an expression for each. Label axes numerically.

10. (a) Repeat part (a) of problem 9 for the functions in Figure P3.10. (b) Integrate each of the functions in part (a). Sketch the resulting functions and write an expression for each.

11. Given the function $f(t) = t + 1$, sketch to scale, in the interval $-2 < t < +2$: (a) $f(t)$, (b) $f(t)u(t)$, (c) $f(t)u(-t)$, (d) $f(t)u(t - 1)$, (e) $f(t - 1)u(t)$, (f) $f(t - 1)u(t - 1)$, (g) $f(t) + u(t)$, and (h) $f(t)u(t + 1)$. (i) Sketch the derivatives of each of the above functions (a) through (h).

12. If $f(t)$ is given by $f(t) = t^2u(t) - tu(t - 2) + u(t - 4)$, evaluate (a) $f(-3)$, (b) $f(1)$, (c) $f(2-)$, (d) $f(2+)$, (e) $f(3)$, and (f) $f(5)$.

figure P3.9

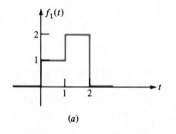

(a)

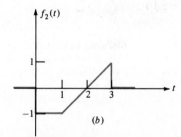

(b)

figure P3.10

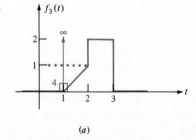

(a)

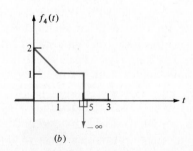

(b)

13. A rectangular voltage pulse $v_1(t) = u(t) - u(t - a)$ is used as input to a system whose output voltage $v_2(t) = v_1(t) - 2v_1(t - a/2)$. (a) Find an expression for $v_2(t)$. (b) Sketch $v_2(t)$ versus t.

14. The voltage $v_1(t) = \frac{1}{2}\delta(t - \frac{1}{2}) + \frac{1}{4}\delta(t - 4)u(t)$ is applied to the input of an electronic integrator. (a) What will the output voltage $v_2(t)$ be? (b) Sketch $v_1(t)$ and $v_2(t)$. (c) What would be the effect of removing the $u(t)$ function from the second term in $v_1(t)$? (d) What is the value of $v_2(100)$?

15. Let $i(t) = 4 - 8[u(t - 2) - u(t - 3)] + tu(t - 2)$ for an interval $0 < t < 5$. For this current, determine the (a) positive and negative peak values, (b) average value, and (c) effective value over this interval.

16. A mad mathematician proposes the following function as being equal to the unit step $u(t)$. Is he correct?

$$u(t) = \lim_{\Delta \to 0} \int_{-\infty}^{t} v_x(t) \, dt$$

where
$$v_x(t) = \begin{cases} \dfrac{\pi}{2\Delta} \sin \dfrac{\pi}{\Delta}(t + \Delta) & -\Delta < t < 0 \\ 0 & t > 0 \end{cases}$$

17. If $dv_1/dt + 3v_1 = 2e^{-t}$ and we know that $v_1(t)$ is of the form $Ae^{-bt} + C$, find $v_1(t)$.

18. Given $i(t) = 3e^{-2t}u(t)$. (a) Find an expression for $v = 4di/dt$. (b) Sketch v versus t for $-1 < t < 2$.

19. Given that $v_1(t) = 4e^{-3t} + 1$ and $v_2 = 2v_1 + dv_1/dt$. (a) Find expressions for $v_2(t)$ and dv_2/dt. (b) Sketch both $v_2(t)$ and dv_2/dt. (c) What is $v_2(\frac{1}{3})$? (d) At what value of t does $v_2(t) = 0$?

20. If, in the previous problem, $v_1(t)$ is multiplied by $u(t)$, find an expression for $v_2(t)$.

21. What is the frequency in hertz of $v(t) = 10 \sin (9t + 30°)$?

22. Convert $10 \cos 3t + 8 \sin 3t$ to the form $A \sin (bt + B)$.

23. Given $v(t) = 125\sqrt{2} \sin (377t + \pi/9)$. (a) What are the values of the maximum and minimum rates of change of $v(t)$? (b) At what value of t does the first $(t > 0)$ maximum slope occur? (c) Repeat (b) for the first minimum slope.

24. Two signals $v_1(t) = 3 \cos 3t$ and $v_2(t) = 4 \cos 4t$ are used as input signals to a circuit, the output voltage of which is $v_3(t) = v_1(t)v_2(t)$. (a) Find an expression for $v_3(t)$ that does not involve products of sinusoidal signals. (b) What is the period of $v_3(t)$?

25. Suppose we have a system whose output voltage $v_2(t) = v_1(t) + \frac{1}{2} \, dv_1/dt$. If $v_1(t) = 2 \cos 2t$, find an expression for $v_2(t)$ in the form $A \cos (\omega t + \theta)$.

26. Repeat problem 25 using $v_1(t) = 2 \cos (2t)u(t)$.

27. At what value of *current* is the 60-Hz waveform $i(t) = 10 \sin (\omega t - 30°)$ changing at the rate of (a) 3265 A/s? (b) 1885 A/s?

28. Find the maximum value of a 60-Hz sinusoidal wave that is changing at 2000 A/s at an instantaneous value 30° from the maximum value of the wave.

29. If $v(t) = 100 \sin (\omega t - 30°)$ and $i(t) = 10 \cos (\omega t - 60°)$, what is the phase angle difference between these two waveforms? Which wave leads?

30. Given a voltage $v(t) = 100 \cos \omega t$. (a) Write an expression using cosine for a current $i(t)$ that has a maximum value of 10 A, the same period as $v(t)$, and which leads $v(t)$ by one-sixth of a cycle. (b) Repeat part (a) using a sine function. (Use radian measure to describe angles in this problem.)

31. (a) Find the instantaneous value of a sinusoidal current, whose maximum value is 90 A, 60° after the current passes through its zero value going positive. (b) Assuming the frequency of the current waveform in (a) is 60 Hz, find the difference in *time* between the 60° value and the 225° value.

32. The current in a particular element is given by $i(t) = 1 + 0.5 \sin 1885t - 0.1 \cos 3770t$. (a) What is the frequency of the sine term? Of the cosine term? (b) What are the maximum and minimum values of $i(t)$? (c) Sketch $i(t)$ over one period.

33. A voltage $v(t) = 100 \cos (\omega t + 60°)$ is impressed across a resistor $R = 10 \, \Omega$. Write an expression for (a) $i(t)$ and (b) $p(t)$, the power delivered to the resistor. (c) What is the maximum value of $p(t)$? (d) What is the minimum value of $p(t)$? (e) What is the average value of $p(t)$ over an integer number of periods of $v(t)$?

34. A voltage $v(t) = 150 \cos 377t$ is applied to a resistor $R = 30 \, \Omega$. (a) Find an expression for the power $p(t)$ delivered to the resistor. (b) What is the frequency of $v(t)$? (c) What is the frequency of $p(t)$? (d) What is the average value of $p(t)$ over an integer number of periods of $v(t)$?

35. A current $i(t) = 5 \sin (110t + 30°)$ exists in a resistor $R = 20 \, \Omega$. (a) Find an expression for $v(t)$, the voltage across the resistor. (b) What is the frequency of $v(t)$? (c) Find an expression for $p(t)$, the power delivered to the resistor. (d) What is the frequency of $p(t)$? (e) What is the average value of $p(t)$ over an integer number of cycles?

36. An electric saw motor draws 20 A for 15 s. Power is then shut off for 45 s, after which this cycle is repeated again and again. (a) If rated full-load current of the motor is 12 A, will it overheat if it is used in this manner? (b) What is the equivalent continuous current that will produce the same average rate of heating?

37. A motor draws 50 A for 10 s after which power is shut off for 20 s. Then the motor draws 60 A for 5 s, after which power is shut off for 1 min. What does the continuous rated current have to be such that the motor will not overheat in repeated usage as described above?

38. A periodic current is described as follows:

$$0 < t < 0.2 \quad \text{and} \quad 1.4 < t < 1.6 \qquad 1 \text{ A}$$

$$0.2 < t < 0.6 \quad \text{and} \qquad 1 < t < 1.4 \qquad 2 \text{ A}$$

$$0.6 < t < 1 \qquad\qquad\qquad\qquad\qquad 3 \text{ A}$$

$$\text{and} \quad i(t + n1.6) = i(t)$$

(a) If this current flows through both a dc ammeter and a true rms ammeter, what will be the readings of the two meters? (b) If the rms meter is a half-wave rectifier instrument that has been calibrated for use with sinusoids, what will it read? What percentage error does this represent?

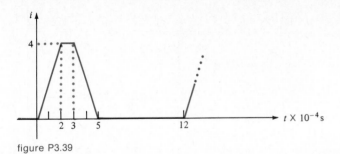

figure P3.39

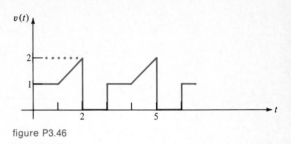

figure P3.46

39. The periodic collector current of a transistor operating as an oscillator is shown in Figure P3.39. What is the (a) frequency of oscillation, (b) average value of this current, and (c) effective value of this current?

40. Suppose a technician wants to measure the rms value of a current that starts at $t = 0$, increases linearly to some value, and then drops abruptly to zero. It then repeats this cycle continuously. The technician mistakenly uses a meter that incorporates a full-wave rectifier and is calibrated for use on sine waves. What percentage error will this result in?

41. A typical cycle of a periodic voltage starts out at 10 V and then decreases linearly to zero before jumping discontinuously to 10 again. What is the form factor of this waveform? The crest factor?

42. Find the rms value of $v(t) = 100 \sin \omega t - 40 \sin 3\omega t$.

43. Two currents flowing toward a node to which three elements are connected are $i_1(t) = 30 \sin (\omega t + 60°)$ and $i_2(t) = 20 \cos (\omega t - 20°)$. Find $i_3(t)$, the current flowing toward the node in the third element.

44. One branch current $i_1(t) = 40 \cos (\omega t - 40°)$ combines with a second branch current to yield a resultant $i(t) = 50 \cos (\omega t + 80°)$. Find an expression in terms of cosine for the second branch current $i_2(t)$. What is the rms value of $i_2(t)$?

45. Find the crest factor of each of the following periodic waveforms (period = T seconds):

(a) $f_a(t) = V_1 \cos \omega t$ where $\omega = 2\pi/T$
(b) $f_b(t) = V_2 \sin \omega t$ for $0 < t < T/2$
     $f_b(t) = 0$ for $T/2 < t < T$
(c) $f_c(t) = (V/T)t$ for $0 < t < T$
(d) $f_d(t) = (2V/T)[t - (T/2)]$ for $0 < t < T$
(e) $f_e(t) = Vu(t)u(0.2T - t)$ for $0 < t < T$

46. Find the effective and average values of the waveform shown in Figure P3.46.

47. Find the average power dissipated in a 4-Ω resistor if the voltage across it is a periodic waveform that is equal to 2 V for 0.1 s and then is zero-valued for 0.2 s before repeating. What is the form factor of such a voltage? The crest factor?

48. Find the rms value of the periodic signal shown in Figure P3.48.

49. The current in a 50-Ω resistor is $3e^{-2t}u(t)$ amperes. Find the average power delivered to the resistor during the time interval: (a) $0 < t < 0.5$, (b) $-0.5 < t < 0.5$, and (c) $0 < t < 5$.

50. Find the value of the average current in the battery charger circuit shown in Figure P3.50. Assume that $R = 0.25$ Ω and $v_s(t) = 20 \sin 377t$.

figure P3.48

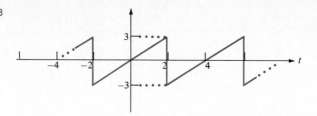

51. The battery in the circuit in Figure P3.50 should not be charged at an average rate exceeding 4 A. If $v_s(t) = 20 \sin 377t$ and R is a variable, at what value should R be set?

52. (*a*) Find the rms value of the signal shown in Figure P3.52. (*b*) Think of $f(t)$ as being the sum of two component signals: its average (dc) value and a square wave whose average value is zero. Sketch both of those two signals versus t. Separately find the ms value (square root of the rms value) of each of these components. Add these two ms values and take the square root. What general statement can be made about signals with nonzero average values?

53. A slide projector lamp is rated at 300 W at 120 V_{rms}, 60 Hz. By placing an ideal diode (no forward voltage drop) in series with this lamp, its brightness will be reduced. A new 300-W lamp is to be designed for use in such a diode circuit. What should be the resistance of the new lamp and its rms current and voltage ratings? How does the new resistance compare with the original lamp resistance?

figure P3.50

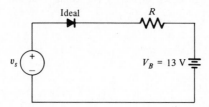

figure P3.52

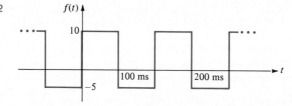

chapter 4
ENERGY-STORAGE ELEMENTS AND ANALOGS

introduction 4.1

In our investigation of circuits and systems we have limited our discussion so far to networks containing only resistors and sources. There are many other types of elements that our analysis techniques will allow us to incorporate in these circuits. In this chapter we investigate several of these other types of elements.

Let us begin by specifying more precisely than we have before what an element is. In the real physical world that we live in, there are *two kinds* of dynamic quantities. One such type consists of the set of all variables that can be described completely without reference to anything else—a kind of "stand alone" variable: A flow of 3 ft^3/s air, 4-A electric current, 3-N force.

The other type of physical variables are those that are meaningful only in reference to something else, for example, any velocity measurement.

When one says simply that the speed of a certain automobile is 60 mi/h, it is presumed to be 60 mi/h *with respect to* the "fixed" road. In reality, of course, the road is attached to the Earth which not only spins on its axis but also revolves around the sun, which in turn is itself flying through space, and so the speed of the car with reference to, say, the sun is *not* 60 mi/h. Temperature is another quantity that must be quantified *with respect to* some other datum or reference level (such as the freezing point of pure water). There are many others. The one we are most concerned with is voltage. We have already seen that the voltage at one point in a circuit must be stated as being so many volts above or below that at some *other point* in the circuit.

The first type of variable we call a *through variable*. The second type is called an *across variable*. We have already seen, in Ohm's law, the relationship between the through variable (current) and the across variable (voltage) that exists in any resistor. The resistor is just a special case of the general element shown in Figure 4.1.1.

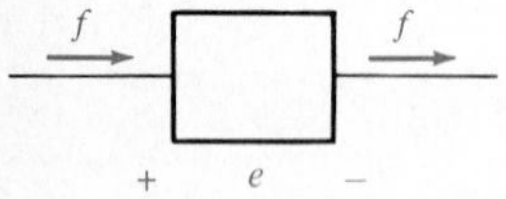

Thus, in general, *an element is any physical entity that defines the relationship between its associated through variable f and its across variable e*. Electric resistors are one of a class of elements wherein

$$e = kf \qquad (4.1.1)$$

In the special case of the resistor:

$$e = \text{voltage } v, \text{ V}$$

$$f = \text{current } i, \text{ A}$$

$$k = \text{resistance } R, \Omega$$

Let us now restate the requirements for linearity in general terms. A linear relationship between any two variables x and y exists if and only if:

1. If $y = g(x)$, then $g(ax) = ay$, where a is a constant. This is called the *homogeneity* property.
2. If $y_1 = g(x_1)$ and $y_2 = g(x_2)$, then $g(x_1 + x_2) = y_1 + y_2$. This is called *additivity*.

These two requirements together are equivalent to saying that a plot of y versus x must be a straight line that passes through the origin (see chapter 1).

figure 4.1.1
The general element and its associated through and across variables.

EXAMPLE 4.1.1

Show that equation (4.1.1) is linear.
ANS.:   Check requirement 1 with

$$y = e \qquad x = f$$

and
$$g(x) = kf$$

Then we can write

$$g(ax) = akf = ag(x) = ay$$

So requirement 1 is satisfied.

Requirement 2 can be checked by writing

$$g(x_1 + x_2) = k(f_1 + f_2)$$
$$= kf_1 + kf_2$$
$$= g(x_1) + g(x_2)$$

So both requirements are satisfied and equation (4.1.1) is shown to be a linear relationship.

EXAMPLE 4.1.2

Show that the integral relationship

$$y(t) = \int_{-\infty}^{t} x(t) \, dt$$

is linear.

ANS.:   We have $$y = g(x),$$

where $$g(x) = \int_{-\infty}^{t} x(t) \, dt.$$

Requirement 1: Does $g(ax) = ay$?

$$g(ax) = \int_{-\infty}^{t} ax(t) \, dt = a \int_{-\infty}^{t} x(t) \, dt$$

So $$g(ax) = ag(x)$$
$$g(ax) = ay$$

Thus requirement 1 is obeyed.

Requirement 2: Does $g(x_1 + x_2) = g(x_1) + g(x_2)$?

$$g(x_1 + x_2) = \int_{-\infty}^{t} x_1(t) + x_2(t) \, dt$$
$$= \int_{-\infty}^{t} x_1(t) \, dt + \int_{-\infty}^{t} x_2(t) \, dt$$

which satisfies requirement 2.

Thus we say that *integration is a linear operation.*

EXAMPLE 4.1.3

Show that differentiation is a linear operation.

ANS.:   Let $y = g(x)$, where $g(x) = dx/dt$.

Requirement 1: Does $g(ax) = ay$?

$$g(ax) = \frac{d(ax)}{dt} = a \frac{dx}{dt} = ag(x) = ay$$

Thus requirement 1 is fulfilled.

Requirement 2: Does $g(x_1 + x_2) = g(x_1) + g(x_2)$?

$$g(x_1 + x_2) = \frac{d}{dt}\left[x_1(t) + x_2(t)\right]$$

$$= \frac{dx_1(t)}{dx} + \frac{dx_2(t)}{dx}$$

$$= g(x_1) + g(x_2)$$

So, yes, requirement 2 is also obeyed and we can say that *differentiation is a linear operation.*

Any element whose across and through variables are linearly related is called a *linear element.*

The product of the instantaneous values of f and e for any element† is the instantaneous value of the power *being delivered to that element.* (What the element does with that power depends on what kind of element it is. Some elements store this power for later use—others immediately dissipate it as heat or light.) Thus we write

$$p(t) = e(t)f(t) \tag{4.1.2}$$

However, we are not at liberty to pick just *any* across variable and multiply it times *any* through variable and then call the result power. For example, force is a through variable, and position is an across variable. Their product is not equal to power. As a matter of fact, we shall specifically choose the major through and across variables that we will be working with so that equation (4.1.2) is obeyed in every possible case. If $p(t)$ turns out to have a negative value at some instant of time, the element is then giving energy back to the system. For some specific kinds of elements (e.g., resistors, mechanical "dashpots") this is impossible. For many other kinds it is quite possible for them to store energy and then give it back at a later time.

We know that the rate at which energy (i.e., ability to do work) is supplied to an element is called power.

$$p(t) = \frac{dW(t)}{dt} \tag{4.1.3}$$

or one can say, equivalently, that the total amount of energy given to an element during the interval $t_1 < t < t_2$ is

$$W = \int_{t_1}^{t_2} e(t)f(t)\,dt \tag{4.1.4}$$

EXAMPLE 4.1.4

a. In translational mechanical systems, power = force × velocity.

$$p = \mathcal{U}f$$

Through variable

Across variable

---

† There are some exceptions to this. For example, in thermal systems, the through variable itself has the units of power and the *ef* product is of no significance.

b. In rotational mechanical systems, power = torque × angular velocity.

$$p = \omega\tau$$

Through variable

Across variable

Later in this chapter we discuss several kinds of different pairs of across and through variables. Let us start by looking at other types of electrical elements.

There are several types of electrical elements each of which has a different relationship between its across variable (voltage) and its through variable (current).

capacitor 4.2

The capacitor is an electrical element wherein the through variable, current, is proportional to the time derivative of the across variable, voltage (see Figure 4.2.1).

$$i = C\frac{dv}{dt} \tag{4.2.1}$$

The proportionality constant C is called the *capacitance* of the capacitor and is measured in farads.[†]

Integrating both sides of equation (4.2.1) from $t = -\infty$ to any arbitrary time t_0, we get

$$v(t_0) = \frac{1}{C}\int_{-\infty}^{t_0} i(t)\,dt \tag{4.2.2}$$

figure 4.2.1
(*a*) The schematic symbol and terminal variables for a capacitor. (*b*) The relationship between the capacitor's terminal variables.

The integral in equation (4.2.2) is the net charge q accumulated on the capacitor at time t_0 due to the current $i(t)$ which has been flowing into the capacitor from $t = -\infty$ until $t = t_0$ that is,

$$q(t_0) = \int_{-\infty}^{t_0} i(t)\,dt \tag{4.2.3}$$

Thus, at any time t, (4.2.2) may be written simply as

$$v(t) = \frac{q(t)}{C} \tag{4.2.4}$$

so that

$$dv(t) = \frac{dq(t)}{C} \tag{4.2.5}$$

The relationship given in equation (4.2.4) satisfies our definition of a linear element in that the q-v plot is a straight line that passes through the origin (see Figure 4.2.2). We have defined a linear element as being one wherein the across and through variables are mathematically linearly related. We note that $q(t)$ is

[†] In honor of Michael Faraday, a famous electrical scientist of the nineteenth century.

figure 4.2.2
The charge-voltage plot for a linear
capacitor. For any value of charge q,
$C = q/v$.

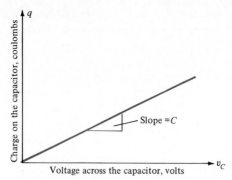

neither the through nor the across variable. However, the variable $q(t)$ is linearly related to the through variable $i(t)$ because the integral/derivative relationship between $q(t)$ and $i(t)$ is a linear one. See Examples 4.1.2 and 4.1.3. Thus, if $v(t)$ and $q(t)$ are linearly related, then so are $v(t)$ and $i(t)$ linearly related, making any capacitor described by equation (4.2.1) or (4.2.2) a linear capacitor.

There are capacitors which are nonlinear as well. For such capacitors equations (4.2.1) and (4.2.2) are not valid.† Note from equation (4.2.4) that the units in which C is measured, farads, are equivalent to coulombs/volt. We see that the capacitor is an element that stores charge. The more stored charge q, the higher the terminal voltage V.

The two plates of any capacitor are electrically separated from each other by a nonconducting layer of dielectric material. So when a constant current flows "through" a capacitor, a positive charge accumulates at a constant rate on the plate into which the current enters. Meanwhile, positive charge is flowing out of the other plate's terminal, and so a net negative charge appears on that plate. The current magnitude is the same at both terminals, so at any instant of time the magnitude of the collected positive charge on one plate is equal to the magnitude of the negative charge accumulated on the other plate. The magnitude of each of these charges in coulombs is $q(t)$ as given by equation (4.2.3).

The separated charge populations, positive on one plate and negative on the other plate, are highly attracted to each other. In fact, work has to have been done (energy delivered) by some source in the circuit in order to achieve this segregation of one type of charge from the other. If the dielectric insulating material were to be removed, these two populations would rush toward each other. This situation is analogous to holding a bowling ball up in the air. The bowling ball and the Earth have an attaction for each other. To raise the ball to, say, shoulder height we must do work on it, i.e., deliver energy to it. At shoulder height, therefore, the ball has (mechanical) potential energy. Similarly, the charged capacitor contains stored (electric) energy. Both these situations are ones wherein the stored (potential) energy can, at any arbitrary time, be released to do work. Such elements are called *conservative* elements. A resistor is an example of a nonconservative element because any energy delivered to a resistor is instantaneously radiated (lost) into its surroundings.

† A general treatment that is applicable to such nonlinear problems will be presented in Chapter 9.

The amount of energy stored in a capacitor at any instant t_0 is the time integral of the power that has been delivered to it:

$$W(t_0) = \int_{-\infty}^{t_0} v(t)i(t)\, dt \tag{4.2.6}$$

We use the expression for current

$$i(t) = \frac{dq}{dt} \tag{4.2.7}$$

to write (4.2.6) as

$$W(t_0) = \int_{q(-\infty)}^{q(t_0)} v(t)\, dq \tag{4.2.8}$$

Then using equation (4.2.5) we get

$$W(t_0) = \int_{v(-\infty)}^{v(t_0)} Cv(t)\, dv \tag{4.2.9}$$

where $v(t_0)$ is the voltage across the capacitor at any arbitrary time t_0 and $v(-\infty)$ was the voltage across it at time $= -\infty$ (say, when it was manufactured). Assuming $v(-\infty) = 0$, the result of performing the integral in (4.2.9) is

$$W(t) = \tfrac{1}{2}C[v(t)]^2 \tag{4.2.10}$$

and taking the time derivative of energy W to get power p:

$$p(t) = Cv\,\frac{dv}{dt} \tag{4.2.11}$$

We could have obtained equation (4.2.11) by simply writing

$$p(t) = v(t)i(t) \tag{4.2.12}$$

and substituting for $i(t)$ from equation (4.2.1).

Consider equation (4.2.8) together with Figure 4.2.2. The area to the left of the $v_c(q)$ curve as plotted in that figure is the energy stored in the capacitor. More specifically, if the charge on the capacitor goes, say, from a value q_1 to a higher value q_2, then the increase in the stored energy is the area to the left of the curve between horizontal lines that intersect the vertical axis at q_1 and q_2. Most importantly, this is true whether the capacitor is linear or *not*.

The actual capacitance of a capacitor and whether or not it is linear is determined by the geometrical shape of the device and the electrical properties of the dielectric material.

The capacitance of a linear, parallel plate capacitor is

$$C = \frac{\varepsilon_r \varepsilon_0 A}{d} \tag{4.2.13}$$

where $\varepsilon_0 = \dfrac{10^{-9}}{36\pi}$ = permittivity of free space

ε_r = relative permittivity of the material between the two plates

A = area of each of the two plates
d = separation distance of the plates

It should be carefully noted that because of equation (4.2.1) the capacitor has the following two properties:

- For an applied sinusoidal voltage, $v_C(t) = V \sin \omega t$, the current is $i_C(t) = C\omega V \cos \omega t$. All else being unchanged, the current will therefore be large for large values of the applied frequency ω and small for low values of ω. In fact for *sufficiently large* ω, any capacitor looks like *a short circuit. For $\omega = 0$* (dc) any capacitor looks like *an open circuit.*
- For the voltage $v_C(t)$ across any capacitor to change instantaneously (that is to say, discontinuously), its time derivative dv/dt must be infinite at the instant of the abrupt change. Equation (4.2.1) indicates that an infinite current, i.e., an impulse, is necessary in any capacitor in order to achieve this. Or, synonymously, we may state that *in the absence of impulses, voltages across capacitors cannot change instantaneously.*

EXAMPLE 4.2.1

In the following, $v_C(t)$, the voltage across the terminals of a 2-F capacitor, is given. Find $i_C(t)$ in each case for all values of t.

a. $v_C(t) = 3 \sin (4t + 45°)$

ANS.:
$$i_C(t) = C \frac{dv_C}{dt}$$
$$= 2(3)(4) \cos (4t + 45°) = 24 \cos (4t + 45°)$$

b. $v_C(t) = 3u(t) \sin (4t + 45°)$

ANS.:
$$i_C(t) = C \frac{dv_C}{dt}$$

Use the chain rule:
$$\frac{d}{dt} [v(t)w(t)] = v(t) \frac{dw}{dt} + w(t) \frac{dv}{dt}$$

Therefore,
$$i_C(t) = 2\{[3u(t)][4 \cos (4t + 45°)] + 3 \, \delta(t) \sin (4t + 45°)\}$$
$$= 24u(t) \cos (4t + 45°) + 4.242 \, \delta(t)$$

c. $v_C(t) = r(t)$
or $v_C(t) = tu(t)$

ANS.:
$$i_C(t) = C \frac{dv}{dt}$$
$$= (2)[t \, \delta(t) + 1u(t)]$$
$$= 2u(t)$$

d. $v_C(t) = u(t)$

ANS.:
$$i_C(t) = C\,\frac{dv}{dt}$$
$$= 2\,\delta(t)$$

e. $v_C(t) = tu(t)u(1-t)$

ANS.:
$$i_C(t) = C\,\frac{dv}{dt}$$
$$= 2[tu(t)][-\delta(t-1)] + 2u(t)[u(1-t)]$$
$$= -2\,\delta(t-1) + 2u(t)u(1-t)$$

EXAMPLE 4.2.2

Given that the current through a 2-F capacitor is $i_C(t) = 4\,\delta(t) - 12e^{-3t}u(t)$, find the corresponding $v_C(t)$.

ANS.:
$$v_C(t) = \frac{1}{C}\int_{-\infty}^{t} i_C(t)\,dt$$
$$= \frac{1}{2}\int_{-\infty}^{t} 4\,\delta(t) - 12e^{-3t}u(t)\,dt$$
$$= 2\int_{0-}^{0+} \delta(t)\,dt - 6\int_{0}^{t} e^{-3t}\,dt \text{ for } t > 0$$
$$= 2 - \frac{6}{-3}\,e^{-3t}\bigg|_{0}^{t} = 2 + 2(e^{-3t} - 1) = 2e^{-3t}$$

For $t < 0$ the integrand (and therefore the integral) is zero. Thus
$$v_C(t) = 2e^{-3t}u(t) \qquad \text{for all } t$$

Very often we are asked to analyze circuits wherein some source, say, for example, a voltage source, is applied at time zero. We are asked to find out what happens afterward, that is, for $t > 0$. The voltage on any given capacitor for any $t > 0$ depends on two things: What current has flowed into the capacitor since $t = 0$? What was the value of v_C at $t = 0$? Mathematically we see this as follows:

INITIAL-CONDITION GENERATORS

$$v_C(t) = \frac{1}{C}\int_{-\infty}^{t} i_C(t)\,dt$$
$$= \frac{1}{C}\int_{-\infty}^{0} i_C(t)\,dt + \frac{1}{C}\int_{0}^{t} i_C(t)\,dt \qquad (4.2.14)$$

The integral in the first term in equation (4.2.14) is the net charge accumulated

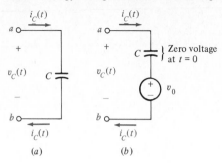

figure 4.2.3
Two completely equivalent circuits (insofar
as terminals *a* and *b* are concerned) for
$t > 0$. (*a*) A capacitor with initial charge, that
is, $v_C(0) = v_o$. (*b*) The capacitor in this circuit
has zero voltage across it at $t = 0$.

over all negative time up until $t = 0$. This is a number, a constant, because the
limits in the integral are constants. Dividing this by C yields the voltage across
the capacitor at $t = 0$. The second term accounts for what happens from $t = 0$
until the arbitrary instant t. This quantity depends on t, is a function of t, because
t is the upper limit of integration. Thus we have:

$$v_C(t) = v_C(0) + \frac{1}{C}\int_0^t i_C(t)\, dt \qquad (4.2.15)$$

or

$$v_C(t) = v_C(0) + \frac{1}{C}\int_{-\infty}^t i_C(t)u(t)\, dt \qquad (4.2.16)$$

Each term in equation (4.2.16) is itself a voltage. This sum of two voltages
describes two things in series: a constant (dc) voltage $v_C(0)$, called the *initial condition*, and a capacitor that has not ever had current in it prior to $t = 0$, that is, is
initially uncharged (at $t = 0$). So we see that at any instant of time after $t = 0$, a
capacitor can be thought of as being equivalent to another capacitor (of equal
capacitance) that has had zero current in it prior to $t = 0$ in series with a constant voltage source equal to the value of the voltage on the original capacitor at
$t = 0$.† (See Figure 4.2.3.)

EXAMPLE 4.2.3

A 4-F capacitor has the following current:

$$i(t) = 2u(t + 1) + u(t) \qquad \text{for all } t$$

See Figure 4.2.4*a* and *b*.

Describe the voltage across this capacitor at any positive instant of time t. Use an initial-condition generator.

ANS.: First find $v_C(t)$ for positive t.

$$v_C(t) = \frac{1}{4}\int_{-\infty}^t i(t)\, dt$$

---

† A second form of this initial-condition generator circuit also exists. It contains an impulse current
source. We will consider this other form in Section 15.6.

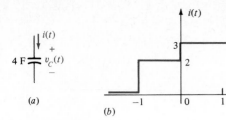

(a)

(b)

figure 4.2.4
The capacitor and waveforms of Example 4.2.3

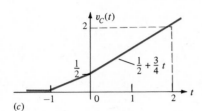

(c)

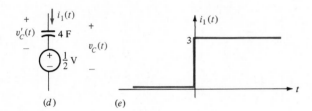

(d)   (e)

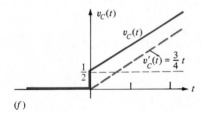

(f)

$$= \frac{1}{4}\left[\int_{-\infty}^{0} i(t)\, dt + \int_{0}^{t} i(t)\, dt\right]$$

$$= \frac{1}{4}\left(\int_{-1}^{0} 2\, dt + \int_{0}^{t} 3\, dt\right)$$

$$= \frac{1}{2} + \frac{3}{4}\int_{0}^{t} dt = \frac{1}{2} + \frac{3}{4}t$$

See Figure 4.2.4c.

The value of $v_C(0) = \frac{1}{2}$ V. Therefore, the equivalent circuit consists of a constant $\frac{1}{2}$-V source in series with an initially *uncharged* 4-F capacitor whose current $i_1(t)$ is the same as the original $i(t)$ but only for positive time—$i(t)$ is zero for negative time:

$$i_1(t) = i(t) \qquad only \text{ for } t > 0$$

that is,

$$i_1(t) = 3u(t)$$

See Figure 4.2.4d and e.

The voltage across an initially uncharged 4-F capacitor due to $i_1(t)$ is

$$v_C'(t) = \frac{1}{4} \int_0^t 3u(t)\, dt$$

$$= \frac{3}{4} \int_0^t dt = \frac{3}{4} t$$

See Figure 4.2.4f.

The equivalent circuits are shown in Figure 4.2.4a and d.

CAPACITORS IN PARALLEL

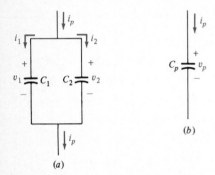

(a)

figure 4.2.5
(a) Two capacitors C_1 and C_2 in parallel. (b)
Their terminal equivalent $C_p = C_1 + C_2$.

The equivalent capacitance C_p of two capacitors C_1 and C_2 in parallel can be easily calculated. See Figure 4.2.5. The total equivalent current is the sum of the individual currents. The voltage across the equivalent capacitor is equal to the voltage across each of the individual capacitors. Thus,

$$i_p = C_p \frac{dv_p}{dt} \tag{4.2.17}$$

$$i_p = i_1 + i_2 = C_1 \frac{dv_1}{dt} + C_2 \frac{dv_2}{dt} \tag{4.2.18}$$

Since $v_p = v_1 = v_2$,

$$i_p = (C_1 + C_2) \frac{dv_p}{dt} \tag{4.2.19}$$

Comparing equations (4.2.17) and (4.2.19) we see that, for two capacitors in parallel, the equivalent single capacitor is:

$$C_p = C_1 + C_2 \tag{4.2.20}$$

Note from Figure 4.2.5a that in connecting two capacitors in parallel, we electrically connect both upper plates together, making an equivalent single larger upper plate (ditto with the bottom plate). Since capacity is directly proportional to plate area, and we have made a single large capacitor whose plate area is the sum of the individual plate areas, equation (4.2.20) seems quite reasonable.

CAPACITORS IN SERIES   The capacitance C_s of two capacitors C_1 and C_2 in series can be similarly obtained. In the series connection the currents in both the individual capacitors and the equivalent single capacitor are all equal. The voltage across the equivalent capacitor is the sum of the voltages across the individual capacitors. See Figure 4.2.6. Thus,

$$v_s(t) = \frac{1}{C_s} \int_{-\infty}^t i_s(t)\, dt \tag{4.2.21}$$

$$v_s(t) = v_1(t) + v_2(t) = \frac{1}{C_1} \int_{-\infty}^t i_1(t)\, dt + \frac{1}{C_2} \int_{-\infty}^t i_2(t)\, dt \tag{4.2.22}$$

$$v_s(t) = \left(\frac{1}{C_1} + \frac{1}{C_2}\right) \int_{-\infty}^{t} i_s(t) \, dt \qquad (4.2.23)$$

Comparing (4.2.21) and (4.2.23), we see that

$$\frac{1}{C_s} = \frac{1}{C_1} + \frac{1}{C_2} \qquad (4.2.24)$$

Alternatively, by putting the right side of (4.2.24) over a common denominator, and then inverting both sides of the result, we can write

$$C_s = \frac{C_1 C_2}{C_1 + C_2} \qquad (4.2.25)$$

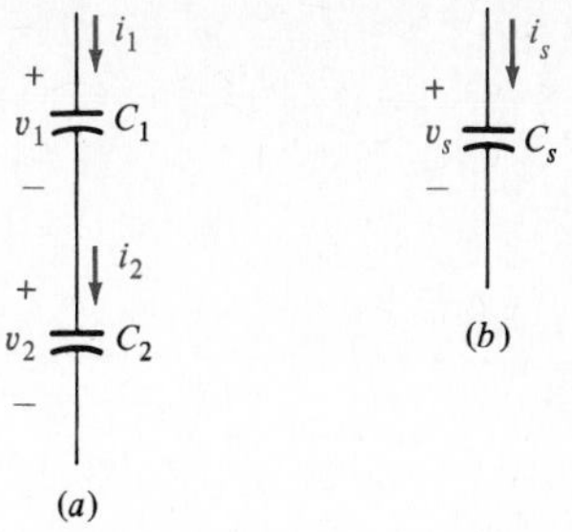

figure 4.2.6
(a) Two capacitors C_1 and C_2 in series.
(b) Their terminal equivalent $C_s = C_1 C_2/(C_1 + C_2)$.

inductor **4.3**

The linear inductor is an electrical element wherein the across variable, voltage, is directly proportional to the time derivative of the through variable, current.

$$v = L \frac{di}{dt} \qquad (4.3.1)$$

The proportionality constant L is called the "inductance" of the inductor and is measured in henrys.[†] See Figure 4.3.1.

Integrating both sides of equation (4.3.1), we get the alternative form:

$$i(t_0) = \frac{1}{L} \int_{-\infty}^{t_0} v \, dt \qquad (4.3.2)$$

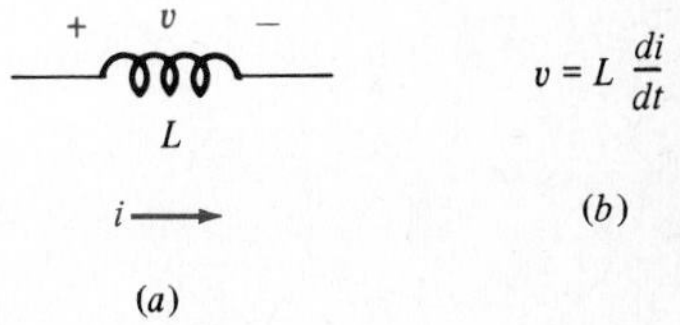

which indicates that the value of the current in an inductor at any time t_0 depends on the entire past history of the voltage that has been applied across it for all time (up to and including time t_0).

We can find whether the inductor is a conservative element or not by solving for the total energy that has been delivered to it. That amount of energy is the time integral of the power delivered to the inductor up until t_0.

$$W(t_0) = \int_{-\infty}^{t_0} v(t) i(t) \, dt \qquad (4.3.3)$$

$$W(t_0) = \int_{-\infty}^{t_0} L \frac{di}{dt} i(t) \, dt \qquad (4.3.4)$$

$$W(t_0) = L \int_{i(-\infty)}^{i(t_0)} i(t) \, di \qquad (4.3.5)$$

Assuming $i(-\infty)$ is zero (because we can always go back, in the case of any

figure 4.3.1
(a) Schematic symbol and terminal variables for the inductor. (b) The relationship between the inductor's terminal variables.

$$v = L \frac{di}{dt}$$

[†] In honor of Joseph Henry of Princeton University (1797–1878).

actual inductor, to the time it rolled off the assembly line and had never had current in it), equation (4.3.5) results in:

$$W(t_0) = \tfrac{1}{2}L[i(t_0)]^2 \qquad (4.3.6)$$

So we observe that in a linear inductor the energy at any instant depends *only* on the magnitude of the current *at that same instant*. Because of the squared value being used, we conclude that the direction (positiveness or negativeness) of the current is unimportant. If the current is at one instant large and at a succeeding instant small, then W gets *smaller*. If the total energy delivered *to* an element goes from a higher value to a lower value with increasing time, then the element must have delivered back some energy to the rest of the circuit. We conclude that the inductor is a conservative element.

Note that equation (4.3.1) implies two very important properties of inductors:

- For sinusoidal currents, $i_L = I \sin \omega t$, the resulting voltage will be $v_L = L\omega I \cos \omega t$. Since the magnitude of v_L is proportional to applied frequency ω, we conclude that *inductors* look like *open circuits for* extremely large values of ω. For $\omega = 0$ (dc), inductors look like *short circuits*.
- In the absence of impulsive voltages *inductor currents cannot change instantaneously.*

THE PHYSICAL BASIS OF INDUCTANCE

Let us look inside a typical inductance (Figure 4.3.2) to see why it operates as it does. The inductance can be thought of as being a coil of resistanceless wire having n turns. If a current i passes through the inductance, a magnetic field Φ (measured in webers) results. There are no sources and sinks of magnetic flux (i.e., lines of Φ have no beginning or end—they form closed contours). This is the familiar electromagnet.

The terminal voltage v across the inductor is equal to the time derivative of the flux linkages $n\Phi$. Thus,

$$v(t) = \frac{d}{dt}(n\Phi) \qquad (4.3.7)$$

or (assuming n is a constant)

$$v(t) = n\frac{d\Phi}{dt} \qquad (4.3.8)$$

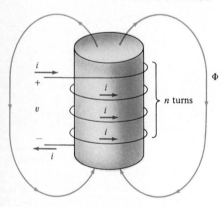

figure 4.3.2
A typical inductor showing its terminal variables v and i and its magnetic field Φ.

This indicates that applying a (positive) voltage v to the terminals of an inductor will result in a growth in the magnetic field. Similarly, if we know the magnetic field is decreasing quickly at some instant, then we know that v will have a large negative value at that instant.

The value of the magnetic field Φ generally depends on (is some function of) the current $i(t)$. For linear inductors the relationship between i and Φ is

$$\Phi = \left(\frac{L}{n}\right)i \qquad (4.3.9)$$

where the directions (and therefore the algebraic signs) of Φ and i are as shown in

Figure 4.3.2.† The orientation of the directions of Φ and i is shown in Figure 4.3.2 constitutes what is called the *right-hand rule*. If the inductor is grasped in the right hand, with the fingers pointing in the direction of the current, then the thumb indicates the direction of the flux Φ. Similarly, if you grasp any wire with your right hand, thumb in the direction of the current i, then the resulting flux Φ surrounding the wire will be in the direction of the fingers.

Substituting equation (4.3.9) into (4.3.8) yields

$$v = L \frac{di}{dt} \tag{4.3.10}$$

which is the same as equation (4.3.1) and from which equation (4.3.2) can be obtained by integration.

Solving equation (4.3.9) for the current i yields:

$$i = \frac{1}{L}(n\Phi) \tag{4.3.11}$$

Comparing equations (4.3.2) and (4.3.11) we see that the integral in (4.3.2) is equal to the flux linkages $n\Phi$ at the instant t_0.

$$\int_{-\infty}^{t_0} v \, dt = n\Phi \tag{4.3.12}$$

Thus the entire past history prior to t_0 of the voltage applied to a linear inductor determines the magnetic flux that exists at time t_0.

The value of the inductance of any inductor depends on its physical shape and size. For example, the inductance of a uniformly wound torus (doughnut shape) of n turns is (see Figure 4.3.3)

$$L = \frac{\mu_r \mu_0 n^2 A}{2\pi R} \tag{4.3.13}$$

where $\mu_0 = 4\pi \times 10^{-7}$ = permeability of free space
μ_r = relative permeability of the material of the torus
A = cross-sectional area of the torus (perpendicular to flux vector)
R = radius of the torus

The expression for the inductance shown in equation (4.3.13) is for a special case (geometrical shape), but it is fairly typical of the results that are usually obtained for inductors of reasonable shape. That is, inductance is proportional to the square of the number of turns n and to the cross-sectional area A through which the flux flows. It is inversely proportional to the length of the path l ($= 2\pi R$ in this case) that the flux takes.

Substituting

$$L = \frac{\mu n^2 A}{l} \tag{4.3.14}$$

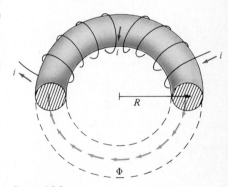

figure 4.3.3
An inductor wound on a toroidal core.

---

† Since $L = n\Phi/i$, we note that an inductor is a flux-linkage maker. For a given current, a larger inductor will have more flux linkages than a smaller inductor.

into (4.3.11) yields

$$\Phi = \mathcal{R}(ni) \qquad (4.3.15)$$

where

$$\mathcal{R} = \frac{\mu A}{l} \qquad (4.3.16)$$

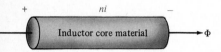

figure 4.3.4
The across variable *ni* and the through variable Φ in a magnetic circuit.

and is called the *reluctance* of the magnetic path along which Φ flows.

Equation (4.3.15) is, thus, a kind of Ohm's law for magnetic circuits, the through variable being magnetic flux Φ and the across variable being *ampere-turns ni* (Figure 4.3.4). If the material that makes up the inductor (through which Φ flows) is air or a vacuum, then $\mathcal{R}$ is a constant and (4.3.15) represents a linear relationship between Φ and *ni*. But, if iron or other ferromagnetic material is used, then μ (and therefore $\mathcal{R}$) is a highly sensitive function of Φ and so the relationship between Φ and *ni* is nonlinear. In such cases the relationship between $v(t)$ and $i(t)$ is *not* given by equation (4.3.1). However, if the Φ versus i relationship is linear (as it is in air or vacuum) then, since Faraday's law

$$v = \frac{dn\Phi}{dt} \qquad (4.3.17)$$

is a linear operation, we see that v and i are linearly related.

Whether or not the inductor is linear, equations (4.3.3) and (4.3.17) are valid. Substituting (4.3.17) into (4.3.3) yields

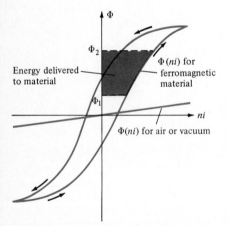

Energy delivered to material

$\Phi(ni)$ for ferromagnetic material

$\Phi(ni)$ for air or vacuum

figure 4.3.5
The Φ versus *ni* plots of a typical ferromagnetic material and for air (or vacuum).

$$W(t_0) = \int_{-\infty}^{t_0} \frac{d(n\Phi)}{dt} i(t)\, dt$$

$$= \int_{\Phi(-\infty)}^{\Phi(t_0)} ni\, d\Phi \qquad (4.3.18)$$

So the energy delivered to an inductor as the flux goes from one value Φ_1 to another Φ_2 is given by the area shown in Figure 4.3.5. Certain ferromagnetic materials exhibit different Φ versus *ni* curves depending on whether the applied *ni* is increasing (energy being stored) or decreasing (energy being delivered back out to the circuit). This phenomenon is shown in Figure 4.3.5 and is called *hysteresis*. If a balanced periodic current, say a sinusoid, is applied to an inductor made from a hysteresis-type material, the entire hysteresis loop is traced out each cycle. One consequence of equation (4.3.18) is that the total area inside the hysteresis loop represents the energy lost (delivered to, but not returned by, the inductor) per cycle. This energy loss heats up the magnetic material from which the inductor core is made.

EXAMPLE 4.3.1

The field structure of a large 1000-turn electromagnet has an inductance of 3 H and carries a current of 6.4 A (dc). Assuming a linear system:

a. Determine the magnetic flux (in webers).
b. How much energy is stored in the magnetic field?

c. Find the expression for the necessary applied voltage if the current is to increase *linearly with time* from 0 to 6.4 A in 0.2 s and then remain constant.

ANS.:

(a) Using equation (4.3.11),

$$\Phi = \frac{Li}{n} = \frac{(3)(6.4)}{1000} = 0.0192 \text{ Wb}$$

(b) Using equation (4.3.6),

$$W = \tfrac{1}{2}Li^2 = \tfrac{3}{2}(6.4)^2 = 61.44 \text{ J}$$

(c) Using equation (4.3.1),

$$v = L\frac{di}{dt}$$

We note that for $t < 0$ and $t > 0.2$ the current is constant (zero in the first interval and 6.4 A in the second), and so during those times $v = 0$. For $0 < t < 0.2$, $i(t) = (6.4/0.2)t$. Therefore we have

$$v = (3)\frac{(6.4)}{0.2}$$

$$= 96 \text{ V} \qquad \text{for} \quad 0 \le t \le 0.2$$

$$v = 0 \qquad \text{for} \qquad t > 0.2$$

The current $i(t)$ in an inductor L at any positive value of time t is given by

$$i(t) = \frac{1}{L}\int_{-\infty}^{t} v(t)\ dt \qquad\qquad (4.3.19)$$

INITIAL-CONDITION
GENERATORS

Or, we may write this as the sum of two integrals:

$$i(t) = \frac{1}{L}\int_{-\infty}^{0} v(t)\ dt + \frac{1}{L}\int_{0}^{t} v(t)\ dt \qquad\qquad (4.3.20)$$

$$i(t) = i(0) + \frac{1}{L}\int_{0}^{t} v(t)\ dt \qquad\qquad (4.3.21)$$

The first term in equation (4.3.21) is a constant. It is the value of the current in the inductor at $t = 0$. The second term may be interpreted as being the current through an inductor having the same inductance value as the original inductor, but which has never experienced current nor voltage prior to $t = 0$. After $t = 0$ it sees the same $v(t)$ as does the original inductor.

Since equation (4.3.21) is the sum of two currents, these two currents can be thought of as being in parallel. Thus, equation (4.3.21) describes the circuit of Figure 4.3.6.[†]

---

† A second form of this initial-condition generator circuit also exists. It contains an impulse voltage source. We will consider this other form in Section 15.6.

figure 4.3.6
(*a*) An inductor with voltage *v*(*t*) and current
i(*t*), where both these terminal variables are
defined to exist over all values of *t*. (*b*) An
equivalent circuit, valid for *t* > 0,
incorporating an initial-condition generator.

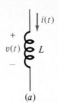

(*a*)

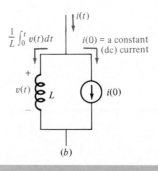

(*b*)

EXAMPLE 4.3.2

A 2-H inductor experiences a *v*(*t*) given in Figure 4.3.7*a*. Find an equivalent circuit valid for $t \geq 0$ that incorporates an initial-condition generator.

ANS.:   For $t > 0$
$$i(t) = \frac{1}{L} \int_0^t v(t)\, dt + i(0)$$

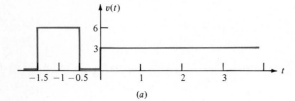

(*a*)

figure 4.3.7
(*a*) Voltage *v*(*t*) applied to the 2-H inductor of Example 4.3.2. (*b*) The equivalent circuit for $t \geq 0$. (*c*) The voltage across the equivalent circuit.

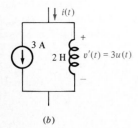

(*b*)

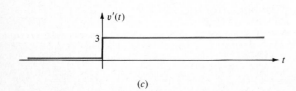

(*c*)

where
$$i(0) = \frac{1}{L} \int_{-\infty}^{0} v(t)\, dt$$

$$= \frac{1}{2} \int_{-1.5}^{-0.5} 6\, dt$$

$$= 3[-0.5 - (-1.5)] = 3 \text{ A}$$

Therefore the equivalent circuit that is valid for $t \geq 0$ is as shown in Figure 4.3.7b, where the voltage across the inductor is $v'(t) = 3u(t)$ as shown in Figure 4.3.7c.

INDUCTORS IN SERIES

We can find the equivalent inductance L_s of two inductors L_1 and L_2 connected in series simply by setting the sum of the voltage across each of them equal to the total voltage across the combination; for example, with reference to Figure 4.3.8:

$$v_1 = L_1 \frac{di_1}{dt} \quad \text{and} \quad v_2 = L_2 \frac{di_2}{dt}$$

where
$$i_1 = i_2 = i_s$$

Thus,
$$v_s = v_1 + v_2$$

$$= L_1 \frac{di_1}{dt} + L_2 \frac{di_2}{dt} = (L_1 + L_2) \frac{di_s}{dt}$$

Since
$$v_s = L_s \frac{di_s}{dt}$$

we see that
$$L_s = L_1 + L_2$$

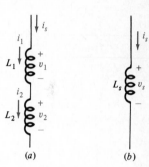

figure 4.3.8
(a) Two inductors in series. (b) Their equivalent inductance.

INDUCTORS IN PARALLEL

The equivalent inductance L_p of two inductors L_1 and L_2 in parallel may be found as follows. With reference to Figure 4.3.9:

$$i_1 = \frac{1}{L_1} \int_{-\infty}^{t} v(t)\, dt$$

$$i_2 = \frac{1}{L_2} \int_{-\infty}^{t} v(t)\, dt$$

and
$$i_p = i_1 + i_2$$

Therefore,
$$i_p = \left(\frac{1}{L_1} + \frac{1}{L_2} \right) \int_{-\infty}^{t} v(t)\, dt$$

Since
$$i_p = \frac{1}{L_p} \int_{-\infty}^{t} v(t)\, dt$$

we see that
$$\frac{1}{L_p} = \frac{1}{L_1} + \frac{1}{L_2}$$

figure 4.3.9
(a) Two inductors in parallel. (b) Their equivalent inductance.

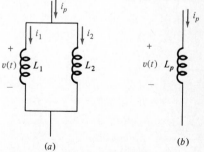

or
$$L_p = \frac{L_1 L_2}{L_1 + L_2}$$

4.4 coupled inductors

Suppose two inductors are situated such that at least some fraction of the magnetic flux produced by each coil links the other coil (Figure 4.4.1). Define:

$$\Phi_{jk} = \text{flux linking winding } j \text{ due to current } i_k \qquad (4.4.1)$$

and also define:

$$L_{jk} = \frac{n_j \Phi_{jk}}{i_k} \qquad (4.4.2)$$

Multiplying through by i_k and taking the derivative, it follows that

$$L_{jk} \frac{di_k}{dt} = n_j \frac{d\Phi_{jk}}{dt} \qquad (4.4.3)$$

Applying equation (4.3.8) of Section 4.3 to coil 1 of Figure 4.4.1 yields

$$v_1 = n_1 \frac{d}{dt} \text{ (total flux linking winding 1)}$$

$$= n_1 \frac{d}{dt} (\Phi_{11} + \Phi_{12})$$

$$= n_1 \frac{d\Phi_{11}}{dt} + n_1 \frac{d\Phi_{12}}{dt}$$

Using equation (4.4.3) above,

$$v_1 = L_{11} \frac{di_1}{dt} + L_{12} \frac{di_2}{dt} \qquad (4.4.4)$$

Similarly for coil 2 we get

$$v_2 = L_{21} \frac{di_1}{dt} + L_{22} \frac{di_2}{dt} \qquad (4.4.5)$$

It turns out that the values of L_{12} and L_{21} for any given geometry are *not* independent of one another. We see this as follows.

Consider the path taken by the flux that links both coils. Φ_{12} passes through this reluctance and Φ_{21} passes through this *same* reluctance. Therefore, using the expression for reluctance, equation (4.3.15) in Section 4.3:

$$\frac{n_1 i_1}{\Phi_{21}} = \frac{n_2 i_2}{\Phi_{12}} \qquad (4.4.6)$$

or, cross multiplying,

$$\frac{n_1 \Phi_{12}}{i_2} = \frac{n_2 \Phi_{21}}{i_1} \qquad (4.4.7)$$

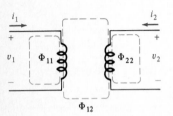

figure 4.4.1
A pair of coupled inductors and the paths of their associated magnetic fields.

Thus, using equation (4.4.2),

$$L_{12} = L_{21} \tag{4.4.8}$$

So call this quantity M, the *mutual inductance*. Thus (4.4.4) and (4.4.5) may be written as:

$$v_1 = L_1 \frac{di_1}{dt} + M \frac{di_2}{dt} \tag{4.4.9}$$

and

$$v_2 = M \frac{di_1}{dt} + L_2 \frac{di_2}{dt} \tag{4.4.10}$$

L_1 and L_2 are always positive quantities, but M may be positive or negative. The *coefficient of coupling* is defined as

$$k = \frac{|M|}{\sqrt{L_1 L_2}} \tag{4.4.11}$$

The largest possible value for k is unity. Thus,

$$k \leq 1 \tag{4.4.12}$$

When $k = 1$, all the flux links both windings, and the two inductors are perfectly coupled.

EXAMPLE 4.4.1

Consider Figure 4.4.2. Two mutually coupled coils have inductances as indicated. A voltage source is connected to coil 1 and coil 2 terminals are left open so no current flows in coil 2. Find the peak voltage $V_{2_{pk}}$ induced in coil 2.

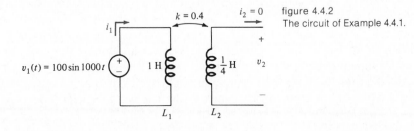

figure 4.4.2
The circuit of Example 4.4.1.

ANS.:

$$v_1 = L_1 \frac{di_1}{dt} + M \frac{di_2}{dt}$$

$$v_2 = M \frac{di_1}{dt} + L_2 \frac{di_2}{dt}$$

$i_2 = 0$ for all t; therefore, $di_2/dt = 0$.

$$v_1 = L_1 \frac{di_1}{dt}$$

$$v_2 = M \frac{di_1}{dt}$$

$$\frac{M}{\sqrt{L_1 L_2}} = k = 0.4 = \frac{M}{\sqrt{1 \times 0.25}}$$

Therefore $M = 0.2$.

$$\frac{v_1}{L_1} = \frac{v_2}{M}$$

or

$$v_2 = \frac{M v_1}{L_1} = \frac{0.2}{1} 100 \sin 1000t$$

thus,

$$V_{2_{\mathrm{pk}}} = 20 \text{ V}$$

DOT NOTATION   Recall that the mathematical form of Faraday's law

$$V = n \frac{d\Phi}{dt} \tag{4.4.13}$$

gives the magnitude, but *not* the polarity of the voltage v. To specify the polarity we need a picture such as Figure 4.4.3.

Similarly, with coupled coils we need a picture of the geometry of the coils in order to tell whether the mutual inductance M is positive or negative. Such a picture might be as shown in Figure 4.4.4.

If v_2 is applied as shown in Figure 4.4.4 such that i_2 is an increasing current, a flux growth $d\Phi/dt$ in the direction shown will result. At the other coil this $d\Phi/dt$ will produce a positive voltage v_1 with the indicated polarity. (See Figure 4.4.3.) Therefore, in this situation M is positive.

An alternative way of indicating this is via a diagram such as Figure 4.4.5. The dots indicate that an applied voltage with its $+$ on the dotted terminal will generate an output voltage with its $+$ on its dotted terminal. (This is true whether v_1 is applied and v_2 is the output or vice versa.) Note that a convenient method for finding the proper dot notation is to use the fact that currents flowing *in* at dotted terminals all *aid* the flux, Φ_1, that links both windings.

In the event that more than two inductors are coupled together a new pair of

figure 4.4.3
The relative polarities (directions) of v and $d\Phi/dt$ as given by Faraday's law.

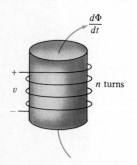

figure 4.4.4
The direction of the increase in magnetic flux due to the application of voltage v_2.

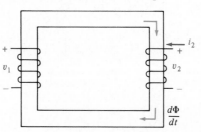

figure 4.4.5
Dot notation used to show the polarity of the output voltage given the polarity of an input voltage.

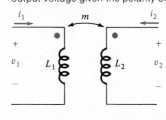

dots (or squares or some other notation) are needed for *each pair* of inductors so coupled.

EXAMPLE 4.4.2

Assign the proper dot notation to the three mutually coupled inductors shown in Figure 4.4.6a.

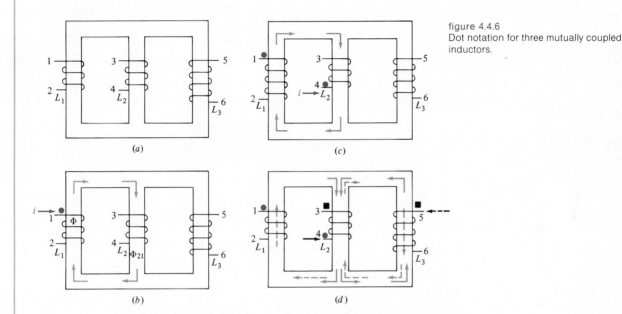

figure 4.4.6
Dot notation for three mutually coupled inductors.

ANS.: First assign dots to L_1 and L_2 as follows:

1. Arbitrarily put a dot on one winding. We choose terminal 1 as shown in Figure 4.4.6b.
2. A current flowing into this dotted terminal will produce a flux Φ as shown in that same figure, upward in leg 1 and down in 2.
3. A current flowing *in* at terminal 4 will *aid* the flux, that links L_1 and L_2, therefore, that is the properly dotted terminal of L_2. See Figure 4.4.6c.

One might be tempted, similarly, to place a third dot on the upper terminal of L_3 (terminal 5) because a current in at 5 will also produce a flux upward in leg A. However, currents flowing into the dotted terminals of L_2 and L_3 do *not* produce fluxes that aid each other in leg B. (See Figure 4.4.6d for proper notation.)

The ideal transformer is a limiting case idealization of the actual transformers that one can get off the shelf of electrical supply houses (Figure 4.4.7). It is ideal (unrealistic) in three specific ways: **IDEAL TRANSFORMER**

1. It does not dissipate energy.
2. It has unity coupling coefficient, $k = 1$ (i.e., no leakage flux).
3. The self-inductances L_1 and L_2 are both infinite.

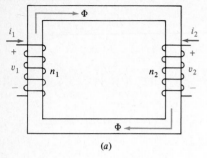

(a)

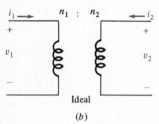

Ideal

(b)

figure 4.4.7
(a) An ideal transformer and *(b)* its
schematic symbol.

Using Faraday's law at both pairs of terminals,

$$v_1 = n_1 \frac{d\Phi_1}{dt} \quad (4.4.14)$$

$$v_2 = n_2 \frac{d\Phi_2}{dt} \quad \left. \right\} \Phi_1 = \Phi_2 \quad (4.4.15)$$

so

$$\boxed{\frac{v_1}{v_2} = \frac{n_1}{n_2}} \quad (4.4.16)$$

Remember that in a magnetic element flux Φ is the through variable and ampere-turns ni is the across variable. Applying this to Figure 4.4.7*a*,

$$n_1 i_1 + n_2 i_2 = \Phi \mathscr{R} \quad (4.4.17)$$

where the reluctance, $\mathscr{R}$, is zero in the ideal case. Thus, $n_1 i_1 + n_2 i_2 = 0$, or

$$\boxed{\frac{i_1}{i_2} = \frac{-n_2}{n_1}} \quad (4.4.18)$$

in Figure 4.4.7.

The fact that L_1 and L_2 are both infinite in the ideal transformer can be shown as follows:

Recall that, in general,

$$v_1 = L_1 \frac{di_1}{dt} + M \frac{di_2}{dt} \quad (4.4.19)$$

$$v_2 = M \frac{di_1}{dt} + L_2 \frac{di_2}{dt} \quad (4.4.20)$$

and try to measure L_1. To do that we might set $di_2/dt = 0$ to eliminate the second term in equation (4.4.19) and apply voltage v_1 from a voltage source. Equation (4.4.18), however, says that therefore di_1/dt will also have to be zero. Clearly, the only way we can have nonzero v_1 is for L_1 to be $= \infty$. A similar argument can be made for L_2 via the second equation above.

It can easily be shown that a consequence of equations (4.4.16) and (4.4.18) is that the ideal transformer dissipates no power. Rearrange the two equations, respectively, as

$$v_1 n_2 = v_2 n_1 \quad (4.4.21)$$

and

$$\frac{i_1}{n_2} = \frac{-i_2}{n_1} \quad (4.4.22)$$

Multiply one by the other:

$$v_1 n_2 \frac{i_1}{n_2} = v_2 n_1 \frac{-i_2}{n_1}$$

$$v_i i_1 = \underbrace{v_2(-i_2)}$$ (4.4.23)

↑ Power *out* of winding 2

Power *in* to winding 1

EXAMPLE 4.4.3

Consider Figure 4.4.8. For $v_1(t) = 10 \sin 377t$, find v_2, i_2, and the power $p(t)$ dissipated in the resistor. The transformer is ideal. The turns ratio is $10:1$.

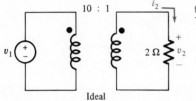

figure 4.4.8
The circuit of Example 4.4.3.

ANS.:

$$\frac{v_2}{n_2} = \frac{v_1}{n_1}$$

$$v_2 = v_1\left(\frac{n_2}{n_1}\right) = \tfrac{10}{10} \sin 377t$$

$$i_2 = \frac{v_2}{R} = \tfrac{1}{2} \sin 377t$$

$$p(t) = [i(t)]^2 R = \tfrac{2}{4} \sin^2 377t = \tfrac{1}{2}[\tfrac{1}{2}(1 - \cos 754t)]$$

$$p(t) = \tfrac{1}{4}(1 - \cos 754t)$$

EXAMPLE 4.4.4

Find $i_a(t)$ if $v_1(t)$ is the same as in Example 4.4.3. See Figure 4.4.9.

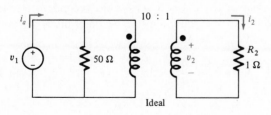

figure 4.4.9
(a) The circuit of Example 4.4.4. (b) The equivalent circuit with R_2 reflected *back* into the input side of the ideal transformer.

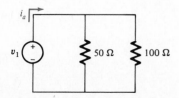

ANS.:

$$\frac{v_2}{v_1} = \frac{n_2}{n_1}$$

$$\frac{i_2}{i_1} = \frac{n_1}{n_2}$$

So

$$\frac{v_2/i_2}{v_1/i_1} = \frac{n_2^2}{n_1^2}$$

$$\frac{R_2}{R_1} = \left(\frac{n_2}{n_1}\right)^2$$

whence

$$R_1 = R_2\left(\frac{n_1}{n_2}\right)^2$$

$$= 1(10)^2 = 100 \ \Omega$$

Thus

$$R_{eq} = \frac{50(100)}{50 + 100} = 33.3 \ \Omega$$

and

$$i_a = \frac{v_1}{R_{eq}}$$

$$= \frac{10}{33.3} \sin 337t = 0.3 \sin 377t$$

Notice that in an ideal transformer one side is the high-voltage, high-turns, low-current side and the other is the low-voltage, low-turns, high-current side. The first of these (high voltage, low current) is called the high-resistance side, and the second is called the low-resistance side. This is because the ratio of a high voltage to a low current is a high resistance and the ratio of a low voltage to a high current is a low resistance. We talk about the value of a resistance as it appears to the circuit on the *other* side of the transformer as its *reflected value*. So, for example, if an R-ohm resistor is connected across the secondary terminals (v_2) of an ideal transformer as shown in Figure 4.4.10a, we compute i_1 as follows. Because of the polarities of i_2 and v_2 we write:

$$-i_2 = \frac{v_2}{R} \tag{4.4.24}$$

From equation (4.4.18),

$$i_1 = \frac{-n_2}{n_1} i_2 \tag{4.4.25}$$

Thus,

$$i_1 = \frac{n_2}{n_1} \frac{v_2}{R} \tag{4.4.26}$$

and from (4.4.16),

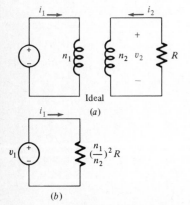

figure 4.4.10
(a) A circuit containing an ideal transformer.
(b) The equivalent input-side circuit showing R reflected back into that side.

$$v_2 = v_1\left(\frac{n_2}{n_1}\right) \tag{4.4.27}$$

Substituting (4.4.27) into (4.4.26) yields

$$i_1 = \frac{v_1}{R}\left(\frac{n_2}{n_1}\right)^2 \tag{4.4.28}$$

or

$$\frac{v_1}{i_1} = R\left(\frac{n_1}{n_2}\right)^2 \tag{4.4.29}$$

Therefore, insofar as source v_1 is concerned, it sees an effective reflected resistance which is $(n_1/n_2)^2$ times the actual R. See Figure 4.4.10*b*.

EXAMPLE 4.4.5

Given the circuit of Figure 4.4.11, find the turns ratio that will result in maximum possible power being delivered to the 4-Ω resistor.

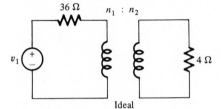

figure 4.4.11
The circuit of Example 4.4.5.

ANS.: The reflected value of the 4-Ω resistor, in order to achieve maximum power transfer from the Thevenin circuit to the left of the transformer, should be 36 (equal to the Thevenin-output resistance). Since

$$R_{eq} = \left(\frac{n_1}{n_2}\right)^2 4$$

and

$$R_{eq} = 36$$

we solve for n_1/n_2:

$$\left(\frac{n_1}{n_2}\right)^2 = \frac{36}{4}$$

$$\frac{n_1}{n_2} = 3$$

ENDtest 2

mechanical translational systems **4.5**

In mechanical translational systems we are interested in the movement of masses and the compression of springs and shock absorbers. The through variable is force and the across variable is velocity. In the standard international (SI) system of dimensions, force is measured in newtons and velocity in meters per second.

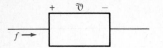

figure 4.5.1
The general mechanical translational element with its associated through variable f = force, and across variable $\mathscr{V}$ = velocity.

Velocity is an across variable because it is an inherently *relative* measure. We cannot speak about the absolute velocity of something. We can only say that something is going a quantified amount faster (and in some specified direction) than some other thing. Perhaps the "other thing" (called the reference or *datum*) is the Earth's surface. But, for example, in the case of an insect flying around inside the cabin of a jet airliner, the reference coordinate system might more reasonably be chosen as being the airframe of the plane.

Power delivered to a mechanical translational element is the product of the across variable times the through variable. Power equals force times velocity in these systems (Figure 4.5.1).

MASS

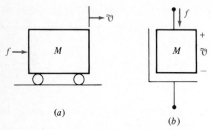

(a)

(b)

figure 4.5.2
(a) The pictorial diagram for a mass M supported by frictionless rollers. (b) The schematic symbol describing the velocity of the mass as being the relative velocity of the upper terminal with respect to the lower (datum) terminal.

The dimensions of mass are kilograms. The mechanical pictorial diagram for a mass in translational motion is shown in Figure 4.5.2a. The mechanical schematic symbol is shown in Figure 4.5.2b. The schematic symbol reminds us that the velocity of a mass must be measured relative to a datum coordinate system (bottom terminal). Thus the *lower terminal of every mass* must be *connected to* the *ground* or reference node in the mechanical circuit.

The defining equation for mass M is given by Newton's law:

$$f = Ma \qquad (4.5.1)$$

or

$$f = M\frac{d\mathscr{V}}{dt} \qquad (4.5.2)$$

where mass is measured in kilograms and

$$a = \frac{d\mathscr{V}}{dt} \qquad (4.5.3)$$

is the acceleration of the mass M. Integrating equation (4.5.2) yields

$$\mathscr{V}(t_0) = \frac{1}{M}\int_{-\infty}^{t_0} f(t)\,dt \qquad (4.5.4)$$

Immediately we see the analogies with the electric capacitor:

$$f \sim i \qquad (4.5.5)$$

$$\mathscr{V} \sim v \qquad (4.5.6)$$

$$M \sim C \qquad (4.5.7)$$

SPRING

The defining equation for the translational spring is

$$x = Kf \qquad (4.5.8)$$

(see Figure 4.5.3) where x is the amount of compression (or expansion) from equilibrium in meters, f is the force of compression or expansion in newtons, and K is the *compliance*† of the spring. In terms of the across variable $\mathscr{V}$, equation (4.5.8) is written as

† Many authors use the *stiffness* of the spring rather than compliance. These are, of course, inverses of one another. Thus, stiffness is analogous to $1/L$.

$$f(t_0) = \frac{1}{K} \int_{-\infty}^{t_0} \mathcal{V} \, dt \tag{4.5.9}$$

In other words, the force being applied to the spring is proportional to the time integral of the velocity of one end of the spring relative to the other end of the spring. This integral, of course, is the distance the spring has been compressed. Taking the derivative of (4.5.9) yields

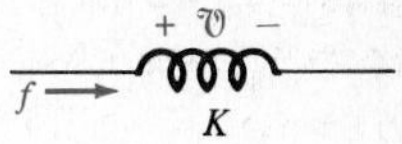

$$\mathcal{V} = K \frac{df}{dt} \tag{4.5.10}$$

figure 4.5.3
The translational spring.

Equations (4.5.9) and (4.5.10) remind us of the defining equations for the electric inductor:

$$f \sim i \tag{4.5.11}$$

$$\mathcal{V} \sim v \tag{4.5.12}$$

$$K \sim L \tag{4.5.13}$$

Damper

The defining equation for the damper (or *dashpot* or *shock absorber*) is

$$f = D\mathcal{V} \tag{4.5.14}$$

where D is the *damping coefficient* of the damper and is measured in newton-seconds per meter—which should be obvious from equation (4.5.14). The schematic symbol for the dashpot is shown in Figure 4.5.4.

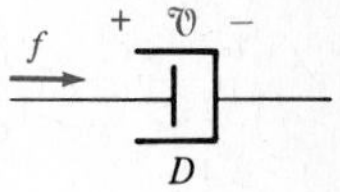

If a compression force is applied to a damper, it will compress with a *velocity* that is proportional to the applied force. If the force is then removed, the overall length of the damper will not change. It will not return to its original length. (That is what a spring would do—not a damper.) A damper may be thought of as a leaky piston moving within a cylinder filled with viscous fluid.

figure 4.5.4
The translational damper.

The damper is often used to model friction. That is when, say, an object is moving through air or other fluid, or when a mass is sliding on a plane. In such cases we assume that the frictional force is proportional to velocity, so the faster the object moves, the more force is needed to overcome the frictional effect. Such friction is called *viscous friction*. It is a linear relationship between force and velocity.

Unfortunately, in many real situations this linear assumption is not completely valid. There are many nonlinear forms of friction that are observed. Two that occur often are *stiction* and *coulomb friction*. Stiction is the name given to the fact that, in order to get a mass that is sitting still to move, a force has to be applied to break the stickiness between the mass and the plane. Once the mass begins to move, however slowly, the stiction force drops to zero. Coulomb friction is a frictional force that seems to be unaffected by the object's velocity. So, when the frictional force is a constant, independent of the object's velocity, it is called coulomb friction.

A LEVER

The lever shown in Figure 4.5.5 is analogous to the electric ideal transformer (and is similarly useful for maximim power transfer applications). To see this analogy, consider the angle θ:

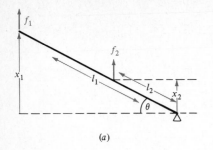

(a)

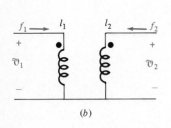

(b)

figure 4.5.5
(a) A lever and (b) its schematic symbol.

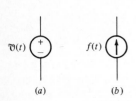

(a)   (b)

figure 4.5.6
(a) An independent velocity source. (b) An
independent force source.

$$\sin \theta = \frac{x_1}{l_1} = \frac{x_2}{l_2} \tag{4.5.15}$$

or

$$x_1 l_2 = x_2 l_1 \tag{4.5.16}$$

Taking the derivative of (4.5.16),

$$\mathcal{V}_1 l_2 = \mathcal{V}_2 l_1 \tag{4.5.17}$$

or, finally,

$$\boxed{\frac{\mathcal{V}_1}{\mathcal{V}_2} = \frac{l_1}{l_2}} \tag{4.5.18}$$

If the lever is lossless (no friction in the joint), then whatever power is delivered to the lever must come out. Thus we can say the sum of all powers in must equal zero. Thus

$$p_{\text{in}} = f_1 \mathcal{V}_1 + f_2 \mathcal{V}_2 = 0 \tag{4.5.19}$$

so

$$\frac{\mathcal{V}_1}{\mathcal{V}_2} = -\frac{f_2}{f_1} \tag{4.5.20}$$

or

$$\boxed{\frac{f_1}{f_2} = -\frac{l_2}{l_1}} \tag{4.5.21}$$

Equations (4.5.18) and (4.5.21) are analogous to the current and voltage equations in the ideal transformer with

$$l_1 \sim n_1 \tag{4.5.22}$$

$$l_2 \sim n_2 \tag{4.5.23}$$

Finally, force, and velocity sources are analogous to current and voltage sources, respectively (Figure 4.5.6).

We have not yet reached the point of solving electric circuits consisting of R's, L's, and C's for node voltages and element currents. However, we will soon be doing just that. All the techniques that we used in analyzing resistive circuits (Kirchhoff's laws, Thevenin, etc.) will be useful to us. If we can reduce a mechanical pictorial diagram to a schematic diagram with nodes and meshes, etc., then the *same techniques* that we will use to solve *RLC* circuits will enable us to solve mechanical systems for their velocities and internal forces.

Let us then investigate how to get the mechanical schematic diagram for any translational system. The method for doing this is quite simple:

1. First draw all masses with one terminal connected to the ground (reference) node.
2. Connect all other elements to each upper mass terminal that exert a force on that mass.

EXAMPLE 4.5.1

Obtain the mechanical schematic for the system shown in Figure 4.5.7a.

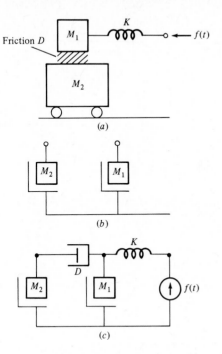

figure 4.5.7
(a) The mechanical system of Example 4.5.1. (b) First one terminal of every mass is connected to the datum. (c) Connecting the remaining elements.

ANS.:   First draw in both masses with one terminal of each connected to the datum (Figure 4.5.7b).
    Since the two masses are connected by the friction element D, connect the top terminals of the two masses by a damper element D. The force source $f(t)$ acts on one end of spring K and the other end of K is connected to M_1. See Figure 4.5.7c.

mechanical rotational systems   **4.6**

In rotational systems the across variable is angular velocity ω. The through variable is torque τ. The SI units for torque are newton-meters (N · m), and for angular velocity, radians per second. The elements inertia J, spring K, and damper D and their schematic symbols are shown in Figure 4.6.1. The defining relationships are:

* Inertia $\tau = J \dfrac{d\omega}{dt}$ or $\omega(t_0) = \dfrac{1}{J} \displaystyle\int_{-\infty}^{t_0} \tau \, dt$ (4.6.1)

* Spring $\omega = K \dfrac{d\tau}{dt}$ or $\tau(t_0) = \dfrac{1}{K} \displaystyle\int_{-\infty}^{t_0} \omega \, dt$ (4.6.2)

* Damper $\tau = D\omega$ or $\omega = \dfrac{1}{D} \tau$ (4.6.3)

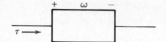

figure 4.6.1
The rotational mechanical elements: (a)
inertia, (b) spring, and (c) damper.

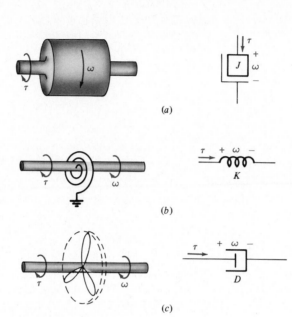

(a)

(b)

(c)

GEARS The gear train is the mechanical rotational equivalent of the ideal electric transformer (see Figure 4.6.2). From geometry we see that the gear with the smaller
number of teeth (smaller diameter) must rotate faster than the big gear. Specifically,

$$\frac{\omega_1}{\omega_2} = \frac{n_2}{n_1} \tag{4.6.4}$$

Since we are assuming idealized gears (no mass to accelerate, no friction to over-

figure 4.6.2
(a) The pictorial diagram of a gear train and
(b) its schematic symbol.

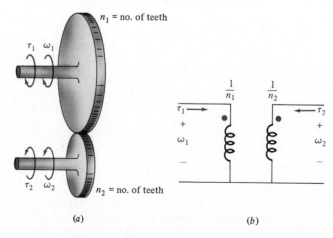

(a)

(b)

come, etc.), no mechanical power is lost in this gear train. Therefore the sum of the powers in equals zero:

$$\tau_1 \omega_1 + \tau_2 \omega_2 = 0 \qquad (4.6.5)$$

or

$$\frac{\tau_1}{\tau_2} = -\frac{\omega_2}{\omega_1} \qquad (4.6.6)$$

Inserting (4.6.4) into (4.6.5) yields

$$\frac{\tau_1}{\tau_2} = -\frac{n_1}{n_2} \qquad (4.6.7)$$

Equations (4.6.4) and (4.6.7) define the ideal gear train. The common datum reference for both the primary and secondary sides of the schematic symbol shown in Figure 4.6.2b implies that both ω_1 and ω_2 are measured with respect to the *same* fixed reference system.

TRANSDUCERS

A system that transforms one kind of energy into another kind (often for measurement purposes) is called a *transducer*. For example, one type of speedometer (see Figure 4.6.3) transforms rotational mechanical energy into a sinusoidal electric voltage, the magnitude of which can be directly measured by an ac voltmeter (the face has been changed to read "mph" instead of "volts"). The generation of this voltage can be described as follows.

From Faraday's law

$$v = \frac{d(n\Phi)}{dt} \qquad (4.6.8)$$

we obtain

$$v = n \frac{d\Phi}{d\theta} \frac{d\theta}{dt} \qquad (4.6.9) \quad \omega = \frac{d\theta}{dt}$$

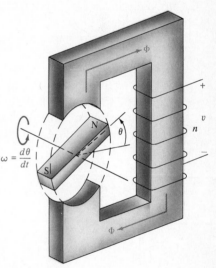

figure 4.6.3
An electromechanical transducer.

where θ is the instantaneous angle of the rotating permanent magnet whose flux is Φ_m, measured from the horizontal. If the rotational speed ω is a constant, then as a first-order approximation we can write

$$\Phi = \Phi_m \sin \theta \qquad (4.6.10)$$

Substituting

$$\omega = \frac{d\theta}{dt} \qquad (4.6.11)$$

and (4.6.10) into (4.6.8) we have

$$v = n\omega\Phi_m \cos \theta \qquad (4.6.12)$$

then using

$$\theta = \omega t \qquad (4.6.13)$$

we have

$$v = n\omega\Phi_m \cos \omega t \qquad (4.6.14)$$

DC MOTOR

The ideal dc electric motor is a device that takes in power in electrical form and emits that power in mechanical form. So with reference to Figure 4.6.4,

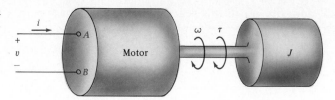

figure 4.6.4
The system of Example 4.6.1.

$$vi = \tau\omega \tag{4.6.15}$$

where
$$v = k\omega \tag{4.6.16}$$

and
$$ik = \tau \tag{4.6.17}$$

The constant k is the same in both equations (4.6.16) and (4.6.17).

EXAMPLE 4.6.1

An ideal motor (lossless) acts as an electromechanical transducer, interconnecting an electrical system and a mechanical element $J = 0.0678$ kg · m². When 120 V is applied to the motor, it runs at a speed of 1800 rpm (188.5 rad/s).
a. Find the electromagnetic constant k.
b. If the shaft torque produced by the motor is 6.366 N · m, find the input current i in amperes.
c. Write an expression relating τ, ω, and J.
d. The J will be reflected back through the transducer, and it along with the electromechanical constant k will control the relationship between v and i. Derive the expression relating v, i, J, and k.
e. A single electric element which gives the same v versus i relationship may be substituted for everything to the right of terminals A and B. What kind of an element will this be and what will be its numerical value? Show units.

ANS.:

(a) $v = k\omega$

$$k = \frac{v}{\omega} = \frac{120}{188.5} = 0.6366 \text{ V/rad}$$

(b) $\tau = ki \quad i = \frac{\tau}{k} = \frac{6.366}{0.6366} = 10 \text{ A}$

(c) $\tau = J\frac{d\omega}{dt}$

(d) $ki = J\frac{d}{dt}\left(\frac{v}{k}\right) \quad i = \frac{J}{k^2}\frac{dv}{dt}$

(e) $i = C\frac{dv}{dt}$

The electric element is a capacitor. Its value is $C = J/k^2 = 0.0678/(0.6366)^2 = 0.1673$ F

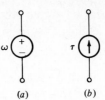

figure 4.6.5
Independent rotational sources: (*a*) velocity and (*b*) torque.

(*a*)   (*b*)

The independent sources in mechanical rotating systems are of the form shown in Figure 4.6.5, and the procedure for constructing schematic diagrams is the same as for translational systems.

liquid systems 4.7

In liquid systems we are interested in the flow of incompressible fluids through pipes, tanks, etc. The through variable is the volume flow Q of the fluid in cubic meters per second (m³/s). The across variable is pressure p in newtons per square meter (N/m²).† The flow of fluids is generally a highly nonlinear mechanism, and so our attempts to model these systems with linear elements are simply a first-order approximation of the real thing.

The defining equation for fluid resistance is

$$p = RQ \qquad (4.7.1)$$

LIQUID RESISTANCE

Although the actual driving force in fluid systems is a difference in pressure, $p_1 - p_2$ (in newtons per square meter), it is more convenient to talk about *pressure head h* (in meters) (Figure 4.7.1). The pressure head h that corresponds to a given pressure p is the height of fluid (in an open container) that would produce pressure p at its bottom (see Figure 4.7.2). To convert from one measure to the other, we need the relationship between p and h. Consider the total downward

† In actuality, the officially correct unit for pressure is the *pascal*.

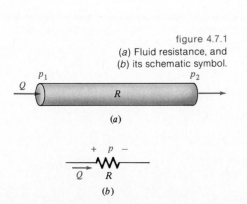

figure 4.7.1
(*a*) Fluid resistance, and
(*b*) its schematic symbol.

(*a*)

(*b*)

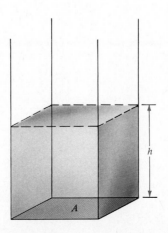

figure 4.7.2
A column of liquid with pressure *head ḣ*.

force f on the bottom of the column in Figure 4.7.2. This total force is the product of the pressure (in newtons per square meter) times the area A of the bottom.

$$f = Ap \tag{4.7.2}$$

Newton's law tells us that force f is equal to the total mass of the fluid times the acceleration due to gravity g. Thus,

$$f = Mg \tag{4.7.3}$$

Combining (4.7.2) and (4.7.3), we have

$$Ap = Mg \tag{4.7.4}$$

The total mass M can be written as the product of the volume occupied (in cubic meters) times the density ρ (in kilograms per cubic meter):

$$Ap = Ah\rho g \tag{4.7.5}$$

where Ah is the volume. Therefore,

$$p = \rho gh \tag{4.7.6}$$

or
$$h = \frac{1}{\rho g}\, p \tag{4.7.7}$$

It turns out that, even though p and h are both equally valid choices for the across variable in fluid systems, using the pressure head h leads to certain simplifications.

LIQUID CAPACITANCE   The volumic flow Q of any liquid (by analogy to the definition of electric current as being the time derivative of charge—coulombs per second) into an open storage tank (capacitor) is

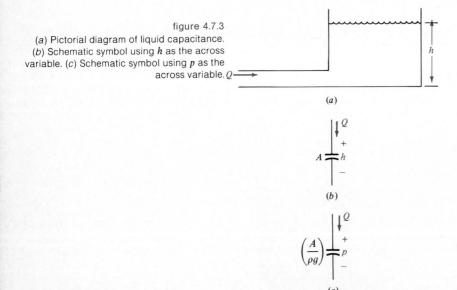

figure 4.7.3
(a) Pictorial diagram of liquid capacitance.
(b) Schematic symbol using h as the across variable. (c) Schematic symbol using p as the across variable.

$$Q = \frac{d}{dt}(\text{volume}) \tag{4.7.8}$$

$$= \frac{d}{dt}(Ah) \tag{4.7.9}$$

which, assuming a constant cross-sectional area A, is

$$Q = A\frac{dh}{dt} \tag{4.7.10}$$

Thus the constant relating the through variable to the time derivative of the across variable is[†] the

$$\text{Liquid capacitance} = A \tag{4.7.11}$$

(see Figure 4.7.3).

In order to accelerate a mass of fluid, we must apply a force to it. We can do this by making the pressure at one end of a slug of the fluid different from what it is at the other end (see Figure 4.7.4). In other words, Newton's law

$$f = Ma \tag{4.7.12}$$

becomes

$$p_1 A - p_2 A = M\frac{d\mathcal{V}}{dt} \tag{4.7.13}$$

where

$$M = \rho Al \tag{4.7.14}$$

and since

$$Q = A\mathcal{V} \tag{4.7.15}$$

we can write

$$\frac{dQ}{dt} = A\frac{d\mathcal{V}}{dt} \tag{4.7.16}$$

so (4.7.13) becomes

$$(p_1 - p_2)A = \rho Al\frac{1}{A}\frac{dQ}{dt} \tag{4.7.17}$$

or

$$p_1 - p_2 = \left(\frac{\rho l}{A}\right)\frac{dQ}{dt} \tag{4.7.18}$$

Letting $p = p_1 - p_2$, the pressure drop across the inertial mass of length l, we have

$$p = \left(\frac{\rho l}{A}\right)\frac{dQ}{dt} \tag{4.7.19}$$

Using equation (4.7.6) in (4.7.19) results in

$$h = \left(\frac{l}{Ag}\right)\frac{dQ}{dt} \tag{4.7.20}$$

LIQUID INERTIA

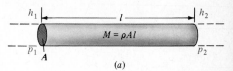

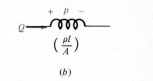

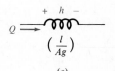

figure 4.7.4
(a) Pictorial diagram of liquid inertial element. (b) Schematic symbol using across variable p. (c) Schematic symbol using across variable h.

[†] If p (newtons per square meter) is used as the across variable, then since $p = h\rho g$, $Q = (A/\rho g)(dp/dt)$ and the capacitance is a function of the density ρ of the fluid.

so equation (4.7.20) implies an analogy between liquid inertia and electric inductance L.

$$J = \left(\frac{l}{Ag}\right) \sim L \tag{4.7.21}$$

LIQUID SYSTEM TRANSFORMERS

There are several forms in which the liquid system analog of the ideal transformer can be found. For example, consider the twin-turbine system of Figure 4.7.5. In this device (which we assume to be lossless) we assume identical propellers that are perfectly coupled to the surrounding fluid. From this geometry we deduce

$$\frac{Q_1}{n_2} = \frac{-Q_2}{n_1} \tag{4.7.22}$$

Because the unit is lossless,

$$Q_1 p_1 + Q_2 p_2 = 0 \tag{4.7.23}$$

so

$$\frac{Q_1}{-Q_2} = \frac{p_2}{p_1} = \frac{h_2}{h_1} = \frac{n_2}{n_1} \tag{4.7.24}$$

or

$$\frac{p_1}{p_2} = \frac{h_1}{h_2} = \frac{n_1}{n_2} \tag{4.7.25}$$

Another liquid system transformer is of the form shown in Figure 4.7.6. In this device

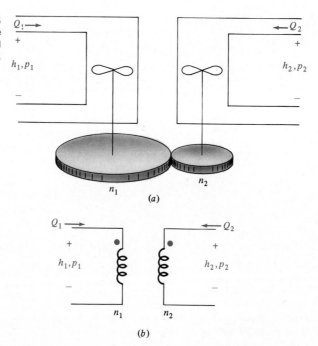

figure 4.7.5
A twin-turbine liquid system (analog to the ideal transformer): (a) pictorial diagram and (b) schematic symbol.

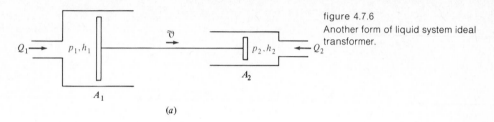

figure 4.7.6
Another form of liquid system ideal
transformer.

(a)

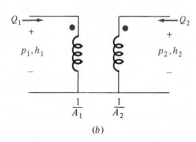

(b)

$$Q_1 = A_1 \mho \qquad (4.7.26)$$

and
$$-Q_2 = A_2 \mho \qquad (4.7.27)$$

where $\mho$ is the velocity of the connected pistons. Assuming no losses, the sum of the powers delivered is zero.

$$p_1 Q_1 + p_2 Q_2 = 0 \qquad (4.7.28)$$

so
$$\frac{-Q_2}{Q_1} = \frac{p_1}{p_2} \qquad (4.7.29)$$

Substituting (4.7.26) and (4.7.27) into (4.7.29) we have

$$\frac{-Q_2}{Q_1} = \frac{A_2}{A_1} \qquad (4.7.30)$$

and
$$\frac{p_1}{p_2} = \frac{A_2}{A_1} \qquad (4.7.31)$$

or
$$\frac{p_1}{1/A_1} = \frac{p_2}{1/A_2} \qquad (4.7.32)$$

Thus $1/A_1$ is analogous to n_1, and $1/A_2$ is analogous to n_2. See Figure 4.7.6b.

EXAMPLE 4.7.1

A mechanical-liquid-mechanical machine is shown in Figure 4.7.7a. Assuming a lossless system, find the mechanical schematic diagram of this device.

ANS.:
$$Q = \mho_1 A_1 \qquad (4.7.33)$$

and
$$Q = -\mho_2 A_2 \qquad (4.7.34)$$

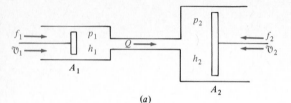

figure 4.7.7
(a) Pictorial diagram of the device in Example 4.7.1. (b) The equivalent schematic diagram.

(a)

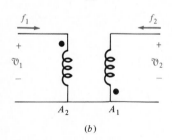

(b)

Therefore,
$$\mathcal{V}_1 A_1 = -\mathcal{V}_2 A_2 \tag{4.7.35}$$

or
$$\frac{-\mathcal{V}_2}{\mathcal{V}_1} = \frac{A_1}{A_2} \tag{4.7.36}$$

Since the sum of all powers entering the device must be zero,
$$f_1 \mathcal{V}_1 + f_2 \mathcal{V}_2 = 0 \tag{4.7.37}$$

or
$$\frac{\mathcal{V}_2}{\mathcal{V}_1} = \frac{f_1}{-f_2} \tag{4.7.38}$$

and using (4.7.36) we can write (4.7.38) as

$$\frac{f_1}{f_2} = \frac{A_1}{A_2} \tag{4.7.39}$$

Equations (4.7.38) and (4.7.39) are analogous to the voltage and current relationships in an ideal electric transformer. The mechanical schematic equivalent is shown in Figure 4.7.7b. A_2 is analogous to the turns on the primary winding n_1, and A_1 is analogous to n_2.

EXAMPLE 4.7.2

An open-top tank is filled with water at a rate Q (cubic meters per minute) specified in Figure 4.7.8. Assuming the tank is empty at $t = 0$,
a. Sketch the h versus t curve from $t = 0$ to $t = 6$ min.
b. What will be the maximum h of the water level? What will be the water level at $t = 6$ min?
c. If the tank height is reduced to 2 m, the water will overflow during a portion of the 6-min interval. Find the water level at $t = 6$ min for this shorter tank.

ANS.:   (a) For $0 < t < 2$ $\qquad\qquad\qquad\qquad Q = a + bt$

For $t = 0$ $\qquad\qquad\qquad\qquad Q = 0 = a + b(0) \qquad a = 0$

For $t = 2$ $\qquad\qquad\qquad\qquad Q = +1 = 0 + b(2) \qquad b = 0.5$

So $\qquad\qquad\qquad\qquad\qquad Q_{0-2} = 0.5t \quad$ for $\; 0 < t < 2$

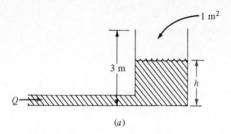

1 m²

3 m

h

Q

(a)

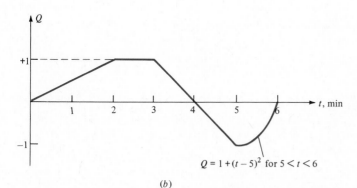

$Q = 1 + (t - 5)^2$ for $5 < t < 6$

(b)

$$Q_{2-3} = +1 \qquad \text{for} \quad 2 < t < 3$$

and

$$Q_{3-5} = a + bt$$

where, for $t = 3$

$$Q = 1 = a + b(3) \tag{4.7.40}$$

for $t = 5$

$$Q = -1 = a + b(5) \tag{4.7.41}$$

Subtracting (4.7.41) from (4.7.40),

$$+2 = 0 + b(-2)$$

Thus,

$$b = -1$$

For $t = 3$

$$Q = +1 = a + (-1)(3)$$

$$a = +4$$

$$Q_{3-5} = 4 - t \qquad \text{for} \quad 3 < t < 5$$

And, as given,

$$Q_{5-6} = -1 + (t - 5)^2 \qquad \text{for} \quad 5 < t < 6$$

Capacitance of the tank $= A = 1 \text{ m}^2$. For $0 < t < 2$

$$h = \frac{1}{C} \int_0^t Q \, dt + h_0 = 1 \int_0^t 0.5t \, dt + 0$$

$$= \frac{0.5t^2}{2} \bigg|_0^t = 0.25t^2$$

At $t = 2$

$$h = 0.25(2)^2 = 1 \text{ m}$$

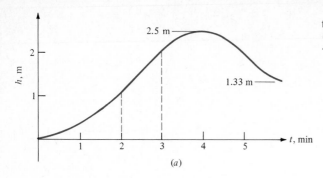

figure 4.7.9
(*a*) The height *h* (pressure head) of the column in Example 4.7.2. (*b*) Height *h* when the tank is only 2 m high.

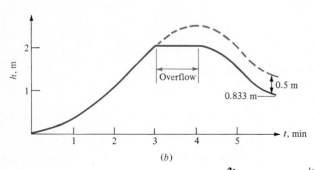

For $2 < t < 3$

$$h = \int_2^t 1\, dt + h_2 = t\Big|_2^t + 1$$

$$= t - 2 + 1 = t - 1 \qquad \text{(straight line)}$$

At $t = 3$
$$h_3 = 3 - 1 = 2 \text{ m}$$

For $3 < t < 5$
$$h = \int_3^t (4 - t)\, dt + h_3 = \left(4t - \frac{t^2}{2}\right)\Big|_3^t + 2$$

$$h = 4t - \frac{t^2}{2} - 12 + 4.5 + 2 = 4t - \frac{t^2}{2} - 5.5$$

$$Q = C\frac{dh}{dt}$$

(*b*) When $Q = 0$, the slope of the *h* curve will be 0 and *h* will be at a peak.

At $t = 4$
$$Q = 0; \qquad h = 4(4) - \frac{4(4)}{2} - 5.5 = 2.5 \text{ m}$$

At $t = 5$
$$h = 20 - \tfrac{25}{2} - 5.5 = 2.0 \text{ m}$$

For $5 < t < 6$
$$h = \int_5^t [-1 + (t - 5)^2]\, dt + h_5$$

$$= \int_5^t (-1 + t^2 - 10t + 25)\, dt + 2$$

$$= \left(-t + \frac{t^3}{3} - \frac{10t^2}{2} + 25t\right)\Big|_5^t + 2$$

$$= \frac{t^3}{3} - 5t^2 + 24t - 34.67$$

When $t = 6$
$$h = \frac{6^3}{3} - 5(6)^2 + 24(6) - 34.67$$

$$= 1.33 \text{ m}$$

(c) If the tank is only 2 m tall, it will start to overflow when $h = 2$ m at $t = 3$ min. It will continue to run over at a constant $h = 2$ m as long as Q is positive. When Q turns negative or water starts to run back out the pipe, h will start to decrease at the same rate at which it decreased before.

The last part of the curve from $t = 4$ to $t = 6$ will just be decreased 0.5 m. Thus $h_6 = 1.333 - 0.5 = 0.833$ m.

thermal systems **4.8**

In thermal systems the across variable is temperature θ measured in degrees kelvin (K) which is the absolute version of the Celsius or centigrade temperature scale (that is, $0°C = 273$ K). The through variable is heat flow in joules per second (J/s). If an object contains heat, its molecules are vibrating around their geometric average positions. The larger the amplitude of these vibrations, the hotter the temperature of the object. Heat is transferred in three different ways: conduction, convection, and radiation.

✳ Conduction When two objects touch each other, molecular vibrational energy from the hotter body is transferred to the cooler one. This effect is nicely approximated by a linear relationship, i.e., the flow Q of heat energy is proportional to the temperature difference θ. Thus we have (see Figure 4.8.1)

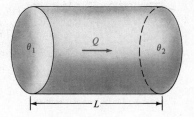

$$Q = \frac{kA}{L} \theta \qquad (4.8.1)$$

where k is the *thermal conductivity*, A is the cross-sectional area, and L is the length of the thermal path. So

figure 4.8.1
Conductive heat transfer.

$$\theta = \frac{L}{kA} Q \qquad (4.8.2)$$

is a linear relationship analogous to Ohm's law except that it is valid for this type of heat-transfer mechanism. The constant of proportionality is the thermal resistance R:

$$R = \frac{L}{kA} \qquad (4.8.3)$$

This expression is identical† to the corresponding one for electric resistance except, of course, in the electrical case it is the electric (rather than the thermal) conductivity of the material that is used.

† In fact, as a rule of thumb, materials that are good thermal conductors are most often also good electric conductors.

✳ Convection   Convection is the mechanism whereby heat is transferred by rising hot air or other fluid. A linear relationship similar to (4.8.2) is also valid as a first-order approximation for this type of heat flow. The proportionality constant R is given by

$$R = \frac{1}{hA}$$

where h is the *convection coefficient*.

✳ Radiation   The transfer of energy by the transmission of infrared (lower in frequency than the lowest-frequency red visible light) or other electromagnetic waves is called *radiation*. Hot objects all emit, to some extent, such radiation. No physical material is needed in the path in order to transfer energy by this technique. It is the mechanism by which the sun warms the Earth across millions of miles of vacuum. The relationship between θ and Q for this mechanism is highly nonlinear.

$$Q = \sigma A \varepsilon (\theta_1^4 - \theta_2^4)$$

where $\sigma = 5.67 \times 10^{-8}$ (the Stefan-Boltzman constant)
ε = emissivity of the surface $(0 < \varepsilon < 1)$

THERMAL CAPACITANCE   Generally, objects that are made from dense materials have the ability to store heat. In the ideal case,

$$Q = C \frac{d\theta}{dt}$$

where C = thermal capacitance
$C = \mathrm{M}c_p$
M = mass (kg) of material
c_p = specific heat at constant pressure of the material

Caution: Especially in thermal systems we must carefully note any assumptions that are made. Often, our assumption of a *lumped* rather than a *distributed* system may be a poor one in thermal system work. This is because the velocity with which the effects of temperature changes propagate through many thermal systems is extremely slow. On the other hand, in electrical systems the analogous effects move along throughout the system at almost the speed of light. Distributed systems require, for their analysis, the use of *partial differential equations* the solution of which is an art that is far beyond the scope of this text.

There is one other important difference between thermal systems and all other types: Heat flow corresponds directly (and dimensionally, joules per second) to power. Therefore, the product of the across variable times the through variable is meaningless in thermal systems. The through variable is, itself, power.

Finally, there is no mechanism that corresponds to the electric inductor or the mechanical spring in thermal systems.

EXAMPLE 4.8.1

A refrigerator has a total thermal resistance (inside to outside) of 0.8 (K · s/J).

a. If the outside of the box is maintained at 300 K and the inside is maintained at 270 K, what is the heat flow into the box?

b. If the refrigeration unit is 20 percent efficient and runs one-third of the time, how much electric power does it use when it is running?

ANS.: (a) Thermal resistance R relates heat flow Q and temperature difference by:

$$\theta = QR$$

$$\theta = 300 - 270 = 30 \text{ K}$$

$$Q = \frac{\theta}{R} = \frac{30 \text{ K}}{0.8 \text{ K} \cdot \text{s/J}} = 37.5 \text{ J/s} = 37.5 \text{ W}$$

(b) 37.5 W is needed continuously by the box.

$$\frac{37.5}{0.20} \text{ W} = 187.5 \text{ W of electric power needed continuously by the refrigeration unit}$$

$$187.5 \times 3 = 562.5 \text{ W needed while unit is running (one-third duty cycle operation)}$$

summary 4.9

In this chapter we have presented descriptions of the two other basic two-terminal electrical elements: the inductor and the capacitor. We saw that their fundamental definitions indicate an integral/derivative relationship between their across (voltage) and through (current) variables. The ideal transformer is a perfectly coupled set of inductors each having infinite values of inductance.

The notion was introduced that other systems (in other areas of science and engineering) can be analyzed via the same techniques that we use for RLC electrical systems. This is because they have mathematical relationships between their across and through variables that are analogous to those of R's, L's, and C's. There is no fundamental reason why electrical engineers should not be able to do vibration analysis of mechanical systems, nor any reason why mechanical engineers should not understand the solution of RLC circuits! Because of their resultant *wide applicability*, the importance of the analysis techniques we will present in the remainder of this text is increased by an order of magnitude over what it would be if they were only applicable to some small, narrow area of engineering.

problems

1. Given $i_C(t) = (0.5t - 1)[u(t - 2) - u(t - 4)]$ in a 1-F capacitor, what is $v_C(5)$?

2. The current through a 1-F capacitor is $i_C(t) = \delta(t - 1)$. What is $v_C(t)$?

3. The voltage across a 1-F capacitor is $v_C(t) = r(t) = tu(t)$. What is $i_C(t)$?

4. Given a $\frac{1}{2}$-F capacitor with $i_C(t) = 10e^{-3t}u(t)$, if $v_C(0-) = 10$ V, find $v_C(t)$ for $t > 0$.

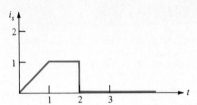

figure P4.11

5. Sketch the voltage $v_C(t)$ across a 1-F capacitor when $i_C(t) = u(t - 1) - 2u(t - 2) - \delta(t - 3)$.

6. Write a single expression for $v_C(t)$ in problem 5.

7. The voltage across a 1-F capacitor is $v_C(t) = 3 + 2u(t)$. (a) Evaluate $v_C(0-)$ and $v_C(0+)$, the value of the voltage just before and just after $t = 0$. (b) Find $i_C(t)$.

8. What is the maximum amount of energy stored in a 1-μF capacitor if $v_C(t) = 100\sqrt{2} \cos 5t$?

9. The current in a 0.5-F capacitor consists of a sequence of two pulses: $i_C(t) = u(t - 1) - u(t - 2) + 2[u(t - 3)u(4 - t)]$. Sketch the voltage across the capacitor. Specifically, what are the values of $v_C(3)$ and $v_C(5)$?

10. Find and sketch $v_C(t)$ across a 2-F capacitor if $v_C(0-) = 0$ and $i_C(t) = tu(t) - (t - 1)u(t - 1)$. Label each segment of the plot with a valid mathematical expression. Also sketch $i_C(t)$.

11. For the capacitive circuit shown in Figure P4.11, sketch the voltage $v_C(t)$ from $t = 0$ to $t = 3$ s. What is the value of v_C at $t = 1, 2,$ and 3 s?

12. Find the value of the voltage across a 1-F capacitor at $t = 7.4$ s if $i_C(t) = [u(t)u(\pi - t)] \sin t$.

13. Given a 2-F capacitor, find the resulting current if $v_C(t)$ is (a) $2u(t)$, (b) $2u(t - 1)$, (c) $4 \cos (377t + 30°)$, (d) $4 u(t) \cos (377t + 30°)$, and (e) $(6t - 10)u(t - 2)$.

14. Repeat problem 13 for a 2-H inductor (all v_C's become v_L's).

15. The voltage across a 0.2-F capacitor is given as $v_C(t) = 10r(t) - 20r(t - 1) + 10r(t - 3)$. Sketch the resulting current, power, and stored energy curves versus time. Show maximum values on each plot. Label axes, and show curvatures carefully.

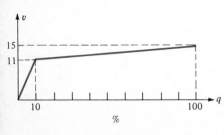

figure P4.16

16. A 45 A·h lead-acid storage battery has a terminal voltage versus charge characteristic as shown in Figure P4.16. (a) Design an equivalent circuit for this battery that is valid for the range 0 to 10 percent of full charge. (b) Repeat part (a) for the range 10 to 100 percent. (c) How much energy is stored in the battery when it contains 10 percent of a full charge? (d) How much energy is stored when the battery is fully charged? (e) If energy is removed from this battery at a constant rate of 25 W, how long will it take to go from being fully charged down to the 10 percent level?

17. If $i_L(t) = u(t) - u(t - 3)$ in a 1-H inductor, what is $v_L(t)$ for all t? Sketch it.

18. Given a $\frac{1}{2}$-H inductor with $i_L(t) = 10e^{-3t}u(t)$, find $v_L(t)$ for all t.

19. A current $i(t)$ is given as $i(t) = t[u(t) - u(t - 2)]$. (a) Sketch this waveform. (b) What voltage will appear across a 2-H inductor if it carries this $i(t)$? Sketch it.

20. The current in a series connection of a 10-Ω resistor and a 3-H inductor is $i(t) = (1 - t^2)[u(t + 1) - u(t - 1)]$. Find an expression for the total voltage across the series combination.

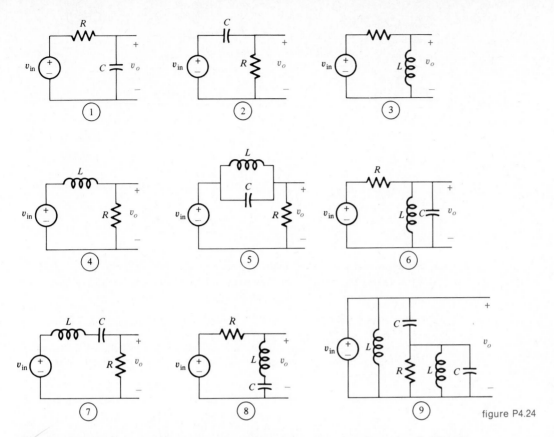

figure P4.24

21. The current in an inductor is $i_L(t) = -r(t) + 2r(t-1) - r(t-2) + u(t-3)$. Sketch this current. Label maximum and minimum values.

22. Assuming the value of the inductor in problem 21 is 6 H, find a single closed-form expression for $v_L(t)$.

23. If $L = 1$ H and $i_L(t) = (t^2 + 1)[u(t) - u(t-5)]$, find $v_L(t)$.

24. For each of the circuits in Figure P4.24, assume that the input voltage $v_{in}(t)$ is a sinusoid having a fixed magnitude but a variable frequency. State whether the output voltage magnitude for each circuit is (a) largest at high frequency (high-pass filter), (b) largest at low frequencies (low-pass filter), (c) large at both high and low frequency, and small at medium frequency (band-stop), (d) large at medium, and small at high and low frequencies (band-pass), or (e) the same for all frequencies (all-pass).

25. The current that goes through a 1-F capacitor and a 1-H inductor in series is given by $i(t) = t^2[u(t) - u(t-2)] + u(t)$. Find and sketch $v_C(t)$, $v_L(t)$, and the total voltage.

26. Repeat problem 25 if the 1-H inductor is replaced by a 1-Ω resistor.

27. In the circuit shown in Figure P4.27, $R_1 = R_2 = 10$ Ω, $R_3 = 5$ Ω, $R_4 = 1$ Ω, $C_1 = C_2 = 1$ F, and $L = 30$ mH. If $V = 10$ V (dc), what is the value of the current in R_1? In R_4?

28. The voltage waveform shown in Figure P4.28 is impressed across a 1-H inductor. Prior to $t = 0$, $v_L = 0$. (a) Find the value of the current in the inductor. Sketch $i(t)$ versus t. What is the slope of $i(t)$ at $t = 4-$? At $t = 4+$? (b) Find a single closed-form expression

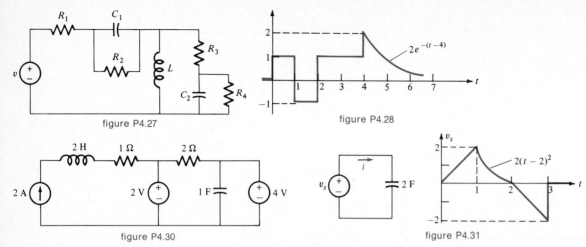

figure P4.27

figure P4.28

figure P4.30

figure P4.31

for $i(t)$. (c) Suppose, at $t = -1000$ s, a unit-impulse voltage had been applied to the terminals of this inductor, and then the waveform shown above was applied as in part (a). What difference, if any, would the impulse make in the current $i(t)$ that flows for $t > 0$?

29. A linear 100-turn inductor is wound on a metal core. When a constant 2-A current exists in the coil, the central 60 turns are linked by a magnetic flux of 0.03 mWb and the two end sections are each linked by 0.015 mWb. (a) What is the inductance of this inductor? (b) If the current drops linearly from 5 A to zero in 0.01 s, what is the value of the voltage v_L during this interval?

30. In the circuit in Figure P4.30, find (a) the power dissipated in each resistor, (b) the power delivered by each source, and (c) the total energy stored in the circuit.

31. For the circuit and voltage waveform shown in Figure P4.31, sketch $i(t)$ versus t. (a) Label each piecewise segment of this current waveform with a valid mathematical expression. (b) Sketch the energy $w(t)$ stored in the capacitor. Label all segments of the curve mathematically and indicate the values of each extremum (max or min). (c) Sketch the power $p(t)$ delivered to the capacitor. Find this both by taking the derivative of $w(t)$ and by taking the product of $v(t)$ and $i(t)$.

32. The energy stored in a 4-H inductor is given by

$$w(t) = 32t^4[u(t) - u(t-1)] + 32u(t-1) - 24u(t-2) - 8u(t-3)$$

You are told that the current in the inductor is positive from $t = 0$ to $t = 1$. It then reverses direction at 1-s intervals. Sketch the inductor voltage that results from the current and stored energy described above.

33. A voltage source impressed across a series R-L circuit ($R = 0.5$ Ω, $L = 0.5$ H) produces a periodic sawtooth current waveshape as follows: The value of the current rises linearly from zero at $t = 0$ to 1 A at $t = 2$, and then drops linearly to 0 A at $t = 3$. It then repeats indefinitely. Sketch the waveshape of the voltage source. Carefully label axes and important points on the waveform.

34. Find the Norton equivalent circuit of the circuit shown in Figure P4.34.

35. A 100-turn inductor is excited by a 60-Hz sinusoidal current whose maximum value is 160 mA. The magnetic flux versus ni curve of the core is shown in Figure P4.35. (a) How much power is lost in the core under these conditions? (b) An approximate model of this

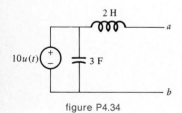

figure P4.34

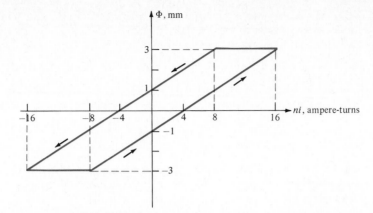

inductor consists of the parallel connection of an ideal inductor and a resistor. Determine the values of each (use the maximum extremes of the hysteresis loop to determine L and the magnitude of the voltage across this linear model).

36. Two inductors ($L_1 = 8$ H, $L_2 = 9$ H) are placed near each other. When the peak value of the voltage across L_1, $v_1 = 12$ V, and $i_2 = 0$, it is found that the peak value of the voltage across the other inductor, $v_2 = 10$. What is the value of the coupling coefficient k?

37. During a certain interval of time, voltages $v_1 = 2$ V and $V_2 = 2$ V are impressed on the terminals of a set of coupled coils. Assume linear operation; $L_1 = 8$ H, $L_2 = 2$ H, and $M = +2$ H. (a) Find the rates of growth of the currents in each coil, di_1/dt and di_2/dt. (b) What is the value of the coupling coefficient?

38. Given a pair of coupled coils with $L_1 = L_2 = 2$ H and $M = +1$ H. What open-circuit output voltage $v_2(t)$ results if $v_1(t) = 10 \sin (3t)$?

39. For the pair of coils in problem 38, find $v_1(t)$ and $v_2(t)$ when $i_1(t) = 2 \sin (2t)$ and $i_2(t) = 0$.

40. In an ideal transformer $n_1 = 10$ and $n_2 = 40$. If a 100-Ω resistor is connected to the output (n_2 side), find the resulting input resistance, $R_{in} = V_1/I_1$.

41. An ideal transformer is connected as follows: The input side is driven by a Thevenin circuit ($V_{Th} =$ a 200-rms V, 60-Hz sinusoid. $R_{Th} = 2 \Omega$), and a 2-Ω load resistor is connected to the transformer's output terminals. If $n_1 = 600$ and $n_2 = 300$, find the magnitude of the secondary voltage, $v_2(t)$.

42. In the ideal transformer circuit shown in Figure P4.42, the turns ratio $n_1/n_2 = 10/1$. (a) Which output terminal should be dotted? (b) For $v_1(t) = 10 \sin (377t)$, find $v_2(t)$, $i_2(t)$, and the power $p(t)$ dissipated in the resistor.

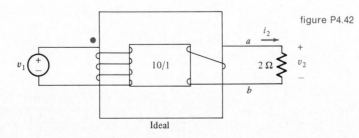

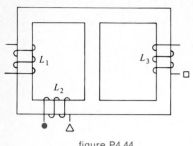

figure P4.44

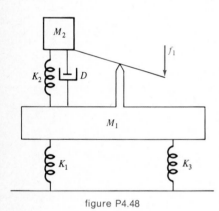

figure P4.48

43. Find the constant of proportionality k_s that relates di_2/dt to v_1 in a nonideal transformer when v_2 is set to zero by short-circuiting the output terminals. Assume that $L_1 = L_2 = 1$ H and $M = +\frac{1}{2}$.

44. Complete the dot notation in Figure P4.44. Use the dot for the M_{12} relationship and the triangle for M_{23}.

45. Two 1-H coils are taped together in an attempt to get maximum mutual coupling. Electrically they are connected in series and the equivalent inductance is measured at 2.5 H. (a) What coefficient of coupling was achieved? (b) If perfect coupling could be achieved, what equivalent inductance would be obtained?

46. Two 60-mH coils are to be connected in series to obtain a 75-mH equivalent inductor. As a first approximation, we can assume the coupling coefficient k equal to the cosine of the angle α between the axes of these coils. Find the proper value for α and sketch the series connection in a way that shows this geometric relationship.

47. $W_{\text{kinetic}} = \frac{1}{2}M\upsilon^2$ and $W_{\text{electric}} = \frac{1}{2}Cv^2$. What is the electrical analog of momentum, $M\upsilon$?

48. In Figure P4.48, a force f_1 is applied to the lever to try to lift M_2. Sketch the mechanical schematic.

49. Draw the schematic diagrams for the systems whose pictorial diagrams are shown in Figure P4.49.

50. Pictorial sketches of a number of mechanical systems are shown in Figure P4.50. Draw mechanical schematics for each.

51. The damping of a linear dashpot is $D = 20$ N·s/m. A sinusoidal velocity with a peak amplitude of 20 m/s and period $= 1$ s is applied to it. How much energy is dissipated during one period of this velocity waveform?

52. A 22,000-lb boat is moving through the water with a velocity of 12 ft/s. It is powered by a diesel engine rated at 30 hp at 2600 rpm that drives its propeller through a 2/1 speed reduction gear. At $t = 0$, the helmsman, attempting to stop, throws the engine into full maximum reverse. Assuming a linear system (no prop cavitation, etc.), (a) how much torque τ_p does the propeller shaft transmit to the propeller? (b) Assume the propeller translates this torque τ_p proportionally into the force $f = 5\tau_p$ that decelerates the boat. How much time does the boat take to come to a stop? (c) How far does it travel during that maneuver? (1 hp $= 746$ W, 1 kG $= 2.2$ lb.)

53. Two ideal (massless) linear springs with compliance $= 0.01$ m/N are connected to an overhead beam in series (one hangs from the end of the other). The upper spring is different from the lower one in that it has a cover that prevents it from stretching more than

figure P4.49

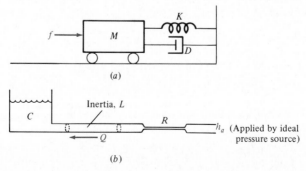

(a)

(b)

figure P4.50

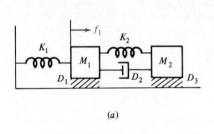

(a)

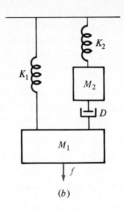

(b)

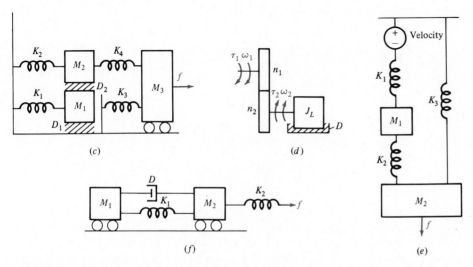

(c)   (d)

(f)   (e)

10 cm. (a) If a 20-N downward force is applied to the bottom end of the bottom spring, how far will that point displace downward? (b) How much force stretches the upper spring? (c) The lower spring? (d) What is the electrical analog of a spring that has a constraint like the one in this problem?

54. An initially empty, open tank (cross-sectional area $= 10 \text{ m}^2$) is filled at a rate $Q = t^2 u(t)$ cubic meters per second. The tank is 7.2 m high. How long does it take before the tank overflows?

55. The head (height) of the fluid in the tank in problem 54 varies as shown in Figure P4.55. Plot to scale the Q versus time that is responsible for this variation. Label axes and give numerical values to important points on the resulting curve.

56. An oil-fired furnace is rated at 130,000 btu/h (38,000 J/s) continuous duty. It needs to run 50 percent of the time to hold the temperature inside a house at 68°F (20°C) when the

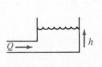

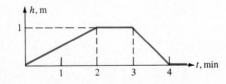

figure P4.55

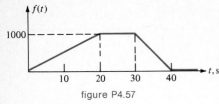

figure P4.57

outdoor temperature is 32°F (0°C) with no wind. Find the thermal resistance of the walls and roof of the house. If the thermostat setting is reduced to 59°F (15°C) with the same outside conditions, find the percentage of time the furnace needs to run to hold this lower inside temperature.

57. A 1000-kG space vehicle is accelerated starting at $t = 0$ by a rocket engine force of $f(t)$ newtons as specified in Figure P4.57. Draw a sketch of velocity versus time for $0 < t < 40$ s. What is the maximum velocity reached during this time?

58. A refrigerator has a total thermal resistance (outside to inside) of 0.8 K·s/J. If the outside of the box is at a constant 300 K and the inside is maintained at 270 K, what is the heat flow into the box? If the refrigeration unit is 20 percent efficient and runs one-third of the time, how much electric power does it use when it is running?

59. (a) Thevenize the circuit to the left of terminals a-b in Figure P4.59. (b) What value of turns ratio n_2/n_1 will maximize the power delivered to the load?

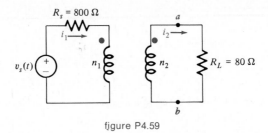

figure P4.59

chapter 5
SYSTEM EQUATIONS

introduction 5.1

When elements are connected together, the result is called a *network*. Electric networks are usually called *circuits*, and mechanical networks are usually called *machines*. In any event, if we connect together the kinds of elements discussed so far in this text, the result is a network to which we can apply Kirchoff's laws:

KVL (Kirchoff's voltage law) The sum of all across variables around any closed path is zero.

KCL (Kirchoff's current law) The sum of all through variables into any node is zero.

Either of these summations will generally lead to an *integrodifferential* system equation, an equation that relates one of the system's variables and the source(s) in the system. We can then do one of two things: (1) solve the system equation(s) and thereby find expressions for one or more of the voltages and currents (or forces or fluid pressure heads, etc.) in the network, or (2) simply look at the form of the equations and the coefficients that result, and from them learn something about the properties of the network. In this chapter we shall concentrate on obtaining system equations for systems that contain all sorts of elements.

EXAMPLE 5.1.1

Find the integrodifferential equation for the single unknown node voltage in Figure 5.1.1.

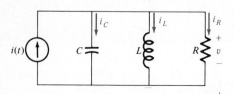

figure 5.1.1
The circuit of Example 5.1.1.

ANS.:   From KCL we have

$$i = i_C + i_L + i_R \tag{5.1.1}$$

where

$$i_C = C \frac{dv}{dt} \tag{5.1.2}$$

$$i_R = \frac{1}{R} v \tag{5.1.3}$$

$$i_L = \frac{1}{L} \int_{-\infty}^{t} v(t) \, dt \tag{5.1.4}$$

Substituting (5.1.2) to (5.1.4) into (5.1.1) gives

$$C \frac{dv}{dt} + \frac{1}{R} v + \frac{1}{L} \int_{-\infty}^{t} v(t) \, dt = i(t) \tag{5.1.5}$$

If a network has several loops and/or nodes, we get several integrodifferential equations by applying Kirchhoff's laws to these several nodes and/or loops. Such sets of equations are called *simultaneous* or *coupled* integrodifferential equations because they must be solved together, as a package, in order to find expressions for the dependent variables (the node voltages and loop currents). In this chapter we study how to write these sets of equations and to manipulate them in several different useful ways. Then in Chapters 6 and 7 we shall learn how to solve them.

You have already learned how to solve resistive circuits, which are simply a special, simpler case of the general circuits we now deal with. When we include energy-storage elements like capacitors, inductors, masses, or springs in a network, we get system equations that contain integrals and derivatives rather than the easily solved sets of simultaneous algebraic equations we dealt with earlier. But the general methods (e.g., Kirchhoff's laws) we used successfully to obtain the system equations of resistive circuits work every bit as well for general *RLC* circuits and their analogs in mechanical, fluid, etc., systems.

EXAMPLE 5.1.2

As a review, write equations that are directly solvable for the unknown node voltages in Figure 5.1.2.

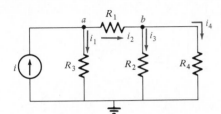

figure 5.1.2
The circuit of Example 5.1.2.

ANS.: At node *a*

$$i(t) = i_1 + i_2 \tag{5.1.6}$$

Hence

$$\frac{1}{R_3} v_a + \frac{1}{R_1}(v_a - v_b) = i(t) \tag{5.1.7}$$

At node *b*

$$\frac{1}{R_1}(v_a - v_b) = \frac{1}{R_2} v_b + \frac{1}{R_4} v_b \tag{5.1.8}$$

The two algebraic equations for the resistive circuit in Example 5.1.2 can be solved for $v_a(t)$ and $v_b(t)$ by many different methods. One way would be to solve the first equation for v_a and then substitute that expression into the second equation and thereby solve for v_b. In Example 5.1.3 we do this with numbers.

EXAMPLE 5.1.3

In Figure 5.1.2, $R_1 = 4\ \Omega$, $R_2 = 3\ \Omega$, $R_3 = 9\ \Omega$, $R_4 = 6\ \Omega$, and $i = 5$ A. Find v_a and v_b.
ANS.: At node *a*

$$5 = \frac{v_a}{9} + \tfrac{1}{4}(v_a - v_b) \tag{5.1.9}$$

At node *b*

$$\frac{v_a - v_b}{4} = \frac{v_b}{3} + \frac{v_b}{6} \tag{5.1.10}$$

Solve the first equation for v_a (by multiplying by 36)

$$180 = 4v_a + 9(v_a - v_b) \tag{5.1.11}$$

$$\frac{180 + 9v_b}{13} = v_a \tag{5.1.12}$$

Now substitute into equation (5.1.10)

$$\frac{1}{4}\left(\frac{180}{13} + \frac{9v_b}{13} - v_b\right) = v_b(\tfrac{1}{3} + \tfrac{1}{6}) \tag{5.1.13}$$

whence (after some simplification) $v_b = 6$, and thus $v_a = 18$.

Consider now a circuit similar in form to that of Examples 5.1.2 and 5.1.3 except that it contains energy-storage elements.

EXAMPLE 5.1.4

Write the simultaneous integrodifferential equations that can be solved for node voltages v_a and v_b in the circuit of Figure 5.1.3.

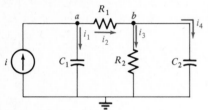

figure 5.1.3
The circuit of Example 5.1.4.

ANS.:   At node a

$$i = i_1 + i_2$$

$$C_1 \frac{dv_a}{dt} + \frac{1}{R_1}(v_a - v_b) = i(t) \tag{5.1.14}$$

At node b

$$i_2 = i_3 + i_4$$

$$\frac{1}{R_1}(v_a - v_b) = \frac{1}{R_2}v_b + C_2\frac{dv_b}{dt} \tag{5.1.15}$$

Equations (5.1.14) and (5.1.15) are simultaneous integrodifferential equations because they must be solved together in order to find $v_a(t)$ and $v_b(t)$.

Note that it is much easier to solve simultaneous algebraic equations than simultaneous integrodifferential equations. In Example 5.1.4, however, we could use the same general method as in Example 5.1.3 and solve the second expression for v_a and then substitute it into the first equation. We would then have a single equation in a single unknown which we shall soon be able to solve. Try it. The answer, after much algebra, is

$$R_1 C_1 C_2 \frac{d^2 v_b}{dt^2} + \frac{(R_1 + R_2)C_1 + R_2 C_2}{R_2} \frac{dv_b}{dt} + \frac{v_b}{R_2} = i \qquad (5.1.16)$$

Alert readers may have noted that the circuit of Example 5.1.4 had *two* energy-storage elements. It also turns out in this example that the highest-order derivative in the system equation is a *second* derivative. This is not a coincidence. In general, each additional energy-storage element adds another derivative to the system equation. We call a system having n energy-storage elements (in which therefore the highest-order derivative in the system equation is an nth-order derivative) an *nth-order system.*† We also say the system has *n degrees of freedom.*

DRILL PROBLEM

A series combination of $L = 2$ H and $C = 3$ F is driven by a voltage source $v_s(t)$. Find the differential equation that relates $v_C(t)$ to $v_s(t)$.

ANS.: $6(d^2 v_C / dt^2) + v_C = v_s$.

---

operational notation 5.2

It would be convenient to be able to *manipulate* integrodifferential equations (although not solve them) in the same way that we do algebraic ones. In order to do this we define the *operator p* as

$$p = \frac{d}{dt} \qquad (5.2.1)$$

Thus

$$pf = \frac{df}{dt} \qquad (5.2.2)$$

and

$$p^n f = \frac{d^n f}{dt^n} \qquad (5.2.3)$$

We also define $1/p$ to mean integration such that

$$\frac{1}{p} pf = p \frac{f}{p} = f \qquad (5.2.4)$$

must be able to integrate & differentiate f

For this to be the case, we must define

---

† Actually, the number of energy-storage elements is the *upper bound* on n, the order of the system. Certain specific ways of interconnecting the energy-storage elements within the circuit can result in n being less than the number of energy-storage elements.

$$\frac{1}{p}f = \int_{-\infty}^{t} f(t)\, dt \tag{5.2.5}$$

under the assumption that $f(t) = 0$ for $t < 0$. Therefore,

$$\boxed{\frac{1}{p}f = \int_{0-}^{t} f(t)\, dt} \tag{5.2.6}$$

EXAMPLE 5.2.1

If $f(t) = \cos t\, u(t)$, find pf and $(1/p)pf$.

ANS.:
$$pf = -\sin t\, u(t) + 1\delta(t)$$

and
$$\frac{1}{p}pf = \int_{0-}^{t} [-\sin t\, u(t) + \delta(t)]\, dt$$

$$= \cos t \Big|_{0-}^{t} + 1 = \cos t - 1 + 1$$

$$= \cos t$$

EXAMPLE 5.2.2

If $f(t) = -\sin t\, u(t)$, find $(1/p)f$ and sketch the result versus t.

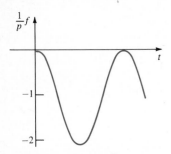

figure 5.2.1
The plot of $(1/p)f$ in Example 5.2.2.

ANS.:   See Figure 5.2.1.

$$\frac{1}{p}f = \int_{-\infty}^{t} -\sin t\, u(t)\, dt = \int_{0-}^{t} -\sin t\, dt = +\cos t \Big|_{0-}^{t} = \cos t - 1$$

EXAMPLE 5.2.3

Write the following integrodifferential equation in p-operator notation

$$\frac{d^2 y}{dt^2} + 4\frac{dy}{dt} + 8y + 3\int_{0}^{t} y\, dt = 8\cos t\, u(t)$$

ANS.: $p^2 y + 4py + 8y + 3\dfrac{1}{p} y = 8 \cos t$ for $t > 0$.

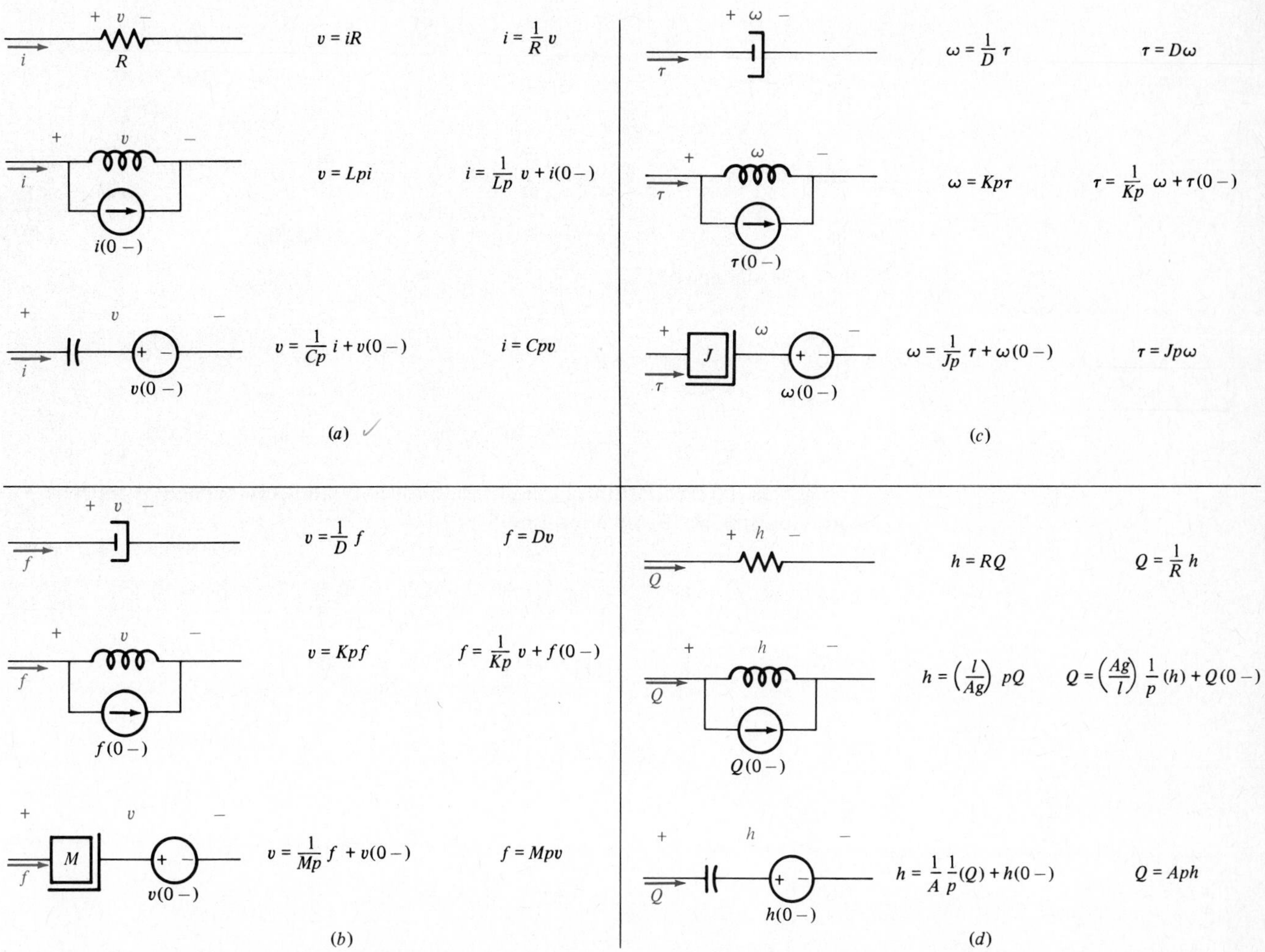

$v = iR$	$i = \dfrac{1}{R} v$	
$v = Lpi$	$i = \dfrac{1}{Lp} v + i(0-)$	
$v = \dfrac{1}{Cp} i + v(0-)$	$i = Cpv$	

(a)

$\omega = \dfrac{1}{D} \tau$	$\tau = D\omega$
$\omega = Kp\tau$	$\tau = \dfrac{1}{Kp} \omega + \tau(0-)$
$\omega = \dfrac{1}{Jp} \tau + \omega(0-)$	$\tau = Jp\omega$

(c)

$v = \dfrac{1}{D} f$	$f = Dv$
$v = Kpf$	$f = \dfrac{1}{Kp} v + f(0-)$
$v = \dfrac{1}{Mp} f + v(0-)$	$f = Mpv$

(b)

$h = RQ$	$Q = \dfrac{1}{R} h$
$h = \left(\dfrac{l}{Ag}\right) pQ$	$Q = \left(\dfrac{Ag}{l}\right)\dfrac{1}{p}(h) + Q(0-)$
$h = \dfrac{1}{A}\dfrac{1}{p}(Q) + h(0-)$	$Q = Aph$

(d)

Defining equations for several kinds of systems are given in p-operator form in Figure 5.2.2. Note that in the light of the definition of the integral operator in equation (5.2.6) we must use initial-condition generators to indicate any nonzero energy-storage conditions at time equal to zero. The p operator is not an algebraic variable but a symbol that denotes a certain operation (differentiation with respect to time) to be performed on the function which immediately *follows* it.

Earlier we defined a linear element as one with two special properties: additivity and homogeneity. Recall their meaning:

＊ Additivity If i_1 produces v_1 and i_2 produces v_2, then $i_1 + i_2$ will produce $v_1 + v_2$.

figure 5.2.2
Defining equations in p-operator form for (a) electrical systems, (b) mechanical translational systems, (c) mechanical rotational systems, and (d) fluid systems.

✳ Homogeneity   If i_1 produces v_1, then ai_1 will produce av_1.

✳ Does the p operator have these two properties? For additivity, if

$$\frac{di_1}{dt} = v_1 \qquad \text{or} \qquad pi_1 = v_1$$

and

$$\frac{di_2}{dt} = v_2 \qquad \text{or} \qquad pi_2 = v_2$$

then

$$\frac{d}{dt}(i_1 + i_2) = \frac{di_1}{dt} + \frac{di_2}{dt}$$

or

$$p(i_1 + i_2) = v_1 + v_2 \qquad \text{checks}$$

✳ For homogeneity, if

$$\frac{di_1}{dt} = v_1 \qquad \text{or} \qquad pi_1 = v_1$$

then (for $a = \text{const}$)

$$\frac{d(ai_1)}{dt} = a\frac{di_1}{dt} \qquad \text{or} \qquad p(ai_1) = api_1 = av_1 \qquad \text{checks}$$

Just as differentiation has both the additivity and homogeneity properties, so does integration. From equation (5.2.6), if

$$v_1 = \frac{1}{p}i_1 = \int_{0-}^{t} i_1\, dt \qquad \text{and} \qquad v_2 = \frac{1}{p}i_2 = \int_{0-}^{t} i_2\, dt$$

then

$$\int_{0-}^{t}(i_1 + i_2)\, dt = \int_{0-}^{t} i_1\, dt + \int_{0-}^{t} i_2\, dt$$

$$\frac{1}{p}(i_1 + i_2) = \frac{1}{p}i_1 + \frac{1}{p}i_2 = v_1 + v_2 \qquad \text{additivity}$$

and, if

$$v_1 = \frac{1}{p}i_1$$

then

$$\frac{1}{p}(ai_1) = \int_{0-}^{t} ai_1\, dt = a\int_{0-}^{t} i_1\, dt = a\frac{1}{p}i_1$$

$$= av_1 \qquad \text{homogeneity}$$

We therefore call the p operator a *linear operator*. The p operator also has the following properties. For addition

$$(p^m + p^n)f = p^m f + p^n f \qquad\qquad \text{commutative}$$

$$[p^l + (p^m + p^n)]f = [(p^l + p^m) + p^n]f \qquad \text{associative}$$

✳ and for multiplication

$$(p^m \cdot p^n)f = (p^n \cdot p^m)f = p^{m+n}f \qquad \text{commutative}$$

$$p^l(p^m \cdot p^n)f = (p^l \cdot p^m)p^n f \qquad\qquad \text{associative}$$

$$p^l(p^m + p^n)f = (p^{l+m} + p^{l+n})f \qquad\qquad \text{distributive}$$

Because the p operator has all the above (algebraic) properties, any expression that contains p operators can be rearranged according to the rules of algebra.

Thus the two equations in Example 5.1.4 can be manipulated as follows. The original equations are

$$C_1 \frac{dv_a}{dt} + \frac{1}{R_1} (v_a - v_b) = i(t) \tag{5.2.7}$$

$$\frac{1}{R_1} (v_a - v_b) = \frac{1}{R_2} v_b + C_2 \frac{dv_b}{dt} \tag{5.2.8}$$

Letting $p = d/dt$, we have

$$C_1 p v_a + \frac{1}{R_1} (v_a - v_b) = i(t) \tag{5.2.9}$$

and

$$\frac{1}{R_1} (v_a - v_b) = \frac{1}{R_2} v_b + C_2 p v_b \tag{5.2.10}$$

First solving (5.2.9) for v_a and then (5.2.10) for v_b yields

$$v_a = \frac{1}{R_1 C_1 p + 1} v_b + \frac{R_1}{R_1 C_1 p + 1} i(t) \tag{5.2.11}$$

and

$$v_b = \frac{R_2 v_a}{R_1 R_2 C_2 p + R_1 + R_2} \tag{5.2.12}$$

Inserting (5.2.11) into (5.2.12) and simplifying, we get, after a few steps of algebra

$$\left[R_1 C_1 C_2 p^2 + \frac{(R_1 + R_2)C_1 + R_2 C_2}{R_2} p + \frac{1}{R_2} \right] v_b = i \tag{5.2.13}$$

or

$$R_1 C_1 C_2 \frac{d^2 v_b}{dt^2} + \frac{(R_1 + R_2)C_1 + R_2 C_2}{R_2} \frac{dv_b}{dt} + \frac{1}{R_2} v_b = i \tag{5.2.14}$$

which is identical to equation (5.1.16).

DRILL PROBLEM

Write the following in p-operator notation:

a.
$$\frac{2d^3 v}{dt^3} + \frac{3d^2 v}{dt^2} + \frac{4dv}{dt} + 5 = \frac{6di_1}{dt} + \frac{9dv_s}{dt}$$

b.
$$v_L(t) = \frac{4di}{dt}$$

c.
$$v_C(t) = \frac{1}{3} \int_{0-}^{t} i_C \, dt + 4$$

ANS.: (a) $2p^3 v + 3p^2 v + 4pv + 5 = 6pi_1 + 9pv_s$; (b) $v_L(t) = 4pi(t)$; (c) $v_C(t) = (1/3p)i_C + 4$.

5.3 generalized impedances

Kirchhoff's laws are just as valid and useful in the analysis of networks that contain energy-storage elements as they are in solving resistive circuits. The sum of all across variables around any closed path is equal to zero (KVL), and the sum of all through variables into any node is zero (KCL). Applying these laws to a particular network yields the mesh and/or node equations for that network. Using *p*-operator notation is simply a way of avoiding writer's cramp and achieving a neat and compact way of writing the resulting system equations.

EXAMPLE 5.3.1

For Figure 5.3.1 find *p*-operator expressions that relate v_a, v_b, and i_4 to $i(t)$ for $t > 0$. Assume that all voltages and currents are zero-valued before $t = 0$.

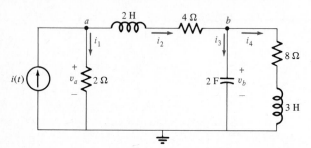

figure 5.3.1
The circuit of Example 5.3.1.

ANS.:   Use the definitions of Figure 5.2.2*a* and the algebraic properties of the *p* operator. Summing voltages around the rightmost mesh gives

$$(8 + 3p)i_4 = v_b \tag{5.3.1}$$

so that

$$i_4 = \frac{1}{8 + 3p}\, v_b \tag{5.3.2}$$

In the 2-F capacitor we have

$$i_3 = 2pv_b \tag{5.3.3}$$

At node *b*

$$i_2 = i_3 + i_4 = \left(2p + \frac{1}{8 + 3p}\right)v_b = \frac{6p^2 + 16p + 1}{8 + 3p}\, v_b \tag{5.3.4}$$

Around the central mesh

$$v_a = v_b + (2p + 4)i_2 = v_b + \frac{12p^3 + 56p^2 + 66p + 4}{8 + 3p}\, v_b \tag{5.3.5}$$

or

$$v_a = \frac{12p^3 + 56p^2 + 69p + 12}{8 + 3p}\, v_b \tag{5.3.6}$$

and so

$$i = i_2 + i_1 = \left(\frac{6p^2 + 16p + 1}{8 + 3p} + \frac{12p^3 + 56p^2 + 69p + 12}{16 + 6p}\right)v_b \tag{5.3.7}$$

or
$$i = \frac{12p^3 + 68p^2 + 101p + 14}{16 + 6p} v_b \qquad (5.3.8)$$

so that
$$v_b(t) = \frac{6p + 16}{12p^3 + 68p^2 + 101p + 14} i(t) \qquad (5.3.9)$$

From (5.3.1) and (5.3.9) we get
$$i_4(t) = \frac{1}{8 + 3p} v_b = \frac{2}{12p^3 + 68p^2 + 101p + 14} i(t) \qquad (5.3.10)$$

and from (5.3.6)
$$v_a(t) = \frac{(12p^3 + 56p^2 + 69p + 12)(6p + 16)i(t)}{(3p + 8)(12p^3 + 68p^2 + 101p + 14)} \qquad (5.3.11)$$

$$v_a(t) = \frac{24p^3 + 112p^2 + 138p + 24}{12p^3 + 68p^2 + 101p + 14} i(t) \qquad (5.3.12)$$

So that (5.3.9) yields
$$12 \frac{d^3 v_b}{dt^3} + 68 \frac{d^2 v_b}{dt^3} + 101 \frac{dv_b}{dt} + 14 v_b = 6 \frac{di(t)}{dt} + 16 i(t) \qquad (5.3.13)$$

(5.3.10) yields
$$12 \frac{d^3 i_4}{dt^3} + 68 \frac{d^2 i_4}{dt^2} + 101 \frac{di_4}{dt} + 14 i_4 = 2 i(t) \qquad (5.3.14)$$

and (5.3.12) yields

• $$12 \frac{d^3 v_a}{dt^3} + 68 \frac{d^2 v_a}{dt^2} + 101 \frac{dv_a}{dt} + 14 v_a = 24 \frac{d^3 i}{dt^3} + 112 \frac{d^2 i}{dt^2} + 138 \frac{di}{dt} + 24 i \qquad (5.3.15)$$

In Example 5.3.1 the variable $i(t)$ and the variables $v_b(t)$, $i_4(t)$, and $v_a(t)$ can be thought of respectively as (1) the *cause* and *effects* or (2) the *forcing function* and three of its *responses* or (3) the *input* and three *outputs* of this system. Figure 5.3.2 is a *block diagram* showing how the three different operators of equations (5.3.9),

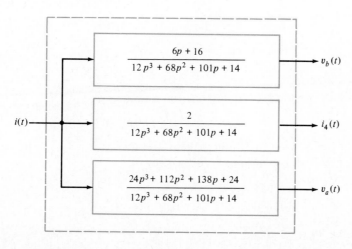

figure 5.3.2
The various p operators that relate the input $i(t)$ to the three output variables v_b, i_4, and v_a in Example 5.3.1.

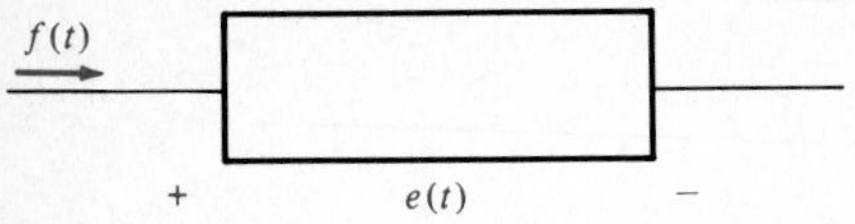

figure 5.3.3.
The across- and through-variable polarities
used in defining driving-point immittances
(impedances and/or admittances).

(5.3.10), and (5.3.12) *multiply* the input $i(t)$, each producing a different output. To describe this, we use the following nomenclature:

Transfer-function operator $H(p)$ × input function = output function

Special cases include the following:

1. If the output is an across variable and the input is a through variable, $H(p)$ is called an *impedance operator*.
2. If the output is a through variable and the input is an across variable, $H(p)$ is called an *admittance operator*.
3. In addition to being either an impedance or admittance operator, if $H(p)$ relates a through variable *and its corresponding* across variable (as defined in Figure 5.3.3), it is called a *driving-point* or *input* admittance or impedance operator.

For example, in Example 5.3.1 and Figure 5.3.2 the operator relating $v_a(t)$ to $i(t)$ is a driving-point impedance. For impedance operators the symbol $Z(p)$ is used; for admittance operators $Y(p)$ is used. If one wishes to talk about either an impedance operator or an admittance operator (or perhaps both) without being specific, one can use *immittance* operator, which means either $Z(p)$ or $Y(p)$. If an immittance is not a driving-point immittance, it is called a *transfer* immittance. Thus in Example 5.3.1 the operator that relates $v_b(t)$ and $i(t)$ is termed a *transfer impedance* (impedance rather than admittance because the operator multiplies a current in order to produce a voltage). The operator that relates $i_4(t)$ to $i(t)$ is not an immittance and is simply called a *transfer function*. Since both the input and output are currents, we may call it a *current transfer function*.

Using this nomenclature, we can write the p-operator driving-point impedance $Z(p)$ and driving-point admittance $Y(p)$ for any of the linear elements we have studied so far. With reference to Figure 5.2.2 and assuming *zero initial conditions*, we can make at least a partial listing of them (Table 5.2.1).

It appears that in every case

$$Z(p) = \frac{1}{Y(p)}$$

This can be easily proved from the definitions

$$Y(p)e(t) = f(t) \tag{5.3.16}$$

and

$$Z(p)f(t) = e(t) \tag{5.3.17}$$

Substituting (5.3.17) into (5.3.16), we get

$$Y(p)Z(p)f(t) = f(t) \tag{5.3.18}$$

from which it must follow that

$$Y(p)Z(p) = 1 \tag{5.3.19}$$

or

$$Z(p) = \frac{1}{Y(p)} \tag{5.3.20}$$

for any element.

Linear element	$Z(p)$	$Y(p)$
Electric	$v = Zi$	$i = Yv$
Resistor	R	$\dfrac{1}{R}$
Inductor	Lp	$\dfrac{1}{Lp}$
Capacitor	$\dfrac{1}{Cp}$	Cp
Mechanical translational	$\mathcal{V} = Zf$	$f = Y\mathcal{V}$
Dashpot	$\dfrac{1}{D}$	D
Spring	Kp	$\dfrac{1}{Kp}$
Mass	$\dfrac{1}{Mp}$	Mp
Mechanical rotational	$\omega = Z\tau$	$\tau = Y\omega$
Dashpot	$\dfrac{1}{D}$	D
Spring	Kp	$\dfrac{1}{Kp}$
Mass	$\dfrac{1}{Jp}$	Jp
Thermal	$\theta = ZQ$	$Q = Y\theta$
Resistance	R	$\dfrac{1}{R}$
Capacitance	$\dfrac{1}{Cp}$	Cp

table 5.2.1

Driving-point impedance $Z(p)$ and admittance $Y(p)$

It can be shown that $Z(p)$ operators obey all the rules that R's do. For example, for any two elements in series (Figure 5.3.4)

$$e = e_1 + e_2 \tag{5.3.21}$$

$$e_1 = Z_1(p)f \tag{5.3.22}$$

$$e_2 = Z_2(p)f \tag{5.3.23}$$

$$e = [Z_1(p) + Z_2(p)]f \tag{5.3.24}$$

So that $Z_1(p) + Z_2(p)$ is the equivalent operator that relates e and f. Thus

$$Z_{eq}(p) = Z_1(p) + Z_2(p) \tag{5.3.25}$$

Similarly, in the case of two elements in parallel (Figure 5.3.5)

$$f = f_1 + f_2 \tag{5.3.26}$$

$$f = Y_1(p)e + Y_2(p)e \tag{5.3.27}$$

$$f = [Y_1(p) + Y_2(p)]e \tag{5.3.28}$$

or

$$f = Y_{eq}(p)e \tag{5.3.29}$$

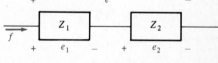

figure 5.3.4
Two impedances in series.

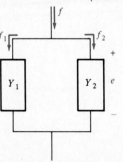

figure 5.3.5
Two admittances in parallel.

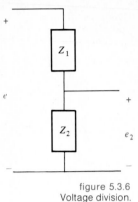

figure 5.3.6
Voltage division.

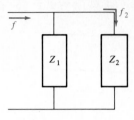

figure 5.3.7
Current division.

where
$$Y_{eq}(p) = Y_1(p) + Y_2(p) \qquad (5.3.30)$$

We realize at this point that the parallel combination of resistors

$$\frac{1}{R_1} + \frac{1}{R_2} = \frac{1}{R_{eq}} \qquad (5.3.31)$$

$$G_1 + G_2 = G_{eq} \qquad (5.3.32)$$

is a special case of

$$Y_{eq}(p) = Y_1(p) + Y_2(p) \qquad (5.3.33)$$

Similarly, we can show that voltage division (Figure 5.3.6) is given by

$$e_2 = \frac{Z_2(p)}{Z_1(p) + Z_2(p)} e \qquad (5.3.34)$$

and current division (Figure 5.3.7) by

$$f_2 = \frac{Z_1(p)}{Z_1(p) + Z_2(p)} f \qquad (5.3.35)$$

EXAMPLE 5.3.2

Use p-operator immittances to obtain a differential equation for $v_L(t)$ in terms of the other elements and input $v_s(t)$ in the circuit shown in Figure 5.3.8.

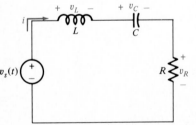

figure 5.3.8
The circuit of Example 5.3.2.

ANS.:   We sum the voltages around the loop

$$v_R + v_C + v_L = v_s \qquad (5.3.36)$$

omit 181-5

or, in terms of the current i,

$$iR + \frac{1}{Cp} i + Lpi = v_s \tag{5.3.37}$$

and thus

$$\left(R + \frac{1}{Cp} + Lp \right) i = v_s \tag{5.3.38}$$

(Note that this equation is consistent with the idea that impedances in series can be added to get an equivalent impedance operator.)

We know that

$i_R = \frac{v_R}{R}$

$i_c = C\frac{dv}{dt}$ $i = i_R = i_c = i_c$

$$i_L = \frac{1}{Lp} v_L \tag{5.3.39}$$

and in this problem $i = i_L$; therefore we can substitute (5.3.39) into (5.3.38) and get

$$\left(R + \frac{1}{Cp} + Lp \right) \frac{1}{Lp} v_L = v_s \tag{5.3.40}$$

or

$$\left(\frac{R}{Lp} + \frac{1}{LCp^2} + 1 \right) v_L = v_s \tag{5.3.41}$$

Putting the left-hand side over a common denominator gives

$$\frac{RCp + 1 + LCp^2}{LCp^2} v_L = v_s \tag{5.3.42}$$

Multiplying through by the denominator and writing the terms in decreasing order of derivative leads to

$$(LCp^2 + RCp + 1)v_L = LCp^2 v_s \tag{5.3.43a}$$

or

$$\left(p^2 + \frac{R}{L} p + \frac{1}{LC} \right) v_L = p^2 v_s \tag{5.3.43b}$$

This is the differential equation that relates $v_L(t)$ to $v_s(t)$. It is written in p-operator notation. We can now write equations (5.3.43) in the usual notation by inspection

$$\frac{d^2 v_L}{dt^2} + \frac{R}{L} \frac{dv_L}{dt} + \frac{1}{LC} v_L = \frac{d^2}{dt^2} v_s \tag{5.3.44}$$

Typically we are told what the expression for source $v_s(t)$ is; it is then a simple matter to write the right-hand side of (5.3.44) immediately as a function of time. Solving for $v_L(t)$ is another matter, which we address in Chapters 6 and 7. At this point we realize that a *system equation* is a *differential equation* such as (5.3.44) that *relates* one or more of the *dependent variables* (ones we would like to solve for) to the *source(s)* in the system. We obtain these system equations by using generalized p-operator impedances and Kirchhoff's laws. Equation (5.3.44) is written in an easily readable form: the terms are in decreasing order of derivative and the highest-order coefficient is unity. You should try to write all system equations in that form.

EXAMPLE 5.3.3

Obtain system equations that relate the velocities of each mass to the force source f_b in the machine shown in Figure 5.3.9a.

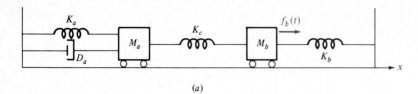

(a)

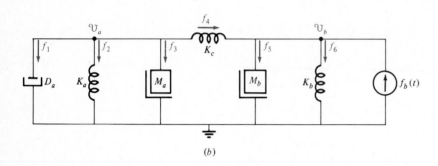

(b)

figure 5.3.9
The mechanical system of Example 5.3.3:
(a) pictorial diagram and (b) schematic diagram.

ANS.:   First draw the schematic diagram. Connect one terminal of each mass to the datum (ground) reference and then connect to the other terminal of each mass those elements and sources which are connected to it in the pictorial. The schematic is shown in Figure 5.3.9b. Then arbitrarily assign a through variable (force) to each element. There are two unknown velocities, the velocity of each of the two masses, $\mathcal{V}_a$ and $\mathcal{V}_b$. Since force $f_b(t)$ helps to increase $\mathcal{V}_b$ in the positive direction (toward the right) it is directed *into* node b.

At node a, KCL tells us that

$$f_1 + f_2 + f_3 + f_4 = 0 \tag{5.3.45}$$

or

$$D_a \mathcal{V}_a + \frac{1}{K_a p} \mathcal{V}_a + M_a p \mathcal{V}_a + \frac{1}{K_c p} (\mathcal{V}_a - \mathcal{V}_b) = 0 \tag{5.3.46}$$

and similarly at node b:

$$f_4 - f_5 - f_6 + f_b = 0 \tag{5.3.47}$$

or

$$\frac{1}{K_c p} (\mathcal{V}_a - \mathcal{V}_b) - M_b p \mathcal{V}_b - \frac{1}{K_b p} \mathcal{V}_b + f_b = 0 \tag{5.3.48}$$

Collecting terms in both (5.3.46) and (5.3.48) gives

$$\left(D_a + \frac{1}{K_a p} + M_a p + \frac{1}{K_c p} \right) \mathcal{V}_a - \frac{1}{K_c p} \mathcal{V}_b = 0 \tag{5.3.49}$$

$$-\frac{1}{K_c p} \mathcal{V}_a + \left(\frac{1}{K_c p} + M_b p + \frac{1}{K_b p} \right) \mathcal{V}_b = f_b \tag{5.3.50}$$

It should be obvious that in the usual form equations (5.3.49) and (5.3.50) are

$$D_a \mathcal{V}_a + \frac{1}{K_a} \int_{0-}^{t} \mathcal{V}_a \, dt + M_a \frac{d\mathcal{V}_a}{dt} + \frac{1}{K_c} \int_{0-}^{t} \mathcal{V}_a \, dt - \frac{1}{K_c} \int_{0-}^{t} \mathcal{V}_b \, dt = 0 \tag{5.3.51}$$

and
$$-\frac{1}{K_c} \int_{0-}^{t} \mathcal{V}_a \, dt + \frac{1}{K_c} \int_{0-}^{t} \mathcal{V}_b \, dt + M_b \frac{d\mathcal{V}_b}{dt} + \frac{1}{K_b} \int_{0-}^{t} \mathcal{V}_b \, dt = f_b \tag{5.3.52}$$

These are actually a pair of *coupled integrodifferential* equations (since each contains more than one dependent variable, they must be solved together, simultaneously). To get rid of the integrals and have ordinary differential equations we can simply multiply (5.3.49) and (5.3.50) through by p. The equivalent operation with (5.3.51) and (5.3.52) is to take d/dt of both sides of each, giving

$$\left[M_a p^2 + D_a p + \left(\frac{1}{K_a} + \frac{1}{K_c} \right) \right] \mathcal{V}_a - \frac{1}{K_c} \mathcal{V}_b = 0 \tag{5.3.53}$$

$$-\frac{1}{K_c} \mathcal{V}_a + \left[M_b p^2 + \left(\frac{1}{K_c} + \frac{1}{K_b} \right) \right] \mathcal{V}_b = p f_b \tag{5.3.54}$$

Since the p operator can be manipulated as if it were an algebraic quantity, the simultaneous differential equations (5.3.53) and (5.3.54) in p-operator form can be subjected to Cramer's rule, back substitution, or matrix inversion to obtain two separate differential equations, one in $\mathcal{V}_a$ and one in $\mathcal{V}_b$. For example, if all element values are unity, equations (5.3.53) and (5.3.54) become

$$(p^2 + p + 2)\mathcal{V}_a - \mathcal{V}_b = 0 \tag{5.3.55}$$

and
$$-\mathcal{V}_a + (p^2 + 2)\mathcal{V}_b = p f_b \tag{5.3.56}$$

From Cramer's rule

$$\Delta = (p^2 + p + 2)(p^2 + 2) - 1 = p^4 + p^3 + 4p^2 + 2p + 3$$

$$\mathcal{V}_a = \frac{\begin{vmatrix} 0 & -1 \\ p f_b & p^2 + 2 \end{vmatrix}}{\Delta} = \frac{p f_b}{\Delta}$$

$$(p^4 + p^3 + 4p^2 + 2p + 3)\mathcal{V}_a = p f_b$$

and, similarly, $(p^4 + p^3 + 4p^2 + 2p + 3)\mathcal{V}_b = (p^3 + p^2 + 2p)f_b$

The use of KCL and KVL together with generalized p-operator impedances is not limited to systems that are totally electrical or totally mechanical. The systems electrical engineers deal with are often electromechanical, electrochemical, etc. The following example concerns a sample electroacoustical system that we use every day.

EXAMPLE 5.3.4

For the loudspeaker in Figure 5.3.10 find system equations that relate:
a. Cone velocity $\mathcal{V}$ to input voltage v_s
b. Current i to voltage v_s
The mass of the speaker cone is M. The frictional coupling to the air is given by D. The effect of movement of the speaker cone on the electric circuit is given by the velocity-dependent voltage source. The mechanism

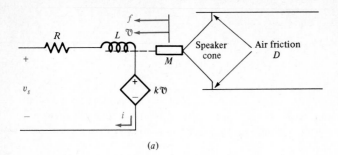

figure 5.3.10
The loudspeaker of Example 5.3.4: (a) overall system and (b) mechanical schematic.

(a)

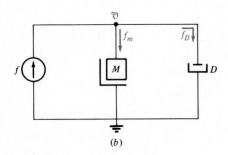

(b)

that moves the cone is the electromagnetic force f produced by the inductor's magnetic field acting on a piece of ferrous metal attached to the cone;

$$f = ki \tag{5.3.57}$$

where k is the same constant as in the dependent source.

ANS.:   (a) In the mechanical schematic we note that

$$f_m + f_D = f \tag{5.3.58}$$

Thus

$$Mp\mho + D\mho = ki \tag{5.3.59}$$

From the electric circuit we have

$$i = \frac{v_s - k\mho}{R + Lp} \tag{5.3.60}$$

Inserting (5.3.60) into (5.3.59) yields

$$Mp\mho + D\mho = \frac{kv_s - k^2\mho}{R + Lp} \tag{5.3.61}$$

Multiplying through by the denominator and collecting terms, we get

$$(R + Lp)(Mp + D)\mho + k^2\mho = kv_s$$

$$[LMp^2 + (LD + RM)p + (RD + k^2)]\mho = kv_s$$

or

$$\left(p^2 + \frac{LD + RM}{LM}p + \frac{RD + k^2}{LM}\right)\mho = \frac{kv_s}{LM} \tag{5.3.62}$$

and finally

$$\frac{d^2\mathcal{V}}{dt^2} + \frac{LD + RM}{LM}\frac{d\mathcal{V}}{dt} + \frac{RD + k^2}{LM}\mathcal{V} = \frac{kv_s}{LM} \tag{5.3.63}$$

(b) To get a relationship involving the current i and voltage v_s, solve (5.3.59) for $\mathcal{V}$ and substitute it into (5.3.60). Thus (5.3.59) becomes

$$\mathcal{V} = \frac{ki}{Mp + D} \tag{5.3.64}$$

and then (5.3.60) yields

$$(R + Lp)i = v_s - \frac{k^2 i}{Mp + D} \tag{5.3.65}$$

Collecting terms and seeking a common denominator leads to

$$\left(p^2 + \frac{LD + MR}{LM}p + \frac{DR + k^2}{LM}\right)i = (Mp + D)v_s\frac{1}{LM} \tag{5.3.66}$$

or

$$\frac{d^2 i}{dt^2} + \frac{LD + MR}{LM}\frac{di}{dt} + \frac{DR + k^2}{LM}i = \frac{1}{L}\frac{dv_s}{dt} + \frac{Dv_s}{LM} \tag{5.3.67}$$

omit ↑

Because we write mesh equations and node equations with $Z(p)$ and $Y(p)$ immittances, using exactly the same techniques as with resistive circuits, the resulting equations have exactly the same symmetry properties as described earlier.

EXAMPLE 5.3.5

Write the set of *mesh* equations for the circuit shown in Figure 5.3.11.

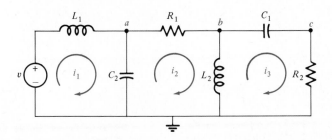

figure 5.3.11
The circuit of Examples 5.3.5 and 5.3.6.

ANS.: Use impedances $Z = Lp$, $1/Cp$, and R and sum voltages around each of the three meshes. For mesh 1

$$v - L_1 p i_1 - \frac{1}{C_2 p} i_1 + \frac{1}{C_2 p} i_2 = 0 \tag{5.3.68}$$

for mesh 2

$$-\left(\frac{1}{C_2 p} + R_1 + L_2 p\right)i_2 + \frac{1}{C_2 p} i_1 + L_2 p i_3 = 0 \tag{5.3.69}$$

and for mesh 3

$$-\left(L_2 p + \frac{1}{C_1 p} + R_2\right)i_3 + L_2 p i_2 = 0 \tag{5.3.70}$$

Collecting terms and rearranging, we get the standard form

$$+\left(L_1 p + \frac{1}{C_2 p}\right)i_1 - \frac{1}{C_2 p}i_2 = v \tag{5.3.71}$$

$$-\frac{1}{C_2 p}i_1 + \left(\frac{1}{C_2 p} + R_1 + L_2 p\right)i_2 - (L_2 p)i_3 = 0 \tag{5.3.72}$$

$$-(L_2 p)i_2 + \left(L_2 p + \frac{1}{C_1 p} + R_2\right)i_3 = 0 \tag{5.3.73}$$

EXAMPLE 5.3.6

Write the *node* equations for the circuit of Figure 5.3.11.
ANS.: The unknown node voltages are v_a, v_b, and v_c. We use admittances $Y(p) = 1/R$, $1/Lp$, and Cp in KCL. At node a

$$\frac{1}{L_1 p}(v_a - v) + C_2 p v_a + \frac{1}{R_1}(v_a - v_b) = 0 \tag{5.3.74}$$

At node b

$$\frac{1}{R_1}(v_b - v_a) + \frac{1}{L_2 p}v_b + C_1 p(v_b - v_c) = 0 \tag{5.3.75}$$

At node c

$$C_1 p(v_c - v_b) + \frac{1}{R_2}v_c = 0 \tag{5.3.76}$$

Collecting terms and rearranging, we get

$$+\left(\frac{1}{L_1 p} + C_2 p + \frac{1}{R_1}\right)v_a - \frac{1}{R_1}v_b = \frac{1}{L_1 p}v \tag{5.3.77}$$

$$-\frac{1}{R_1}v_a + \left(\frac{1}{R_1} + \frac{1}{L_2 p} + C_1 p\right)v_b - C_1 p v_c = 0 \tag{5.3.78}$$

$$-(C_1 p)v_b + \left(C_1 p + \frac{1}{R_2}\right)v_c = 0 \tag{5.3.79}$$

Note the symmetry that occurs in the coefficients of both the mesh equations, (5.3.71) to (5.3.73), and the node equations, (5.3.77) to (5.3.79).

DRILL PROBLEM

Write a set of node equations and obtain the differential equation relating v_a to i_1 and i_2 in the circuit in Figure 5.3.12.

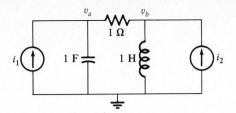

figure 5.3.12
The circuit of the drill problem.

ANS.: $(p^2 + p + 1)v_a = (p + 1)i_1 + pi_2$.

Thevenin and Norton circuits with p operators 5.4

The methods for finding Thevenin and Norton equivalents for circuits that contain energy-storage elements are the same as those we used with simple resistive circuits. The only difference for circuits containing energy-storage elements is that we must use the p-operator impedance of each element. Therefore, the open-circuit voltage, the short-circuit current, and the output impedance all become functions of the p operator. We shall give several examples and discuss the uses of these equivalent circuits.

When we first discussed the Thevenin and Norton terminal equivalents, we saw how to plot the *volt-ampere terminal characteristic*, which is simply a plot of the equation that relates v and i, the terminal variables. If we actually find this equation for any circuit (with or without energy-storage elements), we can look at it and simply write down, by inspection, the expressions for $V_{oc} = V_{Th}$, $I_{sc} = I_N$, and $Z_o(p)$, giving us another method for finding the Thevenin or Norton equivalent circuit. More important is the fact that since KCL and KVL always enable us to solve for $v(i)$ and/or $i(v)$, this demonstrates that we can always find the Thevenin or Norton equivalent for any circuit.

Consider any linear network of the form shown in Figure 5.4.1, where, although we use the familiar electrical symbols v and i, the network could be mechanical, thermal, etc. We can always write a node equation at the upper terminal to obtain an expression in the form

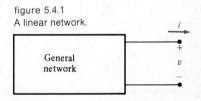

figure 5.4.1
A linear network.

$$v = \sum_{j=1}^{n} [(\text{operator } j)(\text{internal source } j)] - (\text{operator } \phi)i(t) \qquad (5.4.1)$$

where n is the number of internal sources.

EXAMPLE 5.4.1

In the circuit in Figure 5.4.2 solve for v in the form of equation (5.4.1).
ANS.: From KCL we write

$$i_1 - i_2 - i = 0 \qquad (5.4.2)$$

Therefore

$$\left(\frac{1}{R} + \frac{1}{Lp}\right)v - \frac{1}{R}v_s = -i \qquad (5.4.3)$$

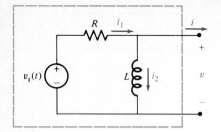

figure 5.4.2
The circuit of Example 5.4.1.

Since

$$\frac{1}{R} + \frac{1}{Lp} = \frac{Lp + R}{RLp}$$

solving for v gives

$$v = \frac{Lp}{Lp + R} v_s - \frac{RLp}{Lp + R} i \qquad (5.4.4)$$

$$\underset{\text{operator 1}}{} \qquad \underset{\text{operator } \phi}{}$$

where v_s is the only internal source.

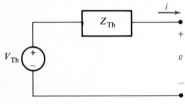

figure 5.4.3
The general Thevenin terminal equivalent
circuit.

Being able to write the voltage v in the form shown in equation (5.4.1) means that this equation also describes the Thevenin terminal equivalent circuit in Figure 5.4.3. In this equivalent circuit

$$v = V_{\text{Th}} - Z_{\text{Th}}(p)i \qquad (5.4.5)$$

and hence we have a method for finding the Thevenin circuit; i.e., solve for v in terms of i and all the internal sources and then with reference to equation (5.4.5) recognize V_{Th} and $Z_{\text{Th}}(p)$ by inspection. For instance, in Example 5.4.1 [see equation (5.4.4)] we recognize that whatever multiplies i must be Z_{Th}. The other term must be V_{Th}. Therefore,

$$V_{\text{Th}} = \frac{Lp}{Lp + R} v_s \qquad (5.4.6a)$$

and

$$Z_{\text{Th}}(p) = \frac{RLp}{Lp + R} \qquad (5.4.6b)$$

We have already said that the rules we learned for finding the Thevenin equivalent for any resistive circuit still apply in the general case, so that to find the Thevenin (open-circuit) voltage in Example 5.4.1 we could have used voltage division. Voltage v is a fraction of $v_s(t)$. That fraction is the impedance of the inductor divided by the sum of the impedance of the resistor plus the impedance of the inductor

$$V_{\text{Th}} = \frac{Lp}{Lp + R} v_s(t) \qquad (5.4.7)$$

The Thevenin (output) impedance is found by first setting the independent source equal to zero (replacing it by a short circuit) and then calculating the impedance seen when looking back into the resulting circuit from the terminals. Since the resistor and inductor are in parallel,

$$Z_{\text{Th}} = \frac{RLp}{R + Lp} \tag{5.4.8}$$

Compare (5.4.6) with (5.4.7) and (5.4.8).

Either method [writing an expression for v in the form of equation (5.4.1) or finding the V_{oc}, I_{sc}, and/or R_{Th}] also works if there are multiple internal sources in the circuit.

EXAMPLE 5.4.2

In the circuit shown in Figure 5.4.4a solve for the voltage v in the form of equation (5.4.1). Show how to find V_{Th} and Z_{Th} from this expression.

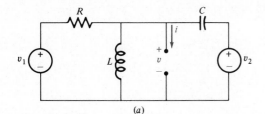

(a)

figure 5.4.4

(a) The circuit of Example 5.4.2. (b) The same circuit rearranged to expose the terminals.

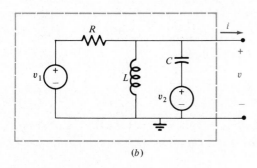

(b)

ANS.:   First redraw the circuit to expose the terminals as in Figure 5.4.4b. At the upper-right-hand node we write a node equation

$$\left(\frac{1}{R} + \frac{1}{Lp} + Cp\right)v - \frac{1}{R}v_1 - Cpv_2 = -i \tag{5.4.9}$$

Since

$$\frac{1}{R} + \frac{1}{Lp} + Cp = \frac{Lp + R + RLCp^2}{RLp}$$

we have

$$v = \frac{RLp}{RLCp^2 + Lp + R}\left(\frac{v_1}{R} + Cpv_2 - i\right) \tag{5.4.10}$$

or

$$v = \left(\frac{Lp}{RLCp^2 + Lp + R}v_1 + \frac{RLCp^2}{RLCp^2 + Lp + R}v_2\right) - \frac{RLp}{RLCp^2 + Lp + R}i \tag{5.4.11}$$

Thevenin (open-circuit) voltage source

Thevenin (output) impedance

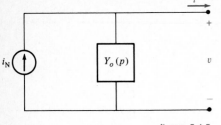

figure 5.4.5
The Norton equivalent circuit.

What about the general Norton equivalent circuit? Can we obtain it by a similar technique? For the circuit shown in Figure 5.4.5 we note that

$$i = i_N - Y_o(p)v \qquad (5.4.12)$$

Therefore if we manipulate the expression relating terminal variables i and v for any circuit into the form of equation (5.4.12), the Norton current i_N and output admittance $Y_o(p)$ are easily obtained by inspection. For example, equation (5.4.4) can be rewritten

$$\frac{RLp}{Lp + R} i = \frac{Lp}{Lp + R} v_s - v \qquad (5.4.13)$$

or

$$i = \frac{1}{R} v_s - \frac{Lp + R}{RLp} v \qquad (5.4.14)$$

So immediately, with reference to equation (5.4.12), we recognize that

$$i_N = \frac{1}{R} v_s \qquad (5.4.15)$$

and

$$Y_o(p) = \frac{Lp + R}{RLp} \qquad (5.4.16)$$

Note that (5.4.16) and (5.4.6b) are reciprocals of each other, as they should be (Y is the reciprocal of Z and vice versa).

Also remember from our earlier work that the relationship

$$V_{Th} = I_N R_{Th} \qquad (5.4.17)$$

which is now

$$V_{Th} = I_N Z_o(p) \qquad (5.4.18)$$

must hold true. We use it to check our work in Example 5.4.1. Equation (5.4.7) gives us the left-hand side of (5.4.18) and equations (5.4.8) and (5.4.15) give us the right-hand side. Thus

$$\frac{Lp}{Lp + R} v_s(t) = \frac{1}{R} v_s(t) \frac{RLp}{R + Lp} \qquad (5.4.19)$$

and our work checks.

Sometimes the internal elements in a Thevenin-Norton equivalent circuit may look strange, as in the next example, where the voltage source in the Thevenin circuit has a *different waveform* from the current source in the Norton circuit. Never mind: remember that these two circuits are only equivalent insofar as their *terminal variables v and i* are concerned. What occurs internally is not important.

EXAMPLE 5.4.3

Find the Thevenin equivalent for the circuit in Figure 5.4.6a.
ANS.:   Write a node equation at the upper node

$$i = 5u(t) - i_C = 5u(t) - 2pv \qquad (5.4.20)$$

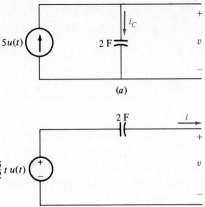

figure 5.4.6
(a) The circuit of Example 5.4.3. (b) The Thevenin equivalent of the circuit in (a).

and solve for v

$$2pv = 5u(t) - i$$

$$v = \frac{5}{2p} u(t) - \frac{1}{2p} i \qquad (5.4.21)$$

With reference to equation (5.4.5) we recognize that

$$V_{\text{Th}} = \frac{5}{2p} u(t)$$

$$= \frac{5}{2} \int_{0-}^{t} u(t)\, dt \qquad \text{for } t > 0$$

$$= \tfrac{5}{2} t\, u(t) = \tfrac{5}{2} r(t) \qquad (5.4.22)$$

and

$$Z_{\text{Th}}(p) = \frac{1}{2p} \qquad (5.4.23)$$

The Thevenin circuit is shown in Figure 5.4.6b.

It may at first seem that these two circuits are unlikely to be equivalent since one contains a ramp-function source and one a step-function source. To check on this equivalence we can write the equation relating v and i in each case. If they turn out to be the same equation, we know that v and i will be the same regardless of which equivalent circuit is used. For the Norton equivalent circuit

$$5u(t) = i + Y_C(p)v$$

$$5u(t) - i = 2pv$$

$$v = \frac{5}{2p} u(t) - \frac{i}{2p}$$

or

$$v = \tfrac{5}{2} t\, u(t) - \frac{1}{2p} i(t) \qquad (5.4.24)$$

For the Thevenin equivalent circuit

$$V_{\text{Th}} - v_C = v$$

$$V_{\text{Th}} - Z_C(p)i(t) = v$$

$$v = \tfrac{5}{2}t\, u(t) - \frac{1}{2p}\, i(t) \tag{5.4.25}$$

Since the resulting v versus i equations, (5.4.24) and (5.4.25), are identical, it is true that although their *internal* sources have different waveforms, these two circuits are *terminally* equivalent.†

The major use of Thevenin-Norton terminal equivalent circuits is to obtain an equation that can be solved for the voltage across (or the current in) whatever element is externally connected to those terminals.

EXAMPLE 5.4.4

Given the network shown in Figure 5.4.7a, first find a Norton and/or Thevenin equivalent circuit for the circuit to the left of terminals a and b. Then find the (differential) equation for the voltage across the 1-Ω resistor.

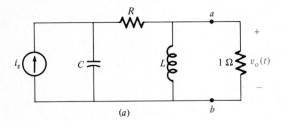

(a)

figure 5.4.7
(a) The circuit of Example 5.4.4. (b) The same circuit with the 1-Ω resistor removed, exposing terminals a and b. (c) The Thevenin terminal equivalent of (b).

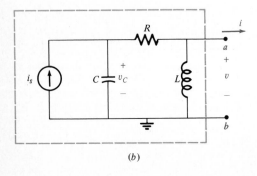

(b)

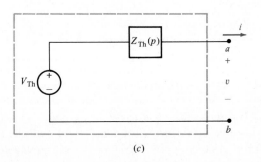

(c)

ANS.:   Let us obtain the Thevenin equivalent circuit for the network in this problem two different ways: (1) by finding the volt-ampere characteristic by writing node equations and (2) by I_{sc}, V_{oc}, and Z_{Th}.

First Method
See Figure 5.4.7b. From KCL

† Some texts call this a *source transformation* from the Thevenin to the Norton form and vice versa.

$$\left(Cp + \frac{1}{R}\right)v_C - \frac{1}{R}\,v = i_s \tag{5.4.26}$$

$$-\frac{1}{R}\,v_C + \left(\frac{1}{R} + \frac{1}{Lp}\right)v = -i \tag{5.4.27}$$

are the node equations. Solving (5.4.27) for v_C

$$Ri + \frac{Lp + R}{Lp}\,v = v_C \tag{5.4.28}$$

and substituting into (5.4.26) yields

$$\frac{RCp + 1}{R}\left(Ri + \frac{Lp + R}{Lp}\,v\right) - \frac{1}{R}\,v = i_s$$

$$(RCp + 1)i + \frac{(RCp + 1)(Lp + R) - Lp}{RLp}\,v = i_s \tag{5.4.29}$$

or

$$v = \frac{(1/C)p}{p^2 + (R/L)p + 1/LC}\,i_s - \frac{Rp(p + 1/RC)}{p^2 + (R/L)p + 1/LC}\,i \tag{5.4.30}$$

Note that equation (5.4.30) equally well describes the Thevenin equivalent circuit of Figure 5.4.7c:

$$v = V_{\text{Th}} - Z_{\text{Th}}(p)i \tag{5.4.31}$$

Comparing (5.4.30) and (5.4.31), we see that

$$V_{\text{Th}} = \frac{(1/C)p}{p^2 + (R/L)p + 1/LC}\,i_s \tag{5.4.32}$$

and

$$Z_{\text{Th}} = \frac{Rp(p + 1/RC)}{p^2 + (R/L)p + 1/LC} \tag{5.4.33}$$

Second Method
Find I_{sc} in Figure 5.4.8a

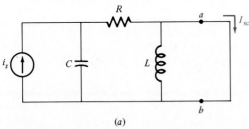

(a)

figure 5.4.8
(a) Finding the short-circuit current I_{sc}. (b) Finding the output impedance Z_o. (c) Using the Thevenin equivalent to solve for the output voltage across the 1-Ω resistor.

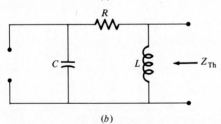

(b)

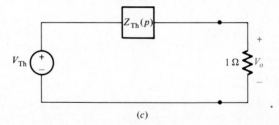

(c)

$$I_{sc} = i_s \frac{Z_C(p)}{Z_C(p) + Z_R(p)} = \frac{1/Cp}{1/Cp + R} i_s \tag{5.4.34}$$

Find Z_{Th}. Remember to set $i_s = 0$ (see Figure 5.4.8b).

$$Z_{Th} = \frac{[Z_C(p) + Z_R(p)]Z_L(p)}{Z_C(p) + Z_R(p) + Z_L(p)} = \frac{(1/Cp + R)Lp}{1/Cp + R + Lp} = \frac{p^2(R + 1/Cp)}{p^2 + (R/L)p + 1/LC} \tag{5.4.35}$$

Note that this agrees with equation (5.4.33).

We can find V_{oc} from $V_{oc} = I_{sc} Z_{Th}$:

$$V_{oc} = \frac{1/Cp}{1/Cp + R} i_s \frac{p^2(R + 1/Cp)}{p^2 + (R/L)p + 1/LC}$$

hence

$$V_{Th} = \frac{(1/C)pi_s}{p^2 + (R/L)p + 1/LC} \tag{5.4.36}$$

Note that this agrees with equation (5.4.32).

Finally the voltage v_o is found as in Figure 5.4.8c after reconnecting the 1-Ω resistor to the Thevenin equivalent circuit

$$v_o(t) = V_{Th}[Z_{Th}(p) + 1]^{-1} = \frac{(1/C)pi_s}{p^2 + (R/L)p + 1/LC}\left[\frac{p^2(R + 1/Cp)}{p^2 + (R/L)p + 1/LC} + 1\right]^{-1}$$

$$v_o(t) = \left[\frac{(1/C)p}{p^2(R + 1) + p(R/L + 1/C) + 1/LC}\right]i_s(t) \tag{5.4.37}$$

From which the system differential equation is obtained by multiplying through by the denominator and letting $p = d/dt$

$$(R + 1)\frac{d^2v_o}{dt^2} + \left(\frac{R}{L} + \frac{1}{C}\right)\frac{dv_o}{dt} + \frac{1}{LC} v_o = \frac{1}{C}\frac{d}{dt} i_s(t) \tag{5.4.38}$$

The term in brackets in equation (5.4.37) is the transfer function that *operates on the input* current $i_s(t)$ *to produce the output* voltage $v_o(t)$. The transfer function is a fundamental concept in electrical engineering of which much use will be made in the remainder of this text.

DRILL PROBLEM

Suppose the 1-Ω load resistor in the circuit in Figure 5.4.7a is replaced by a 2-H inductor. Assume the other elements in the circuit are all unity-valued; $R = L = C = 1$. What differential equation would then relate $v_o(t)$ to $i_s(t)$?

ANS.: $(2p^2 + 3p + 3)v_o = 2pi_s$.

duality 5.5

omit
p195-7

Consider the two circuits shown in Figure 5.5.1 together with their system equations. Notice that the system equations are exactly the same in structure, the only difference being that the variables and constants are opposites of each other. In other words, voltage replaces current, R replaces G, and C replaces L. Thus the following quantities are said to be *duals*:

Voltage	Current
Capacitance	Inductance
Resistance	Conductance
Series	Parallel
Short circuit	Open circuit
Velocity	Force
Mass	Compliance $1/k$
D	$1/D$

Thanks to the nature of the principle of duality, if you have the solution of one network, you also have the solution to its dual. Note, however, that some dual networks have no physical meaning (e.g., there is no thermal element that is the dual of thermal capacity), that in mechanical schematics masses must have one terminal connected to ground, and that only planar networks have duals. To obtain the dual network:

1. Place a node inside each mesh (window) and one in the outside region. Label each with a letter or number.
2. From each such node draw a line to another node if and only if it will pass through exactly one element. Continue to do this until a line has been drawn through every element in the original circuit.
3. Replace each line by the dual element.

The question of how (in which polarity) the sources should be connected into the dual circuit is answered as follows. Draw clockwise mesh currents in

figure 5.5.1
Two dual networks.

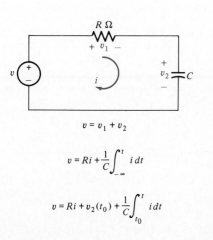

$$v = v_1 + v_2$$

$$v = Ri + \frac{1}{C}\int_{-\infty}^{t} i\, dt$$

$$v = Ri + v_2(t_0) + \frac{1}{C}\int_{t_0}^{t} i\, dt$$

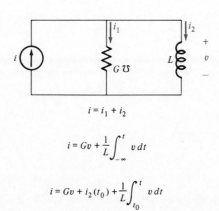

$$i = i_1 + i_2$$

$$i = Gv + \frac{1}{L}\int_{-\infty}^{t} v\, dt$$

$$i = Gv + i_2(t_0) + \frac{1}{L}\int_{t_0}^{t} v\, dt$$

the original circuit (each such mesh current corresponds to a node voltage in the dual); a source in the original circuit that *aids* a given mesh current becomes the opposite type in the dual and drives charge *into* the corresponding node.

EXAMPLE 5.5.1

Find the dual of the network shown in Figure 5.5.2a.

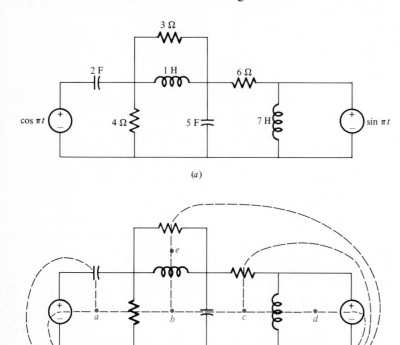

figure 5.5.2
Obtaining the dual of a network.

(a)

(b)

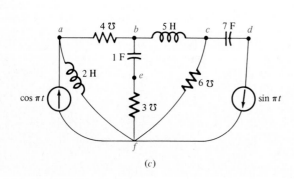

(c)

ANS.: Place dots in each mesh and one outside. Connect them with dashed lines each of which passes through only one element (see Figure 5.5.2b). Then replace each dashed line with the dual of the element the line intersects in the original circuit. For example, since the line from *a* to *b* intersects a 4-Ω resistor, in the final circuit nodes *a* and *b* are connected by a 4-℧ conductance. The dashed line from *b* to *c* goes through a 5-F capacitor in the original circuit, and so *b* and *c* are connected by a 5-H inductor in the dual. The voltage source in mesh *d* *bucks* the assumed clockwise mesh current; therefore in the dual the corresponding current source *removes* charge from node *d*.

The dual of the dual circuit is the original circuit. You should prove this to yourself in the previous example.

DRILL PROBLEM

Find the dual of the circuit shown in Figure 5.5.3*a*.

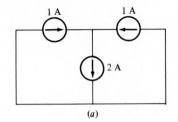

(*a*)

figure 5.5.3
(*a*) The circuit of the drill problem and (*b*) its dual.

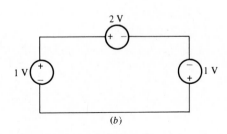

(*b*)

ANS.: See Figure 5.5.3*b*.

summary **5.6**

In this chapter we have generalized the technique of Kirchhoff's laws to be applicable to circuits containing inductances and capacitances as well as resistors. The ideas of *p*-operator notation and generalized impedance expedite the writing and manipulation of system (differential) equations. The transfer function of a system was introduced as the operator applied to one variable in order to produce another variable of the system. Finally we discussed the dual of a system and the general ideas of duality between voltage and current, inductance and capacitance, etc. We have now covered many of the basic definitions and techniques of circuit analysis and know how to obtain equations that relate the sources to the output variables. In Chapter 6 we shall begin to solve for these output signals in systems that contain an energy-storage element.

problems

1. (a) A resistor R and a capacitor C connected in series form a single closed mesh. Write the differential equation for $i(t)$, the current in the mesh. (b) Insert a voltage source $v_s(t)$ into the mesh such that $i(t)$ leaves from its positive terminal and repeat part (a).

2. (a) A mass M hangs on an ideal spring K from an overhead beam. A force $f(t)$ is applied to M in the upward (positive) direction. Write a differential equation for the velocity of the mass. (b) Two 1-kG masses, M_1 and M_2, connected by a dashpot $D = 1 \text{ N} \cdot \text{s/m}$ ride on frictionless rollers on a flat floor. A force $f(t)$ is applied to M_1 in the positive direction. Find the differential equation that relates the velocity of M_1 to $f(t)$.

3. A mass M hangs from the bottom of an ideal spring K. The top end of the spring is connected to an overhead beam by an ideal velocity source whose positive terminal is at the beam. Derive a differential equation, the solution of which will be $\mathcal{V}_M(t)$, the velocity of the mass.

4. (a) What is the net equivalent impedance operator of the parallel combination of a 3-H inductor and a 2-F capacitor? (b) A 1-Ω resistor is in series with the parallel combination described in part (a). A voltage source $v_s(t)$ is connected across the resulting two terminals. Find a differential equation that relates $v_C(t)$ to $v_s(t)$

5. Write the single differential equation needed to solve for $v_a(t)$ in the circuit shown in Figure P5.5.

6. Write the mesh equations for the circuit of Figure P5.6, collecting all terms in i_1, i_2, and i_3 on the left and putting all independent sources on the right.

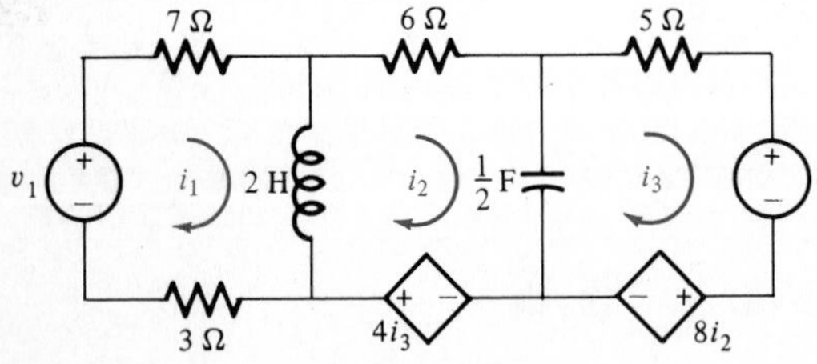

7. A field-effect transistor (FET) used as a high-frequency common-source amplifier has the equivalent circuit shown in Figure P5.7. In p-operator notation write the node equations necessary to solve for all the unknown node voltages.

8. Consider the design of a box in which a canister of plutonium is to be shipped. The canister is extremely heavy (mass M) and is placed inside the box, whose mass is negligible

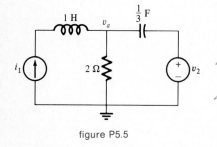

figure P5.5

figure P5.6

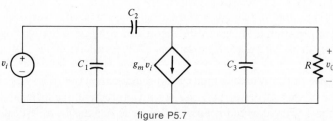

figure P5.7

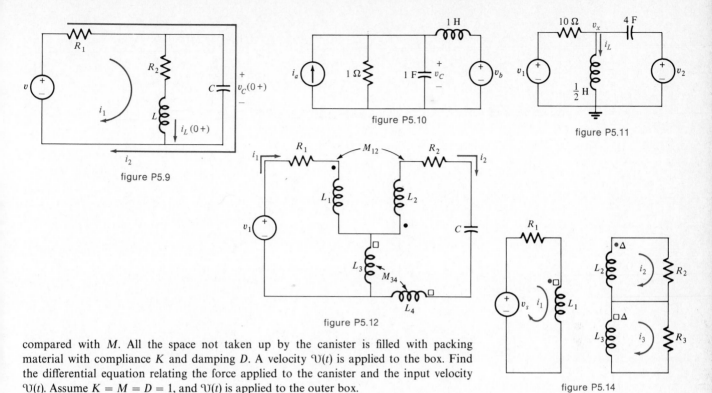

figure P5.9

figure P5.10

figure P5.11

figure P5.12

figure P5.14

compared with M. All the space not taken up by the canister is filled with packing material with compliance K and damping D. A velocity $\mathcal{V}(t)$ is applied to the box. Find the differential equation relating the force applied to the canister and the input velocity $\mathcal{V}(t)$. Assume $K = M = D = 1$, and $\mathcal{V}(t)$ is applied to the outer box.

9. In the circuit shown in Figure P5.9 the initial current through the inductor is $i_L(0+)$, and the initial voltage across the capacitor is $v_C(0+)$. Their direction and polarity are shown in the diagram. Write KVL equations for the two loop currents i_1 and i_2. Put sources on the right and collect terms in i_1 and i_2 on the left.

10. For the circuit in Figure P5.10 write the differential equation that relates $v_C(t)$ to the two sources.

11. (a) For the circuit in Figure P5.11 write the differential equation that relates $v_x(t)$ to the two sources. (b) Repeat part (a) for the current $i_L(t)$.

12. For the circuit in Figure P5.12 write the two mesh equations.

13. The mesh equations of an electric network are as follows. Draw the network indicating the size and type of each element. Is this the only possible network?

$$\left(\frac{4}{p+1}\right)i_1 - 5pi_2 \qquad -i_3 = 10$$

$$-5pi_1 + (5p+12)i_2 \qquad = 0$$

$$-i_1 \qquad\qquad +7i_3 = 15$$

14. (a) Write the mesh equations for the circuit in Figure P5.14. Include the effects of M_{12}, M_{13}, and M_{23}. (b) If resistors R_2 and R_3 are replaced by capacitors C_1 and C_2, respectively, repeat part (a).

15. A series RLC circuit is driven by a voltage source $v_s(t)$. (a) Find the differential equation for the voltage $v_C(t)$ across the capacitor. Repeat part (a) for (b) $v_L(t)$ and (c) $v_R(t)$.

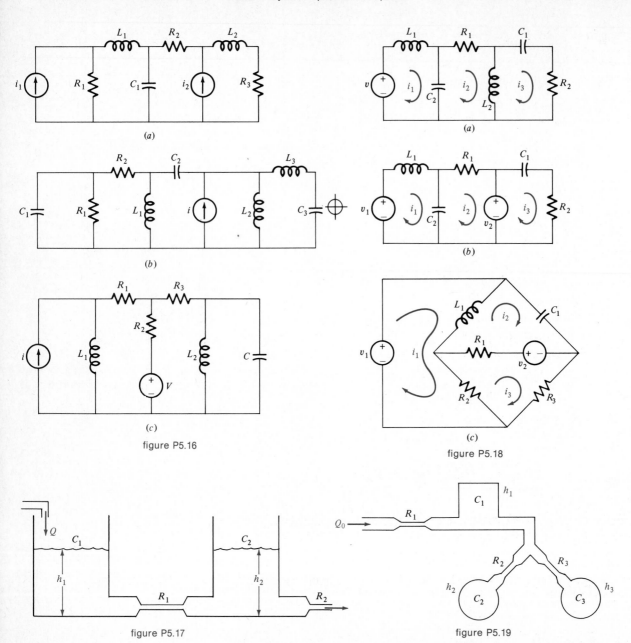

figure P5.16

figure P5.18

figure P5.17

figure P5.19

16. Write a set of node equations for each of the systems shown in Figure P5.16.

17. (a) Draw the schematic diagram and write a set of system differential equations in the variables h_1 and h_2 for the fluid system shown in Figure P5.17. (b) Repeat part (a) if the fluid resistor R_2 leads out to the left from C_1 rather than to the right from C_2 as shown. Assume that Q still goes into C_1.

18. Write the mesh equations for each of the systems shown in Figure P5.18.

19. A model of a patient who is connected to a respirator is shown in Figure P5.19, where

C_1 is the capacity of the machine tube and the patient's mouth and trachea and C_2 and C_3 are the right and left lungs. (a) Draw the schematic diagram and write all necessary KCL (node) equations. (b) Let the value of all fluid resistances and capacitances be unity. Find the single differential equation that relates h_3 to Q_0.

20. In the circuit of Figure P5.20 choose node a as the reference. (a) Write the two p-operator node equations necessary to solve this network. All element values are unity. (b) Use Cramer's rule or any other convenient method to determine from the results in part (a) the single differential equation that relates v_b to the voltage source.

21. Repeat both parts of problem 20 for the circuit in Figure P5.21.

22. For the simple circuit shown in Figure P5.22: (a) write the mesh equations; (b) solve them for two separate differential equations, one in i_1 and one in i_2 ; (c) use the results of part (b) to find a differential equation in $v_{ab}(t)$; and (d) check your answer to part (c) by writing one node equation.

23. (a) Write the p-operator node equations for the circuit shown in Figure P5.23. (b) Find the differential equation in standard d/dt form that relates v_b to the sources. (c) Write the mesh equations in p-operator notation, then write v_b in terms of i_1 and i_2, and finally, determine the expression for v_b in terms of v_o and i_s.

24. For the circuit in Figure P5.24 label nodes, assign current directions, and write the necessary node equations.

25. Redraw the circuit in Figure P5.24, assign loop currents (all in the same direction), and write the necessary mesh equations.

26. Given the circuit of Figure P5.26, (a) write mesh equations and (b) use them to determine the differential equation that relates mesh current i_2 to the inputs $v_1(t)$ and $v_2(t)$. (c) Write the result of part (b) as a transfer-function operator.

27. Two masses M_1 and M_2 are sitting on a flat floor with friction D_1 and D_2, respectively, between each mass and the floor. The masses are connected to each other by the parallel combination of spring K and dashpot D_3. Force $f_s(t)$ is applied in the positive direction to M_1. Obtain simultaneous differential equations in $\mathfrak{V}_1$ and $\mathfrak{V}_2$, the velocity of each mass.

28. Mass M_1 slides on a frictionless floor when force $f(t)$ is applied to it, but it is attached to a nearby wall by a spring K_1. Mass M_2 sits on top of M_1 and is connected to it by friction D. M_2 is connected to the wall by spring K_2. Obtain simultaneous differential equations in $\mathfrak{V}_1$ and $\mathfrak{V}_2$.

29. Three masses are connected as follows: M_1 hangs from an overhead beam by spring K_1, and M_2 is hung from the same beam by K_2. These two masses are in close proximity

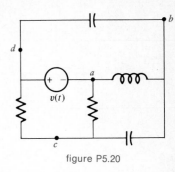

figure P5.20

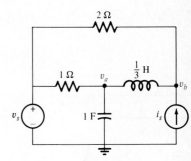

figure P5.21

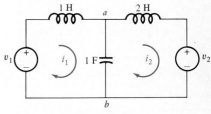

figure P5.22

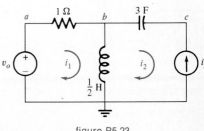

figure P5.23

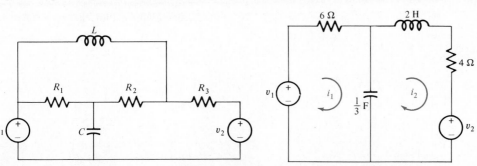

figure P5.24   figure P5.26

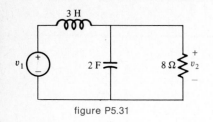

figure P5.31

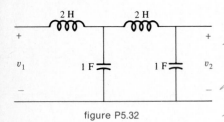

figure P5.32

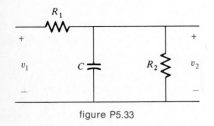

figure P5.33

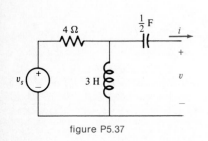

figure P5.37

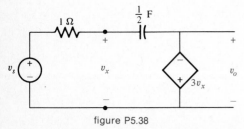

figure P5.38

and are thus connected by air friction D. M_3 hangs from M_2 by spring K_3 and is connected to the floor by a series connection of spring K_4 and dashpot D_2. Use nodal methods (KCL) to obtain the complete set of necessary simultaneous differential equations.

30. An angular-velocity source $\omega_s(t)$ is connected to a drive gear by a lightweight, thin, twistable shaft modeled by compliance K_1. The second gear (which is meshed with the drive gear) has n teeth for every tooth on the driver. By another twistable shaft (compliance K_2) the second gear drives an inertia J, which is supported at its outer end by a bearing having angular friction D. Write three simultaneous differential equations in the following variables: torque transmitted by the drive shaft from the velocity source; torque in the second shaft; torque that overcomes bearing friction.

31. Determine the p-operator transfer function that relates output $v_2(t)$ to input $v_1(t)$ in the circuit in Figure P5.31.

32. Find the transfer function for the circuit in Figure P5.32 that relates the input and output voltages; $v_2(t)$ is the output.

33. Find the p-operator transfer function $H(p) = v_2/v_1$ for the circuit in Figure P5.33.

34. Find the Norton equivalent circuit of a two-terminal network that consists of a unit-ramp-function voltage source $r(t)$ in series with a 3-F capacitor. Find the Thevenin equivalent of your answer. Is it the original circuit?

35. A voltage source $v(t)$ in series with a 1-F capacitor is placed inside a black box, and the two ends of this combination are led to terminals that extend through one wall of the box. Just inside the box a 3-H inductor is connected across the terminals. (a) Find the differential equation in d/dt form that relates $v_1(t)$ to the open-circuit output voltage $v_2(t)$ that can be measured outside the box. (b) Find the Thevenin equivalent circuit. (c) Solve for $v_2(t)$ in the form

$$v_2(t) = (\text{operator } 1 \times \text{source}) - (\text{operator } 2 \times i)$$

(d) Repeat parts (a) to (c) changing the 3-H inductor to a 3-F capacitor.

36. A Thevenin circuit consists of a unit-ramp-function voltage source and a 3-H inductor. A 4-Ω load resistor is connected to the output terminals. (a) Find an expression for the voltage across the resistor $v_R(t)$ as a function of the source voltage and a transfer-function operator. (b) Find the Norton equivalent for the Thevenin circuit (not including the resistor). (c) Use this Norton equivalent to determine an expression for $v_R(t)$ in terms of a transfer-function operator and $i_N(t)$.

37. Sketch the Norton equivalent for the circuit shown in Figure P5.37, carefully labeling the elements with their p-operator descriptions.

38. In the circuit shown in Figure P5.38 find the differential equation that relates $v_o(t)$ to $v_s(t)$.

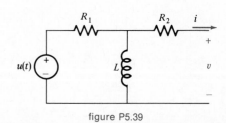

figure P5.39

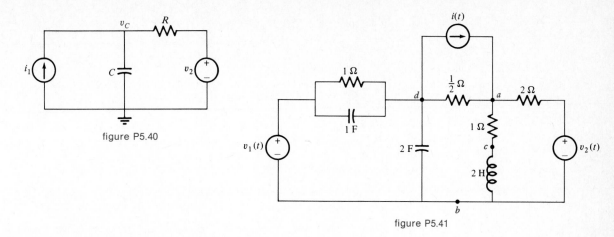

figure P5.40

figure P5.41

39. In the circuit of Figure P5.39, find $v(t)$ in the form

$$v(t) = (\text{operator } 1 \times \text{internal sources}) + [\text{operator } 2 \times i(t)]$$

40. In the circuit in Figure P5.40 use superposition to solve for v_C as a function of $i_1(t)$ and $v_2(t)$. Check your answer by writing a node equation.

41. In the circuit of Figure P5.41, find the p-operator differential equation relating v_{ab} to the sources (a) by successive Thevenin-Norton transformations and then combining elements in series and parallel and (b) by superposition.

42. (a) Write the mesh equations for the circuit shown in Figure P5.42. (b) Sketch the circuit that is the dual of the one in part (a) and write its node equations.

43. In problem 15 you were asked to develop differential equations for the voltage across each element in a voltage-driven series RLC circuit. Using these results, write the differential equations for the current in each element of a current-driven parallel RLC circuit.

44. Write the mesh equations for the circuit shown in Figure P5.44. Sketch the dual circuit and write its node equations.

45. Sketch the dual of the system in (a) part (b) of problem 2; (b) part (b) of problem 4; (c) problem 5; and (d) problem 16.

46. Two mutually coupled inductors L_1 and L_2 are connected in parallel. Their dotted terminals are connected together. The combination is excited by a voltage source $v_s(t)$. Mesh current i_1 leaves the plus terminal of the source and passes through L_1. Mesh current i_2 opposes i_1 in L_1 and also flows through L_2. (a) Write the mesh equations in p-operator notation and (b) solve them for pi_1 as a function of $v_s(t)$. (c) From the result in part (b) find the equivalent inductance seen by $v_s(t)$.

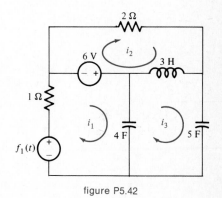

figure P5.42

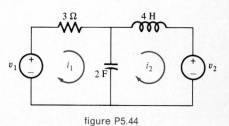

figure P5.44

chapter 6
FIRST-ORDER SYSTEMS

introduction 6.1

If a circuit containing one or more energy-storage elements is excited by a source that abruptly changes its value (e.g., a battery that is connected to the circuit at $t = 0$ by the closing of a switch), the node voltages and element currents in that circuit will take time to reach their final steady-state values. To find expressions (functions of time) for any output variable (node voltage or element current) we must solve the differential system equation that relates that variable to the input source. We should therefore write a differential equation which describes the system during the time interval of interest to us (usually for $t > 0$).

EXAMPLE 6.1.1

For the circuit in Figure 6.1.1 write a system equation that can be solved for $v_C(t)$, for $t > 0$.

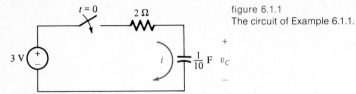

figure 6.1.1
The circuit of Example 6.1.1.

ANS.:   By KVL

$$iR + v_C = 3 \qquad \text{where} \qquad i = C\frac{dv_C}{dt}$$

Thus
$$C\frac{dv_C}{dt}R + v_C = 3 \qquad \text{and} \qquad \frac{dv_C}{dt} + \frac{1}{RC}v_C = \frac{3}{RC}$$

and finally
$$\frac{dv_C}{dt} + 5v_C = 15 \qquad \text{for } t > 0$$

EXAMPLE 6.1.2

Write a differential equation that describes the system in Figure 6.1.2 after the switch is thrown from a to b.

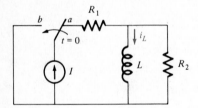

figure 6.1.2
The circuit of Example 6.1.2.

ANS.:   After $t = 0$, i_L flows only through R_2. Thus by KVL around the right-hand mesh

$$L\frac{di_L}{dt} + i_L R_2 = 0$$

$$\frac{di_L}{dt} + \frac{R_2}{L}i_L = 0 \qquad \text{for } t > 0$$

EXAMPLE 6.1.3

Write a differential equation valid for $t > 0$ in the circuit of Figure 6.1.3.
ANS.:   By KCL at node a

$$\frac{v_a - v}{R_1} + i_L + \frac{v_a}{R_2} = 0$$

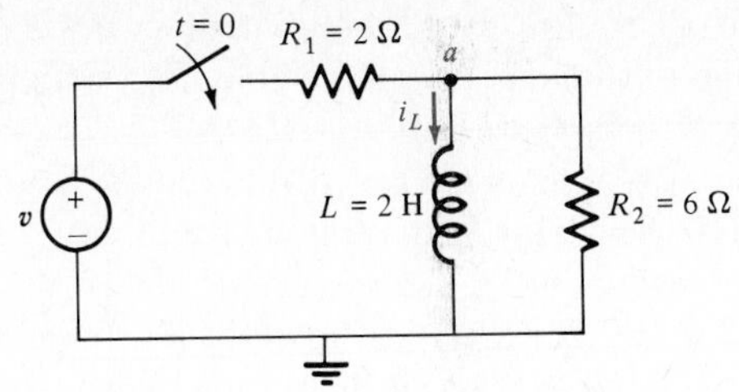

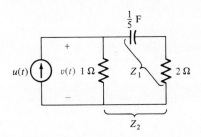

figure 6.1.3
The circuit of Example 6.1.3.

or

$$+v_a \frac{R_1 + R_2}{R_1 R_2} + i_L = \frac{v}{R_1}$$

Since $v_a = L(di_L/dt)$,

$$L \frac{R_1 + R_2}{R_1 R_2} \frac{di_L}{dt} + i_L = \frac{v}{R_1}$$

$$\frac{di_L}{dt} + \frac{R_1 R_2}{L(R_1 + R_2)} i_L = \frac{R_2}{L(R_1 + R_2)} v$$

$$\frac{di_L}{dt} + \tfrac{3}{4} i_L = \tfrac{3}{8} v \qquad \text{for } t > 0$$

EXAMPLE 6.1.4

Write a differential equation in terms of $v(t)$ for $t > 0$ in Figure 6.1.4. Use p-operator methods.

figure 6.1.4
The circuit of Example 6.1.4.

ANS.:   Using the generalized impedance ideas of Chapter 5, we find that

$$Z_1 = 2 + \frac{1}{Cp} = 2 + \frac{5}{p} = \frac{2p + 5}{p}$$

and Z_2 is equal to Z_1 in parallel with $1\ \Omega$:

$$Z_2 = \frac{Z_1(1)}{Z_1 + 1} = \frac{(2p + 5)/p}{(2p + 5)/p + 1} = \frac{2p + 5}{3p + 5}$$

Since $v = iZ_2$,

$$(3p + 5)v(t) = (2p + 5)i(t)$$

$$\frac{3dv}{dt} + 5v = 2\delta(t) + 5u(t) = 5 \qquad \text{for } t > 0$$

Once we have written down the differential equation that describes a given system, we must solve it for the value of the dependent variable. For example, in Example 6.1.4 we would like to solve for the expression for $v(t)$.

Some simple systems have no sources and give rise to differential equations that are easy to solve. We shall solve these easy equations first and then extend our method to systems with one or more sources. In any event, our goal in this chapter is to be able to obtain the correct mathematical function of time for any voltage or current in any place in any circuit that has only one energy source.

DRILL PROBLEM

In the circuit of Figure 6.1.3 replace the 2-H inductor with a 2-F capacitor. Write a differential equation for $v_C(t)$ that is valid for $t > 0$. Use p operators.

ANS.: $(3p + 1)v_C = \frac{3}{4}v.$

6.2 natural response of first-order systems

Suppose we have a system containing no sources that are active during positive values of time. We want to find expressions for the variables for all $t > 0$. To do this we first write the equilibrium equation valid for that system for $t > 0$. Such equations will generally be of the form

$$\frac{dx}{dt} + \alpha x = 0$$

Systems in which the source(s) are switched out of the network give rise to

---

EXAMPLE 6.2.1

Write the equilibrium equation for the system in Figure 6.2.1 for $t > 0$.

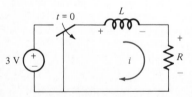

figure 6.2.1
An *RL* circuit.

ANS.:   Use KVL. For $t > 0$

$$L\frac{di}{dt} + iR = 0$$

Hence

$$\frac{di}{dt} + \frac{R}{L}\,i = 0$$

differential equations having no sources or *forcing functions*. Such differential equations are called *homogeneous* differential equations, and their solutions are called the *source-free response* or *natural response* of the system. One way the homogeneous differential equation found in Example 6.2.1

$$\frac{di}{dt} + \frac{R}{L} i = 0 \qquad \text{for } t > 0 \tag{6.2.1}$$

can be solved for $i(t)$ is as follows. For simplicity, let the constant R/L be called k. Then

$$\frac{di}{dt} + ki = 0 \qquad \text{for } t > 0 \tag{6.2.2}$$

which says that we must seek a function $i(t)$ such that its derivative when added to itself (multiplied by k) totals zero for all t. This means that the function $i(t)$ and its derivative di/dt must have the same waveshape. If $i(t)$ were a parabola, its derivative would be a ramp and there is no way their sum could come out zero for all values of t. But what about the exponential? The function

$$i(t) = Ae^{st} \tag{6.2.3}$$

where A and s are constants, has the derivative

$$\frac{di}{dt} = sAe^{st} \tag{6.2.4}$$

Inserting (6.2.3) and (6.2.4) into (6.2.2), we get

$$sAe^{st} + kAe^{st} = 0 \tag{6.2.5}$$

or

$$s + k = 0 \tag{6.2.6}$$

Thus

$$s = -k \tag{6.2.7}$$

and the solution is

$$i(t) = Ae^{-kt} \tag{6.2.8}$$

We still have not evaluated the unknown constant A, but we shall deal with that problem shortly. The main point is that we know the general mathematical expression for the response. The value of k is determined by the values of the system parameters.

Equation (6.2.6) is called the *characteristic equation* or *eigenequation*. The solution, equation (6.2.7), of the characteristic equation is called the *characteristic value* or *eigenvalue*.

The solution to a source-free system is called its *natural* response because this is what the system does naturally—without being forced to. The system acts in this manner because of its inherent structure, not because of any sources connected to it, driving it.

Equation (6.2.8) can be plotted as shown in Figure 6.2.2. Note that when $t = 1/k$, the value of i has decayed to e^{-1} times its initial value at $t = 0 + \dagger$; that is, at $t = 1/k$

---

† Time $t = 0+$ is understood to be the instant of time just after $t = 0$.

figure 6.2.2
The natural response of the circuit in Figure
6.2.1 ($k = R/L$).

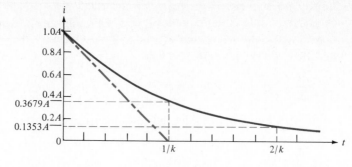

$$i\left(\frac{1}{k}\right) = Ae^{-k(1/k)} = Ae^{-1} = 0.3679A \qquad (6.2.9)$$

where $1/k$ is the *time constant* of the system. Similarly, at two time constants after $t = 0$ the response is e^{-2} times what it was at $t = 0+$. The time constant is the negative reciprocal of the characteristic value.

It should also be remembered from Chapter 3 that if the initial slope of $i(t)$ at $t = 0+$ is extended toward the right, it will intersect the t axis at $t =$ one time constant. The time constant is usually given the symbol τ; so in equation (6.2.8) we say

$$\frac{1}{k} = \tau$$

or $$i(t) = Ae^{-t/\tau} \qquad (6.2.10)$$

EXAMPLE 6.2.2

Solve the circuit in Figure 6.2.3 for $v_C(t)$.

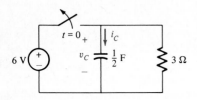

figure 6.2.3
The circuit of Example 6.2.2.

ANS.: For $t > 0$

$$v_C + i_C R = 0 \qquad v_C + RC\frac{dv_C}{dt} = 0 \qquad v_C + (3)\left(\frac{1}{2}\right)\frac{dv_C}{dt} = 0 \qquad \frac{dv_C}{dt} + \tfrac{2}{3}v_C = 0 \qquad (6.2.11)$$

Assume a solution of the form

$$v_C(t) = Ae^{st} \qquad (6.2.12)$$

and write the derivative

$$\frac{dv_C}{dt} = \frac{d}{dt}Ae^{st} = sAe^{st} \qquad (6.2.13)$$

Substitute (6.2.12) and (6.2.13) into (6.2.11)

$$sAe^{st} + \tfrac{2}{3}Ae^{st} = 0$$

Thus the characteristic equation is

$$s + \tfrac{2}{3} = 0$$

and its solution, the characteristic value, is $s = -\tfrac{2}{3}$. Hence from (6.2.12)

$$v_C(t) = Ae^{-(2/3)t} \qquad \text{for } t > 0 \tag{6.2.14}$$

Note that the time constant is $\tau = \tfrac{3}{2}$. The constant A will be evaluated in Sections 6.3 and 6.4.

EXAMPLE 6.2.3

In Figure 6.2.4 solve for $i(t)$ for $t > 0$.

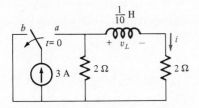

figure 6.2.4
The circuit of Example 6.2.3.

ANS.:   For $t > 0$, by KVL,

$$v_L + i(2 + 2) = 0 \tag{6.2.15}$$

$$L\frac{di}{dt} + 4i = 0 \tag{6.2.16}$$

$$\frac{di}{dt} + 40i = 0 \tag{6.2.17}$$

Thus

$$i(t) = Ae^{st} \tag{6.2.18}$$

and

$$\frac{di}{dt} = sAe^{st} \tag{6.2.19}$$

whence equation (6.2.17) becomes

$$sAe^{st} + 40Ae^{st} = 0 \tag{6.2.20}$$

and the characteristic equation is therefore

$$s + 40 = 0 \tag{6.2.21}$$

The characteristic value is

$$s = -40 \tag{6.2.22}$$

so that

$$i(t) = Ae^{-40t} \qquad t > 0$$

The time constant is $\tau = \tfrac{1}{40}$.

In general, then, any source-free system that contains only a single energy-storage element can be solved for its natural response as follows:

1. Obtain the differential equation that is valid for the time interval of interest. (This equation will be homogeneous, i.e., equal to zero.)
2. Assume a solution of the form $x(t) = Ae^{st}$.
3. Insert this $x(t)$ and its derivative into the differential equation. Divide by Ae^{st} and get the characteristic equation as a result.
4. Solve for the characteristic value and substitute it back into the solution written in step 2.

This is part of a general method that we shall expand later to enable us to solve systems with sources and with more than one energy-storage element.

FINDING THE TIME CONSTANT DIRECTLY

The value of the time constant in any first-order system is equal either to RC (if the energy-storage element is a capacitor) or to L/R (if the energy-storage element is an inductor). In both cases R is the Thevenin equivalent resistance of the rest of the network (as seen by the energy-storage element).

EXAMPLE 6.2.4

Find the time constant of the circuit in Example 6.2.1 without writing the differential equation.

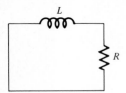

figure 6.2.5
The circuit of Figure 6.2.1 for $t > 0$.

ANS.:   For $t > 0$ the circuit is as shown in Figure 6.2.5. Therefore the time constant τ is L/R.

EXAMPLE 6.2.5

Find the time constant in Example 6.2.2 without writing the differential equation.

figure 6.2.6
The circuit of Figure 6.2.3 for $t > 0$.

ANS.:   For $t > 0$ the circuit is shown in Figure 6.2.6. Therefore the time constant $\tau = RC = \frac{3}{2}$ s.

EXAMPLE 6.2.6

Find the time constant of the circuit in Example 6.2.3.

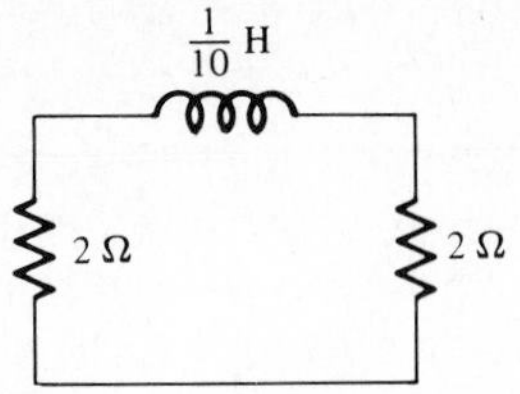

figure 6.2.7
The circuit of Figure 6.2.4 for $t > 0$.

ANS.:   The circuit for $t > 0$ is shown in Figure 6.2.7. The equivalent resistance seen by the $L = \frac{1}{10}$ H inductor is $R = 2 + 2 = 4\ \Omega$. Thus, the time constant is

$$\frac{L}{R} = \frac{\frac{1}{10}}{4} = \frac{1}{40}\ \text{s}$$

For all *first-order* natural responses we can use the following methods.

Method 1
The natural response is of the form

$$x(t) = Ae^{-t/\tau}$$

where $\tau = L/R$ or RC and R is the Thevenin equivalent resistance of the remainder of the circuit *as seen by the energy-storage element*. We shall see that this shortcut method of finding the natural response works for any first-order circuit whether it contains sources during $t > 0$ or not. We discuss this further in Section 6.4.

Method 2
If the differential equation is written in *p*-operator notation, substituting the *algebraic variable s* for the *operator p* yields the characteristic equation.† For example, equation (6.2.17) in *p*-operator notation is

$$pi + 40i = 0$$

Substituting *s* for *p* yields (after canceling *i*)

$$s + 40 = 0$$

which is equation (6.2.21). Thus, *computationally*, we can always jump from the differential equation directly to the characteristic equation; see equations (6.2.18) to (6.2.20).

DRILL PROBLEM

A 3-H inductor and a 4-Ω resistor are connected in series to form a single closed mesh. Write the differential equation in $i(t)$ and solve it to obtain an expression for $i(t)$ that is valid for $t > 0$. Do not evaluate the unknown coefficient.

ANS.: $i(t) = Ae^{-(4/3)t}$.

---

† This is also true for higher-order equations (see Section 7.2).

6.3 initial conditions

In Section 6.2 we learned how the forms of the solutions of single energy-storage (first-order) source-free systems are obtained from their homogeneous differential equations. We now consider how to evaluate the unknown coefficient A that appears in the natural response. The natural response of all first-order systems is

$$x(t) = Ae^{st} \tag{6.3.1}$$

where x is the voltage or current of interest and s is the eigenvalue, which depends on the values of the elements in the system. This function has a plot versus time as shown in Figure 6.3.1. Note that the yet-to-be-determined constant A is the magnitude of $x(t)$ at $t = 0+$; that is, just after $t = 0$

$$x(0+) = Ae^0 \tag{6.3.2}$$

$$x(0+) = A \tag{6.3.3}$$

This is therefore a way to find the value of A because the numerical value of $x(0+)$ can be determined by analyzing the system at the instant $t = 0+$.

figure 6.3.1
The natural response of any first-order system ($\tau = -1/s$).

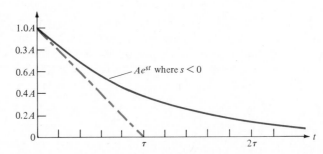

EXAMPLE 6.3.1

In the circuit shown in Figure 6.3.2 find the natural response and the unevaluated constant. The switch opens at $t = 0$ and v_s is constant in time.

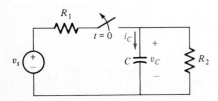

figure 6.3.2
The circuit of Example 6.3.1.

ANS.:   Find the natural response as in the previous section. For $t > 0$ KVL states that

$$v_C + i_C R_2 = 0 \qquad \text{and} \qquad \frac{dv_C}{dt} + \frac{1}{R_2 C} v_C = 0$$

If we assume the form of $v_C(t)$ to be exponential, $v_C = Ae^{st}$ yields the characteristic function

$$sAe^{st} + \frac{1}{R_2 C} Ae^{st} = 0$$

and the characteristic value

$$s = -\frac{1}{R_2 C}$$

so that

$$v_C(t) = Ae^{-t/R_2 C}$$

Now we can find the value of A by means of initial conditions, in other words by recognizing that A is the value of v_C when $t = 0+$.

Since v_s is a constant in this problem and the switch remains closed until $t = 0$, the capacitor acts like an open circuit for $t < 0$. Thus

$$v_C(t) = \frac{R_2}{R_1 + R_2} v_s \qquad \text{for } t < 0$$

so at $t = 0-$

$$v_C(0-) = \frac{R_2}{R_1 + R_2} v_s$$

Since no impulse sources are present, the capacitor voltage will not change instantaneously. Therefore,

$$v_C(0+) = v_C(0-) = \frac{R_2}{R_1 + R_2} v_s$$

and thus the natural response is

$$v_C(t) = \frac{R_2 v_s}{R_1 + R_2} e^{-t/R_2 C}$$

EXAMPLE 6.3.2

Find the natural response for i_L in the circuit shown in Figure 6.3.3. The switch moves from position a to position b at $t = 0$.

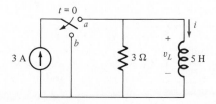

figure 6.3.3
The circuit of Example 6.3.2.

ANS.:   For $t > 0$

$$v_L + iR = 0 \tag{6.3.4}$$

$$\frac{di}{dt} + \frac{R}{L} i = 0 \tag{6.3.5}$$

Assuming that

$$i = Ae^{st} \tag{6.3.6}$$

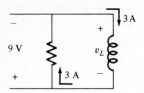

figure 6.3.4
The circuit of Figure 6.3.3 at the instant $t = 0+$.

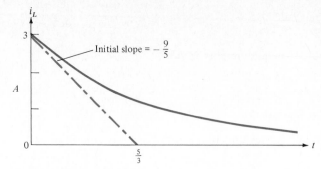

figure 6.3.5
The natural response of the circuit in Figure 6.3.3.

and substituting (6.3.6) back into (6.3.5), we get the characteristic equation

$$s + \frac{R}{L} = 0 \tag{6.3.7}$$

and the characteristic value is therefore $s = -R/L$ or

$$s = -\tfrac{3}{5} \tag{6.3.8}$$

Thus
$$i(t) = Ae^{-3t/5} \tag{6.3.9}$$

For $t < 0$ all current from the 3-A source passes through the inductor. Since the inductor current i_L cannot change instantaneously if v_L remains bounded (does not contain impulses), we have

$$i(0-) = 3 = i(0+) \tag{6.3.10}$$

Substituting equation (6.3.10) into equation (6.3.9) with $t = 0$ gives $A = 3$, and thus the natural response is

$$i(t) = 3e^{-3t/5} \qquad \text{for } t > 0$$

It is always instructive to sketch the circuit at the instant $t = 0+$ and label as many voltages and currents at that time as we can. In Figure 6.3.4 note that $i_L(0+) = 3$ flows through the 3-Ω resistor. We can see that it must follow that at $t = 0+$, $v_L = -9$. But since we know that it must be true (for all t) that

$$v_L = L \frac{di}{dt}$$

we can solve for the value of the slope of the i versus t plot at the initial instant $t = 0+$

$$-9 = 5 \frac{di}{dt}$$

or
$$\left. \frac{di}{dt} \right|_{0+} = -\tfrac{9}{5} \text{ A/s}$$

which is read "the value of the derivative evaluated at time zero plus." The response is plotted in Figure 6.3.5. Note that the initial slope, if extended, does pass through the time axis at one time constant.

Let us solve some other examples of finding initial conditions.

EXAMPLE 6.3.3

Find the numerical value of $i(0+)$ and its derivative di/dt evaluated at $t = 0+$. The switch closes at $t = 0$ (Figure 6.3.6).

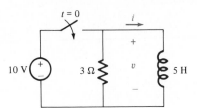

figure 6.3.6
The circuit of Example 6.3.3.

ANS.: Assume that the switch has been open for a long time before $t = 0$; then $i(0-) = 0 = i(0+)$. Closing the switch impresses the 10-V source across the resistor and the inductor. Thus

$$10 = v_R(0+) = v_L(0+) = L\frac{di}{dt}\bigg|_{0+}$$

and so

$$\frac{di}{dt}\bigg|_{0+} = \frac{10}{5} = 2 \text{ A/s}$$

EXAMPLE 6.3.4

In Figure 6.3.7 the switch moves from a to b at $t = 0$. Find the values of i_3, v_4, di_3/dt, and dv_4/dt at time $t = 0+$.

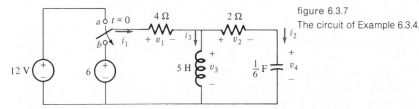

figure 6.3.7
The circuit of Example 6.3.4.

ANS.: For $t < 0$ the switch is in position a. Because the source is constant (dc) in time, the inductor acts like a short circuit and the capacitor like an open circuit. Thus, just before $t = 0$, $v_1(0-) = 12$ and $v_3(0-) = 0$; hence $i_1(0-) = i_3(0-) = \frac{12}{4} = 3$ A. [Also $i_2(0-) = 0$; therefore $v_2(0-) = 0$.] Since $v_3(0-) = v_2(0-) = 0$, KVL around the right-hand mesh tells us that $v_4(0-) = 0$. Immediately after the switch is thrown to position b (at $t = 0+$) the values of i_3 and v_4 will be exactly as they were at $t = 0-$, namely,

$$i_3(0+) = i_3(0-) = 3 \quad \text{and} \quad v_4(0+) = v_4(0-) = 0$$

The state of the circuit at the instant $t = 0+$ is shown in Figure 6.3.8. Thus at $t = 0+$, $i_1 = 3 + i_2$,

$$\frac{6 - v_3}{4} = 3 + \frac{v_3}{2}$$

$$6 - v_3 = 12 + 2v_3$$

$$-2 = v_3$$

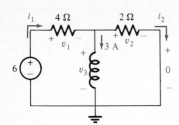

figure 6.3.8
The circuit of Example 6.3.4 at $t = 0+$.

and since $v_4(0+) = 0$, we have $v_3(0+) = v_2(0+) = -2$. Therefore

$$i_1(0+) = \frac{6 - (-2)}{4} = 2 \text{ A} \qquad \text{so that} \qquad v_1(0+) = 8 \text{ V}$$

$$i_2(0+) = \frac{v_3}{2} = \frac{-2}{2} = -1 \text{ A}$$

Since $C(dv_4/dt) = i_2$,

$$\frac{1}{6} \left. \frac{dv_4}{dt} \right|_{0+} = -1 \qquad \text{and} \qquad \left. \frac{dv_4}{dt} \right|_{0+} = -6$$

Also since $v_3 = L(di/dt)$,

$$v_3(0+) = L \left. \frac{di}{dt} \right|_{0+} \qquad -2 = 5 \left. \frac{di}{dt} \right|_{0+}$$

and

$$\left. \frac{di}{dt} \right|_{0+} = -\tfrac{2}{5} \text{ A/s}$$

Even though this last example is not a single-storage element circuit, it should be clear that all initial conditions are determined simply by the application of Ohm's and Kirchhoff's laws at the instant $t = 0+$ and the facts that (1) for constant excitations capacitors act like open circuits and inductors act like short circuits and (2) capacitor voltages and inductor currents cannot change instantaneously in the absence of impulse sources.

**INITIAL CONDITIONS
AT $t = t_0$**

Sometimes initial conditions are given for times other than $t = 0$. In such cases we know the value of the variable of interest, say v, at time t_0 ; in other words we know $v(t_0)$ rather than $v(0+)$. In this event the response is of the form†

$$v(t) = v(t_0)e^{-(t-t_0)/\tau}$$

EXAMPLE 6.3.5

Find $i_L(t)$ for all t in the circuit shown in Figure 6.3.9.

† This is known as the *time-invariance property* of linear systems. The usual response is time-shifted by a translation of axes to the new initial time t_0.

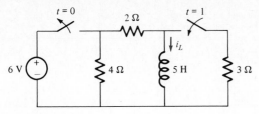

figure 6.3.9
The circuit of Example 6.3.5.

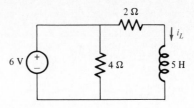

figure 6.3.10
The circuit of Example 6.3.5 for $t < 0$.

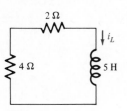

figure 6.3.11
The circuit of Example 6.3.5 for
$0 < t < 1$.

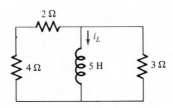

figure 6.3.12
The circuit of Example 6.3.5 for $t > 1$.

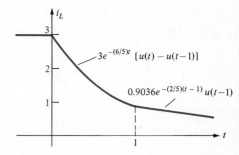

figure 6.3.13
Response of $i_L(t)$ for the circuit shown in Figure
6.3.9.

ANS.:   For $t < 0$ we have the effective circuit shown in Figure 6.3.10. Since the inductor behaves like a short circuit, i_L is determined by the 6-V source and the 2-Ω resistor

$$i_L(t) = 3 \text{ A} \qquad \text{for } t < 0$$

Thus
$$i_L(0-) = 3 \text{ A}$$

For $0 < t < 1$ the circuit is as shown in Figure 6.3.11. Thus, if v_L remains bounded, $i_L(0+) = i(0-) = 3$ A, and so

$$i_L(t) = 3e^{-6t/5} \qquad 0 < t < 1$$

At $t = 1-$

$$i_L(1-) = 3e^{-6/5} = 3(0.3012) = 0.9036$$

and if v_L remains bounded at $t = 1$,

$$i_L(1+) = i_L(1-) = 0.9036$$

Hence for $t > 1$ the circuit is as shown in Figure 6.3.12.
   The time constant during $t > 1$ is L/R, where $L = 5$ H and $R = 6(3)/(6 + 3) = 2 \Omega$; hence $\tau = \frac{5}{2}$ s, so that (see Figure 6.3.13)

$$i_L(t) = 0.9036e^{-(2/5)(t-1)} \qquad \text{for } 1 < t$$

EXAMPLE 6.3.6

Find the natural response $v_C(t)$ of the circuit in Figure 6.3.14.

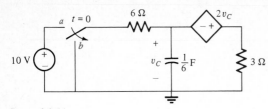

figure 6.3.14
The circuit of Example 6.3.6.

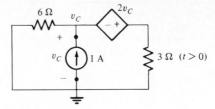

figure 6.3.15
Finding the R_{eq} seen by the capacitor in Figure 6.3.14.

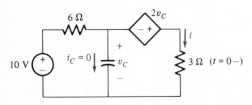

figure 6.3.16
The circuit of Example 6.3.6 at the instant $t = 0 -$
(just before ($t = 0$).

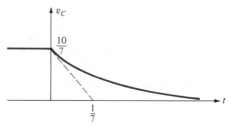

figure 6.3.17
The response $v_C(t)$ of the circuit of Example 6.3.6.

ANS.:   The natural response is of the form

$$v_C(t) = Ae^{-t/\tau} \qquad \text{for } t > 0 \tag{6.3.11}$$

where
$$\tau = R_{eq} C \tag{6.3.12}$$

We can find R_{eq}, the Thevenin equivalent resistance of the rest of the circuit as seen by the capacitor, as follows (see Figure 6.3.15). Remove the capacitor, set all independent sources equal to zero, drive the terminals with a 1-A current source, and calculate the resulting terminal voltage. KCL at the center node yields

$$\frac{v_C}{6} - 1 + \frac{3v_C}{3} = 0$$

$$\tfrac{7}{6}v_C = 1$$

$$v_C = \tfrac{6}{7} \text{ V} \tag{6.3.13}$$

and therefore
$$R_{eq} = \tfrac{6}{7} \, \Omega \tag{6.3.14}$$

The time constant τ is therefore

$$\tau = RC = \tfrac{6}{7}(\tfrac{1}{6}) = \tfrac{1}{7} \tag{6.3.15}$$

To evaluate the undetermined coefficient A we must find the initial condition $v_C(0+)$. For a long time before $t = 0$ and therefore at time $t = 0-$, the circuit is as shown in Figure 6.3.16 with $i_C = 0$. Thus KVL around the outer loop yields

$$3i - 2v_C + 6i = 10 \tag{6.3.16}$$

and around the right-hand mesh

$$v_C = 3i - 2v_C \tag{6.3.17}$$

or
$$3v_C = 3i$$
$$i = v_C \qquad (6.3.18)$$

Substituting (6.3.18) into (6.3.16) yields
$$3v_C - 2v_C + 6v_C = 10$$
$$7v_C = 10$$

or
$$v_C = \tfrac{10}{7} \qquad (6.3.19)$$

Thus we use
$$v_C(0+) = v_C(0-) = \tfrac{10}{7} \qquad (6.3.20)$$

as the initial condition under the assumption that i_C remains bounded. Substituting (6.3.15) and (6.3.20) into (6.3.11), we get
$$v_C(t) = \tfrac{10}{7}e^{-7t} \qquad \text{for } t > 0 \qquad (6.3.21)$$

A sketch of $v_C(t)$ is shown in Figure 6.3.17. This waveform contains no discontinuities, so its derivative will contain no impulses. Thus $i_C = C\,dv/dt$ will remain bounded, and our earlier assumption is shown to be valid.

DRILL PROBLEM

A Thevenin source $[v_{\text{Th}}(t) = 12 \text{ V and } R_{\text{Th}} = 6\,\Omega]$ drives a parallel combination consisting of a 3-Ω resistor and a $\tfrac{1}{2}$-F capacitor. At $t = 0$ a switch between the Thevenin source and the parallel RC combination opens. What is the value of $v_C(0+)$; $i_C(0+)$; dv_C/dt evaluated at $t = 0+$?
ANS.:   4 V, $-\tfrac{4}{3}$ A, and $-\tfrac{8}{3}$ V/s.

---

complete response of first-order systems   6.4

We now consider a method that will enable us to find the complete response for any linear system. Although in this chapter we apply it to single energy-storage (first-order) systems, we shall find later that this same method also works for higher-order systems (ones with more than one energy-storage element). There are other shortcut methods that work if the circuit has certain properties, but it is reassuring to have a general method in our analytical toolbox that always gives the right answer for any linear system. The method is based on the principle of superposition. Earlier in this chapter we learned to solve first-order systems with no sources after $t = 0$. Now we solve a first-order system having a non-zero-valued source for $t > 0$, starting with a specific example.

EXAMPLE 6.4.1

If $v_C(0-) = 0$ and $v_s(t) = 6$ V in the circuit in Figure 6.4.1, write the differential equation for $v_C(t)$ for $t > 0$.
ANS.:   By KCL we have for $t > 0$

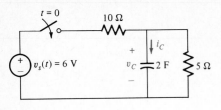

figure 6.4.1
A circuit containing a nonzero source for $t > 0$.

$$\frac{6 - v_C}{10} = i_C + \frac{v_C}{5}$$

$$\frac{6 - v_C}{10} = C\frac{dv_C}{dt} + \frac{v_C}{5}$$

$$2\frac{dv_C}{dt} + v_C(\tfrac{1}{10} + \tfrac{1}{5}) = \tfrac{6}{10} \tag{6.4.1}$$

$$\frac{dv_C}{dt} + \tfrac{3}{20}v_C = \tfrac{3}{10} \qquad \text{for } t > 0 \tag{6.4.2}$$

This is not a homogeneous differential equation because it has a nonzero *forcing function* ($\tfrac{3}{10}$) on the right. We solve this differential equation as follows. Note that a zero can be added to the right-hand side of equation (6.4.2) with no loss of generality; i.e.,

$$\frac{dv_C}{dt} + \tfrac{3}{20}v_C = \tfrac{3}{10} + 0 \tag{6.4.3}$$

As a matter of fact, this is precisely the equation we would get if a second but zero-valued voltage source $v_o(t) = 0$ V were added to the circuit in series with $v_s(t)$. Such a source has no electrical effect on the circuit, of course, since it simply acts like a perfectly conducting piece of wire.

Superposition tells us that if we solve for $v_C(t)$ due to source v_s alone and then for $v_C(t)$ due to source v_o alone, the sum of these component responses is the complete response. Therefore, to solve equation (6.4.3) we proceed as follows. Set the first term on the right-hand side of equation (6.4.3) to zero and find the response due to this zero-valued forcing function. This is just the natural response of the circuit, which we already know how to find; call the natural response v_n. Then assume a solution of the form

$$v_n(t) = Ae^{st} \tag{6.4.4}$$

Substituting (6.4.4) into the source-free differential equation

$$\frac{dv_n}{dt} + \tfrac{3}{20}v_n = 0 \tag{6.4.5}$$

yields
$$sAe^{st} + \tfrac{3}{20}Ae^{st} = 0 \tag{6.4.6}$$

which gives the characteristic equation

$$s + \tfrac{3}{20} = 0 \tag{6.4.7}$$

from which we get the characteristic value

$$s = -\tfrac{3}{20} \qquad\qquad (6.4.8)$$

Therefore
$$v_n(t) = Ae^{-(3/20)t} \qquad\qquad (6.4.9)$$

Alternatively we could have written equation (6.4.4) and evaluated $s = -1/\tau$, where $\tau = RC$ and $R = 10(5)/(10 + 5)$. In any event, the natural response is the first of the two components of the complete solution. It is the response due to zero and is one part of our total answer.

Now return to equation (6.4.3) according to the superposition principle, ignore the second (zero-valued) forcing function and find a time function $v_p(t)$ that will satisfy

$$\frac{dv_p}{dt} + \tfrac{3}{20}v_p = \tfrac{3}{10} \qquad\qquad (6.4.10)$$

Such a $v_p(t)$ is called the *particular response* because it depends on the *particular source* used.

This function $v_p(t)$ is found by recognizing that what is required in equation (6.4.10) is a function whose time derivative when added to itself (multiplied by $\tfrac{3}{20}$) equals a constant ($\tfrac{3}{10}$) for all $t > 0$. Try some of the usual time functions.

First Attempt
Try $v_p(t) = A \sin t$. Then

$$\frac{dv_p}{dt} = A \cos t$$

Substituting into the differential equation yields

$$A \cos t + \tfrac{3}{20}A \sin t = \tfrac{3}{10} \qquad ??$$

Since the left-hand side cannot be a constant in time no matter what value of A we use, this one will not work.

Second Attempt
Try $v_p(t) = Ae^{-at}$. Then

$$\frac{dv_p}{dt} = -aAe^{-at}$$

Substituting into the differential equation yields

$$-aAe^{-at} + \tfrac{3}{20}Ae^{-at} = \tfrac{3}{10} \qquad ??$$

Again, no nonzero values of A or a can be chosen such that the sum on the left-hand side totals to $\tfrac{3}{10}$ (independent of t).

Third Attempt
Suppose we choose a constant $v_p(t) = k$. Then

$$\frac{dv_p}{dt} = 0$$

Substituting into the differential equation, we get

$$0 + \tfrac{3}{20}k = \tfrac{3}{10} \quad \text{and} \quad k = \tfrac{3}{10}\left(\tfrac{20}{3}\right) = 2$$

Thus the solution is

$$v_p(t) = 2 \tag{6.4.11}$$

We note from the circuit that this forced response is precisely the voltage that appears across the 5-Ω resistor if we assume that the capacitor acts like an open circuit. Capacitors do indeed act like open circuits, as we know, in networks that contain only constant voltage and current sources; so this result follows from the fact that the behavior of any energy-storage elements in the particular response is determined by the kind of source driving the network. Superposition tells us that *the complete response is the sum of the natural and the particular responses*, equations (6.4.9) and (6.4.11). Thus

$$v_C(t) = v_n(t) + v_p(t) = Ae^{-(3/20)t} + 2 \tag{6.4.12}$$

It remains to evaluate the coefficient A, which is done by making use of *initial conditions* (what is known about the circuit at $t = 0+$). That is, we were told that

$$v_C(0-) = 0 \tag{6.4.13}$$

In the absence of impulse currents, capacitor voltages cannot change instantaneously. If i_C is bounded, it would follow that

$$v_C(0+) = v_C(0-) = 0 \tag{6.4.14}$$

Therefore from (6.4.12)

$$0 = Ae^{(0)} + 2 \quad \text{and} \quad A = -2$$

so that the complete response is obtained by inserting this value for A into equation (6.4.12)

$$v_C(t) = -2e^{-(3/20)t} + 2 = 2(1 - e^{-(3/20)t}) \quad \text{for } t > 0$$

and
$$v_C(t) = 0 \quad \text{for } t < 0$$

or
$$v_C(t) = 2(1 - e^{-(3/20)t})u(t) \quad \text{for all } t$$

We check on the boundedness of $i_C(t)$ by solving

$$i_C = C \frac{dv_C}{dt}$$

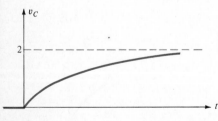

figure 6.4.2
The complete response $v_C(t)$ for the circuit in
Figure 6.4.1.

$$i_C(t) = 2 \frac{d}{dt} 2(1 - e^{-(3/20)t})u(t)$$

$$= 4[(1 - e^{-(3/20)t})\delta(t) + \tfrac{3}{20}e^{-(3/20)t}u(t)]$$

In the first term of this expression it may look as though there were an impulse. However, the coefficient of the impulse is zero-valued at $t = 0$. Thus

$$i_C(t) = \tfrac{3}{5}e^{-(3/20)t}u(t) \quad \text{for all } t$$

which is bounded for all t; see Figure 6.4.2.

Note carefully the order in which things are done in this example:

1. Write the differential equation valid for $t > 0$.
2. Solve the homogeneous (source-free) equation for the natural response by:
  a. Assuming $x_n(t) = Ae^{st}$.
  b. Substituting $x_n(t)$ into the differential equation and solving for the value of s. (But A remains unknown.)
3. Solve the forced differential equation for the particular response $x_p(t)$ by assuming a linear sum of the forcing function and all its derivatives. (More about this in the remainder of this section.)
4. Sum the results found in steps 2 and 3 to get the complete response.
5. Evaluate the constant A by means of initial conditions.

A common error is to do step 5 immediately after step 2. Don't! It will lead to an unavoidable error.

EXAMPLE 6.4.2

Solve for the complete response $v_C(t)$ for $t > 0$ for the same circuit as in Example 6.4.1, again given that $v_C(0-) = 0$ but now with $v_s(t) = 60t^2$; see Figure 6.4.3.

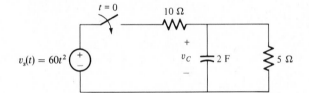

figure 6.4.3
  Another example of finding the complete response.

ANS.: Again we write the differential equation for $t > 0$ for v_C by using KCL

$$\frac{dv_C}{dt} + \tfrac{3}{20}v_C = \frac{v_s(t)}{20}$$

Thus
$$\frac{dv_C}{dt} + \tfrac{3}{20}v_C = 3t^2 \qquad \text{for } t > 0 \tag{6.4.15}$$

Since the natural response does not depend on the particular source driving the circuit, $v_C(t)$ is the same as found in equation (6.4.2). Therefore we simply write

$$v_n(t) = Ae^{-(3/20)t} \tag{6.4.16}$$

as before. As the particular solution try *a linear sum of the forcing function and all its nonzero derivatives*

$$v_p(t) = k_1 t^2 + k_2 t + k_3 \tag{6.4.17}$$

Therefore
$$\frac{dv_p}{dt} = 2k_1 t + k_2 \tag{6.4.18}$$

Substituting (6.4.17) and (6.4.18) back into (6.4.15) gives

$$2k_1 t + k_2 + \tfrac{3}{20}(k_1 t^2 + k_2 t + k_3) = 3t^2 \tag{6.4.19}$$

Note that there is only one term in t^2 on the left; therefore, in order to balance the equation we set like terms equal to like terms. In t^2 we have

$$\tfrac{3}{20}k_1 t^2 = 3t^2 \tag{6.4.20}$$

Thus
$$\tfrac{3}{20}k_1 = 3$$

and
$$k_1 = 20 \tag{6.4.21}$$

From (6.4.19), balancing terms in t, we have
$$2k_1 t + \tfrac{3}{20}k_2 t = 0$$

so that
$$2k_1 + \tfrac{3}{20}k_2 = 0 \tag{6.4.22}$$

and, using (6.4.21),
$$k_2 = -2k_1(\tfrac{20}{3}) = -\tfrac{800}{3} \tag{6.4.23}$$

Finally, balancing the constant terms in equation (6.4.19)
$$k_2 + \tfrac{3}{20}k_3 = 0$$

or
$$k_3 = -k_2(\tfrac{20}{3}) = \frac{16{,}000}{9}$$

so that the particular response is
$$v_p(t) = 20t^2 - \tfrac{800}{3}t + \frac{16{,}000}{9}$$

The complete response is therefore
$$v_C(t) = Ae^{-(3/20)t} + 20t^2 - \tfrac{800}{3}t + \frac{16{,}000}{9}$$

Now using the initial condition that $v_C(0-) = 0$ and assuming bounded $i_c(t)$, we have
$$0 = Ae^0 + 0 - 0 + \frac{16{,}000}{9} \qquad A = -\frac{16{,}000}{9}$$

and so the complete response is
$$v_C(t) = 20t^2 - \tfrac{800}{3}t + \frac{16{,}000}{9}(1 - e^{-(3/20)t}) \qquad \text{for } t > 0$$

EXAMPLE 6.4.3

The switch in Figure 6.4.4 is thrown from a to b at $t = 0$. Find $i_L(t)$ for $t > 0$ given $i_s(t) = 2e^{-5t}$ for all t.

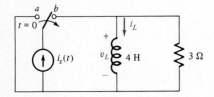

figure 6.4.4
The circuit for Example 6.4.3.

ANS.:   First we note that if the switch has been in position a for a long while (since we have not been told otherwise, it is safe to assume it has been), the 4-H inductor and 3-Ω resistor will both have zero-valued currents and voltage at time $t = 0-$. Then, from KCL for $t > 0$,

$$i_L + \frac{v_L}{R} = 2e^{-5t} \tag{6.4.24}$$

and since

$$v_L = L \frac{di}{dt} \tag{6.4.25}$$

$$\frac{L}{R} \frac{di}{dt} + i_L = 2e^{-5t}$$

and

$$\frac{di}{dt} + \tfrac{3}{4}i_L = \tfrac{3}{2}e^{-5t} \tag{6.4.26}$$

The natural response is obtained by substituting a zero forcing function and

$$i_n(t) = Ae^{st} \tag{6.4.27}$$

into (6.4.26), thus obtaining the characteristic equation

$$s + \tfrac{3}{4} = 0 \tag{6.4.28}$$

and characteristic value

$$s = -\tfrac{3}{4} \tag{6.4.29}$$

Hence

$$i_n(t) = Ae^{-(3/4)t} \tag{6.4.30}$$

The particular response is obtained as follows. Assume that $i_p(t)$ is a linear sum of the forcing function $\tfrac{3}{2}e^{-5t}$ and all its derivatives. Thus

$$i_p(t) = k_1 e^{-5t} + k_2(-5)e^{-5t} + k_3(25)e^{-5t} + \cdots$$
$$= (k_1 - 5k_2 + 25k_3 - 125k_4 + \cdots)e^{-5t} = k_0 e^{-5t} \tag{6.4.31}$$

We quickly see from this equation that the exponential forcing function is an easy one to work with. The forced response is simply a single term because all derivatives of the exponential function have the same time dependence as the original forcing function.

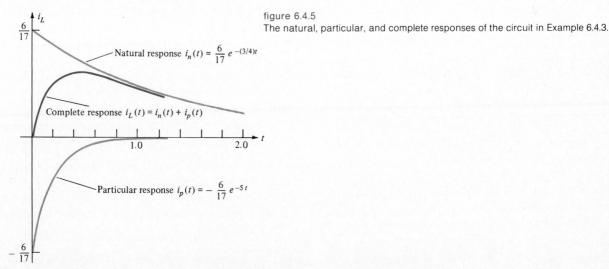

figure 6.4.5
The natural, particular, and complete responses of the circuit in Example 6.4.3.

Natural response $i_n(t) = \frac{6}{17} e^{-(3/4)t}$

Complete response $i_L(t) = i_n(t) + i_p(t)$

Particular response $i_p(t) = -\frac{6}{17} e^{-5t}$

Substituting (6.4.31) into (6.4.26) and setting like terms equal yields

$$-5k_0 e^{-5t} + \tfrac{3}{4}k_0 e^{-5t} = \tfrac{3}{2}e^{-5t}$$

$$(-5 + \tfrac{3}{4})k_0 = \tfrac{3}{2}$$

$$k_0 = -\tfrac{6}{17} \tag{6.4.32}$$

or
$$i_p(t) = -\tfrac{6}{17}e^{-5t} \quad \text{for } t > 0 \tag{6.4.33}$$

Summing (6.4.30) and (6.4.33) yields the complete response

$$i_L(t) = Ae^{-(3/4)t} - \tfrac{6}{17}e^{-5t}$$

and since $i_L(0) = 0$,

$$0 = Ae^0 - \tfrac{6}{17}e^0 \quad \text{and} \quad A = \tfrac{6}{17}$$

Thus
$$i_L(t) = \tfrac{6}{17}(e^{-(3/4)t} - e^{-5t})$$

See Figure 6.4.5.

In the special case† where the exponential forcing function used is identical to the exponential term in the natural response, the particular response has a slightly more complicated form, which can be derived as follows:

$$\frac{dx}{dt} + \alpha x = k_1 e^{-\alpha t} \tag{6.4.34}$$

$$e^{\alpha t}\frac{dx}{dt} + \alpha e^{\alpha t}x = k_1 \tag{6.4.35}$$

or
$$\frac{d}{dt}xe^{\alpha t} = k_1 \tag{6.4.36}$$

The indefinite integral of both sides yields

$$\int \frac{d}{dt}xe^{\alpha t}\, dt = \int k_1\, dt + k_2$$

$$\int d(xe^{\alpha t}) = k_1 t + k_2$$

$$xe^{\alpha t} = k_1 t + k_2$$

or finally
$$\boxed{x(t) = x_p(t) = e^{-\alpha t}(k_1 t + k_2)} \tag{6.4.37}$$

It is not necessary to rederive equation (6.4.37) in this manner each time an equation of the form of (6.4.34) arises; instead simply write down the particular response directly in the form of equation (6.4.37).

Thus, for example, if the forcing function in equation (6.4.26) is changed to $\tfrac{3}{2}e^{-(3/4)t}$,

† We discuss this special case and others in Section 7.7 and return to it again in Sections 11.5 and 15.7.

$$i_p(t) = (k_1 + k_2 t)e^{-(3/4)t}$$

Substitution of this into (6.4.26) reveals the requirement that $k_2 = \frac{3}{2}$ and there is no restriction on k_1. Thus

$$i_p(t) = (k_1 + \tfrac{3}{2}t)e^{-(3/4)t}$$

and, as before, $\qquad\qquad i_n(t) = Ae^{-(3/4)t}$

so that, from the initial condition $i_L(0+) = 0$ we find $k_1 + A = 0$ and $i_L(t) = \frac{3}{2}te^{-(3/4)t}$

DRILL PROBLEM

In the circuit of Figure 6.4.4 replace the 4-H inductor with a 4-F capacitor. Solve for $v_C(t)$ for $t > 0$. Assume, as before, that the switch is thrown at $t = 0$ from a (where it has been for a long time) to b and that $i_s(t) = 2e^{-5t}$. What is the value of $i_C(0+)$?

ANS.: $v_C(t) = \frac{6}{59}(e^{-(1/12)t} - e^{-5t})$ and $i_C(0+) = 2$ A.

---

zero-state and zero-input responses of first-order systems   6.5

We have seen that the complete response of any linear system is the sum of its natural response and its particular response (also called forced response or response due to the source). Thus the complete response for any system contains these two components. Alternatively, we may think of the complete response as being made up of two other very different component responses: (1) the complete response of the variable of interest due to any *nonzero initial states* of the energy-storage elements with *no sources active*, which is called the *zero-input response*, and (2) the complete response of the variable with the source(s) connected but with *all initial conditions set to zero*, which is called the *zero-state response*.

EXAMPLE 6.5.1

Given $v_s(t) = 2t\,u(t)$ and $v_C(0-) = 5$ V:
a. Find the complete response $v_C(t)$ for $t > 0$ for the circuit of Figure 6.5.1.
b. Break this complete response into its zero-input and zero-state components.

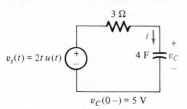

figure 6.5.1
The circuit for Example 6.5.1.

ANS.:   (a) By KVL,

$$iR + v_C = 2t \qquad \text{for } t > 0$$

$$C \frac{dv}{dt} R + v = 2t \qquad \text{and} \qquad \frac{dv}{dt} + \frac{1}{RC} v = \frac{1}{RC} 2t$$

$$\frac{dv}{dt} + \tfrac{1}{12}v = \frac{t}{6} \qquad \text{for } t > 0$$

Natural Response

$$v_n(t) = Ae^{st} \qquad \frac{dv}{dt} + \tfrac{1}{12}v = 0 \qquad s = -\tfrac{1}{12}$$

$$v_n(t) = Ae^{-(1/12)t} \tag{6.5.1}$$

Particular Response

$$v_p(t) = k_1 t + k_2 \qquad \frac{dv_p}{dt} = k_1 \qquad k_1 + \tfrac{1}{12}(k_1 t + k_2) = \tfrac{1}{6}t \qquad k_1 = 2$$

$$k_1 + \tfrac{1}{12}k_2 = 0 \qquad k_2 = -24 \qquad v_p(t) = 2t - 24 \tag{6.5.2}$$

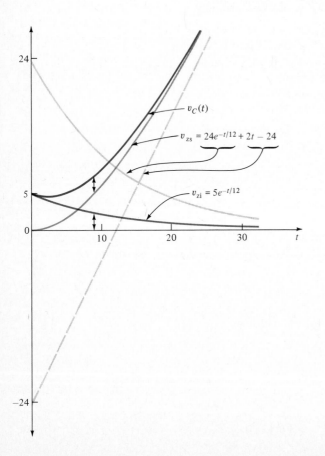

figure 6.5.2
The zero-state and zero-input components of the complete
response.

Complete Response

$$v_C(t) = Ae^{-(1/12)t} + 2t - 24$$

$$5 = A - 24 \qquad A = 29$$

$$v_C(t) = 29e^{-(1/12)t} + 2t - 24 \tag{6.5.3}$$

(b) The zero-input response is the complete response when all sources are zero-valued. This is simply the natural response because the particular response to a zero-valued source is zero. Thus

$$v_{zi} = Ae^{-(1/12)t}$$

from initial conditions $5 = Ae^0$ and thus

$$v_{zi} = 5e^{-(1/12)t} \tag{6.5.4}$$

The zero-state response is the complete response obtained when all initial conditions are zero. Since the particular and natural responses are as they were,

$$v_{zs} = Ae^{-(1/12)t} + 2t - 24$$

From the new initial conditions, $0 = Ae^0 + 0 - 24$, $A = 24$. Therefore

$$v_{zs} = 24e^{-(1/12)t} + 2t - 24 \tag{6.5.5}$$

Note that the sum of the zero-input response and the zero-state response is the complete response (Figure 6.5.2).

EXAMPLE 6.5.2

In Figure 6.5.3:
a. Find $i_L(t)$ given that $i_{s_1}(t) = (t + 3)u(t)$ and $i_{s_2}(t) = 2u(-t)$.
b. Find the zero-state and zero-input responses.

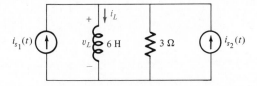

figure 6.5.3
The circuit of Example 6.5.2.

ANS.:   (a) For $t < 0$ $i_{s_1} = 0$, $i_{s_2}(t) = 2$ A $= i_L(0-)$. For $t > 0$ $i_{s_1} = t + 3$ and $i_{s_2} = 0$. Therefore

$$i_L + \frac{v_L}{3} = t + 3 \qquad \text{for } t > 0$$

$$i_L + \tfrac{1}{3}\left(L\,\frac{di_L}{dt}\right) = t + 3$$

$$\frac{di_L}{dt} + \tfrac{1}{2}i_L = \frac{t}{2} + \frac{3}{2} \tag{6.5.6}$$

Natural Response

$$\frac{di_L}{dt} + \tfrac{1}{2}i_L = 0 \qquad i_n(t) = Ae^{st}$$

$$sAe^{st} + \tfrac{1}{2}Ae^{st} = 0 \qquad s = -\tfrac{1}{2} \qquad i_n(t) = Ae^{-(1/2)t} \tag{6.5.7}$$

Particular Response

$$\frac{di_L}{dt} + \tfrac{1}{2}i_L = \frac{t}{2} + \frac{3}{2} \tag{6.5.8}$$

Assume that $i_p(t) = k_1 t + k_2$; then

$$\frac{di_p}{dt} = k_1$$

Substituting back into (6.5.8) leads to

$$k_1 + \frac{k_1}{2}t + \frac{k_2}{2} = \frac{t}{2} + \frac{3}{2}$$

and balancing terms in t gives

$$k_1 = 1$$

so that

$$k_1 + \frac{k_2}{2} = \tfrac{3}{2} \qquad \text{or} \qquad k_2 = 1$$

Therefore the particular response is

$$i_p(t) = t + 1 \tag{6.5.9}$$

Complete Response

$$i_L(t) = i_n(t) + i_p(t) = Ae^{-(1/2)t} + t + 1 \tag{6.5.10}$$

Using the initial condition that $i_L(0+) = 2$ in the complete response, we can evaluate A as

$$2 = Ae^0 + 0 + 1 \qquad A = 2 - 1 = 1$$

figure 6.5.4
The complete response and its two components: the natural and particular responses.

figure 6.5.5
The complete response and its two components: the zero-input and zero-state responses.

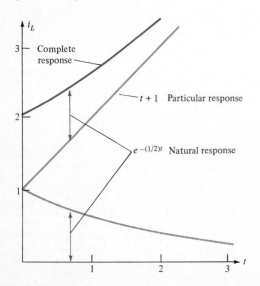

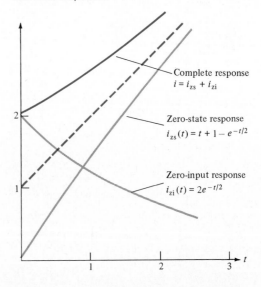

and so
$$i_L(t) = e^{-(1/2)t} + t + 1 \qquad \text{for } t > 0 \tag{6.5.11}$$

See Figure 6.5.4.

(b) The zero-input response is obtained from the natural response by using the initial condition $i_L(0-) = i_L(0+) = 2$

$$i_{zi}(t) = Ae^{-(1/2)t} \qquad 2 = Ae^0 \qquad A = 2$$

therefore
$$i_{zi}(t) = 2e^{-(1/2)t} \qquad \text{for } t > 0 \tag{6.5.12}$$

The zero-state response can be found as usual by evaluating the complete response, equation (6.5.10), assuming zero initial conditions, i.e., from (6.5.10)

$$0 = Ae^{-(1/2)t\theta} + t^0 + 1 \qquad A = -1 \qquad i_{zs}(t) = t + 1 - e^{-(1/2)t} \tag{6.5.13}$$

Of course in this case, since we have already found the complete response and the zero-input response, we could have simply subtracted one from the other. In other words, since

$$\underset{\text{complete}}{i(t)} = \underset{\text{zero-state}}{i(t)} + \underset{\text{zero-input}}{i(t)}$$

then

$$\underset{\text{zero-state}}{i(t)} = \underset{\text{complete}}{i(t)} - \underset{\text{zero-input}}{i(t)}$$

$$= (e^{-(1/2)t} + t + 1) - 2e^{-(1/2)t}$$

$$= t + 1 - e^{-(1/2)t} \qquad \text{for } t > 0 \tag{6.5.14}$$

which is identical to the result in (6.5.13); see Figure 6.5.5.

The sum of $i_{zs}(t)$ and $i_{zi}(t)$ is the complete response; i.e., addition of equations (6.5.12) and (6.5.14) gives

$$i_L(t) = t + 1 + e^{-(1/2)t} \qquad \text{for } t > 0$$

which is identical to equation (6.5.11). You should compare Figures 6.5.4 and 6.5.5 carefully. When solving differential equations for complete responses as we have been doing in this chapter, it may seem easier to think about natural and particular responses. However, there are many times when it is more convenient to find the zero-state and zero-input components.

DRILL PROBLEM

In the circuit of Figure 6.5.3 replace the 6-H inductor with a 6-F capacitor and solve for the zero-input and zero-state responses of $v_C(t)$. Find the complete response. What is the initial slope of this voltage at $t = 0+$?

ANS.: $3t - 45 + 51e^{-t/18}$; $6e^{-t/18}$; $\frac{1}{6}$ V/s.

---

unit-step and unit-impulse responses of first-order systems 6.6

The *unit-step response* of a network is defined as being the *zero-state response* to a *unit-step-function* input, or the complete response of the designated output variable of interest in that network when all initial conditions are zero (i.e., no energy is stored at $t = 0$) to a $u(t)$ source (either voltage or current). Often the unit-step

UNIT-STEP RESPONSE

response is called $w(t)$ as long as it is clear what variable is referred to. Some older texts call the unit-step response the *indicial response*. To find the unit-step response, set the initial condition equal to zero and find the complete response due to a $u(t)$ source.

EXAMPLE 6.6.1

Find the unit-step response for $v_C(t)$ in the circuit shown in Figure 6.6.1, where $v_C(t)$ is the output variable.

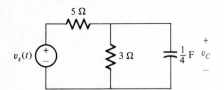

figure 6.6.1
The circuit of Example 6.6.1.

ANS.: By KCL at node v_C for $t > 0$

$$\frac{1 - v_C}{5} = \frac{v_C}{3} + \frac{1}{4}\frac{dv_C}{dt}$$

Multiplying by 4 and collecting terms in v_C gives

$$\frac{dv_C}{dt} + \tfrac{32}{15}v_C = \tfrac{4}{5} \tag{6.6.1}$$

The natural response is thus

$$v_n(t) = Ae^{-(32/15)t} \tag{6.6.2}$$

The particular response to the constant forcing function is

$$v_p(t) = k \tag{6.6.3}$$

Substituting (6.6.3) into (6.6.1) leads to

$$0 + \tfrac{32}{15}k = \tfrac{4}{5}$$

$$k = \tfrac{15}{32}(\tfrac{4}{5}) = \tfrac{3}{8} = v_p(t)$$

Note that v_p is simply the voltage that appears across the $\frac{1}{4}$-F capacitor due to a constant (for all t) source $v_s = 1$ V, for which the capacitor acts like an open circuit. The complete unit-step response is then found to be

$$v_C(t) = v_p(t) + v_n(t) = \tfrac{3}{8} + Ae^{-(32/15)t}$$

where by the initial condition $v_C(0-) = v_C(0+) = 0$ we find

$$0 = \tfrac{3}{8} + Ae^0 \qquad \text{and} \qquad A = -\tfrac{3}{8}$$

Thus

$$v_C(t) = \begin{cases} \tfrac{3}{8}(1 - e^{-(32/15)t}) & \text{for } t > 0 \\ \tfrac{3}{8}(1 - e^{-(32/15)t})u(t) & \text{for all } t \end{cases}$$

Another example will demonstrate the idea that the particular response component of the total unit-step response is the voltage (current) produced across (through) the element of interest by a constant dc 1-V or 1-A source.

EXAMPLE 6.6.2

Find the unit-step response for i_L in the circuit of Figure 6.6.2.

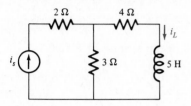

figure 6.6.2
The circuit of Example 6.6.2.

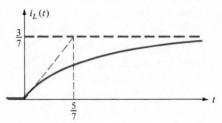

figure 6.6.3
The unit-step response i_L
of the circuit in
Example 6.6.2.

ANS.:   It is implied that $i_s(t) = u(t)$ and $i_L(0-) = 0$.

Natural Response

$$i_{L_n}(t) = Ae^{-(R/L)t}$$

where $L = 5$ H and R is the Thevenin resistance of the circuit to the left of L. Setting $i_s = 0$ (replace it by an open circuit), we see that

$$R = 4 + 3 = 7 \ \Omega$$

Thus

$$i_{L_n}(t) = Ae^{-(7/5)t}$$

Particular Response
For a constant 1-A source i_s, the inductor is an effective short circuit. Thus the 1 A divides between the 3-Ω and 4-Ω resistors

$$i_{L_p}(t) = \frac{3}{3 + 4}\,(1) = \tfrac{3}{7}$$

The complete unit-step response is therefore

$$i_L(t) = \tfrac{3}{7} + Ae^{-(7/5)t}$$

and since $i_L(0-) = i_L(0+) = 0,$ $0 = \tfrac{3}{7} + Ae^0$ and $A = -\tfrac{3}{7}$

Therefore $i_L(t) = \tfrac{3}{7}(1 - e^{-(7/5)t})$ for $t > 0$

or $w(t) = i_L(t) = \tfrac{3}{7}(1 - e^{-(7/5)t})u(t)$ for all t

See Figure 6.6.3.

A SHORTCUT FOR FIRST-ORDER SYSTEMS WITH CONSTANT SOURCES For first-order systems having only sources that are constant for $t > 0$ we note that the response starts at a value dictated by the initial (energy-storage) condition, either $v_C(0-)$ or $i_L(0-)$. Since the natural-response term is a negative exponential that decays away to zero as t gets large, the *final* or so-called *steady-state* value of the response is determined by the source and the resistors (any capacitor being an equivalent open circuit and any inductor a short). The response gets from its initial value to its final value exponentially, with $\tau = RC$ or $\tau = L/R$.

EXAMPLE 6.6.3

The switch in Figure 6.6.4 is thrown from a to b at $t = 0$. Find the complete response $v_C(t)$ for all t.

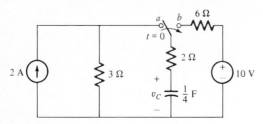

figure 6.6.4
The circuit of Example 6.6.3.

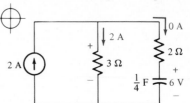

figure 6.6.5
The circuit of Example 6.6.3 for $t < 0$.

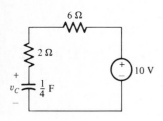

figure 6.6.6
The circuit of Example 6.6.3 for $t > 0$.

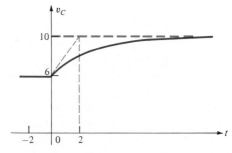

figure 6.6.7
The complete response of $v_C(t)$ in the circuit of Figure 6.6.4.

ANS.: For $t < 0$ we have Figure 6.6.5. Since the capacitor acts like an open circuit for constant excitations, no i_C current flows and therefore there is no drop across the 2-Ω resistor. Thus, $v_C(0-) = 6$ V. For positive values of time ($t > 0$) the circuit looks like that in Figure 6.6.6. The time constant τ is therefore

$$\tau = RC$$

where

$$C = \tfrac{1}{4} \text{ F}$$

$$R = 2 \ \Omega + 6 \ \Omega = 8 \ \Omega$$

so that

$$\tau = \tfrac{1}{4}(8) = 2 \text{ s}$$

For sufficiently large values of t (after the negative exponential term has decayed to negligible magnitude), $v_C = 10$ V and $i_C = 0$ because the capacitor is an open circuit for constant excitations. We therefore say that the *final value* is $v_C = 10$ V. Even without finding the closed-form mathematical expression for $v_C(t)$ we can sketch it, as in Figure 6.6.7.

A general expression for the complete response of any variable in a first-order system having only constant sources is

$$x(t) = X_f - (X_f - X_i)e^{-t/\tau} \qquad \text{for } t > 0$$

where X_f = final ($t \gg 0$) value of variable x
X_i = initial ($t = 0+$) value of variable x
τ = time constant ($t > 0$)

Since for this example $v_i = 6$, $v_f = 10$, and $\tau = 2$, we get

$$v_C(t) = 10 - (10 - 6)e^{-t/2} = 10 - 4e^{-t/2} \qquad \text{for } t > 0$$

or
$$w(t) = v_C(t) = 6 + 4(1 - e^{-t/2})u(t) \qquad \text{for all } t$$

In Section 6.3 we noted that if a switch is thrown or some other change is initi- **TIME INVARIANCE** ated at t_0, a time different from $t = 0$, the response is simply translated in time from $t = 0$ to the time t_0. Suppose, for example, in Figure 6.6.4 that the switch is thrown back to position a at time $t = 1$ s. At $t = 1-$ we calculate the initial value of v_C as

$$v_C(1-) = 6 + 4(1 - e^{-1/2}) = 6 + 4(0.393) = 7.57 \text{ V} = v_C(1+)$$

The final value of v_C is, of course, $(2 \text{ A})(3 \text{ }\Omega) = 6$ V. Therefore

$$v_C(t) = 6 - (6 - 7.57)e^{-(4/5)(t-1)} = 6 + 1.57e^{-(4/5)(t-1)} \qquad \text{for } t > 1$$

The zero-state response (the *complete response* of a linear system that has *zero* **UNIT-IMPULSE RESPONSE** *stored energy* at $t = 0$) to a *unit impulse* $\delta(t)$ is called the *unit-impulse response* of that system, usually written $h(t)$.

EXAMPLE 6.6.4

Find the unit-impulse response $h(t)$ for $v_C(t)$ in Figure 6.6.8.

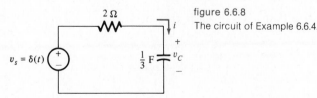

figure 6.6.8
The circuit of Example 6.6.4.

ANS.: Use KVL to find the differential equation for $iR + v_C = v_s$

$$RC\frac{dv_C}{dt} + v_C = v_s \qquad \frac{dv_C}{dt} + \frac{1}{RC}v_C = \frac{1}{RC}v_s$$

$$\frac{dv_C}{dt} + \tfrac{3}{2}v_C = \tfrac{3}{2}\delta(t) \qquad \text{for all } t$$

Now
$$\frac{dv_C}{dt} + \tfrac{3}{2}v_C = 0 \qquad \text{for } t > 0$$

This is a homogeneous equation satisfied by

$$v_C(t) = Ae^{-(3/2)t} \qquad \text{for } t > 0$$

Thus we see that the complete impulse response is equal to the natural response because the particular response for $t > 0$ to an impulse source is zero. The impulse source simply puts nonzero initial conditions (at $t = 0+$) onto the energy-storage element.

In other words, for the circuit in Figure 6.6.8 at time $t = 0$ the source v_s looks as though it might split (by voltage division), some fraction falling across the $\tfrac{1}{3}$-F capacitor and the rest across the resistor. However, the voltage across a capacitor can never be an impulse. For that to happen the *energy* stored in the capacitor would have to be infinite; i.e., all the energy in the universe would be stored in the capacitor. Since $W_C = \tfrac{1}{2}Cv_C^2$, for $v_C(t) = \delta(t)$ we have

$$W_C = \tfrac{1}{2}C[\delta(t)]^2 = \infty^2 \qquad \text{at } t = 0$$

Infinite capacitor currents and/or infinite inductor voltages can be at least approximated in actuality, but since infinite energy is impossible, a capacitor voltage cannot be an impulse. Similarly, inductor currents can never be impulses because $W_L = \tfrac{1}{2}Li^2$. In Figure 6.6.8 the impulse voltage must *all* fall across $R = 2\ \Omega$, making

$$i = \frac{v}{R} = \frac{\delta(t)}{2}$$

This impulse (height $= \infty$, area $= \tfrac{1}{2}$) of current represents a charge of $\tfrac{1}{2}$ C, so that at time $t = 0+$, immediately after the impulse of current has occurred,

$$v_C(0+) = \frac{q}{C} = \frac{\tfrac{1}{2}}{\tfrac{1}{3}} = \frac{3}{2}\ \text{V}$$

that is
$$v_C(0+) = \frac{1}{C}\int_{-\infty}^{0+} \tfrac{1}{2}\delta(t)\,dt = (3)\,\frac{1}{2}\int_{0-}^{0+}\delta(t)\,dt = 3(\tfrac{1}{2})(1) = \tfrac{3}{2}\ \text{V}$$

Thus, the impulse response is

$$v_C(t) = \tfrac{3}{2}e^{-(3/2)t} \qquad \text{for } t > 0$$

or
$$h(t) = v_C(t) = \tfrac{3}{2}e^{-(3/2)t}u(t) \qquad \text{for all } t$$

This example showed that if a capacitor current is $i_C = \delta(t)$, then v_C increases instantaneously by $1/C$ volts. In other words, for a capacitor with $i_C(t) = \delta(t)$,

$$v_C(t) = \frac{1}{C}\int_{0-}^{t} i_C(t)\,dt + v_C(0-)$$

or
$$v_C(0+) - v_C(0-) = \frac{1}{C}\int_{0-}^{0+}\delta(t)\,dt = \frac{1}{C}$$

Similarly, in an inductor $v_L = \delta(t)$ produces a discontinuity of $1/L$ amperes in i_L, that is, with $v_L(t) = \delta(t)$,

$$i_L(t) = \frac{1}{L}\int_{0-}^{t} v_L(t)\,dt + i_L(0-)$$

$$i_L(0+) - i_L(0-) = \frac{1}{L} \int_{0-}^{0+} \delta(t) \, dt = \frac{1}{L}$$

EXAMPLE 6.6.5

In the circuit of Figure 6.6.9 find the unit-impulse response $i_L(t)$.

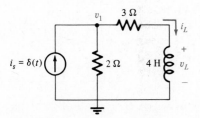

figure 6.6.9
The circuit of Example 6.6.5.

ANS.:   The natural response for $t > 0$ is

$$i_L(t) = Ae^{-t/\tau}$$

where
$$\tau = \frac{L}{R} = \frac{4}{2+3} = \frac{4}{5} \quad \text{and} \quad i_L(t) = Ae^{-(5/4)t}$$

The impulse source sets initial conditions on L at $t = 0+$ that were not there at $t = 0-$.

The impulse-current source does not split although it looks as if it might, some going through the 2- and some through the 3-Ω resistor. Any fraction of $\delta(t)$, say $\frac{2}{5}\delta(t)$, is still an impulse, and i_L can never be an impulse. Therefore all $i_s(t) = \delta(t)$ must go through the 2-Ω resistor, producing by Ohm's law $v_1(t) = 2\delta(t)$ volts. Thus, $v_L = 2\delta(t)$. The resulting inductor current at time $t = 0+$, immediately after the impulse occurs, is therefore

$$i_L(0+) = \frac{1}{L} \int_{-\infty}^{0+} v_L(t) \, dt = \frac{1}{4} \int_{0-}^{0+} 2\delta(t) \, dt = \tfrac{1}{4}(2)(1) = \tfrac{1}{2} \text{ A}$$

Hence the complete impulse response is found by

$$i_L(t) = Ae^{-(5/4)t} \qquad i_L(0+) = \tfrac{1}{2} = Ae^0 \qquad A = \tfrac{1}{2}$$

$$i_L(t) = \begin{cases} \tfrac{1}{2}e^{-(5/4)t} & \text{for } t > 0 \\ \tfrac{1}{2}e^{-(5/4)t}u(t) & \text{for all } t \end{cases}$$

In a linear system if the input source $x(t)$ is known and the corresponding zero-state response $y(t)$ is known, when the source is replaced by its derivative dx/dt, it follows that the zero-state response will be the derivative dy/dt of the original zero-state response. This means that the zero-state response $h(t)$ to a unit impulse is the time derivative of the zero-state response $w(t)$ to a unit step; or, as we say, the unit-impulse response is the derivative of the unit-step response.

EXAMPLE 6.6.6

Find the unit-step response $w(t) = v_C(t)$ of the circuit in Figure 6.6.10.

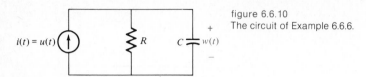

figure 6.6.10
The circuit of Example 6.6.6.

ANS.:

$$v_C(t) = V_f - (V_f - V_i)e^{-t/RC} = R - (R - 0)e^{-t/RC}$$

or

$$v_C(t) = w(t) = R(1 - e^{-t/RC})u(t) \qquad \text{for all } t$$

EXAMPLE 6.6.7

Find $h(t)$ for the circuit in Example 6.6.6 (Figure 6.6.11).

figure 6.6.11
The circuit of Example 6.6.7.

ANS.:   Again we seek $v_C(t)$ for $t > 0$, but when $i(t) = \delta(t)$, where does the $i = \delta(t)$ flow? Can all or part of it flow into R? No, because that would make $v_R = R\delta(t) \to \infty$ at $t = 0$. And since $v_R = v_C$, the energy stored in $C \to \infty$ (impossible). We conclude that all the $i(t) = \delta(t)$ must flow into C. Thus $v_C(0+)$ becomes $= 1/C$ volts, from which, with $x_f - (x_f - x_i)e^{-t/RC}$, we get

$$v_C(t) = h(t) = 0 - \left(0 - \frac{1}{C}\right)e^{-t/RC}$$

or

$$h(t) = \frac{1}{C} e^{-t/RC}u(t) \qquad \text{for all } t$$

Note that the time derivative of the step response in Example 6.6.6 is the unit-impulse response in Example 6.6.7, that is,

$$\frac{dw}{dt} = R(1 - e^{-t/RC})\delta(t) + \frac{R}{RC} e^{-t/RC}u(t) = \frac{1}{C} e^{-t/RC}u(t) \qquad \text{for all } t$$

EXAMPLE 6.6.8

Find the unit-step and unit-impulse responses for $i(t)$ in the circuit of Figure 6.6.12. Check that $dw/dt = h(t)$.
ANS.:   For $v_0(t) = u(t)$ we have $i_f = 0$ and, with $v_C(0-) = 0$, $i_i = v_0/R = 1/R$. Thus

$$w(t) = 0 - \left(0 - \frac{1}{R}\right)e^{-t/RC} = \frac{1}{R} e^{-t/RC}u(t) \qquad \text{for all } t \tag{6.6.4}$$

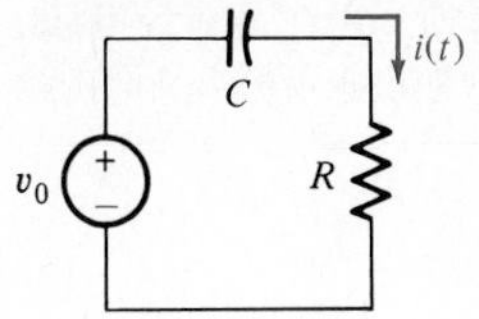

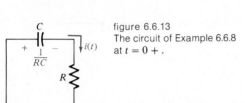

figure 6.6.12
The circuit of Example 6.6.8.

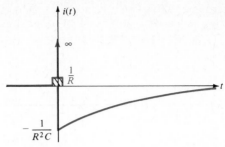

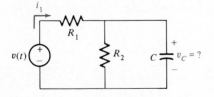

figure 6.6.13
The circuit of Example 6.6.8
at $t = 0+$.

figure 6.6.14
The unit-impulse response of the circuit of Example 6.6.8.

For $v_0(t) = \delta(t)$ we recognize that no fraction of $\delta(t)$ can appear across the C. All of $\delta(t)$ goes across R, causing an impulsive current $i(t) = (1/R)\delta(t)$ to occur at $t = 0$. This puts a voltage of $(1/C)(1/R)$ volts on the capacitor at $t = 0+$, so that we have the situation shown in Figure 6.6.13. Therefore a current flows in the negative i direction with instantaneous magnitude (at $t = 0+$) of $(1/RC)/R$ amperes. Thus

$$i_i(0+) = -\frac{1}{R^2C} \quad \text{and} \quad i(t) = -\frac{1}{R^2C} e^{-t/RC} \quad \text{for } t > 0$$

or

$$h(t) = i(t) = \frac{1}{R}\delta(t) - \frac{1}{R^2C} e^{-t/RC} u(t) \quad \text{for all } t \tag{6.6.5}$$

A sketch of the solution is shown in Figure 6.6.14. Take the time derivative of the unit-step response. Does it equal the unit-impulse response? From equation (6.6.4) we have

$$w(t) = \frac{1}{R} e^{-t/RC} u(t) \quad \text{for all } t$$

whence

$$\frac{dw}{dt} = \frac{1}{R}\left[e^{-t/RC}\delta(t) + \left(-\frac{1}{RC}\right)e^{-t/RC}u(t) \right] \quad \text{for all } t \tag{6.6.6}$$

which is identical to equation (6.6.5).

EXAMPLE 6.6.9

Find and sketch first the unit-impulse response and then the unit-step response of the circuit shown in Figure 6.6.15.

figure 6.6.15
The circuit of Example 6.6.9.

ANS.:   See Figure 6.6.16. Impulse response:

$$v(t) = \delta(t)$$

Since v_C cannot be ∞, all $\delta(t)$ falls across R_1, producing $i_1(t) = (1/R_1)\delta(t)$. Since $v_C(0-) = 0$, all of this i_1 flows into C [also if any of $\delta(t)$ flowed into R_2, it would force $v_C \to \infty$, which is impossible]. The current $i = \delta(t)/R_1$ in C changes v_C from 0 to $1/R_1C$ volts instantaneously, so that

$$v_C(0+) = \frac{1}{R_1C} \quad \text{volts}$$

For $t > 0$ the source $v(t) = 0$ (short circuit), so that the charge on C leaks off through R_1 and R_2 in parallel. Using $x(t) = X_f - (X_f - X_i)e^{-t/RC}$, we have

$$v_C(t) = 0 - \left(0 - \frac{1}{R_1C}\right)e^{-t/RC}$$

so that $\qquad h(t) = \dfrac{1}{R_1C} e^{-t/RC}u(t) \qquad$ where $R = \dfrac{R_1R_2}{R_1 + R_2}$

In the unit-step response $v(t) = u(t)$, and since

$$v_{C_i} = 0 \quad \text{and} \quad v_{C_f} = \frac{R_2}{R_1 + R_2}$$

we have $\qquad w(t) = v_C(t) = \dfrac{R_2}{R_1 + R_2} (1 - e^{-t/RC})u(t)$

In Example 6.6.9 we could have found the Norton equivalent of the source $v(t)$ and R_1 (see Figure 6.6.17); thus the system becomes as shown in Figure 6.6.18. If we wish to find the unit-impulse response in the original network, we can solve the circuit in Figure 6.6.18. Again all $(1/R_1)\delta(t)$ flows into C. Thus $v_C(0+) = 1/R_1C$, and for $t > 0$ the circuit is as shown in Figure 6.6.19 and, as before,

figure 6.6.16
(a) The unit-impulse response and (b) the unit-step response of the circuit of Figure 6.6.15.

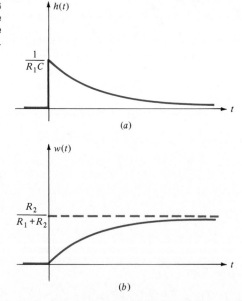

(a)

(b)

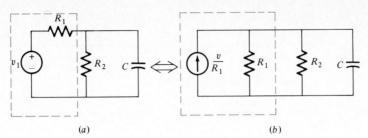

figure 6.6.17
Source transformation in the circuit of Figure 6.6.15.

(a)   (b)

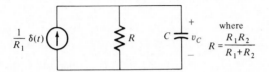

figure 6.6.18
Equivalent circuit to that shown in Figure 6.6.17b.

$$v_C(t) = h(t) = \frac{1}{R_1 C} e^{-t/RC} u(t) \qquad \text{for all } t$$

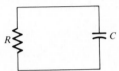

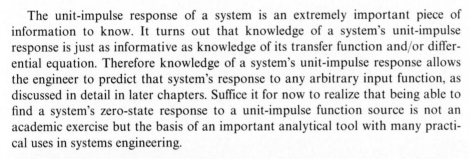

figure 6.6.19
The circuit of Figure 6.6.18 when $t > 0$.

The unit-impulse response of a system is an extremely important piece of information to know. It turns out that knowledge of a system's unit-impulse response is just as informative as knowledge of its transfer function and/or differential equation. Therefore knowledge of a system's unit-impulse response allows the engineer to predict that system's response to any arbitrary input function, as discussed in detail in later chapters. Suffice it for now to realize that being able to find a system's zero-state response to a unit-impulse function source is not an academic exercise but the basis of an important analytical tool with many practical uses in systems engineering.

DRILL PROBLEM

Find the unit-step and unit-impulse responses $w(t)$ and $h(t)$ for the voltage across the 3-Ω resistor in the circuit of Figure 6.6.2. Use shortcuts, do not solve any differential equations and check that $h(t) = d/dt[w(t)]$.

ANS.: $w(t) = \frac{3}{7}(4 + 3e^{-(7/5)t})u(t)$ and $h(t) = 3\delta(t) - \frac{9}{5}e^{-(7/5)t}u(t)$.

complete response by direct integration **6.7**

Nonhomogeneous first-order differential equations like those typically arising in circuit and system theory are generally of the form

$$\frac{dx}{dt} + \alpha x = f(t) \tag{6.7.1}$$

and can always be solved directly for $x(t)$ as follows. Multiply both sides by $e^{\alpha t}$

$$e^{\alpha t}\frac{dx}{dt} + e^{\alpha t}\alpha x = e^{\alpha t}f(t) \tag{6.7.2}$$

Note that the left-hand side of this equation can be written

$$e^{\alpha t}\frac{dx}{dt} + \alpha e^{\alpha t}x = \frac{d}{dt}\,xe^{\alpha t} \tag{6.7.3}$$

so that equation (6.7.2) becomes

$$\frac{d}{dt}\,xe^{\alpha t} = e^{\alpha t}f(t) \tag{6.7.4}$$

Integrating both sides from $-\infty$ to (any time) $t > 0$, we get

$$xe^{\alpha t} = \int_{-\infty}^{t} e^{\alpha \tau}f(\tau)\,d\tau = \int_{-\infty}^{0-} e^{\alpha \tau}f(\tau)\,d\tau + \int_{0-}^{t} e^{\alpha \tau}f(\tau)\,d\tau \tag{6.7.5}$$

The first term on the right is a constant K because both limits in that integral are constants; thus

$$xe^{\alpha t} = K + \int_{-0}^{t} e^{\alpha \tau}f(\tau)\,d\tau \tag{6.7.6}$$

Multiplying by $e^{-\alpha t}$ gives the general solution

$$x(t) = Ke^{-\alpha t} + \int_{-0}^{t} e^{-\alpha(t-\tau)}f(\tau)\,d\tau \tag{6.7.7}$$

The value of the constant K is then determined by initial conditions. The two terms in equation (6.7.7) are the zero-input and zero-state responses, respectively.

EXAMPLE 6.7.1

Find the step response of the circuit in Example 6.6.9 by direct integration.
ANS.: See Figure 6.6.15. By KCL we find the differential equation

$$\frac{v(t) - v_C}{R_1} = \frac{v_C}{R_2} + i_C$$

$$\frac{C\,dv_C}{dt} + \left(\frac{1}{R_1} + \frac{1}{R_2}\right)v_C = \frac{1}{R_1}\,v(t)$$

$$\frac{dv_C}{dt} + \frac{R_1 + R_2}{R_1 R_2 C}\,v_C = \frac{1}{R_1 C}\,u(t) \tag{6.7.8}$$

From equation (6.7.7), with $\alpha = (R_1 + R_2)/R_1 R_2 C$,

$$v_C(t) = Ke^{-\alpha t} + \int_{0-}^{t} e^{-\alpha t}e^{\alpha \tau}\,\frac{1}{R_1 C}\,u(\tau)\,d\tau \qquad \text{for } t > 0$$

$$= Ke^{-\alpha t} + \frac{1}{R_1 C}\,e^{-\alpha t}\int_{0-}^{t} e^{\alpha \tau}(1)\,d\tau = Ke^{-\alpha t} + \frac{1}{R_1 C}\,e^{-\alpha t}\left[\frac{1}{\alpha}\,(e^{\alpha t} - 1)\right]$$

$$= Ke^{-\alpha t} + \frac{1}{R_1 C\alpha}\,(1 - e^{-\alpha t}) = Ke^{-\alpha t} + \frac{R_2}{R_1 + R_2}\,(1 - e^{-\alpha t})$$

Since $v_C(0-) = v_C(0+) = 0$, we see that $K = 0$, and so

$$w(t) = v_C(t) = \frac{R_2}{R_1 + R_2}(1 - e^{\alpha t}) \qquad \text{for } t > 0$$

which is identical to the result of Example 6.6.9.

Solving first-order systems by direct integration is perhaps the most tedious and student-error-producing method we have discussed. It is presented here simply to show that a method for solving first-order systems exists which does not require us first to guess at the correct answer. It is also presented to introduce the idea that zero-state responses can be obtained directly by the integration process of the second term in equation (6.7.7). This process, called *convolution*, will be discussed in detail in Chapter 7.

summary 6.8

In this chapter we have presented a method that enables us to solve for any voltage or current in any first-order system. The concepts of natural, particular, zero-input, and zero-state responses have been discussed in detail and methods presented for obtaining each of them. All these notions will be used again in Chapter 7 when we consider higher-order systems.

We have seen that the exponential source is easy to find the particular response for because there is only one term in the linear sum of this forcing function and all its derivatives. We return to this idea in future chapters. Sources that are not so easily handled can be forced into the exponential form, making our task much easier (more about that later).

At this point you should be able simply to write down the answer to any problem involving first-order circuits with only constant sources after time zero. You should be able to find the unit-impulse response of any first-order system both as the time derivative of the unit-step response and by treating it as a problem in its own right—by knowing where in the circuit impulses can and cannot be found. Also you know that impulse sources set initial conditions on energy-storage elements instantaneously. Thus the complete impulse response is just the natural response.

In Chapter 7 we shall extend and apply all these ideas to the solution of systems with more than one energy-storage element.

problems

1. In the differential equation $4\,dv/dt + 3v = 0$ we can assume a solution of the form Ae^{st}. Find s.

2. Find $v_C(0+)$ and $i_L(0+)$ in the circuit in Figure P6.2. Note the switches open at $t = 0$.

3. The switch in Figure P6.3 opens at $t = 0$. At the instant $t = 0+$ find the values of the following variables: (a) i_L and v_C; (b) di_L/dt and dv_C/dt; (c) i_L due to the 10-V source acting alone (the 18-V source is shut off); (d) i_L due to the 18-V source acting alone (the 10-V source is shut off). (e) Does superposition apply in this case?

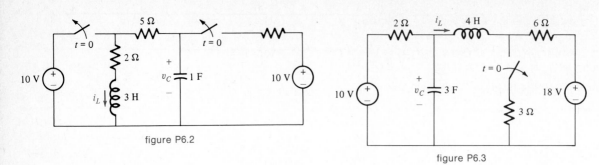

figure P6.2

figure P6.3

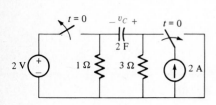

figure P6.4

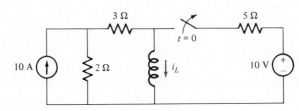

figure P6.5

4. In the circuit in Figure P6.4 find $v_C(0+)$, dv_C/dt evaluated at $t = 0+$, $v_C(\infty)$, and the energy stored in the capacitor $W_C(0+)$. Both switches are thrown in the directions shown at time zero.

5. In the circuit shown in Figure P6.5 the switch opens at $t = 0$. Find (a) $i_L(0-)$; (b) $i_L(0+)$; (c) $i_L(\infty)$; (d) di_L/dt evaluated at $t = 0-$; and (e) di_L/dt evaluated at $t = 0+$.

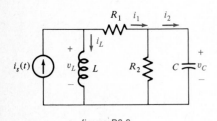

figure P6.6

6. For the circuit in Figure P6.6 the value of the current source $i_s(t) = 8e^{-10t}u(t)$; also $L = \frac{1}{6}$ H, $R_1 = 3$ Ω, $R_2 = 10$ Ω, and $C = \frac{1}{3}$ F. Given that at time $t = 0+$ the initial conditions are $v_C(0+) = 5$ V and $i_L(0+) = 5$ A, find the values of (a) $i_1(0+)$; (b) $i_2(0+)$; (c) $v_L(0+)$; (d) di_L/dt evaluated at $t = 0-$; and (e) dv_C/dt evaluated at $t = 0+$.

7. In the circuit shown in Figure P6.7, $I_s = 9$ A, $R_1 = 3$ Ω, $R_2 = 6$ Ω, and $C = \frac{1}{8}$ F. (a) What is the value of the voltage v_C across the capacitor immediately after the switch is opened at $t = 0$? (b) What is the numerical value of the time constant for $t > 0$? (c) Find $i_C(0+)$. (d) Find dv_C/dt evaluated at $t = 0+$.

8. The switch in Figure P6.8 is thrown from position a to position b at $t = 0$; $V_s = 8$ V, $I_s = 4$ A, $R_1 = 3$ Ω, $R_2 = 10$ Ω, $R_3 = 2$ Ω, $L = 5$ H, and $C = \frac{1}{6}$ F. Find the value of the following for time *less than zero*: (a) v_1; (b) i_1; (c) i_3; (d) v_4; (e) i_4; (f) di_3/dt evaluated at $t = 0+$; and (g) dv_4/dt evaluated at $t = 0+$.

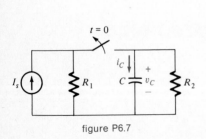

figure P6.7

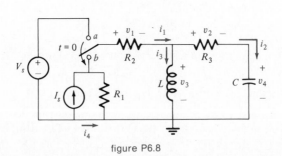

figure P6.8

9. In problem 8 the Norton equivalent circuit connected between node b and ground is replaced by a single, constant, ideal voltage source and everything else is left unchanged. What should the value of this new voltage source be for di_3/dt to equal zero at time $0+$? What will be the value of dv_4/dt at $t = 0+$ under these conditions?

10. In the circuit of Figure P6.10 find the value of d^2i/dt^2 at $t = 0+$.

11. (a) In the circuit of Figure P6.11 find the value of i_1, i_L, i_R, i_C, v_L, and v_C at $t = 0-$. (b) Repeat part (a) for $t = 0+$.

12. A step-function current source is in parallel with both a 10-kΩ resistor and a 50-μF capacitor. (a) Find the natural response with undetermined coefficient(s) for the capacitor voltage. (b) Repeat part (a) but put the capacitor and resistor in series rather than parallel.

13. In Figure P6.13 find the current in the inductance and the voltage at node A for the instant after the switch opens. Then connect an ideal diode from A to ground, as shown, and recalculate the inductor current and the voltage at A for the same instant after the switch opens. Comment on the practical significance of your results.

14. Given a circuit described by the differential equation $dv/dt + \frac{7}{10}v = 6$ for $t > 0$. If $v(0+) = 2$ V, find $v(t)$ for $t > 0$.

15. A voltage source $v_1(t) = 12\,t\,u(t)$ is placed across the series combination of a 3-Ω resistor and a 2-F capacitor. Assume that $v_C(0-) = 3$ V [opposing $v_1(t)$]. Find $v_C(t)$.

16. In the circuit shown in Figure P6.16, $v_s(t) = t^2$. (a) Write the differential equation that describes the capacitor current $i_C(t)$ for $t > 0$. (b) Solve this equation for $i_C(t)$. $R_1 = 6\ \Omega$, $R_2 = 4\ \Omega$, and $C = 2$ F.

17. A known velocity $\mathcal{V}_A(t) = u(t)$ is applied to one end of an ideal dashpot D_1; the other end of the dashpot is connected to a mass M which rests on a flat floor. There is friction D_2 between the mass and the floor. (a) Write the differential equation relating the velocity of the mass $\mathcal{V}_M(t)$ to $\mathcal{V}_A(t)$. (b) Assume $D_1 = D_2 = M = 1$. Solve for $\mathcal{V}_M(t)$ for $t > 0$. (c) Sketch $\mathcal{V}_M(t)$ vs t.

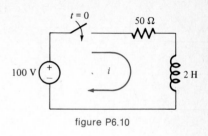

figure P6.10

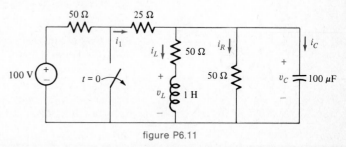

figure P6.11

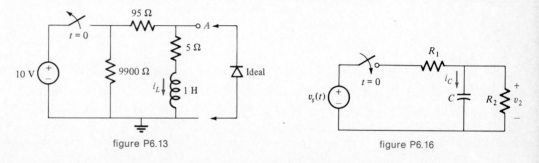

figure P6.13

figure P6.16

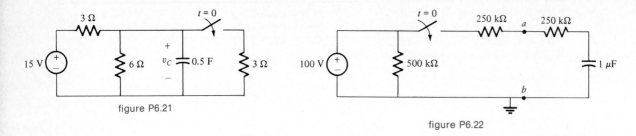

figure P6.21

figure P6.22

18. A series combination of a 20-Ω resistor and a 0.025-F capacitor [with an initial value $v_C(0+) = 4$ V] is driven by a voltage source $v_s(t) = 10e^{-2t}u(t)$. The negative sides of v_s and v_C are connected to ground. Find $v_C(t)$ for all t.

19. A 0.5-F capacitor, a 3-Ω resistor, and a 6-V voltage source are connected in parallel. At $t = 0$ the voltage source is disconnected from the other two elements. (a) Find $v_C(t)$ for all t. (b) Find $i_C(t)$ for all t.

20. A parallel combination of an initially uncharged 2-F capacitor and a 5-Ω resistor is in series with a 10-Ω resistor. This entire circuit is driven by a voltage source $v_s(t) = e^{-t}u(t)$. Find $v_C(t)$ for all t.

21. In the circuit in Figure P6.21 find an expression for $v_C(t)$ for all t. Sketch v_C versus t. What is the value of $v_C(0.5)$?

22. What is the value of $v_{ab}(1)$ in the circuit in Figure P6.22?

23. The switch in the circuit in Figure P6.23 has been open for a long time and closes at $t = 0$. Find $v_C(t)$ for all t. Sketch v_C versus t. Evaluate $v_C(1)$.

24. The thermal resistance (of walls and roof) and capacitance of a building are $R = 0.1$ K·s/J and $C = 10^4$ J/K. On a given day it is found that if the heating system runs continuously, the inside temperature will stay at 305 K (about 90°F). If the inside temperature is initially at 283 K (about 50°F), how warm will it be 30 min after the heating system turns on and stays on?

25. A water tank with capacity A fills through a valve that operates as follows. The flow rate Q is proportional to the difference between the height of the water in the tank h and the maximum water level H; that is, $Q = k(H - h)$. But as the tank becomes almost full and the flow therefore slows, the valve snaps completely off when Q drops to one-tenth its maximum value (which occurs at $t = 0+$ when $h = 0$). How long after filling begins does the valve shut off? How high is the water level at that time?

26. Given that $f(t) = u(t) - u(t - 2)$. (a) In the circuit in Figure P6.26a find $v_C(t)$ and sketch it versus t $[v_C(0-) = 0]$. (b) In the circuit in Figure P6.26b find $i(t)$ and sketch it versus t $[i_L(0-) = 0]$.

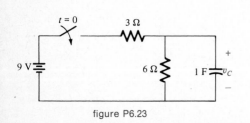

figure P6.23

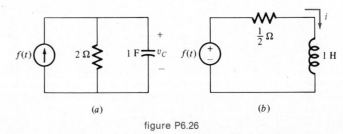

(a)

(b)

figure P6.26

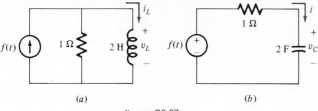

figure P6.27

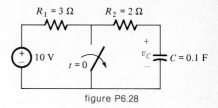

figure P6.28

27. Given that $f(t) = u(t) - u(t-2)$. (a) In the circuit in Figure P6.27a find $v_L(t)$ and sketch it versus t $[i_L(0-) = 0]$. (b) In the circuit in Figure P6.27b find $i(t)$ and sketch it versus t $[v_C(0-) = 0]$.

28. The switch in the circuit of Figure P6.28, which is normally closed, opens at $t = 0$ and then recloses at $t = 0.4$ s and remains closed thereafter. Find $v_C(0.7)$.

29. A voltage source $v_s(t) = 100u(t)$ is across the series combination of a 10-kΩ resistor and an initially uncharged capacitor C. At $t = 0.0139$ s, the voltage across the capacitor has risen to 50 V. Find the value of C.

30. The switch in the circuit in Figure P6.30 has been closed for a long time. Find the value of $v_L(0+)$. Find an expression for $v_L(t)$ for all t. Sketch v_L versus t.

31. In the circuit in Figure P6.31 the switch is thrown from position 1 to position 2 at $t = 0$, a steady-state current having previously been established. Find the complete response for the current $i(t)$.

32. The switch in the circuit in Figure P6.32 has been in position a for a long time. At $t = 0$ the switch is instantaneously moved from a to b. Find $v_2(t)$.

33. The network of Figure P6.33 reaches a steady state with the switch open. At $t = 0$, the switch is closed. Find $i(t)$ and sketch it versus t.

34. In the circuit in Figure P6.34a the switch is closed at $t = 0$. Just before the switch closes the capacitor voltage v_C is 0.5 V. Determine $v_C(t)$ for $t > 0$.

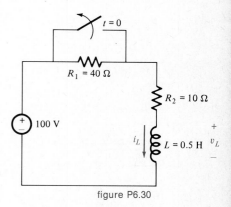

figure P6.30

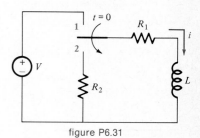

figure P6.31

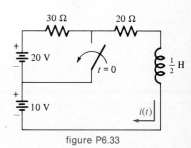

figure P6.33

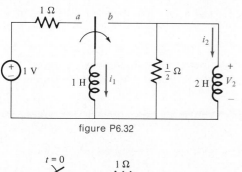

figure P6.32

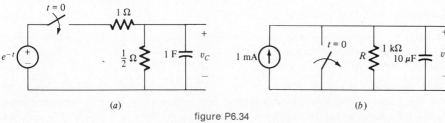

figure P6.34

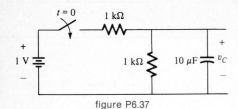

figure P6.37

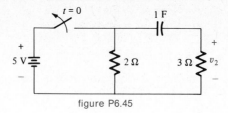

figure P6.45

35. Given the circuit shown in Figure P6.34b find the initial voltage across the capacitor $v_C(0+)$. At what time will the voltage across the capacitor reach (a) 50 percent, (b) 63.2 percent, and (c) 90 percent of its final value?

36. Voltage source $v_s(t)$ is a square wave that switches back and forth between 10 and 0 V. It has a 50 percent duty cycle (half the time it is 10 V and half the time 0 V). Its period is 20 ms and it is 10 V for $0 < t < 10$ ms. This source is applied through a switch that closes at $t = 0$, to the series combination of an initially uncharged 10-μF capacitor and a 1-kΩ resistor. Find $v_C(40 \times 10^{-3})$.

37. Thanks to a previous charging interval, the capacitor voltage in the circuit of Figure P6.37 is $+0.4$ V at $t = 0$, when the switch closes. Find $v_C(t)$ for $t > 0$.

38. A linear system is described by $di/dt + 3i = 2e^{-t}u(t)$, and the initial condition $i(0+) = 10$ A is given. Find the *zero-state response* of this system. Evaluate all constants.

39. Given a circuit described by $di/dt + \frac{7}{20}i = 0.25e^{-t}$ for $t > 0$ and $i(0+) = 2$ A. (a) Find the zero-state response for $i(t)$. (b) Find the zero-input response for $i(t)$. (c) Find the complete response for $i(t)$.

40. Consider a parallel connection of a 1-Ω resistor, a 1-H inductor, and a current source $i_s(t) = r(t)$. The output variable is the resulting voltage. Find (a) the zero-input response, (b) the zero-state response, (c) the complete response, and (d) the unit-impulse response.

41. The parallel combination of a resistor and an inductor is in series with another resistor (all values are unity). An input voltage is applied across this entire circuit. The output variable is the inductor voltage. What is the unit-impulse response $h(t)$?

42. The parallel combination of a 6-Ω resistor and a $\frac{1}{2}$-F capacitor is in series with a 3-Ω resistor. A voltage source $v_s(t) = r(t) = tu(t)$ is impressed across the resulting terminals. Find the zero-state response for $v_C(t)$.

43. An input voltage source $v_s(t)$ is impressed across the series combination of a 4-Ω resistor and a 3-H inductor. The output variable of interest is the resulting current $i(t)$. Find the unit-impulse response.

44. Find the zero-state response in problem 43 if the voltage source is a unit-ramp function $v_s(t) = r(t) = tu(t)$.

45. In the circuit in Figure P6.45 find $v_2(t)$ for all t. The switch has been closed until $t = 0$ and then opens and remains so for all $t > 0$.

46. Applying a voltage source $e^{-3t}u(t)$ to a network yields a particular response of $1u(t)$ amperes. What particular response will result from an application of a voltage source equal to $-3e^{-3t}u(t)$?

47. In the circuit in Figure P6.47, $i_L(0-) = v_C(0-) = 0$. (a) Find $i_L(0+)$ and $v_C(0+)$. (b) How much energy $W_L(0+)$ is stored in the inductor at time $t = 0+$? (c) How much energy $W_C(0+)$ is in the capacitor at $t = 0+$?

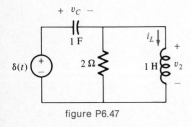

figure P6.47

48. The parallel combination of a 3-H inductor and a 4-Ω resistor is driven by a current source $2u(-t)$. Find (a) $i_L(0+)$; (b) the initial slope di_L/dt evaluated at $t = 0+$; (c) the energy stored in the inductor at time $t = 0+$; and (d) the instantaneous power $p_L(t)$ delivered to the inductor.

49. A series combination of a $\frac{1}{2}$-F capacitor and a 3-Ω resistor is driven by an input voltage source $v_s(t)$. Find the unit-impulse response of the output variables: (a) $v_R(t)$; (b) $v_C(t)$; and (c) $i(t)$. (d) to (f) Find the unit-step response of each of these variables.

50. In Figure 6.6.17 replace the capacitor C with a 4-H inductor. Let $R_1 = 3\ \Omega$ and $R_2 = 6\ \Omega$. (a) Find the unit-impulse response of $v_L(t) = h(t)$. (b) Sketch this $h(t)$ versus t. (c) Find the unit-step response $v_L(t) = w(t)$. (d) Sketch this $w(t)$ versus t.

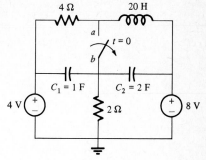

figure P6.52

51. A 1-F capacitor C_1 containing a 1-V initial charge is connected through an initially open switch to a second 1-F capacitor C_2 that initially contains no energy. At $t = 0$ the switch is closed. (a) How much energy is stored in the circuit for $t < 0$? (b) Assuming no charge is lost from the circuit, what is the final voltage across C_1 and C_2? (c) How much total energy is stored in the circuit after the switch closes? (d) How did this happen? *Hint*: Assume an R-ohm resistor is placed in series with the switch and repeat parts (a) and (b). In the limit as R gets vanishingly small, what happens?

52. The switch in Figure P6.52 opens at $t = 0$ after having been closed for a long time. (a) Find an expression for $v_{ab}(t)$, the voltage across the switch for $t > 0$. (b) What is the maximum voltage that appears across this switch? When does it occur?

chapter 7
SECOND-ORDER SYSTEMS

introduction 7.1

In this chapter we shall concern ourselves with obtaining the solutions to higher-order system equations, i.e., system equations whose highest-order derivative is second or greater. Although the method we discussed for solving first-order equations also works well for these higher-order systems, we shall see that in solving an nth-order system equation there are n unevaluated constants to be evaluated (by means of initial conditions) rather than just one. Other than this single complication and some additional terminology, there is nothing about solving higher-order equations that is really different from the method we used to solve first-order systems.

We shall again be concerned with the now familiar natural and particular responses, zero-state and zero-input responses, and unit-step and unit-impulse responses. We start by examining natural (source-free) responses and then concentrate on the three different types of natural responses observed in second-order systems.

7.2 natural response: overdamped case

Suppose we want to find the natural response of a system with more than one energy-storage element. Again we assume that $i_n(t) = Ae^{st}$ and solve for the values of s (eigenvalues) that make this valid.

EXAMPLE 7.2.1

Find the eigenvalues for the circuit in Figure 7.2.1.

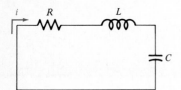

figure 7.2.1
The circuit of Example 7.2.1.

ANS.:   Summing the voltage rises around the loop gives

$$iR + Lpi + \frac{1}{Cp}\, i = 0 \tag{7.2.1}$$

$$\left(R + Lp + \frac{1}{Cp}\right)i = 0 \tag{7.2.2}$$

$$\frac{RCp + LCp^2 + 1}{Cp}\, i = 0 \tag{7.2.3}$$

$$\left(p^2 + \frac{R}{L}\, p + \frac{1}{LC}\right)i = 0 \tag{7.2.4}$$

or

$$\frac{d^2 i}{dt^2} + \frac{R}{L}\frac{di}{dt} + \frac{1}{LC}\, i = 0 \tag{7.2.5}$$

Since the exponential function worked so well on first-order differential equations, let us try it again.

Let us assume that the answer is of the form

$$i_n(t) = Ae^{st} \tag{7.2.6}$$

Is this consistent with equation (7.2.5)? To substitute it in and see we need the first two derivatives of i_n as well as i_n itself. These are

$$\frac{di}{dt} = sAe^{st} \tag{7.2.7}$$

and

$$\frac{d^2 i}{dt^2} = s^2 Ae^{st} \tag{7.2.8}$$

Inserting (7.2.6) to (7.2.8) into (7.2.5) yields

$$s^2 Ae^{st} + \frac{R}{L}\, sAe^{st} + \frac{1}{LC}\, Ae^{st} = 0 \tag{7.2.9}$$

or

$$s^2 + \frac{R}{L}s + \frac{1}{LC} = 0 \qquad (7.2.10)$$

This is the characteristic equation. Solving this by the quadratic formula yields

$$s = \frac{-R/L \pm \sqrt{(R/L)^2 - 4/LC}}{2} = -\frac{R}{2L} \pm \sqrt{\left(\frac{R}{2L}\right)^2 - \frac{1}{LC}} \qquad (7.2.11)$$

If $(R/2L)^2$ is greater than $1/LC$, we have two real eigenvalues (solutions of this characteristic equation). If $1/LC$ is greater than $(R/2L)^2$, we get two complex numbers for the eigenvalues. We shall deal with this eventuality later in the chapter; for now we simply note that a *second-order* system gives rise to *two* eigenvalues.

You should also notice that we could have written equation (7.2.10) directly from (7.2.4) by simply replacing the *operator p* with the *algebraic* variable *s*.† This useful computational shortcut is valid for any *n*th-order linear system.

EXAMPLE 7.2.2

Find the eigenvalues of the system equation

$$\frac{d^2i}{dt^2} + 3\frac{di}{dt} + 2i = 0$$

ANS.: Assume that

$$i(t) = Ae^{st} \qquad \frac{di}{dt} = sAe^{st} \qquad \frac{d^2i}{dt^2} = s^2Ae^{st}$$

$$s^2Ae^{st} + 3sAe^{st} + 2Ae^{st} = 0$$

$$s^2 + 3s + 2 = 0 \qquad \text{characteristic equation}$$

Solving, we get

$$s = \frac{-3 \pm \sqrt{9-8}}{2} = -2, -1$$

In more concise form, we can imagine the differential equation in *p*-operator notation as being

$$(p^2 + 3p + 2)i = 0$$

and then substitute *s* for *p* (or directly for *d/dt* in the original form) to obtain the characteristic equation

$$s^2 + 3s + 2 = 0$$

from which, as before, the eigenvalues are found to be

$$s = -2, -1$$

† See the discussion at the end of Section 6.2 of this technique applied to first-order systems.

Remember that what the eigenvalues are telling us is this: Your guess that the natural solution is $i_n(t) = Ae^{st}$ was generally correct but not for any arbitrary value of s—only for $s = -2$ or $s = -1$. Therefore, in Example 7.2.2 we conclude that either

$$i(t) = Ae^{-2t} \qquad \text{or} \qquad i(t) = Ae^{-t}$$

is the natural response, i.e., the response to a zero-valued input. If a zero-valued input produces Ae^{-2t} and a zero-valued input *also* produces Ae^{-t}, superposition tells us that the sum of those inputs (zero + zero) will produce $A_1 e^{2t} + A_2 e^{-t}$. We can check that this is indeed so by substituting the two-term solution back into the differential equation. Let us do this as follows.

We consider a differential equation in the general form

$$\frac{d^2y}{dt^2} + k_1 \frac{dy}{dt} + k_2 y = 0$$

Assume that the solutions (obtained by finding the zeros of the characteristic polynomial) are $y_1 = A_1 e^{s_1 t}$ and $y_2 = A_2 e^{s_2 t}$. Then form $y_3 = y_1 + y_2$ and substitute it back into the differential equation

$$p^2 y_3 + k_1 p y_3 + k_2 y_3 = 0$$
$$p^2(y_1 + y_2) + k_1 p(y_1 + y_2) + k_2(y_1 + y_2) = 0$$
$$p^2 y_1 + p^2 y_2 + k_1 p y_1 + k_1 p y_2 + k_2 y_1 + k_2 y_2 = 0$$
$$(p^2 y_1 + k_1 p y_1 + k_2 y_1) + (p^2 y_2 + k_1 p y_2 + k_2 y_2) = 0$$
$$0 + 0 = 0$$

Thus, the sum of the two solutions is also a solution. Therefore in Example 7.2.2 we conclude for eigenvalues $s_1 = -2$ and $s_2 = -1$ that the actual solution is $i(t) = A_1 e^{-2t} + A_2 e^{-t}$. The constants A_1 and A_2 must now be evaluated by means of initial conditions.

EXAMPLE 7.2.3

In Example 7.2.2 if we know that $i(0+) = 2$ and di/dt evaluated at $t = 0+$ is 3, find A_1 and A_2 in the natural response

$$i(t) = A_1 e^{-2t} + A_2 e^{-t} \tag{7.2.12}$$

ANS.: Find the derivative of $i(t)$

$$\frac{di}{dt} = -2A_1 e^{-2t} - A_2 e^{-t} \tag{7.2.13}$$

At $t = 0+$ evaluate (7.2.12) and (7.2.13) and set them equal to their given numerical values at that instant

$$i(0+) = A_1 + A_2 = 2 \tag{7.2.14}$$

$$\left.\frac{di}{dt}\right|_{0+} = -2A_1 - A_2 = 3 \tag{7.2.15}$$

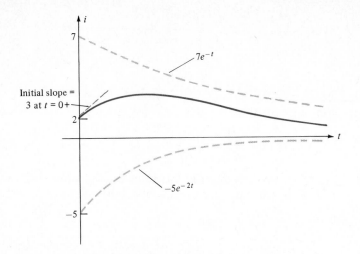

figure 7.2.2
The source-free response of the circuit of Figure 7.2.1.

whence, from (7.2.14) and (7.2.15), $A_1 = -5$ and $A_2 = +7$ so that

$$i(t) = -5e^{-2t} + 7e^{-t} \tag{7.2.16}$$

A sketch of this solution is shown in Figure 7.2.2.

Initial conditions are often presented to us in a form we cannot use, and we must derive the initial conditions we need. This may be true whether the eigenvalues are real or complex. When it is necessary to derive the needed initial conditions from those given, it is often helpful to sketch the circuit and label it with the numerical value of all v's and i's in the circuit at the instant $t = 0+$. Then KVL, KCL, Ohm's law, etc., can be used to find the initial conditions that must be used to evaluate the constants.

For example, we just saw that the solution for the source-free series RLC circuit of Examples 7.2.1 and 7.2.2 with $R = 3 \ \Omega$, $L = 1$ H, $C = \frac{1}{2}$ F is $i(t) = A_1 e^{-2t} + A_2 e^{-t}$. To evaluate A_1 and A_2 we needed to know $i(0+)$ and di/dt evaluated at $t = 0+$. Suppose we are told only that $v_C(0-) = -9$ V and $i_L(0-) = 2$ A; we could then find the needed value of di/dt evaluated at $t = 0+$ as follows. First we realize that, in the absence of impulses, $v_C(0-) = v_C(0+)$ and $i_L(0-) = i_L(0+)$. Then we recall that

$$L \frac{di}{dt} = v_L \tag{7.2.17}$$

Then at $t = 0+$ (see Figure 7.2.3)

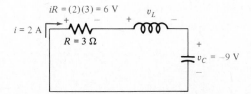

figure 7.2.3
The circuit of Examples 7.2.1 and 7.2.3 at $t = 0+$. Since $v_C = -9$ V and $iR = 6$ V, v_L must equal 3 V.

$$v_L + iR + v_C = 0 \tag{7.2.18}$$

and

$$v_L(t) = -i(t)R - v_C(t)$$

Thus, at $t = 0+$

$$v_L(0+) = -i(0+)R - v_C(0+) = -2(3) + 9 = 3 \tag{7.2.19}$$

Substituting back into (7.2.17) with $L = 1$ gives

$$\left.\frac{di}{dt}\right|_{0+} = 3 \tag{7.2.20}$$

The coefficients A_1 and A_2 are then evaluated as shown in equations (7.2.13) to (7.2.16). In some cases it is convenient to sketch and label the circuit at *both* $t = 0-$ and at $t = 0+$.

EXAMPLE 7.2.4

Consider the circuit shown in Figure 7.2.4a. Solve for $i(t)$ for $t > 0$. Both switches are thrown at $t = 0$ and have been in their previous positions for a long time.

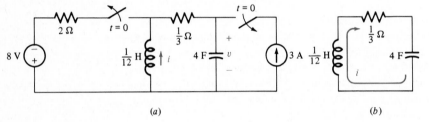

(a)   (b)

figure 7.2.4
(a) The circuit of example 7.2.4. The switches are thrown at the instant $t = 0$. (b) The circuit for time $t > 0$. (c) The circuit at the instant $t = 0-$. (d) The circuit at the instant $t = 0+$.

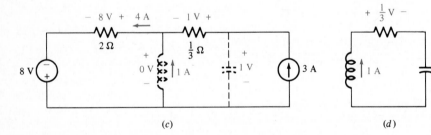

(c)   (d)

ANS.:   For time $t > 0$ the circuit is as shown in Figure 7.2.4b, from which

$$Lpi + iR + \frac{1}{Cp} i = 0 \tag{7.2.21}$$

$$p^2 i + \frac{R}{L} pi + \frac{1}{LC} i = 0$$

or

$$(p^2 + 4p + 3)i = 0 \tag{7.2.22}$$

Thus the characteristic equation is

$$s^2 + 4s + 3 = 0 \tag{7.2.23}$$

and the resulting eigenvalues are

$$s = -3, -1 \tag{7.2.24}$$

so that

$$i_n(t) = A_1 e^{-t} + A_2 e^{-3t} \tag{7.2.25}$$

and

$$\frac{di}{dt} = -A_1 e^{-t} - 3A_2 e^{-3t} \tag{7.2.26}$$

At time $t = 0-$ the circuit is as shown in Figure 7.2.4c. The capacitor is an effective open circuit (for dc sources) and for similar reasons the inductor forms a short circuit. The 8-V source voltage falls across the 2-Ω resistor, making it draw 4 A from the right. The 3-A source pushes 3 A into the top center node from the right. Therefore (via KCL) the inductor must have 1 A flowing up into the center top node. Immediately after the switches are thrown (at $t = 0+$) the inductor current and capacitor voltage remain as they were at $t = 0-$ because no impulse sources are present. The circuit at $t = 0+$ is as shown in Figure 7.2.4d. Therefore (by KVL) at $t = 0+$

$$v_L = -\frac{4}{3} = \frac{1}{12}\frac{di}{dt}$$

so that

$$\left.\frac{di}{dt}\right|_{0+} = -16 \tag{7.2.27}$$

Also

$$i(0+) = 1 \tag{7.2.28}$$

Substituting (7.2.27) and (7.2.28) into (7.2.25) and (7.2.26) evaluated at $t = 0+$ yields

$$1 = A_1 + A_2 \quad \text{and} \quad -16 = -A_1 - 3A_2$$

whence

$$A_1 = -\tfrac{13}{2} \quad \text{and} \quad A_2 = \tfrac{15}{2} \tag{7.2.29}$$

so that

$$i(t) = -\tfrac{13}{2}e^{-t} + \tfrac{15}{2}e^{-3t} \tag{7.2.30}$$

You are encouraged to (1) check that equation (7.2.30) actually does obey the initial conditions [equations (7.2.27) and (7.2.28)] and (2) carefully plot the two terms (separately) in equation (7.2.30) and then plot the $i(t)$ function as the algebraic sum of those two components. Find the maximum value and the time of its occurrence.

natural response: underdamped case   **7.3**

If the eigenvalues turn out to be complex numbers rather than real numbers, we proceed as in the following example.

EXAMPLE 7.3.1

Solve for $y(t)$

$$\frac{d^2y}{dt^2} + 8\frac{dy}{dt} + 25y = 0 \tag{7.3.1}$$

ANS.:   Try $y(t) = Ae^{st}$. This results in the characteristic equation

$$s^2 + 8s + 25 = 0 \tag{7.3.2}$$

whence

$$s = -4 \pm j3 \tag{7.3.3}$$

where†

$$j = \sqrt{-1} \tag{7.3.4}$$

so that

$$y(t) = A_1 e^{(-4+j3)t} + A_2 e^{(-4-j3)t} \tag{7.3.5}$$

Since§

$$e^{j\theta} = \cos\theta + j\sin\theta \tag{7.3.6}$$

and

$$e^{-j\theta} = \cos\theta - j\sin\theta \tag{7.3.7}$$

it follows that

$$y(t) = e^{-4t}[A_1(\cos 3t + j\sin 3t) + A_2(\cos 3t - j\sin 3t)]$$
$$= e^{-4t}[(A_1 + A_2)\cos 3t + j(A_1 - A_2)\sin 3t] \tag{7.3.8}$$

Defining

$$B_1 = A_1 + A_2 \quad \text{and} \quad B_2 = j(A_1 - A_2) \tag{7.3.9}$$

gives‡

$$y(t) = e^{-4t}(B_1 \cos 3t + B_2 \sin 3t) \tag{7.3.10}$$

The form of the quadratic formula used to solve equation (7.3.2) for the eigenvalues ensures that any *complex* eigenvalues will be *conjugates* of each other; i.e., they have equal real parts, and their imaginary parts are of opposite algebraic sign.

Equation (7.3.10) is a solution to the differential equation (7.3.1). In fact, from now on when we see that the eigenvalues are complex conjugates, we can feel free to write down a solution of the form of equation (7.3.10) immediately. The real part of the eigenvalues appears in the leading exponential term, and the imaginary part of the eigenvalues is the radian frequency of both sinusoidal terms. This is as far as we can proceed, however, unless we know the initial conditions. Since there are two constants to evaluate, we need two initial conditions; i.e., we need to know $y(0+)$ and dy/dt evaluated at $t = 0+$.

Suppose in Example 7.3.1 we are told that $y(0+) = 4$ and that dy/dt evaluated at $t = 0+$ is equal to -1. We can use this information to compute B_1 and B_2 as follows. Evaluate (7.3.10) when $t = 0+$

$$y(0+) = e^0(B_1 \cos 0 + B_2 \sin 0) \quad \text{or} \quad 4 = B_1$$

Thus

$$y(t) = e^{-4t}(4\cos 3t + B_2 \sin 3t)$$

Now compute dy/dt and evaluate it at $t = 0+$:

$$\frac{dy}{dt} = e^{-4t}[-4(3)\sin 3t + 3B_2\cos 3t] + (-4)e^{-4t}(4\cos 3t + B_2\sin 3t)$$

† Since electrical engineers use the letter i to describe electric current, they use the letter j for the square root of -1.

‡ Because $y(t)$ is a real quantity (e.g., voltage or current), it follows that B_1 and B_2 must be real. There is no such requirement on A_1 and A_2; see equation (7.3.9).

§ See page 403.

$$\left.\frac{dy}{dt}\right|_{0+} = e^0(-12 \sin 0 + 3B_2 \cos 0) - 4(4 \cos 0 + B_2 \sin 0)$$

$$-1 = 3B_2 - 16 \qquad \text{and} \qquad B_2 = 5$$

Therefore the solution is

$$y(t) = e^{-4t}(4 \cos 3t + 5 \sin 3t) \qquad (7.3.11)$$

We can get this result in slightly different form by using the trigonometric identity

$$\cos (a + b) = \cos a \cos b - \sin a \sin b \qquad (7.3.12)$$

which we can use to write the solution in the form

$$y(t) = e^{-4t}[A \cos (3t + \theta)] \qquad (7.3.13)$$

by noting that, with $a = 3t$ and $b = \theta$, the bracketed term in equation (7.3.13) can be written

$$A \cos (3t + \theta) = A \cos 3t(\cos \theta) - A \sin 3t(\sin \theta) \qquad (7.3.14)$$

The right-hand side of this equation is of the same form as the term in parentheses in equation (7.3.11) provided that

$$A \cos \theta = 4 \qquad (7.3.15)$$

and

$$-A \sin \theta = 5 \qquad (7.3.16)$$

Therefore

$$-\frac{A \sin \theta}{A \cos \theta} = \frac{5}{4} \qquad (7.3.17)$$

$$\tan \theta = -\tfrac{5}{4}$$

$$\theta = -51.34° \qquad (7.3.18)$$

Substituting this into equation (7.3.15) yields

$$A = \frac{4}{\cos (-51.34°)} = 6.40 \qquad (7.3.19)$$

so that (7.3.13) becomes

$$y(t) = 6.4e^{-4t} \cos (3t - 51.34°) \qquad (7.3.20)$$

Alternatively, we could have realized as soon as the complex eigenvalues $s = -4 \pm j3$ were obtained that the general form

$$y(t) = e^{-4t}(B_1 \cos 3t + B_2 \sin 3t) \qquad (7.3.21)$$

is also equivalent to

$$y(t) = Ae^{-4t} \cos (3t + \theta) \qquad (7.3.22)$$

and directly evaluated the two unknown coefficients A and θ in equation (7.3.22) from knowledge of the initial conditions. We know $y(0+) = 4$. Thus

$$4 = Ae^0 \cos (0 + \theta) = A \cos \theta$$

or

$$A = \frac{4}{\cos \theta} \qquad (7.3.23)$$

figure 7.3.1
A plot of $y(t)$ versus time.

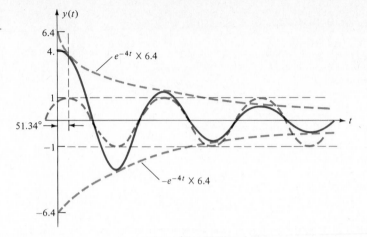

Also we know that dy/dt evaluated at $t = 0+$ equals -1; thus

$$\frac{dy}{dt} = A[-e^{-4t}3 \sin (3t + \theta) - 4e^{-4t} \cos (3t + \theta)] \qquad (7.3.24)$$

evaluated at $t = 0+$ is

$$-1 = A(-3 \sin \theta - 4 \cos \theta) \qquad (7.3.25)$$

Substituting for A from equation (7.3.23) leads to

$$-1 = \frac{4}{\cos \theta} (-3 \sin \theta - 4 \cos \theta) = -12 \tan \theta - 16$$

$$-\tfrac{15}{12} = -\tfrac{5}{4} = \tan \theta$$

$$\theta = -51.34° \qquad (7.3.26)$$

Substituting this into (7.3.23), we find $A = 6.40$. Thus

$$y(t) = 6.40e^{-4t} \cos (3t - 51.34°) \qquad (7.3.27)$$

as found previously in equation (7.3.20).

A sketch of the $y(t)$ waveform is shown in Figure 7.3.1. The t-axis intercepts of $y(t)$ are equally spaced. Since the value of ω in the sinusoidal component is 3, we see that the period $T = 1/f$ of the sinusoidal component can be determined as

$$\omega = 3 \text{ rad/s} \qquad \text{or} \qquad 2\pi f = 3$$

$$\frac{2\pi}{3} = \frac{1}{f} = T$$

So the axis-crossing interval is

$$\frac{T}{2} = \frac{\pi}{3} \qquad \text{s}$$

Notice that the decaying *envelope* produced by the exponential term multiplies the sinusoidal term to produce the *damped sinusoid* of Figure 7.3.1. The maxima of the damped sinusoid do not occur at the same instants as the maxima of the pure sinusoidal component.

EXAMPLE 7.3.2

In the circuit of Figure 7.3.2, at $t = 0+$, $i_L = 0$ and $v_C = 5$ V. Find the expression for i for $t > 0$. Find the time and magnitude of the first maximum, the first zero crossover, the first negative extremum, and the second zero crossover. Plot the current and the envelope from 0 to 2 s.

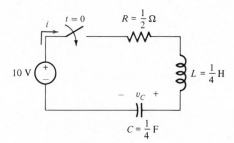

figure 7.3.2
The circuit of Example 7.3.2.

ANS.: First write the system equation

$$\left(R + Lp + \frac{1}{Cp} \right) i = 10 \tag{7.3.28}$$

$$\left(p^2 + \frac{R}{L} p + \frac{1}{LC} \right) i = \frac{p}{L} 10 = 0 \tag{7.3.29}$$

Then the natural response is found by writing the characteristic equation

$$s^2 + 2s + 16 = 0 \tag{7.3.30}$$

the solution of which is

$$s = \frac{-2 \pm \sqrt{4 - 4(16)}}{2} = -1 \pm j3.873 \tag{7.3.31}$$

Thus

$$i_n(t) = Ae^{-t} \cos (3.873t + \theta) \tag{7.3.32}$$

The particular response is zero because the forcing function in (7.3.29) is zero. Therefore the total response is simply the natural response. We use the initial conditions as follows (see Figure 7.3.3b). At $t = 0+$

$$v_L = L \frac{di}{dt}$$

$$\frac{5}{1/4} = 20 = \frac{di}{dt} = A\{e^{-t}[-3.873 \sin (3.8t + \theta)] + [-e^{-t} \cos (3.8t + \theta)]\}$$

$$20 = A(-3.873 \sin \theta - \cos \theta)$$

Also $i_n(0+) = 0$; so from equation (7.3.32), $0 = A \cos \theta$. Thus, $\theta = 90°$ so that $A \neq 0$ is possible:

$$20 = A(-3.873) \tag{7.3.33}$$

$$-5.164 = A \tag{7.3.34}$$

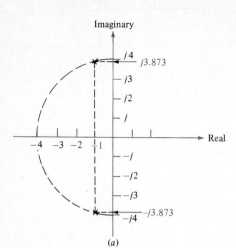

(a)

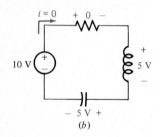

(b)

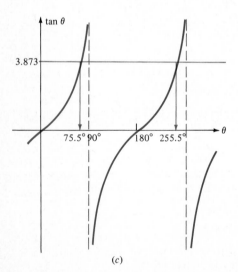

(c)

figure 7.3.3
(a) The complex eigenvalues $s = -1 \pm j3.873$. (b) The circuit at $t = 0+$. (c) The solutions of $3.873 = \tan(3.873t)$.

$$i_n(t) = -5.16e^{-t}\cos(3.873t + 90°) = +5.16e^{-t}\sin 3.873t \qquad (7.3.35)$$

Extrema (maxima and minima) are found by setting the slope equal to zero

$$\frac{di}{dt} = 5.16[e^{-t}(3.873\cos 3.873t) - e^{-t}\sin 3.873t] = 0$$

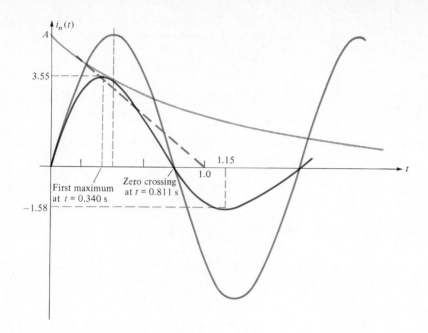

$$3.873 \cos 3.873t = \sin 3.873t$$

$$3.873 = \tan 3.873t \qquad 3.873t = 75.5° = 1.318 \text{ rad} \qquad \text{and} \qquad t = 0.340 \text{ s}$$

The second solution occurs at $75.5° + 180°$ or $t = 4.450/3.873 = 1.151$ s. The first maximum is $i_n(0.34) = 3.55$ A; the first minimum is $i_n(1.151) = -1.58$ A. Zero crossings (Figure 7.3.4) are found from

$$2\pi f = \omega = 3.873 \qquad \frac{1}{f} = T = 1.622 \qquad \frac{T}{2} = 0.811 \text{ s}$$

Crossings are multiples of this.

natural response: critically damped case **7.4**

Suppose the eigenvalues turn out to be real and *equal* to each other. In other words, when we substitute

$$x(t) = Ae^{st} \qquad (7.4.1)$$

into a system equation of the form

$$\frac{d^2x}{dt^2} + 2\alpha \frac{dx}{dt} + \alpha^2 x = 0 \qquad (7.4.2)$$

we get a characteristic equation that looks like

$$s^2 + 2\alpha s + \alpha^2 = 0 \qquad (7.4.3)$$

the solutions of which are

$$s = -\alpha, \; -\alpha \tag{7.4.4}$$

Therefore it looks as though

$$x(t) = Ae^{-\alpha t} \tag{7.4.5}$$

were the solution, but this cannot be correct. We have two initial conditions to satisfy: $x(0+)$ and dx/dt evaluated at $t = 0+$, which cannot be done with only one undetermined coefficient.

We can derive the correct solution as follows. The original system equation is

$$\frac{d^2x}{dt^2} + 2\alpha \frac{dx}{dt} + \alpha^2 x = 0 \tag{7.4.6}$$

Thus we can write it as

$$\frac{d^2x}{dt^2} + \alpha \frac{dx}{dt} + \alpha \frac{dx}{dt} + \alpha^2 x = 0 \tag{7.4.7}$$

$$\frac{d}{dt}\left(\frac{dx}{dt} + \alpha x\right) + \alpha\left(\frac{dx}{dt} + \alpha x\right) = 0 \tag{7.4.8}$$

Define

$$y = \frac{dx}{dt} + \alpha x \tag{7.4.9}$$

so that equation (7.4.8) becomes

$$\frac{dy}{dt} + \alpha y = 0 \tag{7.4.10}$$

Immediately we recognize that the solution for $y(t)$ is

$$y(t) = A_1 e^{-\alpha t} \tag{7.4.11}$$

and therefore, from the definition of y,

$$\frac{dx}{dt} + \alpha x = A_1 e^{-\alpha t} \tag{7.4.12}$$

We can solve for the complete response of the first-order differential equation (7.4.12) by the method we already know from Chapter 6.

Natural Response

$$x_n(t) = A_2 e^{-\alpha t}$$

Particular Response
Since this is an example of the special case where the forcing function is identical to the natural-response exponential term, we know that the particular response has the form

$$x_p(t) = (k_1 + k_2 t)e^{-\alpha t} \tag{7.4.13}$$

and so $\dfrac{dx}{dt} = (k_1 + k_2 t)(-\alpha e^{-\alpha t}) + k_2 e^{-\alpha t}$

$$= (k_2 - \alpha k_1)e^{-\alpha t} - \alpha k_2 t e^{-\alpha t} = (k_2 - \alpha k_1 - \alpha k_2 t)e^{-\alpha t} \tag{7.4.14}$$

Substituting (7.4.13) and (7.4.14) into (7.4.12) gives

$$(k_2 - \alpha k_1 - \alpha k_2 t) + \alpha k_1 + \alpha k_2 t = A_1$$

t terms: $0 = 0$ no information

Constant terms: $k_2 - \alpha k_1 + \alpha k_1 = A_1$

$$k_2 = A_1$$

Since there is evidently no requirement on k_1, let $k_1 = 0$.

Complete Response

$$x(t) = x_n(t) + x_p(t) = A_2 e^{-\alpha t} + A_1 t e^{-\alpha t} = e^{-\alpha t}(A_1 t + A_2) \qquad (7.4.15)$$

This complete response of equation (7.4.12) is actually the natural response $x(t)$ of equation (7.4.6). The constants A_1 and A_2 are the two unevaluated coefficients in the natural response of the second-order system described by equation (7.4.6). Such a situation, where the *characteristic values* are *real and equal*, is called a *critically damped* case. It represents the boundary point between the underdamped (oscillatory) case and the overdamped case (two different exponentials). Any infinitesimal change in the value of any element in a critically damped circuit will result in one of these other types of natural responses rather than the critically damped case.

EXAMPLE 7.4.1

In the circuit shown in Figure 7.4.1a the initial conditions are $v(0+) = 2$ V and $i_L(0+) = \frac{5}{9}$ A. Solve for $v(t)$ for $t > 0$; $R = 3\ \Omega$, $C = \frac{1}{18}$ F, and $L = 2$ H.

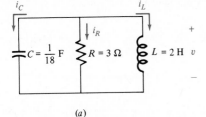

(a)

figure 7.4.1
(a) The circuit of Example 7.4.1. (b) The situation at the instant $t = 0+$.

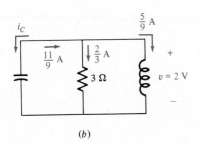

(b)

ANS.:   KCL gives us the system equation by

$$i_C + i_R + i_L = 0 \qquad (7.4.16)$$

$$\left(Cp + \frac{1}{R} + \frac{1}{Lp}\right)v = 0 \tag{7.4.17}$$

$$\left(p^2 + \frac{1}{RC}p + \frac{1}{LC}\right)v = 0 \tag{7.4.18}$$

Inserting numerical element values gives

$$(p^2 + 6p + 9)v = 0 \tag{7.4.19}$$

Assuming $v_n(t) = Ae^{st}$ yields the characteristic equation

$$s^2 + 6s + 9 = 0 \tag{7.4.20}$$

(Remember how simply we can always go from the p-operator form of the system equation to the characteristic equation.) Solving (7.4.20) yields

$$s = \frac{-6 \pm \sqrt{(6)^2 - 4(9)}}{2} = -3, -3 \tag{7.4.21}$$

so that

$$v_n(t) = e^{-3t}(A_1 t + A_2) \tag{7.4.22}$$

and

$$\frac{dv_n}{dt} = A_1(-3te^{-3t} + e^{-3t}) - 3A_2 e^{-3t} \tag{7.4.23}$$

Since to find A_1 and A_2 we need $v(0+)$ and dv/dt evaluated at $t = 0+$, we draw and label the circuit variables as they are at the instant $t = 0+$ (see Figure 7.4.1b). If $v(0+) = 2$, then $i_R = \frac{2}{3}$ A. Therefore the capacitor current i_C must be given by

$$i_C = -(i_R + i_L) \tag{7.4.24}$$

so that

$$C\frac{dv}{dt} = -(\tfrac{2}{3} + \tfrac{5}{9})$$

or

$$\frac{dv}{dt} = -22 \tag{7.4.25}$$

Therefore (7.4.22) and (7.4.23) become (at $t = 0+$)

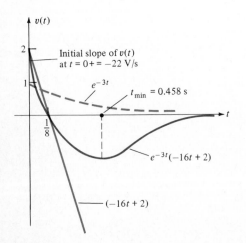

figure 7.4.2
The plot of $v(t)$ versus time for Example 7.4.1.

$$v(0+) = 2 = A_1(0)e^0 + A_2 e^0$$

$$2 = A_2 \qquad\qquad (7.4.26)$$

and $\qquad \dfrac{dv}{dt}\bigg|_{0+} = -22 = -3A_1 t + [A_1 - 3(2)]e^0 \qquad -22 = A_1 - 6$

$$-16 = A_1 \qquad\qquad (7.4.27)$$

Inserting (7.4.26) and (7.4.27) into (7.4.22) yields

$$v(t) = (-16t + 2)e^{-3t}$$

the geometry of the s plane 7.5

The eigenvalues that arise from any second-order system can be plotted on the complex plane, i.e., an orthogonal pair of axes of which the horizontal one is the *real* and the vertical one the *imaginary* axis (Figure 7.5.1). Since we have used the letter s for the eigenvalues, we call this complex plane the *s plane*.

To clarify certain geometrical properties of the s-plane plot of the eigenvalues let us define the two quantities ζ (zeta), the *damping factor*, and ω_n, the *undamped natural frequency*. Any second-order source-free differential equation can be written in the form

$$\frac{d^2 x}{dt^2} + 2\zeta\omega_n \frac{dx}{dt} + \omega_n^2 = 0 \qquad\qquad (7.5.1)$$

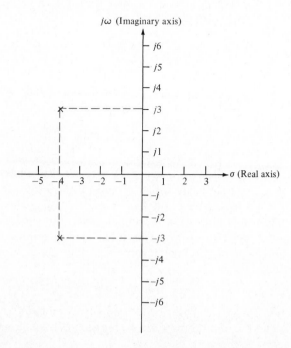

figure 7.5.1
The s plane. A plot of typical eigenvalues $s = -4 \pm j3$ from an underdamped system.

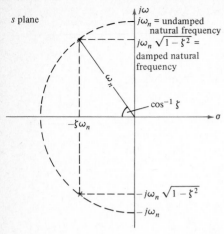

s plane

$j\omega_n$ = undamped natural frequency

$j\omega_n\sqrt{1-\zeta^2}$ = damped natural frequency

$\cos^{-1}\zeta$

$-\zeta\omega_n$

$-j\omega_n\sqrt{1-\zeta^2}$

$-j\omega_n$

figure 7.5.2
A pair of complex eigenvalues showing the geometrical interpretation of ζ and ω_n.

(Why we want to do so will become clear in a moment.) The characteristic equation would then be

$$s^2 + 2\zeta\omega_n s + \omega_n^2 = 0 \qquad (7.5.2)$$

Solving, we get the eigenvalues

$$s = -\zeta\omega_n \pm \omega_n\sqrt{\zeta^2 - 1} \qquad (7.5.3)$$

The case where $\zeta < 1$ is the underdamped case, where the eigenvalues are complex numbers

$$s = -\zeta\omega_n \pm j\omega_n\sqrt{1 - \zeta^2} \qquad (7.5.4)$$

The two quantities ζ and ω_n have great significance when we plot these complex eigenvalues, equation (7.5.4), as shown in Figure 7.5.2. Since the natural response of any system having eigenvalues like those of equation (7.5.4) and Figure 7.5.2 is

$$y_n(t) = Ae^{-(\zeta\omega_n)t}\cos(\omega_n\sqrt{1 - \zeta^2}t + \phi) \qquad (7.5.5)$$

we can immediately see the effect of the damping factor ζ. If $\zeta = 0$, the exponential decaying factor disappears from equation (7.5.5) and the natural response is a pure sinusoid with radian frequency ω_n. The response does not decay with time: it is *undamped*; and it is for this reason that ω_n is called the *undamped natural frequency*.

When ζ is between zero and unity, the exponential damping term is present in the natural response. In such cases the *damped* natural frequency ω_d is

$$\omega_d = \omega_n\sqrt{1 - \zeta^2} \qquad (7.5.6)$$

Note these important points:

1. The radial distance of the eigenvalues from the origin is ω_n, as we can show by the pythagorean theorem

$$(-\zeta\omega_n)^2 + (\omega_n\sqrt{1 - \zeta^2})^2 = \text{radius}^2$$

$$\omega_n = \text{radius} \qquad (7.5.7)$$

2. The damping factor ζ is the cosine of the angle between the radius to either eigenvector and the negative real axis. Let that angle be called θ; then

$$\cos\theta = \frac{\zeta\omega_n}{\omega_n} = \zeta \qquad (7.5.8)$$

3. If ζ is allowed to approach zero, all else remaining unchanged, the eigenvalues move along the circumference of the semicircle shown in Figure 7.5.2 toward the vertical axis, finally reading $s = \pm j\omega_n$ when $\zeta = 0$.

4. If ζ is allowed to approach unity, the eigenvalues work their way down the semicircle toward each other, finally coming to the same location on top of each other on the negative real axis at $s = -\omega_n$.

EXAMPLE 7.5.1

For the series *RLC* circuit shown in Figure 7.5.3 assume that $i(0+) = 1$ and di/dt evaluated at $t = 0+$ is zero and find and plot the eigenvalues and the natural response $i(t)$ for each of the following values of R:

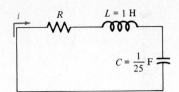

figure 7.5.3
The circuit of Example 7.5.1.

a. $12.5\ \Omega$ **b.** $10\ \Omega$ **c.** $6\ \Omega$ **d.** $0\ \Omega$

ANS.: Since the system is

$$\left(R + \frac{1}{Cp} + Lp\right)i = (LCp^2 + RCp + 1)i = 0$$

$$\left(p^2 + \frac{R}{L}p + \frac{1}{LC}\right)i = 0 \tag{7.5.9}$$

the characteristic equation is

$$s^2 + \frac{R}{L}s + \frac{1}{LC} = 0$$

or with $C = \frac{1}{25}$ F and $L = 1$ H,

$$s^2 + Rs + 25 = 0 \tag{7.5.10}$$

The general form of the characteristic equation

$$s^2 + 2\zeta\omega_n s + \omega_n^2 = 0 \tag{7.5.11}$$

leads us to note that $\omega_n^2 = 25$ and $2\zeta\omega_n = R$, from which

$$\omega_n = 5 \tag{7.5.12}$$

and

$$\zeta = \frac{R}{10} \tag{7.5.13}$$

(a) $R = 12.5\ \Omega$. From (7.5.13), $\zeta = 1.25$, the overdamped case. From (7.5.10) the eigenvalues are found to be

$$s^2 + 12.5s + 25 = 0$$

$$s = \frac{-12.5 \pm \sqrt{(12.5)^2 - 4(25)}}{2} = -10, -2.5 \tag{7.5.14}$$

The natural response is

$$i_n(t) = Ae^{-10t} + Be^{-2.5t} \tag{7.5.15}$$

and

$$i_n(0+) = A + B = 1 \tag{7.5.16}$$

Also

$$\frac{di_n}{dt} = -10Ae^{-10t} - 2.5Be^{-2.5t} \tag{7.5.17}$$

$$\left.\frac{di_n}{dt}\right|_{0+} = 0 = -10A - 2.5B \tag{7.5.18}$$

Solving (7.5.16) and (7.5.18) together gives

$$A = -\frac{1}{3} \quad \text{and} \quad B = \frac{4}{3}$$

so that
$$i_n(t) = -\tfrac{1}{3}e^{-10t} + \tfrac{4}{3}e^{-2.5t} \qquad (7.5.19)$$

(b) $R = 10\ \Omega$. From (7.5.13), $\zeta = 1.0$, the critically damped case. From (7.5.10) the eigenvalues are found to be

$$s^2 + 10s + 25 = 0 \qquad s = -5, -5 \qquad (7.5.14a)$$

The natural response is

$$i_n(t) = (At + B)e^{-5t} \qquad (7.5.15a)$$

and
$$i_n(0+) = B = 1 \qquad (7.5.16a)$$

$$\frac{di_n}{dt} = -(At + B)5e^{-5t} + Ae^{-5t} \qquad (7.5.17a)$$

$$\left.\frac{di_n}{dt}\right|_{0+} = -(B)5 + A = 0 \qquad (7.5.18a)$$

Inserting (7.5.16a) into (7.5.18a) yields $A = 5$. Thus

$$i_n(t) = (5t + 1)e^{-5t} \qquad (7.5.19a)$$

(c) $R = 6\ \Omega$. From (7.5.13), $\zeta = 0.6$, the underdamped case. From (7.5.10) the eigenvalues are found to be

$$s^2 + 6s + 25 = 0$$

$$s = \frac{-6 \pm \sqrt{6^2 - 4(25)}}{2} = -3 \pm j4 \qquad (7.5.20)$$

The natural response is

$$i_n(t) = Ae^{-3t}\cos(4t + \phi) \qquad (7.5.21)$$

and
$$i_n(0+) = A\cos\phi = 1 \qquad (7.5.22)$$

$$\frac{di_n}{dt} = A[-4e^{-3t}\sin(4t + \phi) - 3e^{-3t}\cos(4t + \phi)] \qquad (7.5.23)$$

$$\left.\frac{di_n}{dt}\right|_{0+} = 0 = -4\sin\phi - 3\cos\phi$$

Thus
$$\frac{\sin\phi}{\cos\phi} = -\tfrac{3}{4} = \tan\phi \qquad (7.5.24)$$

and therefore
$$\phi = -36.9°$$

and from (7.5.22)

$$A = \frac{1}{\cos\phi} = 1.25 \qquad (7.5.25)$$

Thus $i_n(t) = 1.25e^{-3t}\cos(4t - 36.9°)$. See Figure 7.5.4.

(d) $R = 0\ \Omega$. From (7.5.13), $\zeta = 0$, the undamped case. The eigenvalues are found to be

$$s^2 + 0s + 25 = 0 \quad s^2 + 25 = 0 \qquad s^2 = -25 \quad s = \pm j5 \qquad (7.5.26)$$

Since the natural response is

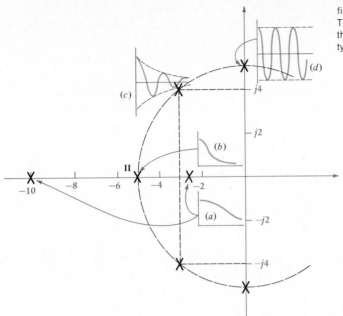

figure 7.5.4
The eigenvalues of Example 7.5.1 plotted on the s plane. Note the form of the natural response associated with each such typical eigenvalue location.

$$i_n(t) = A \cos (5t + \phi) \tag{7.5.27}$$

$$i_n(0+) = A \cos \phi = 1 \tag{7.5.28}$$

$$\frac{di_n}{dt} = -5A \sin (5t + \phi) \tag{7.5.29}$$

$$\left.\frac{di_n}{dt}\right|_{0+} = -5A \sin \phi = 0 \tag{7.5.30}$$

it follows that $\sin \phi = 0$ and

$$\phi = 0 \tag{7.5.31}$$

Inserting (7.5.31) into (7.5.28), we find

$$A = 1 \tag{7.5.32}$$

Thus from (7.5.27)

$$i_n(t) = \cos 5t \tag{7.5.33}$$

EXAMPLE 7.5.2

Given the approximate model of an automobile suspension system in Figure 7.5.5, where $M = $ (mass of car)$/4 = 350$ kg, $1/K = 22{,}880$ N/m is the stiffness of the spring, and D is the effect of the shock absorber:
a. What value for D gives critical damping?
b. What value for D gives $\zeta = 0.707$?
c. If we remove the shock absorber, what will the natural response of the car be to an initial displacement?

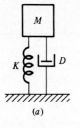

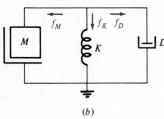

figure 7.5.5
(a) Pictorial diagram of the suspension system of a single automobile wheel. (b) The schematic diagram.

(a)

(b)

ANS.: $\quad\quad\quad f_M = f_K + f_D = 0 \quad\quad$ where $\quad\quad f_M = M\dfrac{dv}{dt} \quad\quad f_K = \dfrac{1}{Kp}v \quad\quad f_D = Dv$

$$M\frac{dv}{dt} + Dv + \frac{1}{K}\int v\,dt = 0$$

$$\frac{d^2v}{dt^2} + \frac{D}{M}\frac{dv}{dt} + \frac{1}{KM}v = 0$$

The characteristic equation is

$$s^2 + \frac{D}{M}s + \frac{1}{KM} = 0 \quad\quad \text{where} \quad\quad 2\zeta\omega_n = \frac{D}{M} \quad\quad \text{and} \quad\quad \omega_n^2 = \frac{1}{KM}$$

Therefore we solve for D:

$$D = 2\zeta\omega_n M = 2\zeta\sqrt{\frac{M}{K}}$$

(a) $D = 2(1)\sqrt{22{,}880(350)} = 5660$ N·s/m.
(b) $D = 2(0.707)\sqrt{22{,}880(350)} = 0.707(5660) = 4002$ N·s/m.
(c) If we remove the shock absorber, the natural response of the car will be

$$y(t) = A\,\cos\,(\omega_n t + \phi)$$

where A is the displacement, $\phi = 0$, and $\omega_n = 1/\sqrt{KM} = \sqrt{22{,}880/350} = 8.085$ rad/s or (since $\omega_n = 2\pi f_n$)

$$f_n = \frac{8.085}{2\pi} = 1.29 \text{ Hz}$$

Systems of higher than second order have more than two eigenvalues. When they are plotted on the *s* plane, it is called the *eigenvalue constellation*, by analogy to a constellation of stars. The eigenvalues closest to the origin, called the *dominant eigenvalues*, give rise to terms in the natural response that have longer time constants (take longer to die away) than the others. Sometimes a complicated higher-order system can be reasonably well described by just its dominant eigenvalues; i.e., we might approximate a fifth-order system having eigenvalues at -100, -50, -20, -2, and -1, with a second-order system with eigenvalues at $s = -1$, -2. Strictly speaking, this introduces error into the response we calculate, but as a first approximation it may be quite good.

complete responses of higher- 7.6
order systems

As with first-order systems, the complete response of higher-order systems is the sum of the natural response and the particular response. The particular response is found exactly as in first-order systems as a linear sum of the forcing function (source) and all its derivatives.†

EXAMPLE 7.6.1

In the circuit shown in Figure 7.6.1a the voltage source is $v_s(t) = (t^2 + 1)u(t)$. Also the initial conditions are given as $i_L(0-) = \frac{1}{9}$ A and $v_C(0-) = \frac{1}{2}$ V. Find $i_L(t)$ for all $t > 0$.

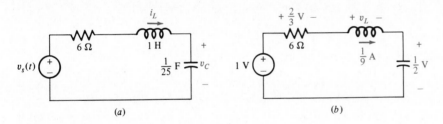

figure 7.6.1
(a) The circuit of Example 7.6.1; $v_s(t) = (t^2 + 1)u(t)$. (b) The circuit at the instant $t = 0+$.

(a)   (b)

ANS.:   Writing a KVL summation gives

$$\left(\frac{1}{Cp} + R + Lp\right)i(t) = v_s(t) \qquad \left(Lp^2 + Rp + \frac{1}{C}\right)i(t) = pv_s(t)$$

$$\left(p^2 + \frac{R}{L}p + \frac{1}{LC}\right)i(t) = p\frac{v_s(t)}{L} = p\frac{t^2 + 1}{1}$$

Hence the system equation that is valid for t greater than zero is

$$(p^2 + 6p + 25)i(t) = 2t \tag{7.6.1}$$

Natural Response

$$s^2 + 6s + 25 = 0 \tag{7.6.2}$$

$$s = -3 \pm j4 \qquad \text{or} \qquad \omega_n = 5 \qquad \text{and} \qquad \zeta = \tfrac{3}{5} \tag{7.6.3}$$

and so

$$i_n(t) = e^{-3t}(A_1 \cos 4t + A_2 \sin 4t) \tag{7.6.4}$$

---

† Except for two special cases (see Sections 6.4, 7.7, 11.5, and 15.7).

Particular Response

$$i_p = k_1 t + k_2 \tag{7.6.5}$$

$$\frac{di}{dt} = k_1 \tag{7.6.6}$$

$$\frac{d^2 i}{dt^2} = 0 \tag{7.6.7}$$

Therefore, substituting (7.6.5) to (7.6.7) into (7.6.1) gives

$$0 + 6k_1 + 25k_1 t + 25k_2 = 2t$$

$$25k_1 = 2$$

$$k_1 = 0.08 \tag{7.6.8}$$

$$6k_1 + 25k_2 = 0$$

$$k_2 = -0.0192 \tag{7.6.9}$$

Hence the complete response is

$$i(t) = 0.08t - 0.0192 + e^{-3t}(A_1 \cos 4t + A_2 \sin 4t) \tag{7.6.10}$$

and

$$\frac{di}{dt} = 0.08 - 3e^{-3t}(A_1 \cos 4t + A_2 \sin 4t) + e^{-3t}(-4A_1 \sin 4t + 4A_2 \cos 4t) \tag{7.6.11}$$

Evaluation of A_1 and A_2

From Figure 7.6.1*b* we can write that at $t = 0+$

$$v_L = L \frac{di}{dt} \tag{7.6.12}$$

and

$$\tfrac{2}{3} + v_L + \tfrac{1}{2} = 1 \tag{7.6.13}$$

therefore

$$v_L = -\tfrac{1}{6} \tag{7.6.14}$$

and

$$\frac{di}{dt} = -\tfrac{1}{6} \tag{7.6.15}$$

Evaluating equations (7.6.10) and (7.6.11) at $t = 0$ leads to

$$\tfrac{1}{9} = -0.0192 + A_1$$

$$A_1 = 0.13 \tag{7.6.16}$$

and

$$-\tfrac{1}{6} = 0.08 - 3A_1 + 4A_2$$

$$A_2 = 0.0358 \tag{7.6.17}$$

Thus

$$i(t) = 0.08t - 0.0192 + e^{-3t}(0.13 \cos 4t + 0.0358 \sin 4t) \tag{7.6.18}$$

or equivalently

$$i(t) = 0.08t - 0.0192 + e^{-3t}[0.13 \cos (4t - 15.4°)] \tag{7.6.19}$$

Recall that the characteristic equation (7.6.2) can be written in the general form

$$s^2 + 2\zeta\omega_n s + \omega_n^2 = 0 \qquad (7.6.20)$$

Hence

$$\omega_n^2 = 25$$

$$\omega_n = 5 \qquad (7.6.21)$$

and

$$2\zeta\omega_n = 6$$

$$\zeta = 0.6 \qquad (7.6.22)$$

We can also compute ω_d as a check on our work

$$\omega_d = \omega_n\sqrt{1 - \zeta^2} = 5\sqrt{1 - 0.6^2} = 4 \text{ rad/s} \qquad (7.6.23)$$

as before.

EXAMPLE 7.6.2

The switch in the circuit of Figure 7.6.2, which has been closed for a long time, is opened at $t = 0$.
a. Find the initial conditions.
b. Find the numerical expression for v_C for $t > 0$.

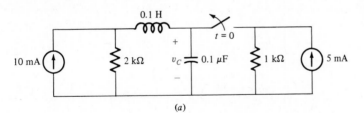

(a)

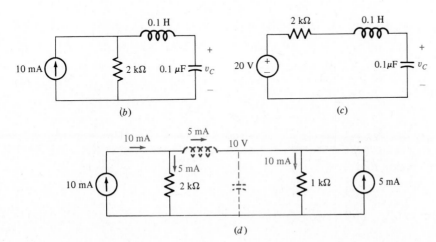

(b)   (c)

(d)

figure 7.6.2
(a) The circuit of Example 7.6.2. (b) For $t > 0$.
(c) An equivalent circuit for $t > 0$. (d) The
circuit at $t = 0-$.

ANS.:   For time $t > 0$ the circuit is as shown in Figure 7.6.2b. The necessary p-operator system equation in terms of v_C could be obtained from two node equations, but let us perform a Norton-Thevenin source transformation, as shown in Figure 7.6.2c. Then simple voltage division yields

$$v_C = \frac{1/Cp}{1/Cp + R + Lp} \, 20 = \frac{1/LC}{p^2 + (R/L)p + 1/LC} \, 20$$

where $\qquad \dfrac{R}{L} = \dfrac{2 \times 10^3}{0.1} = 2 \times 10^4 \quad$ and $\quad \dfrac{1}{LC} = \dfrac{1}{0.1(0.1 \times 10^{-6})} = 10^8$

Thus $\qquad\qquad\qquad (p^2 + 2 \times 10^4 p + 10^8)v_C = 20 \times 10^8 \qquad\qquad\qquad$ (7.6.24)

The characteristic equation is therefore

$$s^2 + 2 \times 10^4 s + 10^8 = 0 \quad \text{with} \quad \omega_n^2 = 10^8 \qquad \omega_n = 10^4 \qquad (7.6.25)$$

and $\qquad\qquad\qquad\qquad 2\zeta\omega_n = 2 \times 10^4$

$$\zeta = 1 \qquad \text{critical damping}$$

Therefore $\qquad\qquad\qquad\qquad s = -10^4, \, -10^4$

and the natural response is

$$v_{Cn}(t) = (A_1 t + A_2)e^{-10^4 t} \qquad\qquad\qquad (7.6.26)$$

Since the particular response to the constant forcing function 20×10^8 is

$$v_{Cp} = k \qquad\qquad\qquad (7.6.27)$$

it follows that

$$\frac{dv_{Cp}}{dt} = \frac{d^2 v_{Cp}}{dt^2} = 0 \qquad\qquad\qquad (7.6.28)$$

and substituting (7.6.27) and (7.6.28) into (7.6.24) yields

$$10^8 k = 20 \times 10^8 \qquad \text{and} \qquad k = 20 = v_{Cp}$$

Hence the complete response is

$$v_C(t) = v_{Cp}(t) + v_{Cn}(t) = 20 + (A_1 t + A_2)e^{-10^4 t} \qquad\qquad\qquad (7.6.29)$$

In order to evaluate A_1 and A_2 we need $v_C(0+)$ and dv_C/dt evaluated at $t = 0+$. Sketch the circuit at the instant $t = 0-$ (see Figure 7.6.2d). Note that with the inductor replaced by a short and the capacitor by an open circuit the voltage across the parallel combination of resistors and therefore across the capacitor, $v_C(0-)$, is 10 V (15 mA into an effective $\frac{2}{3}$ kΩ). Thus

$$v_C(0-) = 10 \text{ V} \qquad\qquad\qquad (7.6.30)$$

and $\qquad\qquad\qquad\qquad i_L(0-) = 5 \text{ mA} \qquad\qquad\qquad (7.6.31)$

In the absence of impulse sources we realize that

$$i_L(0-) = i_L(0+) \qquad\qquad\qquad (7.6.32)$$

and $\qquad\qquad\qquad\qquad v_C(0-) = v_C(0+) \qquad\qquad\qquad (7.6.33)$

Thus from (7.6.30) and (7.6.33)

$$v_C(0+) = 10 \qquad\qquad\qquad (7.6.34)$$

and from Figure 7.6.2b or c and equations (7.6.31) and (7.6.32)

$$i_L(0+) = 5 \text{ mA} = i_C(0+)$$

$$5 \times 10^{-3} = C \left. \frac{dv_C}{dt} \right|_{0+}$$

$$\left. \frac{dv_C}{dt} \right|_{0+} = \frac{5 \times 10^{-3}}{0.1 \times 10^{-6}} = 50 \times 10^{+3} \tag{7.6.35}$$

Using (7.6.34) and (7.6.35) in (7.6.29) evaluated at $t = 0+$, we get

$$10 = 20 + A_2 \qquad \text{or} \qquad A_2 = -10$$

and

$$\frac{dv_C}{dt} = (A_1 t + A_2)(-10^4 e^{-10^4 t}) + e^{-10^4 t} A_1 \tag{7.6.36}$$

At $t = 0+$, (7.6.35) and (7.6.36) yield

$$50 \times 10^3 = 10 \times 10^4 + A_1 \qquad \text{or} \qquad A_1 = -50 \times 10^3$$

so that finally

$$v_C(t) = 20 - (10 + 5 \times 10^4 t) e^{-10^4 t} \tag{7.6.37}$$

two special cases 7.7

In general the particular response is obtained in a straightforward way as the linear sum of the source and all its derivatives, as shown in Table 7.7.1, but two special cases need slightly different techniques:

1. If one eigenvalue is zero, i.e., if the characteristic polynomial has a *zero root* $s = 0$, the assumed form of the particular response should be the *integral* of that which is normal.
2. If the source contains a term which also appears in the natural response, this should be treated the same as the critically damped case and the normal particular response multiplied by $k_1 t + k_2$.

Source	Particular response
v_o (= const)	k (= const)
$3t^n$ (n = positive integer)	$k_1 t^n + k_2 t^{n-1} + \cdots + k_{n-1} t + k_n$
$4e^{\gamma t}$	$k e^{\gamma t}$
$10 \cos 3t$	$k_1 \cos 3t + k_2 \sin 3t$
$15 \sin 3t$	$k_1 \sin 3t + k_2 \cos 3t$

table 7.7.1
Particular responses to typical sources

EXAMPLE 7.7.1

For the first special case, given

$$\frac{d^2 y}{dt^2} + \frac{dy}{dt} = 2t \qquad \text{for } t > 0$$

Find $y(t)$ for $t > 0$.

ANS.:

Natural Response

$$\frac{d^2y}{dt^2} + \frac{dy}{dt} = 0$$

The characteristic equation is $s^2 + s = 0$ or $s(s + 1) = 0$, so that

$$s = 0, \, -1$$

and therefore

$$y_n(t) = A_1 e^{0t} + A_2 e^{-t} \qquad \text{or} \qquad y_n(t) = A_1 + A_2 e^{-t}$$

Particular Response
Try $y_p(t) = k_1 t + k_2$ as usual

$$\dot{y}_p(t) = k_1 \qquad \text{and} \qquad \ddot{y}_p(t) = 0$$

Substituting back into $d^2y/dt^2 + dy/dt = 2t$ yields

$$0 + k_1 = 2t$$

There is no k_1 for which this is true. Realizing that the differential equation can be written in the equivalent form

$$p(p + 1)y = 2t$$

$$(p + 1)y = \frac{1}{p}(2t) = t^2 + k_0$$

we try

$$y_p(t) = k_1 t^2 + k_2 t + k_3 \qquad \dot{y}_p(t) = 2k_1 t + k_2 \qquad \ddot{y}(t) = 2k_1$$

Substituting into $d^2y/dt^2 + dy/dt = 2t$ yields

$$2k_1 + 2k_1 t + k_2 = 2t$$

whence $2k_1 + k_2 = 0$ and $2k_1 = 2$, which lead to

$$k_1 = 1 \qquad \text{and} \qquad k_2 = -2$$

and so

$$y_p(t) = t^2 - 2t$$

We can choose $k_3 = 0$ since there is no requirement on it from this set of equations (and its contribution of a constant will be taken care of by A_1). Thus

$$y(t) = y_n(t) + y_p(t)$$

$$y(t) = A_1 + A_2 e^{-t} + t^2 - 2t$$

is the complete response. The constants A_1 and A_2 are to be evaluated by means of the initial conditions.

EXAMPLE 7.7.2

For the second special case solve for $y(t)$

$$\frac{d^2y}{dt^2} - y = e^{-t} \qquad \text{for } t > 0$$

ANS.:

Natural Response

The characteristic equation is $s^2 - 1 = 0$, and so the eigenvalues are $s = +1, -1$. Thus,

$$y_n(t) = A_1 e^t + A_2 e^{-t}$$

Particular Response

Try

$$y_p(t) = ke^{-t} \text{ as usual}$$

$$\dot{y}_p(t) = -ke^{-t} \qquad \ddot{y}_p(t) = ke^{-t}$$

Substituting back into the system equation yields

$$ke^{-t} - ke^{-t} = e^{-t} \qquad 0 = e^{-t}$$

There is no value of k for which this can be true. Try (as in the critically damped case)

$$y_p(t) = (k_1 + k_2 t)e^{-t}$$

$$\dot{y}_p(t) = -k_1 e^{-t} + k_2(-te^{-t} + e^{-t}) = (k_2 - k_1)e^{-t} - k_2 te^{-t}$$

and

$$\ddot{y}(t) = -(k_2 - k_1)e^{-t} - k_2(-te^{-t} + e^{-t}) = (k_1 - 2k_2)e^{-t} + k_2 te^{-t}$$

Substituting this into the system equation yields

$$(k_1 - 2k_2)e^{-t} + k_2 te^{-t} - k_1 e^{-t} - k_2 te^{-t} = e^{-t}$$

from which, setting like coefficients equal, we get

$$k_1 - 2k_2 - k_1 = 1 \qquad \text{and} \qquad k_2 = -\tfrac{1}{2}$$

We can choose k_1 arbitrarily to be zero since (1) there is no requirement placed on it by the above equation and (2) the effect of $k_1 e^{-t}$ will be taken care of in the complete solution by the component in the natural response $A_2 e^{-t}$. Thus the complete response is given by

$$y(t) = A_1 e^t + A_2 e^{-t} - \tfrac{1}{2}te^{-t} = A_1 e^t + (A_2 - \tfrac{1}{2}t)e^{-t}$$

where the A's are to be evaluated by means of initial conditions.

zero-state and zero-input responses 7.8

Just as for single energy-storage systems, the zero-state response is the complete response when all initial conditions are zero. The zero-input response is the complete response when all sources are set equal to zero.

EXAMPLE 7.8.1

In the circuit of Figure 7.8.1a, $v(0+) = 4$ V, $i_L(0+) = 1$ A, and $i_s(t) = t^2 u(t)$; $R = 4\ \Omega$, $L = 3$ H, and $C = \tfrac{1}{24}$ F.
a. Find the complete response for $t > 0$.
b. Find the zero-state response.
c. Find the zero-input response.
d. Show that the sum of the answers to (b) and (c) is the complete response obtained in (a).

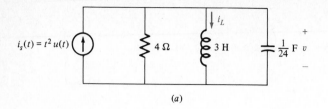

figure 7.8.1
(a) The circuit of Example 7.8.1. (b) The circuit at $t = 0+$ with $i_L(0+) = 1$ A and $v_C(0+) = 4$ V. (c) The circuit at $t = 0+$ with $i_L(0+) = v_C(0+) = 0$.

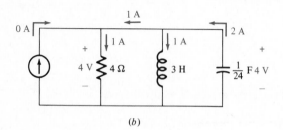

(b)

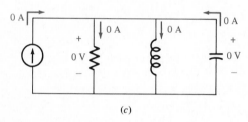

(c)

ANS.: (a)

Complete Response
From KCL at the upper node

$$\left(\frac{1}{R} + \frac{1}{Lp} + Cp\right)v = i_s(t)$$

$$(p^2 + 6p + 8)v = 24pi_s(t) = 24(2t)$$

$$(p^2 + 6p + 8)v = 48t \tag{7.8.1}$$

Natural Response

$$s^2 + 6s + 8 = 0$$

$$s = \frac{-6 \pm \sqrt{36 - 32}}{2} = -2, -4 \qquad \text{overdamped}$$

Therefore $$v_n(t) = A_1 e^{-2t} + A_2 e^{-4t} \tag{7.8.2}$$

Particular Response
From equation (7.8.1) we see that the forcing function in the system equation is $48t$. Thus we let

$$v_p(t) = k_1 t + k_2 \tag{7.8.3}$$

$$\frac{dv_p}{dt} = k_1 \qquad \text{and} \qquad \frac{dv_p^2}{dt^2} = 0 \tag{7.8.4}$$

Inserting (7.8.3) and (7.8.4) into (7.8.1) yields

$$0 + 6k_1 + 8(k_1 t + k_2) = 48t$$

Thus

$$8k_1 = 48 \qquad \text{and} \qquad k_1 = 6$$

and

$$6k_1 + 8k_2 = 0$$

$$8k_2 = -36$$

$$k_2 = -\tfrac{9}{2}$$

so that

$$v_p(t) = 6t - \tfrac{9}{2} \tag{7.8.5}$$

The complete response is therefore of the form

$$v(t) = 6t - \tfrac{9}{2} + A_1 e^{-2t} + A_2 e^{-4t} \tag{7.8.6}$$

and

$$\frac{dv}{dt} = 6 - 2A_1 e^{-2t} - 4A_2 e^{-4t} \tag{7.8.7}$$

Using the initial conditions $v_C(0+) = 4$ V and $i_L(0+) = 1$ A (Figure 7.8.1b), we find dv/dt evaluated at $t = 0+$ is -48. Then evaluating (7.8.6) and (7.8.7) at $t = 0+$ gives

$$4 = 0 - \tfrac{9}{2} + A_1 + A_2 \tag{7.8.8}$$

$$-48 = 6 - 2A_1 - 4A_2 \tag{7.8.9}$$

Solving (7.8.8) and (7.8.9) simultaneously yields

$$A_1 = -10 \qquad A_2 = \tfrac{37}{2} \tag{7.8.10}$$

Thus (7.8.6), the complete response, is

$$v(t) = 6t - \tfrac{9}{2} - 10e^{-2t} + \tfrac{37}{2}e^{-4t} \qquad \text{for } t > 0 \tag{7.8.11}$$

(b) The zero-state response contains the same particular response component as before, equation (7.8.5); but with $i_L(0+) = 0$ and $v_C(0+) = 0$ we have $dv/dt = 0$ (see Figure 7.8.1c). Rewriting (7.8.8) and (7.8.9) accordingly, we get

$$0 = 0 - \tfrac{9}{2} + A_1 + A_2 \tag{7.8.12}$$

$$0 = 6 - 2A_1 - 4A_2 \tag{7.8.13}$$

Solving for A_1 and A_2 gives

$$A_1 = 6 \qquad \text{and} \qquad A_2 = -\tfrac{3}{2}$$

Thus the zero-state response is

$$v_{zs}(t) = 6t - \tfrac{9}{2} + 6e^{-2t} - \tfrac{3}{2}e^{-4t} \qquad \text{for } t > 0 \tag{7.8.14}$$

(c) Since the zero-input response is simply the natural response with the coefficients evaluated by means of the actual initial conditions, we write

$$v_{zi}(t) = A_1 e^{-2t} + A_2 e^{-4t} \tag{7.8.15}$$

$$\frac{dv_{zi}}{dt} = -2A_1 e^{-2t} - 4A_2 e^{-4t}$$

Again using $v(0+) = 4$ and $dv/dt = -48$ evaluated at $t = 0+$, as in part (a), we have

$$4 = A_1 + A_2$$

$$-48 = -2A_1 - 4A_2$$

whence $\qquad\qquad A_1 = -16 \quad\text{and}\quad A_2 = 20$

Thus $\qquad\qquad v_{zi}(t) = -16e^{-2t} + 20e^{-4t} \qquad \text{for } t > 0 \qquad\qquad (7.8.16)$

(d) Note that the sum of (7.8.14) and (7.8.16) equals (7.8.11); that is,

$$v(t) = v_{zs}(t) + v_{zi}(t) \qquad\qquad (7.8.17)$$

as is always the case.

7.9 unit-step and unit-impulse responses

* Unit-Step Response   The unit-step response $w(t)$ of a system is the complete response of the desired (output) variable of that system to a unit-step function (input) $u(t)$ when all initial conditions are zero.

* Unit-Impulse Response   The unit-impulse response $h(t)$ of a system is the complete response of the desired (output) variable of that system to a unit impulse (input) $\delta(t)$ when all initial conditions are zero.

Therefore, we note that the unit-step and the unit-impulse responses are simply the respective zero-state responses to those inputs. Since the two inputs $u(t)$ and $\delta(t)$ are the integral and derivative of each other, from the homogeneity property it follows that $w(t)$ and $h(t)$ are also related by that same integral-derivative relationship.

EXAMPLE 7.9.1

Find the unit-step response of the system shown in Figure 7.9.1. The output is the capacitor voltage $v_C(t)$, and the input is $v_s(t)$.

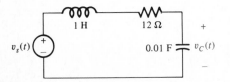

figure 7.9.1
The circuit of Examples 7.9.1 and 7.9.2.

ANS.:   From voltage division we have

$$v_C = \frac{1/Cp}{Lp + R + 1/Cp}\, v_s \quad \text{or} \quad v_C = \frac{1/LC}{p^2 + (R/L)p + 1/LC}\, v_s$$

Thus, with $R = 12 \ \Omega$, $L = 1$ H, and $C = 0.01$ F, the system equation in v_C is

$$(p^2 + 12p + 100)v_C = 100v_s(t) \qquad\qquad (7.9.1)$$

and with $v_s(t) = u(t)$ we have

$$(p^2 + 12p + 100)v_C = 100 \qquad t > 0 \tag{7.9.2}$$

Natural Response

$$s^2 + 12s + 100 = 0 \tag{7.9.3}$$

$$s = -6 \pm j8 \qquad \text{Underdamped case} \tag{7.9.4}$$

Also

$$2\zeta\omega_n = 12 \qquad \text{and} \qquad \omega_n^2 = 100$$

so that

$$\omega_n = 10 \qquad \text{and} \qquad \zeta = 0.6 \tag{7.9.5}$$

As a result

$$v_{Cn}(t) = e^{-6t}(A \cos 8t + B \sin 8t) \tag{7.9.6}$$

Particular Response

The particular response to the forcing function 100 in equation (7.9.2) is

$$v_{Cp}(t) = k \tag{7.9.7}$$

and thus

$$\frac{dv_{Cp}}{dt} = \frac{d^2 v_{Cp}}{dt^2} = 0 \tag{7.9.8}$$

Inserting (7.9.7) and (7.9.8) into (7.9.2) yields

$$(0 + 0 + 100)k = 100$$

$$k = 1 \tag{7.9.9}$$

and so

$$v_{Cp} = 1$$

The complete unit-step response is thus

$$v_C(t) = v_{Cp}(t) + v_{Cn}(t) = 1 + e^{-6t}(A \cos 8t + B \sin 8t) \tag{7.9.10}$$

and

$$\frac{dv_C}{dt} = e^{-6t}(-8A \sin 8t + 8B \cos 8t) - 6e^{-6t}(A \cos 8t + B \sin 8t) \tag{7.9.11}$$

Evaluating (7.9.10) at $t = 0+$ with zero initial conditions gives

$$0 = 1 + (1)[A(1) + 0]$$

or

$$A = -1 \tag{7.9.12}$$

and using this in (7.9.11) at $t = 0+$ with zero initial conditions gives

$$0 = 1(0 + 8B) - 6(1)(-1 + 0)$$

$$B = -\tfrac{3}{4} \tag{7.9.13}$$

Hence

$$w(t) = v_C(t) = 1 - e^{-6t}(\cos 8t + \tfrac{3}{4} \sin 8t) \tag{7.9.14}$$

or

$$w(t) = v_C(t) = 1 - 1.25e^{-6t} \cos(8t - 36.9°) \tag{7.9.15}$$

The unit-step response in Example 7.9.1 is shown with several others on the normalized plot in Figure 7.9.2. Note that on this plot the horizontal axis is

figure 7.9.2
Normalized step response of a second-order
system in Example 7.9.1.

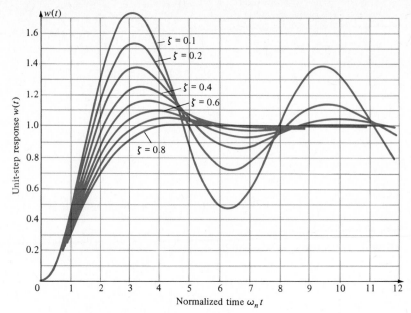

normalized time $\omega_n t$ and the parameter identifying each curve is the damping factor ζ. The unit-impulse response $h(t)$ of any system is the time derivative of the unit-step response $w(t)$. Thus the unit-impulse response for $v_C(t)$ in Example 7.9.1 would be†

$$h(t) = v_C(t) = \tfrac{25}{2}e^{-6t}\sin 8t \qquad (7.9.16)$$

However, impulse responses can be calculated directly.

We recall that unit-impulse sources change inductor currents, capacitor voltages, spring forces, etc., instantaneously (discontinously), as in Table 7.9.1.

table 7.9.1

Element	$\delta(t)$	Discontinuous	Amount
Inductor	v_L	i_L	L^{-1}
Capacitor	i_C	v_C	C^{-1}
Spring	$\mathcal{V}$	f_K	K^{-1}
Mass	f_m	$\mathcal{V}$	M^{-1}

EXAMPLE 7.9.2

Find the unit-impulse response $h(t)$ for the capacitor voltage in the circuit of Figure 7.9.1.

ANS.:   For $v_s(t) = \delta(t)$ we note that no fraction of this input can appear across the capacitor (infinite energy storage would result). Similarly, no fraction of $v_s(t) = \delta(t)$ can appear across the resistor because it would create $i_R = i_L$, an impulse current, which would store infinite energy in the inductor. Consequently all $v_s(t) = \delta(t)$ must fall across the inductor, so that at $t = 0+$

$$i_L(0+) = \frac{1}{L} \qquad \text{amperes} \qquad (7.9.17)$$

Since this current flows into the capacitor,

$$i_C(0+) = \frac{1}{L} = C\frac{dv_C}{dt}\Big|_{0+}$$

† You are encouraged to verify this result by taking the time derivative of equation (7.9.15).

and thus
$$\left.\frac{dv_C}{dt}\right|_{0+} = \frac{1}{LC} = 100 \qquad (7.9.18)$$

which together with
$$v_C(0+) = 0 \qquad (7.9.19)$$

gives the necessary initial conditions on $v_C(t)$. Since for $t > 0$ the impulse source equals zero, the complete unit-impulse response is simply the natural response, equation (7.9.6), with the constants evaluated by means of (7.9.18) and (7.9.19); that is, from (7.9.6)

$$0 = (1)[A(1) + 0]$$

so that
$$A = 0 \qquad (7.9.20)$$

Taking the time derivative of equation (7.9.6) leads to

$$\frac{dv}{dt} = e^{-6t}(-8A \sin 8t + 8B \cos 8t) - 6e^{-6t}(A \cos 8t + B \sin 8t) \qquad (7.9.21)$$

Evaluating (7.9.21) at $t = 0+$ and using (7.9.18) and (7.9.20), we have

$$100 = 1(0 + 8B) - 6(1)(0 + 0)$$

$$B = \tfrac{25}{2} \qquad (7.9.22)$$

so that
$$h(t) = v_C(t) = \tfrac{25}{2}e^{-6t} \sin 8t \qquad (7.9.23)$$

Compare equations (7.9.23) and (7.9.16).

EXAMPLE 7.9.3

Find the unit-impulse response for the capacitor voltage v_C in the circuit in Figure 7.9.3. Then find the unit-step response.

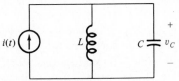

figure 7.9.3
The circuit of Example 7.9.3.

ANS.:   Impulse response $h(t)$

$$v_C(t) = \frac{(1/Cp)(Lp)i}{1/Cp + Lp} = \frac{(1/C)p}{p^2 + 1/LC}\, i$$

$$\left(p^2 + \frac{1}{LC}\right)v_C(t) = \frac{1}{C}\, pi(t) \qquad t > 0 \qquad (7.9.24)$$

Natural Response
The characteristic equation is

$$s^2 + \frac{1}{LC} = 0 \qquad (7.9.25)$$

$$s = \pm j \frac{1}{\sqrt{LC}} \tag{7.9.26}$$

$$v_{Cn}(t) = A_1\left(\cos \frac{1}{\sqrt{LC}} t\right) + A_2\left(\sin \frac{1}{\sqrt{LC}} t\right) \tag{7.9.27}$$

and
$$\frac{dv_{Cn}}{dt} = -\frac{A_1}{\sqrt{LC}}\left(\sin \frac{1}{\sqrt{LC}} t\right) + \frac{A_2}{\sqrt{LC}}\left(\cos \frac{1}{\sqrt{LC}} t\right) \tag{7.9.28}$$

Initial Conditions
All $i(t) = \delta(t)$ must flow into C since $i_L = \infty$ is an impossibility. Thus $v_C(0+) = 1/C$. Also, $C\ dv/dt = i_C$ at $t = 0+$ equals 0 because $i(0+) = i_L(0+) = 0$. Therefore $dv_C/dt = 0$ at $0+$, and (7.9.27) and (7.9.28) become

$$\frac{1}{C} = A_1 \cos 0 + A_2 \sin 0 \tag{7.9.29}$$

$$0 = -\frac{A_1}{\sqrt{LC}} \sin 0 + \frac{A_2}{\sqrt{LC}} \cos 0 \tag{7.9.30}$$

Consequently $A_1 = 1/C$ and $A_2 = 0$, and for $i(t) = \delta(t)$ we have

$$v_C(t) = h(t) = \frac{1}{C}\left(\cos \frac{1}{\sqrt{LC}} t\right)u(t)$$

We integrate $h(t)$ to find $w(t)$

$$w(t) = \frac{\sqrt{LC}}{C}\left(\sin \frac{1}{\sqrt{LC}} t\right)u(t)$$

You are urged to verify this by directly finding v_C when $i(t) = u(t)$.

7.10 convolution

SAMPLING   When we discussed the impulse for the first time, we defined it as the limiting case of a pulse Δt seconds long and $1/\Delta t$ units high as $\Delta t \to 0$ (see Figure 7.10.1). How different is the zero-state response of any given linear system to a unit impulse from its response to the approximate impulse? It is not very different if Δt is much much less than the shortest time constant in the system.

figure 7.10.1
The approximate unit impulse. As $\Delta t \to 0$, $\tilde{\delta}(t) \to \delta(t)$.

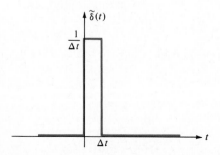

EXAMPLE 7.10.1

In the circuit of Figure 7.10.2 all initial conditions are zero, i.e., all voltages and currents equal zero at $t = 0-$ (zero state).

a. Find the unit-impulse response for v_2.

b. Find the response due to a short, high pulse of area $= 1$ (an approximate impulse).

ANS.:   (a)

$$v_2 = \frac{(1/Cp)v_1}{R + 1/Cp} = \frac{1/RC}{p + 1/RC}\,v_1 \qquad (7.10.1)$$

and so

$$v_2(t) = Ae^{-t/RC}$$

If $v_1(t) = \delta(t)$, the impulse voltage falls across R. Thus, $i = (1/R)\delta(t)$. Since this goes into C,

$$v_2(0+) = \frac{1}{C}\int_{0-}^{0+} i\,dt = \frac{1}{RC} \qquad \text{volt}$$

Therefore

$$v_2(t) = h(t) = \frac{1}{RC}\,e^{-t/RC} \qquad (7.10.2)$$

(b) Let

$$v_1(t) = u(t) - u(t-1) \qquad (7.10.3)$$

(see Figure 7.10.3). Then

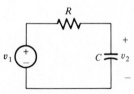

figure 7.10.2
The circuit of Example 7.10.1

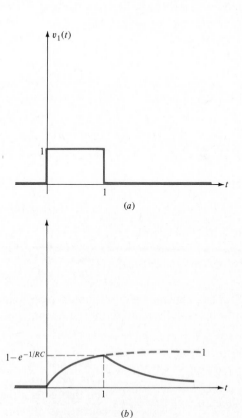

(a)

(b)

figure 7.10.3
(a) The input function $v_1(t) = u(t) - u(t-1)$. (b) The corresponding output.

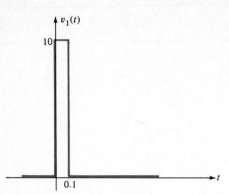

figure 7.10.4
A better approximation to a unit-impulse input function than
the waveform in Figure 7.10.3.

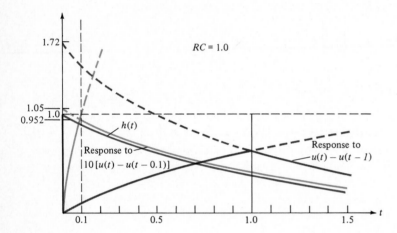

figure 7.10.5
A short-duration, high-amplitude pulse can
be used as input to determine a system's
approximate impulse response. The time
duration of the pulse should be less than
one-tenth the shortest time constant of the
system.

$$v_2(t) = v_f - (v_f - v_i)e^{-t/RC} \qquad \text{for } 0 < t < 1$$

and
$$v_2(t) = \begin{cases} 1 - e^{-t/RC} & \text{for } 0 < t < 1 \\ (1 - e^{-1/RC})e^{-(t-1)/RC} & \text{for } t > 1 \end{cases}$$

(7.10.4)

(7.10.5)

Now let

$$v_1(t) = 10[u(t) - u(t - 0.1)]$$

(7.10.6)

(see Figure 7.10.4). This leads to

$$v_2(t) = \begin{cases} 10(1 - e^{-t/RC}) & \text{for } 0 < t < 0.1 \\ 10(1 - e^{-0.1/RC})e^{-(t-0.1)/RC} & \text{for } t > 0.1 \end{cases}$$

(7.10.7)

(7.10.8)

Figure 7.10.5 summarizes these responses. Note that, for $\Delta t < 0.1RC$, the zero-state response to the approximation $\tilde{\delta}(t)$ is essentially equal to the actual impulse response $h(t)$.

Suppose we have as input to a linear system some arbitrarily shaped continuous waveform $f_{in}(t)$, as shown in Figure 7.10.6a. Let us approximate this $f_{in}(t)$ by
one that equals the original signal at the times $t = kt_0$ only (see Figure 7.10.6b);
call it $\hat{f}_{in}(t)$. If this approximate waveform $\hat{f}_{in}(t)$ were used as input to a linear

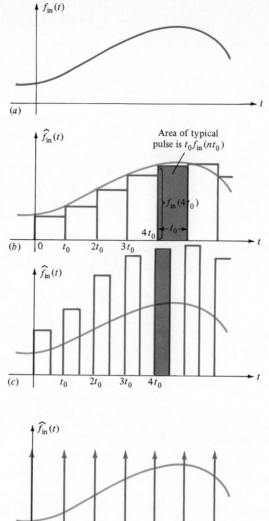

figure 7.10.6

(a) An arbitrary input function $f_{in}(t)$. (b) An approximation of $f_{in}(t)$ consisting of pulses $f_{in}(nt_0)$ high and t_0 wide. (c) Another approximation. Each pulse is proportionally higher than its counterpart in (b) but narrower, so that its area is the same as it was in (b): $f_{in}(nt_0)t_0$. (d) The limiting approximation. Each pulse has become an impulse of the same area as the corresponding pulses in (b) and (c).

system, would the resulting zero-state output closely resemble the zero-state output produced by the original $f_{in}(t)$? The answer is that by forcing t_0 to approach closer and closer to zero we can approximate the original waveform better and better, so that the response of $\hat{f}_{in}$ will approximate the response of f_{in}.

Think of a single one of the approximating pulses. It is $f_{in}(kt_0)$ high and lasts from $t = kt_0$ to $t = (k + 1)t_0$. If we, say, double its height and at the same time shorten its length to $t_0/2$, the resulting response, as in Example 7.10.1, should be relatively unchanged (provided that $\Delta t = t_0$ was much less than the shortest time constant in the system to start with). We can continue to shorten the pulse width and raise the heights accordingly until the $\hat{f}_{in}(t)$ becomes a train of impulses, each having an area equal to the magnitude of f_{in} at that time multiplied by t_0. The value of the *area* of a typical impulse in Figure 7.10.6d at, say, $t = \tau$ seconds is $f_{in}(\tau)t_0$ or $f_{in}(\tau)\,\Delta\tau$. If such numerical values are transmitted every t_0 seconds, we have a *digitized signal*.

CONVOLUTION

If we use a digitized (impulse-train) signal as input to a linear system, each impulse produces a response. Consider the impulse that occurs at any arbitrary time $t = \tau$:

$$\text{Input} = f_{in}(\tau)\delta(t - \tau)\,\Delta\tau \qquad \text{Output} = f_{in}(\tau)h(t - \tau)\,\Delta\tau \qquad (7.10.9)$$

The total input is the sum of all the impulses in the train. (It is the *sum* because where one impulse occurs, all others are equal to zero.) Superposition tells us that we can add all the responses to get the total response. Therefore,

$$\text{Input} = \sum_\tau f_{in}(\tau)\delta(t - \tau)\,\Delta\tau \qquad \text{Output} = \sum_\tau f_{in}(\tau)h(t - \tau)\,\Delta\tau \quad (7.10.10)$$

Finally, to make our approximation as good as possible, we let $t_0 \to 0$. In the limit, $\Delta\tau \to d\tau$ and $\sum \to \int$. Thus

$$\text{Input} = f_{in}(t) = \int_{-\infty}^{\infty} f_{in}(\tau)\delta(t - \tau)\,d\tau \qquad \text{Output} = \int_{-\infty}^{\infty} f_{in}(\tau)h(t - \tau)\,d\tau$$

$$(7.10.11)$$

We see that the arbitrary input $f_{in}(t)$ produces the *zero-state response*.†

$$y(t) = \int_{-\infty}^{\infty} f_{in}(\tau)h(t - \tau)\,d\tau \qquad (7.10.12)$$

The output is termed the *convolution* of the input with the impulse response and is written

$$y(t) = f_{in}(t) * h(t) \qquad (7.10.13)$$

We can get an equivalent form with a simple substitution of variables. In equation (7.10.12) let $t - \tau = \lambda$; then $\tau = t - \lambda$ and $d\tau = -d\lambda$ (remember that in this expression t is a constant)

$$y(t) = \int_{\infty}^{-\infty} f_{in}(t - \lambda)h(\lambda)(-d\lambda)$$

or

$$y(t) = \int_{-\infty}^{\infty} f_{in}(t - \lambda)h(\lambda)\,d\lambda \qquad (7.10.14)$$

Thus we see that the arguments of f_{in} and h are completely interchangeable inside the integral. Since convolution is a process that concerns two functions, just like the multiplication, division, addition, and subtraction of two functions, we talk about the *convolution process*.

† Because it is derived from the impulse response of the system, which is itself a zero-state response.

A *causal system* is one that cannot possibly begin to respond before it receives an input signal. Because we deal only with causal systems, $h(t)$ is zero for $t < 0$. Also, $f_{in}(t)$ usually is zero-valued for $t < 0$. Thus, the limits of integration are usually 0 to t.

EXAMPLE 7.10.2

Find the zero-state response to $f_{in}(t) = u(t)$ if $h(t) = e^{-t}u(t)$.

ANS.:   Use equation (7.10.14)

$$y(t) = \int_{-\infty}^{\infty} f_{in}(t - \lambda)h(\lambda) \, d\lambda$$

where

$$f_{in}(t - \lambda) = u(t - \lambda) \qquad \text{and} \qquad h(\lambda) = e^{-\lambda}u(\lambda)$$

so that

$$y(t) = \int_{0}^{t} (1)e^{-\lambda} \, d\lambda = \frac{e^{-\lambda}}{-1}\bigg|_{0}^{t} = \begin{cases} 1 - e^{-t} & \text{for } t > 0 \\ 0 & \text{for } t < 0 \end{cases}$$

or

$$y(t) = (1 - e^{-t})u(t)$$

EXAMPLE 7.10.3

Find the output $y(t)$ given the input $f_{in} = tu(t)$ into a system whose impulse response is $h(t) = e^{-t}u(t)$. The initial conditions are zero.

ANS.:   Again use equation (7.10.14)

$$y(t) = \int_{-\infty}^{\infty} f_{in}(t - \lambda)h(\lambda) \, d\lambda$$

where

$$f_{in}(t - \lambda) = (t - \lambda)u(t - \lambda) \qquad \text{and} \qquad h(\lambda) = e^{-\lambda}u(\lambda)$$

so that

$$y(t) = \int_{-\infty}^{\infty} (t - \lambda)u(t - \lambda)e^{-\lambda}u(\lambda) \, d\lambda = \int_{0}^{t} (t - \lambda)e^{-\lambda} \, d\lambda \qquad \text{if and only if } t > 0$$

$$= \int_{0}^{t} te^{-\lambda} \, d\lambda - \int_{0}^{t} \lambda e^{-\lambda} \, d\lambda \qquad \text{do the second integral by parts}$$

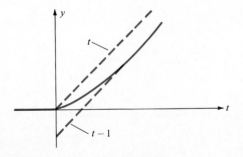

figure 7.10.7
The output function $y(t)$ in Example 7.10.3.

$$= t \left. \frac{e^{-\lambda}}{-1} \right|_0^t - \left[\frac{e^{-\lambda}}{+1}(-\lambda - 1) \right]\Bigg|_0^t = -t(e^{-t} - 1) + \left[e^{-\lambda}(\lambda + 1) \right]\Big|_0^t = -te^{-t} + t + e^{-t}(t + 1) - 1$$

$$= \begin{cases} t - 1 + e^{-t} & \text{for } t > 0 \\ 0 & \text{for } t < 0 \end{cases}$$

$$y(t) = (t - 1 + e^{-t})u(t)$$

See Figure 7.10.7.

Note that the convolution of the input with the impulse response yields the complete response only if all $v_C(0-) = i_L(0-) = 0$ (that is, convolution yields the *zero-state* response).

CONVOLUTION DONE GRAPHICALLY   Given an input $f_{in}(t)$ and an impulse response $h(t)$, if we plot the various functions involved in accomplishing $f_{in} * h$, we can gain a better understanding of the process.

EXAMPLE 7.10.4

Plot the functions involved in solving Example 7.10.2.

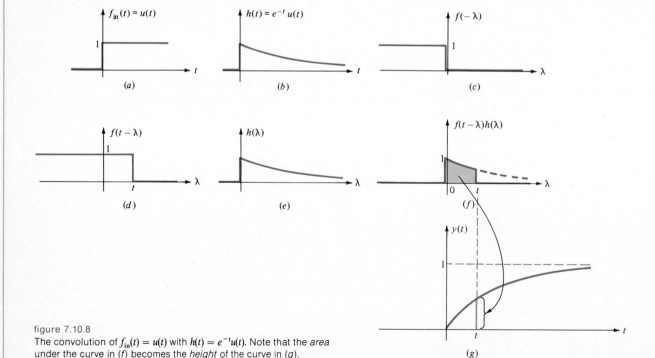

figure 7.10.8
The convolution of $f_{in}(t) = u(t)$ with $h(t) = e^{-t}u(t)$. Note that the *area* under the curve in (*f*) becomes the *height* of the curve in (*g*).

ANS.: See Figure 7.10.8. Multiply the two curves in Figure 7.10.8d and e point by point to get the integrand function shown in Figure 7.10.8f. Then

$$y(t) = \int_0^t f(t - \lambda)h(\lambda) \, d\lambda = \text{area in Figure 7.10.8}f$$

As t increases, the area changes. We plot this *area versus t* in Figure 7.10.8g.

EXAMPLE 7.10.5

For $h(t) = u(t)$ and $f_{in}(t) = 2[u(t) - u(t - 2)]$, find the zero-state output $y(t)$.
ANS.: Do it graphically (see Figure 7.10.9).

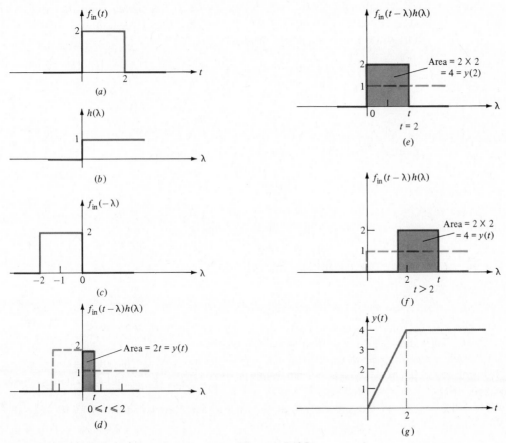

figure 7.10.9 The graphical convolution process of Example 7.10.5.

The observation could be made that if the expression

$$y(t) = \int_{-\infty}^{\infty} f_{in}(t - \lambda)h(\lambda) \, d\lambda$$

figure 7.10.10
Graphical convolution shows, that if $f_{in}(t) = \delta(t)$, then $f_o(t) = h(t)$.

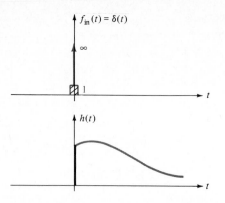

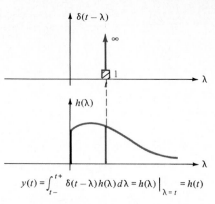

$$y(t) = \int_{t-}^{t+} \delta(t - \lambda)h(\lambda)\,d\lambda = h(\lambda)\Big|_{\lambda = t} = h(t)$$

and its equivalent $y(t) = \int_{-\infty}^{\infty} f_{in}(\lambda)h(t - \lambda)\,d\lambda$ are correct (and they *are*), when we set $f_{in} = \delta(t)$, we should find that $y(t) = h(t)$. Let us try. Using the first form of the convolution integral, above, we have

$$y(t) = \int_{-\infty}^{\infty} \delta(t - \lambda)h(\lambda)\,d\lambda$$

—— impulse occurs at $\lambda = t$

Therefore $\qquad\qquad\qquad y(t) = h(t) \qquad$ checks

Similarly, with the second (equivalent) expression,

$$y(t) = \int_{-\infty}^{\infty} \delta(\lambda)h(t - \lambda)\,d\lambda$$

—— impulse occurs at $\lambda = 0$

Therefore $\qquad\qquad\qquad y(t) = h(t - 0) = h(t) \qquad$ checks

Or do it graphically (see Figure 7.10.10).

EXAMPLE 7.10.6

Find the zero-state response $v_2(t)$ by convolution when $i(t) = u(t) - u(t - 1)$ (see Figure 7.10.11a).
ANS.: First find $h(t)$. If $i(t) = \delta(t)$ and $v_C(0-) = 0$, then at $t = 0-$ the two resistors are essentially in parallel, so that $i_2 = \frac{1}{2}\delta(t)$. The impulse current i_2 flowing into the capacitor produces an instantaneous jump in the value of v_C from 0 to $\frac{1}{2}$ V. For $t > 0$, $i_2(t) = -\frac{1}{2}(\frac{1}{2})e^{-t/2}$. Therefore, $i_2(t) = \frac{1}{2}\delta(t) - \frac{1}{4}e^{-t/2}$ and $v_2(t) = h(t) = \frac{1}{2}\delta(t) + \frac{1}{4}e^{-t/2}$ (see Figure 7.10.11b).

The zero-state output voltage is found by convolution as the area under the curve formed by forming the product of $h(t - \lambda)$ and $i(\lambda)$ (see Figure 7.10.11c to e)

$$v_2(t) = \text{area} = \frac{1}{2} + \int_0^t \frac{1}{4}e^{-x/2}\,dx = \frac{1}{2} + \frac{1}{2}(1 - e^{-t/2}) \qquad 0 < t < 1$$

For $t > 1$ the impulse has shifted beyond $\lambda = 1$, and so the area is (see Figure 7.10.11f)

$$\int_{t-1}^t \frac{1}{4}e^{-x/2}\,dx = \frac{e^{1/2} - 1}{2}\,e^{-t/2} \qquad t > 1$$

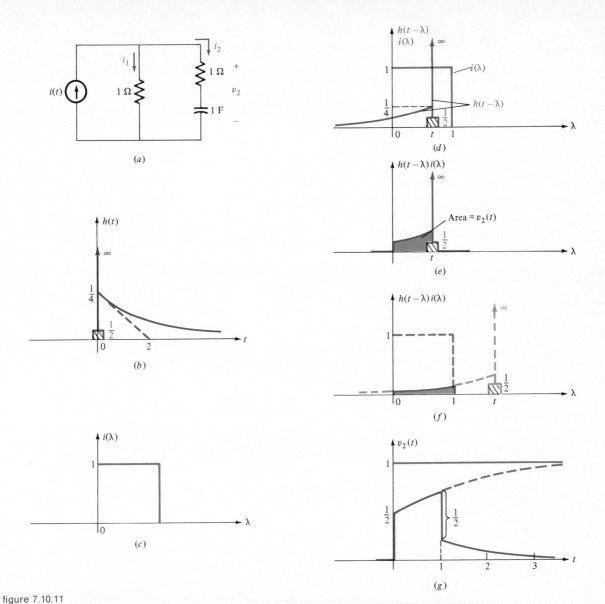

figure 7.10.11
(a) The circuit of Example 7.10.7.   (b) $h(t) = \frac{1}{2}\delta(t) + \frac{1}{4}e^{-t/2}u(t)$.   (c) $i(\lambda) = u(\lambda) - u(\lambda - 1)$.   (d) $h(t - \lambda)$ and $i(\lambda)$ when $0 < t < 1$.
(e) The area under (time integral of) the product $h(t - \lambda)i(\lambda)$.   (f) The area as in (e) except for $t > 1$.   (g) Plot of $v_2(t)$ = the area for all t.

EXAMPLE 7.10.7

If we excite the system in Figure 7.10.12a with an input velocity exactly the same as its impulse response, what will happen?

ANS.:   Since the impulse response is simply the natural response (with coefficients evaluated using initial conditions) this is an example of special case 2. We are using as input a function exactly like one of the terms in the natural response. Thus, the normal particular response is multiplied by $k_1 t + k_2$. However, let us

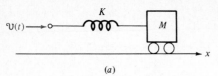

(a)

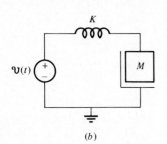

(b)

figure 7.10.12
The convolution process of Example 7.10.6.

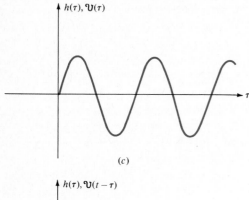

(c)

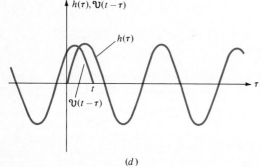

(d)

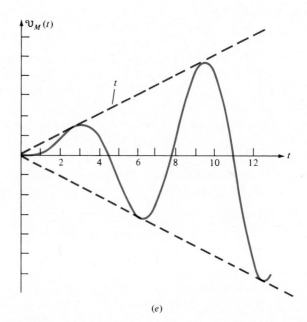

(e)

answer this using convolution. When we get the final answer, we can check to see if it is the usual particular response $h(t) = \mho_M(t)$ multiplied by t.

First we need the impulse response. The mechanical schematic is shown in Figure 7.10.12b. We find $h(t)$ by velocity division using

$$f_M = Mp\mho \tag{7.10.15}$$

$$f_K = \frac{1}{Kp}\,\mho \tag{7.10.16}$$

Now

$$Z_M(p) = \frac{1}{Mp} \tag{7.10.17}$$

and
$$Z_K(p) = Kp \tag{7.10.18}$$

$$\mathcal{U}_M = \frac{1/Mp}{1/Mp + Kp}\,\mathcal{U} = \frac{1/KM}{p^2 + 1/KM}\,\mathcal{U} \tag{7.10.19}$$

or
$$\left(p^2 + \frac{1}{KM}\right)\mathcal{U}_M = \frac{1}{KM}\,\mathcal{U} \tag{7.10.20}$$

The eigenvalues are obtained from the characteristic equation $s^2 + 1/KM = 0$ as

$$s = \pm j\sqrt{\frac{1}{KM}} \qquad \text{undamped case} \tag{7.10.21}$$

and the natural response contains no exponential factor

$$\mathcal{U}_{Mn}(t) = A\cos\frac{t}{\sqrt{KM}} + B\sin\frac{t}{\sqrt{KM}} \tag{7.10.22}$$

Now we need the values of $\mathcal{U}_M(0+)$ and $d\mathcal{U}_M/dt$ evaluated at $t = 0+$ in order to evaluate A and B. We know by definition that

$$\mathcal{U}_M(0-) = f_K(0-) = 0 \tag{7.10.23}$$

The impulse-velocity source

$$\mathcal{U}(t) = \delta(t) \tag{7.10.24}$$

cannot fall across M since $W = \tfrac{1}{2}M\mathcal{U}_M^2$; so

$$\mathcal{U}_K(t) = \delta(t) \tag{7.10.25}$$

and thus
$$f_K(0+) = \frac{1}{Kp}\,\delta(t) = \frac{1}{K} = f_M(0+) \tag{7.10.26}$$

i.e., the initial force on the mass at $t = 0+$ is $1/K$ newton, and this force accelerates the mass

$$\frac{1}{K} = Mp\mathcal{U}_M = f_M$$

or
$$\left.\frac{d\mathcal{U}_M}{dt}\right|_{0+} = \frac{1}{KM} \tag{7.10.27}$$

Using (7.10.23) in (7.10.22), we get $A = 0$; thus

$$\mathcal{U}_{Mn}(t) = B\sin\frac{t}{\sqrt{KM}} \tag{7.10.28}$$

Taking the time derivative, we get

$$\frac{d\mathcal{U}_{Mn}}{dt} = \frac{B}{\sqrt{KM}}\cos\frac{t}{\sqrt{KM}} \tag{7.10.29}$$

Inserting (7.10.27) into (7.10.29) evaluated at $t = 0+$ leads to

$$\frac{1}{KM} = \frac{B}{\sqrt{KM}}$$

so that $B = 1/\sqrt{KM}$ and

$$\mathcal{U}_{Mn}(t) = h(t) = \frac{1}{\sqrt{KM}} \sin \frac{t}{\sqrt{KM}}$$

For simplicity let us call

$$1/\sqrt{KM} = \omega_0 \qquad\qquad (7.10.30)$$

Thus

$$h(t) = \omega_0 \sin \omega_0 t \qquad\qquad (7.10.31)$$

Since in this example we have been asked to use as input a waveform identical to $h(t)$,

$$\mathcal{U}(t) = \omega_0 \sin \omega_0 t \, u(t) \qquad\qquad (7.10.32)$$

By convolution the output is found to be

$$h(\tau) = \omega_0 \sin \omega_0 \tau \, u(\tau)$$

$$\mathcal{U}(t - \tau) = \omega_0 \sin \omega_0(t - \tau) \, u(t - \tau)$$

$$h(\tau)\mathcal{U}(t - \tau) = \omega_0^2(\sin \omega_0 \tau)[\sin(-\omega_0 \tau + \omega_0 t)]u(\tau)u(t - \tau)$$

$$= -\omega_0^2(\sin \omega_0 \tau)[\sin(\omega_0 \tau - \omega_0 t)]u(\tau)u(t - \tau)$$

Using $\cos(a \pm b) = \cos a \cos b \mp \sin a \sin b$, we get

$$\tfrac{1}{2}[\cos(a + b) - \cos(a - b)] = -\sin a \sin b$$

with

$$a = \omega_0 \tau - \omega_0 t \qquad \text{and} \qquad b = \omega_0 \tau$$

Then

$$h(\tau)\mathcal{U}(t - \tau) = \frac{\omega_0^2}{2}\left[\cos(2\omega_0 \tau - \omega_0 t) - \cos(-\omega_0 t)\right] \qquad (7.10.33)$$

(see Figure 7.10.12c and d. The interval of integration is $0 < \tau < t$, and

$$\mathcal{U}_M(t) = \frac{\omega_0^2}{2}\int_0^t \cos(2\omega_0 \tau - \omega_0 t) - \cos \omega_0 t \, d\tau = \frac{\omega_0^2}{2}\frac{1}{2\omega_0}\left[\sin(2\omega_0 \tau - \omega_0 t)\right]_0^t - \frac{\tau\omega_0^2}{2}\cos \omega_0 t\Big|_0^t$$

$$= \frac{\omega_0}{4}\left[\sin \omega_0 t - \sin(-\omega_0 t)\right] - t\frac{\omega_0^2}{2}\cos \omega_0 t$$

$$= \frac{\omega_0}{2}(\sin \omega_0 t - \omega_0 t \cos \omega_0 t) \qquad\qquad (7.10.34)$$

This result is plotted in Figure 7.10.12e for $\omega_0 = 1/\sqrt{KM} = 1$.

In this example the particular response is multiplied by t, as predicted at the beginning of the answer. If the system eigenvalues had been located in the left-half s plane rather than on the imaginary axis, they would have given rise to a natural response containing an exponential damping term (the farther out the eigenvalue is to the left the faster the damping occurs). Multiplying such responses by t creates a term of the form te^{-at}, which approaches zero as t gets large. Therefore we conclude that if a system has eigenvalues $\pm j\omega_0$ on the imaginary axis of the s plane and we use as input a pure sinusoid for $t > 0$ that has the same frequency ω_0 as the natural response, the response will increase without bound as time goes on. (The system will go nonlinear eventually—or explode.)

If a system with eigenvalues *within* the s plane but not on the imaginary axis is excited by a function identical to one of the terms in its natural response, nothing disastrous will happen: a damped type of response that eventually dies out will occur.

instantaneous power $p(t)$ 7.11

As discussed earlier, the instantaneous product of the across and through variables for any element (one exception thermal systems) yields the instantaneous power being delivered to that element (Figure 7.11.1). If $v(t)$ and $i(t)$ are of such polarities that $p(t)$ is *positive* at some time t, power is being *absorbed* by the element (this goes for the source too).

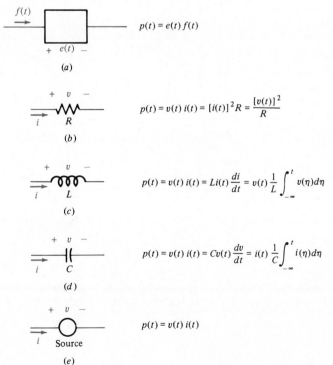

figure 7.11.1
(a) Power delivered *to* a generalized element.
(b) to (e) Power delivered *to* specific electrical elements.

EXAMPLE 7.11.1

Show that at any instant $t > 0$ the power *out* of the source equals the sum of the powers being absorbed by the two passive elements in Figure 7.11.2.

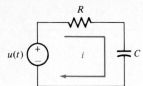

figure 7.11.2
The circuit of Example 7.11.1.

ANS.:   Assuming $v_C(0-) = 0$, $v_C(t) = 1 - e^{-t/RC}$, and $i(t) = (1/R)e^{-t/RC}$, so that $v_R(t) = e^{-t/RC}$, we get

$$p_C(t) = Cv\frac{dv}{dt} = C(1 - e^{-t/RC})\left(\frac{1}{RC}e^{-t/RC}\right) = \frac{e^{-t/RC}}{R} - \frac{e^{-2t/RC}}{R}$$

$$p_R(t) = \frac{v_R^2}{R} = \frac{e^{-2t/RC}}{R}$$

and the power delivered by (out of) the source is

$$-p_s(t) = v_s i = \frac{e^{-t/RC}}{R} = p_C(t) + p_R(t) \qquad \text{for any } t > 0$$

EXAMPLE 7.11.2

In a parallel LC circuit (Figure 7.11.3) initially 1 J of energy is stored in the capacitor and none in the inductor. Determine and sketch $W_C(t)$, $W_L(t)$, $p_C(t)$, and $p_L(t)$.

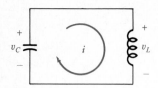

figure 7.11.3
The circuit of Example 7.11.2

ANS.:   At $t = 0-$

$$W_C = 1\text{ J} = \tfrac{1}{2}Cv^2 \qquad v_C(0-) = \sqrt{\frac{2}{C}} = v_C(0+)$$

Thus,

$$W_L = 0 = \tfrac{1}{2}Li^2 \qquad \text{and} \qquad i(0-) = 0 = i(0+)$$

$$v_L(0+) = v_C(0+) = \sqrt{\frac{2}{C}} = L\frac{di}{dt}\bigg|_{0+} \qquad \text{so} \qquad \frac{di}{dt}\bigg|_{0+} = \frac{\sqrt{2}}{L\sqrt{C}}$$

$$\frac{1}{C}\int_{-\infty}^{t} i\,dt + L\frac{di}{dt} = 0 \qquad \text{for } t > 0$$

$$\frac{d^2i}{dt^2} + \frac{1}{LC}i = 0$$

$$s^2 + \frac{1}{LC} = 0 \qquad s = \pm j\frac{1}{\sqrt{LC}}$$

$$i_n(t) = A \cos \frac{1}{\sqrt{LC}} t + B \sin \frac{1}{\sqrt{LC}} t$$

and

$$\frac{di}{dt} = -\frac{A}{\sqrt{LC}} \sin \frac{1}{\sqrt{LC}} t + \frac{B}{\sqrt{LC}} \cos \frac{1}{\sqrt{LC}} t$$

Since

$$i_n(0+) = 0 \quad \text{and} \quad \frac{di}{dt}\bigg|_{0+} = \frac{\sqrt{2}}{L\sqrt{C}}$$

$$A = 0 \quad B = \sqrt{\frac{2}{L}}$$

we have

$$i_n(t) = \sqrt{\frac{2}{L}} \sin \frac{1}{\sqrt{LC}} t$$

and

$$W_L(t) = \tfrac{1}{2} Li^2 = \sin^2 \frac{1}{\sqrt{LC}} t = \frac{1 - \cos(2t/\sqrt{LC})}{2}$$

Similarly

$$v_C(t) = E \cos \frac{1}{\sqrt{LC}} t + D \sin \frac{1}{\sqrt{LC}} t$$

so that

$$\frac{dv_C}{dt} = -\frac{E}{\sqrt{LC}} \sin \frac{t}{\sqrt{LC}} + \frac{D}{\sqrt{LC}} \cos \frac{t}{\sqrt{LC}}$$

Since

$$v_C(0+) = \sqrt{\frac{2}{C}} \quad \text{and} \quad i(0+) = -C \frac{dv}{dt}\bigg|_{0+} = 0$$

figure 7.11.4
Plots of (a) energy stored in the inductor, power delivered to the inductor, and the current $i(t)$ and (b) energy stored in the capacitor, power delivered to the capacitor, and the voltage $v_C(t)$. The amplitudes of the instantaneous power waveforms are both $1/\sqrt{LC}$.

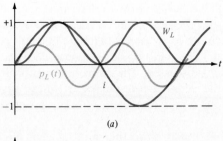

(a)

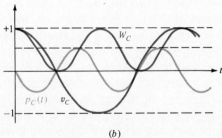

(b)

we find

$$v_C(t) = \sqrt{\frac{2}{C}} \cos \frac{t}{\sqrt{LC}}$$

so that

$$W_C(t) = \tfrac{1}{2}Cv^2 = \cos^2 \frac{t}{\sqrt{LC}} = \frac{1 + \cos(2t/\sqrt{LC})}{2}$$

Consequently $W_C + W_L = 1$ for any $t > 0$ (Figure 7.11.4). As energy is leaving the inductor, it is being absorbed by the capacitor at an equal rate

$$p_L(t) = \frac{dW_L}{dt} = \frac{1}{\sqrt{LC}} \sin \frac{2}{\sqrt{LC}} t$$

and

$$p_C(t) = \frac{dW_C}{dt} = -\frac{1}{\sqrt{LC}} \sin \frac{2}{\sqrt{LC}} t$$

so that

$$p_C(t) = -p_L(t) \quad \text{or} \quad p_C(t) + p_L(t) = 0$$

No power leaves or enters the circuit.

7.12 summary

We have seen in this chapter that the same technique used to solve first-order differential system equations can be extended to solve higher-order equations as well. You are now able to solve for the natural response of a system (overdamped, underdamped, and critically damped cases). You know what eigenvalues (characteristic values) are, how to determine them, and how to interpret their locations on the complex s plane. You can use initial conditions to evaluate unknown coefficients in the response. We showed how to determine the particular response, the complete response, and the zero-input and zero-state responses. Two special cases that you must watch out for when evaluating a system's particular response were discussed in detail. The unit-impulse response and unit-step response of a system were defined and discussed. The convolution process was presented and used to determine the zero-state responses of linear systems. Finally we discussed the instantaneous-power and energy-storage properties of higher-order systems.

The material that we have been looking at up to this point is what might be termed "the basics" of circuit and system analysis. Now we are ready to try our wings a bit. In extending these ideas and techniques in the chapters to come we shall be discussing other ways of designing systems and other ways of modeling them mathematically that have certain very nice advantages. We shall develop some shortcut techniques to make our analytical work easier and some additional techniques to enable us to handle more complicated problems. For now, if you have thoroughly understood all we have said so far, you have a good basic knowledge of the fundamental notions and techniques of circuit and systems analysis.

problems

1. In the circuit shown in Figure P7.1, find $i_L(0+)$, $v_C(0+)$, and di_L/dt, dv_C/dt, and di_C/dt, all evaluated at $t = 0+$.

2. Both switches in the circuit in Figure P7.2, which have been closed for a long time, open at $t = 0$. (a) Find $i_L(0+)$ and $v_C(0+)$. (b) Find dv_C/dt and di_L/dt both evaluated at $t = 0+$.

3. Find an expression for the current $i(t)$ in a series RLC circuit containing $R = 30\ \Omega$, $L = 5$ H, and $C = 0.2$ F. Assume that $v_C(0-) = 10$ V and that the initial current, which flows into the positive terminal of the capacitor, is $i_L(0-) = 1$ A.

4. A series RLC circuit ($R = 5\ \Omega$, $L = 0.5$ H, and $C = 0.125$ F) is driven by a voltage source $v_s(t) = 12u(t)$. There is no energy stored in the circuit at $t = 0-$. Find $i(t)$ for all t.

5. Find $v_b(t)$ for $t > 0$ in the circuit shown in Figure P7.5.

6. In the circuit shown in Figure P7.6 the switch closes at $t = 0$. (a) Find the current through and the voltage across the capacitor at $t = 0+$, the instant after the switch closes. (b) Solve for $v_C(t)$ for $t > 0$.

7. A source-free parallel circuit is composed of the elements $L = 10$ H, $R = 320\ \Omega$, and $C = \frac{125}{8}\ \mu$F. Given that $v_C(0+) = -160$ V and i_C, the corresponding capacitor current, is $i_C(0+) = 0.7$ A, (a) find the initial energy stored in the inductor and capacitor. (b) At what time t_0 is $v_C = 0$? (c) At what time t_m is v_C a positive maximum?

8. The circuit shown in Figure P7.8 has been in the configuration shown (with the switch closed) for a long time. At time $t = 0$ the switch opens. Find $i_L(t)$ for all $t > 0$.

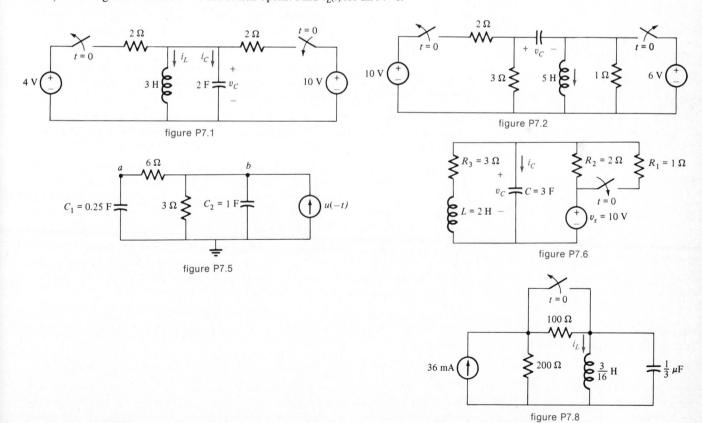

figure P7.1

figure P7.2

figure P7.5

figure P7.6

figure P7.8

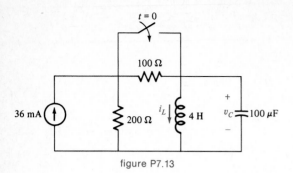

figure P7.13

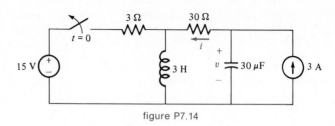

figure P7.14

9. Find the *natural* response for $v(t)$ but do not evaluate unknown coefficients in

$$\frac{d^2v}{dt^2} + \frac{10dv}{dt} + 169v = u(t)$$

10. Find the *complete* response for $v(t)$ in problem 9 given that $v(0-) = 0$ and dv/dt evaluated at $t = 0+$ is equal to $\frac{17}{169}$.

11. (*a*) Find the eigenvalues of the system described by

$$(p^2 + 2p + 5)v = (t + 1)u(t)$$

(*b*) What is the form of the natural response? (Write the expression but do not attempt to evaluate the coefficients.) What is the particular response (numerical values)?

12. Given a series *RLC* circuit ($R = 6\ \Omega$, $L = 1$ H, and $C = 0.04$ F) driven by a voltage source $10u(t)$. At $t = 0+$, $i_L = v_C = 0$. Find the time $t_0 > 0$ at which the first zero crossing occurs in $i(t)$.

13. The circuit shown in Figure P7.13 has been in the configuration shown (with the switch open) for a long time. At $t = 0$ the switch closes. (*a*) Find $v_C(0+)$ and dv_C/dt_0 evaluated at $t = 0+$. (*b*) Find the complete response $v_C(t)$ for all t.

14. In the circuit shown in Figure P7.14 the switch opens at $t = 0$. (*a*) Find the natural component of i, the current through the 30-Ω resistor. Do not evaluate coefficients. (*b*) Find the forced component of the voltage across the 30-μF capacitor.

15. A Thevenin equivalent circuit ($V_{\text{Th}} = 10$ V, $R_{\text{Th}} = 1\ \Omega$) connected by a *normally closed* switch to a parallel *RLC* circuit ($R = 500\ \Omega$, $L = \frac{50}{13}$ H, $C = 100\ \mu$F) is being used in a student laboratory experiment. Seeing a maximum source voltage of only 10 V, student A carelessly puts her hands across the inductor just as her lab partner opens the switch. What is the maximum voltage to which student A may be subjected?

16. A zero-state parallel-connected *RLC* circuit ($R = 25\ \Omega$, $L = \frac{1}{16}$ H, and $C = 16\ \mu$F) is driven by a step-function current source. At $t = 0+$ the current in the resistor is growing at a rate of 25 A/s. (*a*) Is this circuit underdamped, overdamped, or critically damped? (*b*) What is the magnitude of the current source?

17. In the circuit shown in Figure P7.17 (*a*) what value of resistance must be placed in series with the 25-Ω resistor to create critical damping? (*b*) With this new resistor in place, find $v(t)$ for $t > 0$.

18. Given that $v_C(0-) = 2$ V and $i(0-) = 3$ A in the circuit shown in Figure P7.18, find $i(t)$ for all t.

19. A $\frac{1}{3}$-kg mass that slides on a frictionless floor is connected to a wall by a spring whose compliance is $K = \frac{1}{3}$ m/N. The mass is released at $t = 0$ from a fixed position

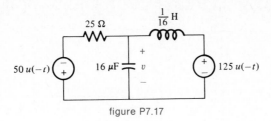

figure P7.17

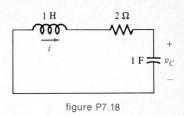

figure P7.18

$[\mathcal{V}(0-)=0]$ after the spring has been initially compressed so that it is pushing on the mass, in the positive direction, with a force of 6 N. (*a*) Find $v(t)$ of the mass for $t > 0$. (*b*) When will the mass first reach its maximum velocity? (*c*) How far *from where it is released* will the mass move in the positive direction before it stops and begins to move back? (*d*) What amount of friction D between the mass and the floor will cause critical damping?

20. The initial energy stored in the inductance in a critically damped source-free parallel *RLC* circuit is zero. Assume $L = C = 1$. At $t = 2$ s the energy stored in the capacitor is 2 J. Find the value of (*a*) the energy stored in the capacitor at $t = 0$ and (*b*) the energy stored in the inductor at $t = 2$ s. (*c*) At what rate is energy leaving the circuit at $t = 1$ s?

21. Given a system described by $d^2i/dt^2 + 3di/dt + 9i = 4dv_s/dt$, where $v_s(t)$ is the input and $i(t)$ is the output. (*a*) Find the eigenvalues of the system, the damping coefficient, ω_n, and ω_d. (*b*) What is the form of the zero-state response?

22. A system has the differential equation $(p^3 + 4p^2 + 5p + 2)y = (p + 3)x$. Given that one of the terms in the natural response is Ae^{-t}, find the form of the entire natural response.

23. A series closed loop consists of the series connection of $L = 1$ H, $C = 1$ F, and a variable resistor R. Given $i(0-) = 1$ and $v_C(0-) = 0$, find the zero-input response $i(t)$ and carefully sketch it versus t for $0 < t < 5$ for (*a*) $R = 1.9\ \Omega$; (*b*) $2\ \Omega$; (*c*) $2.1\ \Omega$.

24. A source-free parallel *RLC* circuit in which $C = \frac{1}{16}$ F and $L = 1$ H has $v_C(0-) = i_L(0-) = 1$. If the damping coefficient ζ is 0.5, (*a*) what is the value of R? What is the value of (*b*) the natural *damped* frequency and (*c*) the natural *undamped* frequency? (*d*) What time will elapse between zero crossings of the $y(t)$ response? (*e*) If the value of the resistance is decreased, will the answers to (*b*) and/or (*c*) change?

25. A circuit consisting of a single closed mesh contains a 10-V source, a switch, a 4-Ω resistor, a 0.25-H inductor, and a 0.25-F capacitor. At $t = 0$ the value of the current (which leaves the positive terminal of the source) is zero; at that same time the voltage across the capacitor is 3 V (the plus sign toward the incoming current). The switch has been open for a long time and then closes at $t = 0$. (*a*) Find an expression for $i(t)$ valid for $t > 0$. Sketch i versus t. Find the maximum value of i and the time at which it occurs. (*b*) Find the zero-state response. (*c*) Find the zero-input response.

26. A series connection of $R = \frac{6}{5}\ \Omega$, $L = \frac{1}{5}$ H, and $C = \frac{1}{5}$ F is driven by a voltage source $v_s(t) = tu(t)$. Find the complete response $i(t)$ if $i(0+) = 0$ and there is a $\frac{4}{5}$-C charge in the capacitor tending to oppose the voltage source.

27. Find the *natural* response of each of the following; do *not* evaluate the unknown coefficients: (*a*) $d^2i/dt^2 + 10di/dt + 21i = 0$; (*b*) $(p^2 + 4p + 16)y(t) = 5t$; (*c*) $2d^2v/dt^2 + 4dv/dt + 2v = 3$; (*d*) $(p^2 + 10p + 169)y = 4e^{-2t}u(t)$.

28. Find the *particular* response (response due to the source) for each of the differential equations in problem 27.

figure P7.30

29. Given the differential equation $(p^2 + 2p + 4)y = f(t)$, find the form of the particular response to each of the following inputs; do not evaluate unknown coefficients: (a) $10e^{-9t}$; (b) $10e^{-t}$; (c) $10e^{-t} \cos \sqrt{3}t$; (d) t^3; (e) $t^3 + t^2 + t + 1$.

30. In the circuit shown in Figure P7.30 find (a) the forced component of the voltage across C and (b) dv_C/dt at $t = 0+$. (c) Is the circuit overdamped, underdamped, or critically damped?

31. (a) Find the *natural* response of the following differential equation. Do not evaluate unknown coefficients

$$\frac{d^2i}{dt^2} + \frac{di}{dt} + \tfrac{5}{4}i = 10t \qquad \text{for } t > 0$$

(b) Find the *particular* (forced) response of

$$\frac{d^2v}{dt^2} + \frac{8dv}{dt} + 7v = 10t \qquad \text{for } t > 0$$

Evaluate all coefficients in this particular response.

32. A linear second-order system's particular response to the input $3tu(t)$ is $i_p(t) = (t - \tfrac{4}{3})u(t)$. What is the system's differential equation?

33. A series RLC circuit ($R = L = C = 1$) is driven by a voltage source $v(t) = \sin t$. The source current $i(t)$ is defined as being the output. What is the value of (a) ζ, (b) ω_n, (c) ω_d? Find (d) the form of the natural response; (e) the particular response; (f) the complete response assuming no initial energy storage in the circuit at $t = 0$.

34. A $\tfrac{1}{6}$-F capacitor is placed in series with the parallel combination of $R = 2\ \Omega$ and $L = 1$ H. The entire circuit is driven by a voltage source $v_s(t) = \cos 2t$. The output variable is the inductor voltage. (a) Find the value of the damping coefficient, the value of the natural undamped frequency, and the value of the natural damped frequency. (b) Find the zero-state response.

35. A 6-℧ conductance, a 0.25-H inductor, and a 2-F capacitor are connected in parallel. One of the two nodes is grounded. A current source $e^{-2t}u(t)$ is also connected in parallel with these other elements so that it forces positive current to flow *into* the ground node. At $t = 0+$ the voltage v at the ungrounded node is 2 V, and its slope is $+3$ V/s. (a) Find $v(t)$ for $t > 0$. (b) Sketch $v(t)$ versus time. (c) At time $t = 0+$ what is the value of the current flowing toward ground in the inductor?

36. For each of the circuits shown in Figure P7.36 find the numerical values of the capacitor voltage $v_C(0+)$, the inductor current $i_L(0+)$, and the slopes of each of these time functions evaluated at $t = 0+$. Assume no energy is stored in either circuit before $t = 0$.

37. A 2-Ω resistor is in series with the parallel combination of $L = 3$ H and $C = \tfrac{1}{6}$ F. A

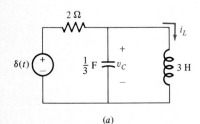

(a)

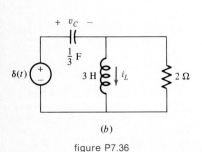

(b)

figure P7.36

unit-impulse voltage source is impressed across the resulting terminals. If $v_C(0-) = i_L(0-) = 0$, find $v_C(0+)$, $i_L(0+)$, dv_C/dt and di_L/dt both evaluated at $t = 0+$, and the total energy $W(0+)$ stored in the circuit at time $t = 0+$.

38. A series combination of $L = 1$ H and $C = 1$ F is connected in parallel with $R = 1\ \Omega$ and $v_s(t)$, a unit-impulse voltage source. If the system stores zero energy at $t = 0-$, (a) what are the values of $v_C(0+)$ and $i_L(0+)$? (b) Find an expression for $i_R(t)$ valid for all t.

39. A series RLC circuit ($R = 6\ \Omega$, $L = \frac{1}{2}$ H, $C = \frac{1}{50}$ F) is connected to a voltage source $v_s(t)$ at time $t = 0$. (a) Find the unit-impulse response of this circuit. The current is considered to be the output. (b) Find the unit-step response. Is the result consistent with your answer to part (a)?

40. For a system whose unit-impulse response is $h(t) = u(t) - u(t - 1)$, sketch and label the zero-state output if the input is $u(t) - u(t - 1)$.

41. The input to the system described in problem 40 is changed to $2[u(t) - u(t - 2)]$. Sketch the zero-state response to this input signal. Label the sketch with valid mathematical expressions that describe the output signal in a piecewise manner. Write a single expression for the output that is valid for all t.

42. A given circuit's unit-impulse response is $h(t) = e^{-3t}u(t)$, and its input is $x(t) = tu(t) - tu(t - 4) - 4u(t - 4)$. Write an expression that will yield the value of the zero-state output $y(t)$ at any time $t > 4$ s. Your answer should be in the form of a single integral whose integrand contains no steps, ramps, or impulses and whose limits do not contain t. Do not evaluate the integral.

43. For a given two-terminal network that is being driven by a current source $i(t)$, the unit-impulse response $h(t)$ of the voltage is the sum of a unit-impulse and a unit-step function. Find the zero-state output voltage that would be generated by the input $i(t) = \frac{3}{2}[u(t) - u(t - 2)]$. Use convolution and make several sketches that clearly demonstrate your solution.

44. Using the convolution process, determine and sketch the zero-state voltage response $v_o(t)$ of a certain voltage amplifier whose unit-impulse response is $h(t) = t[u(t) - u(t - 1)]$ if $v_{in}(t) = u(t) - u(t - 1)$ is applied to its input terminals.

45. A linear circuit is put into a black box and two terminals are brought out and are driven by an ideal voltage source. When no energy is stored initially in the circuit, the current flowing in response to a unit-impulse voltage input is $i(t) = 6e^{-t}u(t)$. (a) What will be the zero-state response $i(t)$ to an input-voltage pulse that is equal to 2 V for $0 < t < 2$ and is zero at all other times? (b) Sketch a possible schematic diagram of this circuit. Label the elements numerically.

46. The return signal to a given radar receiver is $v_{in}(t) = e^{-2t}[u(t) - u(t - 2)]$. If the unit-impulse response of this receiver is $h(t) = e^{-t}u(t)$, find the resulting zero-state output.

47. A linear electrical system controls the left-right angle of the main thruster rocket engines on a space vehicle in response to the input signal generated by the pilot's joystick. The unit-impulse response of this system is $h(t) = e^{-2t}u(t)$. (a) Find an expression for the angle of the thruster if the input control signal is of the form $v_{in}(t) = Ae^{-2t}u(t)$. (b) When does the maximum value of the output angle occur?

48. The unit-impulse response of a network is found to be $7u(t)$. Find the output that this system will generate if the input is $f_{in}(t) = t[u(t)u(1 - t)] - u(t - 2)$. Use convolution. What must be true about the network in order for this method to yield a valid answer?

49. A voltage pulse having unity amplitude and a duration of 1 s serves as input to a

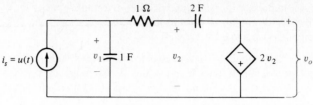

figure P7.52

system whose unit-impulse response is $h(t) = \cos(\pi t/2)u(t)$. What is the amplitude of the zero-state response at $t = 2$ s?

50. A unit-step function input $v_1(t) = u(t)$ is applied to an RC circuit in which no energy is stored initially. The response to this input is $v_2(t) = e^{-t/2}u(t)$. (*a*) What would be the zero-state response of this system to the input $v_1(t) = u(t)u(1 - t)$? (*b*) Can you find a simple RC network which fits this description?

51. The following differential equation describes a certain system for all $t > 0$. Given that $v(0+) = dv/dt$ evaluated at $t = 0+$ is zero, find $v(t)$ for $t > 0$. $d^2v/dt^2 + dv/dt = 4$.

52. Solve for the unit-step response $v_o(t) = w(t)$ of the circuit shown in Figure P7.52.

53. A linear, second-order circuit has a unit-step response $w(t) = (t + 1 - e^{-t})u(t)$. (*a*) Find the *particular* response of this system for $t > 0$ to the input: $\cos t\, u(t)$. Evaluate any unknown coefficients. (*b*) After all transients have died out, what terms may remain significant in the *complete* response to the input given in part (*a*)? (Simply describe any such terms; you do not have to evaluate them numerically.)

chapter 8
OPERATIONAL AMPLIFIERS

introduction 8.1

The operational amplifier (op amp) in its modern form is an integrated-circuit chip with a pair of input terminals leading to a very high resistance. The voltage placed across these input terminals is amplified greatly and appears at the output terminals (one of which is ground). Note from Figure 8.1.1 that if voltage $v_a - v_b$ is positive (that is, $v_a > v_b$), the output voltage v_o will be negative. In practice R_{in} is usually well in excess of 1 MΩ, A is at least 10^5, and R_o is a few tens of ohms.

An ideal op amp would have infinite input resistance, $R_{in} = \infty\,\Omega$ (an open circuit so as not to draw any input current); it would also have zero output resistance $R_o = 0\ \Omega$ (so that the output voltage will be independent of whatever load current we may ask the amplifier to put out); and $A = \infty$. In other words, an ideal op amp is simply a voltage-dependent voltage source (VDVS) that has infinite gain.

figure 8.1.1
Equivalent circuit of an op amp.

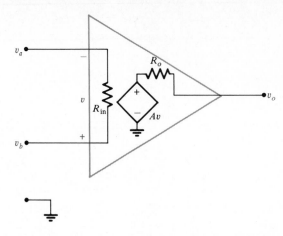

figure 8.1.1
Equivalent circuit of an op amp.

The op amp has been around for a long time. Originally it was constructed of vacuum tubes; then discrete transistors, resistors, wire, etc., were used. Now it comes in a tiny $\frac{1}{4}$-in totally integrated package costing less than 35 cents apiece. At the start the *op* amp was used to perform the *operations* of addition, subtraction, integration, and differentiation in *analog computers*, an application we discuss briefly in Section 8.6. Today, however, thanks to its small size, ready availability, and low cost, the op amp is more often used in a number of different ways directly in electric circuits. Time and space permit us to discuss only a few, but the op amp's properties and capabilities are such that its usefulness is limited only by the cleverness and imagination of the circuit designer. See Figure 8.1.2.

The op amp, like any other amplifier, needs an external source of power. Although the two additional terminals on every op amp are almost never shown explicitly in the schematic diagram because they do not enter into the input-output signal flow in any way, they must be connected to a balanced power supply (e.g., ± 15 V) or the amplifier will not function. A common student error in the laboratory is to leave these terminals disconnected, which is like forgetting to plug your hi-fi amplifier into the 120-V wall socket. When shown, these terminals are usually labeled $\pm V_{cc}$.

8.2 amplifiers (buffers)

There are two basic ways to connect the op amp in order to make it into a practical simple amplifier. In addition there are many variations and combinations of op-amp circuits for use in a wide variety of applications.

Let us develop an expression for the overall gain of each of the two basic ideal operational voltage-amplifier (buffer) circuits in Figure 8.2.1. Consider for a moment the very large (ideally infinite) value of A, the *open-loop gain* of the amplifier. It is called open-loop gain because it is the ratio of output voltage to input voltage *without* any feedback resistor (like R_2 in Figure 8.2.1). If A is, say, 1

signetics

μA741

LINEAR INTEGRATED CIRCUITS

figure 8.1.2
Specification sheet of a commercially available op amp. (© 1985 Signetics Corp. All Rights Reserved.)

The μA741 is a high performance operational amplifier with high open loop gain, internal compensation, high common mode range and exceptional temperature stability. The μA741 is short-circuit protected and allows for nulling of offset voltage.

- INTERNAL FREQUENCY COMPENSATION
- SHORT CIRCUIT PROTECTION
- OFFSET VOLTAGE NULL CAPABILITY
- EXCELLENT TEMPERATURE STABILITY
- HIGH INPUT VOLTAGE RANGE
- NO LATCH-UP

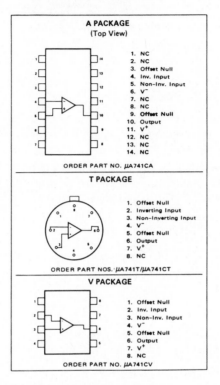

A PACKAGE
(Top View)

1. NC
2. NC
3. Offset Null
4. Inv. Input
5. Non-Inv. Input
6. V⁻
7. NC
8. NC
9. Offset Null
10. Output
11. V⁺
12. NC
13. NC
14. NC

ORDER PART NO. μA741CA

T PACKAGE

1. Offset Null
2. Inverting Input
3. Non-Inverting Input
4. V⁻
5. Offset Null
6. Output
7. V⁺
8. NC

ORDER PART NOS. μA741T/μA741CT

V PACKAGE

1. Offset Null
2. Inv. Input
3. Non-Inv. Input
4. V⁻
5. Offset Null
6. Output
7. V⁺
8. NC

ORDER PART NO. μA741CV

	μA741C	μA741
Supply Voltage	±18V	±22V
Internal Power Dissipation (Note 1)	500mW	500mW
Differential Input Voltage	±30V	±30V
Input Voltage (Note 2)	±15V	±15V
Voltage between Offset Null and V⁻	±0.5V	±0.5V
Operating Temperature Range	0°C to +70°C	-55°C to +125°C
Storage Temperature Range	–65°C to +150°C	–65°C to +150°C
Lead Temperature (Solder, 60 sec)	300°C	300°C
Output Short Circuit Duration (Note 3)	Indefinite	Indefinite

Notes
1. Rating applies for case temperatures to 125°C; derate linearly at 6.5mW/°C for ambient temperatures above +75°C.
2. For supply voltages less than ±15V, the absolute maximum input voltage is equal to the supply voltage.
3. Short circuit may be to ground or either supply. Rating applies to +125°C case temperature or +75°C ambient temperature.

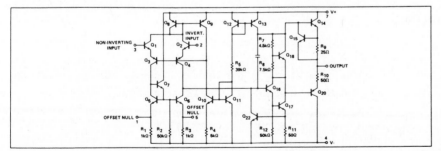

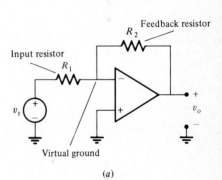

(a)

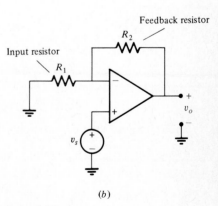

(b)

figure 8.2.1
(a) The inverting amplifier. (b) The noninverting amplifier.

million, this does *not* mean that a 1-V input will produce 1 million volts at the output terminals. The largest voltage the op amp can put out is approximately equal† to its power-supply voltage $\pm V_{cc}$ (typically ± 12 V or ± 15 V). A gain of $A = 10^6$ does mean that, say, 10 V will appear at the output when the difference between v_a and v_b (the net input voltage) is 10 μV. The ideal case $A = \infty$ makes it

† In practice, the maximum saturated output voltage is always somewhat less than the corresponding V_{cc} (supply) voltage.

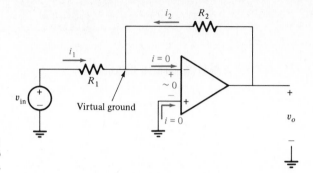

figure 8.2.2
Defining voltages and currents used to
analyze the inverting amplifier.

appear that we could get an output voltage when $v_a - v_b = 0$! Not quite but
almost. When we analyze any op-amp circuit, as a first approximation we can
assume that $v_a - v_b$ is virtually zero. The value of the overall voltage gain will be
determined by the external resistors R_1 and R_2. Let us see how.

Consider the ideal inverting amplifier. We say *ideal* because we shall assume
$A = \infty$, $R_{in} = \infty$ (so that the input current is 0), and $R_o = 0$. Since the input
terminals are at virtually the same voltage,† both input terminals are virtually
(almost) at ground potential. Therefore in Figure 8.2.2

$$i_1 + i_2 = 0 \tag{8.2.1}$$

$$\frac{v_{in}}{R_1} + \frac{v_o}{R_2} = 0 \tag{8.2.2}$$

$$\frac{v_o}{v_{in}} = -\frac{R_2}{R_1} \tag{8.2.3}$$

Consequently, although the open-loop gain A is infinite, the closed-loop gain
(with feedback resistor from output to input) is equal to the finite value given by
equation (8.2.3). The minus sign indicates that a positive voltage v_{in} will create a
negative voltage v_o. The overall gain of the circuit is thus completely determined
by the external resistor values, as we said before.

EXAMPLE 8.2.1

An inverting op-amp buffer like that in Figure 8.2.2 has $R_1 = 10$ kΩ, $R_2 = 100$ kΩ, and $v_{in} = 0.75$ V. Find v_o.
ANS.:   From equation (8.2.3)

$$v_o = v_{in}\left(-\frac{R_2}{R_1}\right) = 0.75\left(-\frac{100 \text{ k}\Omega}{10 \text{ k}\Omega}\right) = -7.5 \text{ V}$$

This assumes the power-supply voltage is greater than ± 7.5 V.

If $R_1 = R_2$, the overall circuit voltage gain is -1. Such a circuit is called an
inverter: $v_o = -v_{in}$. The noninverting amplifier of Figure 8.2.3a, redrawn in

---

† The one marked with a minus sign, the *inverting terminal*, is of course slightly higher than the other
but only by the smallest amount ε.

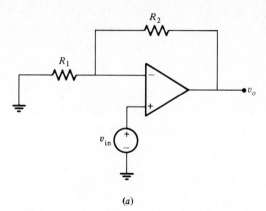

figure 8.2.3
(*a*) The noninverting amplifier. (*b*) The same
circuit redrawn; note the reversal of the input
terminals.

(*a*)

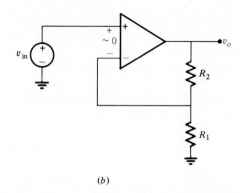

(*b*)

Figure 8.2.3*b*, can be analyzed almost as simply. Recall again in Figure 8.2.3*b*
that the two input terminals of the op amp are almost at the same potential; in
this case that means they are both at voltage v_{in}. Then by the obvious voltage-
divider network

$$v_{in} = \frac{R_1}{R_1 + R_2} v_o \qquad (8.2.4)$$

or

$$v_o = \frac{R_1 + R_2}{R_1} v_{in} \qquad (8.2.5)$$

$$v_o = \left(1 + \frac{R_2}{R_1}\right) v_{in} \qquad (8.2.6)$$

so that again we have an amplifier whose gain is determined solely by the value
of the external resistors. Note that although the magnitude of the gain of the
inverting amplifier can range from zero to a very large value, the magnitude of
the gain of the noninverting buffer cannot be less than unity.

As a matter of fact, we get the limiting case of unity gain for the noninverting
buffer when $R_2 = 0$. The value of R_1 is then arbitrary. Since it serves only to load

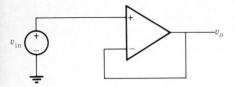

figure 8.2.4
The voltage follower, a special case of the noninverting amplifier (with R_2 set $= 0$).

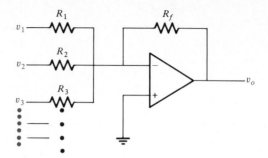

figure 8.2.5
The summer, a special case of the inverting amplifier. The output is a linear sum of the input voltages.

down the output of the buffer, we can choose $R_2 = \infty\ \Omega$ and so discard it. This special case of the noninverting amplifier is called a *voltage follower* (see Figure 8.2.4). Such an amplifier is a good buffer between two other circuits, say A and B. The buffer draws no current from stage A. Any such current would create a voltage drop across the Thevenin output resistance of stage A, thereby lowering its terminal output voltage. But this does not occur. The voltage follower itself has low output resistance and so can deliver output current to stage B without suffering any drop in its own terminal output voltage.

A special case of the inverting amplifier is the *adder* or *summer* circuit (Figure 8.2.5), which is simply an inverting amplifier with several inputs, each with its own input resistor. At the node where these resistors all join we write by KCL:

$$\frac{v_1}{R_1} + \frac{v_2}{R_2} + \frac{v_3}{R_3} + \cdots + \frac{v_o}{R_f} = 0 \tag{8.2.7}$$

where the subscript f stands for feedback. Thus

$$v_o = -\left(\frac{R_f}{R_1} v_1 + \frac{R_f}{R_2} v_2 + \frac{R_f}{R_3} v_3 + \cdots\right) \tag{8.2.8}$$

We shall look at a few examples of the many interesting and extremely useful circuits that can be made up using the inverting amplifier, the noninverting amplifier, or both.

EXAMPLE 8.2.2

A *current-to-voltage converter* can be made from the inverting amplifier simply by setting $R_1 = 0$. Suppose we want to measure the current i_R in Figure 8.2.6a but have only a voltmeter. We add the op amp in Figure 8.2.6b. Since terminal a of the op amp is at virtual ground, i_R has the same value in both circuits. In other words, the addition of the op amp does not in any way disturb or change the circuit made up of v and R. The output voltage v_o is calculated simply by realizing that i_R must flow from left to right through R_2. Thus it flows back into the output terminal of the op amp and then internally to ground, with the result that

$$i_R = \frac{0 - v_o}{R_2} \tag{8.2.9}$$

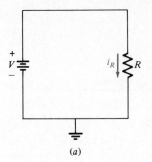

figure 8.2.6
(a) A circuit in which we want to measure the value of
i_R. (b) A method for converting i_R into a voltage so
that we can read its value with a voltmeter.

(a)

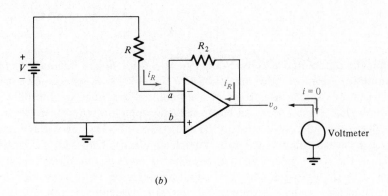

(b)

or
$$v_o = -i_R R_2 \qquad (8.2.10)$$

R_2 provides a convenient scale switch. If, for example, our meter reads in volts and i_R is in milliamperes, we can choose $R_2 = 1000 \ \Omega$.

EXAMPLE 8.2.3

A *differential amplifier* is simply an op amp connected as *both* an inverting and noninverting amplifier (see Figure 8.2.7). Find its output voltage in terms of the input voltages v_1 and v_2.

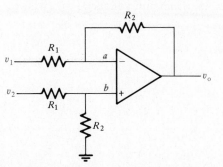

figure 8.2.7
A differential amplifier.

ANS.: Writing an expression for v_a as the sum of v_1 and the voltage rise from left to right across R_1, we have

$$v_a = v_1 + \frac{(v_o - v_1)R_1}{R_1 + R_2} \qquad (8.2.11)$$

By simple voltage division

$$v_b = v_2 \frac{R_2}{R_1 + R_2} \qquad (8.2.12)$$

Equating (8.2.11) and (8.2.12) because of the assumed infinite gain of the op amp yields

$$(v_o - v_1)R_1 + (R_1 + R_2)v_1 = v_2 R_2$$

$$v_o R_1 = -v_1 R_2 + v_2 R_2$$

or

$$v_o = \frac{R_2}{R_1}(v_2 - v_1) \qquad (8.2.13)$$

Note that the differential amplifier does not pass any voltage fluctuations that appear simultaneously on both inputs. Such signals are called *common-mode* signals. An example of the usefulness of a differential amplifier is in amplifying low-level biological signals (e.g., an electrocardiographic voltage sensed by two electrodes placed at different points on the chest). Stray 60-Hz line-voltage interference in such measurements generally arises from electric fields due to lighting, wiring, etc., in the room where the measurement is being made. Since this stray interference appears almost equally and simultaneously on both electrode leads, its effect is canceled out by the differential amplifier rejection of common-mode signals. Only the electrocardiographic signal is amplified because it is a differential signal.

If the same input voltage is used for both v_1 and v_2, the output voltage of the differential amplifier should be zero. This of course assumes that the two pairs of resistors R_1 and R_2 are perfectly matched.

A circuit that is useful for measuring bending deflections and other movements by means of a strain gauge is discussed in the next example.

EXAMPLE 8.2.4

A *strain gauge* is simply a resistor that changes its value slightly when it is bent or twisted. The circuit of Figure 8.2.8 is a differential amplifier both of whose inputs are connected to a common voltage V_s and whose

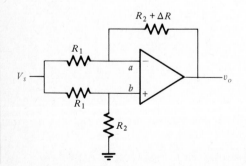

figure 8.2.8
A wide deviation strain-gauge amplifier.

inverting feedback resistor is a strain gauge with a value of $R_2 + \Delta R$. Find an expression for the output voltage.

ANS.:   Exactly as in Example 8.2.3, we write expressions for the voltage at points a and b and then set them equal.

$$v_a = \frac{(v_o - V_s)R_1}{R_2 + R_2 + \Delta R} + V_s \tag{8.2.14}$$

$$v_b = \frac{V_s R_2}{R_1 + R_2} \tag{8.2.15}$$

Equating (8.2.14) and (8.2.15), we have

$$\frac{(v_o - V_s)R_1}{R_1 + R_2 + \Delta R} = \frac{V_s R_2}{R_1 + R_2} - V_s \tag{8.2.16}$$

Collecting terms in V_s and simplifying leads to

$$v_o = V_s\left(-\frac{\Delta R}{R_1 + R_2}\right) \tag{8.2.17}$$

or in terms of γ, the fractional change in R_2,

$$\gamma \equiv \frac{\Delta R}{R_2} \tag{8.2.18}$$

$$v_o = -V_s \frac{R_2}{R_1 + R_2} \gamma \tag{8.2.19}$$

In our discussion of resistive circuits in Section 2.6 we examined the three-terminal wye and delta circuits. It turns out that these circuits are of great use in designing practical, high-voltage-gain amplifiers.

EXAMPLE 8.2.5

a. Find the voltage gain of the inverting-amplifier circuit shown in Figure 8.2.9a if $R_a = R_b = R_4 = 12$ kΩ and $R_c = 120$ Ω.

Find the effect of using this T (wye) circuit (rather than a single feedback resistor) on:

b. The input resistance as seen by source v_s

c. The current loading on the output of the amplifier

d. The overall reliability and performance of the amplifier

ANS.:   (a) Performing a wye-delta transform on the feedback circuit in Figure 8.2.9a yields the circuit of Figure 8.2.9b with values given by

$$R_1 = \frac{R_a R_b + R_b R_c + R_c R_a}{R_a} = \frac{R_{sp}}{R_a}$$

where sp stands for the sum of products, and

$$R_1 = \frac{144 \times 10^6 + 144 \times 10^4 + 144 \times 10^4}{12 \times 10^3} = \frac{14,688 \times 10^4}{12 \times 10^3} = 12.24 \text{ k}\Omega$$

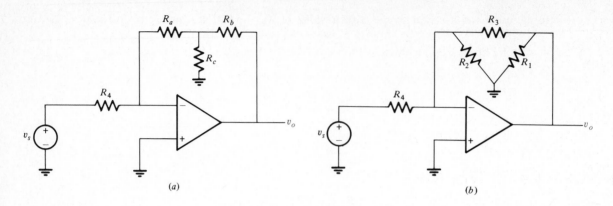

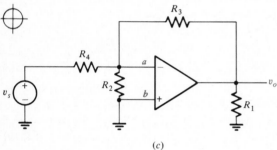

figure 8.2.9
The circuit of Example 8.2.5.

Similarly $$R_2 = \frac{R_{sp}}{R_b} = 12.24 \text{ k}\Omega \quad \text{and} \quad R_3 = \frac{R_{sp}}{R_c} = 1.224 \text{ M}\Omega$$

Hence the voltage gain is

$$\frac{v_o}{v_s} = -\frac{R_3}{R_4} = \frac{-1.224 \times 10^6}{12 \times 10^3} = -102$$

(b) The right-hand end of R_4 is connected to terminal a, which (assuming infinite gain A of the op amp) is at virtual ground. The addition, in parallel, of R_2 with the op amp's internal input resistance changes nothing insofar as the virtual ground potential of terminal a is concerned. The resistance seen by v_s is still R_4.

(c) By a similar argument (the virtual ground potential of node a) we see that R_3 and R_1 are in parallel to ground. The op amp must now supply current to R_1.

(d) The use of the lower-valued resistors in the T circuit rather than a single 1.2-MΩ feedback resistor reduces the noise and stray capacitance effects associated with extremely high resistance values.

We sometimes find useful the fact that an op amp cannot put out a voltage whose magnitude is any greater than that of the power-supply voltage $\pm V_{cc}$. Consider the primitive circuit shown in Figure 8.2.10a. Assume that a real op amp having a gain of, say, 10^6 is connected in this manner to the two voltage sources v_a and v_b and that the power-supply voltages are ± 15 V. If the voltage v_a

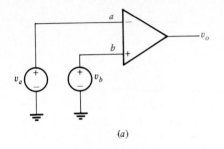

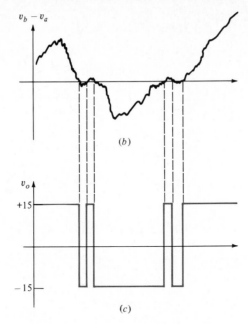

(a)

(b)

(c)

figure 8.2.10
(a) A voltage comparator and typical (b)
input and (c) output voltages.

exceeds the voltage v_b by more than 15 μV, v_o will *saturate* at -15 V. If v_a is
more than 15 μV less than v_b, then v_o will saturate at $+15$ V. For this reason the
circuit of Figure 8.2.10a is called a comparator: it compares the relative magnitudes of v_a and v_b. Essentially, if v_a is more positive than v_b, then v_o is negative
(and vice versa). This circuit has many uses. For example, if $v_a = 0$, then v_o gives
the algebraic sign of v_b. Note in Figure 8.2.10c that there may be chatter in the
output voltage if there is any noise or high-frequency fluctuation in the voltage
difference $v_b - v_a$ which excites the comparator.

A variation on the voltage comparator that eliminates the chatter caused by
minor fluctuations in the input is called the *Schmitt-trigger circuit* (Figure 8.2.11).
Assume that v_o is at its positive V_{cc} saturation level. The voltage v_b at the positive
input terminal is fixed by the voltage divider at some point called the *upper
threshold point* (UTP) between the positive-saturated output voltage v_o and the
fixed reference voltage V_{ref}. If and when v_s gets larger than v_b, v_o will jump to its
negative saturation level, where it will stay until v_s becomes less than the new
value of v_b, called the *lower threshold point* (LTP), which is between V_{ref} and
$-V_{cc}$. The UTP is thus

$$\text{UTP} = V_{ref} + [(v_o)_{max} - V_{ref}] \frac{R_1}{R_1 + R_2} \qquad (8.2.20)$$

and the LTP is

$$\text{LTP} = V_{ref} + [(v_o)_{min} - V_{ref}] \frac{R_1}{R_1 + R_2} \qquad (8.2.21)$$

We can solve for the value of V_{ref} and the voltage-divider fraction
$R_1/(R_1 + R_2)$ by first subtracting (8.2.21) from (8.2.20), which yields

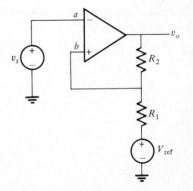

figure 8.2.11
A Schmitt trigger.

$$\text{UTP} - \text{LTP} = V_{\text{ref}} + [(v_o)_{\max} - V_{\text{ref}}]\frac{R_1}{R_1 + R_2} - V_{\text{ref}} - [(v_o)_{\min} - V_{\text{ref}}]\frac{R_1}{R_1 + R_2}$$

$$= [(v_o)_{\max} - (v_o)_{\min}]\frac{R_1}{R_1 + R_2}$$

or
$$\frac{R_1}{R_1 + R_2} = \frac{\text{UTP} - \text{LTP}}{(v_o)_{\max} - (v_o)_{\min}} \qquad (8.2.22)$$

Then adding (8.2.20) and (8.2.21) gives

$$\text{UTP} + \text{LTP} = 2V_{\text{ref}} + [(v_o)_{\max} - V_{\text{ref}} + (v_o)_{\min} - V_{\text{ref}}]\frac{R_1}{R_1 + R_2}$$

$$= 2V_{\text{ref}}\left(1 - \frac{R_1}{R_1 + R_2}\right) + [(v_o)_{\max} + (v_o)_{\min}]\frac{R_1}{R_1 + R_2}$$

where
$$1 - \frac{R_1}{R_1 + R_2} = \frac{R_1 + R_2 - R_1}{R_1 + R_2}$$

$$V_{\text{ref}} = \left\{\text{UTP} + \text{LTP} - [(v_o)_{\max} + (v_o)_{\min}]\frac{R_1}{R_1 + R_2}\right\}\frac{R_1 + R_2}{2R_2} \qquad (8.2.23)$$

EXAMPLE 8.2.6

Design a Schmitt trigger using a power supply $\pm V_{cc} = \pm 15$ V such that UTP $= 2$ V and LTP $= -1$ V.
ANS.: Using equation (8.2.22), we see that

$$\frac{R_1}{R_1 + R_2} = \frac{2 - (-1)}{15 - (-15)} = \frac{3}{30} = 0.1$$

If we choose $R_1 = 1$ kΩ, then $R_2 = 9$ kΩ and, from (8.2.23),

$$V_{\text{ref}} = (2 - 1 - 0)(\tfrac{5}{9}) = 0.56 \text{ V}$$

8.3 integrator-differentiator

In the op-amp circuit of Figure 8.3.1, which is actually just a generalization of the inverting amplifier, the feedback element is specified by its admittance operator $y_f(p)$ and the input element is $y_{\text{in}}(p)$. Summing currents into node a (and remembering that for the ideal op amp $R_{\text{in}} = \infty$, $A = \infty$, and $R_o = 0$), we have

$$v_s y_{\text{in}}(p) + v_o y_f(p) = 0$$

$$v_o = \frac{-y_{\text{in}}(p)}{y_f(p)} v_s \qquad (8.3.1)$$

For the inverting amplifier $y_{\text{in}}(p) = 1/R_1$ and $y_f(p) = 1/R_2$. Thus from (8.3.1)

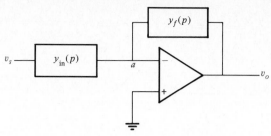

figure 8.3.1
The generalized inverting op-amp circuit.

we have, as before,

$$v_o = -\frac{R_2}{R_1} v_s \qquad (8.3.2)$$

If we choose the input element to be a resistor R and the feedback element to be a capacitor C, then

INTEGRATOR

$$y_{in}(p) = \frac{1}{R} \qquad \text{and} \qquad y_f(p) = Cp$$

and so

$$v_o(t) = -\frac{1}{RCp} v_s = -\frac{1}{RC} \int_{0-}^{t} v_s \, dt \qquad (8.3.3)$$

This is called an *ideal integrator* (Figure 8.3.2).

If we choose $y_{in}(p) = Cp$ and $y_f(p) = 1/R$, we get

DIFFERENTIATOR

$$v_o(t) = -RCpv_s(t) = -RC \frac{dv_s}{dt} \qquad (8.3.4)$$

and we have the *ideal differentiator* (Figure 8.3.3).

Although the integrator circuit of Figure 8.3.2 is used extensively in practice, the differentiator of Figure 8.3.3 is hardly used at all for the following reason. In reality, all signals have some corrupting noise associated with them as, for example, in Figure 8.3.4b. The magnitude of this additive noise may be quite low,

figure 8.3.2
The ideal integrator.

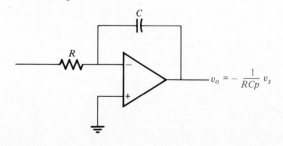

figure 8.3.3
The ideal differentiator.

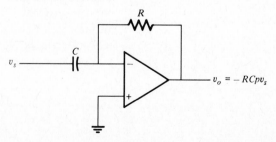

figure 8.3.4
(a) An ideal signal $v(t)$. (b) A typical actual signal. All signals contain at least some corrupting noise.

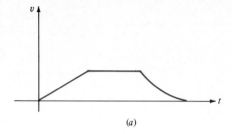

(a)

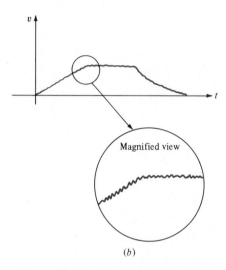

Magnified view

(b)

but the output voltage of the differentiator is proportional not to the magnitude of the input voltage but to its *slope*. Since the slope of the noise may be large indeed, differentiators are notorious for their noisy outputs and are to be avoided whenever possible in good electric-circuit design.

LOSSY INTEGRATOR

Another type of circuit is the lossy integrator, shown in Figure 8.3.5. The ideal integrator will put out a voltage proportional to the area under the $v_s(t)$ curve from $t = 0$ (when the capacitor has no charge on it) until time t. If at time t the integrator output is, say, v_o and from then on the input $v_s(t)$ is zero, the output of the ideal integrator will stay at v_o forever. But at times it is useful to have a lossy integrator—one that will *forget* with increasing time what its output has been. If its input has been zero for a time, such an integrator will have a zero-valued output regardless of the fact that in the distant past it was (and so still should be) putting out a constant value v_o. We obtain this lossy integrator by making the feedback capacitor lossy, i.e., by shunting it with a resistor.

We calculate the voltage-transfer function as follows:

$$\frac{v_o}{v_s} = -\frac{y_{\text{in}}(p)}{y_f(p)} = -\frac{1/R_1}{Cp + 1/R_2} = -\frac{1/R_1 C}{p + 1/R_2 C} \tag{8.3.5}$$

or

$$\left(p + \frac{1}{R_2 C}\right)v_o = -\frac{1}{R_1 C} v_s \tag{8.3.6}$$

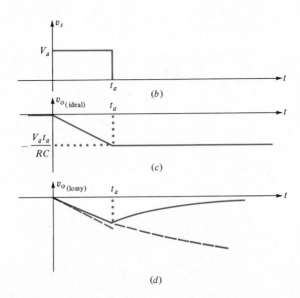

figure 8.3.5
(a) A lossy integrator. (b) The input signal of Example 8.3.1. (c) The response of an ideal integrator to the input signal shown in (b). (d) The response of the lossy integrator in (a) to the input (b).

$$\frac{v_o}{v_s} = \frac{1/R_1 C}{p + 1/R_2 C}$$

(a)

(b)

(c)

(d)

EXAMPLE 8.3.1

Find the zero-state response to the input signal $v_s(t) = V_a[u(t) - u(t - t_a)]$ (see Figure 8.3.5b) of:

a. An ideal integrator

b. The lossy integrator of Figure 8.3.5a

ANS.: (a) From equation (8.3.3) we get

$$v_o(t) = -\frac{1}{RC} \int_{0-}^{t} v_s(t)\, dt$$

from which

$$v_o(t) = \begin{cases} -\dfrac{1}{RC} \displaystyle\int_{0-}^{t} V_a\, dt = -\dfrac{V_a}{RC}\, t & \text{for } 0 < t < t_a \\[4mm] -\dfrac{1}{RC}\left(\displaystyle\int_{0-}^{t_a} V_a\, dt + \displaystyle\int_{t_a}^{t} 0\, dt \right) = -\dfrac{V_a t_a}{RC} & \text{for } \quad t > t_a \end{cases}$$

This response is plotted in Figure 8.3.5c for all t.

(b) For the lossy integrator of Figure 8.3.5a we have from equation (8.3.6)

$$\frac{dv_o}{dt} + \frac{1}{R_2 C} v_o = \frac{-1}{R_1 C} v_s \tag{8.3.7}$$

For $0 < t < t_a$ we have

$$\frac{dv_o}{dt} + \frac{1}{R_2 C} v_o = -\frac{V_a}{R_1 C}$$

Natural Response

$$s + \frac{1}{R_2 C} = 0 \qquad s = -\frac{1}{R_2 C}$$

$$v_{on}(t) = A e^{-t/R_2 C} \tag{8.3.8}$$

Particular Response

$$v_{op}(t) = k$$

$$0 + \frac{1}{R_2 C} k = \frac{-V_a}{R_1 C}$$

$$k = -\frac{R_2}{R_1} V_a \tag{8.3.9}$$

Note that capacitors are open circuits for dc sources. Thus for dc inputs the lossy integrator becomes a simple inverting amplifier with gain $-R_2/R_1$.

Complete Response
Adding (8.3.8) and (8.3.9), we get

$$v_o(t) = v_{on}(t) + v_{op}(t) = A e^{-t/R_2 C} - \frac{R_2}{R_1} V_a \tag{8.3.10}$$

and with $v_o(0) = 0$

$$0 = A e^{(0)} - \frac{R_2}{R_1} V_a$$

$$A = \frac{R_2}{R_1} V_a \tag{8.3.11}$$

so that

$$v_o(t) = -\frac{R_2}{R_1} V_a (1 - e^{-t/R_2 C}) \qquad 0 < t < t_a \tag{8.3.12}$$

Note that the time derivative is

$$\frac{dv_o}{dt} = -\frac{V_a}{R_1 C} e^{-t/R_2 C} \tag{8.3.13}$$

and so at $t = 0$, $v_o(t)$ has the same slope as the ideal (lossless) integrator of part (a). However, if R_2 is large, the final asymptotic value that $v_o(t)$ approaches as time goes on may be quite large indeed and in this case the time constant $R_2 C$ will be large. This means that the response will take a long time to reach its final value, $-(R_2/R_1)V_a$. If t_a is less than $R_2 C$, this response is a reasonably good approximation to the output of the ideal integrator (compare Figure 8.3.5c and d for $0 < t < t_a$). For $t > t_a$ the lossy integrator's response is, from equation (8.3.7),

$$\frac{dv_o}{dt} + \frac{1}{R_2 C} v_o = 0 \qquad t > t_a \tag{8.3.14}$$

Thus the complete response is most easily found from the natural response in the delayed form

$$v_o(t) = B e^{-(t-t_a)/R_2 C}$$

where B is found from the value of $v_o(t)$ evaluated at time $t = t_a$. From (8.3.12)

$$v_o(t_a) = -\frac{R_2}{R_1} V_a (1 - e^{-t_a/R_2 C}) = B e^0$$

and so

$$v_o(t) = -\left[\frac{R_2}{R_1} V_a (1 - e^{-t_a/R_2 C})\right] e^{-(t-t_a)/R_2 C} \qquad t > t_a \tag{8.3.15}$$

which approaches a value of zero as $t \to \infty$. The rate of this decay is governed by the time constant $R_2 C$.

If the ideal-integrator or lossy-integrator circuits have multiple input resistors (as the summer does), they can also perform the summation function first and then integration (or lossy integration) of that sum. In a later chapter we shall use circuits similar to the integrator (both lossy and ideal), the inverter, and the summer to make *filters*, i.e., circuits that pass sinusoids having certain frequencies but not others.

Many circuits use energy-storage elements, resistors, and op amps to perform varying tasks. Consider the circuit of Figure 8.3.6, in which the resistors R have equal values. The operators $y_1(p)$, $y_2(p)$, $y_3(p)$, and $y_4(p)$ are arbitrary. Assuming infinite-gain op amps having infinite input resistances, we simply write a KCL equation at each inverting input. For the first op amp

$$v_s y_1 + v_o y_4 + \frac{v_1}{R} = 0 \tag{8.3.16}$$

and for the second

figure 8.3.6
A useful two op-amp circuit.

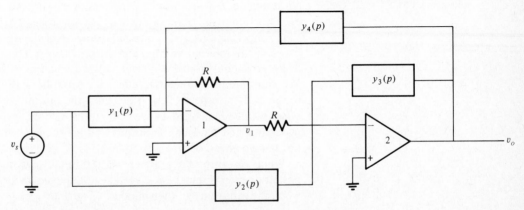

$$\frac{v_1}{R} + v_o\,y_3 + v_s\,y_2 = 0 \tag{8.3.17}$$

Solving both (8.3.16) and (8.3.17) for v_1/R and setting them equal gives

$$v_s\,y_1 + v_o\,y_4 = v_o\,y_3 + v_s\,y_2$$

Collecting terms, we have

$$v_o(y_4 - y_3) = v_s(y_2 - y_1)$$

so that

$$\frac{v_o}{v_s} = \frac{y_2 - y_1}{y_4 - y_3} \tag{8.3.18}$$

or, multiplying numerator and denominator both by -1,

$$\frac{v_o}{v_s} = \frac{y_1 - y_2}{y_3 - y_4} \tag{8.3.19}$$

This circuit can have widely varying responses (eigenvalues, etc.) depending on how the admittances are chosen.

8.4 some practical matters

When one actually begins to work with real op amps in circuits, it becomes obvious that in certain ways they differ from the ideal op amp we have been describing. We now discuss some of these differences.

SLEW RATE   If an actual op amp is connected as a noninverting amplifier with a closed-loop gain, say, of 10 and we apply an input voltage of $v_s(t) = u(t)$, we expect an output $v_o(t) = 10u(t)$. In actuality, the output will not jump discontinuously to its final 10-V level but will take some time to rise to it, because of the op amp's inherent internal time constants. Each different op-amp design has a different capability insofar as its maximum rate of change of output voltage is concerned. This figure of merit, called the *slew rate*, is measured in volts per microsecond.

INPUT OFFSET VOLTAGE   Although ideally we would expect to get 0 V out when we put 0 V into an op-amp buffer, in practice such is not the case. There are almost always some slight imperfections in the internal elements of the op amp, which result in imbalances in the internal voltage and current levels. These internal imbalances result in an erroneous nonzero output voltage when both inputs are shorted together. We can counterbalance this effect by carefully raising the voltage of one input terminal with respect to the other until the output voltage goes to zero. The *input offset voltage* is defined as the differential voltage *measured at the input terminals* needed to just balance the output voltage to zero. The polarity of this voltage is not known ahead of time.

Some op amps have a pair of terminals to which the designer connects a potentiometer (Figure 8.4.1), which is adjusted to obtain a zero output voltage when the inputs are both grounded.

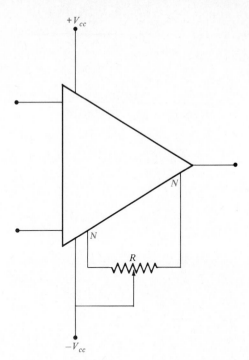

figure 8.4.1
An operational amplifier showing the
power-supply connections $\pm V_{cc}$ and the
potentiometer for balancing out the effect of
the input offset voltage. Terminals N-N are
the offset null terminals.

INPUT BIAS CURRENT

The input transistors of most op amps do, unfortunately, have to be supplied with a small current in order to function. This low-level constant current, called the *input bias current* I_b, is typically between 0.2 and 80 μA. Although I_b is an extremely small current, its effect on the output voltage can be quite noticeable at times. Consider the inverting amplifier of Figure 8.4.2a. Because of the extremely high inherent gain of the op amp we can consider terminals a and b to be at the same voltage. Since b is grounded, we say that terminal a is at virtual ground potential. Therefore, there cannot be any current in R_1. Since all I_b flowing into terminal a must come through R_2, the output voltage must be (from Ohm's law)

$$v_o = I_b R_2 \tag{8.4.1}$$

Hence if R_2 is, say, 100 kΩ, then with $I_b = 0.2$ μA we see from (8.4.1) that we get an erroneous output voltage (with no input signal) of

$$v_o = (2 \times 10^{-7} \text{ A})(10^5 \text{ }\Omega) = 20 \text{ mV}$$

This might or might not be bothersome, depending on the particular application in which the circuit is used, but if the input bias current is as high as 80 μA, it would produce an output voltage error (with R_2 of 100 kΩ) equal to 8 V!

The circuit of Figure 8.4.2b avoids this problem as long as we pick the correct value of R_3, which can be determined as follows. Since I_b flows into terminal b through R_3, we have

$$v_b = -I_b R_3 \tag{8.4.2}$$

Assuming $A \approx \infty$, it is true that

figure 8.4.2
(a) An inverting amplifier showing input bias current I_b in both input terminals. (b) The same amplifier with R_3 added to avoid the effect (on the output voltage) of the input bias current I_b which enters both input terminals.

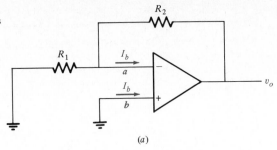

(a)

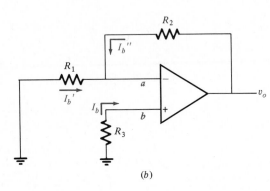

(b)

$$v_a = v_b \tag{8.4.3}$$

so that
$$v_a = -I_b R_3 \tag{8.4.4}$$

This value can be used to calculate I_b' and I_b'', the currents in R_1 and R_2, respectively,

$$I_b' = -\frac{v_a}{R_1} = \frac{I_b R_3}{R_1} \tag{8.4.5}$$

and for the desired value $v_o = 0$,

$$I_b'' = -\frac{v_a}{R_2} = \frac{I_b R_3}{R_2} \tag{8.4.6}$$

The current flowing into terminal a of the op amp is thus

$$I_b' + I_b'' = I_b R_3 \left(\frac{1}{R_1} + \frac{1}{R_2} \right) = I_b R_3 \frac{R_2 + R_1}{R_1 R_2} \tag{8.4.7}$$

For $I_b' + I_b''$, the current flowing into terminal a, to equal the required value I_b, the same as that going into terminal b, we choose

$$R_3 = \frac{R_1 R_2}{R_1 + R_2} \tag{8.4.8}$$

Then (8.4.7) becomes

$$I_b' + I_b'' = I_b \tag{8.4.9}$$

as required and v_o is equal to zero. Note that this is true for any value of I_b.

In effect what we are doing is to make the value of the resistance to ground equal at both input terminals. R_1 and R_2 are in parallel to ground (if $v_o = 0$) from terminal a.

EXAMPLE 8.4.1

In the circuit of Figure 8.4.2, $R_1 = 2$ kΩ, $R_2 = 50$ kΩ, and $I_b = 20$ μA. Find the values of v_a, v_b, and v_o that occur (with both signal inputs grounded as shown) first without and then with R_3.

ANS.: With $R_3 = 0$,

$$v_a = v_b = 0$$

and

$$v_o = I_b R_2 = (20 \times 10^{-6})(50 \times 10^3) = 1 \text{ V}$$

With R_3 present, i.e.,

$$R_3 = \frac{R_1 R_2}{R_1 + R_2} = 1.92 \text{ k}\Omega$$

we have

$$v_a = v_b = -I_b R_3 = -(20 \times 10^{-6})(1.92 \times 10^3) = -38.4 \text{ mV}$$

Therefore $I_b' = 19.2$ μA, $I_b'' = 0.768$ μA, and $v_o = 0$.

In real op amps the bias currents flowing into both terminals a and b are almost never equal. The *magnitude* of the difference between their numerical values is called the *input offset current*

INPUT OFFSET CURRENT

$$|I_{ba} - I_{bb}| = I_{\text{offset}}$$

EXAMPLE 8.4.2

In the circuit of Figure 8.4.2 with $R_1 = 2$ kΩ, $R_2 = 50$ kΩ, $I_{ba} = 21$ μA, and $I_{bb} = 19$ μA, find the resulting output voltage when the input voltage is zero.

ANS.: As in Example 8.4.1, we use $R_3 = 1.92$ kΩ, so that

$$v_b = -I_{bb} R_3 = -(19 \times 10^{-6})(1.93 \times 10^3) = -36.7 \text{ mV}$$

and since $v_a \approx v_b$, we get a current I_b' in R_1 of

$$I_b' = \frac{0 - v_a}{R_1} = \frac{+36.7 \times 10^{-3}}{2 \times 10^3} = 18.4 \text{ } \mu\text{A}$$

Since the current into terminal a of the op amp is given as $I_{ba} = 21$ μA, the current I_b'' must be

$$I_b'' = I_{ba} - I_b' = 21 \times 10^{-6} - 18.4 \times 10^{-6} = 2.6 \text{ } \mu\text{A}$$

This creates a voltage across R_2 such that the output voltage v_o is

$$v_o = v_a + I_b'' R_2 = -36.7 \text{ mV} + (2.6 \times 10^{-6})(50 \times 10^3) = 93.3 \text{ mV}$$

The input offset current causes an erroneous output voltage even when we have done our best to counteract bias-current effects (by inserting the proper value of R_3). The input-offset-current specification of an op amp determines the upper bound on the resistor values R_1 and R_2 that can be used.

So far, in discussing op amps we have assumed that the actual open-loop voltage gain A (the ratio v_o/v_{in} if feedback resistor R_2 is disconnected) is infinitely large. This assumption is almost always acceptable in practice and usually gives rise to insignificantly small errors in the calculated values that result. However, we can obtain the exact expressions for voltage gain, input resistance, and output (Thevenin) resistance for both the inverting and noninverting amplifiers by the basic methods of Chapters 1 and 2.

EXACT FORMULAS FOR INVERTING AMPLIFIER

Voltage Gain

Consider the circuit of Figure 8.4.3a. Under the assumption that the input resistance of the op amp is very high, we have

$$i_1 + i_2 = 0 \tag{8.4.10}$$

and so

$$\frac{v_s - v_a}{R_1} + \frac{v_o - v_a}{R_2} = 0 \tag{8.4.11}$$

figure 8.4.3
(a) The inverting amplifier. (b) Circuit used to determine input resistance. (c) To determine output resistance a unit current is driven into the output terminals and the resulting terminal voltage is measured. (d) The Thevenin equivalent circuit for the procedure described in (c).

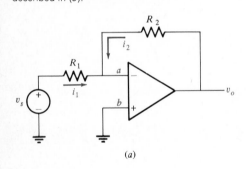

(a)

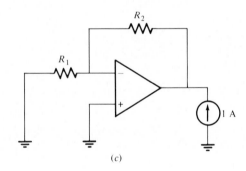

(c)

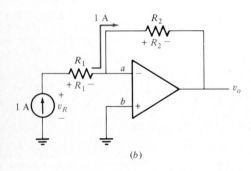

(b)

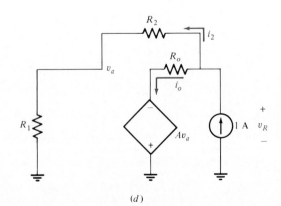

(d)

The output voltage v_o is, by definition,

$$v_o = -A(v_a - v_b) \tag{8.4.12}$$

and $v_b = 0$, so that

$$-v_a = \frac{v_o}{A} \tag{8.4.13}$$

Inserting (8.4.13) into (8.4.11) yields

$$\frac{v_o + v_o/A}{R_2} + \frac{v_s + v_o/A}{R_1} = 0 \tag{8.4.14}$$

which, after collecting terms in v_o, results in

$$\boxed{\frac{v_o}{v_s} = \frac{-A}{1 + (R_1/R_2)(1 + A)}} \tag{8.4.15}$$

In the limit, as A approaches infinity (we use l'Hospital's rule),

$$\lim_{A \to \infty} \frac{v_o}{v_s} = \frac{-1}{R_1/R_2} = -\frac{R_2}{R_1} \tag{8.4.16}$$

which is what we developed before under this assumption of infinite open-loop gain.

Input Resistance

We can find the resistance presented, to the left of R_1, to any input voltage source v_s by the fundamental method used in Chapter 2. Insert a 1-A current into the input terminals as shown in Figure 8.4.3b. Calculate the resulting voltage v_R; this is the numerical value of the input resistance because

$$\frac{v_R}{I_{\text{in}}} = R_{\text{input}} = \frac{v_R}{1} \tag{8.4.17}$$

From Figure 8.4.3b we see that

$$v_R - R_1 = v_a \tag{8.4.18}$$

$$v_o = -Av_a \tag{8.4.19}$$

and

$$v_o = v_a - R_2 \tag{8.4.20}$$

Combining (8.4.19) and (8.4.20) gives

$$-Av_a = v_a - R_2$$

$$R_2 = v_a(1 + A) \tag{8.4.21}$$

Then substituting (8.4.18) into (8.4.21), we get

$$R_2 = (v_R - R_1)(1 + A)$$

$$\frac{R_2}{1 + A} = v_R - R_1$$

or finally

$$R_{\text{input}} = \frac{R_2}{1 + A} + R_1 \tag{8.4.22}$$

In the limit as A gets large we see that the first term becomes insignificant and the resistance seen by the input-voltage source v_s is R_1. This is indeed logical because v_a is virtual ground.

Output Resistance

Use the same technique as above but at the output terminals, as shown in Figure 8.4.3c or d, in which the Thevenin output circuit of the op amp itself is shown. We see that

$$i_o = 1 - i_2 \tag{8.4.23}$$

$$i_2 = \frac{v_R}{R_1 + R_2} \tag{8.4.24}$$

and

$$v_a = i_2 R_1 \tag{8.4.25}$$

Inserting (8.4.24) into (8.4.23) and (8.4.25) gives

$$i_o = 1 - \frac{v_R}{R_1 + R_2} = \frac{R_1 + R_2 - v_R}{R_1 + R_2} \tag{8.4.26}$$

and

$$v_a = \frac{R_1 v_R}{R_1 + R_2} \tag{8.4.27}$$

respectively. Writing KVL around the lower right-hand mesh in Figure 8.4.3d leads to

$$v_R = -A v_a + i_o R_o \tag{8.4.28}$$

Substituting (8.4.26) and (8.4.27) into (8.4.28) yields

$$v_R = \frac{-A R_1 v_R + R_o (R_1 + R_2 - v_R)}{R_1 + R_2} \tag{8.4.29}$$

Collecting terms in v_R and rearranging gives the output resistance

$$R_{\text{output}} = \frac{R_o}{1 + (A R_1 + R_o)/(R_1 + R_2)} \tag{8.4.30}$$

which in the limit, as $A \to \infty$, becomes

$$R_{\text{output}} = \frac{R_o (1 + R_2/R_1)}{A} \tag{8.4.31}$$

$$= 0 \tag{8.4.32}$$

The voltage gain, input resistance, and output resistance of the noninverting amplifier can be found with the same techniques as the inverting amplifier. The only difference is that in calculating the input resistance the actual internal resistance R_{internal} of the op amp between terminals a and b must be included (since there is no other path for the 1-A input current).

Voltage Gain

$$\boxed{\frac{v_o}{v_s} = \frac{A}{1 + R_1 A/(R_1 + R_2)}} \tag{8.4.33}$$

which in the limit as $A \to \infty$ becomes, as before,

$$\frac{v_o}{v_s} = 1 + \frac{R_2}{R_1} \tag{8.4.34}$$

Input Resistance

$$\boxed{R_{\text{input}} = \frac{R_{\text{internal}} R_1(1 + A) + R_2(R_1 + R_{\text{internal}})}{R_1 + R_2}} \tag{8.4.35}$$

which becomes (as $A \to \infty$)

$$R_{\text{input}} \approx \frac{R_{\text{internal}} A}{1 + R_2/R_1} \to \infty \tag{8.4.36}$$

Output Resistance

$$\boxed{R_{\text{output}} = \frac{R_o(R_1 + R_2)}{R_1 + R_2 + AR_1 + R_o}} \tag{8.4.37}$$

which becomes (as $A \to \infty$)

$$R_{\text{output}} = \frac{R_o(1 + R_2/R_1)}{A} \to 0 \tag{8.4.38}$$

block diagrams 8.5

When we first discussed p operators in this text we briefly mentioned the block diagram. Generally, a block diagram consists of a box that contains a p operator, an input signal, and an output signal. The two signals are shown as arrows directed into and out of the box, respectively. The contents of the box operate on (multiply) the input, and the result is the output. Some examples are shown in Figure 8.5.1. Note that the operations in Figure 8.5.1*a*, *c*, *e*, and *f* are all easily performed by op-amp circuits. By interconnecting such blocks we can construct an equivalent op-amp type of system for any given mechanical, hydraulic, or *other electric* system.† Of course, we could get an equivalent electric circuit using

---

† Most important of all, we can use op amps to build electric networks that have almost any specified linear *p*-operator transfer functions. Such op-amp networks are called *active filters* (see Section 12.9).

figure 8.5.1
Some useful block diagrams and their
associated inputs, outputs, and p-operator
transfer functions.

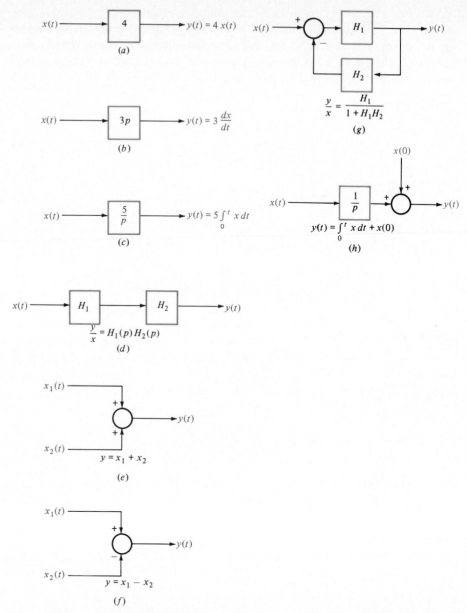

the analog methods of Chapter 4, but it is often impossible to obtain electrical
elements with the required values; a 250-kg mass would mean having a 250-F
capacitor! The analog method works well if all we want is a mathematical solution, but if we actually want to build an electrical model of some other system we
must use op amps. The method is straightforward: build a system of interconnected op-amp integrators, amplifiers, and adders that will have the same p-
operator transfer function as the actual system. Typical building blocks are

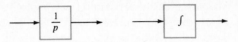

figure 8.5.2
Two equivalent symbols for the block
diagram of an ideal integrator.

shown in Figure 8.5.1. The block diagrams we obtain serve as the basis for understanding another method of describing circuits and systems called *state-variable
analysis.*

The simple ideal integrator with transfer function $H(p) = 1/p$ has the alternate
symbol shown in Figure 8.5.2. We can write the block diagram for more and
more complicated transfer functions by using this basic integrator and the rules
of block-diagram algebra shown in Figure 8.5.1*d* to *g*. Figure 8.5.1*e* and *f* shows
two forms of what is generally called the *summing junction.* This function is easily
performed by the op-amp adder circuit. Figure 8.5.1*d* shows the *cascade* connection of two systems. The overall transfer function $H = y/x$ of such a connection is
equal to the product of each of the individual transfer functions; i.e., if $H_2 = y/q$
and $H_1 = q/x$, then

$$H = \frac{y}{x} = \frac{q}{x}\frac{y}{q} = H_1 H_2$$

Figure 8.5.1*g* represents the interconnection of two systems in *negative feedback.* The signal y is *fed back,* operated on by H_2, and then subtracted from the
input. This difference (sometimes called the *error signal*) is what is operated on by
H_1 in order to produce the output $y(t)$. The expression for the output in Figure
8.5.1*g* is easily derived as follows. Call the signal entering the H_1 block $e(t)$. Then
we can write

$$y = H_1 e \qquad (8.5.1)$$

and

$$e = x - H_2 y \qquad (8.5.2)$$

Substituting (8.5.2) into (8.5.1), we have

$$y = H_1(x - H_2 y) = H_1 x - H_1 H_2 y$$

so that

$$y(1 + H_1 H_2) = H_1 x$$

and

$$\frac{y}{x} = \frac{H_1}{1 + H_1 H_2} \qquad (8.5.3)$$

EXAMPLE 8.5.1

Sketch the block diagram for

$$H(p) = \frac{1}{p + 15} = \frac{1/p}{1 + 15/p}$$

ANS.: Using Figure 8.5.1*g* with $H_1 = 1/p$ and $H_2 = 15$, we get the diagram
in Figure 8.5.3.

figure 8.5.3
The block diagram for
$H(p) = 1/(p + 15)$ in Example 8.5.1.

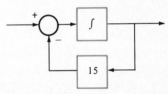

CONTROLLER FORM

Vannevar Bush, the famous scientist-engineer, developed a specifically structured block diagram that can model any given transfer function $H(p)$. His method is called *direct programming*, and the resulting block diagram is called the *controller form* of the system. Suppose we have the general transfer function

$$H(p) = \frac{p^m + b_1 p^{m-1} + b_2 p^{m-2} + \cdots + b_m}{p^n + a_1 p^{n-1} + a_2 p^{n-2} + \cdots + a_n} \qquad (8.5.4)$$

Once we have derived the controller form of the block diagram that models this general $H(p)$ we need not derive it again each time we use it: we simply fill in the blocks with the proper numbers. To develop it once from scratch we first work with a transfer function having the same denominator as that in equation (8.5.4), but with the numerator 1. Call this transfer function $H_d(p)$.

figure 8.5.4
Development of the general form of the
denominator of $H(p)$ in the controller form.

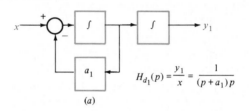

$$H_{d_1}(p) = \frac{y_1}{x} = \frac{1}{(p + a_1)p}$$

(a)

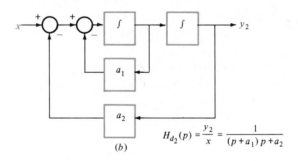

$$H_{d_2}(p) = \frac{y_2}{x} = \frac{1}{(p + a_1)p + a_2}$$

(b)

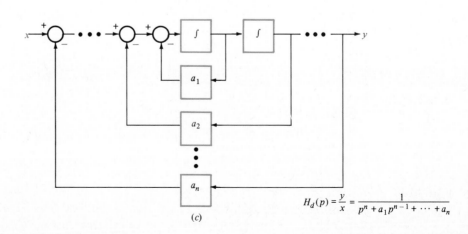

$$H_d(p) = \frac{y}{x} = \frac{1}{p^n + a_1 p^{n-1} + \cdots + a_n}$$

(c)

$$H_d(p) = \frac{1}{p^n + a_1 p^{n-1} + a_2 p^{n-2} + \cdots + a_n} \qquad (8.5.5)$$

Write the denominator of $H_d(p)$ in nested form

$$H_d(p) = \frac{1}{\{[(p + a_1)p + a_2]p + a_3\}p + \cdots + a_n} \qquad (8.5.6)$$

Such a transfer function has a simple block-diagram form. For instance, we know that

$$H_{d1}(p) = \frac{1}{(p + a_1)p} \qquad (8.5.7)$$

can be modeled as in Figure 8.5.4a. Then if we add to this another closed negative-feedback loop containing a_2, as in Figure 8.5.4b, we get (see Figure 8.5.1g)

$$H_{d2}(p) = \frac{1/(p + a_1)p}{1 + a_2/(p + a_1)p} = \frac{1}{(p + a_1)p + a_2} \qquad (8.5.8)$$

Continuing to add cascaded integrators and feedback loops, we find that the general form of $H_d(p)$, equation (8.5.6), is as shown in Figure 8.5.4c.

EXAMPLE 8.5.2

Find the controller form of the block diagram for

$$H(p) = \frac{1}{p^4 + 15p^3 + 71p^2 + 105p}$$

ANS.: First put the transfer function into the form

$$H(p) = \frac{1}{\{[(p + 15)p + 71]p + 105\}p}$$

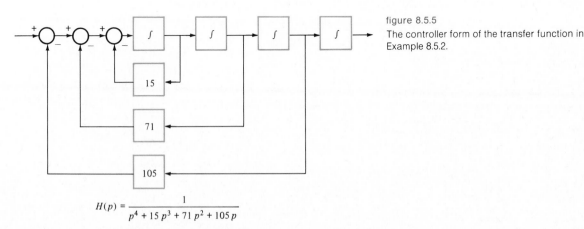

figure 8.5.5
The controller form of the transfer function in Example 8.5.2.

$$H(p) = \frac{1}{p^4 + 15\,p^3 + 71\,p^2 + 105\,p}$$

Starting with a diagram similar to Figure 8.5.4b and alternately adding cascaded integrators and feedback gains, we proceed as shown in Figure 8.5.4. After enough experience, we can simply jump to Figure 8.5.5.

What if the numerator in Example 8.5.2 is not unity? In the next example we develop the general block diagram of the controller form.

EXAMPLE 8.5.3

Develop a block diagram in controller form for

$$H(p) = \frac{p^3 + 12p^2 + 44p + 48}{p^4 + 15p^3 + 71p^2 + 105p} \tag{8.5.9}$$

ANS.:   Write this term by term as

$$H(p) = \frac{48}{p^4 + 15p^3 + 71p^2 + 105p} + \frac{44p}{p^4 + 15p^3 + 71p^2 + 105p}$$
$$+ \frac{12p^2}{p^4 + 15p^3 + 71p^2 + 105p} + \frac{p^3}{p^4 + 15p^3 + 71p^2 + 105p} \tag{8.5.10}$$

Note that each of the four terms in this equation can be realized by the technique demonstrated in Example 8.5.2. For instance, the term

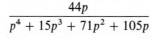

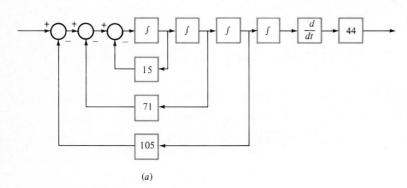

(a)

figure 8.5.6
(a) A realization of

$$\frac{44p}{p^4 + 15p^3 + 71p^2 + 105p}$$

(b) A more efficient realization.

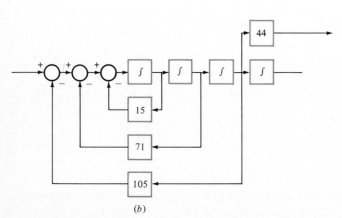

(b)

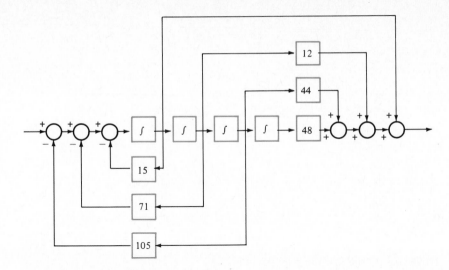

figure 8.5.7
The controller-form realization of the
transfer function of Example 8.5.3

$$H(p) = \frac{p^3 + 12p^2 + 44p + 48}{p^4 + 15p^3 + 71p^2 + 105p}$$

can be built as shown in Figure 8.5.6a. Since the input signal to an integrator is the derivative of the output, the need for the differentiator in Figure 8.5.6a is eliminated if we take the signal from the input of the last integrator rather than the output (Figure 8.5.6b). The complete realization of the transfer function is shown in Figure 8.5.7.

In the controller form of the system block diagram the *input enters each integrator either directly or by way of some previous integrations* but not through other linear operations.

Another interesting block-diagram form of any transfer function is the so called observer form, which, given $H(p)$, we can derive as follows. As with the controller form, once we have done this for a particular $H(p)$, we can jump to the final form of the block diagram. We shall develop the method using a third-order system since it is a simple matter to extrapolate our answer to higher-order systems. Suppose

OBSERVER FORM

$$H(p) = \frac{b_1 p^2 + b_2 p + b_3}{p^3 + a_1 p^2 + a_2 p + a_3}$$

or, in terms of input $x(t)$ and output $y(t)$,

$$(p^3 + a_1 p^2 + a_2 p + a_3)y = (b_1 p^2 + b_2 p + b_3)x$$

Divide through by the highest power of p

$$(1 + a_1 p^{-1} + a_2 p^{-2} + a_3 p^{-3})y = (b_1 p^{-1} + b_2 p^{-2} + b_3 p^{-3})x$$

Take all terms except y itself to one side

$$
\begin{aligned}
y &= -a_1 p^{-1}y - a_2 p^{-2}y - a_3 p^{-3}y + b_1 p^{-1}x + b_2 p^{-2}x + b_3 p^{-3}x \\
&= p^{-1}(b_1 x - a_1 y) + p^{-2}(b_2 x - a_2 y) + p^{-3}(b_3 x - a_3 y) \\
&= p^{-1}\{b_1 x - a_1 y + p^{-1}[b_2 x - a_2 y + p^{-1}(b_3 x - a_3 y)]\}
\end{aligned}
$$

figure 8.5.8
The observer form of

$$H(p) = \frac{b_1 p^2 + b_2 p + b_3}{p^3 + a_1 p^2 + a_2 p + a_3}$$

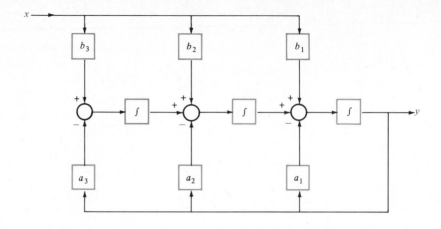

This suggests the form of Figure 8.5.8.

EXAMPLE 8.5.4

Find an observer-form realization of the $H(p)$ used in Example 8.5.3, namely,

$$H(p) = \frac{p^3 + 12p^2 + 44p + 48}{p^4 + 15p^3 + 71p^2 + 105p}$$

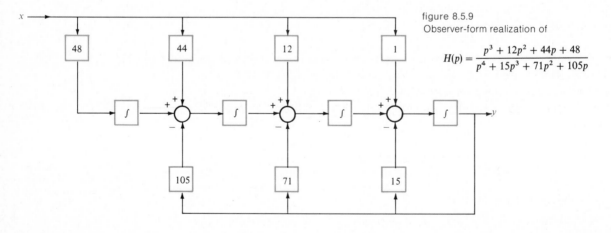

figure 8.5.9
Observer-form realization of

$$H(p) = \frac{p^3 + 12p^2 + 44p + 48}{p^4 + 15p^3 + 71p^2 + 105p}$$

ANS.: See Figure 8.5.9.

Notice that in the observer form of the system block diagram the *output of every integrator affects the output either directly* or through other integrators. No linear sums of integrator outputs contribute to the output in this block diagram.

Transfer functions can be expanded into a *partial-fraction expansion*. When the denominator has only simple factors (real eigenvalues), this expansion is of the form

PARALLEL (SUM, DIAGONAL, FOSTER'S) FORM

$$H(p) = \frac{k_1}{p + a_1} + \frac{k_2}{p + a_2} + \cdots$$

where the k_i are constants and the $p + a_i$ are factors of the denominator of $H(p)$.

EXAMPLE 8.5.5

Construct a parallel-form realization for the $H(p)$ used in Examples 8.5.3 and 8.5.4, namely,

$$H(p) = \frac{p^3 + 12p^2 + 44p + 48}{p^4 + 15p^3 + 71p^2 + 105p} \tag{8.5.11}$$

ANS.:   Since the roots of the denominator (eigenvalues) are $0, -3, -5,$ and $-7,$

$$H(p) = \frac{k_1}{p} + \frac{k_2}{p + 3} + \frac{k_3}{p + 5} + \frac{k_4}{p + 7} \tag{8.5.12}$$

To find k_i, multiply through by $p + a_i$ and then evaluate each term at $p = -a_i$. In other words, for k_1 we multiply equation (8.5.12) by p and get

$$pH(p) = k_1 + \frac{pk_2}{p + 3} + \frac{pk_3}{p + 5} + \frac{pk_4}{p + 7} \tag{8.5.13}$$

Now evaluate equation (8.5.13) at $p = 0$. All terms on the right except for the one with k_1 in it will go to zero

$$\left. \frac{\not{p}(p^3 + 12p^2 + 44p + 48)}{\not{p}(p + 3)(p + 5)(p + 7)} \right|_{p=0} = k_1 + 0 + 0 + 0$$

Thus

$$k_1 = \tfrac{48}{105} \tag{8.5.14}$$

Similarly, for k_2,

$$\left. \frac{(p + 3)(p^3 + 12p^2 + 44p + 48)}{p(p + 3)(p + 5)(p + 7)} \right|_{p=-3} = 0 + k_2 + 0 + 0$$

$$k_2 = \frac{-27 + 12(9) + 44(-3) + 48}{-3(2)(4)} = \frac{-3}{-24} = \frac{1}{8} \tag{8.5.15}$$

and for k_3

$$\left. \frac{(p + 5)(p^3 + 12p^2 + 44p + 48)}{(p + 3)(p + 5)(p + 7)} \right|_{p=-5} = 0 + 0 + k_3 + 0$$

$$\frac{-125 + 12(25) + 44(-5) + 48}{-5(-2)(2)} = k_3$$

$$k_3 = \tfrac{3}{20} \tag{8.5.16}$$

and finally for k_4

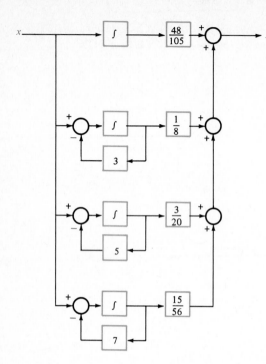

figure 8.5.10
The parallel-form realization of the $H(p)$ transfer function of Example 8.5.5.

$$\frac{(p+7)(p^3 + 12p^2 + 44p + 48)}{p(p+3)(p+5)(p+7)}\bigg|_{p=-7} = 0 + 0 + 0 + k_4$$

$$\frac{-343 + 12(49) + 44(-7) + 48}{-7(-4)(-2)} = k_4$$

$$k_4 = \tfrac{15}{56} \qquad (8.5.17)$$

Thus, inserting (8.5.14) to (8.5.17) into (8.5.13), we get

$$H(p) = \frac{48/105}{p} + \frac{1/8}{p+3} + \frac{3/20}{p+5} + \frac{15/56}{p+7} \qquad (8.5.18)$$

the realization of which is shown in Figure 8.5.10.

Repeated roots in the denominator polynomial can be handled using a special case of the parallel form known as *Jordan's form*.

EXAMPLE 8.5.6

Find Jordan's form for

$$H(p) = \frac{p+4}{(p+3)^2(p+7)} \qquad (8.5.19)$$

ANS.: We use a partial-fraction expansion similar to equation (8.5.12) except that there is a *separate term for every power* of the repeated factor.

$$H(p) = \frac{k_1}{(p+3)^2} + \frac{k_2}{p+3} + \frac{k_3}{p+7} \tag{8.5.20}$$

We can find k_1 and k_3 as before, but k_2 presents a different problem. Let us find k_1 first by multiplying through equation (8.5.20) by $(p+3)^2$:

$$\frac{(p+4)\cancel{(p+3)^2}}{\cancel{(p+3)^2}(p+7)} = k_1 + (p+3)k_2 + \frac{(p+3)^2}{p+7}k_3 \tag{8.5.21}$$

Now evaluate (8.5.21) at $p = -3$:

$$\tfrac{1}{4} = k_1 + 0 + 0$$

or

$$k_1 = \tfrac{1}{4} \tag{8.5.22}$$

Now to find k_2, differentiate (8.5.21) with respect to p, using the rule for a ratio of functions

$$\frac{(p+7)(1) - (p+4)(1)}{(p+7)^2} = k_2 + \frac{(p+7)2(p+3) - (p+3)^2(1)}{(p+7)^2}k_3 \tag{8.5.23}$$

Evaluating equation (8.5.23) at $p = -3$, we get

$$\frac{4-1}{4^2} = k_2 + 0$$

or

$$\tfrac{3}{16} = k_2 \tag{8.5.24}$$

Then, as usual, we find k_3 from (8.5.20) by multiplying by $p + 7$

$$\frac{(p+4)\cancel{(p+7)}}{(p+3)^2\cancel{(p+7)}} = \frac{k_1(p+7)}{(p+3)^2} + \frac{k_2(p+7)}{p+3} + k_3\frac{\cancel{p+7}}{\cancel{p+7}}$$

and evaluating at $p = -7$, we get

$$\frac{-3}{(-4)^2} = 0 + 0 + k_3$$

or

$$k_3 = -\tfrac{3}{16} \tag{8.5.25}$$

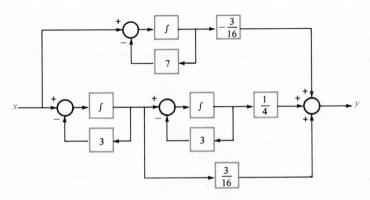

figure 8.5.11
A Jordan's-form realization of a transfer function having repeated eigenvalues.

Inserting (8.5.22), (8.5.24), and (8.5.25) into (8.5.20) gives

$$H(p) = \frac{1/4}{(p+3)^2} + \frac{3/16}{p+3} + \frac{-3/16}{p+7} \tag{8.5.26}$$

The realization is as shown in Figure 8.5.11.

Note that in both the observer form and the controller form the coefficients in the transfer function appear directly in the block diagrams. Unfortunately this is not the case in the parallel form where it is the coefficients in the partial-fraction expansion that appear. Although this adds another step of work, it means that the eigenvalues appear (in the local feedback loops) in the block diagram, which is very useful in certain applications.

If the *eigenvalues* are *complex conjugates*, we resort to a clever subterfuge to get a parallel form. Suppose we have a transfer function $H(p)$

$$H(p) = \frac{ap + b}{cp^2 + dp + e} \tag{8.5.27}$$

where a, b, c, d, and e are real constants such that $d^2 - 4ce < 0$ and the eigenvalues are complex. We complete the square in the denominator, obtaining $H(p)$ in the form

$$H(p) = \frac{\alpha p + \beta}{(p + \gamma)^2 + \delta^2} \tag{8.5.28}$$

This is all we have to do whenever we have a numerical example, because now we can jump to the final form of the answer. Since we do not yet know what that form is, let us derive it as follows. Divide through the top and bottom of equation (8.5.28) by $(p + \gamma)^2$

$$H(p) = \frac{(\alpha p + \beta)/(p + \gamma)^2}{1 + [\delta/(p + \gamma)]^2} \tag{8.5.29}$$

Now a partial-fraction expansion on the *numerator* of equation (8.5.29) gives

$$\frac{\alpha p + \beta}{(p + \gamma)^2} = \frac{k_1}{(p + \gamma)^2} + \frac{k_2}{p + \gamma} \tag{8.5.30}$$

Find k_1 (as usual) by multiplying (8.5.30) through by $(p + \gamma)^2$ and evaluating the result at $p = -\gamma$, that is,

$$\alpha p + \beta = k_1 + \frac{(p + \gamma)^2 k_2}{p + \gamma} \tag{8.5.31}$$

Setting $p = -\gamma$ leads to

$$\alpha(-\gamma) + \beta = k_1 + 0$$

and so

$$k_1 = \beta - \alpha\gamma \tag{8.5.32}$$

Taking the derivative of (8.5.31) with respect to p yields

$$\alpha = k_2 \tag{8.5.33}$$

Inserting (8.5.32) and (8.5.33) into (8.5.30) yields

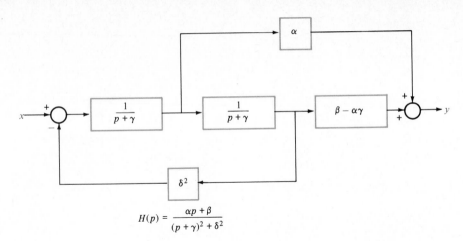

figure 8.5.12
The form for complex eigenvalues.

$$H(p) = \frac{\alpha p + \beta}{(p + \gamma)^2 + \delta^2}$$

$$\frac{\alpha p + \beta}{(p + \gamma)^2} = \frac{\beta - \alpha \gamma}{(p + \gamma)^2} + \frac{\alpha}{p + \gamma} \qquad (8.5.34)$$

Finally, inserting (8.5.34) into (8.5.29) leads to

$$H(p) = \frac{(\beta - \alpha \gamma)/(p + \gamma)^2}{1 + \delta^2/(p + \gamma)^2} + \frac{\alpha/(p + \gamma)}{1 + \delta^2/(p + \gamma)^2}$$

which can be realized by the form shown in Figure 8.5.12.

EXAMPLE 8.5.7

Develop a parallel-form realization for

$$H(p) = \frac{4p^2 + 21p + 38}{p^3 + 7p^2 + 18p + 12}$$

ANS.:   First factor the denominator

$$H(p) = \frac{4p^2 + 21p + 38}{(p^2 + 6p + 12)(p + 1)} = \frac{N(p)}{p^2 + 6p + 12} + \frac{k}{p + 1}$$

Now find k as usual by multiplying through by $p + 1$

$$\frac{4p^2 + 21p + 38}{(p^2 + 6p + 12)\cancel{(p+1)}} \cancel{(p+1)} = \frac{N(p)(p + 1)}{p^2 + 6p + 12} + k$$

and evaluating at $p = -1$

$$\frac{4(1) - 21 + 38}{1 - 6 + 12} = 0 + k \qquad k = \frac{21}{7} = 3$$

We find $N(p)$ by subtracting

$$\frac{4p^2 + 21p + 38}{(p^2 + 6p + 12)(p + 1)} - \frac{3}{p + 1} = \frac{N(p)}{p^2 + 6p + 12}$$

and multiplying through by the quadratic term

$$N(p) = \frac{(4p^2 + 21p + 38) - 3(p^2 + 6p + 12)}{p + 1} = \frac{p^2 + 3p + 2}{p + 1}$$

which is found by long division

$$
\begin{array}{r}
p + 2 = N(p) \\
p + 1\overline{)p^2 + 3p + 2} \\
\underline{p^2 + p} \\
2p + 2 \\
\underline{2p + 2} \\
0
\end{array}
$$

so that

$$H(p) = \frac{p + 2}{p^2 + 6p + 12} + \frac{3}{p + 1}$$

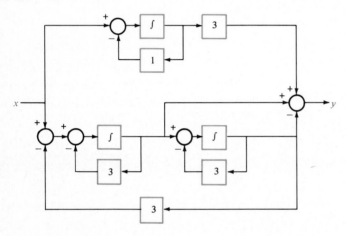

figure 8.5.13
The parallel-form realization of the transfer function of Example 8.5.7.

Completing the square on the first-term denominator gives

$$H(p) = \frac{p + 2}{(p^2 + 6p + 9) + 3} + \frac{3}{p + 1} = \frac{p + 2}{(p + 3)^2 + (\sqrt{3})^2} + \frac{3}{p + 1}$$

By analogy with equation (8.5.28) we note that

$$\alpha = 1 \qquad \beta = 2 \qquad \gamma = 3 \qquad \delta = \sqrt{3}$$

and we have the realization of Figure 8.5.13.

8.6 simulation

When we actually interconnect a set of op-amp integrators, adders, and amplifiers in one of the forms discussed in the previous section, it is called an *analog-computer model* of the actual system. In Chapter 9 we shall see that these various forms of block-diagram models are of paramount importance in obtaining *digital-computer models* of systems.

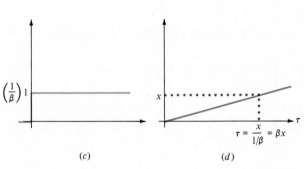

figure 8.6.1
(*a*) Function to be integrated. (*b*) Its integral. (*c*) The function that will be integrated if integrator inputs are multiplied by 1/β. (*d*) The output from such an integrator.

In real life things often happen very fast. System variables in many electric, mechanical, hydraulic, etc., systems vary rapidly in time. It would be convenient in our analog-computer model to be able to slow time down—to make the model work in slow motion.

Such time scaling is simply accomplished by multiplying the input to every integrator by the factor 1/β, where

$$\tau = \beta t \qquad (8.6.1)$$

in which τ is the model time and t the real time, both in seconds. Hence $\beta > 1$ (or $1/\beta < 1$) for slowing down a problem, and $\beta < 1$ (or $1/\beta > 1$) for speeding up a problem. The effect of multiplying the integrator input by 1/β is shown in Figure 8.6.1. In real time if a unit-step function is integrated, the integrator output reaches a value x at time $t = x$. If the input is multiplied by 1/β before being integrated, the output will reach that same value x at time $t = \beta x$. Of course, if $\beta > 1$, that will be a *later* time than before and we have slowed the model down.

summary **8.7**

In this chapter we have seen what an op amp is and examined its varied uses as a circuit element. Although its two basic forms are as inverting and noninverting voltage amplifiers, the variations on these basic themes are almost limitless. Current-to-voltage converters, voltage followers, Schmitt triggers, and high-gain differential amplifiers (to name only a few of the applications discussed) are all extremely useful in designing circuits for processing signals. Ideal and/or lossy integrators are also useful in active filter design; we shall make use of them as well as the summing amplifier when we discuss that topic in Chapter 12.

The functions provided by op amps are concisely and neatly described by block diagrams and the laws governing their interconnection (called block-diagram algebra). Such diagrams are used to describe systems, to design filters, and to determine the state equations of a network. We shall use block diagrams to do all these things: as a matter of fact our first use of them occurs in the very next chapter.

problems

1. In Figure P8.1 find v_o as a function of v_1, v_2, v_3, and the resistor values.

2. Find a single expression for $v_o(t)$ in Figure P8.2 in terms of ramps, steps, and impulse functions. Sketch $v_o(t)$ versus t.

3. In Figure P8.3 find an expression for $v_o(t)$ in terms of the two input voltages ($v_{in(1)}$ and $v_{in(2)}$).

4. (a) In the circuit of Figure P8.4 find the output voltage v_o in terms of v_1 and v_2. (b) If $v_1 = v_2$ and $R_1 = R_2 = C = 1$, what is v_o in terms of the input voltage?

5. Determine the p-operator transfer function of the network shown in Figure P8.5.

6. Suppose a variable-resistance strain gauge is used to replace the feedback resistor in a normal inverting amplifier (with infinite open-loop gain). Find an expression for the

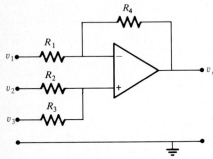

figure P8.1

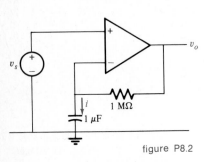

figure P8.2

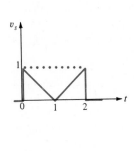

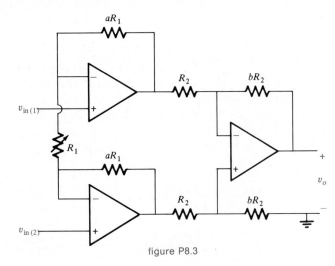

figure P8.3

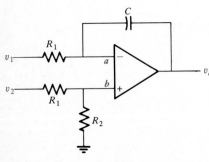

figure P8.4

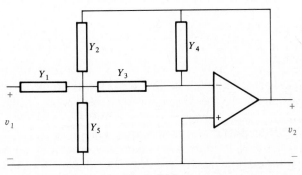

figure P8.5

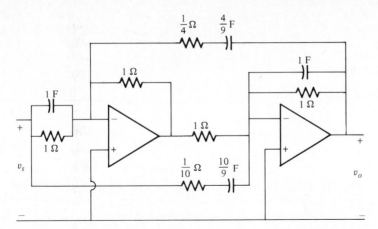

figure P8.13

output voltage v_o in terms of $\gamma = \Delta R_2/R_2$. Why might the strain-gauge amplifier described in Example 8.2.4 be preferable to this one?

7. A T network of resistors is used in the feedback loop of an inverting amplifier, as shown in Figure 8.2.9; $R_a = R_b = R_4 = 10$ kΩ, and $R_c = 100$ Ω. If all these resistors have a ± 10 percent tolerance, what is the maximum possible variation in the magnitude of the gain of the amplifier?

8. Design an inverting amplifier whose overall gain is -200 and whose input resistance (as seen by the external source) is 10 kΩ. Use 10- and 50-kΩ resistors and a 2-kΩ variable resistor. Assume an ideal op amp. (a) What is the correct setting for the variable resistor? (b) If the overall amplifier is putting out 10 V to an open circuit, what is the value of the op amp's output current?

9. Use op amps to design a Schmitt-trigger circuit that will produce a dead zone of ± 1 V. The only available voltages are $+14$ and -9 V. Assume that the op-amp's output voltage can reach $\pm V_{cc}$. Use a 22-kΩ resistor for R_2 in Figure 8.2.11. Use another op-amp circuit to obtain v_{ref}.

10. Sketch v_{in} versus v_o of an ideal Schmitt trigger and by including arrows show in what direction the plot is being traversed. Label UTP, LTP, $+v_{\text{sat}}$, and $-v_{\text{sat}}$. What is the general name given to this shape of input-output characteristic?

11. In Figure 8.3.5a, $R_1 = \frac{1}{2}\,\Omega$, $R_2 = 1\,\Omega$, and $C = 1$ F. If $v_1(t) = u(t)$, find the zero-state response $v_o(t)$.

12. A resistor, $R_1 = 2\,\Omega$, is in series with the parallel combination of $R_2 = 3\,\Omega$ and $C = \frac{1}{2}$ F. This overall circuit is driven by a step-function voltage source whose magnitude is 10 V. Using only one op amp, design an analog simulator whose output voltage will be equal to the voltage across the capacitor. Use a 10-μF capacitor in the simulator.

13. Find the p-operator transfer function of the op-amp network in Figure P8.13.

14. Derive the open-circuit voltage gain [equation (8.4.33)] for a noninverting amplifier whose op amp has finite gain.

15. Derive the input resistance [equation (8.4.35)] for a noninverting amplifier whose op amp has finite gain.

16. Derive the output resistance [equation (8.4.37)] for a noninverting amplifier whose op amp has finite gain.

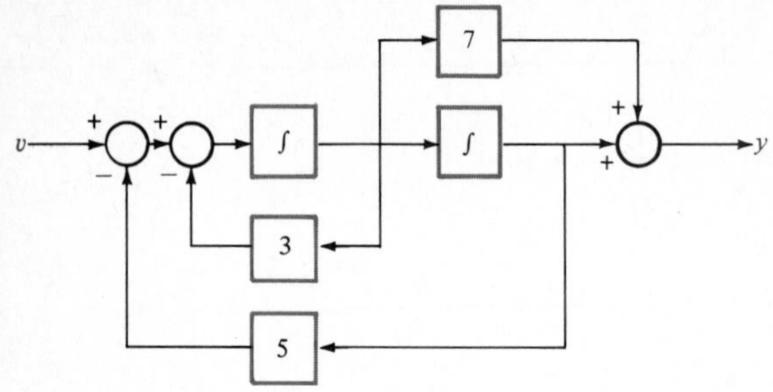

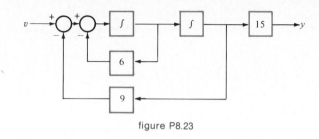

figure P8.22

figure P8.23

17. The op amp in an inverting single-input amplifier (Figure 8.2.2) has finite open-circuit gain A. Find an expression for the current i_2 in the feedback resistor as a function of A and input-voltage source v_i. What is the necessary restriction on R_2 if the magnitude of i_2 is to be independent of the value of R_2?

18. A noninverting amplifier (Figure 8.2.3) incorporates an op amp with finite open-loop gain A. What is the restriction on the value of feedback resistor R_2 if the current in R_2 is to be independent of R_2?

19. An inverting amplifier uses an infinite-gain op amp whose input bias current is 50 μA. A 220-kΩ feedback resistor produces an overall voltage gain of 10. (a) What value of resistance should be placed in series with the *positive* input terminal? (b) If the action suggested in (a) is not taken, what dc offset level will be observed at the output terminals? (c) If the compensating resistor *is* used and the input offset current is 4 μA (so that $I_a = 52 \mu$A), what will the value of the resulting quiescent output voltage be?

20. (a) Find the overall voltage gain of the finite-open-loop-gain inverting amplifier of Figure 8.4.3a. Include the effect of the possibly finite input resistance R_{ab} of the op amp itself. (b) If $A = 10^5$, $R_1 = 10$ kΩ, $R_2 = 100$ kΩ, and $R_{ab} = 100$ kΩ, find the resulting overall voltage gain. (c) Suppose that in part (b) A is infinite and $R_{ab} = 1$ kΩ. Find the overall gain.

21. Find a block diagram of the system described by the differential equation. $dy/dt + 4y = v$, where v is input and y output. Include in your diagram *only* pure integrators, pure gains, and summing junctions.

22. Find the single differential equation, in d/dt form, that describes the system shown in Figure P8.22.

23. Leaving coefficients undetermined, find the form of the natural response $y_n(t)$ in the system shown in Figure P8.23.

24. (a) Sketch a controller-form block diagram for the circuit shown in Figure P8.24. Consider $i_1(t)$ to be the input and $v_1(t)$ to be the output. (b) Repeat part (a) with $v_C(t)$ as the output. Can you combine parts (a) and (b) on a single diagram with two outputs?

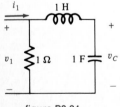

figure P8.24

25. (a) Given the transfer function

$$H(p) = \frac{4p}{p^2 + 2p + 4}$$

find a controller-form block diagram that will model this system. (b) Repeat part (a) but find an observer form.

26. Find the transfer function $H(p)$ of the block diagram in Figure P8.26.

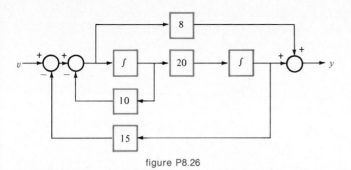

figure P8.26

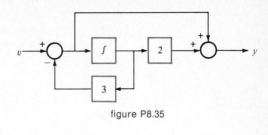

figure P8.35

27. Design an op-amp circuit containing only inverting op amps that will perform the function of the block diagram shown in Figure 8.5.3.

28. For the circuit described in the drill problem at the end of Section 5.1, find (a) the controller-form block-diagram realization and (b) the observer form. (c) What can you say about these two forms of this system? Anything unusual?

29. Find the controller form of the system described by Figure 5.3.2. Include each of the three outputs on one diagram.

30. Consider the mechanical system described in Example 5.3.3. With input $f_b(t)$ the differential equation was found to be $(p^4 + p^3 + 4p^2 + 2p + 3)\mathcal{V}_b = (p^3 + p^2 + 2p)f_b$. Find (a) the controller form and (b) the observer form of the block diagram.

31. Find the parallel form of the block diagram for

$$H(p) = \frac{1000p}{(p + 1)(p + 10)(p + 100)}$$

32. The differential equation that describes the loudspeaker of Figure 5.3.10 is given in equation (5.3.67). If, to minimize computational complexity, we allow the values of all the elements in that problem to be 1, the equation becomes

$$(p^2 + p + 1)i = (p + 1)v_s$$

where $i(t)$ is the current that flows into the loudspeaker in response to the applied voltage $v_s(t)$. Find a block diagram for this system.

33. Find a parallel-form block diagram for the transfer function

$$H(p) = \frac{3p^2 + 14p + 51}{p^3 + 7p^2 + 31p + 25}$$

Hint: The system's natural response contains the term Ae^{-t}.

34. Find a block diagram for the transfer function

$$H(p) = \frac{(p + 2)^2}{(p + 1)^3}$$

35. What is the *p*-operator transfer function $H(p)$ of the block diagram in Figure P8.35, which is called *Guillemin's iterative form*? (b) Using this approach, develop a block diagram for the transfer function

$$H(p) = \frac{(p + 2)(p + 4)(p + 6)}{(p + 3)(p + 5)(p + 7)}$$

36. A system is described by the following pair of simultaneous differential equations, for which $v(t)$ is the input, $y(t)$ is the output, and $x(t)$ is an internal intermediate variable. Develop a block diagram that models this system

$$(p^2 + 5)y = 3x$$

$$(p^2 + 7)x = 5y + 3v$$

37. Given the transfer function

$$H(p) = \frac{Kp}{(p + a)(p + b)}$$

(a) Find a parallel-form block diagram for the system. (b) Show in the block diagram how to speed up the response of this network by a factor of 1000.

chapter 9
STATE-VARIABLE ANALYSIS

introduction 9.1

In this chapter we discuss another method of describing a circuit or system. We have already discussed how to write system differential equations and how to solve them; we know about complete responses and transfer functions, about zero-input responses and convolution. Why do we need another method to describe systems and their behavior? The answer is that the methods we have looked at so far (1) are not applicable to nonlinear systems, (2) are not easily adapted to computer solution, and (3) do not yield much information about the system unless the solution is carried out completely and in detail. A method with all these missing properties, called *state-variable analysis*, affords the engineer insight into the behavior of a system that earlier methods do not. It is readily applied to nonlinear and time-varying systems and results in a set of equations that can be easily programmed into a digital computer.

The essence of state-variable analysis is that it is a technique for determining what state a system is in, where the word *state* means the same as when a physician reports that the state of a patient is poor, or the President makes his annual speech on the state of the Union. An airplane in a tailspin is in a very different state from that of normal flight.

Recall that a system is a collection of interconnected devices or elements designed to accomplish certain objectives. If the energy stored in any one of those physical elements changes, the system is in a different state. For electric systems, the amount of energy stored at any instant in each type of element is given by

$$W_R = 0 \qquad W_L = \tfrac{1}{2}Li^2 \qquad W_C = \tfrac{1}{2}Cv^2$$

This suggests that we could choose the current in each inductor and the voltage across each capacitor in the network as the set of variables to describe the state of the electric circuit (and, by analogy, forces in springs and velocities of masses, etc.).

In general there is more than one possible set of such *state variables* for any given system. We must choose a set of variables in such a way that if we know the value of each variable, we can calculate the amount of energy stored in the given system and how this energy is distributed among the network elements at any instant. At any instant of time, a given distribution of stored energy in the system's elements uniquely determines the location of a point in a coordinate system. Each axis of this coordinate system is labeled with a different state variable. As time goes on and the amount of energy stored in each storage element changes, the point moves, describing a path, locus, or *trajectory* in the coordinate system. The coordinate system is called a *state space* because the instantaneous location of a point in that space specifies exactly how much energy is stored in each element at that instant and thus describes the state of the system. A vector that always reaches from the origin to the point is called the *state vector*. As the point moves in the state space, the state vector changes in both magnitude and direction.

EXAMPLE 9.1.1

Given the circuit shown in Figure 9.1.1, plot the state-space trajectory.

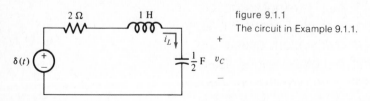

figure 9.1.1
The circuit in Example 9.1.1.

ANS.: The solutions for inductor current i_L and capacitor voltage v_C are

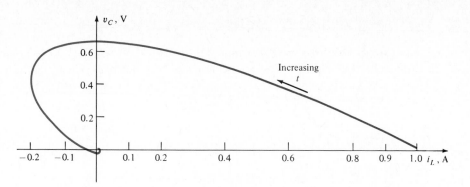

figure 9.1.2
State-space trajectory for the
circuit in Example 9.1.1.

$$i_L(t) = \sqrt{2}e^{-t}\cos\left(t + \frac{\pi}{4}\right) \qquad \text{and} \qquad v_C(t) = 2e^{-t}\sin t$$

Plotting the instantaneous value of v_C versus the corresponding value of i gives the trajectory of Figure 9.1.2.

Note that the trajectory in Figure 9.1.2 comes to rest at the origin. For any source-free system, a point where the time derivatives of every state variable are simultaneously zero is called an *equilibrium point*. An equilibrium point is said to be *unstable* if a slight displacement of the system vector away from that point results in the system vector's moving still farther away on a new trajectory. An equilibrium point is called *stable* if the system vector stays in the neighborhood of the equilibrium point (or returns to it) after being displaced slightly. In any source-free *linear system, the origin is the only equilibrium point.* Nonlinear systems can have more than one equilibrium point, the location and description of which are always of great interest to the systems engineer.

In this chapter we discuss how to choose state variables and set up state equations in terms of them. We shall see how to solve a set of state equations numerically (by computer). This method works just as well for rather complicated nonlinear systems as for linear circuits. Although finding closed-form solutions of the state equations of a system is always desirable, for nonlinear systems it is often not possible. Even for linear systems the easiest methods for solution involve higher mathematical techniques than the student has probably been exposed to at this point.† The major advantages of the state-variable formulation of a system are as follows:

1. We can learn a great deal about a system without solving the state equations—just by examining certain properties of those equations.
2. The state-variable method is easily applied to higher-order systems and nonlinear systems.
3. State equations lend themselves easily to algorithmic (computer) methods of solution.

† We discuss one such method in Chapter 15.

9.2 writing a set of state equations

State equations follow a specific format. On the left-hand side each state equation has the derivative of one (and only one) of the state variables. On the right-hand side is a mathematical function involving any or all of the state variables and the sources. No derivatives appear on the right-hand side.

EXAMPLE 9.2.1

Write the state equations for the circuit shown in Figure 9.2.1.

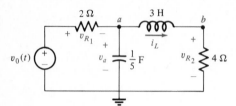

figure 9.2.1
The circuit in Example 9.2.1.

ANS.:   We choose i_L and v_a as state variables. Solve for di_L/dt as follows:

$$v_L = L\frac{di_L}{dt} = v_a - v_b$$

$$L\frac{di_L}{dt} = v_a - 4i_L$$

$$\frac{di_L}{dt} = -\tfrac{4}{3}i_L + \tfrac{1}{3}v_a \tag{9.2.1}$$

and

$$i_C = C\frac{dv_a}{dt} = \frac{v_o(t) - v_a}{2} - i_L$$

$$\frac{dv_a}{dt} = -5i_L - \tfrac{5}{2}v_a + \tfrac{5}{2}v_o(t) \tag{9.2.2}$$

Equations (9.2.1) and (9.2.2) are the state equations for the circuit of Example 9.2.1. Using matrix notation, we can write them as

$$\frac{d}{dt}\begin{bmatrix} i_L \\ v_a \end{bmatrix} = \begin{bmatrix} -4/3 & 1/3 \\ -5 & -5/2 \end{bmatrix}\begin{bmatrix} i_L \\ v_a \end{bmatrix} + \begin{bmatrix} 0 \\ 5/2 \end{bmatrix}v_o(t) \tag{9.2.3}$$

where the scalar d/dt operation on a matrix implies taking the derivative of every element in the matrix.

We are at liberty to specify any variables as the *output variable(s)*, i.e., variables though not necessarily state variables that we are most interested in knowing. Let us arbitrarily choose the two resistor voltages v_{R_1} and v_{R_2}. We call the set of output variables **y**

$$y = \begin{bmatrix} v_{R_1} \\ v_{R_2} \end{bmatrix}$$

and write expressions for them in terms of the state variables and the input sources

$$v_{R_1} = v_o(t) - v_a \qquad v_{R_2} = i_L(4)$$

$$y = \begin{bmatrix} 0 & -1 \\ 4 & 0 \end{bmatrix} \begin{bmatrix} i_L \\ v_a \end{bmatrix} + \begin{bmatrix} 1 \\ 0 \end{bmatrix} v_o(t) \qquad\qquad (9.2.4)$$

These are called the *output equations.*

Any linear system can be characterized by equations such as (9.2.3) and (9.2.4), which in general form† are

$$\boxed{\dfrac{d}{dt}\, \mathbf{x} = \mathbf{A}\mathbf{x} + \mathbf{B}\mathbf{u}} \qquad\qquad (9.2.5)$$

and

$$\boxed{\mathbf{y} = \mathbf{C}\mathbf{x} + \mathbf{D}\mathbf{u}} \qquad\qquad (9.2.6)$$

Equations (9.2.3) and (9.2.4) are in the general form of (9.2.5) and (9.2.6), where

$$\mathbf{A} = \begin{bmatrix} -4/3 & 1/3 \\ -5 & -5/2 \end{bmatrix} \quad \mathbf{B} = \begin{bmatrix} 0 \\ 5/2 \end{bmatrix} \quad \mathbf{C} = \begin{bmatrix} 0 & -1 \\ 4 & 0 \end{bmatrix}$$

$$\mathbf{D} = \begin{bmatrix} 1 \\ 0 \end{bmatrix} \quad \mathbf{x} = \begin{bmatrix} i_L \\ v_a \end{bmatrix} \quad \mathbf{u} = v_o(t)$$

EXAMPLE 9.2.2

Write the state equations for the circuit shown in Figure 9.2.2.

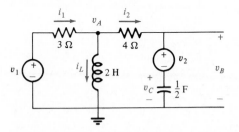

figure 9.2.2
The circuit in Example 9.2.2.

ANS.: There are two outputs, voltages v_A and v_B:

$$\mathbf{x} = \begin{bmatrix} i_L \\ v_C \end{bmatrix} \qquad \mathbf{y} = \begin{bmatrix} v_A \\ v_B \end{bmatrix}$$

---

† The notation here has become standard. The symbol **u** simply denotes the inputs, whatever functions of time they may be. The use of this letter does *not* imply a unit-step function.

First solve for di_L/dt by means of

$$v_L = L\frac{di}{dt} = v_1 - 3i_1 = v_1 - 3(i_L + i_2)$$

$$= v_1 - 3\left[i_L + \frac{v_L - (v_2 + v_C)}{4}\right] = v_1 - 3i_L - \tfrac{3}{4}v_L + \tfrac{3}{4}v_2 + \tfrac{3}{4}v_C$$

Collecting terms in v_L, we have

$$v_L(1 + \tfrac{3}{4}) = -3i_L + \tfrac{3}{4}v_C + v_1 + \tfrac{3}{4}v_2$$

$$v_L = 2\frac{di}{dt} = -\tfrac{12}{7}i_L + \tfrac{3}{7}v_C + \tfrac{4}{7}v_1 + \tfrac{3}{7}v_2 \tag{9.2.7}$$

and the first state equation is

$$\frac{di}{dt} = -\tfrac{6}{7}i_L + \tfrac{3}{14}v_C + \tfrac{2}{7}v_1 + \tfrac{3}{14}v_2 \tag{9.2.8}$$

And for dv_C/dt we solve for

$$i_C = C\frac{dv_C}{dt} = i_2 = i_1 - i_L = \frac{v_1 - v_L}{3} - i_L$$

Substituting equation (9.2.7) for v_L gives

$$i_C = \frac{v_1}{3} - \tfrac{1}{3}(-\tfrac{12}{7}i_L + \tfrac{3}{7}v_C + \tfrac{4}{7}v_1 + \tfrac{3}{7}v_2) - i_L$$

collecting terms yields

$$i_C = -\tfrac{3}{7}i_L - \tfrac{1}{7}v_C + \tfrac{1}{7}v_1 - \tfrac{1}{7}v_2$$

and dividing through by $C = \tfrac{1}{2}$ F leads to

$$\frac{dv_C}{dt} = -\tfrac{6}{7}i_L - \tfrac{2}{7}v_C + \tfrac{2}{7}v_1 - \tfrac{2}{7}v_2$$

$$\begin{bmatrix} \dfrac{di_L}{dt} \\ \dfrac{dv_C}{dt} \end{bmatrix} = \begin{bmatrix} -6/7 & 3/14 \\ -6/7 & -2/7 \end{bmatrix}\begin{bmatrix} i_L \\ v_C \end{bmatrix} + \begin{bmatrix} 2/7 & 3/14 \\ 2/7 & -2/7 \end{bmatrix}\begin{bmatrix} v_1 \\ v_2 \end{bmatrix}$$

Thus

$$\mathbf{x} = \begin{bmatrix} i_L \\ v_C \end{bmatrix} \qquad \mathbf{A} = \begin{bmatrix} -6/7 & 3/14 \\ -6/7 & -2/7 \end{bmatrix} \qquad \mathbf{B} = \begin{bmatrix} 2/7 & 3/14 \\ 2/7 & -2/7 \end{bmatrix} \qquad \mathbf{u} = \begin{bmatrix} v_1 \\ v_2 \end{bmatrix}$$

and, since $y_1 = v_A = v_L$ and $y_2 = v_B = v_C + v_2$,

$$\begin{bmatrix} y_1 \\ y_2 \end{bmatrix} = \begin{bmatrix} -12/7 & 3/7 \\ 0 & 1 \end{bmatrix}\begin{bmatrix} i_L \\ v_C \end{bmatrix} + \begin{bmatrix} 4/7 & 3/7 \\ 0 & 1 \end{bmatrix}\begin{bmatrix} v_1 \\ v_2 \end{bmatrix}$$

and so

$$\mathbf{C} = \begin{bmatrix} -12/7 & 3/7 \\ 0 & 1 \end{bmatrix} \quad \text{and} \quad \mathbf{D} = \begin{bmatrix} 4/7 & 3/7 \\ 0 & 1 \end{bmatrix}$$

EXAMPLE 9.2.3

Write the state equations for the loudspeaker in Fig. 9.2.3. Include the effects of coil resistance, coil inductance, voltage induced by speaker movement, springiness of the speaker diaphram, and the damping and
mass of the speaker. The electromechanical schematic is shown in Fig. 9.2.3*b*.

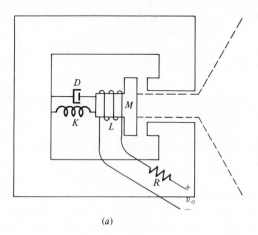

figure 9.2.3
(*a*) A loudspeaker and (*b*) the schematic diagram for
this electromechanical system.

(*a*)

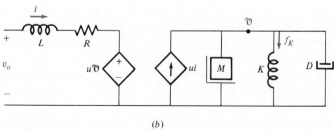

(*b*)

ANS.:   We choose as state variables the current in the inductor, the velocity of the mass, and the force in the
spring:

$$\mathbf{x} = \begin{bmatrix} i \\ \mathcal{V} \\ f_K \end{bmatrix} = \begin{bmatrix} x_1 \\ x_2 \\ x_3 \end{bmatrix}$$

so that

$$v_L = L\frac{di}{dt} = v_o - iR - u\mathcal{V}$$

and

$$\frac{di}{dt} = -i\frac{R}{L} - \frac{u}{L}\mathcal{V} + \frac{v_o}{L} \quad \text{or} \quad \frac{dx_1}{dt} = -\frac{R}{L}x_1 - \frac{u}{L}x_2 + \frac{v_o}{L}$$

Also

$$f_M = M\frac{d\mathcal{V}}{dt} = ui - f_K - D\mathcal{V}$$

$$\frac{d\mathcal{U}}{dt} = +\frac{u}{M}\, i - \frac{D}{M}\, \mathcal{U} - \frac{1}{M}\, f_K \qquad \frac{dx_2}{dt} = \frac{u}{M}\, x_1 - \frac{D}{M}\, x_2 - \frac{1}{M}\, x_3$$

and last

$$\mathcal{U} = K\, \frac{df_K}{dt} \qquad \frac{dx_3}{dt} = \frac{1}{K}\, x_2$$

Hence in matrix form

$$\frac{d}{dt} \begin{bmatrix} x_1 \\ x_2 \\ x_3 \end{bmatrix} = \begin{bmatrix} -\dfrac{R}{L} & -\dfrac{u}{L} & 0 \\ \dfrac{u}{M} & -\dfrac{D}{M} & -\dfrac{1}{M} \\ 0 & \dfrac{1}{K} & 0 \end{bmatrix} \begin{bmatrix} x_1 \\ x_2 \\ x_3 \end{bmatrix} + \begin{bmatrix} \dfrac{1}{L} \\ 0 \\ 0 \end{bmatrix} v_o$$

9.3 selecting different sets of state variables

The location of a moving point in state space can be described by coordinate systems other than the usual cartesian system, in which all axes are orthogonal (at right angles to each other). When we describe a point on a given trajectory by a different cartesian system in which axes are not orthogonal or using spherical or cylindrical coordinates, it remains the same system and the same general trajectory even though the variables differ from one coordinate system to another. Consequently there are many different sets of state variables that can be used to describe any system. A set of state variables is any set of quantities by which the location of the point is uniquely specified in state space. As already mentioned, one set of variables that (almost) always describes any electric network is the set made up of every capacitor voltage and every inductor current.† But for any given system more than one set of state variables is possible. Any set of state variables for a given system has the same number of variables (the degrees of freedom or order of the system), and these *variables are independent* of each other. In other words, *different initial conditions can be assigned separately to each state variable without affecting the initial conditions of any other variable in that set.*

figure 9.3.1
The circuit in Example 9.3.1.

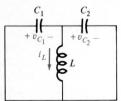

EXAMPLE 9.3.1

How many state variables does it take to uniquely describe the system shown in Fig. 9.3.1?
ANS.: Two different initial conditions cannot be placed on C_1 and C_2 independently. Obviously from KVL around the outside loop $v_{C_1}(0-) = -v_{C_2}(0-)$, and so one possible choice would be to choose v_{C_1} and i_L.

---

† If there is a node to which two or more capacitors and no resistors are connected, or if there is a loop of nothing but inductors, fewer state variables are necessary.

It is convenient to describe the point in state space by the length and direction of a vector from the origin to the point. We can use the notation of a column matrix (vector) to list these coordinate values at any instant t:

$$\mathbf{x}(t) = \begin{bmatrix} x_1(t) \\ x_2(t) \\ x_3(t) \\ \vdots \\ x_n(t) \end{bmatrix}$$

where n equals the order (number of degrees of freedom) of the system.

EXAMPLE 9.3.2

Write a possible set of state variables for the systems shown in Fig. 9.3.2.

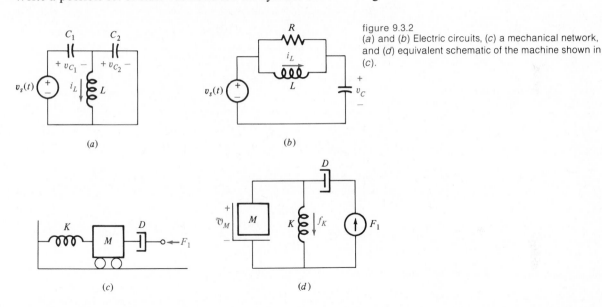

figure 9.3.2
(a) and (b) Electric circuits, (c) a mechanical network, and (d) equivalent schematic of the machine shown in (c).

ANS.:

(a) $\mathbf{x} = \begin{bmatrix} v_{C_2} \\ i_L \end{bmatrix}$ (b) $\mathbf{x} = \begin{bmatrix} i_L \\ v_C \end{bmatrix}$ (c) and (d) $\mathbf{x} = \begin{bmatrix} \mathcal{V}_M \\ f_K \end{bmatrix}$

If we use our knowledge of p-operator notation and the various different methods for block diagraming (see Chapter 8), we can quickly select several different sets of state variables for any given linear system. Each different set of state variables will generate a different $\mathbf{A}$ matrix, some of which have very convenient forms, as we shall see. In order to examine several methods for choosing a set of state variables, suppose that we have found the single linear differential equation that describes the system of interest and then change this differential equation into a transfer function by the p-operator technique.

EXAMPLE 9.3.3

Find the transfer function in p-operator notation given the linear differential equation

$$\frac{d^4y}{dt^4} + 15\frac{d^3y}{dt^3} + 71\frac{d^2y}{dt^2} + 105\frac{dy}{dt} = \frac{d^3u}{dt^3} + 12\frac{d^2u}{dt^2} + 44\frac{du}{dt} + 48u \tag{9.3.1}$$

where $u(t)$ is the input signal and $y(t)$ the output signal.

ANS.:   Using $p \equiv d/dt$ and $p^n \equiv d^n/dt^n$, we can write

$$(p^4 + 15p^3 + 71p^2 + 105p)y = (p^3 + 12p^2 + 44p + 48)u$$

Since the transfer function is defined as the p-operator ratio of output divided by input,

$$G(p) = \frac{y(p)}{u(p)} = \frac{p^3 + 12p^2 + 44p + 48}{p^4 + 15p^3 + 71p^2 + 105p} \tag{9.3.2}$$

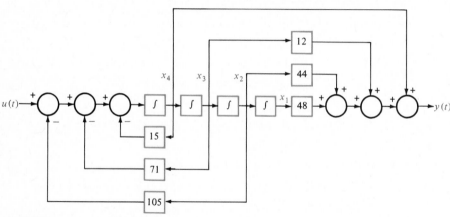

figure 9.3.3
Choosing state variables: Bush's controller form.

To set up a block diagram for this transfer function any of the several forms developed in Chapter 8 will suffice; arbitrarily we pick Bush's controller form (see Example 8.5.3) The result is repeated in Figure 9.3.3.

Select, as state variables, the output of each integrator. In Figure 9.3.3 the state variables are shown. The state and output equations are then simply written down by inspection:

$$\frac{d}{dt}x_1 = x_2 \qquad \frac{d}{dt}x_2 = x_3$$

$$\frac{d}{dt}x_3 = x_4 \qquad \frac{d}{dt}x_4 = -105x_2 - 71x_3 - 15x_4 + u(t)$$

$$y(t) = 48x_1 + 44x_2 + 12x_3 + x_4$$

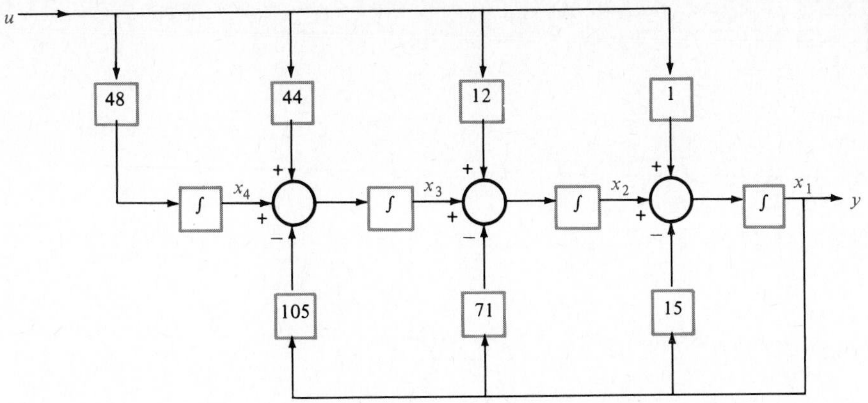

figure 9.3.4
Choosing state variables: the observer form.

which, in matrix notation, is

$$\frac{d}{dt}\mathbf{x} = \mathbf{Ax} + \mathbf{Bu} \qquad y = \mathbf{Cx} + \mathbf{Du} \qquad (9.3.3)$$

where

$$\mathbf{A} = \begin{bmatrix} 0 & 1 & 0 & 0 \\ 0 & 0 & 1 & 0 \\ 0 & 0 & 0 & 1 \\ 0 & -105 & -71 & -15 \end{bmatrix} \qquad \mathbf{B} = \begin{bmatrix} 0 \\ 0 \\ 0 \\ 1 \end{bmatrix} \qquad \mathbf{C} = [48 \quad 44 \quad 12 \quad 1] \qquad \mathbf{D} = \varnothing$$

Another set of state variables can be obtained using a different block-diagram realization of the transfer function, e.g., the *observer form* of equation (9.3.2) in Example 8.5.4. Again defining the output of each integrator to be a state variable (as in Figure 9.3.4) yields

$$\frac{dx_1}{dt} = -15x_1 + x_2 + u \qquad \frac{dx_2}{dt} = -71x_1 + x_3 + 12u$$

$$\frac{dx_3}{dt} = 105x_1 + x_4 + 44u \qquad \frac{dx_4}{dt} = 48u$$

$$y = x_1 \qquad (9.3.4)$$

or

$$\frac{d}{dt}\mathbf{x} = \mathbf{AX} + \mathbf{Bu} \qquad y = \mathbf{Cx} + \mathbf{Du}$$

where

$$\mathbf{A} = \begin{bmatrix} -15 & 1 & 0 & 0 \\ -71 & 0 & 1 & 0 \\ -105 & 0 & 0 & 1 \\ 0 & 0 & 0 & 0 \end{bmatrix} \qquad \mathbf{B} = \begin{bmatrix} 1 \\ 12 \\ 44 \\ 48 \end{bmatrix} \qquad \mathbf{C} = [1 \quad 0 \quad 0 \quad 0] \qquad \mathbf{D} = \varnothing$$

Another realization that yields a useful form of the **A** matrix is Foster's

figure 9.3.5
Choosing state variables: parallel
form

$$G(p) = \frac{p^3 + 12p^2 + 44p + 48}{p^3 + 15p^2 + 71p + 105}$$

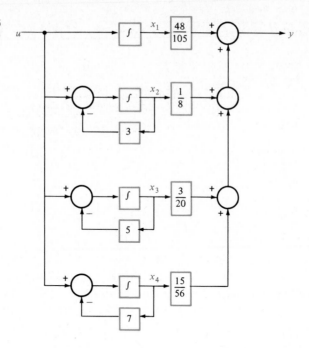

parallel-programming method (see Example 8.5.5). Again choosing the integrator outputs as the state variables (Figure 9.3.5) yields

$$\frac{dx_1}{dt} = u \qquad \frac{dx_2}{dt} = -3x_2 + u$$

$$\frac{dx_3}{dt} = -5x_3 + u \qquad \frac{dx_4}{dt} = -7x_4 + u$$

$$y = \tfrac{48}{105}x_1 + \tfrac{1}{8}x_2 + \tfrac{3}{20}x_3 + \tfrac{15}{56}x_4$$

or

$$\frac{d}{dt}\mathbf{x} = \mathbf{Ax} + \mathbf{Bu} \qquad y = \mathbf{Cx} + \mathbf{Du}$$

where

$$\mathbf{A} = \begin{bmatrix} 0 & 0 & 0 & 0 \\ 0 & -3 & 0 & 0 \\ 0 & 0 & -5 & 0 \\ 0 & 0 & 0 & -7 \end{bmatrix} \qquad \mathbf{B} = \begin{bmatrix} 1 \\ 1 \\ 1 \\ 1 \end{bmatrix}$$

$$\mathbf{C} = [48/105 \quad 1/8 \quad 3/20 \quad 15/56] \qquad \mathbf{D} = \varnothing$$

This simplest form of the $\mathbf{A}$ matrix has several unique uses. Although parallel programming is clearly an important technique for selecting a set of state variables, unfortunately it results only in this simple form of the $\mathbf{A}$ matrix for unrepeated factors in the denominator polynomial. For repeated factors we must use Jordan's form.

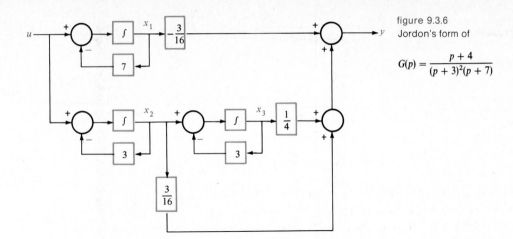

figure 9.3.6
Jordon's form of

$$G(p) = \frac{p + 4}{(p + 3)^2(p + 7)}$$

In Chapter 8 we saw (see Example 8.5.6) that Jordan's form of

$$G(p) = \frac{p + 4}{(p + 3)^2(p + 7)}$$

is as shown in Figure 9.3.6. Choosing integrator outputs as the state variables yields

$$\frac{dx_1}{dt} = -7x_1 + u \qquad \frac{dx_2}{dt} = -3x_2 + u \qquad \frac{dx_3}{dt} = x_2 - 3x_3$$

$$y = -\tfrac{3}{16}x_1 + \tfrac{3}{16}x_2 + \tfrac{1}{4}x_3$$

Therefore, in matrix notation

$$\mathbf{A} = \begin{bmatrix} -7 & 0 & 0 \\ 0 & -3 & 0 \\ 0 & 1 & -3 \end{bmatrix} \quad \mathbf{B} = \begin{bmatrix} 1 \\ 1 \\ 0 \end{bmatrix} \quad \mathbf{C} = [-3/16 \quad 3/16 \quad 1/4] \quad \mathbf{D} = \varnothing$$

Finally we saw in Figure 8.5.12 that to handle the general case of complex-conjugate factors in the denominator of any transfer function we complete the square. Example 8.5.7 was a specific example of a system with one real and two complex-conjugate eigenvalues. Again choosing as the state variables the output of each integrator (Figure 9.3.7) we have

$$\frac{dx_1}{dt} = -x_1 + u \qquad \frac{dx_2}{dt} = -3x_2 + x_3 \qquad \frac{dx_3}{dt} = -3x_2 - 3x_3 + u$$

$$y = 3x_1 - x_2 + x_3$$

or

$$\frac{d}{dt}\mathbf{x} = \mathbf{Ax} + \mathbf{Bu} \qquad y = \mathbf{Cx} + \mathbf{Du}$$

where

$$\mathbf{A} = \begin{bmatrix} -1 & 0 & 0 \\ 0 & -3 & 1 \\ 0 & -3 & -3 \end{bmatrix} \quad \mathbf{B} = \begin{bmatrix} 1 \\ 0 \\ 1 \end{bmatrix} \quad \mathbf{C} = [3 \quad -1 \quad 0] \quad \mathbf{D} = \varnothing$$

figure 9.3.7
Choosing state variables: parallel form with
complex conjugate eigenvalues.

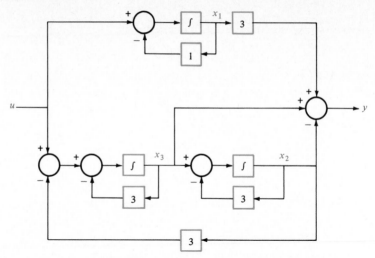

figure 9.3.7
Choosing state variables: parallel form with
complex conjugate eigenvalues.

9.4 eigenvalues and eigenvectors

DECOUPLED SYSTEMS
Suppose an overall circuit consists of three separate meshes, as shown in Figure 9.4.1. Obviously we can solve for each of the three state variables separately because none of these variables affects the others in any way. Such a system is called a *decoupled system*. We write the state equations for the system of Figure 9.4.1 as

$$\frac{dx_1}{dt} = -1x_1 \qquad \frac{dx_2}{dt} = -2x_2 \qquad \frac{dx_3}{dt} = -6x_3 + v_s \qquad (9.4.1)$$

or

$$\frac{d}{dt}\mathbf{x} = \mathbf{A}\mathbf{x} + \mathbf{B}\mathbf{u}$$

where

$$\mathbf{x} = \begin{bmatrix} x_1 \\ x_2 \\ x_3 \end{bmatrix} \qquad \mathbf{A} = \begin{bmatrix} -1 & 0 & 0 \\ 0 & -2 & 0 \\ 0 & 0 & -6 \end{bmatrix} \qquad \mathbf{B} = \begin{bmatrix} 0 \\ 0 \\ 1 \end{bmatrix} \qquad u = v_s$$

Suppose we want to obtain the natural (zero-input) response of this system. Setting $v_s = 0$, we can solve the equations (9.4.1) by inspection, finding

$$x_1(t) = x_1(0)e^{-t}$$

figure 9.4.1
A decoupled system.

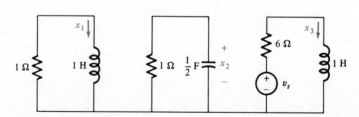

$$x_2(t) = x_2(0)e^{-2t}$$

$$x_3(t) = x_3(0)e^{-6t}$$

In matrix notation, this is

$$\begin{bmatrix} x_1(t) \\ x_2(t) \\ x_3(t) \end{bmatrix} = \begin{bmatrix} e^{-t} & 0 & 0 \\ 0 & e^{-2t} & 0 \\ 0 & 0 & e^{-6t} \end{bmatrix} \begin{bmatrix} x_1(0) \\ x_2(0) \\ x_2(0) \end{bmatrix}$$

or

$$\mathbf{x}(t) = e^{[A]t}\mathbf{x}(0) \tag{9.4.2}$$

where

$$\begin{bmatrix} e^{-t} & 0 & 0 \\ 0 & e^{-2t} & 0 \\ 0 & 0 & e^{-6t} \end{bmatrix} \equiv e^{[A]t} = [\Phi(t)] \tag{9.4.3}$$

This is called the *transition matrix*. For a decoupled system the **A** matrix and hence the transition matrix are diagonal matrices. In general, the transition matrix is a square but not necessarily diagonal matrix.

The scalar exponential function is defined by the infinite series

$$e^{at} = 1 + a\frac{t}{1!} + a^2\frac{t^2}{2!} + a^3\frac{t^3}{3!} + \cdots$$

In similar fashion the transition matrix can be written out as

$$[\Phi(t)] = [I] + [A]\frac{t}{1!} + [A]^2\frac{t^2}{2!} + [A]^3\frac{t^3}{3!} + \cdots \tag{9.4.4}$$

EXAMPLE 9.4.1

Find $[\Phi(t)]$ by using equation (9.4.4) for

$$\mathbf{A} = \begin{bmatrix} -1 & 0 & 0 \\ 0 & -2 & 0 \\ 0 & 0 & -6 \end{bmatrix}$$

ANS.:

$$\mathbf{A}^2 = \begin{bmatrix} 1 & 0 & 0 \\ 0 & 4 & 0 \\ 0 & 0 & 36 \end{bmatrix} \qquad \mathbf{A}^3 = \begin{bmatrix} -1 & 0 & 0 \\ 0 & -8 & 0 \\ 0 & 0 & -216 \end{bmatrix}$$

and so

$$[\Phi(t)] = \begin{bmatrix} 1 & 0 & 0 \\ 0 & 1 & 0 \\ 0 & 0 & 1 \end{bmatrix} + \begin{bmatrix} -t & 0 & 0 \\ 0 & -2t & 0 \\ 0 & 0 & -6t \end{bmatrix} + \begin{bmatrix} \dfrac{t^2}{2} & 0 & 0 \\ 0 & 4\dfrac{t^2}{2} & 0 \\ 0 & 0 & 36\dfrac{t^2}{2} \end{bmatrix} + \begin{bmatrix} -\dfrac{t^3}{6} & 0 & 0 \\ 0 & -8\dfrac{t^3}{6} & 0 \\ 0 & 0 & -216\dfrac{t^3}{6} \end{bmatrix} + \cdots$$

$$
= \begin{bmatrix} 1 - t + \dfrac{t^2}{2} - \dfrac{t^3}{6} + \cdots & 0 & 0 \\[2ex] 0 & 1 - 2t + 4\dfrac{t^2}{2} - 8\dfrac{t^3}{6} + \cdots & 0 \\[2ex] 0 & 0 & 1 - 6t + 36\dfrac{t^2}{2} - 216\dfrac{t^3}{6} + \cdots \end{bmatrix}
$$

$$
= \begin{bmatrix} e^{-t} & 0 & 0 \\ 0 & e^{-2t} & 0 \\ 0 & 0 & e^{-6t} \end{bmatrix}
$$

We note from equation (9.4.2) that one way to determine the transition matrix $[\Phi(t)]$ of a linear system is to set the initial conditions of all but one state variable to zero and evaluate the response of each $x(t)$, repeating this procedure until a nonzero initial condition has been placed [and the resultant response of all the $x(t)$'s recorded] on each state variable. In other words, a typical element Φ_{ij} of the transition matrix is the response of the ith state variable due to an initial condition on only the jth state variable.

In the special case of a decoupled system the resultant trajectories are straight lines coincident with the axes. As the point moves along each straight line, the time dependency of its distance from the origin is given by the single corresponding exponential term in the transition matrix. For example, with zero initial conditions on x_1 and x_3, we find that the natural response of x_2 is

$$
x_2(t) = x_2(0)e^{-2t}
$$

In Chapter 6 it was pointed out that the constant in the exponent term (-2 in this case) is the eigenvalue. Comparing equation (9.4.3) with the **A** matrix in equation (9.4.1) demonstrates that the eigenvalues of a decoupled system can be determined by inspection from its diagonal **A** matrix.

Section 6.7 showed how to get the complete response of a first-order differential equation [see equation (6.7.7)] by direct integration. Since state equations are nothing more than several simultaneous such first-order equations, they have a general solution as follows. In matrix notation

$$
\mathbf{x}(t) = e^{[A]t}\mathbf{x}(0) + \int_{0-}^{t} e^{[A](t-\tau)}\mathbf{u}(\tau)\,d\tau \tag{9.4.5}
$$

where $u(t)$ is the vector of input time functions. At this point we shall not attempt to perform the convolution process called for in equation (9.4.5) because there are easier ways to find the complete solution, but we shall return to this problem when we consider the Laplace transform in Chapter 15.

EIGENVALUES   Since the eigenvalues of a decoupled system are the elements of its diagonal **A** matrix, they can be determined easily; but what if a set of state equations does not have an **A** matrix that is diagonal? We now consider the general situation

$$\frac{d}{dt}\mathbf{x} = \mathbf{Ax} \qquad (9.4.6)$$

where **A** is not necessarily diagonal. Let us write the right-hand side of (9.4.6) as

$$\mathbf{Ax} = \mathbf{z} \qquad (9.4.7)$$

We interpret this as a transformation of the vector **x** into the vector **z**. In general **z** is in a different direction from **x** and has a different magnitude.

EXAMPLE 9.4.2

Find the vector **z** given

$$\mathbf{x} = \begin{bmatrix} 1 \\ 1 \\ 1 \end{bmatrix} \qquad \mathbf{A} = \begin{bmatrix} 2 & -1 & 3 \\ 4 & 2 & 6 \\ 1 & 0 & -1 \end{bmatrix} \qquad \mathbf{z} = \mathbf{Ax}$$

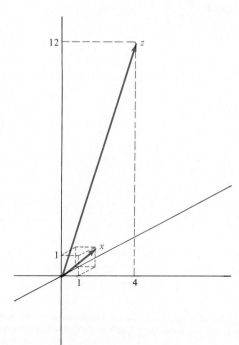

figure 9.4.2
Vectors **x** and **z** in Example 9.4.2.

ANS.:   See Figure 9.4.2

$$\mathbf{z} = \begin{bmatrix} 2 & -1 & 3 \\ 4 & 2 & 6 \\ 1 & 0 & -1 \end{bmatrix} \begin{bmatrix} 1 \\ 1 \\ 1 \end{bmatrix} = \begin{bmatrix} 2 - 1 + 3 \\ 4 + 2 + 6 \\ 1 + 0 - 1 \end{bmatrix} = \begin{bmatrix} 4 \\ 12 \\ 0 \end{bmatrix}$$

Is there a specific vector **x** (or possibly more than one) for which the transformation (9.4.6) does not produce a change in direction—only a change in magnitude? In other words,

$$\mathbf{Ax} = \lambda \mathbf{x} \qquad (9.4.8)$$

where λ is just a scalar multiplier. Or, with reference to equation (9.4.6) what we are really asking is: Are there any directions for the vector **x** in which it can grow or shrink in magnitude but *not* simultaneously *change* in *direction*? This can be seen by substituting (9.4.8) into (9.4.6)

$$\frac{d}{dt}\mathbf{x} = \lambda \mathbf{x} \qquad (9.4.9)$$

which indicates that the direction of the *growth* of the state vector is *coincident with the vector itself* and would produce a straight-line trajectory in the state space.

To see whether this is possible we must determine if there are any values of λ for which (9.4.8) is true. We write equation (9.4.8) in expanded form

$$\begin{aligned}
a_{11}x_1 + a_{12}x_2 + a_{13}x_3 + \cdots + a_{1n}x_n &= \lambda x_1 \\
a_{21}x_1 + a_{22}x_2 + \cdots \qquad\qquad + a_{2n}x_n &= \lambda x_2 \\
\cdots\cdots\cdots\cdots\cdots\cdots\cdots\cdots\cdots\cdots\cdots\cdots\cdots \\
a_{n1}x_1 + \cdots \qquad\qquad\qquad + a_{nn}x_n &= \lambda x_n
\end{aligned} \qquad (9.4.10)$$

Now, take the terms on the left over to the right.

$$\begin{aligned}
0 &= (\lambda - a_{11})x_1 - a_{12}x_2 - a_{13}x_3 - \cdots - a_{1n}x_n \\
0 &= -a_{21}x_1 + (\lambda - a_{22})x_2 - a_{23}x_3 - \cdots - a_{2n}x_n \\
&\cdots\cdots\cdots\cdots\cdots\cdots\cdots\cdots\cdots\cdots\cdots\cdots\cdots \\
0 &= -a_{n1}x_1 - a_{n2}x_2 - \cdots + (\lambda - a_{nn})x_n
\end{aligned}$$

In matrix notation this is

$$[\lambda \mathbf{I} - \mathbf{A}]\mathbf{x} = 0 \qquad (9.4.11)$$

and this can have a nonzero result for **x** if and only if the determinant of $[\lambda \mathbf{I} - \mathbf{A}]$ is equal to zero (see Appendix A). Thus, setting det $[\lambda \mathbf{I} - \mathbf{A}] = 0$, we solve for λ.

EXAMPLE 9.4.3

Find values for λ for the system whose **A** matrix is

$$\mathbf{A} = \begin{bmatrix} 0 & 3 \\ -1 & -4 \end{bmatrix}$$

ANS.:   Form

$$\det [\lambda \mathbf{I} - \mathbf{A}] = \det \left[\lambda \begin{bmatrix} 1 & 0 \\ 0 & 1 \end{bmatrix} - \begin{bmatrix} 0 & 3 \\ -1 & -4 \end{bmatrix} \right] = \det \left[\begin{bmatrix} \lambda & 0 \\ 0 & \lambda \end{bmatrix} - \begin{bmatrix} 0 & 3 \\ -1 & -4 \end{bmatrix} \right] = \det \begin{bmatrix} \lambda & -3 \\ 1 & \lambda+4 \end{bmatrix}$$

$$= \lambda(\lambda + 4) - (1)(-3)$$

Setting this equal to 0 gives

$$\lambda^2 + 4\lambda + 3 = 0 \qquad \lambda = -1, -3$$

The equation obtained by setting

$$\det \left[\lambda \mathbf{I} - \mathbf{A} \right] = 0 \tag{9.4.12}$$

is the eigenequation, and the values of λ that result from its solution are the eigenvalues of the system. Eigenvalues are either real or complex-conjugate pairs. Any eigenvalue with a positive real part indicates the presence of a natural response which grows as time increases, indicating that the system is unstable. Circuits made up only of independent sources, R's, L's, and C's, cannot be unstable; but since many other types of systems (e.g., circuits with dependent sources and some mechanical systems) can be unstable, it is always of interest to find the system eigenvalues to check on this detail.

EXAMPLE 9.4.4

Develop state equations for the circuit shown in Figure 9.4.3 and solve for the eigenvalues.

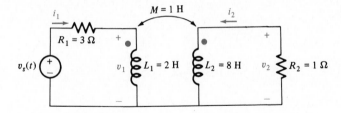

figure 9.4.3
The circuit in Example 9.4.4.

ANS.:   Choose as the set of state variables the inductor currents

$$\mathbf{x} = \begin{bmatrix} i_1 \\ i_2 \end{bmatrix}$$

Then

$$v_1 = L_1 \frac{dx_1}{dt} + M \frac{dx_2}{dt} \tag{9.4.13}$$

$$v_2 = M \frac{dx_1}{dt} + L_2 \frac{dx_2}{dt} \tag{9.4.14}$$

Solve equation (9.4.13) for dx_1/dt

$$\frac{dx_1}{dt} = \frac{v_1}{L_1} - \frac{M}{L_1} \frac{dx_2}{dt} \tag{9.4.15}$$

Insert (9.4.15) into (9.4.14) and solve for dx_2/dt

$$\frac{dx_2}{dt} = -\frac{M}{L_1 L_2 - M^2} v_1 + \frac{L_1}{L_1 L_2 - M^2} v_2 \tag{9.4.16}$$

Now insert (9.4.16) into (9.4.15)

$$\frac{dx_1}{dt} = \frac{v_1}{L_1} - \frac{M}{L_1}\left(-\frac{M}{L_1L_2 - M^2}v_1 + \frac{L_1}{L_1L_2 - M^2}v_2\right)$$

$$= \left[\frac{1}{L_1} + \frac{M^2}{L_1(L_1L_2 - M^2)}\right]v_1 - \frac{M}{L_1L_2 - M^2}v_2$$

$$= \frac{L_2}{L_1L_2 - M^2}v_1 - \frac{M}{L_1L_2 - M^2}v_2 \tag{9.4.17}$$

For $R_1 = 3\ \Omega$, $R_2 = 1\ \Omega$, $L_1 = 2$ H, $L_2 = 8$ H, and $M = 1$ H equations (9.4.17) and (9.4.16) become

$$\frac{dx_1}{dt} = 0.533v_1 - 0.066v_2 \tag{9.4.18}$$

and

$$\frac{dx_2}{dt} = -0.066v_1 + 0.133v_2 \tag{9.4.19}$$

From the circuit diagram we see that

$$v_1 = v_s - x_1 R_1 \tag{9.4.20}$$

$$v_2 = -x_2 R_2 \tag{9.4.21}$$

Substituting (9.4.20) and (9.4.21) into (9.4.18) and (9.4.19) gives

$$\frac{dx_1}{dt} = 0.53\bar{3}(v_s - x_1 3) + 0.066x_2 = -1.6x_1 + 0.066\bar{6}x_2 + 0.53\bar{3}v_s$$

$$\frac{dx_2}{dt} = -0.066\bar{6}(v_s - x_1 3) - 0.13\bar{3}x_2 = +0.2x_1 - 0.13\bar{3}x_2 - 0.066\bar{6}v_s$$

We find the eigenvalues to be

$$\begin{vmatrix} \lambda + 1.6 & -0.066\bar{6} \\ -0.2 & \lambda + 0.133 \end{vmatrix} = (\lambda + 1.6)(\lambda + 0.133) - 0.013\bar{3} = 0$$

$$\lambda^2 + 1.73\lambda + 0.2 = 0$$

$$\lambda = \frac{-1.73 \pm \sqrt{3 - 0.8}}{2} = -0.124, \ -1.61 \tag{9.4.22}$$

As a check on this, we can use the classical methods of Chapter 7 to analyze this circuit starting with

$$v_s = I_1 R_1 + L_1 p I_1 + M p I_2 \quad \text{and} \quad v_2 = M p I_1 + L_2 p I_2 \quad \text{where} \quad v_2 = -I_2 R_2$$

This results in

$$I_1 = \frac{0.53(p + 0.125)}{p^2 + 1.73p + 0.2} v_s \tag{9.4.23}$$

The natural response is obtained as usual from the denominator (characteristic polynomial) of equation (9.4.23). Since the resulting equation is identical to (9.4.22), the eigenvalues are shown to be the quantities calculated above.

For each eigenvalue we can go back to equation (9.4.11) and solve for **x**. Such **x**-vector directions are called *eigenvectors*. With reference to equation (9.4.9) we note that these directions in the state space are those directions along which the state vector of a source-free system changes magnitude but not direction. Consequently, if the state of any linear system initially starts out at any point located on the locus defined by the eigenvector, as time goes on, the state of that system will approach the origin *along this eigenvector*. During this time *only one mode* (i.e., a single exponential term) appears in the natural response. Other modes are not excited by such a set of initial conditions. If the eigenvalues of a system are complex conjugates, no real fixed-direction eigenvectors exist.

EXAMPLE 9.4.5

For the system of Example 9.4.3 find the eigenvectors.

ANS.:   Equation (9.4.11) states that

$$[\lambda \mathbf{I} - \mathbf{A}]\mathbf{x} = 0 \qquad \text{for each } \lambda$$

So, for $\lambda_1 = -1$,

$$\left[-1\begin{bmatrix} 1 & 0 \\ 0 & 1 \end{bmatrix} - \begin{bmatrix} 0 & 3 \\ -1 & -4 \end{bmatrix} \right]\begin{bmatrix} x_1 \\ x_2 \end{bmatrix} = 0 \qquad \text{and} \qquad \begin{bmatrix} -1 & -3 \\ 1 & 3 \end{bmatrix}\begin{bmatrix} x_1 \\ x_2 \end{bmatrix} = 0$$

resulting in the simultaneous equations

$$-1x_1 - 3x_2 = 0$$

$$1x_1 + 3x_2 = 0$$

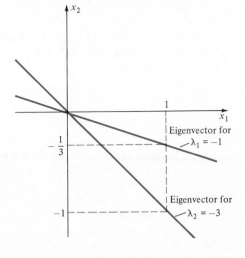

figure 9.4.4
The eigenvalues of Examples 9.4.3 and 9.4.5.

This set of equations does not specify a magnitude for x. Try to solve them. You cannot get a unique solution. But they do specify a direction. Let us pick a value for x_1 and solve for x_2, that is, for

$$x_1 = 1 \qquad x_2 = -\tfrac{1}{3}$$

This point defines one eigenvector. Similarly, for $\lambda_2 = -3$,

$$\left\{-3\begin{bmatrix} 1 & 0 \\ 0 & 1 \end{bmatrix} - \begin{bmatrix} 0 & 3 \\ -1 & -4 \end{bmatrix}\right\}\begin{bmatrix} x_1 \\ x_2 \end{bmatrix} = 0 \quad \text{and} \quad \begin{bmatrix} -3 & -3 \\ 1 & 1 \end{bmatrix}\begin{bmatrix} x_1 \\ x_2 \end{bmatrix} = 0$$

For $x_1 = 1$ this yields $x_2 = -1$, which defines the other eigenvector (Figure 9.4.4). Since the eigenvalues are -1 and -3 the zero-input response of this system must be of the form

$$x_1(t) = Ae^{-t} + Be^{-3t}$$

If the initial conditions are located on one of the eigenvectors, say, at $x_1 = 1$ and $x_2 = -\frac{1}{3}$, using the first state equation leads to

$$\left.\frac{dx_1}{dt}\right|_{0+} = 0x_1 + 3x_2 = -1$$

Hence
$$x_1(0+) = 1 = A(1) + B(1) \quad \text{and} \quad \left.\frac{dx_1}{dt}\right|_{0+} = -1 = -A(1) - 3B(1)$$

from which, by adding, $0 = -2B$ or $B = 0$ and $A = 1$. Therefore, the zero-input response is given by

$$x_1(t) = e^{-t}$$

and the e^{-3t} mode is totally absent.

9.5 numerical solution of linear state equations

Consider the following problem. A car is traveling along a road at 60 mi/h. At $t = 0$ the car is at position $x = 10$. What position will the car be at at $t = 1$ min? We solve it simply by finding out how much farther the car has traveled during that 1-min interval and adding this distance on to the original position. In other words, during the 1-min time span the car will go a distance of 60 mi/h ($=1$ mi/min) times the 1-min time interval which is 1 mi. Mathematically,

$$\Delta x = \frac{dx}{dt}\Delta t = 60(\tfrac{1}{60}) = 1 \text{ mi}$$

Then we find the new position by updating x as follows:

$$x_{\text{new}} = x_{\text{old}} + \Delta x = 10 + 1 = 11$$

But suppose the driver does not maintain a constant speed during that 1-min interval. In that event our answer will not be accurate. We can guard against that possibility by checking the speed more often than once a minute, say, every 0.1 s, and computing the updated position each time. It is highly unlikely that the velocity of the car will change appreciably in so short a time span as 0.1 s. In other words, if we pick a sampling interval shorter than the shortest time constant of the system, our answer should be reasonably accurate. Let us use this method to solve for the values of the state variables as functions of time. After all, the state equations are nothing more than expressions for the velocity dx/dt of the state variables.

Hence we can always solve a set of state equations for an approximate numerical answer; i.e., consider the single state equation obtained from the simple system shown in Figure 9.5.1

$$v_L = L\frac{di}{dt} = v_s - iR$$

$$\frac{di}{dt} = -\frac{R}{L}i + \frac{1}{L}v_s$$

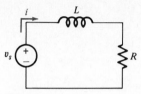

figure 9.5.1
A typical first-order system.

Remembering that

$$\frac{di}{dt} \equiv \lim_{\Delta t \to 0}\frac{i(t + \Delta t) - i(t)}{\Delta t} = \frac{\Delta i}{\Delta t}\Big|_{\Delta t \to 0}$$

we think of time as being broken up into small increments Δt s apart. Call these times $t_0, t_1, t_2, t_3, \dots$. Therefore at time t_0

$$\Delta i = \left[-\frac{R}{L}i(t_0) + \frac{1}{L}v_s(t_0) \right]\Delta t$$

such that at time t_1

$$i(t_1) = i(t_0) + \Delta i$$

At time t_1 we get a new value for Δi by substituting the correct values of i and v_s for that new time

$$\Delta i = \left[-\frac{R}{L}i(t_1) + \frac{1}{L}v_s(t_1) \right]\Delta t$$

whence

$$i(t_2) = i(t_1) + \Delta i$$

and so on. Suppose in Figure 9.5.1 that $L = 1$, $R = 2$, and $v_s(t) = 10u(t)$. Assume that the initial conditions are all zero:

$$\frac{di}{dt} = -2i(t) + 10u(t)$$

Define $t_0 = 0$ and let $\Delta t = 0.01$ s. We then have

$$\Delta i = [-2i(t) + 10](0.01) \qquad \text{for } t > 0$$

Of course any iterative scheme like this is ideal for the digital computer. In general the smaller Δt is made the more accurate the solution (and also the more work involved but as long as it is the computer doing the work we do not have to worry).

t	$i(t)$	Δi
0	0	0.1
0.01	0.1	0.098
0.02	0.198	0.09604
0.03	0.29404	0.0941192
0.04	0.3881592	0.092236816
0.05	0.480396016	0.0903920797

Let us find the exact answer for $i(t)$ in the RL circuit we just discussed. The state equation is

EXACT SOLUTION

$$\frac{di}{dt} = -2i + 10 \qquad \text{or} \qquad \frac{di}{dt} + 2i = 10$$

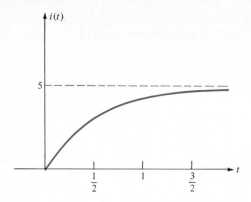

figure 9.5.2
The state variable plotted as a
function of time.

The solution is simply obtained (through the methods of Chapter 6), i.e.,

$$i_n(t) = Ae^{-2t} \quad \text{and} \quad i_p(t) = k$$

whence $\qquad\qquad\qquad\qquad k = 5$

so that $\qquad\qquad\qquad\qquad i(t) = Ae^{-2t} + 5$

With zero initial conditions, $A = -5$; thus, finally for $t > 0$,

$$i(t) = 5(1 - e^{-2t})$$

This is shown in Figure 9.5.2. The error between this exact solution and our numerical approximation at $t = 0.05$ is less than 1 percent.

A system with n degrees of freedom (e.g., a circuit with n energy-storage elements) requires n coupled state equations to describe it. Such equation sets can also be solved numerically by the technique just described.

EXAMPLE 9.5.1

Consider the equations developed in Example 9.2.2:

$$\frac{dx_1}{dt} = -0.8571x_1 + 0.2143x_2 + 0.2857v_1 + 0.2143v_2$$

$$\frac{dx_2}{dt} = -0.8571x_1 - 0.2857x_2 + 0.2857v_1 - 0.2857v_2$$

a. Write a FORTRAN computer program to solve for $x_1(t)$ and $x_2(t)$ for $t > 0$. Assume zero initial conditions and $v_1 = 4$ and $v_2 = 0$.
b. Plot the trajectory.
ANS.: (a) See Figure 9.5.3. (b) The results are plotted in Figure 9.5.4.

```
10 PROGRAM STATE(INPUT,OUTPUT)
20 T=0.
30 DELT=.01
40 X1=0.
50 X2=0.
60 3 CONTINUE
```
⎫
⎬ ←——Initializing statements
⎭

figure 9.5.3

```
70 DO 1 I=1,50
80 V1=4.
90 DELX1=-.8571*X1+.2143*X2+.2857*V1
100 DELX2=-.8571*X1-.2857*X2+.2857*V1
110 T=T+DELT
120 X1=X1+DELX1*DELT
130 1 X2=X2+DELX2*DELT
140 PRINT 2,T,X1,X2
150 2 FORMAT(1X,3F10.3)
160 GO TO 3
170 END
```

lines 90–100 } State equations

lines 110–130 } ←——Updating statements

140 ←——Produces output

160 ←— Start over – – calculate the next point

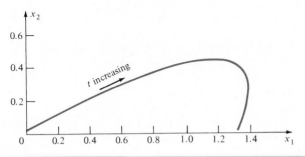

figure 9.5.4
The state-space trajectory of Example 9.5.1.

EXAMPLE 9.5.2

Write a program for one or more of the popular personal computers with high-resolution graphics to plot out several trajectories of the system in Examples 9.4.3 and 9.4.5.

ANS.:   The program in Figure 9.5.5*a* is written for the **IBM PC** (and compatibles such as the Compaq) in Advanced Basic (BASICA). The program in Figure 9.5.5*b*, written for the APPLE II, will plot one trajectory, the initial conditions for which are in lines 60 and 70. The trajectories are shown in Figure 9.5.6. Note that the eigenvectors for this example (originally shown in Figure 9.4.4) are evident in Figure 9.5.6. The trajectories

figure 9.5.5a

```
1 'STATE3
2 'This program plots the state trajectories of the system in
3 'examples 9.4.3 and 9.4.5.  This result is shown in figure 9.5.6.
4 T$="PROGRAM STATE3"
5 L=LEN(T$)
6 SCREEN 2:CLS
7 PRINT TAB(INT((80-L)/2+.5));T$
8 T$="See examples 9.4.3 and 9.4.5."
9 L=LEN(T$)
20 PRINT TAB(INT((80-L)/2+.5));T$
60 LINE(320,0)-(320,199)
70 LINE(0,100)-(639,100),3,,&HAAAA
75 SCALE=2.5
```

figure 9.5.5a (continued)

```
80 INC%=320/SCALE
90 PRESET(0,100)
95 FOR I=1 TO 2*SCALE
100      DRAW "C1 D U"
105      DRAW "BR=INC%;"
110      NEXT
120 PRESET(319,0)
122 INC%=100/SCALE
125 FOR I=1 TO 2*SCALE
130      DRAW "C1 R L"
135      DRAW "BD=INC%;"
140      NEXT
145 FOR X20=-1.5 TO 1.5 STEP 3
150      FOR X10=-2 TO 2 STEP   .5
160              T=0
170              DT=.05
180              X1=X10
190              X2=X20
220              XO=X1/(2*SCALE)*639+320
230              YO=-X2/(2*SCALE)*199+100
240              PRESET(XO,YO)
250              X=X1/(2*SCALE)*639+320
260              Y=-X2/(2*SCALE)*199+100
270              LINE-(X,Y)
300              DX1=3*X2
310              DX2=-X1-4*X2
320              T=T+DT
330              X1=X1+DX1*DT
340              X2=X2+DX2*DT
345              IF((X1^2+X2^2)<.005) GOTO 355 ELSE
350              GOTO 250
355              NEXT X10
357      NEXT X20
360 END
```

figure 9.5.5b

```
10   HGR
20   HCOLOR= 3
22   HPLOT 0,80 TO 279,80
24   HCOLOR= 0
26   HPLOT 140,0
28   HCOLOR= 3
30   HPLOT 140,0 TO 140,159
35   G = 50
40   DT = .01
50   T = 0.
60   XA = 0
70   XB = 1.6
80   F = 0
90   DA = 3 * XB
100  DB =  - 1 * XA - 4 * XB
```

```
110  T = T + DT
120  XA = XA + DA * DT
130  XB = XB + DB * DT
140  PRINT XA,XB
150  HCOLOR= 3
160  XH = G * XA + 140
170  XV = 80. - G * XB
180  HPLOT XH,XV
190  GOTO 80
200  END
```

figure 9.5.5b (continued)

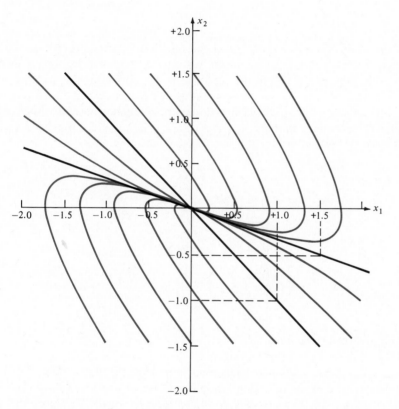

figure 9.5.6
The state-space trajectories of Examples
9.4.3 and 9.4.5:

$$\frac{dx_1}{dt} = 3x_2 \quad \text{and} \quad \frac{dx_2}{dt} = -x_1 - 4x_2$$

begin by following a path that is generally parallel to one eigenvector and then approach zero asymptotic to the other eigenvector.† This is a general rule about how such trajectories behave, so that knowing the eigenvectors in any given problem will often help us in sketching the state-space trajectories.

nonlinear state equations 9.6

One of the major strengths of the state-variable method is the relative ease with

† First the trajectories parallel the eigenvector associated with the fast (shorter time constant) mode (exponential term); then they parallel the slow (longer time constant) eigenvector.

which nonlinear systems can be analyzed. Any set of state equations for a source-free system has the general form

$$\frac{dx_1}{dt} = f_1(x_1, x_2, \ldots, x_n)$$

$$\frac{dx_2}{dt} = f_2(x_1, x_2, \ldots, x_n) \qquad (9.6.1)$$

$$\cdots\cdots\cdots\cdots\cdots\cdots\cdots$$

$$\frac{dx_n}{dt} = f_n(x_1, x_2, \ldots, x_n)$$

where the x_i are the state variables. If the functions f_i are all linear, we say the system is linear. If any f_i is nonlinear, we say the system is nonlinear and we cannot write equation (9.6.1) in matrix form. If we confine the fluctuations of the state variables within a small enough region of the state space, we can approximate the system by a set of linear functions. These functions usually do a reasonable job of describing what will happen to the state variables (but in that local region only). The equations for this *linearized system* are†

$$\frac{d}{dt}\mathbf{x} = \mathbf{A}\mathbf{x}$$

where
$$\mathbf{A} = \begin{bmatrix} \dfrac{\partial f_1}{\partial x_1} & \dfrac{\partial f_1}{\partial x_2} & \dfrac{\partial f_1}{\partial x_3} & \cdots & \dfrac{\partial f_1}{\partial x_n} \\[2mm] \dfrac{\partial f_2}{\partial x_1} & \cdots & \cdots & \cdots & \dfrac{\partial f_2}{\partial x_n} \\[2mm] \cdots\cdots\cdots\cdots\cdots\cdots\cdots\cdots \\[1mm] \dfrac{\partial f_n}{\partial x_1} & \cdots & \cdots & \cdots & \dfrac{\partial f_n}{\partial x_n} \end{bmatrix} \qquad (9.6.2)$$

This is called the *jacobian* matrix.

The partial derivatives in the jacobian, when evaluated at any given point in the state space, yield the linear state equations that are valid at that point. It is particularly useful when analyzing a nonlinear system to (1) find the equilibrium points, i.e., points in the state space where the vector that produces the state trajectory can remain at rest with fixed magnitude and direction‡ (at such points the time derivative of every state variable equals zero) and (2) linearize the system at each such equilibrium point and find the eigenvalues of the linear model that is valid in that neighborhood. Any eigenvalues having positive real parts will produce natural responses that increase with time, the system thus being unstable in the neighborhood of that equilibrium point. Usually these two steps are sufficient to enable the engineer to sketch the general shape of typical trajectories. One additional helpful step is to plot, if possible, the eigenvectors for the linear model that are valid at each equilibrium point. These direction vectors help in

---

† We set the source(s) equal to zero and seek the zero-input response.
‡ Every linear system has one and only one equilibrium point, which is located at the origin of the state space.

sketching typical trajectories. Let us now consider two examples of nonlinear systems.

EXAMPLE 9.6.1

The network in Figure 9.6.1 contains a tunnel diode and is an electrical model of the voltage v_C across the membrane of a nerve cell. Find any equilibrium point(s), linearize the system at any such point(s), find eigenvalues, and plot the state trajectory resulting from an initial condition of $v_C(0-) = 1$ V and $i_L(0-) = 0$. In the model $L = 12.5$ H, $v_s = 0.7$ V, $R = 0.8\ \Omega$, and $C = 1$ F.

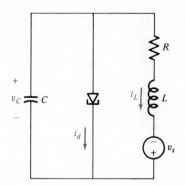

figure 9.6.1
An electrical model of a nerve axon membrane.

ANS.:   The tunnel-diode current i_d is

$$i_d = \frac{v_C^3}{3} - v_C$$

We choose $[x] = \begin{bmatrix} v_C \\ i_L \end{bmatrix}$ and write a pair of state equations

$$i_C = C \frac{dv_C}{dt} = -i_d - i_L$$

$$\frac{dv_C}{dt} = -\frac{v_C^3}{3} + v_C - i_L \tag{9.6.3}$$

and $$v_L = L \frac{di_L}{dt} = v_C + v_s - i_L R$$

$$\frac{di_L}{dt} = \frac{1}{L} v_C - \frac{R}{L} i_L + \frac{1}{L} v_s = 0.08v_C - 0.064i_L + (0.08)(0.7) \tag{9.6.4}$$

We can find the equilibrium points for i_L and v_C by setting equations (9.6.3) and (9.6.4) to zero (i.e., the zero growth rate of all state variables is the definition of *equilibrium*). Thus

$$0 = -\frac{v_C^3}{3} + v_C - i_L \tag{9.6.5}$$

and $$0 = 0.08v_C - 0.064i_L + 0.056 \tag{9.6.6}$$

Both these equations can be plotted in the state space.† Any intersection(s) will thus be the point(s) which satisfy the requirements that $dv_C/dt = di_L/dt = 0$ simultaneously. Alternatively, we can simply solve equations (9.6.5) and (9.6.6) and find by either method $v_C = -1.1994$ and $i_L = -0.6243$. This, then, is the equilibrium point.

Now, find the jacobian matrix of the functions f_1 and f_2 on the right sides of equations (9.6.3) and (9.6.4)

$$f_1 = -\frac{v_C^3}{3} + v_C - i_L \qquad f_2 = 0.08v_C - 0.064i_L + 0.056$$

$$\begin{bmatrix} \dfrac{\partial f_1}{\partial v_C} = -v_C^2 + 1 & \dfrac{\partial f_1}{\partial i_L} = -1 \\ \dfrac{\partial f_2}{\partial v_C} = 0.08 & \dfrac{\partial f_2}{\partial i_L} = -0.064 \end{bmatrix}$$

Evaluating at $v_C = -1.1994$ and $i_L = -0.6243$ gives

$$A = \begin{bmatrix} -0.4386 & -1 \\ 0.08 & -0.064 \end{bmatrix}$$

and so the eigenvalues are found by

$$\det[\lambda I - A] = 0 \qquad \det \begin{bmatrix} \lambda + 0.4386 & 1 \\ -0.08 & \lambda + 0.064 \end{bmatrix} = 0$$

$$\lambda^2 + 0.5026\lambda + 0.0281 + 0.08 = 0$$

$$\lambda^2 + 0.5026\lambda + 0.1081 = 0$$

$$\lambda = \frac{-0.5026 \pm \sqrt{0.2526 - 0.4324}}{2} = -0.251 \pm j0.212$$

$$2\pi f = 0.212$$

$$T = 29.6 \text{ s}$$

Since complex-conjugate eigenvalues indicate an underdamped natural response, the trajectory will spiral in toward the equilibrium point.

figure 9.6.2

```
10   HGR
20   HCOLOR=3
22   HPLOT 0,80 TO 279,80
24   HCOLOR=0
26   HPLOT 140,0
28   HCOLOR=3
30   HPLOT 140,0 TO 140,159
35   G=50
40   DT=.1
50   T=0
60   XA=.3
70   XB=0
80   F=0
90   DA=XA-XB-(XA^3)/3
100  DB=.08*XA-.064*XB
101  DB=DB+.056
110  T=T+DT
120  XA=XA+DA*DT
130  XB=XB+DB*DT
140  PRINT XA,XB
150  HCOLOR=3
160  XH=G*XA+140
170  XV=80-G*XB
180  HPLOT XH,XV
190  GOTO 80
200  END
```

A program for the popular Apple II computer that uses its high-resolution graphics is shown in Figure 9.6.2. Running that program results in the state-space trajectory in Figure 9.6.3. The voltage $v_C(t)$ versus time is shown in Figure 9.6.4.

† Such loci are called *isoclines*, lines of equal slope. Any locus-connecting points where the trajectories have equal slope are termed isoclines. The isoclines referred to here connect points having *zero* slope.

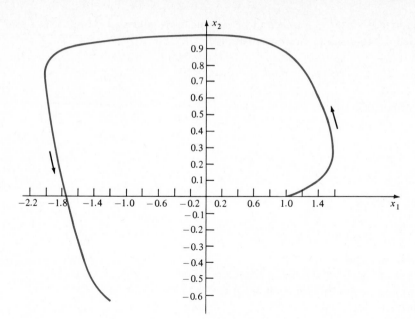

figure 9.6.3
The state-space trajectory of the circuit in
Example 9.6.1

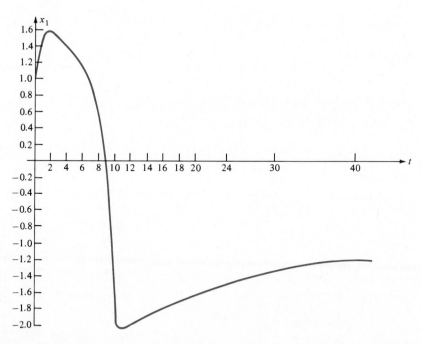

figure 9.6.4
State variable $x_1 = v_C$ plotted as a function
of time.

EXAMPLE 9.6.2

Consider an ecological system which supposedly models the interaction between a population of squid and a
population of herring in a typical acre of seabed in some fishing ground. Squid eat herring as their major food
source and the birthrate depends on being well fed. Both species will start to leave the area if their population

becomes too dense. The state variables are x_1 = number of herring per acre and x_2 = number of squid per acre. The state equations proposed to model these two variables are†

$$\frac{dx_1}{dt} = k_1 x_1 - k_2 x_1^2 - k_3 x_1 x_2 \qquad \frac{dx_2}{dt} = -k_4 x_2 - k_5 x_2^2 + b k_3 x_1 x_2$$

where $k_1 = 1.1$ = reproduction of herring

$k_2 = 10^{-5}$ = herring density factor

$k_3 = 10^{-3}$ = effect of squid feeding

$k_4 = 0.9$ = death from old age of squid

$k_5 = 10^{-4}$ = squid density factor

$b = 0.02$ = birthrate coefficient

a. Find the locations of all equilibrium points.
b. Find an approximate linear model that is valid near each equilibrium point.
c. Find the eigenvalues and any real eigenvectors for each linearized model found in part (b) above.
d. Use the eigenvectors found in part (c) to sketch several typical trajectories.
ANS.:   (a) The state equations are

$$\frac{dx_1}{dt} = 1.1 x_1 - 10^{-5} x_1^2 - 10^{-3} x_1 x_2 \tag{9.6.7}$$

$$\frac{dx_2}{dt} = -0.9 x_2 - 10^{-4} x_2^2 + 2 \times 10^{-5} x_1 x_2 \tag{9.6.8}$$

Setting each of these equal to zero, we can solve for three different equilibrium points; i.e.,

$$0 = 1.1 x_1 - 10^{-5} x_1^2 - 10^{-3} x_1 x_2 \tag{9.6.9}$$

$$0 = -0.9 x_2 - 10^{-4} x_2^2 + 2 \times 10^{-5} x_1 x_2 \tag{9.6.10}$$

yields the following:

Equilibrium Point 1
By inspection, $x_1 = 0$, $x_2 = 0$ satisfy equations (9.6.9) and (9.6.10).

Equilibrium Point 2
Setting $x_2 = 0$ in (9.6.9) yields

$$0 = 1.1 x_1 - 10^{-5} x_1^2 \qquad \text{or} \qquad x_1 = 1.1 \times 10^5$$

Equilibrium Point 3
Solving (9.6.9) for x_2 yields

$$x_2 = 1100 - 10^{-2} x_1 \tag{9.6.11}$$

and (9.6.10) for x_2 yields

$$x_2 = 0.2 x_1 - 0.9 \times 10^4 \tag{9.6.12}$$

When (9.6.11) is set equal to (9.6.12), we get $x_1 = 48{,}095$, and inserting this into either (9.6.11) or (9.6.12) yields

$$x_2 = 619$$

† Equations of this form are sometimes called *Volterra competition equations*.

Equilibrium Point 4
Setting $x_1 = 0$ in (9.6.10) yields $x_2 = -9000$, which has no physical meaning in this problem (what is a negative population density?).

(b) The linearized model is found by using the jacobian matrix with

$$\frac{dx_1}{dt} = f_1(x_1, x_2) = 1.1x_1 - 10^{-5}x_1^2 - 10^{-3}x_1x_2$$

$$\frac{dx_2}{dt} = f_2(x_1, x_2) = 0.9x_2 - 10^{-4}x_2^2 + 2 \times 10^{-5}x_1x_2$$

$$\mathbf{A} = \begin{bmatrix} \dfrac{\partial f_1}{\partial x_1} & \dfrac{\partial f_1}{\partial x_2} \\ \dfrac{\partial f_2}{\partial x_1} & \dfrac{\partial f_2}{\partial x_2} \end{bmatrix} = \begin{bmatrix} 1.1 - 2 \times 10^{-5}x_1 - 10^{-3}x_2 & -10^{-3}x_1 \\ 2 \times 10^{-5}x_2 & -0.9 - 2 \times 10^{-4}x_2 + 2 \times 10^{-5}x_1 \end{bmatrix}$$

At equilibrium point 1 ($x_1 = 0$, $x_2 = 0$):

$$\mathbf{A} = \begin{bmatrix} 1.1 & 0 \\ 0 & -0.9 \end{bmatrix}$$

Thus we have a decoupled system in the neighborhood of ($x_1 = 0$, $x_2 = 0$), that is,

$$\frac{dx_1}{dt} = 1.1x_1 \qquad \text{and} \qquad \frac{dx_2}{dt} = -0.9x_2$$

$$\text{unstable} \qquad\qquad\qquad \text{stable}$$

At equilibrium point 2 ($x_1 = 1.1 \times 10^5$, $x_2 = 0$) we have

$$\mathbf{A} = \begin{bmatrix} 1.1 - (2 \times 10^{-5})(1.1 \times 10^5) - (10^{-3} \times 0) & -10^{-3}(1.1 \times 10^5) \\ 0 & -0.9 + (2 \times 10^{-5})(1.1 \times 10^5) \end{bmatrix}$$

or

$$\mathbf{A} = \begin{bmatrix} -1.1 & -110 \\ 0 & 1.3 \end{bmatrix}$$

(c) Eigenvalues are found from

$$\det [\lambda\mathbf{I} - \mathbf{A}] = 0 \qquad \begin{vmatrix} \lambda + 1.1 & 110 \\ 0 & \lambda - 1.3 \end{vmatrix} = (\lambda + 1.1)(\lambda - 1.3) = 0 \qquad \lambda = 1.3, -1.1$$

We find the eigenvector associated with each eigenvalue by using $[\lambda\mathbf{I} - \mathbf{A}]\mathbf{x} = 0$. For $\lambda = -1.1$

$$\begin{bmatrix} -1.1\begin{bmatrix} 1 & 0 \\ 0 & 1 \end{bmatrix} - \begin{bmatrix} -1.1 & -110 \\ 0 & 1.3 \end{bmatrix} \end{bmatrix}\begin{bmatrix} x_1 \\ x_2 \end{bmatrix} = 0$$

$$\begin{bmatrix} \begin{bmatrix} -1.1 & 0 \\ 0 & -1.1 \end{bmatrix} - \begin{bmatrix} -1.1 & -110 \\ 0 & 1.3 \end{bmatrix} \end{bmatrix}\begin{bmatrix} x_1 \\ x_2 \end{bmatrix} = 0 \qquad \begin{bmatrix} 0 & 110 \\ 0 & -2.4 \end{bmatrix}\begin{bmatrix} x_1 \\ x_2 \end{bmatrix} = 0$$

so $\qquad 110x_2 = 0 \qquad$ and $\qquad -2.4x_2 = 0$

$$x_2 = 0 \qquad\qquad\qquad x_2 = 0$$

and there is no restriction on x_1, that is, the x_1 axis.
For $\lambda = +1.3$,

$$\left[1.3 \begin{bmatrix} 1 & 0 \\ 0 & 1 \end{bmatrix} - \begin{bmatrix} -1.1 & -110 \\ 0 & 1.3 \end{bmatrix} \right] \begin{bmatrix} x_1 \\ x_2 \end{bmatrix} = 0$$

yields

$$\begin{bmatrix} 2.4 & 110 \\ 0 & 0 \end{bmatrix} \begin{bmatrix} x_1 \\ x_2 \end{bmatrix} = 0$$

and so

$$2.4x_1 + 110x_2 = 0 \qquad \text{or} \qquad x_2 = \frac{-2.4}{110} x_1$$

At equilibrium point 3 ($x_1 = 48{,}095$, $x_2 = 619$) setting det $[\lambda \mathbf{I} - \mathbf{A}] = 0$ yields

$$\left| \begin{bmatrix} \lambda & 0 \\ 0 & \lambda \end{bmatrix} - \begin{bmatrix} 1.1 - 0.9619 - 0.619 & -48.095 \\ 0.01238 & -0.9 - 0.1238 + 0.9619 \end{bmatrix} \right| = 0$$

$$\begin{vmatrix} \lambda + 0.4809 & 48.095 \\ -0.01238 & \lambda + 0.0619 \end{vmatrix} = 0$$

$$\lambda^2 + 0.5428\lambda + 0.62518 = 0$$

$$\lambda = 0.271 \pm j0.743$$

which is an underdamped response and leads to no real eigenvectors.

(d) Several trajectories are shown in Figure 9.6.5.

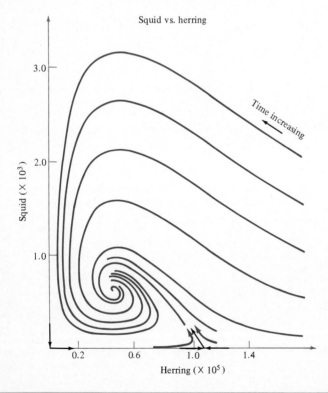

figure 9.6.5
State trajectories of the system in Example
9.6.2. Equilibrium points are at (0, 0),
(110,000, 0), and (48,095, 619). Eigenvectors
are shown at the first two of these
equilibrium points.

time-varying systems 9.7

Certain systems have *elements* that vary with time. This is very different from saying that their input *variables* are functions of time—a very common thing and one that presents no problem in solving for equilibrium points, eigenvalues, etc. A time-varying system is one in which one or more of the **A** or **B** matrix elements change with time. For example, in a certain type of microphone the incoming sound-pressure waves actually vary the spacing between the plates of a small capacitance, so that the equivalent circuit of such a device incorporates a capacitor whose value varies with time.

In general, a linear time-varying system's state equations would look like

$$\frac{d}{dt}\mathbf{x} = \mathbf{A}(t)\mathbf{x} + \mathbf{B}(t)\mathbf{u}$$

Time-varying systems may be linear or nonlinear, and obtaining closed-form solutions for any such system is a difficult problem even when it is possible at all. Not all such systems have closed-form solutions. In any event, the computer method we have discussed for obtaining state-space trajectories of linear and nonlinear time-invariant systems also works equally well with most time-varying systems.

EXAMPLE 9.7.1

Write a FORTRAN computer program that will list values of a set of state variables for the time-varying system described by

$$\frac{d^2v}{dt^2} + L\frac{dv}{dt} + v = 0 \qquad \text{where} \qquad L = \frac{2}{t+1} - 0.1$$

ANS.:   Let $$x_1 = v \qquad \text{and} \qquad x_2 = \frac{dv}{dt}$$

Thus $$\frac{dx_1}{dt} = x_2 \qquad \text{and} \qquad \frac{dx_2}{dt} = -Lx_2 - x_1$$

So we write, as before, the program in Figure 9.7.1. The output is plotted in Figure 9.7.2. An Apple II+ high-resolution graphics program is Figure 9.7.3.

```
100 PROGRAM TIMEVAR(OUTPUT, TAPE6=OUTPUT)       figure 9.7.1

110 DELT=0.1

120 T=0

130 X1=5

140 X2=0

150 1 CONTINUE

160 DO 2 K=1,5

170 D=2./(T+1.)-0.1

180 DELX1=X2

190 DELX2=-D*X2-X1
```

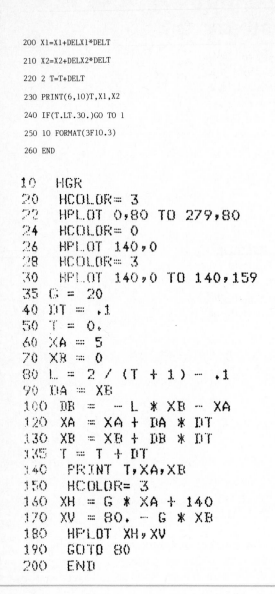

```
200 X1=X1+DELX1*DELT
210 X2=X2+DELX2*DELT
220 2 T=T+DELT
230 PRINT(6,10)T,X1,X2
240 IF(T.LT.30.)GO TO 1
250 10 FORMAT(3F10.3)
260 END
```

```
10    HGR
20    HCOLOR= 3
22    HPLOT 0,80 TO 279,80
24    HCOLOR= 0
26    HPLOT 140,0
28    HCOLOR= 3
30    HPLOT 140,0 TO 140,159
35    G = 20
40    DT = .1
50    T = 0.
60    XA = 5
70    XB = 0
80    L = 2 / (T + 1) - .1
90    DA = XB
100   DB = - L * XB - XA
120   XA = XA + DA * DT
130   XB = XB + DB * DT
135   T = T + DT
140   PRINT T,XA,XB
150   HCOLOR= 3
160   XH = G * XA + 140
170   XV = 80. - G * XB
180   HPLOT XH,XV
190   GOTO 80
200   END
```

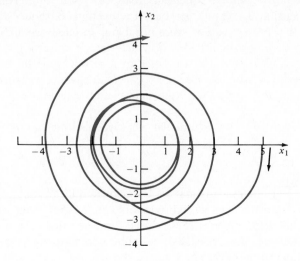

figure 9.7.2
The state-space trajectory of the time-varying system of Example 9.7.1.

9.8 summary

State-variable analysis is the basic method used by modern control engineers to design large automatic systems, e.g., control systems for airplanes and rockets, industrial temperature controllers and complex process controllers, as well as navigation systems for ships and planes. We have been able only to scratch the surface of the method in this short chapter. The main points introduced are as follows:

1. What is meant by the *state* of a system

2. How to choose state variables

3. How to write a set of state equations for a system

4. How to come up with different sets of state variables for any given system

5. How to find the eigenvalues of a linear system

6. How to find the eigenvectors of a linear system and to interpret them geo-metrically

7. How to use the computer to generate numerical solutions of state equations and how to plot these solutions

8. How to linearize nonlinear state equations in order to work with them more easily

If you are pursuing a career in control or communication engineering, you will be seeing a great deal more of this useful technique, to which we return in Chapter 15 when we discuss the Laplace transform.

problems

1. A series LC circuit is driven by a voltage source $v(t)$. The output voltage is taken across the capacitor. Assuming $L = 2$ H and $C = 3$ F, find the **A**, **B**, **C**, and **D** matrices in the state equations for this circuit. Let $i(t)$ be x_1.

2. Write a set of state equations for the circuit in Figure P9.2.

3. An inductor L and a resistor R are in parallel. In series with this combination is a capacitor C. The entire circuit is driven by a voltage source $v_s(t)$. Develop a set of state and output equations for this system. Assume the output variable to be the capacitor voltage.

4. Write state equations for the circuit in Figure P9.4.

5. Write state equations for the circuit in Figure P9.5.

6. Write state equations for the circuit in Figure P9.6.

7. Write state equations for the circuit in Figure P9.7.

8. Write state equations for the circuit in Figure P9.8.

9. Write state equations for the circuit in Figure P9.9.

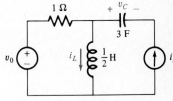

figure P9.2

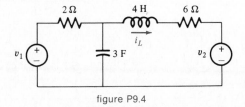

figure P9.4

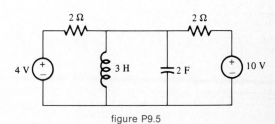

figure P9.5

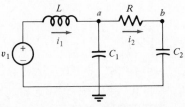

figure P9.6

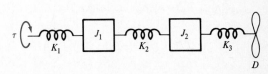

figure P9.7

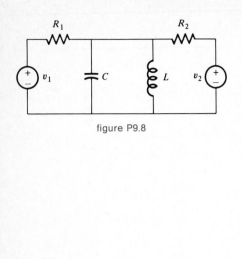

figure P9.8

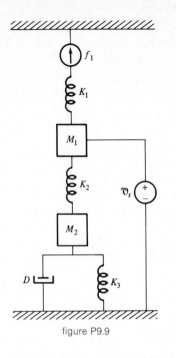

figure P9.9

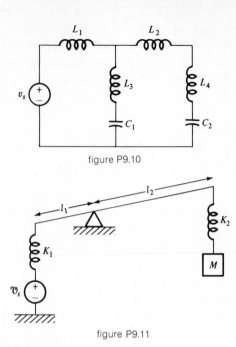

figure P9.10

figure P9.11

10. Write state equations for the circuit in Figure P9.10.

11. Write state equations for the circuit in Figure P9.11.

12. Write state equations for the circuit in Figure P9.12. Show the equations in written-out matrix form (each element in the matrix shown explicitly).

13. Write state equations for the circuit shown in Figure P9.13. The output variable is the voltage $v_o(t)$ across the inductor. Find the **A, B, C,** and **D** matrices.

14. For the two-input circuit in Figure P9.14 use i_L and v_C as state variables. The output is voltage $v_a(t)$. Find **A, B, C,** and **D**.

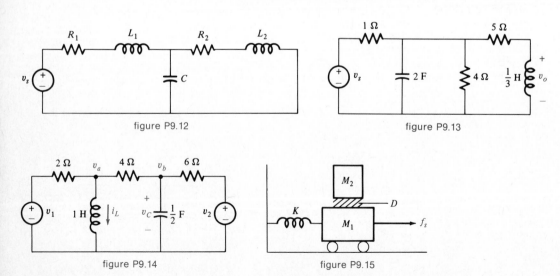

figure P9.12

figure P9.13

figure P9.14

figure P9.15

15. (*a*) Sketch the schematic diagram of the mechanical system in Figure P9.15. (*b*) Write a set of state equations and put them into matrix form.

16. Using the results of problem 8.25, choose an appropriate set of state variables and write the corresponding state and output equations. Write the **A**, **B**, **C**, and **D** matrices.

17. (*a*) Sketch a controller-form block diagram for the single-input–two-output system in Figure P9.17. (*b*) Define a set of state variables consistent with your answer to part (*a*) and write the state and output equations and the **A**, **B**, **C**, and **D** matrices.

18. An attempt is made to stabilize by negative feedback an inherently unstable system whose transfer function is $1/[p(p-1)]$ (Figure P9.18). (*a*) Will this scheme result in an overall stable system? (*b*) Suppose, in addition to the above feedback scheme, we set the input $u(t)$ proportional to the time derivative of the output (so that $u = kpy$). Write the resulting controller-form state equations. (*c*) Can we now achieve a damping factor $\zeta = 0.707$? How?

19. (*a*) Verify the results of Example 9.1.1. (*b*) Write a set of state equations for this circuit and determine the system eigenvalues from the resulting **A** matrix.

20. (*a*) Write the state equations for the block diagram shown in Figure P8.22. (*b*) Write the **A**, **B**, **C**, and **D** matrices. (*c*) Find the system eigenvalues.

21. Find the eigenvalues of the circuit in problem 1 from the **A** matrix.

22. Use the state equations in problem 2 to determine the eigenvalues of that system.

23. Write the state equations for the block diagram in Figure P8.23. Write the **A**, **B**, **C**, and **D** matrices and find the system eigenvalues.

24. Use the results of problem 8.25 to write *two different* sets of state and output equations for the given transfer function. Find the eigenvalues of each of the two resulting **A** matrices.

25. Use the results of problem 8.30 to write *two different* sets of **A**, **B**, **C**, and **D** matrices for the system described in that problem.

26. Use the result of problem 8.33 to develop a set of state equations for the system whose transfer function is given in that problem.

27. Use the result of problem 8.34 to develop a set of state equations for the system whose transfer function is given in that problem.

28. Use Guillemin's iterative form (see problem 8.35) for the transfer function

$$H(p) = \frac{(p+2)(p+4)(p+6)}{(p+3)(p+5)(p+7)}$$

to find a set of state equations for this system. Find the **A**, **B**, **C**, and **D** matrices.

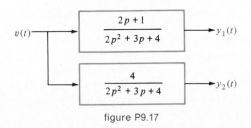

figure P9.17

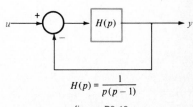

$$H(p) = \frac{1}{p(p-1)}$$

figure P9.18

29. In the squid-versus-herring problem (Example 9.6.2) the units of time are years. For several years after any severe disturbance, the populations in this system will oscillate. (*a*) What is the period of this oscillation? (*b*) Suppose every last squid were removed from this area and all surrounding areas. Would the herring population grow larger without bound? Would there be an upper limit on the number of herring per acre? If so, what is its value? (*c*) If the squid-free condition described in part (*b*) has existed for many years and then a pair of fertile squid is introduced into this area, how many squid will be there 2 years later?

30. A simple pendulum consists of a mass M at the end of a stiff, massless rod of length l which is suspended from a frictionless pivot. A torque (input signal) can be applied to the pendulum at the pivot. The output variable θ is the instantaneous value of the angle of the pendulum from its quiescent (straight-down) position. (*a*) Verify that the equation of motion of this system is

$$p^2\theta + \frac{g}{l}\sin\theta = \frac{\tau_{\text{applied}}}{J}$$

where g is the acceleration due to gravity. *Hint*: The moment of inertia of the system is $J = Ml^2$. (*b*) Develop a block diagram for this system. One of the blocks will contain the operator $(g/l)\sin(\cdot)$. (*c*) Write the state equations for this block diagram. (*d*) Linearize these state equations. (*e*) Find all equilibrium points and evaluate the eigenvalues of the linearized system at each such point. (*f*) Sketch several state-space trajectories.

31. The sizes x_1 and x_2 of two opposing armies fighting for control of a Central American country are described by the two state equations

$$\dot{x}_1 = 5x_1 - 4 \times 10^{-4}x_1 x_2$$
$$\dot{x}_2 = 4x_2 - 4.5 \times 10^{-4}x_1 x_2$$

We note from the first terms of both equations that the first group x_1 can conscript "volunteers" at a more rapid rate than the other group. The second terms of each equation seem to indicate that the second group loses more men when combat occurs between the two sides. Identify all equilibrium points, eigenvalues, eigenvectors, etc. Sketch several trajectories. If army x_1 has 9000 soldiers, how many does x_2 need (minimum) to take over the country?

32. For the nerve-cell model in Example 9.6.1, set the time derivatives of each state variable, as given by the state equations, equal to zero and plot the resulting loci. Comment on the significance of these loci.

33. A series circuit, $R = \frac{1}{3}\,\Omega$, $L = \frac{1}{12}\,$H, and $C = 4\,$F, is driven by a voltage-source input. Consider the current to be the output variable. (*a*) Develop a controller-form set of state equations; find the eigenvalues; find the eigenvectors; sketch a few typical zero-input state trajectories. (*b*) Repeat part (*a*) but use the typical circuit-theory variables i_L and v_c. (*c*) Repeat using the parallel form.

34. The faculty of a certain engineering school is made up of x_1 untenured and x_2 tenured professors. The school has the following personnel policy. Each year a number of new untenured people are hired equal to 10 percent of the entire faculty. Also 10 percent of the untenured faculty are given tenure and 10 percent are laid off. Historical data show that 5 percent of the tenured people retire or leave each year. (*a*) Sketch some typical state-space trajectories of resulting faculty rank populations. (*b*) In the long run, tenured-faculty growth will be what percentage greater than untenured-faculty growth? (*c*) When this policy is *first instituted*, what ratio of tenured to untenured faculty will result *initially* in the most constant, unchanging population of untenured faculty?

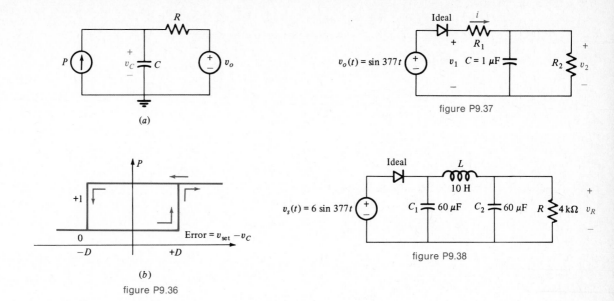

figure P9.37

(a)

(b)

figure P9.36

figure P9.38

35. Use the results of problem 3 to determine the eigenvalues (in terms of the circuit-element values) of that system.

36. The model of a home heating system contains the following elements: P = furnance to supply heat, C = heat capacity of the house, R = insulation of walls and roof, v_o = outside temperature. The thermostat has a dead zone $2D°F$ wide, as shown. Assume $R = 100$, $C = 0.05$, $D = 1°F$, and $v_o = 40°F$. Write a computer program to investigate the operation of this model. The units of time are hours. Use DELT = 0.001. Let the COST of fuel be equal to P*DELT. Investigate the following questions. (a) If the homeowner is going to be away for 8 h, should she set the thermostat down to 65°F from 70°F or does she lose by doing this because the furnace has to run for a long time when she comes home and sets the thermostat to 70°F? (b) Does a 10 percent increase in insulation R yield an average 10 percent reduction in fuel costs? (c) Does a 50 percent increase in furnace capacity (*not efficiency*) yield a saving in fuel cost?

37. Consider the primitive power-supply circuit shown in Figure P9.37 and investigate the advantages and disadvantages to the designer of high (and low) values of both R_1 and R_2. To do this plot v_o, v_2, and i versus t for two cycles of the input waveform $v_o(t) = \sin 377t$ for (a) $R_1 = 100\ \Omega$, $R_2 = 100\ k\Omega$; (b) $R_1 = 100\ \Omega$, $R_2 = 10\ k\Omega$; and (c) $R_1 = 1000\ \Omega$, $R_2 = 10\ k\Omega$.

38. Consider the practical power-supply circuit shown in Figure P9.38. Using state variables $x_1 = v_{C1}$, $x_2 = i_L$, and $x_3 = v_{C2}$, develop a computer program to print out and/or plot the zero-state response when $v_s(t)$ is switched on at $t = 0$. (a) What is the maximum value of v_R and when does it occur? (b) What must the current rating of the diode be? (c) R is changed to 1 kΩ. Repeat (a) and (b).

39. Write a set of state equations for the circuit in Figure P9.39. Use v_C and i_L as the state variables. The voltage source is a unit step, and the initial conditions are zero. Write a computer program to obtain the state trajectory. Use a time increment of 1 μs in your program for (a) $R_1 = R_2 = 1\ k\Omega$ and (b) $R_1 = 3\ k\Omega$, $R_2 = 100\ \Omega$.

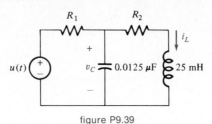

figure P9.39

figure P9.40

40. Write a computer program to plot (or at least print out) the values of currents i_1 and i_2 in Figure P9.40. Plot the results over the interval $t = 0$ to $t = 2$ s.

41. As an example of a time-varying system consider the circuit of problem 3 as the electrical analog of a mechanical system. Given that $L = C = 1$ and that R changes from 10 to 1 Ω according to the following rule. If $f(t) = \sin(4\pi t)$ is positive, then $R = 10 \Omega$; otherwise $R = 1 \Omega$. Suppose we want to increase the velocity of the mass from initial rest to 2 m/s (v_C from 0 to 2 V) as quickly as possible and with a minimum of fluctuation (*hunting*) in the resulting velocity. Develop a computer program that will help you sketch the state trajectories for several different input voltages $v_s(t)$. For example, try, among others, $v_s(t) = 2$ and $v_s(t) = 4 - v_C(t)$. Can you think of some better ways to specify $v_s(t)$? The real answers to such questions lie in the area of electrical engineering called *optimum control theory*.

For each of the four following problems you are now asked to (a) develop state equations, (b) use a computer to solve for the resulting state trajectories, and (c) plot each variable versus time. If you have not done those earlier problems, do them now as a timely review.

42. Problem 7.25.

43. Problem 7.25 with $R = 0.5 \Omega$ and $v_C(0+) = 3$ V.

44. Example 7.6.2 (Figure 7.6.2).

45. Problem 7.35.

chapter 10
SYSTEMS WITH SINUSOIDAL INPUTS

introduction 10.1

The entire power grid of every civilized nation today uses sinusoidal voltages to supply the electrical requirements of homes and industries. You can imagine how complicated the circuit diagram of even a small part of any such national power grid would be. A small town or one fair-sized factory would have a very complicated *RLC* circuit with one or several sinusoidal sources (power-company lines) and many switches, relays, and transformers. All radio, TV, and satellite communications systems use sinusoidal signals. Even the digital signals we hear so much about nowadays are really bursts of sinusoidal signals when they are transmitted by radio or telecommunication links.

Therefore, finding the resulting sinusoidal currents and voltages in some complicated circuit is a recurring problem in electrical engineering. At this point in our study of circuits and systems we know at least one way of solving such problems as long as the network is linear. This method is of course the basic one of finding the complete response by (1) deriving the system differential equation, (2) finding the natural response, (3) finding the particular (forced) response, and (4) adding these together and evaluating unknown coefficients by means of initial conditions.

A useful shortcut† for obtaining the particular response to any sinusoidal input is based on a simple geometrical fact. Consider a point on a plane which is moving in a circle with constant angular velocity. If we observe this motion from a point *in the plane* (and far enough away to be able to ignore any parallax effect), the point appears to be oscillating sinusoidally with time. Have you ever seen a bicycle rider at night in your headlights? The reflective pedals appear to go up and down sinusoidally if the bicycle is moving at a constant velocity.

If we want to talk about sinusoidal waveforms and use them as inputs to systems, it is far simpler and easier mathematically to work with the point that rotates in a circle. The *projection* (shadow) of this point is the sinusoidal waveform we are actually interested in. We shall use a constant-radius vector that rotates in a circle centered at the origin of the complex (real-axis–imaginary-axis) plane. Since this method requires the use of complex numbers to describe this rotating vector (complex) quantity, we begin with a brief review of the algebra of complex numbers.

10.2   complex numbers

A complex number is the *sum* of a *real* number and an *imaginary* number. Since the square root of -1 is defined as the imaginary operator, an imaginary number is the product of $\sqrt{-1}$ times a real number. Let‡ $j = \sqrt{-1}$; then

$$3 = \text{a real number}$$

$$j4 = \text{an imaginary number}$$

and
$$3 + j4 = \text{a complex number}$$

A convenient geometric interpretation of any complex number is that it represents a point on a complex plane (Figure 10.2.1). By geometry we see that such a point can always be described by giving its radial distance from the origin and its mathematical angle measured counterclockwise from the positive real axis. Thus we write $A = 3 + j4 = 5\underline{/53.13°}$, where $3 + j4$ is the *cartesian* or *rectangular form* and $5\underline{/53.13°}$ the *polar* form of the number. We thus have two different but equivalent ways of writing a complex number; we shall soon find a third way.

† This method is due to Charles Proteus Steinmetz (1865–1923) an electrical engineer with General Electric in Schenectady, N.Y.

‡ Mathematicians and physicists use the letter i to denote this quantity, but electrical engineers use j instead to avoid confusion with i denoting current.

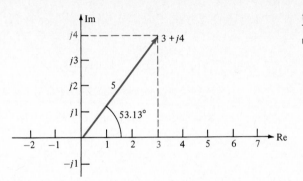

figure 10.2.1
The complex number plane showing the number $3 + j4$.

EXAMPLE 10.2.1

Find the polar form of $A = 9 + j7$.

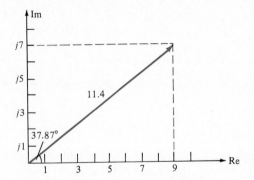

figure 10.2.2
Polar and rectangular forms of $9 + j7 = 11.4\underline{/37.87°}$.

ANS.: The magnitude (hypotenuse) is $\sqrt{9^2 + 7^2}$, and the angle is the angle whose tangent is $\frac{7}{9}$. Thus

$$A = 11.4\underline{/37.87°} = \text{polar form}$$

See Figure 10.2.2.

We add complex numbers by adding the real and imaginary parts separately.

ADDITION OF COMPLEX NUMBERS

EXAMPLE 10.2.2

$$(3 + j4) + (2 + j8) = 5 + j12$$

Addition can be done only after converting all numbers to be added into rectangular form. Note that this definition of complex addition is consistent with two-dimensional vector addition.

EXAMPLE 10.2.3

Find $5\underline{/30°} + (-2 + j2)$.

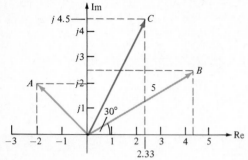

figure 10.2.3
The sum of two complex numbers (vectors). **A** + **B** = **C**.

ANS.:

$$5\underline{/30°} = \quad 4.33 + j2.5$$
$$\underline{\quad -2 \quad + j2 \quad}$$
$$2.33 + j4.5$$

See Figure 10.2.3.

SUBTRACTION OF COMPLEX NUMBERS   Subtraction is accomplished exactly the same way as addition after changing the sign of the real and imaginary parts of the subtrahend.

EXAMPLE 10.2.4

Find $5\underline{/30°} - (-2 + j2)$.

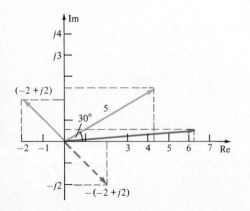

figure 10.2.4
The difference (subtraction process) between two complex numbers; see Example 10.2.4.

ANS.:

$$5\underline{/30^\circ} = 4.33 + j2.5$$
$$\underline{2\quad - j2}$$
$$6.33 + j0.5$$

This is equivalent to rotating $-2 + j2$ through 180° and then adding (see Figure 10.2.4).

Multiplication of complex numbers is defined as

$$(A\underline{/\alpha})(B\underline{/\beta}) = AB\underline{/\alpha + \beta}$$

or, $(a + jb)(c + jd) = ac + jbc + jad + j^2bd = (ac - bd) + j(bc + ad)$

Thus, we see that multiplication can be done in either form, cartesian or polar, but polar is usually much easier.

MULTIPLICATION OF COMPLEX NUMBERS

EXAMPLE 10.2.5

$$(5\underline{/23^\circ})(11\underline{/14^\circ}) = 55\underline{/37^\circ}$$

Division is the inverse of multiplication

$$\frac{AB\underline{/\alpha + \beta}}{B\underline{/\beta}} = A\underline{/\alpha}$$

This is easiest in polar form: divide the magnitudes and subtract the denominator angle from the numerator angle.

DIVISION

EXAMPLE 10.2.6

$$\frac{55\underline{/37^\circ}}{11\underline{/14^\circ}} = 5\underline{/23^\circ}$$

Taking the nth root of a complex number $A = |A|\underline{/\alpha}$ is accomplished by seeking another complex number $B = |B|\underline{/\beta}$ such that $B^n = A$:

$$\underset{n\ entries}{B \cdot B \cdot B \cdots B} = A$$

ROOTS

Therefore the nth root of A is given by $B_1 = \sqrt[n]{|A|}\ \underline{/\alpha/n}$. But remember that $A\underline{/\alpha} = A\underline{/\alpha + 360^\circ} = A\underline{/\alpha + 720^\circ} = \cdots$, therefore another value of B is obtained from

$$B_2 = \sqrt[n]{|A|}\ \underline{\bigg/\frac{\alpha + 360}{n}}$$

and a third is

$$B_3 = \sqrt[n]{|A|} \left| \frac{\alpha + 720°}{n} \right.$$

In general there will be n distinct values for B when we take the nth root of A.

EXAMPLE 10.2.7

Find the square root of $9\underline{/60°}$.
ANS.:   One root is

$$\sqrt{9} \left| \frac{60°}{2} \right. = 3\underline{/30°} \qquad \text{such that} \qquad (3\underline{/30})(3\underline{/30}) = 9\underline{/60°}$$

Another root is

$$\sqrt{9} \left| \frac{60 + 360}{2} \right. = 3\underline{/210°} = -(3\underline{/30°})$$

EXAMPLE 10.2.8

Find the cube root of $9\underline{/60°}$.

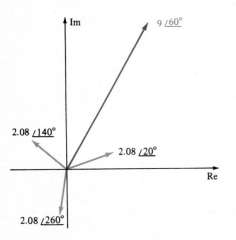

figure 10.2.5.
The quantity $9\underline{/60°}$ and its cube roots.

ANS.: $$(9\underline{/60°})^{1/3} = (9\underline{/420°})^{1/3} = (9\underline{/780°})^{1/3}$$

Therefore the three values are

$$2.08\underline{/20°} \qquad 2.08\underline{/140°} \qquad 2.08\underline{/260°}$$

See Figure 10.2.5.

Any complex number $C\underline{/\theta}$ can be written in *exponential form* $C\underline{/\theta} = Ce^{j\theta}$. We can prove this as follows. Consider the infinite series for e^x

$$e^x = 1 + x + \frac{x^2}{2!} + \frac{x^3}{3!} + \frac{x^4}{4!} + \frac{x^5}{5!} + \cdots \qquad (10.2.1)$$

Letting $x = j\theta$ gives

$$e^{j\theta} = 1 + j\theta - \frac{\theta^2}{2!} - j\frac{\theta^3}{3!} + \frac{\theta^4}{4!} + j\frac{\theta^5}{5!} \cdots \qquad (10.2.2)$$

The infinite series for $\cos\theta$ and $\sin\theta$ are

$$\cos\theta = 1 - \frac{\theta^2}{2!} + \frac{\theta^4}{4!} - \frac{\theta^6}{6!} + \cdots \qquad (10.2.3)$$

and

$$\sin\theta = \theta - \frac{\theta^3}{3!} + \frac{\theta^5}{5!} - \frac{\theta^7}{7!} + \cdots \qquad (10.2.4)$$

From equations (10.2.2) to (10.2.4) we see that†

$$e^{j\theta} = \cos\theta + j\sin\theta \qquad (10.2.5)$$

or, multiplying by the scalar constant C,

$$Ce^{j\theta} = C\cos\theta + jC\sin\theta \qquad (10.2.6)$$

The quantity on the right-hand side of (10.2.6) is a complex number whose magnitude is

$$\sqrt{C^2\cos^2\theta + C^2\sin^2\theta} = C$$

and whose angle is

$$\tan^{-1}\frac{C\sin\theta}{C\cos\theta} = \tan^{-1}(\tan\theta) = \theta \qquad \text{QED}$$

Therefore $e^{j(\cdot)}$ is simply the equivalent of $\underline{/\quad}$.

EXPONENTIAL FORM OF A COMPLEX NUMBER

EXAMPLE 10.2.9

$$4e^{j45°} = 4\underline{/45°}$$

This is consistent with the rule pertaining to products of exponentials $e^a e^b = e^{a+b}$, for example, $e^3 e^{-5} = e^{-2}$ and now

$$e^{j\pi}e^{j\pi/2} = e^{j3\pi/2} = 1\underline{\left/\frac{3\pi}{2}\right.}\text{ rad}$$

Other examples are

$$e^{j40°}e^{j22°} = e^{j62°} = 1\underline{/62°}$$

† Equation 10.2.5 is called *Euler's relationship*.

and
$$e^{-4+j3} = e^{-4}e^{j3} = 0.0183\underline{/3} \text{ rad} = 0.0183\underline{/171.9°}$$

CONJUGATE   The conjugate of a complex number is obtained by changing the algebraic sign of the imaginary part (in rectangular form). If the number is in polar form, change the sign of the angle. The conjugate of complex number C is written C^*.

EXAMPLE 10.2.10

$$C = 3 + j4 \qquad C^* = 3 - j4 \qquad D = 73\underline{/42°} \qquad D^* = 73\underline{/-42°}$$

On the complex plane there is an upper-half-plane–lower-half-plane symmetry between a complex number and its conjugate (Figure 10.2.6).

DRILL PROBLEM

Evaluate

$$\sqrt[3]{\frac{4e^{j\pi/4}(3 + j4)}{6\underline{/72°}}}$$

ANS.: $1.49\underline{/8.71°}$, $1.49\underline{/128.7°}$, $1.49\underline{/248.7°}$.

10.3 phasors

Consider a complex quantity A whose magnitude is unity and whose angle keeps increasing at a *constant rate* ω rad/s. If the angle is $0°$ when $t = 0$, we can write this algebraically as

$$A = 1\underline{/\omega t} = e^{j\omega t} \qquad (10.3.1)$$

This quantity is called a *complex exponential function*. Using Euler's relationship, we can write equation (10.3.1) as

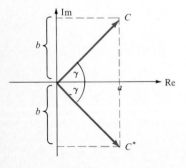

figure 10.2.6
The complex number $C = |C|\underline{/\gamma} = a + jb$, and its complex conjugate $C^* = |C|\underline{/-\gamma} = a - jb$.

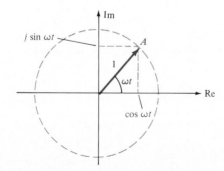

figure 10.3.1
A complex exponential function

$$A = e^{j\omega t} = \cos \omega t + j \sin \omega t$$

$$A = \cos \omega t + j \sin \omega t \qquad (10.3.2)$$

See Figure 10.3.1.

Even though we cannot obtain complex voltage sources for actual use in the laboratory, it is convenient to use them mathematically. Such a voltage could be written as

$$v(t) = V_m e^{j(\omega t + \theta)} = V_m \cos (\omega t + \theta) + jV_m \sin (\omega t + \theta) \qquad (10.3.3)$$

so that

$$V_m \cos (\omega t + \theta) = \text{Re } V_m e^{j(\omega t + \theta)} \qquad (10.3.4)$$

where Re means the *real part of*. The real part of the complex number $A = 3 + j4$ is simply Re $A = 3$. The real part of a complex exponential is obtained in exactly the same way. Graphically, it is the projection of the complex (vector) quantity onto the real axis. In Figure 10.3.2 the real part of the complex exponential function is obtained from simple trigonometry by dropping a perpendicular vertically down from the tip of the complex exponential to the real axis. Thus, equation (10.3.4) is seen to be correct.

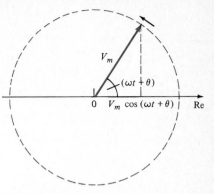

figure 10.3.2
The complex exponential function $v(t) = V_m e^{j(\omega t + \theta)}$ showing its real part to be Re $v(t) = V_m \cos (\omega t + \theta)$.

Rewriting equation (10.3.3), we have

$$v(t) = V_m e^{j\theta} e^{j\omega t} = (V_m \underline{/\theta})e^{j\omega t} \qquad (10.3.5)$$

where the complex coefficient in parentheses is called the *phasor*† for $v(t)$. Equation (10.3.5) simply states that $v(t)$ is a vector on the complex plane that rotates counterclockwise (direction of increasing mathematical angle). At $t = 0$ its value is $V_m \underline{/\theta}$. One way to define a phasor quantity is to say that it is a snapshot (instantaneous still photograph) of the rotating complex exponential taken at $t = 0$.

Whenever we have to deal with a pure *sinusoidal* voltage or current (or any other quantity for that matter), it is usually easier to work with its corresponding phasor, the complex amplitude of the complex exponential whose real part is our real sinusoid. For example, to add two sinusoids we can simply add their corresponding phasors, because the real part of the sum of two complex quantities is the sum of their individual real parts.

Let us seek the sum $s(t)$ of

$$x(t) = X \cos (\omega t + \alpha) \qquad (10.3.6)$$

and

$$y(t) = Y \cos (\omega t + \beta) \qquad (10.3.7)$$

The corresponding complex exponentials are

$$\hat{x}(t) = X e^{j(\omega t + \alpha)} = X e^{j\alpha} e^{j\omega t} = X \underline{/\alpha} e^{j\omega t} \qquad (10.3.8)$$

and

$$\hat{y}(t) = Y e^{j(\omega t + \beta)} = Y e^{j\beta} e^{j\omega t} = Y \underline{/\beta} e^{j\omega t} \qquad (10.3.9)$$

where

$$\text{Re } \hat{x}(t) = X \cos (\omega t + \alpha) = x(t)$$

and

$$\text{Re } \hat{y}(t) = Y \cos (\omega t + \beta) = y(t)$$

The complex sum $\hat{s}(t)$ is

† Phasor quantities are denoted by boldface type, $\mathbf{V} = V_m \underline{/\theta}$.

$$\hat{s}(t) = \hat{x}(t) + \hat{y}(t) = X\underline{/\alpha}e^{j\omega t} + Y\underline{/\beta}e^{j\omega t} = (X\underline{/\alpha} + Y\underline{/\beta})e^{j\omega t}$$
$$= S\underline{/\gamma}e^{j\omega t}$$
$$= Se^{j\gamma}e^{j\omega t}$$
$$= Se^{j(\omega t + \gamma)} \qquad\qquad\qquad (10.3.10)$$

and the real part of $\hat{s}(t)$ is

$$\text{Re } \hat{s}(t) = S \cos (\omega t + \gamma)$$

Note that in computing equation (10.3.10) the term $e^{j\omega t}$ is a common factor in both $\hat{x}(t)$ and $\hat{y}(t)$, which means that $x(t)$ *and* $y(t)$ *must have the same radian frequency* ω. We cannot add the phasors $X\underline{/\alpha}$ and $Y\underline{/\beta}$ if $x(t)$ and $y(t)$ have different frequencies. In equation (10.3.10) we see that

$$Se^{j\gamma} = X\underline{/\alpha} + Y\underline{/\beta} \qquad \text{or} \qquad \mathbf{S} = \mathbf{X} + \mathbf{Y}$$

EXAMPLE 10.3.1

Given

$$v_1(t) = 10 \cos (3t - 76°) \qquad \text{and} \qquad v_2(t) = 5 \cos (3t + 71°)$$

find $v_1(t) + v_2(t)$.

ANS.:   Find the complex exponential voltage, the real part of which is $v_1(t)$. Do the same for $v_2(t)$. Add the complex voltages. Take the real part of the result:

$$10 \cos (3t - 76°) \rightarrow 10e^{j(3t-76°)} \qquad \text{for } t = 0 \rightarrow 10\underline{/-76°} = \mathbf{V}_1$$
$$5 \cos (3t + 71°) \rightarrow 5e^{j(3t+71°)} \qquad \text{for } t = 0 \rightarrow \underline{5\underline{/+71°} = \mathbf{V}_2}$$
$$\text{phasors}$$

$$\mathbf{V}_1 = 10\underline{/-76°} = 2.42 - j9.70$$
$$\mathbf{V}_2 = 5\underline{/+71°} = \underline{1.63 + j4.73}$$

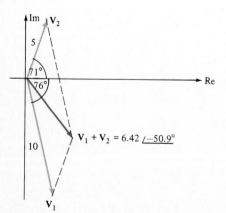

figure10.3.3
The (complex) sum of the phasors in Example 10.3.1.

$$\mathbf{V}_1 + \mathbf{V}_2 = 4.05 - j4.97 = 6.42\underline{/-50.9^\circ}$$

$$6.42\underline{/-50.9^\circ}e^{j3t} = 6.42e^{j(3t-50.9^\circ)}$$

$$v_1(t) + v_2(t) = \mathrm{Re}\,(6.42e^{j(3t-50.9^\circ)}) = 6.42\cos{(3t-50.9^\circ)}$$

This addition process is shown graphically in Figure 10.3.3.

Phasors are also useful in converting from one form of a sinusoidal expression to another. Since the real part of the complex exponential *at any instant of time* is the instantaneous value of the real function, the phasors for cos ωt and for sin ωt must be as shown in Figure 10.3.4. For

$$v_{\cos}(t) = V_m \cos{\omega t}$$

the corresponding complex exponential function is

$$\hat{v}_{\cos}(t) = V_m\,e^{j\omega t} = (V_m\underline{/0^\circ})e^{j\omega t} \qquad (10.3.11)$$

such that $$\mathrm{Re}\,\hat{v}_{\cos}(t) = v_{\cos}(t)$$

Similarly, for

$$v_{\sin}(t) = V_m \sin{\omega t} = V_m \cos{(\omega t - 90^\circ)}$$

the corresponding complex exponential function is

$$\hat{v}_{\sin}(t) = V_m\,e^{j(\omega t - 90^\circ)} = (V_m\underline{/-90^\circ})e^{j\omega t} \qquad (10.3.12)$$

such that $$\mathrm{Re}\,\hat{v}_{\sin}(t) = V_m \cos{(\omega t - 90^\circ)} = V_m \sin{\omega t}$$

Evaluating equations (10.3.11) and (10.3.12) at $t = 0$ yields the phasor quantities

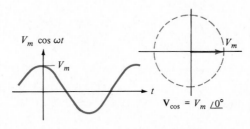

(a)

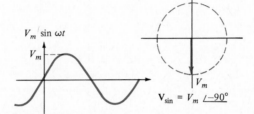

(b)

figure 10.3.4
The waveform, the complex exponential, and phasor for (a) $V_m \cos{\omega t}$ and (b) $V_m \sin{\omega t}$. Both complex exponentials are rotating counterclockwise with angular velocity ω rad/s and are pictured at $t = 0$.

$$\mathbf{V}_{\cos} = V_m \underline{/0^\circ} \quad \text{and} \quad \mathbf{V}_{\sin} = V_m \underline{/-90^\circ}$$

Figure 10.3.4 shows the rotating complex exponential functions as they are positioned at $t = 0$. As time increases from zero, the instantaneous value of $V_m \cos \omega t$ decreases and $V_m \sin \omega t$ increases (from zero). Note that the respective projections of the two complex exponentials onto the real axis do exactly that. As a result of the information in Figure 10.3.4 we see that sinusoids can be easily converted from sine to cosine form. Simply remember that the phasor for sine is at the -90° position and the phasor for cosine is at the 0° position (see the examples in Figure 10.3.5).

EXAMPLE 10.3.2

Find a single term equivalent to $2 \cos 4t + 4 \sin 4t$ (see Figure 10.3.6).
ANS.: $4.472 \cos (4t - 63.4^\circ)$ or $4.472 \sin (4t + 26.6^\circ)$.

DRILL PROBLEM

In Figure 10.3.5 evaluate the sum of the three time functions $x_1(t) + x_2(t) + x_3(t)$.
ANS.: $9.78 \sin (\omega t - 10.9^\circ)$.

---

figure 10.3.5
Three different sinusoids with their phasors and equivalent alternate forms.

figure 10.3.6
Performing the phasor sum of the two sinusoids of Example 10.3.2.

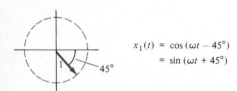

$$x_1(t) = \cos (\omega t - 45^\circ)$$
$$= \sin (\omega t + 45^\circ)$$

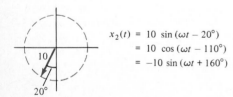

$$x_2(t) = 10 \sin (\omega t - 20^\circ)$$
$$= 10 \cos (\omega t - 110^\circ)$$
$$= -10 \sin (\omega t + 160^\circ)$$

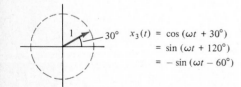

$$x_3(t) = \cos (\omega t + 30^\circ)$$
$$= \sin (\omega t + 120^\circ)$$
$$= -\sin (\omega t - 60^\circ)$$

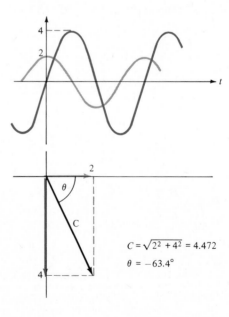

$$C = \sqrt{2^2 + 4^2} = 4.472$$
$$\theta = -63.4^\circ$$

linear systems with sinusoidal inputs 10.4

The real advantage of using complex (phasor) descriptions of sinusoidal voltage, current, etc., sources comes when we use these sources as inputs to linear systems. We know that a sinusoidal input voltage

$$v(t) = V_m \cos(\omega t + \theta) \tag{10.4.1}$$

will produce a sinusoidal *particular-response* current that has the same frequency as $v(t)$

$$i_p(t) = I_m \cos(\omega t + \phi) \tag{10.4.2}$$

in any linear electric network.

EXAMPLE 10.4.1

Find $i_p(t)$ in the circuit of Figure 10.4.1 when $v(t) = 4 \cos 3t$.

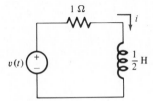

figure 10.4.1
The circuit of Example 10.4.1.

ANS.: Since $di/dt + 2i = 2(4 \cos 3t)$, assume that

$$i_p = k_1 \cos 3t + k_2 \sin 3t$$

Insert this back into the differential equation and get

$$-3k_1 \sin 3t + 3k_2 \cos 3t + 2k_1 \cos 3t + 2k_2 \sin 3t = 8 \cos 3t$$

Balancing this equation term by term gives $k_1 = \frac{16}{13}$ and $k_2 = \frac{24}{13}$, so that

$$i_p(t) = \tfrac{16}{13} \cos 3t + \tfrac{24}{13} \sin 3t = 2.22 \cos(3t - 56.3°)$$

Also, from equations (10.4.1) and (10.4.2), using the *homogeneity* or *multiplicative property* of linearity, we find that the purely imaginary source

$$v(t) = jV_m \sin(\omega t + \theta) \tag{10.4.3}$$

will produce

$$i_p(t) = jI_m \sin(\omega t + \phi) \tag{10.4.4}$$

We conclude that *a real input produces a real output* and *an imaginary input produces an imaginary output*. Therefore (from the additive property of linearity)

$$v(t) = V_m \cos(\omega t + \theta) + jV_m \sin(\omega t + \theta) \tag{10.4.5}$$

will produce

$$i_p(t) = I_m \cos(\omega t + \phi) + j I_m \sin(\omega t + \phi) \qquad (10.4.6)$$

or

$$v(t) = V_m e^{j(\omega t + \theta)} \qquad (10.4.7)$$

will produce

$$i_p(t) = I_m e^{j(\omega t + \phi)} \qquad (10.4.8)$$

The same result could be obtained for a current-source input and/or voltage output or any input-output combination. Thus *a complex exponential source produces a complex exponential output at the same radian frequency* ω.

It turns out to be easier to determine the particular response due to a complex exponential source than that due to a real one. Therefore, the best way to find the particular response if the source is sinusoidal is to use the corresponding complex exponential source as input, find the complex exponential output, and take its real part.

EXAMPLE 10.4.2

Solve the system equation of Example 10.4.1 by means of complex exponentials.
ANS.:

$$\frac{di}{dt} + 2i = 2v(t) \qquad \text{where } v(t) = 4 \cos 3t$$

Use, instead, $\hat{v}(t) = 4e^{j3t}$ to find $\hat{i}_p(t)$. We do this by seeking $\hat{i}_p(t)$ in the form

$$\hat{i}_p(t) = I_m e^{j(3t + \phi)} = (I_m e^{j\phi})e^{j3t}$$

where the quantity in the parentheses is the phasor for i, a complex number

$$\mathbf{I} = I_m \underline{/\phi}$$

Substitute this $\hat{i}_p(t)$ back into the differential equation

$$j3\mathbf{I}e^{j3t} + 2\mathbf{I}e^{j3t} = 8e^{j3t}$$

$$(2 + j3)\mathbf{I} = 8$$

$$\mathbf{I} = \frac{8}{2 + j3} = \frac{8\underline{/0^\circ}}{3.6\underline{/56.3^\circ}} = 2.22\underline{/-56.3^\circ}$$

Therefore

$$\hat{i}_p(t) = (2.22e^{-j56.3^\circ})e^{j3t} = 2.22e^{j(3t - 56.3^\circ)}$$

whose real part, as before, is

$$i_p(t) = 2.22 \cos(3t - 56.3^\circ)$$

EXAMPLE 10.4.3

Find $y_p(t)$ for

$$\frac{d^2y}{dt^2} + \frac{dy}{dt} + \frac{4}{3}y = 6\cos(2t + 60°)$$

ANS.:   Use, instead, the forcing function $(6\underline{/60°})e^{j2t}$. Thus $\hat{y}_p(t)$ is of the form $\mathbf{Y}e^{j2t}$ and

$$p\hat{y}_p(t) = j2\mathbf{Y}e^{j2t} \qquad \text{and} \qquad p^2\hat{y}_p(t) = -4\mathbf{Y}e^{j2t}$$

Substituting into the differential equation gives

$$(-4 + j2 + \tfrac{4}{3})\mathbf{Y}e^{j2t} = (6\underline{/60°})e^{j2t}$$

or

$$\mathbf{Y} = \frac{6\underline{/60°}}{-8/3 + j2} = \frac{6\underline{/60°}}{3.33\dots\underline{/143°}} = 1.8\underline{/-83°}$$

Taking the real part

$$\text{Re }(1.8\underline{/-83°}e^{j2t}) = \text{Re } 1.8e^{j(2t-83°)}$$

gives

$$y_p(t) = 1.8\cos(2t - 83°)$$

We saw in an earlier chapter that a transfer-function operator is the p operator that acts on one system variable (the input) and yields another system variable (the output).

EXAMPLE 10.4.4

Find the transfer-function operator that acts on the input voltage v_1 to produce output v_2 in Figure 10.4.2.

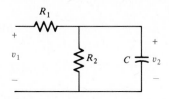

figure 10.4.2
The circuit of Example 10.4.4.

ANS.:

$$(v_2 - v_1)\frac{1}{R_1} + \frac{v_2}{R_2} + C\frac{dv_2}{dt} = 0$$

$$v_2\left(\frac{1}{R_1} + \frac{1}{R_2} + Cp\right) - v_1\left(\frac{1}{R_1}\right) = 0$$

$$v_2(t) = \frac{1/R_1}{1/R_1 + 1/R_2 + Cp}v_1(t) \tag{10.4.9}$$

We can obtain a system differential equation from the p-operator transfer-function equation simply by multiplying through by the denominator polynomial, e.g.,

$$f_o(t) = [H(p)]f_{in}(t) = \frac{a_m p^m + a_{m-1} p^{m-1} + \cdots + a_1 p + a_0}{b_n p^n + b_{n-1} p^{n-1} + \cdots + b_1 p + b_0} f_{in}(t) \quad (10.4.10)$$

$$(b_n p^n + b_{n-1} p^{n-1} + \cdots + b_1 p + b_0) f_o(t)$$

$$= (a_m p^m + a_{m-1} p^{m-1} + \cdots + a_1 p + a_0) f_{in}(t)$$

$$b_n \frac{d^n}{dt^n} f_o + b_{n-1} \frac{d^{n-1} f_o}{dt^{n-1}} + \cdots + b_1 \frac{df_o}{dt} + b_0 f_o$$

$$= a_m \frac{d^m}{dt^m} f_{in} + \cdots + a_0 f_{in}(t) \quad (10.4.11)$$

Let us do this for Example 10.4.4 and then solve for the particular response due to a sinusoidal input.

EXAMPLE 10.4.5

In the circuit in Example 10.4.4, solve for the particular response if $R_1 = 1\ \Omega$, $R_2 = \frac{1}{2}\ \Omega$, $C = 3$ F, and $v_1(t) = 10 \cos 2t$.

ANS.:   Multiply the result of Example 10.4.4 by the denominator

$$\left(\frac{1}{R_1} + \frac{1}{R_2} + Cp \right) v_2(t) = \frac{1}{R_1} v_1(t) \quad (10.4.12)$$

Substitute values for the elements

$$(3 + 3p)v_2(t) = 10 \cos 2t$$

or

$$3 \frac{dv_2}{dt} + 3v_2 = 10 \cos 2t \quad (10.4.13)$$

By the method of complex exponential sources (Example 10.4.2) equation (10.4.13) becomes

$$3 \frac{d\hat{v}_2}{dt} + 3\hat{v}_2 = 10 e^{j2t} \quad (10.4.14)$$

where

$$\hat{v}_2(t) = V_2 \underline{/\theta} e^{j2t} \quad (10.4.15)$$

and

$$\frac{d\hat{v}_2}{dt} = j2 V_2 \underline{/\theta} e^{j2t} \quad (10.4.16)$$

Substitute (10.4.15) and (10.4.16) into (10.4.14)

$$3(j2 V_2 \underline{/\theta} e^{j2t}) + 3 V_2 \underline{/\theta} e^{j2t} = 10 e^{j2t}$$

$$(j6 + 3) V_2 \underline{/\theta} = 10 \underline{/0°}$$

$$V_2 \underline{/\theta} = \frac{1}{3 + j6} 10 \underline{/0°} \quad (10.4.17)$$

$$V_2 \underline{/\theta} = 1.49 \underline{/-63°} \quad (10.4.18)$$

Substituting (10.4.18) into (10.4.15) and taking the real part yields

$$v_2(t) = 1.49 \cos (2t - 63°) \quad (10.4.19)$$

A shortcut is available for reducing the work in Example 10.4.5 and all similar problems. Equation (10.4.17) can be obtained directly from equation (10.4.9) by substituting the numerical value $j\omega$, where ω is the frequency of the input source, for the operator p in the transfer-function equation and substituting the phasor quantities for the time functions. This *converts the operator equation into an algebraic equation*, from which we can solve for the output phasor; i.e., equation (10.4.9) is

$$v_2(t) = \frac{1/R_1}{1/R_1 + 1/R_2 + Cp}\, v_1(t) \qquad (10.4.9)$$

which, with element values inserted is

$$v_2(t) = \frac{1}{3 + 3p}\, v_1(t) \qquad (10.4.20)$$

Since the input frequency $\omega = 2$ rad/s, we substitute

$$\hat{v}_1(t) \to 10\underline{/0°}\,e^{j2t} \qquad \hat{v}_2(t) \to V_2\underline{/\theta}\,e^{j2t} \qquad p \to j\omega = j2 \qquad (10.4.21)$$

into (10.4.20) and obtain

$$V_2\underline{/\theta} = \frac{1}{3 + j6}\, 10\underline{/0°} \qquad (10.4.22)$$

which is identical to equation (10.4.17).

If we had not specified any particular numerical value for ω, the best we could have done would have been to write

$$\mathbf{V}_2 = \left(\frac{1}{3 + j3\omega}\right)\mathbf{V}_1$$

where $\mathbf{V}_1$ and $\mathbf{V}_2$ are phasors. The quantity inside the parentheses is called the *sinusoidal transfer function* $H(j\omega)$. We can plot the magnitude and the angle of this transfer function versus ω. First find $|H(j\omega)|$ and $\underline{/H(j\omega)}$ from $H(j\omega)$:

$$H(j\omega) = \frac{1}{3 + j3\omega} = \frac{1}{3 + j3\omega}\,\frac{3 - j3\omega}{3 - j3\omega}$$

and then

$$H(j\omega) = \frac{3 - j3\omega}{9 + 9\omega^2} = \frac{1}{3 + 3\omega^2} - j\,\frac{\omega}{3 + 3\omega^2} = \frac{1}{3 + 3\omega^2}\,(1 - j\omega)$$

See Figure 10.4.3. Thus the magnitude of $H(j\omega)$ is found by letting $A = 1/(3 + 3\omega^2)$, so that $H(j\omega) = A(1 - j\omega)$, which leads to

$$|H(j\omega)| = \sqrt{A^2 + \omega^2 A^2} = A\sqrt{1 + \omega^2} \qquad |H(j\omega)| = \frac{\sqrt{1 + \omega^2}}{3 + 3\omega^2} = \frac{1}{3\sqrt{1 + \omega^2}}$$

and the angle is

$$\underline{/H(j\omega)} = \tan^{-1}(-\omega) = -\tan^{-1}\omega$$

Choosing several values of ω, we can tabulate $|H(j\omega)|$ and $\underline{/H(j\omega)}$ both versus ω as shown in Table 10.4.1.

Of course, the output magnitude is the input magnitude *times* $|H(j\omega)|$, and the angle of the output is the angle of the input *plus* $\underline{/H(j\omega)}$. We can see how these

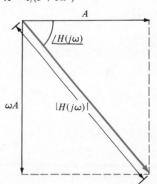

figure 10.4.3
The complex quantity
$H(j\omega) = A(1 - j\omega)$ where
$A = 1/(3 + 3\omega^2)$

table 10.4.1

ω, rad/s	$|H(j\omega)|$	$\underline{/H(j\omega)}$
0	0.333	0°
0.5	0.298	−26.6°
1.0	0.236	−45°
1.5	0.185	−56.3°
2.0	0.149	−63.4°
10.0	0.0332	−84.3°
∞	0	−90°

figure 10.4.4
Plots of (a) $|H(j\omega)|$ and (b) $\underline{/H(j\omega)}$.

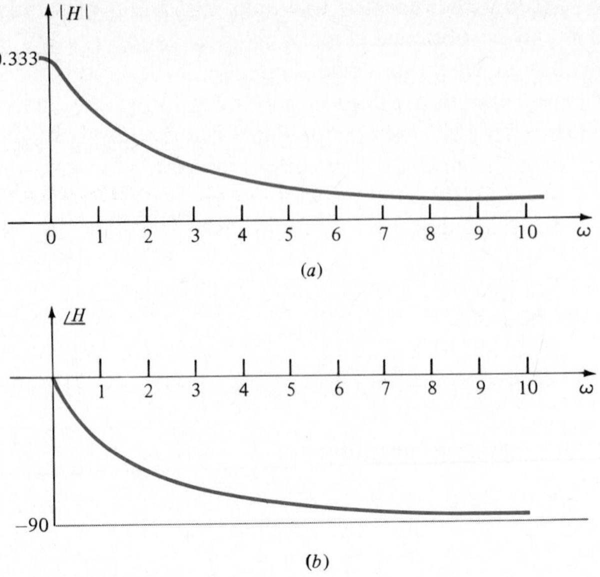

two quantities vary with applied input frequency in Figure 10.4.4. Such plots are extremely informative about a system's behavior under the influence of different inputs. We shall return to this subject in greater detail in a later chapter. It suffices at this point to realize that general conclusions about how the system will treat sinusoidal inputs of different frequency are easily seen from plots such as those in Figure 10.4.4. For example, the system whose transfer function is described in these plots obviously attenuates low frequencies less than it does high frequencies.

DRILL PROBLEM

For all $t < 0$, a series circuit ($R = 24\ \Omega$, $L = 8$ H) is driven by a sinusoidal voltage source $v_s(t) = 10 \cos 2t$. At $t = 0$ exactly the amplitude of the source becomes (and remains forever after)

$$v_s(t) = 30 \cos 2t \qquad \text{for } t > 0$$

Find $i(t)$ for $t > 0$.
ANS.: $i(t) = 1.04 \cos(2t - 33.7°) - 0.576e^{-3t}$.

10.5 sinusoidal impedance and admittance

Consider any single linear electrical element which has a sinusoidal signal impressed on it. What is the general relationship between the voltage across it and the current through it? Consider first the inductor

$$L\frac{di}{dt} = v(t)$$

Let $\hat{v}(t) = \mathbf{V}e^{j\omega t}$. Thus $\hat{i}_p(t) = \mathbf{I}e^{j\omega t}$ and

$$j\omega L\mathbf{I}e^{j\omega t} = \mathbf{V}e^{j\omega t}$$

$$\frac{\mathbf{V}}{\mathbf{I}} = Z_L(j\omega) = j\omega L$$

where $Z_L(j\omega)$ is the *sinusoidal impedance* of the inductor. Remember that $\mathbf{V}$ and $\mathbf{I}$ are both phasors, i.e., complex numbers.

EXAMPLE 10.5.1

A voltage $10\cos(5t + 15°)$ is impressed across a 0.5-H inductor. Find $i_p(t)$.

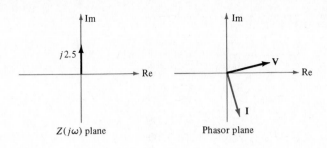

figure 10.5.1
The impedance and phasors of Example 10.5.1.

ANS.:

$$\mathbf{V} = 10\underline{/15°} \qquad Z_L(j\omega) = j\omega L = j(5)(0.5) = j2.5$$

$$\mathbf{V} = \mathbf{I}Z \qquad \mathbf{I} = \frac{\mathbf{V}}{Z} = \frac{10\underline{/15°}}{2.5\underline{/90°}} = 4\underline{/-75°}$$

$$i_p(t) = 4\cos(5t - 75°)$$

See Figure 10.5.1.

We make the following observations relative to Figure 10.5.1:

1. Since Z is in ohms, $\mathbf{V}$ in volts, and $\mathbf{I}$ in amperes, the scale factors volts per inch and amperes per inch need not be the same when we sketch the $\mathbf{V}$ phasor and the $\mathbf{I}$ phasor.
2. The impedance $Z(j\omega)$ is a fixed quantity. It does not vary with time. Both the $\mathbf{V}$ and the $\mathbf{I}$ control the magnitude and relative angle of their corresponding complex exponential rotating functions. These complex exponentials, $\mathbf{V}e^{j\omega t}$ and $\mathbf{I}e^{j\omega t}$, rotate in the counterclockwise direction with angular velocity $\omega = 5$ rad/s.
3. The angle between $\mathbf{V}e^{j\omega t}$ and $\mathbf{I}e^{j\omega t}$ remains fixed at the impedance angle (90° in this case). The current lags the voltage as they both rotate together.

It is therefore correct to define the sinusoidal impedance as follows: *The sinusoidal impedance $Z(j\omega)$ of an element is the ratio of the element's complex voltage phasor divided by its corresponding complex current phasor.*

CAPACITOR   The defining equation for a pure capacitor is

$$C \frac{dv}{dt} = i$$

Let $\hat{i} = \mathbf{I}e^{j\omega t}$; we choose v of the form $\hat{v}_p = \mathbf{V}e^{j\omega t}$ (remember that $\mathbf{I}$ and $\mathbf{V}$ are, in general, complex numbers). Substituting into the defining equation gives

$$Cj\omega \mathbf{V}e^{j\omega t} = \mathbf{I}e^{j\omega t}$$

whence $$\frac{\mathbf{V}}{\mathbf{I}} = \frac{1}{j\omega C} = -j\frac{1}{\omega C} = Z_C(j\omega)$$

EXAMPLE 10.5.2

Given that $i_C(t) = 4 \cos(5t - 75°)$ in a capacitor of $C = 0.05$ F, find $v_C(t)$.

ANS.: $$\mathbf{I} = 4\underline{/-75°} \qquad Z_C(j\omega) = -j\frac{1}{\omega C} = -j\frac{1}{5(0.05)} = -j4$$

$$\mathbf{V} = \mathbf{I}Z = (4\underline{/-75°})(-j4) = (4\underline{/-75°})(4\underline{/-90°}) = 16\underline{/-165°}$$

$$v_C(t) = 16 \cos(5t - 165°) = 16 \sin(5t - 75°)$$

RESISTOR   The defining equation for a pure resistor is

$$\frac{1}{R}v = i$$

Let $\hat{i} = \mathbf{I}e^{j\omega t}$; then $\hat{v}_p = \mathbf{V}e^{j\omega t}$. Therefore the equation yields

$$\frac{1}{R}\mathbf{V}e^{j\omega t} = \mathbf{I}e^{j\omega t} \qquad \text{or} \qquad \frac{\mathbf{V}}{\mathbf{I}} = R = Z_R$$

SINUSOIDAL ADMITTANCE   The ratio of complex voltage divided by complex current is impedance. The inverse is the sinusoidal *admittance* $Y(j\omega)$; that is,

$$Z(j\omega) = \frac{\mathbf{V}}{\mathbf{I}}$$

where $\mathbf{V}$ and $\mathbf{I}$ are, in general, complex, and

$$Y(j\omega) = \frac{\mathbf{I}}{\mathbf{V}}$$

EXAMPLE 10.5.3

List the sinusoidal impedance and admittance of each electrical element.
ANS.:

Inductor: $\qquad Z_L(j\omega) = j\omega L \qquad Y_L(j\omega) = \dfrac{1}{j\omega L} = -j\,\dfrac{1}{\omega L}$

Capacitor: $\qquad Z_C(j\omega) = -j\,\dfrac{1}{\omega C} \qquad Y_C(j\omega) = j\omega C$

Resistor: $\qquad Z_R = R \qquad Y_R = \dfrac{1}{R}$

Assume that we know the value of the sinusoidal current through a series combination of elements (Figure 10.5.2). How can we find the overall voltage? In other words, knowing that $i(t) = I_m \cos(\omega t + \theta)$, how can we find $v(t) = v_1(t) + v_2(t)$? The answer of course is to use phasors. Use $\mathbf{I} = I_m\underline{/\theta}$. Then

$$\mathbf{V} = I_m\underline{/\theta}Z_1(j\omega) + I_m\underline{/\theta}Z_2(j\omega) = [Z_1(j\omega) + Z_2(j\omega)]I_m\underline{/\theta}$$

Thus we can find the overall equivalent impedance of elements in series by summing the individual impedances.

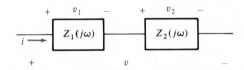

figure 10.5.2
Two sinusoidal impedances in series.

EXAMPLE 10.5.4

Given a series connection of $R = 3\ \Omega$, $C = \frac{1}{30}$ F, and $L = 2$ H carrying $i(t) = 15 \cos(3t + 30°)$, find $v(t)$, the voltage across the entire circuit (see Figure 10.5.3).

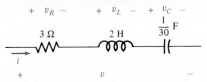

figure 10.5.3
The circuit of Example 10.5.4.

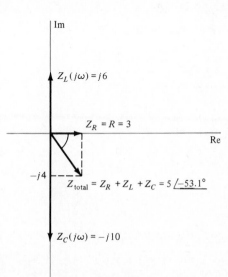

figure 10.5.4
Impedance diagram of the circuit of Example 10.5.4. These complex numbers are all fixed quantities. (They do *not* rotate about the origin. They are *not phasors*.)

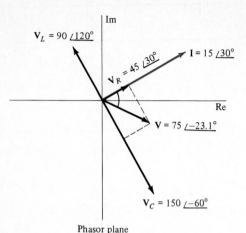

figure 10.5.5
The phasor diagram of Example 10.5.4.

Phasor plane

ANS.:   Use $\mathbf{I} = 15\underline{/30°}$. Then

$$Z_R = 3 \qquad Z_C(j3) = -j\,\frac{1}{3(1/30)} = -j10 \qquad Z_L(j3) = j3(2) = j6$$

$$Z_{\text{total}} = 3 - j10 + j6 = 3 - j4 = 5\underline{/-53.1°}$$

See Figure 10.5.4.

$$\mathbf{V} = \mathbf{I}Z = (15\underline{/30°})(5\underline{/-53.1°}) = 75\underline{/-23°}$$

$$v(t) = 75\cos(3t - 23.1°)$$

See Figure 10.5.5.

EXAMPLE 10.5.5

Given that the inductor voltage is $v_L(t) = 60\cos(5t + 40°)$ in the network shown in Figure 10.5.6, find $v_{\text{in}}(t)$.

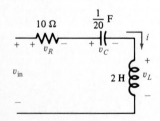

figure 10.5.6
The circuit of Example 10.5.5.

ANS.:

Method 1
Use $\mathbf{V}_L = 60\underline{/40°}$ and $Z_L = j\omega L = j10$ to find

$$\mathbf{I}_L = \frac{\mathbf{V}_L}{Z_L} = \frac{60\underline{/40°}}{10\underline{/90°}} = 6\underline{/-50°}$$

Then
$$\mathbf{V}_R = \mathbf{I}Z_R = (6\underline{/-50°})(10\underline{/0°}) = 60\underline{/-50°} = 38.6 - j46$$

$$\mathbf{V}_C = \mathbf{I}Z_C = (6\underline{/-50°})\left[-j\frac{1}{5(1/20)}\right] = (6\underline{/-50})(4\underline{/-90}) = 24\underline{/-140°}$$

$$\mathbf{V}_L + \mathbf{V}_C = 60\underline{/40°} + 24\underline{/-140°} = 36\underline{/40°} = 27.6 + j23.1$$

$$\mathbf{V}_R = 38.6 - j46 \qquad \mathbf{V}_{in} = 66.2 - j22.9 = 70\underline{/-19°}$$

or
$$v_{in}(t) = 70 \cos(5t - 19°)$$

The phasor diagram is shown in Figure 10.5.7.

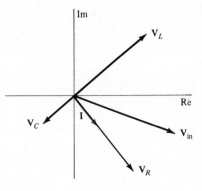

figure 10.5.7
The phasor diagram of Example 10.5.5.

Method 2
Make use of the fact that sinusoidal impedances (like impedance operators in general) obey the same circuit laws in the same way as resistors; i.e.,

$$\mathbf{V}_L = \mathbf{V}_{in}\frac{Z_L}{Z_L + Z_C + Z_R} \qquad \text{voltage division}$$

$$60\underline{/40°} = \mathbf{V}_{in}\frac{j10}{j10 - j4 + 10} = \mathbf{V}_{in}\frac{10\underline{/90°}}{10 + j6} = \mathbf{V}_{in}\frac{10\underline{/90°}}{11.7\underline{/31°}}$$

$$= \mathbf{V}_{in}(0.857\underline{/59°})$$

$$\mathbf{V}_{in} = \frac{60\underline{/40°}}{0.857\underline{/59°}} = 70\underline{/-19°}$$

as before. Hence by either method

$$v_{in}(t) = 70 \cos(5t - 19°)$$

Phasors are therefore the easiest way of solving for the particular response of any linear system with a sinusoidal input. Let us find a complete response.

EXAMPLE 10.5.6

Assuming that all initial conditions are zero, find the complete response $i(t)$ for $t > 0$ in the circuit shown in Figure 10.5.8.

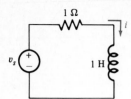

figure 10.5.8
The circuit of Example 10.5.6.

$$v_s(t) = 10 \cos 2t u(t)$$

ANS.:   The system equation is obtained with KVL

$$\frac{di}{dt} + i = 10 \cos 2t \qquad t > 0$$

Natural response:

$$s + 1 = 0 \qquad s = -1 \qquad i_n(t) = Ae^{-t}$$

Particular response:

$$\hat{v}_s = 10\underline{/0°}e^{j\omega t} \qquad \mathbf{I} = \frac{\mathbf{V}}{\mathbf{Z}}$$

$$\mathbf{I} = \frac{10\underline{/0°}}{\mathbf{Z}} = \frac{10\underline{/0°}}{1 + j2} = \frac{10\underline{/0°}}{2.2\underline{/63.4}} = 4.5\underline{/-63.4°}$$

$$i_p(t) = 4.5 \cos(2t - 63.4°) \qquad \text{and} \qquad i(t) = 4.5 \cos(2t - 63.4°) + Ae^{-t}$$

Using the initial conditions gives

$$0 = 4.5 \cos(-63.4°) + A \qquad A = -2.01$$

$$i(t) = 4.5 \cos(2t - 63.4) - 2.01e^{-t} \qquad t > 0$$

REACTANCE AND   ✻ Reactance   The imaginary part of a complex impedance is called the *reac-*
SUSCEPTANCE   *tance* of the impedance. The usual symbol is X.

EXAMPLE 10.5.7

If the impedance $Z(j\omega) = 3 + j4$, then the rectance $X = 4$.

In general, therefore, $Z(j\omega) = R + jX$.

✻ Susceptance   The imaginary part of a complex admittance is called the *sus-ceptance* and is denoted by the symbol B.

In general, therefore, $Y(j\omega) = G + jB$, where G is the *conductance* and B is the susceptance. These relationships are summarized in Figure 10.5.9 and Table 10.5.1.

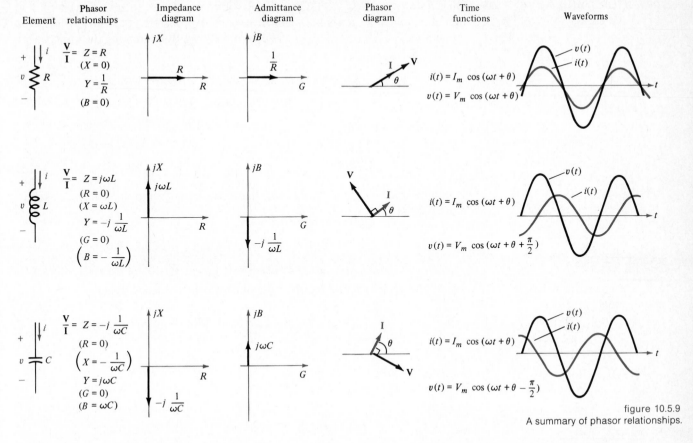

figure 10.5.9
A summary of phasor relationships.

The sinusoidal impedance Z is the ratio of the *complex* exponential voltage divided by the *complex* exponential current. It is *not*, in general, $v(t)/i(t)$. For example,

IMPORTANT REMINDER

$$Z(j\omega) = \frac{\mathbf{V}e^{j\omega t}}{\mathbf{I}e^{j\omega t}} = \frac{V\underline{/\theta}}{I\underline{/\alpha}} = \frac{V}{I}\underline{/\theta - \alpha} \qquad Z(j\omega) = \frac{10 \cos (2t + \theta)}{4 \cos (2t + \alpha)}$$

$$\textbf{(Correct)} \hspace{5cm} \textbf{(Incorrect)}$$

	Inductor	Capacitor	Resistor
Impedance	$Z_L(j\omega) = j\omega L$	$Z_C(j\omega) = -j\dfrac{1}{\omega C}$	$Z_R = R$
Reactance	$X_L = \omega L$	$X_C = \dfrac{-1}{\omega C}$	0
Admittance	$Y_L(j\omega) = \dfrac{1}{Z_L(j\omega)} = -j\dfrac{1}{\omega L}$	$Y_C(j\omega) = \dfrac{1}{Z_C(j\omega)} = j\omega C$	$Y_R(j\omega) = \dfrac{1}{Z_R} = \dfrac{1}{R}$
Conductance G	0	0	$\dfrac{1}{R}$
Susceptance B	$-\dfrac{1}{\omega L}$	ωC	0

table 10.5.1

CAPACITORS AND INDUCTORS AT ZERO AND INFINITE FREQUENCY

Since for a capacitor $i_C(t) = C(dv/dt)$ and for an inductor $v_L(t) = L(di/dt)$, we know that the capacitor acts like an open circuit for constant (dc) voltages and the inductor acts like a short circuit. This is also shown by examining the impedances

$$Z_C(j\omega) = -j\frac{1}{\omega C} \quad \text{and} \quad Z_L(j\omega) = j\omega L$$

Since

$$Z_C = \begin{cases} \infty\underline{/-90°} \\ 0\underline{/-90°} \end{cases} \quad \text{and} \quad Z_L = \begin{cases} 0\underline{/90°} & \text{for } \omega = 0 \\ \infty\underline{/90°} & \text{for } \omega \to \infty \end{cases}$$

at infinite frequency, inductors look like open circuits and capacitors look like short circuits.

EXAMPLE 10.5.8

a. Describe the behavior of the circuit shown in Figure 10.5.10 as a function of frequency, where $v_1(t)$ is sinusoidal.

b. If $v_1(t) = 10 \cos \omega t$, $R = 1$ MΩ, and $C = 0.1$ μF, find $v_2(t)$ for $\omega = 1$, 10, and 50 rad/s.

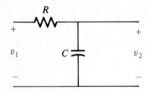

figure 10.5.10
The circuit of Example 10.5.8.

ANS.:

(a) If the frequency of v_1 is very low ($\omega \to 0$), the capacitor looks very much like an open circuit; there is no voltage drop in R, and $v_2 = v_1$. On the other hand, if the frequency of the applied voltage $v_1(t)$ is high ($\omega \to \infty$), the capacitor looks like a short circuit; all v_1 falls across R, $v_2 \to 0$. Since this circuit discriminates against high frequencies, it is a *low-pass filter*.

(b) For $\omega = 1$

$$V_2 = \frac{-j(1/\omega C)}{R - j(1/\omega C)}V_1 = \frac{10\underline{/-90°} \times 10^6}{(1 - j10) \times 10^6}V_1 = \frac{(10\underline{/-90°})(10\underline{/0°})}{10(+)\underline{/-84.3°}} = 10(-)\underline{/-5.7°}$$

$$v_2(t) \approx 10 \cos (t - 5.7°)$$

For $\omega = 10$

$$V_2 = \frac{(1\underline{/-90°})(10\underline{/0°})}{1 - j1} = \frac{10\underline{/-90}}{\sqrt{2}\underline{/-45°}} = 7.07\underline{/-45°}$$

$$v_2(t) = 7.07 \cos (10t - 45°)$$

For $\omega = 50$

$$\mathbf{V}_2 = \frac{-j[10^6/(50 \times 0)]}{10^6 - j[10^6/(50 \times 0.1)]} \mathbf{V}_1 = \frac{-j0.2}{1 - j0.2} \mathbf{V}_1 = \frac{0.2\underline{/-90°}}{1.02\underline{/-11.3°}} 10\underline{/0°} = 0.196\underline{/-78.7°}10\underline{/0°}$$

$$v_2(t) = 1.96 \cos(50t - 78.7°)$$

EXAMPLE 10.5.9

Given that $v_C(t) = 283 \cos(500t + 45°)$ in the circuit shown in Figure 10.5.11, find v_g and i. Draw the phasor diagram.

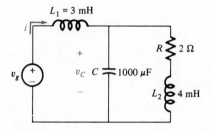

figure 10.5.11
The circuit of Example 10.5.9.

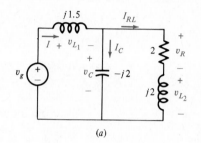

(a)

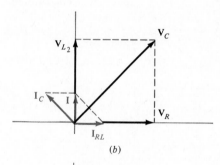

(b)

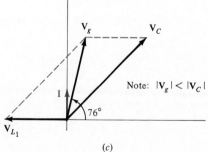

(c)

figure 10.5.12
(a) The circuit with impedance values labeled. (b) The phasors for some of the variables in the circuit. (c) The remaining phasors.

ANS.:   See Figure 10.5.12.

$$Z_{L_1} = j\omega L = j500(3 \times 10^{-3}) = j1.5$$

$$Z_{L_2} = j500(4 \times 10^{-3}) = j2$$

$$Z_C = -j\frac{1}{500 \times 10^{-3}} = -j2$$

$$Y_{RL} = \frac{1}{2 + j2} = \frac{1}{2\sqrt{2}/45°} = 0.354/-45° = 0.25 - j0.25$$

$$Y = Y_C + Y_{RL} = j0.5 + 0.25 - j0.25 = 0.25 + j0.25 = 0.354/45°$$

$$\mathbf{I} = Y\mathbf{V}_C = (0.354/45°)(283/45°) = 100/90°$$

Hence
$$i(t) = 100 \cos (500t + 90°) = -100 \sin 500t$$

$$\mathbf{I}_C = Y_C \mathbf{V}_C = (0.5/90°)(283/45°) = 141/135°$$

$$\mathbf{I}_{RL} = Y_{RL} \mathbf{V}_C = (0.354/-45°)(283/45°) = 100/0°$$

$$\mathbf{V}_R = \mathbf{I}_{RL} R = (100/0°)(2/0°) = 200/0°$$

$$\mathbf{V}_{L_2} = \mathbf{I}_{RL}\mathbf{Z}_{L_2} = (100/0°)(2/90°) = 200/90°$$

$$\mathbf{V}_g = \mathbf{Z}_{L_1}\mathbf{I} + \mathbf{V}_C = (1.5/90°)(100/90°) + 283/45° = 206/76°$$

$$v_g(t) = 206 \cos (500t + 76°)$$

In the circuit shown in Example 10.5.9 one of the branch voltages was actually larger than the source voltage. This is quite common in *RLC* sinusoidally excited circuits. The branch current(s) may also be greater than the current in the source.

EXAMPLE 10.5.10

Find $i(t)$, the sinusoidal source current in the circuit shown in Figure 10.5.13. The amplitude of $v_s(t)$ is 230 V.

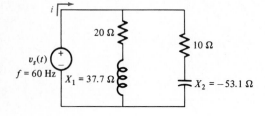

figure 10.5.13
The circuit of Example 10.5.10.

ANS.: $$\mathbf{Z}_1 = 20 + j37.7 \qquad \mathbf{Z}_2 = 10 - j53.1$$

Let $\mathbf{V}_s = 230/0°$. Then

$$\mathbf{I}_1 = \frac{230/0°}{20 + j37.7} = \frac{230/0°}{42.7/62.1°} = 5.41/-62.1° = 2.53 - j4.78$$

$$\mathbf{I}_2 = \frac{230/0°}{10 - j53.1} = \frac{230/0°}{54/-79.3°} = 4.26/79.3° = 0.79 + j4.19$$

$$\mathbf{I} = \mathbf{I}_1 + \mathbf{I}_2 = 3.32 - j0.59 = 3.37/-10.1°$$

$$i(t) = 3.37 \cos (377t - 10.1°)$$

Note that both $|\mathbf{I}_1|$ and $|\mathbf{I}_2|$ are greater than $|\mathbf{I}|$.

Sinusoidal impedances and admittances can be used to write mesh and node equations just as we used resistors and then generalized impedance operators. Simply write a KVL summation around each mesh (through each link in the graph) or KCL summation at each node where the voltage is an independent unknown. The result will be a set of mesh or node equations, respectively.

EXAMPLE 10.5.11

Given the circuit of Figure 10.5.14, in which $i_s(t) = 10 \cos (2t + 30°)$, find $v_3(t)$.

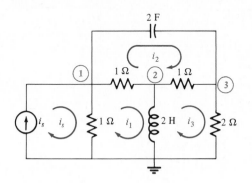

figure 10.5.14
The circuit of Example 10.5.11.

ANS.:   Write node equations:

Node 1:
$$\mathbf{V}_1 + j4(\mathbf{V}_1 - \mathbf{V}_3) + (\mathbf{V}_1 - \mathbf{V}_2) = \mathbf{I}_s$$

Node 2:
$$\frac{1}{j4}\mathbf{V}_2 + (\mathbf{V}_2 - \mathbf{V}_1) + (\mathbf{V}_2 - \mathbf{V}_3) = 0$$

Node 3:
$$\tfrac{1}{2}\mathbf{V}_3 + j4(\mathbf{V}_3 - \mathbf{V}_1) + (\mathbf{V}_3 - \mathbf{V}_2) = 0$$

Rearranging gives

$$(2 + j4)\mathbf{V}_1 - \mathbf{V}_2 - j4\mathbf{V}_3 = \mathbf{I}_s$$

$$-\mathbf{V}_1 + \left(2 + \frac{1}{j4}\right)\mathbf{V}_2 - \mathbf{V}_3 = 0$$

$$-j4\mathbf{V}_1 - \mathbf{V}_2 + (\tfrac{3}{2} + j4)\mathbf{V}_3 = 0$$

$$\mathbf{V}_3 = \frac{\begin{vmatrix} 2 + j4 & -1 & I_s \\ -1 & 2 + \dfrac{1}{j4} & 0 \\ -j4 & -1 & 0 \end{vmatrix}}{\begin{vmatrix} 2 + j4 & -1 & -j4 \\ -1 & 2 + \dfrac{1}{j4} & -1 \\ -j4 & -1 & \tfrac{3}{2} + j4 \end{vmatrix}} = \frac{2 + j8}{6 + j11.25}\,\mathbf{I}_s$$

Since $\mathbf{I}_s = 10\underline{/30°}$, we have $\mathbf{V}_3 = 6.46\underline{/44°}$, or

$$v_3(t) = 6.46 \cos (2t + 44°)$$

We can solve this same problem by mesh equations.

Mesh 1:
$$(\mathbf{I}_1 - \mathbf{I}_s)(1) + (\mathbf{I}_1 - \mathbf{I}_2)(1) + (\mathbf{I}_1 - \mathbf{I}_3)j4 = 0$$

or
$$(2 + j4)\mathbf{I}_1 - (1)(\mathbf{I}_2) - j4(\mathbf{I}_3) = \mathbf{I}_s$$

Mesh 2:
$$-1\mathbf{I}_1 + \left(2 + \frac{1}{j4}\right)\mathbf{I}_2 - (1)(\mathbf{I}_3) = 0$$

Mesh 3:
$$-j4(\mathbf{I}_1) - (1)(\mathbf{I}_2) + (3 + j4)(\mathbf{I}_3) = 0$$

Solving the above three simultaneous equations for $\mathbf{I}_3$ by Cramer's rule gives

$$\mathbf{I}_3 = \frac{\begin{vmatrix} 2+j4 & -1 & \mathbf{I}_s \\ -1 & 2+\frac{1}{j4} & 0 \\ -j4 & -1 & 0 \end{vmatrix}}{\begin{vmatrix} 2+j4 & -1 & -j4 \\ -1 & 2+\frac{1}{j4} & -1 \\ -j4 & -1 & 3+j4 \end{vmatrix}} = \frac{2+j8}{12+j22.5}\mathbf{I}_s$$

Again since, $\mathbf{I}_s = 10\underline{/30°}$, we get

$$\mathbf{I}_3 = 3.24\underline{/44°}$$

Then
$$\mathbf{V}_3 = \mathbf{I}_3(2\underline{/0°}) = 6.46\underline{/44°}$$

or, as before,
$$v_3(t) = 6.46 \cos (2t + 44°)$$

The careful reader will again note the symmetry in both the mesh- and node-equation sets in Example 5.10.11. The node-equation symmetry is inherent, but the mesh-equation symmetry will not occur unless all mesh currents are in the same direction.

EXAMPLE 10.5.12

Find $v_a(t)$ in Figure 10.5.15a using either loop equations or node equations; $\omega = 377$ rad/s.
ANS.: Two loop (mesh) equations are needed

$$(1.4 + j1.6 + 11.66 + j8.75)\mathbf{I}_1 - (11.66 + j8.75)\mathbf{I}_2 = 460\underline{/0°}$$
$$-(11.66 + j8.75)\mathbf{I}_1 + (11.66 + j8.75 + 0.8 + j1)\mathbf{I}_2 = 451\underline{/180°}$$

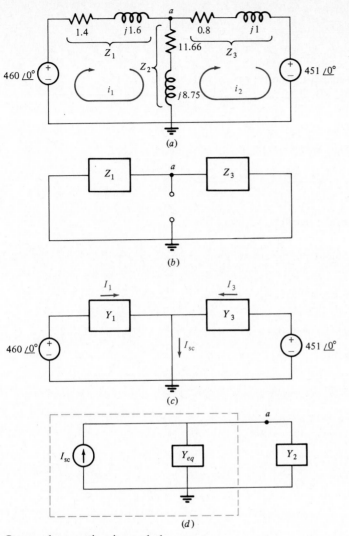

figure 10.5.15
(a) The circuit of Example 10.5.12 (b) Finding Z_{eq}.
(c) Finding I_{sc}. (d) The Norton equivalent circuit.

One node equation is needed

$$\mathbf{V}_a(Y_1 + Y_2 + Y_3) - 460\underline{/0°}\,Y_1 - 451\underline{/0°}\,Y_3 = 0$$

where the various admittances Y_1, Y_2, and Y_3 are

$$Z_1 = 1.4 + j1.6 = 2.126\underline{/48.8°} \qquad Y_1 = 0.470\underline{/-48.8°} = 0.3097 - j0.354$$

$$Z_2 = 11.66 + j8.75 = 14.58\underline{/36.9°} \qquad Y_2 = 0.0686\underline{/-36.9°} = 0.0549 - j0.0412$$

$$Z_3 = 0.8 + j1 = 1.28\underline{/51.34°} \qquad Y_3 = 0.781\underline{/-51.34°} = 0.4878 - j0.6098$$

The single node equation is obviously the best method of solution:

$$\mathbf{V}_a(1.32\underline{/-49.7°}) = 216.2\underline{/-48.8°} + 352.23\underline{/-51.34°}$$

$$= 142.4 - j162.67 + 220.04 - j275.05$$

$$= 362.45 - j437.72 = 568.3\underline{/-50.37°}$$

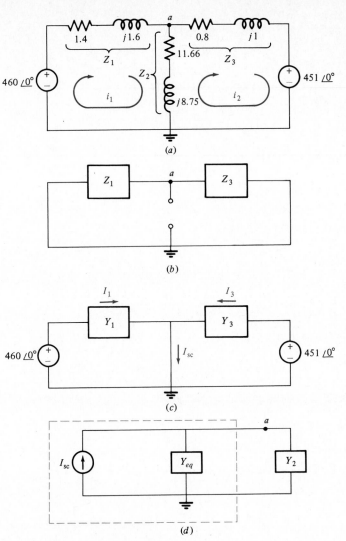

figure 10.5.15
(a) The circuit of Example 10.5.12 (b) Finding Z_{eq}.
(c) Finding I_{sc}. (d) The Norton equivalent circuit.

One node equation is needed

$$\mathbf{V}_a(Y_1 + Y_2 + Y_3) - 460\underline{/0°}\,Y_1 - 451\underline{/0°}\,Y_3 = 0$$

where the various admittances Y_1, Y_2, and Y_3 are

$$Z_1 = 1.4 + j1.6 = 2.126\underline{/48.8°} \qquad Y_1 = 0.470\underline{/-48.8°} = 0.3097 - j0.354$$

$$Z_2 = 11.66 + j8.75 = 14.58\underline{/36.9°} \qquad Y_2 = 0.0686\underline{/-36.9°} = 0.0549 - j0.0412$$

$$Z_3 = 0.8 + j1 = 1.28\underline{/51.34°} \qquad Y_3 = 0.781\underline{/-51.34°} = 0.4878 - j0.6098$$

The single node equation is obviously the best method of solution:

$$\mathbf{V}_a(1.32\underline{/-49.7°}) = 216.2\underline{/-48.8°} + 352.23\underline{/-51.34°}$$

$$= 142.4 - j162.67 + 220.04 - j275.05$$

$$= 362.45 - j437.72 = 568.3\underline{/-50.37°}$$

$$\mathbf{V}_a = \frac{568.3\underline{/-50.37°}}{1.32\underline{/-49.7°}} = 430\underline{/-0.6°} \approx 430\underline{/0°}$$

or
$$v_a(t) = 430 \cos 377t$$

Let us repeat this example using a Norton equivalent circuit. For Z_{eq} (see Figure 10.5.15b)

$$Y_{eq} = Y_1 + Y_3 = 0.3097 - j0.354 + 0.4878 - j0.6098 = 0.7975 - j0.9638$$

For I_{sc} (see Figure 10.5.15c)

$$\mathbf{I}_1 = (460\underline{/0°})(Y_1) = (460\underline{/0°})(0.47\underline{/-48.8°}) = 216.2\underline{/-48.8°} = 142.4 - j162.7$$

$$\mathbf{I}_3 = (451\underline{/0°})(Y_3) = (451\underline{/0°})(0.781\underline{/-51.34°}) = 352.23\underline{/-51.34°} = 220 - j275$$

$$\mathbf{I}_{sc} = \mathbf{I}_1 + \mathbf{I}_3 = 362.4 - j437.7 = 568.3\underline{/-50.38°}$$

Thus the Norton equivalent circuit is as shown in Figure 10.5.15d, and

$$\mathbf{V}_a = \mathbf{I}_{sc}\frac{1}{Y_{eq} + Y_2} = 568.3\underline{/-50.38°}\,\frac{1}{0.7975 - j0.9638 + 0.0549 - j0.0412}$$

$$= 568.3\underline{/-50.38°}\,\frac{1}{0.8524 - j1.005} = 568.3\underline{/-50.38°}\,\frac{1}{1.318\underline{/-49.7°}}$$

$$= (568.3\underline{/-50.38°})(0.759\underline{/49.7°}) = 430\underline{/0°}$$

as before.

DRILL PROBLEM

Solve for the mesh currents $i_1(t)$ and $i_2(t)$ in the circuit of Figure 10.5.16.

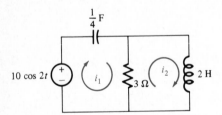

figure 10.5.16
The circuit of the drill problem.

ANS.: $i_1(t) = 5 \cos (2t + 16.2°)$ and $i_2(t) = 3 \cos (2t - 36.9°)$.

10.6 immittance loci

In this section we investigate what happens to impedances and admittances as certain quantities vary from one extreme value to another. For instance, what happens to the impedance of a series connection of R, L, and C when the total reactance X goes from $-\infty$ to $+\infty$? Such a variation in the impedance (see Figure 10.6.1a) could be obtained by allowing the applied frequency ω to vary from zero to infinity or by keeping ω fixed and varying L and C. Since

$$Z = R + jX = R + j\omega L - j\frac{1}{\omega C} \qquad (10.6.1)$$

we can list several typical values as in Table 10.6.1. See Figure 10.6.1b.

What happens to the corresponding admittance $Y(j\omega)$? List in Table 10.6.2 $Y(j\omega) = 1/Z(j\omega)$ using the values of $Z(j\omega)$ from Table 10.6.1 (remember that $Y = G + jB$). The resulting sinusoidal admittance values $Y(j\omega)$ are plotted in Figure 10.6.2.

It turns out that *all* such loci, as we vary R or jX or ω, *are circles* (we include straight lines as being part of the circumference of a circle of infinite radius whose center is at ∞). For example, we can prove this to be true for any $Z(j\omega)$ with a fixed value of R and an $X(\omega)$ that varies:

$$Z(j\omega) = \frac{1}{Y(j\omega)}$$

$$R + jX(\omega) = \frac{1}{G + jB} \qquad (10.6.2)$$

$$R + jX(\omega) = \frac{G - jB}{G^2 + B^2} \qquad (10.6.3)$$

so that $\qquad R = \dfrac{G}{G^2 + B^2} \qquad$ or $\qquad G^2 - \dfrac{G}{R} + B^2 = 0$

Add $(1/2R)^2$ to each side to complete the square on the left

$$\left(G - \frac{1}{2R}\right)^2 + B^2 = \left(\frac{1}{2R}\right)^2 \qquad (10.6.4)$$

This is the equation of a circle of radius $1/2R$ centered at $G = 1/2R$, $B = 0$ on the $Y(j\omega)$ plane. Indeed, it is the equation of the circle shown in Figure 10.6.2. It describes that set of points (G, B), where $Y(j\omega) = G + jB$, which are consistent with an impedance $Z(j\omega) = R + jX$ having a fixed value of R and allowing X to vary from $-\infty$ to $+\infty$. Such a set of points defines a locus in the complex admittance plane that shows where the tip of the $Y(j\omega)$ vector will travel to as $X(\omega)$ varies.

Figure 10.6.3 shows immittance loci for four possible RC and RL circuits as

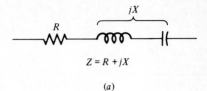

$Z = R + jX$

(a)

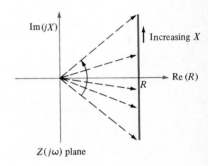

(b)

figure 10.6.1
(a) An impedance. (b) Typical values of the complex quantity $Z(j\omega)$ as ω varies from zero to infinity.

table 10.6.1

ω	jX	$|Z|$	$\underline{/Z}$
0	$-\infty$	∞	$-90°$
	$-jR$	$R\sqrt{2}$	$-45°$
	$0°$	R	$0°$
	$+jR$	$R\sqrt{2}$	$+45°$
∞	∞	∞	$+90°$

table 10.6.2

$|Z|$	$\underline{/Z}$	$|Y|$	$\underline{/Y}$
∞	$-90°$	0	$+90°$
$R\sqrt{2}$	$-45°$	$\dfrac{0.707}{R}$	$+45°$
R	$0°$	$\dfrac{1}{R}$	$0°$
$R\sqrt{2}$	$45°$	$\dfrac{0.707}{R}$	$-45°$
∞	$90°$	0	$-90°$

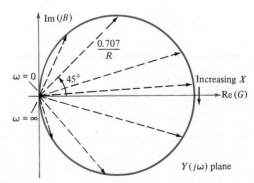

figure 10.6.2
The admittance of a series *RLC* circuit as the applied frequency varies from zero to infinity.

figure 10.6.3
Immittance loci of simple *RL* and *RC* circuits.

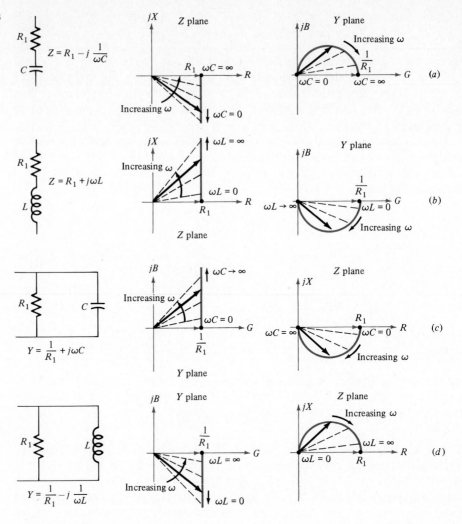

the applied frequency ω varies from zero to infinity. For the series *RC* circuit of Figure 10.6.3a, R has the value R_1, and when $\omega = 0$, the reactance is $X = -1/\omega C = -\infty$. Thus, the impedance vector is infinitely long and is pointing *down* the negative jX axis. As frequency ω increases from zero, the tip of the impedance Z rides up along the vertical straight line that eventually intersects the R axis at R_1. So when ω approaches infinity, the impedance vector approaches the horizontal $Z(j\infty) = R_1$. A typical intermediate value of $Z(j\omega)$ is shown in the impedance-plane sketch in Figure 10.6.3a, and several other values are implied by dashed lines.

As far as the admittance-plane plot is concerned, at $\omega = 0$, when the impedance is infinitely long and at an angle of $-90°$, the reciprocal of this complex number is at $+90°$ and of zero length. As ω increases, the length of Z decreases and its angle increases (gets less negative). Therefore $|Y|$ increases and its angle decreases. For example, when the frequency is such that $Z = R_1 - jR_1 = R_1\sqrt{2}\underline{/-45°}$, then $Y = 1/(R_1\sqrt{2})\underline{/+45°}$. Finally as the frequency gets large without bound, the impedance approaches R_1 (the capacitor looks like a short) and $Y = 1/R_1$. Similar arguments can be made for the three remaining circuits in Figure 10.6.3. You should carefully study the $Z(j\omega)$ and $Y(j\omega)$ loci of those circuits. The construction of any immittance-locus plot is accomplished simply by noting, for the immittance under consideration:

1. Its complex value at one extreme, for example, $\omega = 0$
2. Its complex value at the other extreme, for example, $\omega = \infty$
3. An intermediate value (like $Z = R_1 + jR_1$)
4. That locus must be a circle

It makes no difference whether it is the frequency ω or one of the element values (R or L or C) that is varying; the result is a circle locus for $Z(j\omega)$ or $Y(j\omega)$.

In Figures 10.6.1 and 10.6.2 there is one frequency called the *resonant frequency*   **RESONANCE**
ω_r such that *the impedance $Z(j\omega_r)$ and admittance $Y(j\omega_r)$ are real*; i.e., the imaginary parts of both Z and Y are zero. From equation (10.6.1) we solve for that frequency value by setting

$$\text{Im } Z(j\omega) = X(\omega) = 0 \qquad (10.6.5)$$

$$j\omega L - j\frac{1}{\omega C} = 0$$

$$\omega = \frac{1}{\sqrt{LC}} = \omega_r \qquad (10.6.6)$$

This is the resonant frequency of a series RLC circuit. Other circuits have other resonant frequencies. From equation (10.6.3), if $X(\omega) = 0$, then $B(\omega) = 0$ and $Y(j\omega) = G(\omega)$, which is also a real quantity. Thus, if the input impedance of any circuit, $Z(j\omega)$, is *real*, $Y(j\omega)$ *is also real*. This condition is called *resonance*, and the frequency $\omega = \omega_r$ at which it occurs is the *resonant frequency* of that circuit.

At resonance, a series RLC circuit excited by a fixed-magnitude sinusoidal voltage will experience a current maximum. A parallel RLC circuit will experience a total-input-current minimum at its resonant frequency. This is a *coincidence*. Other combinations of R's, L's, and C's do *not* have current or voltage extrema at their resonant frequencies.

EXAMPLE 10.6.1

Show that the circuit in Figure 10.6.4 can have more than one resonant condition. Assume all values fixed except for C, which is variable.

ANS.:   Draw the phasor diagram. Find the values of C that makes i be in phase with the input voltage. If $V_{in} = 230\underline{/0°}$,

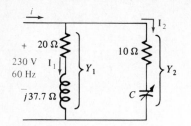

figure 10.6.4
The circuit of Example 10.6.1.

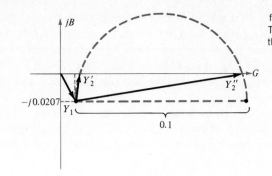

figure 10.6.5
The admittance locus for
the circuit in Example 10.6.1.

$$\mathbf{I}_1 = \frac{230\underline{/0^\circ}}{20 + j37.7} = 5.39\underline{/-62.1^\circ} \tag{10.6.7}$$

Similarly
$$\mathbf{I}_2 = \mathbf{V}_{in}\, Y_2 \tag{10.6.8}$$

and
$$Y_1 = \frac{1}{20 + j37.7} = 0.0234\underline{/-62.1^\circ} \tag{10.6.9}$$

or $Y_1 = 0.011 - j0.0207$ and $Y_2 = \dfrac{1}{10 - jX_C} = \dfrac{10 + jX_C}{100 + X_C^2}$ \hfill (10.6.10)

and so $Y_{in} = Y_1 + Y_2$, where Y_1 is a fixed quantity and Y_2 depends on X_C (see Figure 10.6.3a).

The plot of the total input (driving-point) admittance is the graphical summation of the fixed $Y_1 = 0.011 - j0.0207$ and the semicircular plot of Y_2 shown in Figure 10.6.3a. This result is shown in Figure 10.6.5, and we see that two resonant conditions are possible: $Y_{in} = Y_1 + Y_2'$ or $Y_1 + Y_2''$; since both cause Y_{in} to be a purely real quantity, $i(t)$ will be in phase with $v_{in}(t)$.

What value of C will produce these two different resonances? We can answer this numerically by setting the magnitude of the imaginary part of Y_2 equal to the magnitude of the imaginary part of Y_1

$$\text{Im } Y_2 = -\text{Im } Y_1$$

$$+0.0207 = \frac{X_C}{100 + X_C^2}$$

or
$$0.0207X_C^2 - X_C + 2.07 = 0$$

whence
$$X_C = 46.142,\ 2.167 \tag{10.6.11}$$

and since $X_C = 1/\omega C$, where $\omega = 377$ rad/s,

$$C = \frac{1}{377 X_C} = 57.49\ \mu\text{F},\ 1224\ \mu\text{F} \tag{10.6.12}$$

As a check, insert these values into the expression for Y_2 and see what Y_{in} results in each case

$$Y_2 = \frac{1}{10 - jX_C}$$

For $X_C = 46.14$

$$C = 57.49\ \mu\text{F} \quad \text{and} \quad Y_2' = 0.00449 + j0.0207 \tag{10.6.13}$$

but for $X_C = 2.167$

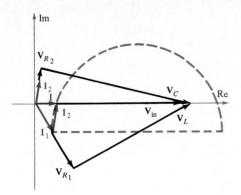

figure 10.6.6
The phasor diagram that results from choosing $C = 57.49$ μF in the circuit shown in Figure 10.6.4.

$$C = 1224 \ \mu F \quad \text{and} \quad Y_2'' = 0.0955 + j0.0207 \tag{10.6.14}$$

Thus
$$Y_{in}' = Y_1 + Y_2' = (0.011 - j0.0207) + (0.00449 + j0.0207) = 0.01549\underline{/0^\circ} \tag{10.6.15}$$

so
$$\mathbf{I} = Y_{in}'\mathbf{V} = 3.563\underline{/0^\circ} \tag{10.6.16}$$

Similarly
$$Y_{in}'' = (0.011 - j0.0207) + (0.0955 + j0.0207) = 0.1065\underline{/0^\circ}$$

so
$$\mathbf{I}'' = 24.5\underline{/0^\circ} \tag{10.6.17}$$

For $C = 57.49 \ \mu F$ and $Y_{in} = Y_{in}' = 0.01549\underline{/0^\circ}$ the resulting phasor diagram is given in Figure 10.6.6. The construction of the phasor diagram for $C = 1224 \ \mu F$ is left as an exercise.

One final conclusion that can be drawn from the admittance and phasor diagrams is that we can use this geometry to compute the value of C resulting in a *minimum input current*. The phasor for this minimum current (Figure 10.6.7a) is perpendicular to the circumference of the circle locus. The admittance diagram is shown in Figure 10.6.7b and an enlarged view in Figure 10.6.7c, from which

$$Y_1 = 0.0234\underline{/-62.1^\circ} = 0.011 - j0.0207 \tag{10.6.18}$$

and
$$\alpha = \tan^{-1}\frac{0.0207}{0.011 + 0.05} = 18.7^\circ \tag{10.6.19}$$

$$\text{Im } Y_2 = 0.05 \sin 18.7^\circ = 0.0161 \tag{10.6.20}$$

and
$$0.05 - \text{Re } Y_2 = 0.5 \cos 18.7^\circ = 0.0473$$

$$\text{Re } Y_2 = 0.05 - 0.0473 = 0.00265 \tag{10.6.21}$$

or
$$Y_2 = 0.00265 + j0.0161 \tag{10.6.22}$$

so that
$$Y_2 = 0.0163\underline{/80.6^\circ} \tag{10.6.23}$$

Therefore
$$Y = Y_1 + Y_2 = (0.011 - j0.0207) + (0.00265 + j0.0161)$$
$$= 0.01365 - j0.0046 = 0.0144\underline{/-18.6^\circ} \tag{10.6.24}$$

Finally
$$\mathbf{I}_{min} = \mathbf{V}Y = (230\underline{/0^\circ})(0.0144\underline{/-18.6^\circ}) \tag{10.6.25}$$

figure 10.6.7
Solving for minimum input current: (*a*) the
resulting phasor diagram; (*b*) the admittance
diagram; (*c*) an enlarged view of the
admittance diagram (not to scale).

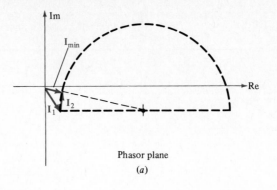

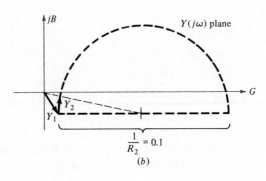

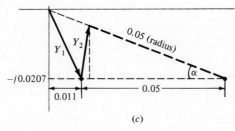

or
$$i_{min}(t) = 3.312 \cos (377t - 18.6°) \qquad (10.6.26)$$

The value of C necessary to produce this minimum value of i can be computed
from (10.6.23)

$$Y_2 = 0.0163\underline{/80.6°} = \frac{1}{Z_2} \qquad (10.6.27)$$

and so
$$Z_2 = \frac{1}{Y_2} = 61.35\underline{/-80.6°} = 10.0 - j60.53$$

where
$$60.53 = -X_C = \frac{1}{\omega C} \qquad (10.6.28)$$

Therefore
$$C = \frac{1}{377(60.53)} = 43.8 \ \mu F \qquad (10.6.29)$$

DRILL PROBLEM

In the circuit shown in Figure 10.6.8, $v(t) = 120 \cos 100t$, $R_2 = 100\ \Omega$, and $C = 100\ \mu$F. If R_1 and L are both variable, what value of R_1 will result in there being only one possible resonant condition? What value of L will then attain this resonance?

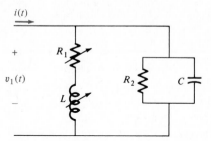

figure 10.6.8
The circuit in the drill problem.

ANS.: $R_1 = 50\ \Omega$, $L = 0.5$ H.

power in sinusoidally excited systems 10.7

For a general impedance $Z = |Z|\underline{/\theta}$ let us find the average power dissipated when a sinusoidal excitation is used (see Figure 10.7.1)

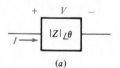

$$v(t) = V_m \cos(\omega t + \theta) \qquad (10.7.1)$$

$$i(t) = I_m \cos \omega t \qquad (10.7.2)$$

Instantaneous power $p(t)$ delivered to the impedance element is given by the product of $v(t)$ and $i(t)$

$$p(t) = v(t)i(t) = V_m I_m \cos(\omega t + \theta)\cos(\omega t)$$

Using

$$\cos(a \pm b) = \cos a \cos b \mp \sin a \sin b$$

we get

$$\cos(a + b) + \cos(a - b) = 2\cos a \cos b$$

Thus

$$p(t) = \frac{V_m I_m}{2}[\cos(2\omega t + \theta) + \cos\theta] \qquad (10.7.3)$$

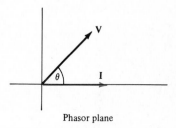

figure 10.7.1
(a) A general sinusoidal impedance with its associated terminal voltage and current. (b) The corresponding phasor diagram.

We note from equation (10.7.3) that electric power is delivered to any impedance in pulsations having a frequency that is *double that* of the applied voltage and current. Thus when we walk past a large power transformer or switchgear and hear it humming, that hum is 120 not 60 Hz. Fluorescent lights blink on and off at a rate of 120 not 60 Hz. Single-phase electric motors produce a torque which is pulsatory and which can be bothersome in such precise applications as the design and construction of hi-fi record turntables where such pulsations are clearly audible.

What is the average value P_{av} (over a long period of time) of $p(t)$? Since every cycle of the instantaneous-power waveform $p(t)$ looks like every other cycle, we can find the average value by dividing the area under $p(t)$ over two cycles by the period of those cycles in seconds. [Since $p(t)$ fluctuates at frequency 2ω, the period of two $p(t)$ cycles is equal to T s, the period of one cycle of $v(t)$ or $i(t)$.] Thus

$$P_{av} = \frac{1}{T} \int_{t_0}^{t_0 + T} p(t) \, dt$$

Choose as the lower limit of integration the instant of time when the argument of $\cos (2\omega t + \theta)$ is zero: $t_0 = -\theta/2\omega$. Thus

$$t_0 + T = -\frac{\theta}{2\omega} + \frac{1}{f} = -\frac{\theta}{2\omega} + \frac{4\pi}{2(2\pi f)} = \frac{4\pi - \theta}{2\omega}$$

$$P_{av} = \frac{1}{T} \int_{-\theta/2\omega}^{(4\pi - \theta)/2\omega} \frac{V_m I_m}{2} [\cos (2\omega t + \theta) + \cos \theta] \, dt$$

$$= \frac{1}{T} \left[\frac{V_m I_m}{2} \frac{\sin (2\omega t + \theta)}{2\omega} \bigg|_{-\theta/2\omega}^{(4\pi - \theta)/2\omega} + \frac{V_m I_m}{2} \cos \theta \frac{4\pi - \theta + \theta}{2\omega} \right]$$

$$= \frac{1}{T} \left[\frac{V_m I_m}{4\omega} (\sin 4\pi - \sin 0) + \frac{V_m I_m}{2} \cos \theta \frac{4\pi}{4\pi f} \right]$$

Since $\sin 4\pi - \sin 0 = 0$,

$$P_{av} = \frac{1}{T} \frac{V_m I_m}{2} (\cos \theta) T = \frac{V_m}{\sqrt{2}} \frac{I_m}{\sqrt{2}} \cos \theta = V_{rms} I_{rms} \cos \theta \qquad (10.7.4)$$

In this equation the product of the rms voltage value and the rms current values $V_{rms} I_{rms}$ is called the *apparent power* and is measured in units of volt-amperes. One would expect the product of volts and amperes to be power. Indeed, the insulation on the wires, switches, etc., is under as much stress and the copper wire is carrying as much current as if that apparent amount of power were being delivered to $Z(j\omega)$. However, the *real amount of power is always less than the apparent power* because of the presence of the $\cos \theta$ factor in equation (10.7.4). This *power factor* (pf) is

$$pf = \cos \theta \qquad \text{where} \qquad 0 \leq \cos \theta \leq 1 \qquad (10.7.5)$$

Therefore the impedance angle also plays a part in determining how much real power gets delivered to any impedance. For a pure resistor, the impedance angle is zero and the power factor is unity (UPF) so that the apparent and real power are equal. But for a pure inductor or pure capacitor, the magnitude of the impedance angle is 90° and the power factor is therefore equal to zero. No real average power is delivered to a purely reactive component no matter how large the apparent power is.

EXAMPLE 10.7.1

Find P_{av} delivered by the source in Figure 10.7.2:

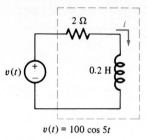

$v(t) = 100 \cos 5t$

figure 10.7.2
The circuit of Example 10.7.1.

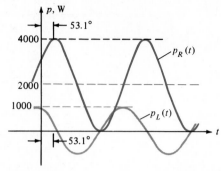

figure 10.7.3
Instantaneous powers $p_R(t)$ and $p_L(t)$ plotted versus t.

a. By finding $p_R(t)$ and $p_L(t)$, adding them together to get $p_z(t)$, and then finding the average
b. By using $V_{rms} I_{rms} \cos \theta$.
ANS.:   (a) First find $i(t)$ using phasors

$$\mathbf{V} = 100\underline{/0} \qquad \mathbf{Z} = 2 + j5(0.2) = 2 + j = 2.24\underline{/26.6°}$$

$$\mathbf{I} = \frac{\mathbf{V}}{\mathbf{Z}} = \frac{100\underline{/0°}}{2.24\underline{/26.6°}} = 44.7\underline{/-26.6°}$$

Therefore $i(t) = 44.7 \cos (5t - 26.6°)$

and the power delivered to the resistor is

$$p_R(t) = i^2 R = 2000(2) \cos^2 (5t - 26.6°) = 4000 \cos^2 (5t - 26.6°) = 2000[1 + \cos (10t - 53.1°)]$$

Note that this quantity is never less than zero. Now, with $v_L(t) = L(di/dt)$, the power delivered to the inductor is

$$p_L(t) = Li\,\frac{di}{dt} = 0.2(44.7) \cos (5t - 26.6°)(44.7)(5)[-\sin (5t - 26.6°)]$$

$$= 2000[\cos (5t - 26.6°)][-\sin (5t - 26.6°)]$$

$$= -\tfrac{2000}{2} \sin (10t - 53.1°)$$

Note that the average value of this quantity is zero. The total $p(t)$ is $p_R + p_L$ and

$$p(t) = 2000[1 + \cos (10t - 53.2°) - \tfrac{1}{2} \sin (10t - 53.1°)]$$

$$= 2000[1 + 1.12 \cos (10t - 26.6°)]$$

so that $P_{av} = 2000 \text{ W}$

Note that the total instantaneous power is sometimes negative. The sketch of $p_R(t)$ and $p_L(t)$ is shown in Figure 10.7.3.

(b) $P_{av} = V_{rms} I_{rms} \cos \theta = \dfrac{100}{\sqrt{2}} \dfrac{44.7}{\sqrt{2}} \cos 26.6° = 2000 \text{ W}$

COMPLEX POWER   Since, from equation (10.7.4)

$$P_{av} = V_{rms} I_{rms} \cos (\theta_v - \theta_I) \qquad (10.7.6)$$

where θ_v and θ_I are the phase angles of the voltage and current phasors, respectively, we can use the Euler identity to write

$$P_{av} = V_{rms} I_{rms} \text{Re } e^{j(\theta_v - \theta_I)}$$

or

$$P_{av} = \text{Re } V_{rms} e^{j\theta_v} I_{rms} e^{-j\theta_I} \qquad (10.7.7)$$

Define two complex quantities

$$\mathbf{V}_{rms} = V_{rms}\underline{/\theta_v} \qquad \text{and} \qquad \mathbf{I}_{rms} = I_{rms}\underline{/\theta_I}$$

which are merely scaled-down versions of the phasor quantities $\mathbf{V} = V\underline{/\theta_v}$ and $\mathbf{I} = I\underline{/\theta_I}$. Then, noting the negative sign in equation (10.7.7), we have

$$P_{av} = \text{Re } \mathbf{V}_{rms} \mathbf{I}^*_{rms} \qquad (10.7.8)$$

Equation (10.7.8) defines a complex quantity whose real part is the average (real) power P_{av}. We call this quantity the *complex power*

$$\boxed{S = P_{av} + jQ = \mathbf{V}_{rms} \mathbf{I}^*_{rms}} \qquad (10.7.9)$$

so that $P_{av} = \text{Re } S$, where Q is the *reactive* or *quadrature* power, measured in *vars* (volt-amperes reactive), S is the *complex power*, measured in *volt-amperes* (its magnitude is the apparent power $V_{rms} I_{rms}$), and P_{av} is, of course, measured in watts.

EXAMPLE 10.7.2

Find S, P_{av}, and Q delivered to a series connection of a 15-Ω resistor and a 100-mH inductor. The voltage across the combination is $v(t) = 120\sqrt{2} \cos (377t + 30°)$.

ANS.:

$$\mathbf{V}_{rms} = 120\underline{/30°} \qquad Z = 15 + j37.7 = 40.6\underline{/68.3°}$$

$$\mathbf{I}_{rms} = \frac{\mathbf{V}_{rms}}{Z} = 2.96\underline{/-38.3°}$$

$$S = \mathbf{V}_{rms} \mathbf{I}^*_{rms} = 355\underline{/30° - (-38.3°)} = 355\underline{/68.3°} = P_{av} + jQ = 131 + j330$$

The power factor is $\cos 68.3° = 0.369$. See Figure 10.7.4.

For any impedance $Z(j\omega) = Z$ (a complex number)

$$\mathbf{V}_{rms} = \mathbf{I}_{rms} Z \qquad (10.7.10)$$

Substituting (10.7.10) into (10.7.9) yields complex power

$$S = P_{av} + jQ = \mathbf{I}_{rms} Z \mathbf{I}^*_{rms} = (\mathbf{I}_{rms} \mathbf{I}^*_{rms})Z \qquad (10.7.11)$$

Because the product of a complex quantity times its conjugate is real and equal to the magnitude squared of the complex quantity, we can write (10.7.11) as

$$S = |\mathbf{I}_{\mathrm{rms}}|^2 Z = I^2_{\mathrm{rms}} Z \qquad (10.7.12)$$

which states that the complex power S and impedance Z have the same angle (since I^2_{rms} is a real quantity). Because the power triangle and the impedance triangle are *similar triangles* in the geometric sense, $P_{\mathrm{av}} = |S| \cos \theta$ and $Q = |S| \sin \theta$, where θ is the impedance angle (for this reason the impedance angle is sometimes called the *power angle*).

For a *resistor*, equation (10.7.12) becomes

$$S = P_{\mathrm{av}} + jQ = I^2_{\mathrm{rms}} R$$

so that

$$P_{\mathrm{av}} = I^2_{\mathrm{rms}} R \qquad (10.7.13)$$

and

$$Q = 0 \qquad (10.7.14)$$

Thus, a resistor absorbs real power from the circuit. For an inductor, equation (10.7.12) becomes

$$S = P_{\mathrm{av}} + jQ = I^2_{\mathrm{rms}}(jX_L) = I^2_{\mathrm{rms}}(j\omega L)$$

so that

$$P_{\mathrm{av}} = 0 \qquad (10.7.15)$$

and

$$Q = I^2_{\mathrm{rms}} X_L \qquad (10.7.16)$$

Thus an inductor absorbs reactive power from the circuit. For a capacitor, equation (10.7.12) becomes

$$S = P_{\mathrm{av}} + jQ = I^2_{\mathrm{rms}}(jX_C) = I^2_{\mathrm{rms}}\left(-j\,\frac{1}{\omega C}\right) = -jI^2_{\mathrm{rms}}|X_C|$$

so that

$$P_{\mathrm{av}} = 0 \qquad (10.7.17)$$

and†

$$Q = -I^2_{\mathrm{rms}}|X_C| \qquad (10.7.18)$$

In equation (10.7.9) we can substitute for $\mathbf{I}^*_{\mathrm{rms}}$ [instead of for $\mathbf{V}_{\mathrm{rms}}$ as we did in developing equation (10.7.11)]

$$\mathbf{I}^*_{\mathrm{rms}} = \frac{\mathbf{V}^*_{\mathrm{rms}}}{Z^*}$$

Thus equation (10.7.9) becomes

$$S = P_{\mathrm{av}} + jQ = \mathbf{V}_{\mathrm{rms}}\frac{\mathbf{V}^*_{\mathrm{rms}}}{Z^*}$$

so that

$$S = P_{\mathrm{av}} + jQ = \frac{V^2_{\mathrm{rms}}}{Z^*} \qquad (10.7.19)$$

Since for a resistor $Z = R$, equation (10.7.19) yields

$$P_{\mathrm{av}} = \frac{V^2_{\mathrm{rms}}}{R} \quad \text{and} \quad Q = 0 \qquad (10.7.20)$$

† Equation (10.7.18) is sometimes interpreted to mean that a capacitor is a *source* of reactive power since it *absorbs* a *negative* amount of Q.

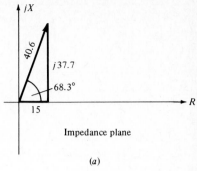

Impedance plane

(a)

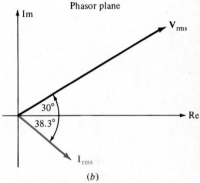

Phasor plane

(b)

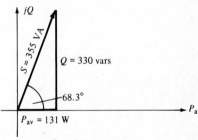

Complex power plane

(c)

figure 10.7.4
(a) The impedance diagram, (b) the phasor diagram, and (c) the complex power diagram for the circuit of Example 10.7.2.

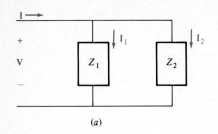

(a)

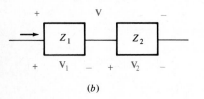

(b)

figure 10.7.5
(a) Two impedances in parallel. (b) Two impedances in series.

For an inductor $Z = jX_L$ and $Z^* = -jX_L$; hence

$$S = \frac{V_{rms}^2}{-jX_L} = +j\frac{V_{rms}^2}{X_L}$$

$$P_{av} = 0 \quad \text{and} \quad Q = \frac{V_{rms}^2}{X_L} \tag{10.7.21}$$

For a capacitor $Z = -j|X_C|$; hence

$$S = \frac{V_{rms}^2}{+j|X_C|} = -j\frac{V_{rms}^2}{|X_C|}$$

$$P_{av} = 0 \quad \text{and} \quad Q = -\frac{V_{rms}^2}{|X_C|} \tag{10.7.22}$$

The total complex power delivered to two impedances in parallel is the sum of the complex powers delivered to each. Proof of this is as follows (see Figure 10.7.5a):

$$S = VI^* = V(I_1 + I_2)^* = V(I_1^* + I_2^*) = VI_1^* + VI_2^*$$

The same holds true for a series connection of two impedances (Figure 10.7.5b):

$$S = VI^* = (V_1 + V_2)I^* = V_1I^* + V_2I^*$$

Does superposition apply to power? What do these equations say about this?

Electric utility companies are always interested in maintaining the power factor as close to unity as possible at every point in the network. If a given transmission line has a fixed maximum current capacity (and is operating at rated voltage), it can carry more power when the power factor is close to unity than it can when the power factor is less. In order to maintain the power factor close to unity value, power companies install banks of capacitors throughout the network, as needed. They also charge extra to industries that have low power factors each month. Power factor is called *leading* when the load impedance is *capacitive* and *lagging* when the load is *inductive*.†

EXAMPLE 10.7.3

Two factories are supplied by one 4400-V (rms) 60-Hz power line. Factory A draws 5 MVA at 0.6 pf lagging, and factory B uses 10 MVA at 0.4 pf lagging.
a. What is the rms value of the transmission-line current?
b. What value of capacitance should be placed across the line in order to attain UPF?
c. What percentage increase in the power-carrying capability of the line does this result in?

ANS.:   (a) $S_A = 5\underline{/\cos^{-1} 0.6} = 5\underline{/53.1°} = 3 + j4$ MVA

and $S_B = 10\underline{/\cos^{-1} 0.4} = 10\underline{/66.4°} = 4 + j9.165$ MVA

The total complex power delivered by the transmission line is

---

† This nomenclature arises from the fact that power engineers often use the line *voltage* as their reference phasor. *RC* impedances then draw *leading* currents and *RL* circuits draw *lagging* currents.

$$S_A + S_B = 7 + j13.165 = 14.9\underline{/62°} \quad \text{MVA}$$

$$P_{av} = V_{rms} I_{rms} \cos \theta$$

$$7 \times 10^6 = 4400 I_{rms} \cos 62° \quad I_{rms} = 3389 \text{ A}$$

(b) We need $Q = -13.165$ megavars and from equation (10.7.22)

$$Q = \frac{-V^2_{rms}}{|X_C|} \quad \text{or} \quad 13.165 \text{ megavars} = \frac{4400^2}{|X_C|}$$

$$|X_C| = 1.47 \ \Omega = \frac{1}{\omega C}$$

where $\omega = 2\pi 60 = 377$ rad/s. Thus $C = 1804 \ \mu$F.

(c) If pf = 1, then $P_{av} = 4400(3389) = 14.9$ MW instead of 7 MW and the real power-carrying capacity of the line would be more than doubled after the addition of the power-factor-correcting capacitor bank.

DRILL PROBLEM

An impedance Z connected across a 100-V rms 1 rad/s line draws $P_{av} = 192$ W at 0.196 pf lagging. What is the value of the capacitor C to be connected in parallel with Z to improve the power factor to 0.98 lagging?
ANS.: $C = 0.0922$ F.

---

three-phase circuits **10.8**

The power grid in the United States and almost every other major nation today uses sinusoidal (ac) voltages. This historical choice was made primarily because sinusoidal voltages can be generated without using the complicated mechanical contact devices called commutators that were necessary in early dc generators and because sinusoidal voltages and currents can be stepped up or down by means of transformers; e.g., high-voltage ac can be converted simply and efficiently into 125-V household ac by power transformers without any moving devices or parts. Transmission of electric power is more efficient at higher voltages because for a given amount of power the current is less. Thus, the $i^2 R_T$ loss in the low but nonzero resistance R_T of the transmission lines is less. This is why electric-power lines are operated at extremely high voltages.†

A sinusoidal voltage is created (see Figure 10.8.1) when a constant magnetic flux is rotated in space in such a way that it links and unlinks a fixed coil of wire sinusoidally with time. The voltage produced is found from Faraday's law to be

$$v(t) = n \frac{d\phi}{dt} \tag{10.8.1}$$

---

† Since ac power lines also have inherent series inductance and shunt capacitance (power leaks to ground) that dc lines do not have, extremely high voltage dc would be the most efficient way to transmit power; but, at present we have no good way of using, i.e., transforming down, such voltages when we get them to the end of the line.

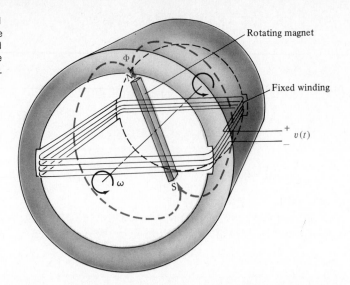

The constant-strength magnetic field in Figure 10.8.1 emanates from the north
pole of the permanent magnet; it then splits, some going one way and some the
other (most of it *outside* the fixed coil of wire) and then recombines to enter the
south pole of the permanent magnet. This magnet is mounted inside a cylinder
called the *rotor* of this machine, which is called an *alternator*. If the rotor spins at
a constant angular velocity of ω rad/s, the flux linking the coil is essentially

$$\phi(t) = \Phi_m \sin \omega t \qquad (10.8.2)$$

and so equation (10.8.1) yields

$$v_A(t) = \omega n \Phi_m \cos \omega t \qquad (10.8.3)$$

figure 10.8.2
Generation of a three-phase voltage set. An
end view of the alternator showing coils 1–2,
3–4, and 5–6. The constant amplitude, but
rotating, magnetic field is Φ.

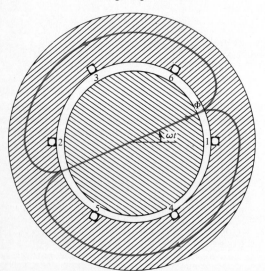

figure 10.8.3
The three voltages $v_A(t)$, $v_B(t)$, and $v_C(t)$.

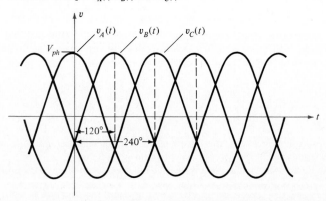

The fixed coil is mounted on the *stator*. If, for example, the rotor is driven at a rotational speed of 3600 rpm, the frequency of the stator voltage is 60 Hz = 377 rad/s. In practice, the constant magnetic field is usually obtained from an electromagnet rather than a permanent magnet, but the principle is the same.

If two more similar coils are wound on the stator of the machine, each indexed physically 120° from the others, three voltages $v_A(t)$, $v_B(t)$, and $v_C(t)$ will be generated (Figures 10.8.2 and 10.8.3). These voltages, which will be 120° out of phase with each other but equal in magnitude, are called the *phase voltages* and have magnitude V_{ph}. The resultant voltages are

$$v_1 - v_2 = v_A(t) = V_{ph} \cos \omega t \qquad (10.8.4)$$

$$v_3 - v_4 = v_B(t) = V_{ph} \cos (\omega t - 120°) \qquad (10.8.5)$$

$$v_5 - v_6 = v_C(t) = V_{ph} \cos (\omega t - 240°) \qquad (10.8.6)$$

If conductors 2, 4, and 6 are connected together and brought out to a terminal (called the *neutral* line) and conductors 1, 3, and 5 are each brought out to separate terminals, the resulting four-terminal system is described by the phasor diagram of Figure 10.8.4a. On the other hand, if conductors 1 and 4, 3 and 6, and

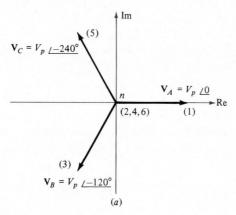

(a)

(b)

figure 10.8.4
(a) A wye-connected three-phase source. (b) A delta-connected three-phase source. (The stator coil connections are shown in parentheses.)

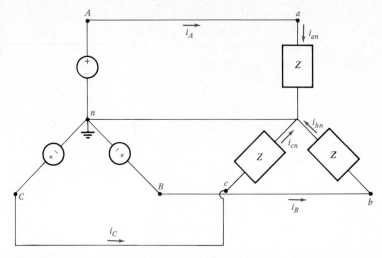

figure 10.8.5
A grounded-neutral, balanced
wye-connected three-phase network.
Identical sinusoidal impedances Z are found
in each of the three phases of the load.

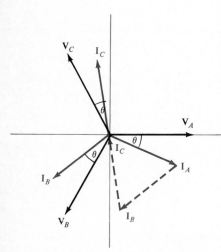

figure 10.8.6
The phasor diagram of the balanced
network of Figure 10.8.5. The transposed
phasors $\mathbf{I}_B$ and $\mathbf{I}_C$ demonstrate that the
neutral current (the sum of $\mathbf{I}_A + \mathbf{I}_B + \mathbf{I}_C$) is
identically zero.

2 and 5 are connected together and these three resulting nodes are brought out separately to three terminals, the resulting system is described by the phasor diagram of Figure 10.8.4b. The first method of interconnecting the stator conductors gives rise to a *wye*-connected three-phase source (we shall call the four terminals A, B, C, and n). The second method produces a *delta*-connected three-phase source.

In the wye connection, the neutral terminal n is almost always the reference (ground) node†, but in the delta connection no terminal suggests itself as a logical choice for reference node‡. The voltages V_A, V_B, and V_C are the phase voltages.

Suppose three equal load impedances are connected separately across the terminals of a wye-connected three-phase source (Figure 10.8.5). When the three load impedances are equal, the three-phase circuit is *balanced* because the three resulting currrents $i_A(t)$, $i_B(t)$, and $i_C(t)$ will all be equal in magnitude and 120° out of phase with each other. The phasor diagram is shown in Figure 10.8.6, where

$$\mathbf{I}_A = \frac{\mathbf{V}_A}{Z} = \frac{V_{\text{ph}}\underline{/0°}}{|Z|\underline{/\theta}} = |I_A|\underline{/-\theta} \qquad (10.8.7)$$

$$\mathbf{I}_B = \frac{\mathbf{V}_B}{Z} = \frac{V_{\text{ph}}\underline{/-120°}}{|Z|\underline{/\theta}} = |I_B|\underline{/-120°-\theta} \qquad (10.8.8)$$

$$\mathbf{I}_C = \frac{\mathbf{V}_C}{Z} = \frac{V_{\text{ph}}\underline{/-240°}}{|Z|\underline{/\theta}} = |I_C|\underline{/-240°-\theta} \qquad (10.8.9)$$

The circuit operates like three separate circuits but with a common ground (neutral or *return*) line. Since $|I_A| = |I_B| = |I_C| = V_{\text{ph}}/|Z|$ and all three of these currents are mutually out of phase by 120°, simple geometry applied to the current phasors in Figure 10.8.6 shows that the current in the neutral wire (sum

† In the wye configuration, the phase voltages are *line-to-neutral* voltages.

‡ In delta-connected loads, a transformer can be used in each phase, and one side of each such output winding can be grounded.

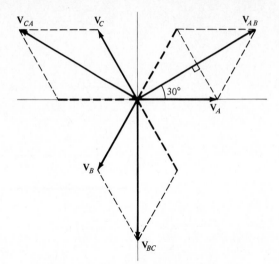

figure 10.8.7
The line-to-line voltages as obtained from
the phase voltages. Note that
$$|V_{\mathrm{LL}}| = \sqrt{3}\,|V_{\mathrm{ph}}|.$$

of $\mathbf{I}_A + \mathbf{I}_B + \mathbf{I}_C$) is zero (and hence that wire can be eliminated from the circuit). In wye-connected loads the phase currents are identical to their corresponding line currents.

In three-phase circuits the source may be connected in either the wye or delta configuration, and the load can also be (independently of the source) connected in either wye or delta. Clearly, there are four possibilities. Sometimes the neutral line is included, in which case we have a four-wire three-phase wye-connected source. If the neutral is omitted, the result is a three-wire transmission line.

An easy voltage measurement to make in practice is the magnitude of the line-to-line voltage, say $|V_{AB}|$. The three line-to-line voltages of the circuit in Figure 10.8.5 are shown in Figure 10.8.7, where we note (from the 30-60-90 triangle) that

$$\frac{|V_{AB}|/2}{\sqrt{3}} = \frac{|V_A|}{2}$$

Thus $|V_{AB}| = \sqrt{3}|V_A|$, or, in general, each of the line-to-line voltages has the magnitude

$$|V_{LL}| = \sqrt{3}|V_{\mathrm{ph}}| \qquad (10.8.10)$$

For example, if the rms value of the phase voltage is 120 V, the line-to-line rms voltage will be $\sqrt{3}(120) = 208$ V. Delta-connected sources can be converted into four-wire wye-connected sources with the use of power transformers.

In the wye-connected load (Figure 10.8.5) the phase currents are all indentical to the corresponding line current

$$i_{an}(t) = i_A(t) \qquad (10.8.11)$$

$$i_{bn}(t) = i_B(t) \qquad (10.8.12)$$

$$i_{cn}(t) = i_C(t) \qquad (10.8.13)$$

The relationships between the phase and line currents in a delta-connected load

figure 10.8.8
A delta-connected balanced load (where each phase contains the identical sinusoidal impedance Z).

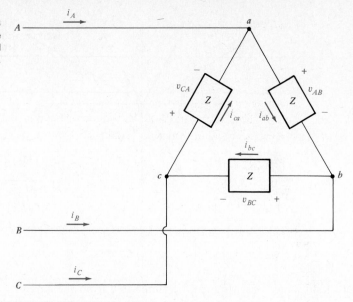

(Figure 10.8.8) are obtained by writing KCL at each node, a, b, and c:

$$i_A(t) = i_{ab}(t) - i_{ca}(t) \qquad (10.8.14)$$

$$i_B(t) = i_{bc}(t) - i_{ab}(t) \qquad (10.8.15)$$

$$i_C(t) = i_{ca}(t) - i_{bc}(t) \qquad (10.8.16)$$

The resulting phasors are shown in Figure 10.8.9. From geometry it is clear that

figure 10.8.9
The phasor diagram for the delta-connected load of Figure 10.8.8. The angle θ is the impedance angle of each of the three load impedances, Z.

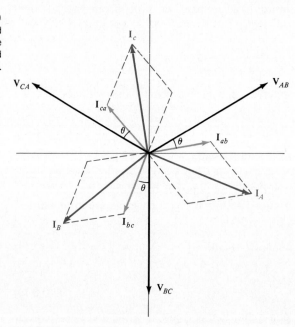

the magnitude of the line current is

$$|I_L| = \sqrt{3}|I_{ph}| \tag{10.8.17}$$

Let us now find the *total instantaneous power* delivered to a three-phase load, working first with a wye-connected load. For the load shown in Figure 10.8.5 and using the corresponding phasor diagram of Figure 10.8.6 we can write

THREE-PHASE POWER

$$v_A(t) = V_{ph} \cos \omega t \tag{10.8.18}$$

$$i_A(t) = I_L \cos(\omega t - \theta) \tag{10.8.19}$$

so that the instantaneous power delivered to the phase A impedance is

$$p_A(t) = v_A(t)i_A(t) = V_{ph} I_L \cos \omega t \cos(\omega t - \theta) \tag{10.8.20}$$

Using the trigonometric identity

$$\cos(a \pm b) = \cos a \cos b \mp \sin a \sin b \tag{10.8.21}$$

we get (with $a = \omega t$ and $b = \omega t - \theta$)

$$p_A(t) = \frac{V_{ph}}{\sqrt{2}} \frac{I_L}{\sqrt{2}} [\cos(2\omega t - \theta) + \cos \theta] \tag{10.8.22}$$

Similarly

$$p_B(t) = \frac{V_{ph}}{\sqrt{2}} \frac{I_L}{\sqrt{2}} [\cos(2\omega t - 240° - \theta) + \cos \theta] \tag{10.8.23}$$

and

$$p_C(t) = \frac{V_{ph}}{\sqrt{2}} \frac{I_L}{\sqrt{2}} [\cos(2\omega t - 480° - \theta) + \cos \theta] \tag{10.8.24}$$

The total instantaneous power is given by the sum of equations (10.8.22) to (10.8.24). Thus (because the balanced three-phase set of quantities in those equations sum to zero)

$$p_T(t) = 3 \frac{V_{ph}}{\sqrt{2}} \frac{I_L}{\sqrt{2}} \cos \theta \tag{10.8.25}$$

which is equal to 3 times the average power delivered to any one load impedance, where $V_{ph}/\sqrt{2}$ and $I_L/\sqrt{2}$ are the rms values of the phase voltage and line (phase) current, respectively. Note that *the total instantaneous power is a constant (independent of time)*. Thus a properly designed three-phase motor is theoretically capable of producing perfectly smooth (nonpulsating) mechanical output power. Since $|V_{LL}| = \sqrt{3}|V_{ph}|$, equation (10.8.25) can be written

$$P_{av} = \sqrt{3}(V_{LL})_{rms}(I_L)_{rms} \cos \theta \tag{10.8.26}$$

Similarly, the total reactive power Q delivered to such a balanced three-phase load is 3 times the reactive power delivered to each phase

$$Q = 3 \frac{V_{ph}}{\sqrt{2}} \frac{I_L}{\sqrt{2}} \sin \theta = \sqrt{3}(V_{LL})_{rms}(I_L)_{rms} \sin \theta \tag{10.8.27}$$

POWER MEASUREMENT IN THREE-PHASE CIRCUITS

One straightforward way of measuring the total average power delivered to a three-phase load (whether balanced or not) is to measure each of the average powers delivered to the three impedances independently. This of course requires three wattmeters if we do not want to time-share a single meter for the three phases, turning the system off and rewiring the meter between measurements (see Figures 10.8.10 and 10.8.11).

A wattmeter is an instrument with two sensing coils, one for voltage and one for current. The current coil is placed in series with the current to be measured, and the voltage coil is placed across the voltage to be measured. Each coil has one terminal marked $\pm$; these terminals are connected together and then into the circuit in such a way that positive current flows into that node. A wattmeter designed to be used with sinusoidal voltages and currents yields a reading equal to $V_{\text{rms}} I_{\text{rms}} \cos \Psi$, where Ψ is the phase angle between V and I.

Although it is obvious that the circuits of Figures 10.8.10 and 10.8.11 will measure total average power, a better way requires only *two wattmeters* (see Figure 10.8.12). Regardless of whether the three-phase load is *balanced or not* it is always true, whether the currents and voltages are sinusoidal or not, that

$$i_A + i_B + i_C = 0 \tag{10.8.28}$$

The total instantaneous power delivered to the load is

$$p_T(t) = v_{An} i_A + v_{Bn} i_B + v_{Cn} i_C \tag{10.8.29}$$

Solving equation (10.8.28) for i_B and inserting that into equation (10.8.29) yields the total instantaneous power in the form

$$p_T(t) = v_{An} i_A + v_{Bn}(-i_A - i_C) + v_{Cn} i_C = i_A(v_{An} - v_{Bn}) + i_C(v_{Cn} - v_{Bn})$$

$$= p(W_1) + p(W_2) \tag{10.8.30}$$

figure 10.8.10
The three-wattmeter method for measuring average power delivered to a wye-connected set of unknown load impedances.

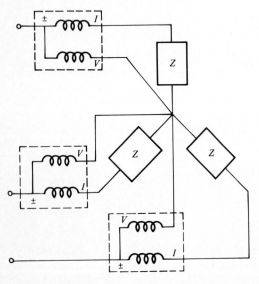

figure 10.8.11
The three-wattmeter method for measuring average power delivered to a delta-connected set of unknown load impedances.

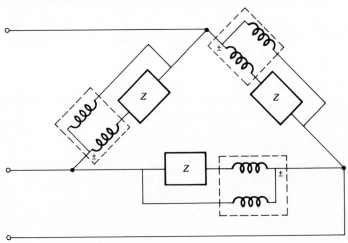

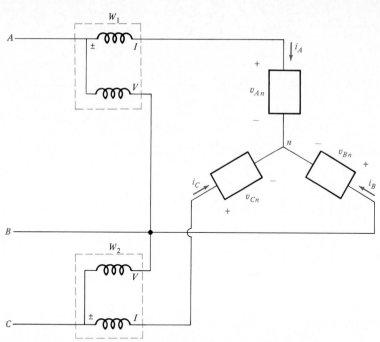

figure 10.8.12
The two-wattmeter method for measuring average power delivered to any three-phase load.

where W_1 and W_2 are the wattmeters. Hence the total of the two instantaneous powers measured by the two wattmeters is the total instantaneous power delivered to the load. Each meter reads the average value of the instantaneous power it senses. Since the average value of a sum is the sum of the individual average values,

$$P_{\text{av}} = \text{algebraic sum of } W_1 \text{ and } W_2 \text{ readings}$$

Any delta-connected load always has a wye equivalent (see Chapter 2). Thus the two-wattmeter method works for delta loads (both balanced and unbalanced) just as well as for wye-connected loads.

In this chapter, however, we limit our discussion to sinusoidally excited, balanced three-phase loads. In such systems we can determine from the readings of W_1 and W_2 not only the total average power delivered to the total load but also the reactive power (and thus the *power factor*, which of course is the same for each phase because the system is balanced).

The phasor diagram of the circuit in Figure 10.8.12 is given in Figure 10.8.13. Wattmeters yield readings equal to the applied rms voltage times the rms current times the cosine of the phase angle between these two quantities. Thus, in this circuit, the two meter readings are

$$W_1 = (V_{AB})_{\text{rms}}(I_A)_{\text{rms}} \cos \underline{/V_{AB}, I_A} \qquad W_2 = (V_{CB})_{\text{rms}}(I_C)_{\text{rms}} \cos \underline{/V_{CB}, I_C}$$

or
$$W_1 = (V_{LL})_{\text{rms}}(I_L)_{\text{rms}} \cos (30° + \theta)$$

$$= (V_{LL})_{\text{rms}}(I_L)_{\text{rms}}(\cos 30° \cos \theta - \sin 30° \sin \theta)$$

$$W_2 = (V_{LL})_{\text{rms}}(I_L)_{\text{rms}} \cos (\theta - 30°)$$

$$= (V_{LL})_{\text{rms}}(I_L)_{\text{rms}}(\cos 30° \cos \theta + \sin 30° \sin \theta)$$

figure 10.8.13
The phasor diagram of the circuit in Figure 10.8.12. The impedance angles seen by the two wattmeters are labeled $\underline{/W_1}$ and $\underline{/W_2}$, respectively.

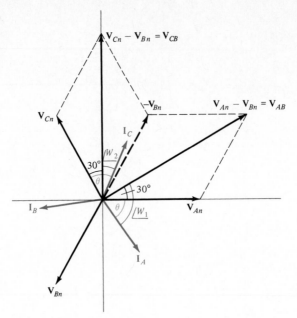

Thus
$$W_1 + W_2 = \sqrt{3}(V_{LL})_{\text{rms}}(I_L)_{\text{rms}} \cos \theta$$
$$= 3(V_{Ln})_{\text{rms}}(I_L)_{\text{rms}} \cos \theta = \text{total power} \qquad (10.8.31)$$

and
$$W_2 - W_1 = (V_{LL})_{\text{rms}}(I_L)_{\text{rms}} \, 2 \sin 30° \sin \theta$$
$$= (V_{LL})_{\text{rms}}(I_L)_{\text{rms}} \sin \theta$$

Using equation (10.8.27), we get
$$Q = \sqrt{3}(W_2 - W_1) \qquad (10.8.32)$$

Thus
$$\text{pf} = \frac{Q}{P} = \sqrt{3}\,\frac{W_2 - W_1}{W_2 + W_1} \qquad (10.8.33)$$

EXAMPLE 10.8.1

A balanced wye-connected load containing impedances of $Z = 30 + j40 \ \Omega$ per phase is driven by a balanced three-phase source such that the line-to-neutral voltage is 120 V rms. Two wattmeters are connected as in Figure 10.8.12. What will be the reading of *each* wattmeter?

ANS.:
$$Z = 30 + j40 = 50\underline{/53.1°}$$

A typical rms line current will be

$$\mathbf{I}_A = \frac{120\underline{/0°}}{50\underline{/53.1°}} = 2.4\underline{/-53.1°}$$

Thus the power per phase P_{ph} is

$$P_{ph} = I_{rms}^2 R = 2.4^2(30) = 172.8 \text{ W}$$

and so

$$P_{total} = 3P_{ph} = 518 \text{ W}$$

Total reactive power is obtained from $Q/P = \tan$ (impedance angle) as $Q = 518 \tan 53.1° = 690$ vars. Thus, from equations (10.8.31) and (10.8.32)

$$W_2 + W_1 = 518 \qquad W_2 - W_1 = \frac{690}{\sqrt{3}}$$

$$2W_2 = 916$$

So that

$$W_2 = 458 \text{ W} \qquad \text{and} \qquad W_1 = 60 \text{ W}$$

If three wires A, B, and C define a balanced three-phase source, the order in which the three possible line-to-line voltages V_{AB}, V_{BC}, and V_{CA} reach their maxima determines the *phase sequence*. There are only two possible sequences, ABC and CBA. A circuit which can be used to determine which phase sequence is occurring is shown in Figure 10.8.14, from which we can write

$$\mathbf{V}_{DB} + \mathbf{I}_{AC} R = \mathbf{V}_{AB}$$

$\mathbf{V}_{AB}$ is a line-to-line voltage defined by the phase voltage sources as

$$\mathbf{V}_{AB} = \mathbf{V}_{An} - \mathbf{V}_{Bn}$$

Current $\mathbf{I}_{AC}$ is determined by line-to-line voltage $\mathbf{V}_{AC}$ and the series RC admittance. Depending on the choice of C and R, the current $\mathbf{I}_{AC}$ could be anywhere from almost in phase with $\mathbf{V}_{AC}$ to $90°$ *ahead* of $\mathbf{V}_{AC}$. In any event,

$$\mathbf{V}_{AC} = \mathbf{I}_{AC} R - \frac{j\mathbf{I}_{AC}}{\omega C}$$

A typical value of $\mathbf{I}_{AC} R$ is shown (along with other phasors of interest in this problem) in Figure 10.8.15. You should examine this diagram until you are fully convinced that the phasor relationships shown there are correct. The phase sequence used in drawing this phasor diagram is ABC. Regardless of the R and C values chosen, this diagram shows that $|V_{DB}| < |V_{LL}|$.

The CBA sequence is shown in Figure 10.8.16. The same phasors as in Figure 10.8.15 now demonstrate that, regardless of the values of R and C, $|V_{DB}| > |V_{LL}|$. This circuit represents a simple and inexpensive method for determining phase sequence. For example, in a three-phase motor, the phase sequence determines in what direction the motor will turn. Needless to say, it is well to know the phase sequence before starting up such a machine.

PHASE SEQUENCE

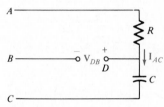

figure 10.8.14
A phase sequence detector. If $|V_{DB}| < |V_{LL}|$, the sequence is ABC. If $|V_{DB}| > |V_{LL}|$, the sequence is CBA.

figure 10.8.15
Phasor diagram for the phase sequence detector of Figure 10.8.14; note the sequence is ABC.

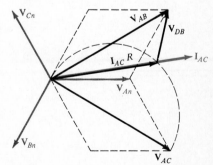

figure 10.8.16
Phasor diagram for the phase sequence
detector of Figure 10.8.14; note the sequence
is *CBA*.

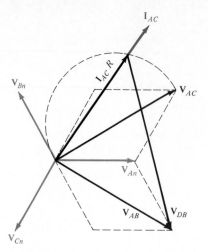

10.9   summary

In this chapter we have investigated methods used by electrical engineers in dealing with sinusoidal waveforms as inputs to linear systems. Basic is the concept of the complex exponential function $f(t) = \mathbf{A}e^{j\omega t}$, where $\mathbf{A}$ is a complex number called the phasor for $f(t)$.

To use phasors we must be adept at the algebra of complex numbers. Perhaps the most important idea presented in this chapter is that every sinusoid has an equivalent complex exponential function whose phasor multiplied by the transfer function $H(j\omega)$ yields the phasor of the particular response of that system.

Sinusoidal impedance and admittance values are used in developing transfer functions. How these complex quantities vary as functions of input frequency or element value was investigated in the discussion of immittance loci.

The amount of power (and the inherent double-frequency nature of that power) delivered to an arbitrary impedance was discussed together with the concept of complex power and power-factor correction. This last topic is a matter of extreme practical importance in commercial electric-power systems. A natural extension of any power-transmission discussion is the three-phase power method used to supply electric energy throughout the civilized world.

problems

1. Convert to polar form:

(a) $2 + j4$ (b) $-2 - j4$ (c) $2 + j0.04$
(d) $1.5 - j2.5$ (e) $2.5 + j1.5$ (f) $-1.5 - j2.5$
(g) to (l) Convert all the above to exponential form.

2. Convert to rectangular form:

(a) $5\underline{/30°}$ (b) $5e^{-j30°}$ (c) $3\underline{/50°}$
(d) $3\underline{/1°}$ (e) $2.5\underline{/160°}$ (f) $1.5\underline{/200°}$ (g) $e^{1+j\pi}$

3. Evaluate for $Z = 1\underline{/25°}$

$$F(Z) = \frac{(Z + 3)(Z + j)(3 + j2)}{3/40° + 3/65°(Z + j2)}$$

4. Given $A = 4 + j3$, $B = 3 - j1$, and $C = -3 + j4$, find the value of

$$D = \frac{AB + BC - CA}{A + B + C}$$

5. Two branches of a network are connected in parallel. One carries current $i_1(t) = 10 \cos (100t + 100°)$, and the other carries $i_2(t) = 20 \cos (100t - 40°)$. Find $i_T(t) = i_1(t) + i_2(t)$.

6. If three currents entering a node are $10 \cos (2t - 40°)$ A, $8 \cos (2t - 100°)$ A, and $15 \sin (2t + 30°)$ A, find the current $i_4(t)$ leaving the node in the fourth and last conductor.

7. The peak value of the voltage across the current source in Figure P10.7 is measured and found to be 26 V. A similar measurement of the inductor voltage yields a value of 10 V. Find the value of R. Draw a complete phasor diagram.

8. Find the particular (forced) response of the variable v in $pv + 2v = 10 \cos (3t + 20°)$.

9. An ideal voltage source $v_s(t) = 10 \cos (10t + 20°)u(t)$ drives the series combination of $R = 3 \ \Omega$ and $C = 0.025$ F. (a) Find the steady-state sinusoidal response $i(t)$. (b) Find the natural response $i_n(t)$. Do not evaluate the unevaluated coefficient. (c) Find the forced (particular) response $i_p(t)$. (d) Assuming all initial conditions to be zero, find the complete response $i(t)$ for all t.

10. Find the zero-state response of $v_C(t)$ in the circuit shown in Figure P10.10.

11. A system is described by the differential equation $(p + 1)v_2 = pv_1$, where $v_1(t)$ is the input and $v_2(t)$ is the output. If $v_1(t) = 6 + 10 \cos 4t$ for *all* t, then $v_2(t) = K + A \cos (t + \theta)$. Find K, A, and θ.

12. Given the equation $d^2v/dt^2 + 10dv/dt + v = 100 \cos 7t$, find the particular response $v_p(t)$ due to the steady-state sinusoidal source. Use the rotating complex exponential function whose real part is $100 \cos 7t$.

13. A single mesh contains an RLC series circuit ($R = 1 \ \Omega$, $L = 1$ H, and $C = 1$ F) and a voltage source $v_s(t) = 10\sqrt{2} \cos t \ u(t)$. Find $i(t)$ for $t > 0$. Assume $v_C(0-) = 0$.

14. The parallel combination of $R = \frac{1}{2} \ \Omega$, $C = 1$ F, and $L = \frac{1}{2}$ H is driven by a pure sinusoidal current source $i(t) = 10 \cos (2t + 30°)$. Find the resulting steady-state voltage $v(t)$.

15. Given the circuit in Figure P10.15 with $R = 5 \ \Omega$, $L = 3$ H, and $Z_1 = 3 - j4$, (a) draw and label impedances Z_1, Z_2, and Z_{in} on the complex Z plane. (b) Demonstrate graphically that $Z_1 + Z_2 = Z_{in}$. (c) Label Z_{in} with its value in polar form. (d) Draw and label the current phasor and all element voltage phasors. Use dotted construction lines to show how KVL applies to this circuit.

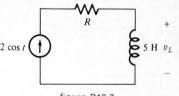

figure P10.7

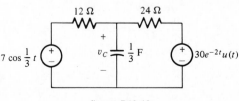

figure P10.10

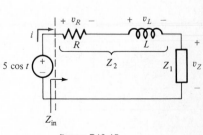

figure P10.15

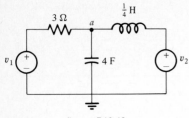

figure P10.19

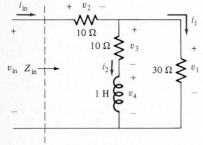

figure P10.20

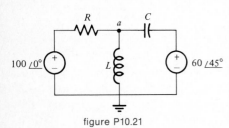

figure P10.21

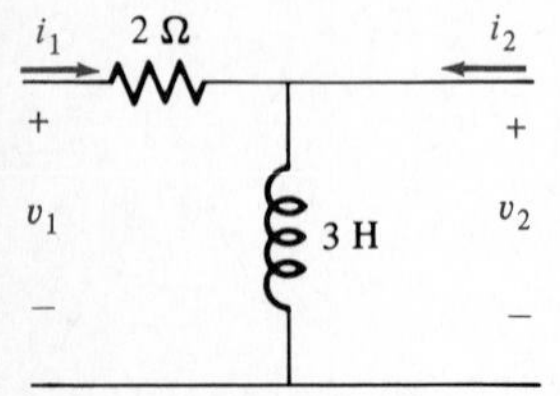

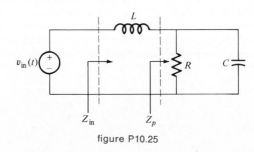

figure P10.23

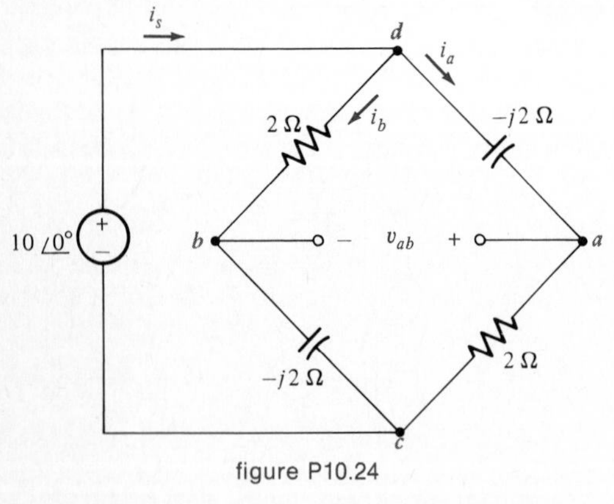

16. A parallel LC combination is in series with a resistor ($R = 8\ \Omega$, $L = 3$ H, and $C = \frac{1}{6}$ F). The resulting two-terminal circuit is driven by $v_s(t) = 10 \cos 2t$. (a) What sinusoidal impedance Z_{in} is seen by the voltage source? (b) Find $i(t)$, the current leaving the positive terminal of the source.

17. A series combination of $R = 40\ \Omega$ and $L = 20$ H is driven by $v_s(t) = 14 \cos (3t + 23°)$. Find $v_L(t)$, the steady-state sinusoidal voltage across the inductor.

18. An unknown element is placed in series with a 1-Ω resistor. This series combination is connected across a 60-Hz sinusoidal voltage source whose magnitude is 10 V. The steady-state sinusoidal current flowing out of the source lags the source voltage by 30°. The voltage across the resistor has a magnitude of 3 V. Find two possible equivalent circuits for the unknown element.

19. Find $v_a(t)$ in the circuit of Figure P10.19 if $v_1(t) = 30 \cos t$ and $v_2(t) = 15 \cos (t + 45°)$.

20. For the circuit shown in Figure P10.20 (a) find the steady-state sinusoidal impedance Z_{in} if the frequency of the applied voltage is 10 rad/s. (b) If $I_{in} = 10\underline{/0°}$, find V_1, V_2, V_3, V_4, V_{in}, I_1, and I_2. Sketch each of them on a phasor diagram. Show phasor summations with dotted construction lines.

21. In the circuit shown in Figure P10.21, $Y_R = 10\ \mho$, $Y_C = j10\ \mho$, and $Y_L = -j10\ \mho$. Find V_a.

22. A series combination of a 2-Ω resistor and a 0.4-H inductor is driven by $v(t) = 10 \cos 10t$. A pair of output terminals are connected, one to each end of the inductor. Find the Thevenin equivalent circuit as seen by any external impedance connected to those terminals.

23. Given the circuit shown in Figure P10.23, (a) what is the differential equation in d/dt notation relating output $v_2(t)$ to the input $v_1(t)$? Assume $i_2(t) = 0$. (b) For $i_2 = 0$ and $v_1 = 10 \cos (2t + 30°)$, find the steady-state current $i_1(t)$. (c) For $i_2 = 0$ and $v_1(t) = 10\ u(t) \cos (2t + 30°)$ find the *particular* response for $i_1(t)$.

24. Draw a phasor diagram for the circuit in Figure P10.24 that includes I_s, I_a, I_b, V_{db}, and V_{ab}.

25. Given the circuit of Figure P10.25, which is being driven by a sinusoidal source $v_{in}(t) = 9 \cos 2t$, (a) find Z_p if $R = 2\ \Omega$ and $Z_C = -j$. (b) What value should be selected for L in order to make Z_{in} purely resistive? (c) Sketch the phasor diagram for the resonant condition specified in part (b). Include all branch currents and voltages.

figure P10.24

figure P10.25

26. Draw the phasor diagram for a single mesh containing $v_s(t) = 10\sqrt{2} \cos 5t$, $R = 5\ \Omega$, $L = 1$ H, and $C = \frac{1}{75}$ F. Show the current phasor and all element voltage phasors.

27. A 60-Ω resistor is in series with a 5-mH inductor. What size capacitor should be placed in parallel with this RL combination for the total input impedance to have zero reactance at 24 krad/s?

28. A 200-Ω resistor, a 0.04-H inductor, and a 0.25-μF capacitor are connected in series. Find the phasor of the voltage across the combination if the current phasor $30\underline{/45°}$ mA is applied at (a) $\omega = 800$ rad/s, (b) $\omega = 10{,}000$ rad/s, and (c) $\omega = 12{,}500$ rad/s.

29. When the voltage $20 \cos (2000t + 60°)$ is applied across a circuit element, the current flowing in it has a magnitude of 0.1 A. (a) If the element is an inductor, find the value of L and write an expression for $i(t)$. (b) If it is a capacitor, find the value of C and write the expression for $i(t)$.

30. A circuit consists of the *series* combination of $R_1 = 3\ \Omega$ and an inductor whose reactance is $1\ \Omega$. Determine the values of resistor R_2 and reactance X_2 which when placed in *parallel* with each other will present the same total input impedance as the first circuit.

31. A circuit is placed inside a black box and two terminals are brought out. Inside the box the circuit consists of two branches connected in parallel across the terminals. One branch is a single 1-μF capacitor. The other branch is a series combination of $v_s(t) = 100 \cos 10^5 t$ and $R = 10\ \Omega$. Find the Thevenin equivalent of this circuit in the box.

32. Given that $v_1(t) = 100 \cos 377t$ in the network shown in Figure P10.32. Find $i_1(t)$, $i_2(t)$, $i_3(t)$, and $v_2(t)$. Sketch a phasor diagram that includes all voltage and current phasors.

33. Calculate the *complete response* of the current flowing in a single closed mesh that contains the series connection of $R = L = C = 1$ and $v_s(t) = 10\sqrt{2} \cos t\, u(t)$. Assume that no initial energy is stored in the circuit at $t = 0-$.

34. In the circuit of Figure P10.34 the remaining two terminals are left open-circuited and $\omega = 1$ krad/s. (a) Find the input impedance at terminals a-b, c-d, and a-c. (b) Find Z_{in} at c-d if a-b are short-circuited.

35. A circuit consists of $R = 1\ \Omega$ and $C = 1$ F connected in parallel, and this combination is in series with $L = 1$ H. This entire circuit is driven by a voltage source $v_s(t)$. If the voltage across the RC combination is $2 \cos (2t - 45°)$, find $v_s(t)$. Draw a phasor diagram that includes all voltages and currents. See Figure P10.35.

36. A series RLC circuit ($R = C = 1$ and $L = 2$ H) is driven by a voltage source $v_s(t)$. The inductor voltage is observed to be $v_L(t) = 2\sqrt{2} \cos (t - 30°)$. Find $v_s(t)$ and draw a phasor diagram that includes all voltages and the current.

37. Find $v(t)$ in the circuit in Figure P10.37.

38. In the circuit of Figure P10.38 $v_s(t)$ is a steady-state sinusoidal voltage source whose magnitude is 4 and whose frequency is 1 rad/s. R is adjusted until $|\mathbf{V}_R| = |\mathbf{V}_1|$, at which time $|\mathbf{V}_2| = \frac{1}{2}|\mathbf{V}_1|$, and the value of R is found to be $10\ \Omega$. Find L_1, M, and $\mathbf{I}_1$. Show all significant voltages and the current on a phasor diagram.

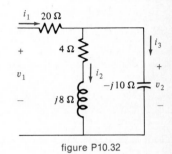

figure P10.32

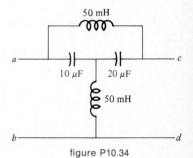

figure P10.34

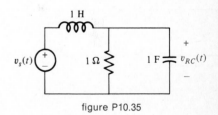

figure P10.35

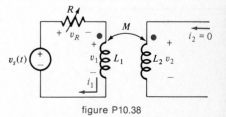

figure P10.38

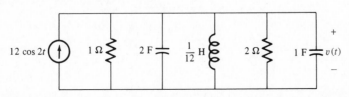

figure P10.37

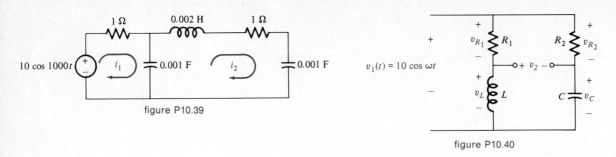

figure P10.39

figure P10.40

39. Find $i_2(t)$ in the circuit of Figure P10.39.

40. If the frequency of the input voltage $v_1(t)$ in Figure P10.40 is adjusted so that $|\mathbf{V}_{R_1}| = |\mathbf{V}_{R_2}|$, what is the phase angle of $v_2(t)$ with respect to $v_1(t)$? *Hint:* Draw the loci of the voltage phasors.

41. In the circuit of Figure P10.41 the input frequency is fixed. (*a*) What value of R_2 will result in only one possible resonant condition? (*b*) If the other elements have values such that only one possible value of L will produce resonance, what must the value of L at resonance be? (*c*) If $X_L = X_C$, what must the values of R_1 and R_2 be in order to achieve resonance?

42. In the circuit of Figure P10.41, set $R_1 = \infty\ \Omega$, $R_2 = 1\ \Omega$, $C = 1$ F, and $L = 1$ H. Now drive the circuit with $i(t) = 6u(t) \cos t$. Find the *particular* response $v_p(t)$.

43. Let the circuit in Figure P10.41 be driven by the input *voltage source* $v_{in}(t) = 10 \cos \omega t$. Also let $R_1 = \infty\ \Omega$, $R_2 = \frac{1}{2}\ \Omega$, $X_L = 0.8\ \Omega$, and $|X_C| = \frac{5}{4}\ \Omega$. Draw and label the phasor diagram. Include $\mathbf{V}_L$, $\mathbf{V}_{R_2}$, $\mathbf{V}_{in}$, $\mathbf{I}_{in}$, $\mathbf{I}_C$, and $\mathbf{I}_{RL}$. Demonstrate on the diagram that $\mathbf{I}_{in} = \mathbf{I}_C + \mathbf{I}_{RL}$ and that $\mathbf{V}_{in} = \mathbf{V}_{R_2} + \mathbf{V}_L$.

44. A capacitor C is in parallel with $R_1 = 20\ \Omega$. Connected in series with this RL combination are $R_2 = 10\ \Omega$ and $L = 0.05$ H. A 60-Hz voltage source ($V_{in} = 150\underline{/0°}$) drives the total resulting impedance Z_{in}. (*a*) Assuming C to be a variable, draw the impedance locus for Z_{in}. (*b*) If C is tuned so that the input current is a maximum, what is the phase angle between $v_{in}(t)$ and $i_{in}(t)$? (*c*) What is the value of the maximum current magnitude achieved in part (*b*)?

45. In the circuit shown in Figure P10.45 assume that R can vary from zero to infinity. (*a*) Sketch the locus of the impedance seen by the current source as R varies over its entire range. (*b*) What value of R will maximize the average power delivered by the source? What will that maximum value of P_{av} be?

46. A 120-Ω resistor and a variable capacitor are connected in parallel. In series with this RC combination is a 50-Ω inductive reactance. What is the power factor presented by this total circuit when C is set so as to maximize the magnitude of the input current drawn from a fixed-frequency sinusoidal voltage source that drives the total circuit?

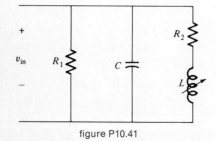

figure P10.41

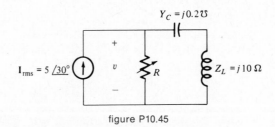

figure P10.45

47. A factory draws 10 kW average power at pf $= 0.5$ lagging from a 440-V rms 60-Hz line. (*a*) What is the apparent power in kilovolt-amperes? (*b*) What value of capacitance should be placed in parallel with the factory to correct the power factor completely (UPF)?

48. A series circuit consisting of $R = 10\ \Omega$, $L = 2$ H, and $C = \frac{1}{10}$ F is driven by a 10-V rms 1 rad/s sinusoidal voltage source. What is the value of the average power delivered by the source?

49. A sinusoidal voltage source $v_{\text{in}}(t)$ drives the series combination of $R = 2\ \Omega$ and an unknown impedance Z. The magnitude of Z is $10\ \Omega$, and it dissipates an average power of 6 W. If the source supplies 8 W, find (*a*) the rms value of the resulting current, (*b*) the rms value of v_{in}, (*c*) the power factor seen by the source, (*d*) the rms value of v_Z, and (*e*) the power factor of Z.

50. In the circuit shown in Figure P10.45, $R = 10\ \Omega$ and the rms voltage phasor is found to be $\mathbf{V}_{\text{rms}} = 22.36\underline{/93.43°}$. (*a*) Find the complex power S absorbed by each of the three passive circuit elements. (*b*) What value of capacitance when placed *in series* with the source current will maximize the power factor?

51. A circuit consisting of a single resistor R and an inductor L is driven by a 25-V rms 60-Hz sinusoidal voltage source. A capacitor is to be placed in parallel with the source to improve the power factor. Given that the average power dissipated in the resistor is 100 W and that the reactive power delivered to the inductor is 75 vars, what value of C will yield a 0.9 lagging power factor as seen by the source?

52. The operating frequency of the circuit shown in Figure P10.52 is 1000 rad/s. Find the real, reactive, and apparent power supplied by the source if $\mathbf{I}_s = 2.6\underline{/12°}$ A rms.

53. A sinusoidal source, 120 V rms at 60 Hz, supplies 2400 VA to a load operating at a 0.707 lagging pf. Find the value of a parallel capacitor (and its kilovolt-ampere rating) such that the power factor will be raised to 0.95 lagging.

54. Three load impedances Z_1, Z_2, and Z_3 are receiving the complex power values $2 + j3$, $3 - j1$, and $1 + j2$ VA, respectively. What total complex power is received if the three loads are (*a*) in series with a 100-V rms voltage source; (*b*) in parallel with a 1200-A rms current source?

55. In the circuit shown in Figure P10.55 each element receives the apparent powers $|S|_{C_1} = 45.25$ VA, $|S|_L = 68$ VA, $|S|_R = 75$ VA, and $|S|_{C_2} = 45$ VA. (*a*) Find the power factor as seen by the source. Tell whether it is leading or lagging. (*b*) By what factor should the capacitance C_1 be multiplied in order to obtain resonance?

56. An equivalent network with $\mathbf{V}_{\text{Th}} = 100\underline{/0°}$ V and $Z_{\text{Th}} = 3 + j4\ \Omega$ has a variable load impedance Z_L connected to its terminals. (*a*) At what impedance value Z_L will maximum power be delivered to this load? (*b*) What is the value of this maximum power? (*c*) Find the voltages across Z_L and Z_{Th} at maximum power conditions and plot a phasor diagram showing these voltages and $\mathbf{I}_L$.

57. In the circuit shown in Figure P10.57 $v_s(t)$ is a 1-kHz sinusoidal source. (*a*) What element values R_L and L should be chosen for the load impedance in order to maximize the power delivered to Z_L? (*b*) Under the maximum power conditions of part (*a*), what effective value of $v_s(t)$ will deliver 1 W to Z_L?

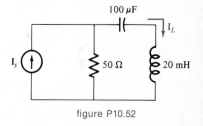

figure P10.52

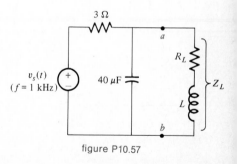

figure P10.57

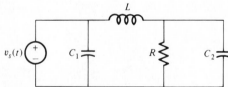

figure P10.55

58. A 440-V rms line-to-line balanced three-phase line supplies a balanced wye-connected load whose per phase impedance is a 10-Ω resistor. Find the rms value of the line current and the total power dissipated in the three-phase load.

59. The line-to-line voltage in a balanced three-phase system is 110 V rms. Find the rms value of the line current, power per phase, and total power delivered if the per phase load impedances are $3 + j4$ connected in (a) wye and (b) delta.

60. A balanced three-phase wye-connected system ($\omega = 100$ rad/s) has in each phase of its load the parallel connection of $R = 500\ \Omega$, $C = 50\ \mu F$, and $L = 1.25$ H. Given that $\mathbf{V}_{An} = 240\underline{/0°}$ rms and $\mathbf{V}_{Bn} = 240\underline{/-120°}$ rms, (a) find $\mathbf{V}_{CA}$. (b) Draw a phasor diagram showing $\mathbf{I}_A$, all phase voltages, and $\mathbf{V}_{CA}$. Label all important angles and lengths. (c) What is the value of the complex power delivered to each phase? (d) What element should be placed *in series* in each line in order to maximize the power factor seen by the source? (e) Repeat part (d) if the added elements are to be placed in delta across the three lines.

61. A balanced three-phase line supplies a balanced load as shown in Figure P10.61. Power being delivered is measured at two points A (at the source) and B (at the load) in both cases by the two-wattmeter method. (a) What will *each* wattmeter read at point A? (b) What will each wattmeter read at point B? (c) What is the efficiency of the line as a power-distribution system?

62. A series RL load is placed across a single-phase 254-V rms line. The readings of a voltmeter, an ammeter, and a wattmeter connected in order to determine the complex power being delivered are 254 V, 5 A, and 635 W, respectively. (a) What is the value of the power factor? (b) How much reactive power is being delivered? (c) What value of C, when connected across the load, will increase the power factor to unity (UPF)? (d) If the RL load with 254 V across it represents the line-to-neutral circuit of one phase of a balanced wye-connected three-phase system, what will each wattmeter read if the two-wattmeter method is used?

63. Each of the three load impedances in a balanced 60-Hz three-phase system consists of a 10-Ω resistor in parallel with a purely reactive element. The system is instrumented according to the standard two-wattmeter method. If one of the wattmeters reads twice the value of the other, what is the value of the reactive elements? If this question has more than one possible nontrivial answer, find them.

64. An RC phase sequence detector like the one shown in Figure 10.8.14 incorporates a 1-kΩ resistor. What value capacitor will maximize the difference between the magnitude of the detector's output voltage V_{DB} and the magnitude of the line-to-line voltage?

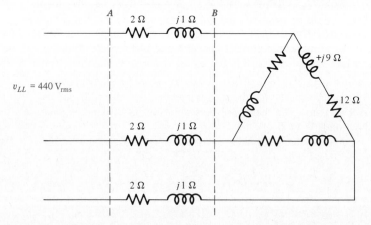

figure P10.61

chapter 11
SYSTEMS WITH COMPLEX EXPONENTIAL INPUTS

introduction 11.1

The method of phasors, sinusoidal impedances $Z(j\omega)$, and admittances $Y(j\omega)$ is applicable to only one type of input, the pure sinusoid. In this chapter we expand and generalize these concepts to enable us to use other input waveforms besides the pure sinusoid. We start by recalling that the particular response of any linear system to any exponential source will also be an exponential and will have the same general shape as the source. The exponential particular response has the same shape (because it has the same exponent) as the source function. This is true regardless of whether the exponent of the source function is real, imaginary, or complex.

In other words, if we have a linear system, driven by source

$$x(t) = Xe^{st} \qquad (11.1.1)$$

the particular response $y_p(t)$ will be of the form

$$y_p(t) = Ye^{st} \qquad (11.1.2)$$

The quantities X, Y, and s can be complex numbers, but the important thing to note is that the quantity s is numerically the same in both equations (11.1.1) and (11.1.2).

For example, when we learned to solve first-order systems (see Example 6.4.3) we saw that a source of, say, $2e^{-5t}$ produces a particular response of $k_0 e^{-5t}$. Also when we use the notions of $Z(j\omega)$ impedances and $Y(j\omega)$ admittances to find steady-state sinusoidal (particular) responses, we are really making use of this same idea, namely that a complex exponential input

$$\hat{x}(t) = X\underline{/\theta}e^{j\omega_0 t} \qquad (11.1.3)$$

where $X\underline{/\theta}$ is a complex number describing the magnitude and phase angle of the input sine wave, will produce a particular response

$$\hat{y}(t) = Y\underline{/\phi}e^{j\omega_0 t} \qquad (11.1.4)$$

Note that $j\omega_0$ is numerically the same in both (11.1.3) and (11.1.4). Those two complex exponential functions have the same exponent.

What, if anything, can we say in general about the relationship between $X\underline{/\theta}$, the complex amplitude of the input, and $Y\underline{/\phi}$, the complex amplitude of the output? Let us look at the general case of a linear system described by its differential equation

$$a_n \frac{d^n y}{dt^n} + \cdots + a_1 \frac{dy}{dt} + a_0 y = b_m \frac{d^m x}{dt^m} + \cdots + b_1 \frac{dx}{dt} + b_0 x \qquad (11.1.5)$$

and suppose we have a complex exponential input

$$x(t) = Xe^{st} \qquad (11.1.6)$$

(X may also be complex). As usual, we seek a particular response of the form

$$y(t) = k_1 e^{st} + k_2 se^{st} + k_3 s^2 e^{st} + \cdots = (k_1 + k_2 s + k_3 s^2 + \cdots)e^{st} \qquad (11.1.7)$$

i.e., a linear sum of the forcing function plus all its derivatives. Because the quantity in the parentheses is just a number (possibly complex), we let the single quantity Y represent it

$$Y = k_1 + k_2 s + k_3 s^2 + \cdots \qquad (11.1.8)$$

and so the particular response is also a complex exponential function

$$y(t) = Ye^{st} \qquad (11.1.9)$$

Compare (11.1.9) and (11.1.6). The e^{st} parts are identical.

How does Y depend on X and s? We answer this by substituting (11.1.6) and (11.1.9) into (11.1.5). This yields

$$a_n s^n Ye^{st} + \cdots + a_1 sYe^{st} + a_0 Ye^{st} = b_m s^m Xe^{st} + \cdots + b_1 sXe^{st} + b_0 Xe^{st} \qquad (11.1.10)$$

Dividing through by the exponential that appears in each term of (11.1.10), we get

$$(a_n s^n + \cdots + a_1 s + a_0)Y = (b_m s^m + \cdots + b_1 s + b_0)X \qquad (11.1.11)$$

which is just the p-operator form of the differential equation with the operator p replaced by the complex number $s = \sigma + j\omega$. Solving for the ratio of Y/X, we get

$$\frac{Y}{X} = \frac{b_m s^m + \cdots + b_1 s + b_0}{a_n s^n + \cdots + a_1 s + a_0} = H(s) \qquad (11.1.12)$$

Thus we see that the ratio of the complex amplitude of the complex exponential output divided by the complex amplitude of the complex exponential input is simply the p-operator transfer function $H(p)$ with the (possibly complex) numerical value of s inserted in place of the operator p everywhere it appears in $H(p)$. This turns $H(p)$, which is an operator with no numerical value, into a numerical quantity $H(s)$. If we multiply $H(s)$ by X, the complex amplitude of the complex exponential input function, we get Y, the complex amplitude of the corresponding complex output function. The variable $s = \sigma + j\omega$ is called the *complex frequency*, where σ is the *neper frequency* and ω is the *radian frequency* (it is equal to $2\pi f$, where f is the frequency in hertz).

Note that the denominator of the transfer function comes from the left (output-variable) side of the system differential equation. The numerator comes from the right (input-variable) side of the differential equation. The terms in each of these polynomials are in one-to-one correspondence with the terms in the differential equation. The power to which s is raised in any transfer-function term is equal to the order of the derivative in the corresponding term in the differential equation.

EXAMPLE 11.1.1

Find the forced (particular) response for $v_o(t)$ in the circuit shown in Figure 11.1.1 if

$$i_1(t) = 2e^{-3t} \qquad t > 0 \qquad (11.1.13)$$

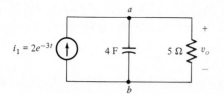

figure 11.1.1
The circuit of Example 11.1.1.

ANS.: Write the system equation via KCL at node a:

$$C\frac{dv_o}{dt} + \frac{v_o}{R} = i_1(t) \qquad (11.1.14)$$

Since in p-operator notation this is

$$\left(Cp + \frac{1}{R}\right)v_o(t) = i_1(t) \qquad (11.1.15)$$

we have

$$H(p) = \frac{v_o(t)}{i_1(t)} = \frac{1}{Cp + 1/R} = \frac{1/C}{p + 1/RC} = \frac{1/4}{p + 1/20} \qquad (11.1.16)$$

and thus

$$H(s) = \frac{1/4}{s + 1/20} \qquad (11.1.17)$$

Since the complex frequency is $s = -3$ for this source, we evaluate H as follows:

$$H(-3) = \frac{1/4}{-3 + 1/20} = -0.0847 \qquad (11.1.18)$$

Hence the coefficient of the particular output response is

$$V = H(-3)(2) = -0.0847(2) = -0.169 \qquad (11.1.19)$$

and the particular response is

$$v_o(t) = Ve^{-3t} = -0.169e^{-3t} \qquad t > 0 \qquad (11.1.20)$$

Having seen that we can use the p-operator transfer function together with any complex exponential input to obtain a numerical quantity $H(s)$, we can use this quantity and the input function to evaluate the particular response. You *must* remember that this method is valid *only* when the input is a *complex exponential function* (of the form Ae^{st}).

The transfer function $H(s)$ is a vitally important and useful thing to know about any linear system. If we know a system's differential equation, we know its $H(p)$ and thus, for any complex exponential input, its $H(s)$.

11.2 using complex exponentials

We have just seen how relatively simple it is to obtain the particular response to any complex exponential source. We now take advantage of this fact by forcing as many other input functions as we can into the Ae^{st} complex exponential form. Consider the following four possible input functions, all of which can be written in the form Ae^{st}:

1. Real exponential, for example, $v(t) = 4e^{-5t}$. Use $A = 4\underline{/0}$, $s = -5$; that is, $v(t) = 4\underline{/0}e^{-5t}$.
2. Sinusoid, for example, $v(t) = 10 \cos(377t + 14°)$. Use $A = 10\underline{/14°}$ and $s = j377$ so $v(t) = 10\underline{/14°}e^{j377t}$. Then we take the real part of the answer, as in Chapter 10.
3. Constant, for example, $v(t) = 13.7$. Use $A = 13.7\underline{/0°}s = 0$ (because $e^0 = 1$); that is, $v(t) = 13.7e^0$.
4. Damped sinusoid, for example, $v(t) = 15e^{-8t}\cos(9t + \pi/4)$. Use $A = 15\underline{/\pi/4}$, $s = -8 + j9$ so that $v(t) = 15\underline{/\pi/4}e^{(-8+j9)t}$. Then take the real part of the answer.

EXAMPLE 11.2.1

Find the particular response for the current $i(t)$ in the circuit of Figure 11.2.1 if $v(t) = 60e^{-4t}$.

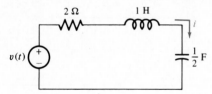

figure 11.2.1
The circuit of Examples 11.2.1 to 11.2.4.

ANS.:   Write the system equation for i from KVL

$$\left(R + Lp + \frac{1}{Cp}\right)i(t) = v(t) = 60e^{-4t} \tag{11.2.1}$$

so that

$$H(p) = \frac{1}{R + Lp + 1/Cp} = \frac{1}{2 + p + 2/p} \tag{11.2.2}$$

and since $s = -4$, we get

$$H(-4) = \frac{1}{2 - 4 - 2/4} = -0.4 \tag{11.2.3}$$

and

$$i(t) = 60(-0.4)e^{-4t} = -24e^{-4t} \tag{11.2.4}$$

EXAMPLE 11.2.2

Repeat Example 11.2.1 with $v(t) = 10 \cos (2t + \pi/4)$.
ANS.:   Use $v = 10\underline{/45^\circ}e^{j2t}$ as input; find the response

$$i(t) = 10\underline{/45^\circ}H(j2)e^{j2t} \tag{11.2.5}$$

where, using (11.2.2) with $p \to s = j2$, we have

$$H(j2) = \frac{1}{2 + j2 + 2/j2} = \frac{1}{2 + j2 - j} = \frac{1}{2.24\underline{/26.6^\circ}} = 0.447\underline{/-26.6^\circ} \tag{11.2.6}$$

Inserting (11.2.6) into (11.2.5) gives

$$i(t) = 10\underline{/45^\circ}(0.447\underline{/-26.6^\circ})e^{j2t} = 4.47\underline{/18.4^\circ}e^{j2t} \tag{11.2.7}$$

Taking the real part of (11.2.7), we have

$$i(t) = 4.47 \cos (2t + 18.4^\circ) = 4.47 \cos (2t + 0.321) \tag{11.2.8}$$

EXAMPLE 11.2.3

Repeat Example 11.2.1 with $v(t) = 10$.
ANS.:   Use $v = 10\underline{/0^\circ}e^0$ as input. Thus

$$i(t) = 10\underline{/0°}\,H(0) \tag{11.2.9}$$

and

$$H(p) = \frac{1}{2 + p + 2/p} = \frac{p}{2p + p^2 + 2} \tag{11.2.10}$$

so that

$$H(0) = \tfrac{0}{2} = 0 \tag{11.2.11}$$

and

$$i(t) = 0 \tag{11.2.12}$$

which does not surprise us very much because the series capacitor blocks dc currents.

EXAMPLE 11.2.4

Repeat Example 11.2.1 with $v(t) = 10e^{-5t}\cos(3t + 60°)$.
ANS.: Use the corresponding complex exponential $10\underline{/60°}\,e^{(-5+j3)t}$ with

$$H(-5+j3) = \frac{1}{2 + (-5+j3) + 2/(-5+j3)} = 0.23\underline{/-139°}$$

Thus

$$i(t) = H(-5+j3)10\underline{/60°}\,e^{(-5+j3)t} = 2.3\underline{/-79°}\,e^{(-5+j3)t}$$

or, taking the real part,

$$i(t) = 2.3e^{-5t}\cos(3t - 79°)$$

EXAMPLE 11.2.5

Repeat Example 11.2.1 if the source $v(t)$ is given as

$$v(t) = 60e^{-4t} + 10 + 10e^{-5t}\cos(3t + 60°)$$

ANS.: Using the *superposition* theorem together with the results of Examples 11.2.1, 11.2.3, and 11.2.4, we find the particular response to be

$$i(t) = -24e^{-4t} + 2.3e^{-5t}\cos(3t - 79°)$$

11.3 transfer functions $H(s)$, $Z(s)$, $Y(s)$

Consider any single inductor L across which is impressed a complex exponential voltage

$$v(t) = Ve^{st} \tag{11.3.1}$$

where V and s are both (possibly) *complex*. The defining relationship is

$$v(t) = L\frac{di}{dt} \tag{11.3.2}$$

or in operator form

$$v(t) = Lpi \tag{11.3.3}$$

Inserting (11.3.1) into (11.3.2) yields

$$\frac{di}{dt} = \frac{1}{L} V e^{st} \tag{11.3.4}$$

a small but perfectly valid differential equation; we find the particular response

$$i(t) = I e^{st} \tag{11.3.5}$$

by substituting (11.3.5) into (11.3.4) and thus obtaining

$$s I e^{st} = \frac{1}{L} V e^{st} \tag{11.3.6}$$

or

$$V = LsI \tag{11.3.7}$$

Defining the ratio of the complex exponential voltage divided by the complex exponential current to be the impedance $Z(s)$, we get

$$Z_L(s) = \frac{V}{I} = Ls \tag{11.3.8}$$

The inverse of $Z_L(s)$ is the admittance $Y_L(s)$

$$Y_L(s) = \frac{I}{V} = \frac{1}{Ls} \tag{11.3.9}$$

Compare equations (11.3.3) and (11.3.7). The former is a time-domain p-operator system equation. The latter is exactly the same equation with the complex frequency variable s substituted for p. Equation (11.3.7)—and therefore (11.3.8) and (11.3.9)—is valid, of course, only for systems with complex exponential inputs. Similarly for any capacitor C we have

$$C \frac{dv_C(t)}{dt} = i_C(t) \tag{11.3.10}$$

or

$$Cp V_C(t) = i_C(t) \tag{11.3.11}$$

For complex exponential v_C and i_C (11.3.10) becomes

$$sCV e^{st} = I e^{st} \tag{11.3.12}$$

Solving this for the ratio V/I yields the impedance

$$\frac{V}{I} = Z_C(s) = \frac{1}{Cs} \tag{11.3.13}$$

where V, I, and Z are all generally complex quantities. Inverting gives

$$Y_C(s) = Cs \tag{11.3.14}$$

Obviously for the sinusoidal case $s = j\omega$ and equations (11.3.8), (11.3.9), (11.3.13), and (11.3.14) become

$$Z_L(s) \rightarrow Z_L(j\omega) = j\omega L \tag{11.3.15}$$

$$Y_L(s) \rightarrow Y_L(j\omega) = -j \frac{1}{\omega L} \tag{11.3.16}$$

$$Z_C(s) \to Z_C(j\omega) = -j\frac{1}{\omega C} \tag{11.3.17}$$

$$Y_C(s) \to Y_C(j\omega) = j\omega C \tag{11.3.18}$$

Equations (11.3.15) to (11.3.18) are the (now familiar) sinusoidal immittances we have already used to obtain particular responses to pure sinusoidal inputs. The $Z(j\omega)$ impedances used in solving those problems are simply special cases of these more general $Z(s)$ impedances. Using $Z(j\omega)$ is valid only if we have pure sinusoidal inputs to linear systems, but $Z(s)$ is applicable for solving for particular responses in circuits having other input waveforms as well as sinusoids. What other types of waveforms? Any that can be written in the complex exponential form Ae^{st}.

We saw earlier, in discussing the p operator, that we can obtain system equations using the techniques of circuit analysis (Ohm, Kirchhoff, etc.) and the laws of algebra. In this section we have noticed that the generalized complex exponential impedance $Z(s)$ can be obtained from $Z(p)$ simply by substituting s for p. It makes no difference whether we obtain the transfer function $H(s)$ of a large circuit by (1) first finding $H(p)$ (using p-operator methods) and then substituting s for p in the final expression or (2) using $Z_L(s) = Ls$, $Z_R = R$, or $Z_C = 1/Cs$ for each element and then combining them into $H(s)$.

EXAMPLE 11.3.1

Let us repeat the last example in Section 11.2 this time making use of generalized impedances $Z(s)$. In other words, let us solve the series RLC circuit of Figure 11.2.1 with $v(t) = 10e^{-5t}\cos(3t + 60°)$ using the fact that when we convert $v(t)$ into a complex exponential,

$$Z_R(s) = R \qquad Z_L(s) = Ls \qquad \text{and} \qquad Z_C(s) = \frac{1}{Cs}$$

ANS.: The amplitude of the complex exponential voltage across each of the three passive elements is

$$V_R = IZ_R(s) \qquad V_L = IZ_L(s) \qquad V_C = IZ_C(s)$$

and the complex exponential voltage $v(t)$ is the sum of each of those components

$$v(t) = Ve^{st} = V_R e^{st} + V_L e^{st} + V_C e^{st} = [IZ_R(s) + IZ_L(s) + IZ_C(s)]e^{st}$$

so that

$$V = [Z_R(s) + Z_L(s) + Z_C(s)]I$$

or

$$I = Y(s)V \qquad \text{where} \qquad Y(s) = [Z_R(s) + Z_L(s) + Z_C(s)]^{-1}$$

Knowing that $s = -5 + j3$, we calculate the total input admittance

$$Y(s) = \left[R + Ls + \frac{1}{Cs}\right]^{-1}$$

or

$$Y(-5 + j3) = \left[2 + 1(-5 + j3) + \frac{2}{-5 + j3}\right]^{-1}$$

which, as before, is

$$Y(-5 + j3) = 0.23\underline{/-139°}$$

Therefore, the particular response

$$i(t) = Ie^{st} = Ie^{(-5+j3)t}$$

has the complex amplitude

$$I = Y(-5 + j3)V = (0.23\underline{/-139°})(10\underline{/60°}) = 2.3\underline{/-79°}$$

so that

$$i(t) = 2.3\underline{/-79°}e^{(-5+j3)t}$$

Taking the real part gives

$$i(t) = 2.3e^{-5t}\cos(3t - 79°)$$

In this last example the transfer function turned out to be an admittance. All admittances and all impedances are transfer functions, but not all transfer functions are either impedances or admittances. If the input and output variables in a system are both voltages, we simply say that the transfer function is a voltage transfer function and call it $H(s)$. Similarly when both the input and output are currents we have a current transfer function $H(s)$. You should also remember that the entire concept of a transfer function $H(s)$ has meaning only for a system for which the assumptions we have made are correct: (1) that we are using complex exponential sources and (2) that the system is linear. Both must be true for the particular output response to be a complex exponential having the same shape (exponent) as the input function.

In the next few sections we shall investigate various properties of $H(s)$ and what we can find out about the system by examining that function. In all of this do not forget that the numerical value of the complex frequency $s = \sigma + j\omega$ is determined in any linear system by the particular complex exponential input function we apply. That is, s is *determined by the source* we pick to excite the system. Its value has nothing to do with the R's, L's, or C's in the circuit. For example, we might choose a sinusoidal source whose frequency is variable. In that event the neper frequency σ would be zero and ω would be a continuous variable.

poles and zeros 11.4

In general, the transfer function $H(s)$ can be written in the form of a rational function (ratio of polynomials) in s

$$H(s) = \frac{b_m s^m + \cdots + b_1 s + b_0}{a_n s^n + \cdots + a_1 s + a_0} \tag{11.4.1}$$

or in factored form

$$H(s) = \frac{K(s - z_1)(s - z_2) \cdots}{(s - p_1)(s - p_2) \cdots} \tag{11.4.2}$$

where the z_i are called *zeros* of $H(s)$ and the p_i are called *poles* of $H(s)$. *The zeros are values of s such that the transfer function is 0 for that value of s. The poles are those values of s for which the H(s) becomes ∞. The poles and zeros of a system are sometimes called its* critical frequencies. *We can plot the poles and zeros on the complex s plane. Such a plot is called the* pole-zero constellation.†

EXAMPLE 11.4.1

Plot the pole-zero constellation of

$$H(s) = \frac{s}{s^2 + 2s + 2} = \frac{s}{(s + 1 - j)(s + 1 + j)}$$

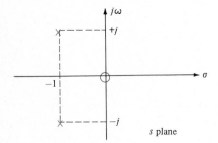

figure 11.4.1
The pole-zero constellation of Example 11.4.1.

ANS.:   See Figure 11.4.1. Zeros are $s = 0$ and $s = \infty$; poles are $s = -1 + j$ and $s = -1 - j$.

Since the locations of complex poles and zeros are determined by the quadratic formula

$$as^2 + bs + c = 0$$

$$s = \frac{-b \pm \sqrt{b^2 - 4ac}}{2a}$$

we note that any complex-valued critical frequency is *always* accompanied by its complex conjugate. Therefore, the pole-zero constellation has an upper-half-plane–lower-half-plane symmetry

---

EXAMPLE 11.4.2

Find the pole-zero constellation (s-plane plot of poles and zeros) for the transfer function that relates the input $i(t) = Ie^{st}$ to the output $v(t) = Ve^{st}$ in the circuit of Figure 11.4.2.
ANS.:   The transfer function in this case is the driving-point impedance because

$$Ve^{st} = H(s)Ie^{st} = Z(s)Ie^{st}$$

---

† In general, every $H(s)$ has the same number of poles as it does zeros, but one or more of them may be located at infinity. We plot the *finite* locations of the critical frequencies on the pole-zero constellation.

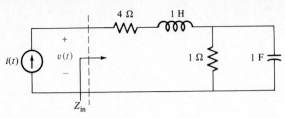

figure 11.4.2
The circuit of Example 11.4.2.

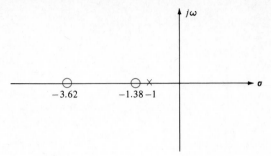

figure 11.4.3
The pole-zero constellation of Example 11.4.2.

Therefore

$$Z(s) = 4 + Ls + \frac{1}{Cs + 1}$$

which, with $L = 1$ H and $C = 1$ F, is

$$Z(s) = 4 + s + \frac{1}{s + 1} = \frac{4(s + 1) + (s + 1)s + 1}{s + 1}$$

$$= \frac{s^2 + 5s + 5}{s + 1} = \frac{(s + 1.38)(s + 3.62)}{s + 1}$$

The pole-zero constellation is shown in Figure 11.4.3.

Plotting the poles and zeros of any transfer function can tell us a great deal about the system without going through the work of solving for the complete response, as we shall see in coming sections. Remember that the pole-zero constellation specifies the transfer function *within a constant*. With reference to equation (11.4.2), we need the value of K as well as the critical frequency locations to specify $H(s)$ fully.

graphical evaluation of $H(s)$ **11.5**

The complex numerical value of the transfer function $H(s)$ can be evaluated for any value of s via a very convenient graphical procedure. Consider, for example, the simple circuit of Figure 11.5.1. Defining $v_1(t)$ as the input and $v_2(t)$ as the output, we have, by voltage division

$$V_2 = \frac{(1/Cs)V_1}{R + 1/Cs} = \frac{1/RCV_1}{s + 1/RC} \qquad (11.5.1)$$

or

$$H(s) = \frac{1/RC}{s + 1/RC} \qquad (11.5.2)$$

and with $R = C = 1$ we have

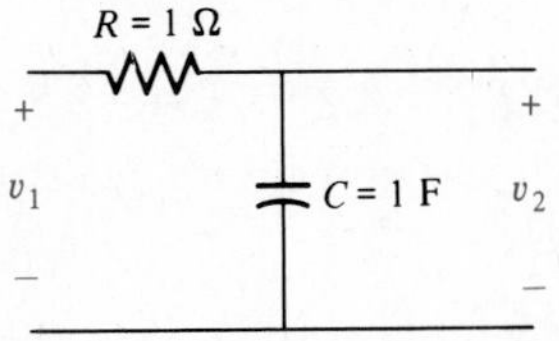

figure 11.5.1
A simple circuit.

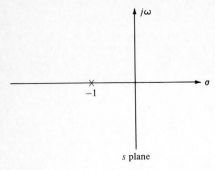

figure 11.5.2
The pole-zero constellation of the transfer
function of the system in Figure 11.5.1.

$$H(s) = \frac{1}{s+1} \tag{11.5.3}$$

Look at the pole-zero constellation of this transfer function (Figure 11.5.2). Choosing a sinusoidal source for $v_1(t)$ is equivalent to setting $s = j\omega$ in $H(s)$. Picking a particular frequency for this source fixes the numerical value of ω. In turn, this defines a single point on the $j\omega$ axis of the s plane. For example, if we choose for the circuit of Figure 11.5.1 a sinusoidal input with radian frequency $\omega = 1$, this specifies the point $s = j1$.

In Figure 11.5.2 think about placing a dot on the imaginary axis at $s = j1$. This point is the same distance up the $j\omega$ axis as the pole is horizontally distant from the origin. Draw a vector from the pole up to the dot. This vector is at a 45° angle and (since this creates a 45–45–90 triangle) its length is equal to $\sqrt{2}$. What is the significance of this vector? To answer this question let us evaluate $H(j1)$ just as we normally would. Substituting $s = j$ into equation (11.5.3), we get

$$H(j) = \frac{1}{j+1} = \frac{1}{1+j} \tag{11.5.4}$$

$$H(j) = \frac{1}{\sqrt{2}\underline{/45°}} \tag{11.5.5}$$

$$H(j) = 0.707\underline{/-45°} \tag{11.5.6}$$

We note that the value of the factor in the denominator of equations (11.5.4) and (11.5.5) is identically the complex number defined by the vector drawn from the pole to the point $s = j$. This is the basis for the following four-step graphical method for evaluating $|H(s)|$ and $\underline{/H(s)}$:

1. Plot the pole-zero constellation of

$$H(s) = \frac{K(s+a_1)(s+a_2)\cdots}{(s+b_1)(s+b_2)\cdots}$$

2. Place a dot at the value of s dictated by the input Ae^{st}.
3. Draw vectors from each finite pole and zero to the point s.
4. Then the magnitude of $H(s)$ is given by

$$|H(s)| = \frac{K\prod(\text{length of vectors from zeros})}{\prod(\text{length of vectors from poles})}$$

and the angle of $H(s)$ will be given by

$$\underline{/H(s)} = \sum(\text{angles of vectors from zeros}) - \sum(\text{angles of vectors from poles})$$

where $\prod$ signifies the product operation and $\sum$ signifies a summation.

EXAMPLE 11.5.1

Given a system with

$$H(s) = \frac{s(s+2)}{s^2 + 2s + 2}$$

and with input = 10 cos 2t, find the output $v(t)$.

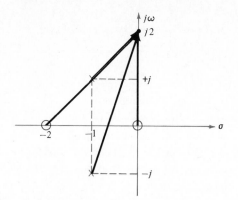

figure 11.5.3
The evaluation of the transfer function of Example 11.5.1 at $s = j2$.

ANS.:   See Figure 11.5.3. (*Step 1*) Draw the pole-zero constellation for

$$H(s) = \frac{s(s + 2)}{(s + 1 - j)(s + 1 + j)}$$

(*Step 2*) Place a dot at $j2$ since the input is Re $10e^{j2t}$. (*Step 3*) Draw vectors from each pole and zero to $j2$ (see Table 11.5.1).

table 11.5.1

Term	Critical frequency	Type	Magnitude	Angle
s	$s = 0$	Zero	2	90°
$s + 2$	-2	Zero	$2\sqrt{2}$	45°
$s + 1 - j$	$-1 + j$	Pole	$\sqrt{2}$	45°
$s + 1 + j$	$-1 - j$	Pole	3.16	71.57°

Then (*Step 4*) $|H(s)| = \dfrac{2(2\sqrt{2})}{\sqrt{2}(3.16)} = 1.26$ $\underline{/H(s)} = (90° + 45°) - (45° + 71.57°) = 18.4°$

As always, the complex phasor output is equal to the complex phasor input times this complex transfer function

Complex output = $1.26\underline{/18.4°}\,10\underline{/0°} = 12.6\underline{/18.4°}$

and taking the real part gives

$$v(t) = 12.6 \cos (2t + 18.4°)$$

This technique for finding $|H(s)|$ and $\underline{/H(s)}$ makes use of the fact that, for a given value of s, each term in the factored form of $H(s)$ is itself a complex (vector) quantity. For example, $s + 4$ is the vector leading from -4 to s because $s + 4 = s - (-4)$. Say $s = +j2$. The graphical construction is shown in Figure 11.5.4.

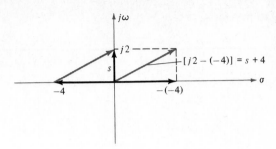

figure 11.5.4
The graphical interpretation of the term $s + 4$ evaluated at $s = +j2$ obtained by adding the vectors $+j2$ and $-(-4)$.

Since $H(s)$ is of the form $K(n_1)(n_2) \cdots/(n_3)(n_4) \cdots$, where the various n's are complex, we follow the rules of complex multiplication and complex division to get the numerical evaluation of $|H|$ and $\underline{/H}$.

MAGNITUDE AND PHASE RESPONSE AS s IS VARIED

The value of $|H|$ and $\underline{/H}$ can both be plotted as a function of s as s varies. For example, if s starts at the origin and progresses up the $+j\omega$ axis, this is equivalent to setting $s = j\omega$ and smoothly varying ω from $\omega = 0$ upward toward $\omega = \infty$. We could also select a discrete set of different values for $0 < \omega < \infty$ and plot† $|H|$ and $\underline{/H}$ at each such value of ω.

EXAMPLE 11.5.2

Consider the circuit in Figure 11.5.5, where

$$H(s) = \frac{V_2}{V_1} = \frac{RLs/(R + Ls)}{RLs/(R + Ls) + 1/Cs} = \frac{s^2}{s^2 + (1/RC)s + 1/LC}$$

Given that $R = 10\ \Omega$, $C = \frac{1}{60}$ F, and $L = 2.4$ H, sketch $|H(s)|$ and $\underline{/H(s)}$ versus ω for $0 < \omega < \infty$.

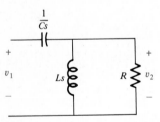

figure 11.5.5
The circuit of Example 11.5.2.

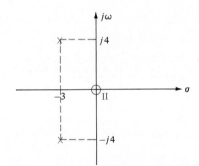

figure 11.5.6
The pole-zero constellation of the circuit of Example 11.5.2.

† The plot of $|H(j\omega)|$ versus ω and that of $\underline{/H(j\omega)}$ versus ω are sometimes called the *magnitude response* and the *phase response* respectively. This can be misleading in that the natural response, particular response, zero-state response, and zero-input response are all *time-function responses* of a system to some specific input function and set of initial conditions. To make matters worse, $|H(j\omega)|$ plotted versus ω is sometimes called the *frequency response*. This terminology makes no sense whatever and should be avoided.

ANS.:   This problem is easily solved by first plotting the pole-zero constellation. Inserting the values of R, L, and C, we obtain

$$H(s) = \frac{V_2}{V_1} = \frac{s^2}{s^2 + 6s + 25} = \frac{s^2}{(s + 3 - j4)(s + 3 + j4)}$$

The pole-zero constellation is shown in Figure 11.5.6. Note that $K = 1$. The roman numeral II next to the zero at the origin indicates there are two zeros on top of each other at that point, a so-called *double-order* zero. For $s = j\omega$ and ω very close to zero but not equal to it:

1. Both vectors from the zeros at the origin up to $j\omega$ have almost zero length; $|H| \to 0$.
2. Since the angles at the poles are almost equal and of opposite sign, their sum is approximately $0°$.
3. The angles of both vectors from the zeros are $90°$ (and will be $90°$ for any $s = +j\omega$); $\underline{/H} = 180°$.

For $\omega \to \infty$ all vectors are very long (approximately equal) and all are at approximately $90°$

$$H = \frac{l^2}{ll} = 1 \quad \text{and} \quad \underline{/H} = (90° + 90°) - (90° + 90°) = 0°$$

For $\omega = 0+$, 1, and 2 rad/s see Figure 11.5.7. Similarly at $\omega = 4$

$$H = \frac{4\underline{/90°}\,4\underline{/90°}}{3\underline{/0°}(3 + j8)} = 0.625\underline{/111°}$$

and at $\omega = 8$

$$H = \frac{8\underline{/90°}\;8\underline{/90°}}{(3 + j4)(3 + j12)} = 1.03\underline{/51°}$$

This last magnitude may seem surprising. $|H|$ is greater than unity! Let us go back and try a value of ω between 4 and 8, say $\omega = \omega_n = 5$; then

$$H(j5) = 0.834\underline{/90°}$$

Plotting the data we have so far suggests the pair of curves in Figure 11.5.8. For practice you should try to locate the frequency of the maximum in $H(j\omega)$ more precisely.

The pole-zero constellation suggests that the maximum in $|H|$ would be more pronounced if the two poles were nearer the $j\omega$ axis because vectors from one of those poles would be very short ($|H|$ large) for a range of s near $j4$. If we reexamine Figure 11.5.7 we can visualize what would happen if we allowed the point s to travel down the $-j\omega$ axis starting from the origin (rather than going up the $+j\omega$ as it did). In such a case ($s = -j\omega$, $0 < \omega < \infty$) the resultant vector lengths are the same but the angles all change sign. With reference to Figure 11.5.7c, for example, we can see that for $s = -j2$

$$H(-j2) = \frac{2\underline{/-90°}\,2\underline{/-90°}}{3.61\underline{/+33.7°}\;6.71\underline{/-63.4°}} = 0.165\underline{/-150°}$$

In general, then, we can say that

$$H(-j\omega) = H^*(j\omega)$$

where the asterisk denotes complex conjugation. Since the square of the magnitude of any complex quantity can be found by multiplying it by its complex conjugate,

$$|H(j\omega)|^2 = H(j\omega)H^*(j\omega) = H(j\omega)H(-j\omega) \tag{11.5.7}$$

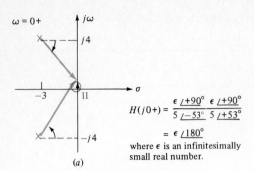

figure 11.5.7

Evaluating $H(s)$, the transfer function in Example 11.5.2, at (a) $\omega = 0+$, (b) $\omega = 1$, and (c) $\omega = 2$ rad/s.

$$H(j0+) = \frac{\epsilon\ \underline{/+90°}\ \epsilon\ \underline{/+90°}}{5\ \underline{/-53°}\ 5\ \underline{/+53°}}$$

$$= \epsilon\ \underline{/180°}$$

where ϵ is an infinitesimally small real number.

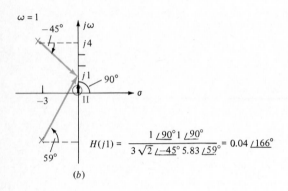

$$H(j1) = \frac{1\ \underline{/90°}\ 1\ \underline{/90°}}{3\sqrt{2}\ \underline{/-45°}\ 5.83\ \underline{/59°}} = 0.04\ \underline{/166°}$$

figure 11.5.8

(a) The magnitude and (b) the angle of the transfer function in Example 11.5.2 plotted versus radian frequency, ω.

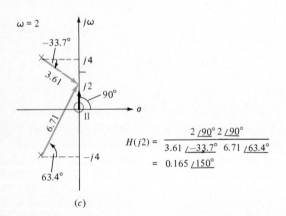

$$H(j2) = \frac{2\ \underline{/90°}\ 2\ \underline{/90°}}{3.61\ \underline{/-33.7°}\ 6.71\ \underline{/63.4°}}$$

$$= 0.165\ \underline{/150°}$$

Later we shall be interested in the magnitude of $H(j\omega)$. One way to get the square of that quantity is with equation (11.5.7).

The example we have just done shows that we can plot $|H|$ and $\underline{/H}$ as s varies (travels along some path in the s plane). This method will work regardless of the locus (path) taken by s.

EXAMPLE 11.5.3

Find $|H|$ and $\underline{/H}$ as s varies along the path $|s| \, \underline{/135°}$ for $H(s) = 1/(s + 2)$ (see Figure 11.5.9).

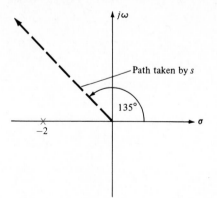

figure 11.5.9
The locus of s in Example 11.5.3.

table 11.5.2

$|s|$	$|s + 2|$	$\underline{/s + 2}$	$|H|$	$\underline{/H}$
0	2	0°	0.5	0°
1	1.47	28.7°	0.680	−28.7°
$\sqrt{2}$	$\sqrt{2}$	45°	0.707	−45°
$2\sqrt{2}$	2	90°	0.5	−90°
∞	∞	135°	0	−135°

ANS.: Table 11.5.2 shows the value of $s + 2$ and $1/(s + 2)$ as s starts at the origin and moves up along this path. We can sketch the two variables $|H|$ and $\underline{/H}$ versus $|s|$.

Suppose we let the path that s takes be the negative real axis of the s plane. This is equivalent to using $e^{\sigma t}$, where $\sigma < 0$, as the input to our system. The output will be, we assume, in the form $H(s)e^{st}$.

$H(s)$ **AS A FUNCTION OF** Re s

EXAMPLE 11.5.4

Given the transfer function

$$H(s) = \frac{s + 1}{(s + 3.7)(s^2 + 4.8s + 8.65)}$$

plot $|H|$ and $\underline{/H}$ for $s = \sigma$, where $-\infty < \sigma < 0$.
ANS.: Factor the denominator quadratic

$$H(s) = \frac{s + 1}{(s + 3.7)(s + 2.4 - j1.7)(s + 2.4 + j1.7)}$$

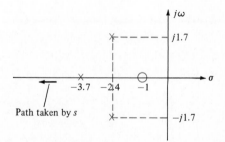

figure 11.5.10
The pole-zero constellation in Example 11.5.4 showing the locus of s along the negative real axis.

Plot the pole-zero constellation and proceed as before except that this time s is on the negative real axis (Figure 11.5.10). Observe that, since complex poles and zeros of $H(s)$ always come in conjugate pairs for linear systems with real coefficients in their differential equations, the angle of $H(\sigma)$ will always be either 0 or 180°; $H(\sigma)$ *will always be real*. We list the vector magnitudes for several values of σ in Table 11.5.3. A sketch of $|H|$ versus σ is shown in Figure 11.5.11.

table 11.5.3

s	$s+1$	$s+2.4-j1.7$	$(s+2.4-j1.7)^2$	$s+3.7$	$\lvert H\rvert$	$\underline{/H}$
0	1	2.94	8.65	3.7	0.0312	0°
-0.5	0.5	2.54	6.5	3.2	0.024	0°
-1	0				0	Undefined
-1.5	0.5	1.92	3.7	2.2	0.0614	180°
-2.4	1.4	1.7	2.89	1.3	0.373	180°
-3.5	2.5	2.02	4.1	0.2	3.05	180°
-3.7	2.7	2.14	4.58	0	∞	Undefined
-4	3	2.33	5.45	0.3	1.83	0°
-10	9	7.79	60.7	6.3	0.024	0°
$-\infty$	∞	∞	∞^2	∞	$\dfrac{1}{\infty^2}=0$	0°

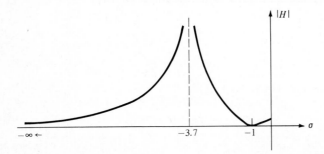

figure 11.5.11
The magnitude of $H(\sigma)$ for $-\infty < \sigma < 0$.

Does the fact that $|H(s)| \to \infty$ at $s = -3.7$ in Example 11.5.4 mean that the response of this system to the input $e^{-3.7t}$ will be ∞? No; the response of any linear system when excited at a negative real pole (at say $s = -\sigma_1$) is $Ate^{-\sigma_1 t}$ not $Ae^{-\sigma_1 t}$. This technique of evaluating $H(s)$ with $s = \sigma$ for input $e^{\sigma t}$ works arbitrarily close to a negative real system pole but not *at* that pole.

Remember that the output is given by $H(s)e^{st}$ under the *assumption* that the output has this same e^{st} form. When we excite a system with a forcing function (source) e^{-at} at a negative real pole $s = \sigma = -a$, the output does *not* have this same form. The output has the form†

$$(k_1 t + k_2)e^{-at} \tag{11.5.8}$$

Therefore, the value of $H(s)$ for that s does not help us to find the particular response. $H(s)$ is not defined, i.e., does not exist, at the system poles; i.e., when the input function $e^{(-1+j)t}$ is used to drive a system whose transfer function is

$$H(s) = \frac{4(s+1)}{(s+1-j)(s+1+j)}$$

† See equation (6.4.15) and Section 7.7.

it does *not* generate a response function $Ae^{-t}e^{jt}$ and thus $Ae^{-t}\cos(t+\theta)$, where $A=\infty$. The particular response *will* be of the form $(A+Bt)e^{-t}e^{jt}$, which results in

$$(A+Bt)e^{-t}\cos(t+\theta) \tag{11.5.9}$$

Note that the functions in both (11.5.8) and (11.5.9) are bounded; i.e., the act of exciting a system at one of its poles (within the left half s plane) does *not* result in a response that goes to infinity because neither of the reasons why it might (the coefficient is ∞ or the form of the function is such that, as time goes on, its value increases without bound) actually occurs.

The case of a $j\omega$ axis pole is different. Using an input of the form $\cos\omega_0 t$ with a system having poles at $s=\pm j\omega_0$ results in a particular response of the form $(A+Bt)\cos(\omega_0 t+\theta)$, which does increase unboundedly with time, as discussed in Section 7.11. Systems with right-half s-plane poles are a different matter altogether, which we investigate in the next section.

natural response from the transfer function 11.6

In Section 11.1 it was pointed out that the denominator polynomial in the transfer function $H(s)$ comes directly in term-by-term correspondence from the left (output-variable) side of the system differential equation.

EXAMPLE 11.6.1

Find the corresponding $H(s)$ for the system differential equation

$$\frac{d^2x}{dt^2}+6\frac{dx}{dt}+25x=2\frac{df}{dt}+36f \tag{11.6.1}$$

ANS.: Assume $f(t)$ to be of the form Fe^{st}; then

$$(s^2+6s+25)Xe^{st}=(2s+36)Fe^{st}$$

and

$$\frac{Xe^{st}}{Fe^{st}}=H(s)=\frac{2(s+18)}{s^2+6s+25} \tag{11.6.2}$$

$$H(s)=\frac{2(s+18)}{(s+3-j4)(s+3+j4)} \tag{11.6.3}$$

To solve for the natural response of the output variable $x(t)$ in Example 11.6.1 we would simply write the homogeneous equation

$$\frac{d^2x}{dt^2}+6\frac{dx}{dt}+25x=0 \tag{11.6.4}$$

assume a solution of the form

$$X_n(t)=Ae^{st} \tag{11.6.5}$$

insert it into (11.6.4) to obtain

$$s^2 + 6s + 25 = 0 \qquad (11.6.6)$$

which is the eigenequation, and thus find

$$s = -3 + j4, \; -3 - j4 \qquad (11.6.7)$$

which are the eigenvalues. Note that the left side of (11.6.6) is the denominator of (11.6.2) and that the eigenvalues of (11.6.7) are the system poles of (11.6.3). Does this surprise you? It should not. They are actually the same quantities.† We simply called them eigenvalues when we wanted to find natural responses and poles when we wanted to find forced responses due to complex exponential inputs. Thus the natural response is

$$X_n(t) = Ae^{-3t}(e^{j4t} + e^{-j4t})$$

or, taking the real part,

$$x_n(t) = e^{-3t}(B_1 \cos 4t + B_2 \sin 4t) \qquad (11.6.8)$$

The natural response can thus be obtained in any linear system by using the poles (eigenvalues) of $H(s)$.

EXAMPLE 11.6.2

Find the natural response $v_n(t)$ of the circuit in Figure 11.6.1.

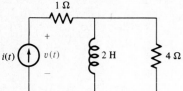

figure 11.6.1
The circuit of Example 11.6.2.

ANS.:

$$Z_{in}(s) = 1 + \frac{2s(4)}{2s + 4} = \frac{5s + 2}{s + 2} = \frac{5(s + 2/5)}{s + 2} \qquad (11.6.9)$$

and

$$V = Z_{in}(s)I \qquad (11.6.10)$$

The natural response is

$$v_n(t) = Ae^{-2t} \qquad (11.6.11)$$

For a nonzero voltage to exist while $I = 0$, it is necessary in equation (11.6.10) that the denominator of $Z_{in}(s)$ also be zero, thus producing an indeterminate form.

---

† System poles are those specific values of s for which we can get a nonzero response ($X \neq 0$) while the input *is* zero ($F = 0$). This is what we have called the natural response.

EXAMPLE 11.6.3

Find the complete solution $i(t)$ of the circuit in Figure 11.6.2, where $i_L(0-) = 0$.

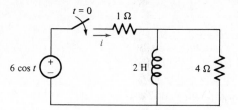

figure 11.6.2
The circuit of Example 11.6.3.

ANS.:   Use $v(t) = 6e^{st}$, where $s = j$; for $i(t) = Ie^{st}$

$$I = H(s)6\underline{/0°} \quad \text{and} \quad H(s) = Y_{\text{in}}(s) = \frac{0.2(s + 2)}{s + 0.4}$$

Thus

$$i_n(t) = Ae^{-0.4t}$$

The particular response is obtained from

$$H(j) = \frac{0.2(j + 2)}{j + 0.4} = \frac{0.2(2.24\underline{/26.6°})}{1.08\underline{/68.2°}} = 0.415\underline{/-41.6°}$$

so that $I_p = (6\underline{/0°})(0.415\underline{/-41.6°})$ and

$$I_p(t) = 2.49 \cos (t - 41.6°)$$

The complete response is

$$i(t) = 2.49 \cos (t - 41.6°) + Ae^{-0.4t}$$

Since $i_L(0-) = i_L(0+) = 0$ (no impulses present), we see that $i(0+) = (6 \text{ V})/(5 \text{ Ω})$

$$\tfrac{6}{5} = 2.49 \cos (-41.6°) + A$$

so that $A = -0.662$ and $i(t) = 2.49 \cos (t - 41.6°) - 0.662e^{-0.4t}$

Now that we recognize that a system's eigenvalues are also its poles, it becomes clear that any system with one or more poles in the right half s plane will have a natural response containing an exponential function with a positive real exponent. Therefore, such natural responses will increase unboundedly with increasing time. Hence any such system is called *unstable* because, even with no input, its output grows larger and larger until either the system self-destructs or in some manner becomes self-limiting. In either event the system is nonlinear and cannot be analyzed by transfer-function methods.

Electric circuits made up only of linear R's, L's, and C's (no dependent sources allowed) cannot have right-half-plane poles and thus cannot be unstable, and the same can be said for the mechanical, hydraulic, etc., analogs of these circuits.

Although a stable system cannot have any *poles* in the right half plane, there is no restriction on the locations of its *zeros*. Let us consider what effect the presence of one or more right-half-plane zeros in $H(s)$ might have on the plots of

**NON-MINIMUM-PHASE
NETWORKS**

figure 11.6.3
Two similar pole-zero constellations, one
with (a) left-half-plane zeros only and the
other with (b) symmetrically located
right-half-plane zeros. (c) The angle plot
produced by the constellation shown in (a).
(d) The angle plot produced by the
constellation shown in (b).

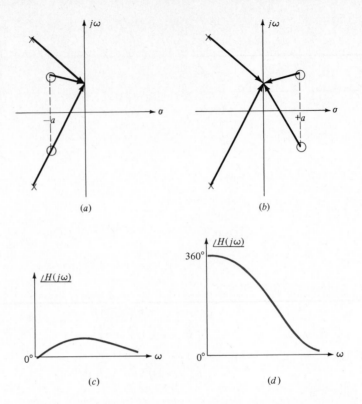

figure 11.6.3
Two similar pole-zero constellations, one with (a) left-half-plane zeros only and the other with (b) symmetrically located right-half-plane zeros. (c) The angle plot produced by the constellation shown in (a). (d) The angle plot produced by the constellation shown in (b).

$|H(j\omega)|$ and $\underline{/H(j\omega)}$ versus ω. Consider the two pole-zero constellations of two different systems shown in Figure 11.6.3. The pole locations are identical in both constellations and the zeros are mirror images of each other; i.e., one pair of zeros is in the left half plane and the other is identically located in the right half plane. For any point $s = j\omega$ we note that the magnitude $|H(j\omega)|$ is the same for both these systems. This is true because the magnitude of $H(j\omega)$ depends only on the *lengths* of the various vectors drawn from the critical frequencies to the point s. The lengths of corresponding vectors in Figure 11.6.3a and b are equal. Therefore the plot of $|H(j\omega)|$ versus ω will be the same for both of these systems.

The angle plot $\underline{/H(j\omega)}$ versus ω is a different story. Consider the constellation in Figure 11.6.3a. For $s = j0$ (dc input) the angles of the vectors from both pairs of critical frequencies cancel out (the angle of the numerator is zero and so is the angle of the denominator). As ω increases (s progresses up the $j\omega$ axis) we note that the angle at the upper-half-plane zero increases faster than the (subtractive) angle at the upper pole. Thus, the total angle of $\underline{/H(j\omega)}$ increases. Eventually, for very large ω values, all the vector angles approach $90°$ and so the overall $\underline{/H(j\omega)}$ approaches zero.

In Figure 11.6.3b the total angle begins (for $\omega = 0$) at zero. The pole angles cancel each other, and the sum of the angles at the zeros is $360° = 0°$. As ω increases, the total of the two angles at the zeros decreases and the total denominator angle (sum of the angles at the poles) increases. Thus the total angle decreases sharply. As ω approaches an extremely large value, all vectors approach $90°$ and thus the total angle approaches zero.

Clearly, the plot of $|H(j\omega)|$ for the system that has right-half-plane zeros varies over a wider range of angles than that of the system having only left-half-plane zeros. In general, the plot of $|H(j\omega)|$ versus ω of a system having only left-half-plane zeros will vary through a smaller range of angles than the plot of $|H(j\omega)|$ versus ω for that same system when one or more of its zeros is switched over symmetrically about the $j\omega$ axis to a right-half-plane location. Systems having only left-half-plane zeros are therefore called *minimum-phase* systems. Systems with one or more right-half-plane zeros are called *non-minimum-phase* systems.

EXAMPLE 11.6.4

Describe qualitatively the properties of a system having the pole-zero constellation shown in Fig. 11.6.4. Assume a sinusoidal input.

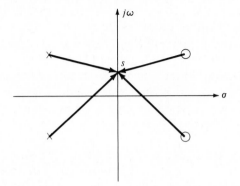

figure 11.6.4
The pole-zero plot of an all-pass network. For this system $|H(s)| = 1$ for any $s = j\omega$.

ANS.:   The vectors from the critical frequencies to any point on the $s = j\omega$ axis cancel out. Thus

$$|H(j\omega)| = 1 \qquad \text{for all } \omega$$

The angle decreases from 360° to 0° (or equivalently, from 0° to −360°) as ω ranges from 0 to ∞. This *all-pass network* is useful in that it can supply any desired amount of phase lag. If, in some system, a signal suffers an unwanted phase lag, we can design an all-pass network that will supply enough *additional* phase lag to bring the total angle to 360° = 0°.

---

rubber-sheet analogy for $|H(s)|$ **11.7**

We have seen that for any arbitrary value of s we can find the value of the magnitude of the transfer function

$$H(s) = \frac{K(s + a)(s + b) \cdots}{(s + c)(s + d) \cdots} \qquad (11.7.1)$$

graphically. Each factor in equation (11.7.1) is a complex number whose magnitude and angle are the magnitude and angle of a vector running from the critical frequency to the point s. If the test point s is extremely close to any one critical frequency, the length of the vector from the critical frequency to the test point

figure 11.7.1
The axes for a three-dimensional plot of
$|H(s)|$ versus σ and $j\omega$.

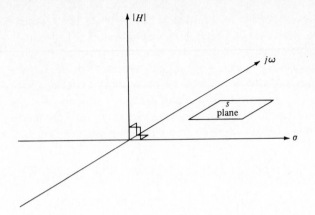

figure 11.7.1
The axes for a three-dimensional plot of
$|H(s)|$ versus σ and $j\omega$.

will be very short. If the critical frequency is a zero, $H(s)$ will be very small (the factor in the numerator approaches 0). For s near a pole, the vector length is in the denominator of $H(s)$, so that $|H(s)|$ will become large.

Think of making a three-dimensional plot of $|H(s)|$ versus s, as shown in Figure 11.7.1. The $s = \sigma + j\omega$ plane is horizontal, $|H|$ being plotted on the vertical axis. Plot the pole-zero constellation on the s plane. Stretch a rubber sheet horizontally so that it intersects the $|H|$ axis at the value of $|H(s)|$ for $s = \infty$. Then thumbtack the sheet down to the s plane at the zeros and place an infinitely long tent pole at each pole location. The sheet then takes on the value of $|H(s)|$ for any s.

For any system undergoing pure sinusoidal excitation the magnitude of the transfer function $|H(s)|$ can be plotted versus ω. Such a plot is the intersection between the rubber sheet and the plane formed by the $|H|$ and $j\omega$ axes. It is the cross section obtained by cutting vertically down to the $j\omega$ axis through the sheet, the blade moving downward along the $|H|$ axis. Similarly the $|H|$-versus-σ plot can be obtained by cutting vertically downward along $|H|$ but with the cutting edge running left to right so that it ends up along the σ axis. Since complex poles and zeros come in conjugate pairs, the lower s plane generates a surface which is the mirror image of that in the upper s plane.

figure 11.7.2
A plot of $|H(s)|$ versus σ and $j\omega$ for a system
having poles at $s = -2 \pm j4$ and a zero at
$s = -3.5$.

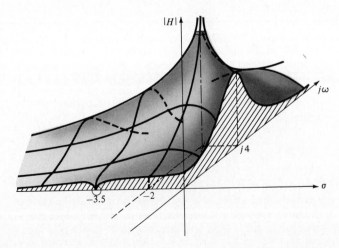

The rubber-sheet analogy is convenient for visualizing the effects of shifting pole and/or zero locations or introducing new poles and/or zeros into an existing constellation. For example, if the complex pair of poles in Figure 11.7.2 were located closer to the $j\omega$ axis, say at $s = -0.5 \pm j4$, the rubber-sheet analogy indicates that the peak in $|H(j\omega)|$ plotted versus ω will be higher and sharper. Although we cannot gain any quantitative information from the analogy, it provides qualitative insight into the general overall behavior of any linear system.

Q **and bandwidth 11.8**

If a pair of complex-conjugate poles occurs in a transfer function, it is clear from the rubber-sheet analogy that for sinusoidal inputs the magnitude of the transfer function will have a peak somewhere in the vicinity of the imaginary part of the pole location. For example, in Figure 11.7.2 $|H(j\omega)|$ appears to have a peak at or near $s = \pm j4$. In this section we examine three questions: Does $|H(j\omega)|$ indeed have a maximum? If so, at what frequency? How sharp a peak is it?

First, let us look at a simple circuit we have worked with before, the *RLC* parallel circuit of Figure 11.8.1. We saw in Chapter 10 that only for certain special circuits does an extremum occur in the input immittance magnitude at the resonant frequency, which, you recall, is the frequency that produces zero phase angle in $Z_{in}(j\omega)$ and $Y_{in}(j\omega)$. For the circuit of Figure 11.8.1 the input is $i(t)$, the output is $v(t)$, and therefore the transfer function is the driving-point impedance $H(j\omega) = Z_{in}(j\omega)$. We find this by summing the individual admittances

$$Y_{in}(s) = \frac{1}{R} + \frac{1}{Ls} + Cs \tag{11.8.1}$$

$$= \frac{s^2 + (1/RC)s + 1/LC}{(1/C)s} \tag{11.8.2}$$

so that

$$Z_{in}(s) = \frac{(1/C)s}{s^2 + (1/RC)s + 1/LC} \tag{11.8.3}$$

From the quadratic formula the pole locations are found to be

$$s = \frac{-1/RC \pm \sqrt{(1/RC)^2 - 4(1/LC)}}{2} \tag{11.8.4}$$

figure 11.8.2
The pole-zero constellation of the parallel *RLC* circuit of Figure 11.8.1.

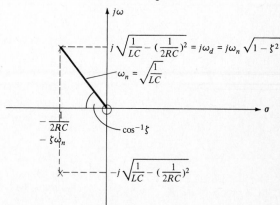

figure 11.8.1
A simple *RLC* circuit with input $i(t)$ and output $v(t)$.

which is (assuming complex conjugates result)

$$s = -\frac{1}{2RC} \pm j\sqrt{\frac{1}{LC} - \left(\frac{1}{2RC}\right)^2} \tag{11.8.5}$$

as shown in Figure 11.8.2. Using the standard nomenclature for complex-eigenvalue locations, we recall that the cosine of the angle from the negative real axis to the pole is equal to ζ, the damping factor. In this parallel RLC circuit we note that

$$\zeta = \frac{1}{2RC\omega_n} \tag{11.8.6}$$

and

$$\omega_n^2 = \frac{1}{LC} \tag{11.8.7}$$

so that, after inserting (11.8.7) into (11.8.6),

$$\zeta = \frac{1}{2R}\sqrt{\frac{L}{C}} \tag{11.8.8}$$

The rubber-sheet analogy indicates that $|Z_{in}(j\omega)|$ should have a peak near $\omega = \omega_d$, but in this particular circuit we already know that this peak occurs when the capacitive reactance equals the inductive reactance; i.e.,

$$Y_{in}(j\omega) = \frac{1}{R} + j\omega C - j\frac{1}{\omega L} \tag{11.8.9}$$

the magnitude of which is minimized (as a function of ω) when

$$j\omega C - j\frac{1}{\omega L} = 0 \tag{11.8.10}$$

or when

$$\omega = \omega_0 = \frac{1}{\sqrt{LC}} \tag{11.8.11}$$

Thus, inserting (11.8.11) into (11.8.9) and inverting gives

$$Z_{max}(j\omega) = R$$

The resonant frequency ω_r is that value of ω for which $Z_{in}(j\omega)$ is real. From (11.8.9) we see that $Y_{in}(j\omega)$ is real if and only if

$$\omega C - \frac{1}{\omega L} = 0 \quad\text{or}\quad \omega_r = \frac{1}{\sqrt{LC}}$$

Hence for this particular circuit the resonant frequency, the frequency of the peak in the magnitude of the transfer function, and the undamped natural frequency are all equal

$$\omega_r = \omega_0 = \omega_n = \frac{1}{\sqrt{LC}} \tag{11.8.12}$$

This is *not true* for most circuits.

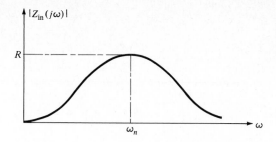

figure 11.8.3
The magnitude of $Z_{in}(j\omega)$ versus ω (showing the resonant peak at $\omega = \omega_n$) for the parallel *RLC* circuit of Figure 11.8.1.

It would be useful to be able to determine quantitatively how wide the resonant peak is. Certainly one answer to this question is that the peak is infinitely wide because the plot in Figure 11.8.3 in fact stretches out with nonzero height all the way to the right (to ω equals infinity). But that statement really tells us little about the shape of the peak in $Z_{in}(j\omega)$. To get a better quantitative description of the shape of this peak we define the *quality factor Q* (which must not be confused with quadrature power *Q*). The quality factor is

$$Q = 2\pi \frac{\text{maximum energy stored}}{\text{total energy lost per period}} \qquad (11.8.13)$$

This definition can be used in conjunction with any *RLC* circuit (not just this parallel connection). The numerator of equation (11.8.13) is the maximum value of energy ever stored in the (energy-storage) elements of the system. The denominator of (11.8.13) is the total energy lost (dissipated) during one period of the input waveform.

A moment's thought will tell us that *Q* is quite probably a function of input frequency. For example, the amount of energy stored in any inductor is

$$W_L(t) = \tfrac{1}{2}L[i(t)]^2 \qquad (11.8.14)$$

and certainly the value of the current $i(t)$ depends on the impedance of the inductor, which in turn depends on the frequency of the input signal. Also, to find the value of the denominator we can integrate the power dissipated in all the resistors in the circuit over an interval of time equal to T, one period of the input signal. If the input frequency is low, we integrate for a longer time and vice versa. So for any given circuit there are an uncountable and infinite number of correct values for *Q*, depending on the value of the input frequency.

Let us agree to use as input frequency the resonant frequency ω_0. The value of *Q* that we calculate for this particularly important frequency is denoted Q_0 (the resonant *Q*). We can calculate the value of Q_0 for our parallel *RLC* circuit as follows. Assume that $i(t) = I_m \cos \omega_0 t$. We calculate Q_0 as

$$Q_0 = 2\pi \frac{[W_L(t) + W_C(t)]_{\text{max}}}{P_{\text{av}} T} \qquad \text{at } \omega = \omega_0 \qquad (11.8.15)$$

At resonance $Y_L + Y_C = 0$, and so

$$Y_{in} = \frac{1}{R} \qquad (11.8.16)$$

and
$$v(t) = I_m R \cos \omega_0 t \tag{11.8.17}$$

Thus
$$W_C(t) = \tfrac{1}{2} C v_C^2 = \frac{C I_m^2 R^2}{2} \cos^2 \omega_0 t \tag{11.8.18}$$

The inductor current is found (by phasors) to be

$$\mathbf{I}_L = \mathbf{V} Y_L = I_m R \underline{/0°} \; \frac{1}{\omega_0 L} \underline{/-90°} \tag{11.8.19}$$

so that
$$i_L(t) = \frac{I_m R}{\omega_0 L} \cos (\omega_0 t - 90°) \tag{11.8.20}$$

or, since $\omega_0 = 1/\sqrt{LC}$,

$$i_L(t) = I_m R \sqrt{\frac{C}{L}} \sin \omega_0 t \tag{11.8.21}$$

Thus
$$W_L(t) = \tfrac{1}{2} L i^2 = \frac{I_m^2 R^2 C}{2} \sin^2 \omega_0 t \tag{11.8.22}$$

The total energy stored is $W(t) = W_L(t) + W_C(t)$, so adding equations (11.8.18) and (11.8.22) we get

$$W(t) = \frac{I_m^2 R^2 C}{2} = \text{const} \tag{11.8.23}$$

The *maximum* energy stored is, therefore, the numerical constant given in equation (11.8.23).

Now for the denominator. The average power dissipated in the resistor is $(I_m/\sqrt{2})^2 R$. The total energy lost each period is therefore

$$W_{\text{diss}} = \frac{I_m^2}{2} RT = \frac{I_m^2 R}{2 f_0} = \frac{2\pi I_m^2 R}{2\omega_0} \tag{11.8.24}$$

Therefore, inserting (11.8.23) and (11.8.24) into (11.8.15) gives

$$Q_0 = \frac{2\pi \tfrac{1}{2} I_m^2 R^2 C}{2\pi I_m^2 R / 2\omega_0} = \omega_0 RC \tag{11.8.25}$$

or, since $-X_C = X_L$ at $\omega = \omega_0$

$$\boxed{Q_0 = \frac{R}{-X_{C_0}} = \frac{R}{X_{L_0}}} \tag{11.8.26}$$

and from
$$\zeta = \frac{1}{2RC\omega_0} \tag{11.8.6}$$

we have
$$\zeta = \frac{1}{2Q_0} \tag{11.8.27}$$

or
$$Q_0 = \frac{1}{2\zeta} \tag{11.8.28}$$

Hence we can write the pole locations in terms of Q_0

$$\text{Re} = -\frac{1}{2RC} = -\zeta\omega_0 = -\frac{\omega_0}{2Q_0} \tag{11.8.29}$$

$$\text{Im} = \omega_0\sqrt{1-\zeta^2} = \omega_0\sqrt{1-\left(\frac{1}{2Q_0}\right)^2} = \omega_d \tag{11.8.30}$$

Let us now try to describe quantitatively how wide the peak in $|Z(j\omega)|$ is. To do this we designate two frequencies ω_1 and ω_2, one higher than the frequency at which the peak occurs and one lower, picking the frequencies where the value of $|Z(j\omega)|$ is $1/\sqrt{2}$ times its peak value. Therefore at each of these frequencies the magnitude of the output signal is 0.707 times what it would be at the frequency where the peak occurs. Since $p = v^2/R$ or i^2R, the power-producing capability of the output signal at either ω_1 or ω_2 is thus half what it is at the resonant frequency. For this reason ω_1 and ω_2 are called the *half-power frequencies* and $\omega_2 - \omega_1$ is called the *bandwidth* of the peak in $|Z(j\omega)|$.

The bandwidth can then be determined as follows. For our parallel *RLC* circuit

$$Y(j\omega) = \frac{1}{R} + j\left(\omega C - \frac{1}{\omega L}\right) \tag{11.8.31}$$

Multiplying through by $\omega_0/\omega_0 = 1$ and $R/R = 1$, we get

$$Y(j\omega) = \frac{1}{R} + j\frac{1}{R}\left(\frac{\omega\omega_0 RC}{\omega_0} - \frac{\omega_0 R}{\omega\omega_0 L}\right) \tag{11.8.32}$$

Noting that, from equation (11.8.26),

$$Q_0 = \omega_0 RC = \frac{R}{\omega_0 L}$$

we can write (11.8.32) as

$$Y(j\omega) = \frac{1}{R}\left[1 + j\left(\frac{\omega Q_0}{\omega_0} - \frac{\omega_0 Q_0}{\omega}\right)\right] = \frac{1}{R}\left[1 + jQ_0\left(\frac{\omega}{\omega_0} - \frac{\omega_0}{\omega}\right)\right] \tag{11.8.32a}$$

[We are happy to see in passing that (11.8.32a) gives $Y(j\omega) = 1/R$ when $\omega = \omega_0$, as it should.] At the half-power frequencies ω_1 and ω_2, $|Z(j\omega)| = Z_{max}/\sqrt{2}$, or

$$|Y(j\omega)| = \frac{\sqrt{2}}{R} \tag{11.8.33}$$

This occurs, according to equation (11.8.32a), when

$$Q_0\left(\frac{\omega}{\omega_0} - \frac{\omega_0}{\omega}\right) = \pm 1 \tag{11.8.34}$$

Solving equation (11.8.34) using the plus sign and setting $\omega = \omega_2$ gives

$$\frac{\omega_2}{\omega_0} - \frac{\omega_0}{\omega_2} = \frac{1}{Q_0} \qquad \frac{\omega_2^2 - \omega_0^2}{\omega_0\omega_2} = \frac{1}{Q_0}$$

$$\omega_2^2 - \frac{\omega_0}{Q_0}\omega_2 - \omega_0^2 = 0$$

Therefore

$$\omega_2 = \frac{\omega_0}{2Q_0} \pm \sqrt{\left(\frac{\omega_0}{2Q_0}\right)^2 + \frac{4\omega_0^2}{4}}$$

$$= \omega_0\left[\frac{1}{2Q_0} + \sqrt{\left(\frac{1}{2Q_0}\right)^2 + 1}\right] \qquad (11.8.35)$$

Similarly from equation (11.8.34) using the minus sign, we get

$$\omega_1 = \omega_0\left[\frac{-1}{2Q_0} + \sqrt{\left(\frac{1}{2Q_0}\right)^2 + 1}\right] \qquad (11.8.36)$$

Thus the bandwidth B is

$$\boxed{B = \omega_2 - \omega_1 = \frac{\omega_0}{Q_0} = 2\zeta\omega_0} \qquad (11.8.37)$$

When electrical engineers use the value of Q_0 to compute bandwidth, it is usually in situations where the resonant peak is sharp, i.e., where Q_0 is a relatively large quantity.† We state this condition formally as

$$\left(\frac{1}{2Q_0}\right)^2 \ll 1 \qquad (11.8.38)$$

or

$$\zeta^2 \ll 1 \qquad (11.8.39)$$

Using (11.8.38) in (11.8.35) and (11.8.36), we see, that *for high-Q_0* circuits and systems (with very low damping)

$$\omega_2 \approx \omega_0\left(\frac{1}{2Q_0} + 1\right) = \omega_0(1 + \zeta) \qquad (11.8.40)$$

† Clever (and arcane) methods exist for extrapolating the Q method presented here to other *RLC* forms, but the next section contains a simpler way to handle any such circuit.

figure 11.8.4
A *high-Q_0* circuit showing the bandwidth B, the resonant frequency ω_0, and the half-power frequencies ω_1 and ω_2.

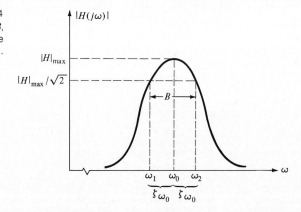

and

$$\omega_1 \approx \omega_0\left(-\frac{1}{2Q_0} + 1\right) = \omega_0(1 - \zeta) \qquad (11.8.41)$$

which simply states that ω_0 sits halfway between ω_1 and ω_2 (see Figure 11.8.4). Note that equations (11.8.40) and (11.8.41) are only approximations. The actual values of ω_1 and ω_2 given by (11.8.35) and (11.8.36) must be used for all systems having appreciable damping (say $Q_0 < 10$).

EXAMPLE 11.8.1

If $C = 0.1$ μF and $L = 10$ mH are placed in parallel, what leakage resistance R will yield a 0.2-kHz bandwidth around the resonant frequency?

ANS.:   For the parallel *RLC* circuit $\omega_0 = 1/\sqrt{LC} = 31.6$ krad/s, so that

$$-X_C = \frac{1}{\omega_0 C} = 316.5\ \Omega \quad \text{and} \quad B = \frac{\omega_0}{Q_0} \rightarrow Q_0 = \frac{31.6 \times 10^3}{200(2\pi)} \approx 25$$

Therefore

$$R = Q_0|X_C| = 25(316.5) = 7912\ \Omega.$$

In Figure 11.8.5 we follow the same line of reasoning as in the parallel case. The input admittance relates the output (current) to the input (voltage). We write

RESONANCE IN *RLC* SERIES CIRCUITS

$$Z_{\text{in}}(s) = R + Ls + \frac{1}{Cs} \qquad (11.8.42)$$

Multiplying by s/L and inverting gives

$$H(s) = Y_{\text{in}}(s) = \frac{(1/L)s}{s^2 + R/Ls + 1/LC} \qquad (11.8.43)$$

Poles (eigenvalues) are located at

$$s = -\frac{R}{2L} \pm j\sqrt{\frac{1}{LC} - \left(\frac{R}{2L}\right)^2} \qquad (11.8.44)$$

Also, from equation (11.8.42) setting $s = j\omega$ gives

$$Z_{\text{in}}(j\omega) = R + j\omega L - j\frac{1}{\omega C} \qquad (11.8.45)$$

figure 11.8.5
The *RLC* series circuit.

or

$$|Z_{\text{in}}(j\omega)| = \sqrt{R^2 + \left(\omega L - \frac{1}{\omega C}\right)^2} \qquad (11.8.46)$$

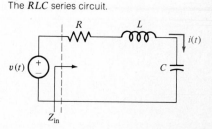

Obviously, the frequency that yields a minimum value of $|Z_{\text{in}}(j\omega)|$ and thus a maximum value of $|Y_{\text{in}}(j\omega)|$ is found from (11.8.46) to be

$$\omega L - \frac{1}{\omega C} = 0$$

$$\omega = \omega_0 = \frac{1}{\sqrt{LC}} \tag{11.8.47}$$

so $|(Y_{in})_{max}| = 1/R$.

The resonant frequency ω_r is that frequency for which the angle of $Z(j\omega)$ [and therefore $Y(j\omega)$] is equal to zero. From (11.8.45) this is found by setting the imaginary part equal to zero.

$$\omega L - \frac{1}{\omega C} = 0$$

$$\omega_r = \frac{1}{\sqrt{LC}} \tag{11.8.48}$$

Also from (11.8.43) or (11.8.44) the undamped natural frequency ω_n is

$$\omega_n = \frac{1}{\sqrt{LC}} \tag{11.8.49}$$

Hence from (11.8.47) to (11.8.49)

$$\omega_0 = \omega_r = \omega_n = \frac{1}{\sqrt{LC}} \tag{11.8.50}$$

Let us use a sinusoidal voltage source with frequency ω_0 and find the value of Q_0 [see equation (11.8.13)]. With

$$v(t) = V_m \cos \omega_0 t \tag{11.8.51}$$

$$\mathbf{V} = V_m\underline{/0°} \quad \text{and} \quad Y_{in} = \frac{1}{R}\underline{/0°}$$

$$\mathbf{I} = \frac{V_m}{R}\underline{/0°} \quad \text{and} \quad i(t) = \frac{V_m}{R} \cos \omega_0 t$$

$$\mathbf{V}_C = \mathbf{I}Z_C = \frac{V_m}{R}\underline{/0°}\,\frac{1}{\omega_0 C}\underline{/-90°} \rightarrow v_C(t) = \frac{V_m}{\omega_0 RC} \sin \omega_0 t$$

so that

$$W_L(t) = \tfrac{1}{2}Li^2 = \omega_0 L \frac{V_m^2}{2R^2\omega_0} \cos^2 \omega_0 t \tag{11.8.52}$$

and

$$W_C(t) = \tfrac{1}{2}CV^2 = \frac{1}{\omega_0 C} \frac{V_m^2}{2R^2\omega_0} \sin^2 \omega_0 t \tag{11.8.53}$$

The total stored energy is the sum of $W_L(t)$ and $W_C(t)$. Thus from (11.8.52) and (11.8.53) the total stored energy (which is also the maximum) is

$$W_{st} = \frac{V_m^2 X_{L0}}{2R^2\omega_0} = \text{const} \tag{11.8.54}$$

On the other hand, the average power being dissipated in R is

$$P_{av} = I_{rms}^2 R = \frac{V_m^2}{2R} \tag{11.8.55}$$

so the energy dissipated during one period T is

$$P_{av} T = \frac{V_m^2}{2R} \frac{1}{f_0} = \frac{2\pi V_m^2}{2R\omega_0} \tag{11.8.56}$$

Thus

$$Q_0 = 2\pi \frac{V_m^2 X_{L0}}{2R^2 \omega_0} \frac{2R\omega_0}{2\pi V_m^2} = \frac{\omega_0 L}{R} = \frac{1}{R\omega_0 C}$$

$$\boxed{Q_0 = \frac{X_{L0}}{R} = \frac{-X_{C0}}{R}} \tag{11.8.57}$$

Note these are the inverse of the relationships we found for the parallel case.

Again writing the pole locations given in equation (11.8.44) in terms of Q_0 and using $Q_0 = \omega_0 L/R$ or $R/2\omega_0 L = 1/2Q_0$ from (11.8.57), we get

$$\text{Re} = -\frac{R}{2L} = -\zeta\omega_0 = -\frac{\omega_0}{2Q_0} \tag{11.8.58}$$

$$\text{Im} = \sqrt{\frac{1}{LC} - \left(\frac{R\omega_0}{2L\omega_0}\right)^2} = \omega_0 \sqrt{1 - \left(\frac{1}{2Q_0}\right)^2} = \omega_d \tag{11.8.59}$$

In order to determine the bandwidth we seek those frequencies where the response i has dropped to $1/\sqrt{2}$ times its resonant (maximum) value. Since V_m is a constant, this will occur when $|Y_{in}(j\omega)| = |Y_{in}|_{max}/\sqrt{2}$ or $|Z_{in}(j\omega)| = |Z_{in}|_{min}\sqrt{2}$. Thus we seek ω_1 and ω_2 as follows:

$$Z(j\omega) = R + j\left(\omega L - \frac{1}{\omega C}\right) = R + jR\left(\frac{\omega_0 \omega L}{\omega_0 R} - \frac{\omega_0}{\omega_0 \omega RC}\right)$$

$$= R\left[1 + jQ_0\left(\frac{\omega}{\omega_0} - \frac{\omega_0}{\omega}\right)\right] \tag{11.8.60}$$

Since we seek the value of ω that makes the magnitude of the quantity in brackets equal to $\sqrt{2}$, we set the imaginary part equal to unity and find, exactly as with equations (11.8.34) to (11.8.37),

$$\boxed{B = \omega_2 - \omega_1 = \frac{\omega_0}{Q_0} = 2\zeta\omega_0} \tag{11.8.61}$$

Equations (11.8.37) and (11.8.61) are equally applicable to series and parallel *RLC* circuits. Only the value of Q_0 is different.

In order to achieve high, sharp peaks in *RLC* circuits it is necessary to remove the damping as much as possible. In the parallel circuit this is done by setting R to as large a value as possible; i.e., damping would be totally removed if $R = \infty$. Thus we can easily remember that in the parallel case we define Q_0 as

$$Q_0 = \frac{R}{|X_0|} \qquad \text{parallel} \tag{11.8.62}$$

In the series case we remove damping by setting R to a low value (zero if possible) in order to achieve high Q_0 values. Thus

$$Q_0 = \frac{|X_0|}{R} \qquad \text{series} \qquad (11.8.63)$$

EXAMPLE 11.8.2

Given a series connection of $R = 2\ \Omega$, $L = 10$ mH, and $C = 100\ \mu$F, find ω_0, Q_0, bandwidth, and the error in assuming that the bandwidth is evenly centered on ω_0.

ANS.:
$$\omega_0 = \frac{1}{\sqrt{LC}} = \frac{1}{\sqrt{0.01(100 \times 10^{-6})}} = 1000 \text{ rad/s}$$

$$X_{L_0} = \omega_0 L = 1000(0.01) = 10\ \Omega \qquad Q_0 = \frac{\omega_0 L}{R} = \frac{10}{2} = 5$$

$$B = \frac{\omega_0}{Q_0} = \frac{1000}{5} = 200 \text{ rad/s}$$

The approximation yields a peak frequency of $\omega_0 = 1000$ rad/s, $\omega_2 = \omega_0 + B/2 = 1100$ rad/s, and $\omega_1 = \omega_0 - B/2 = 900$ rad/s. This is not exactly correct, however. At 900 rad/s we expect the magnitude of the transfer function $|Y(j\omega)|$ to be $1/\sqrt{2}$ times its peak value; i.e., since $|Y(j\omega)|_{\text{peak}} = 1/R = \frac{1}{2}$, we assume

$$|Y(j900)| = \frac{1/2}{\sqrt{2}} = 0.354$$

However, the actual value, from equation (11.8.43), is

$$Y(j900) = \frac{1/(10 \times 10^{-3})(j900)}{(j900)^2 + 2/(10 \times 10^{-3})(j900) + 1/(10 \times 10^{-3})(100 \times 10^{-6})}$$

$$= \frac{100(900\underline{/90^\circ})}{[(-8.1 \times 10^5) + 10^6] + j1.8 \times 10^5} = \frac{9 \times 10^4\underline{/90^\circ}}{261{,}725\underline{/43.45^\circ}}$$

$$= 0.344\underline{/46.55^\circ}$$

so that
$$|Y(j900)| = 0.344$$

We have seen in this section that the notion of Q_0, the resonant quality factor, is particularly useful in determining the bandwidth of highly underdamped (so-called high-Q_0) systems. For more heavily damped systems the exact equations (11.8.13), (11.8.35), and (11.8.36) rather than the approximations (11.8.40) and (11.8.41) must be used. Moreover, when a highly damped RLC circuit of some different form (other than simple series or parallel) is to be worked with, we must also start with (11.8.13) and derive Q_0 for that circuit.

Bode plots 11.9

A general and relatively straightforward method for sketching both $|H(j\omega)|$ and $\underline{/H(j\omega)}$ versus ω, called the *Bode plot*, is based on the following ideas. When we evaluated $H(s)$ by drawing vectors from each pole and zero to the point s, we were actually treating each factor $s + a$ in the numerator and denominator as a separate complex number. At each value of complex frequency s we multiplied the magnitudes and added the angles of these complex numbers to find the value of $H(s)$. We repeated this process at each new value of s. Usually this locus of points s was the $j\omega$ axis, i.e., we were interested in how the magnitude and angle of the transfer function vary with changing sinusoidal frequency.

If we could find out how *each separate factor* of the numerator and denominator of $H(j\omega)$ varies in magnitude and angle over a wide range of ω, and if we could *add* all these contributions together, we could see how the total $H(j\omega)$ varies. But how can we *add* numbers that should be multiplied? Here is how it is done. Taking a simple transfer function first, consider one with real poles and zeros

$$H(s) = \frac{K(s + a)(s + c) \cdots}{(s + b)(s + d) \cdots} \tag{11.9.1}$$

Take the quantities a, b, c, etc., outside the parentheses so that $H(s)$ is in the form

$$H(s) = \frac{Kac \cdots (s/a + 1)(s/c + 1) \cdots}{bd \cdots (s/b + 1)(s/d + 1) \cdots} \tag{11.9.2}$$

Letting $s = j\omega$, we have

$$H(j\omega) = \frac{Kac \cdots (1 + j\omega/a)(1 + j\omega/c) \cdots}{bd \cdots (1 + j\omega/b)(1 + j\omega/d) \cdots} \tag{11.9.3}$$

and its magnitude is

$$|H(j\omega)| = \frac{Kac}{bd} \frac{|1 + j\omega/a| \cdot |1 + j\omega/c| \cdots}{|1 + j\omega/b| \cdot |1 + j\omega/d| \cdots} \tag{11.9.4}$$

How does each factor in (11.9.4) contribute to a plot of $|H(j\omega)|$ versus ω? If we are going to *add* the contribution of each magnitude term in (11.9.4) to the contribution of all other such terms in order to find $H(j\omega)$, we must use a *logarithmic measure* because we can add logarithms of terms that should be multiplied. Let us therefore take the logarithm of equation (11.9.4) and arbitrarily multiply by 20

$$20 \log |H(j\omega)| = 20 \log \frac{Kac \cdots}{bd \cdots} + 20 \log \left| 1 + \frac{j\omega}{a} \right| + 20 \log \left| 1 + \frac{j\omega}{c} \right| + \cdots$$

$$- 20 \log \left| 1 + \frac{j\omega}{b} \right| - 20 \log \left| 1 + \frac{j\omega}{d} \right| - \cdots$$

$$\tag{11.9.5}$$

Since 20 times the logarithm of any quantity is usually called the *decibel* measure of that quantity, we digress to examine briefly the origins of this unit of measurement.

DECIBELS   In the early days of the telephone system, engineers wanted to have a way of measuring the effectiveness of adding power amplifiers into telephone lines in order to make the sound of the transmitted message loud enough to be heard at the receiving end of a long line. Figure 11.9.1*a* shows a simplified schematic of a telephone line without any amplifier, where R_L is the line resistance and R_R is the receiver. The output power P_1 may not be sufficient for intelligible reception of the signal if the line is very long, $R_L \gg R_R$. Figure 11.9.1*b* shows the same line after the installation of an amplifier. Presumably the voltage V_2 is now larger than the V_1 it was before. Certainly that is true if the output power P_2 is greater than its previous value P_1. Since the human ear is a logarithmic device, if P_2 is twice P_1, the message sounds not twice as loud but only slightly louder.

Aware of this the telephone engineers decided to measure the effectiveness of line amplifiers in logarithmic units and named them after Alexander Graham Bell, the inventor of the telephone. A *bel* (B) was defined as the logarithm of the new (amplified) output power divided by the original (unamplified) output power

$$B = \log \frac{P_2}{P_1} \tag{11.9.6}$$

Since $P_2 = V_2^2/R_R$ and $P_1 = V_1^2/R_R$, the effective gain of inserting the amplifier can also be written

$$B = 2 \log \frac{V_2}{V_1} \tag{11.9.7}$$

Since the bel turned out to be a large unit, engineers began to talk about power gains in terms of a unit one-tenth as large as a bel, the *decibel* (dB), and we have

$$dB = 20 \log \frac{V_2}{V_1} \tag{11.9.8}$$

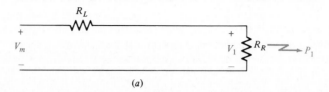

(a)

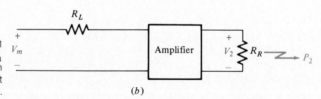

figure 11.9.1
(a) A telephone line with output voltage V_1 and output power P_1. (b) The same system after an amplifier has been inserted so that $P_2 > P_1$ and $V_2 > V_1$.

(b)

Note that in going from (11.9.6) to (11.9.7) the value R_R canceled out. Because of this cancellation we sometimes lose sight of the fact that V_1 and V_2 are supposed to be across the *same value of resistance*. Nonetheless, engineers today seem to measure almost any quantity imaginable by simply taking its common logarithm and multiplying by 20. Although not strictly correct, this is a convenient measure in many situations and the measurement of the magnitudes of the factors of $|H(j\omega)|$ is one of those situations.

The first term on the right of equation (11.9.5) is simply a constant, which we shall call the *gain constant*. Consider the next term. Let us sketch

$$20 \log \left| 1 + \frac{j\omega}{a} \right| \qquad (11.9.9)$$

for all values of ω. First, look at what happens for very small values of ω, that is, $1 \gg \omega/a$. In equation (11.9.9) the log magnitude is approximately $20 \log 1 = 0$ dB. For very large values of ω, that is, $1 \ll \omega/a$, in equation (11.9.9) the log magnitude is $20 \log \omega/a$ dB. We list this one for several large values of ω in Table 11.9.1.

Since every time we multiply the frequency by 10 we say we have gone a *decade* higher in ω, for large ω the log magnitude has a positive slope of $+20$ dB/decade. We plot this using a horizontal logarithmic ω axis in Figure 11.9.2. A line with slope of $+20$ dB/decade can also be described by saying that its slope is $+6$ dB/octave. An octave (eight notes) in music represents a doubling of frequency. Middle A is 440 Hz and the next A up the scale, one octave higher, is 880 Hz. Using the high-frequency approximation dB $= 20 \log (\omega/a)$, if we are at, say, $\omega = a \times 10^4$, then

$$20 \log \frac{a \times 10^4}{a} = 80 \text{ dB}$$

If we double ω (go up an octave), we get

$$20 \log \frac{2a \times 10^4}{a} = 20(\log 2 + \log 10^4) = 20(0.301 + 4) = 6 + 80$$

$$= 86 \text{ dB} \qquad 6 \text{ dB higher}$$

table 11.9.1

ω	ω/a	dB
$a \times 10^2$	10^2	40
$a \times 10^3$	10^3	60
$a \times 10^4$	10^4	80

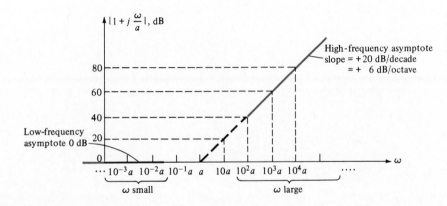

figure 11.9.2
The approximate magnitude in decibels of the factor $1 + j(\omega/a)$ for very large and very small values of ω.

Whenever we are at a high frequency and we double it, the log magnitude, as given by this straight line, goes up 6 dB. Notice that if we extend the straight line valid for high values of ω down into the middle range of ω (where it is *not* valid), it intersects the horizontal ω axis at $\omega = a$. Therefore, we can easily plot both the *low-frequency asymptote* and the *high-frequency asymptote* for any factor having the form $s/a + 1$. For low frequencies it is a straight horizontal line coincident with the horizontal (logarithmic) axis that extends from infinity on the left up to the frequency $\omega = a$. From that point we plot a straight line upward toward the right with slope equal to $+20$ dB/decade (or $+6$ dB/octave).

The actual log-magnitude curve does not make the sharp upward break at $\omega = a$. Let us find its value at $\omega = a$:

$$20 \log\left|1 + \frac{j\omega}{a}\right|\Bigg|_{\omega=a} = 20 \log |1 + j| = 20 \log \sqrt{2}$$

$$= 10 \log 2$$

$$= 3 \text{ dB}$$

So at the *break frequency* or *corner frequency* $\omega = a$ the actual curve is up at $+3$ dB. There is a 3-dB correction at the corner. At one octave *below* the corner ($\omega = a/2$)

$$20 \log \left|1 + \frac{ja/2}{a}\right| = 20 \log |1 + j\tfrac{1}{2}| = 20 \log 1.118$$

$$= +1 \text{ dB}$$

and at one octave *above* the corner ($\omega = 2a$)

$$20 \log \left|1 + \frac{j2a}{a}\right| = 20 \log |1 + j2| = 20 \log 2.236 = +7 \text{ dB}$$

But since at $\omega = 2a$ the high-frequency straight line is already up to $+6$ dB, we note that the *correction* at both $+1$ and -1 octave away from the corner frequency is 1 dB above the asymptotes (see Figure 11.9.3). The *angle contribution* of a term such as $1 + j\omega/a$ is simply

figure 11.9.3
The magnitude contribution of the term
$s/a + 1$.

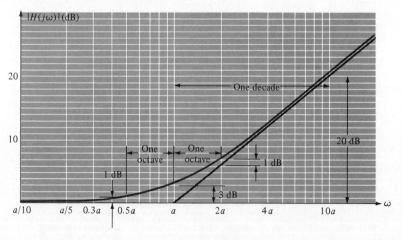

$$\underline{/1 + j\,\frac{\omega}{a}} = \begin{cases} 0° & \text{for } \omega \ll a & (11.9.10) \\ +90° & \text{for } \omega \gg a & (11.9.11) \\ +45° & \text{at } \omega = a & (11.9.12) \end{cases}$$

Also 1 octave away from the break the angle correction is 26.6°, as follows:

$$\arg\left(1 + j\,\frac{\omega}{a}\right)\Bigg|_{\omega = a/2} = \begin{cases} \underline{/1 + j(\tfrac{1}{2})} = 26.6° & \text{at } \omega = a/2 \\ \underline{/1 + j2} = 63.4° & \text{at } \omega = 2a \end{cases}$$

When 63.4° is subtracted from the high-frequency asymptote value 90°, it leaves 26.6°.

A good approximation of this angle plot, *for a first-order term such as this*, is obtained by plotting a straight line through the 45° point at the break frequency ($\omega = a$). This straight line has a slope of 45° per decade (see Figure 11.9.4).

If the term $s/a + 1$ appears in the *denominator* of $H(s)$, its contributions to both the overall magnitude and the overall angle are *negative*. For high values of ω the slope goes downward toward the lower right and the angle asymptote is at $-90°$ at frequencies higher than (to the right of) the corner frequency. The reasons are as follows. A term $s/a + 1$ in the denominator is equivalent to a term $(s/a + 1)^{-1}$ in the numerator. The magnitude in decibels of such a term is

$$20 \log\left|\left(1 + j\,\frac{\omega}{a}\right)^{-1}\right| = -20 \log\left|\left(1 + j\,\frac{\omega}{a}\right)\right| \qquad (11.9.13)$$

which is simply the negative of the curve shown in Figure 11.9.3. The angle contribution of a denominator term to the overall angle is the negative of the one indicated in equations (11.9.10) to (11.9.12) because of the algebra of complex numbers. When we divide by a complex number, we *subtract* its angle from the angle of the numerator. A single, lone s may be found as a factor either in the numerator or denominator. Its effect on the log magnitude, letting $s = j\omega$, is

$$20 \log \omega \qquad \text{dB} \qquad (11.9.14)$$

which is a straight line (when plotted on a horizontal log-frequency axis) with slope $+20$ dB/decade $= +6$ dB/octave. It intersects the horizontal (0-dB) axis at $\omega = 1$. The angle contribution is $+90°$ for all ω.

Any quadratic factor of the form $s^2 + 2\zeta\omega_n s + \omega_n^2$ that results in complex poles (and/or zeros, depending on whether it appears in numerator or denominator) contributes to the overall log magnitude of $H(s)$ as follows. Putting it in the usual form, we have

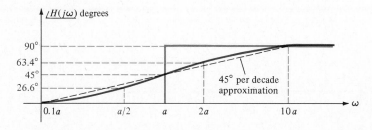

figure 11.9.4
The angle contribution of the term $s/a + 1$.

$$\frac{s^2}{\omega_n^2} + \frac{2\zeta}{\omega_n} s + 1$$

Letting $s = j\omega$ and measuring its magnitude in decibels gives

$$20 \log \left| \left(1 - \frac{\omega^2}{\omega_n^2}\right) + j \frac{2\zeta\omega}{\omega_n} \right| \qquad \text{dB} \tag{11.9.15}$$

For large ω this is approximately given by

$$20 \log \left(\frac{\omega}{\omega_n}\right)^2 = 40 \log \frac{\omega}{\omega_n} \qquad \text{dB} \tag{11.9.16}$$

so that every decade (tenfold) increase in frequency ω results in a $+40$-dB increment in the number of decibels. Also this approximation intersects the ω axis at $\omega = \omega_n$. For small ω, equation (11.9.15) becomes $20 \log 1 = 0$ dB. Thus the high- and low-frequency behavior of the magnitude of this quadratic term, shown in Figure 11.9.5, is similar to the behavior of the simple $s + a$ type of term (Figure 11.9.2) except that the high-frequency-range slope of the quadratic term is twice the slope of the first-order term.

What is the value of the magnitude of the quadratic term for values of ω near ω_n? From (11.9.15) with $\omega = \omega_n$ we have

$$\boxed{20 \log 2\zeta \qquad \text{dB}} \tag{11.9.17}$$

Let us define normalized frequency as $w = \omega/\omega_n$. A plot of the magnitude of this quadratic term versus normalized frequency w is given in Figure 11.9.6. Note that the curve for $\zeta = 0.5$ does indeed pass through 0 dB at $\omega = \omega_n$, as predicted by equation (11.9.17). You should study Figure 11.9.6 carefully, noticing, among other things, that:

- The extremum in $|H|$ gets larger and larger as the damping factor ζ decreases. This is consistent with the rubber-sheet analogy: as ζ approaches zero, the critical frequency approaches the $j\omega$ axis more closely. When $\zeta = 0$, the zero is on the $j\omega$ axis and the magnitude of the term is zero (or $-\infty$ dB).
- The smallest value of ζ that does not result in an extremum (dip) in $|H|$ is $\zeta = 0.707$.

figure 11.9.5
The approximate magnitude of the quadratic term

$$\left. \left(\frac{s^2}{\omega_n^2} + \frac{2\zeta s}{\omega_n} + 1\right) \right|_{s = j\omega}$$

for very large and very small values of ω.

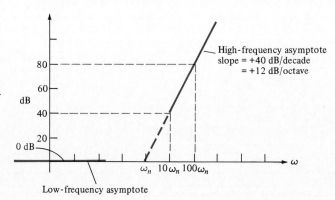

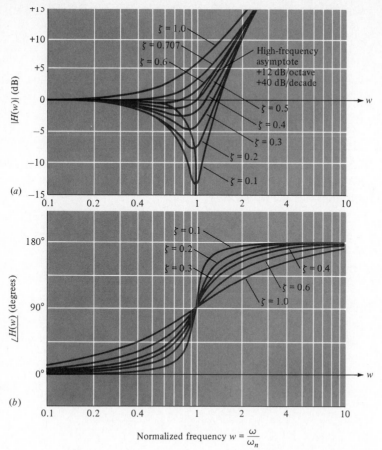

figure 11.9.6
(a) The magnitude of a second-order term

$$\frac{s^2}{\omega_n^2} + \frac{2\zeta s}{\omega_n} + 1$$

versus normalized frequency $w = \omega/\omega_n$.
(b) The angle contribution of this term.

- The correction is 0 dB for $\zeta = 0.5$; that is, at $\omega = \omega_n$, $|H(\omega_n)| = 1$ when $\zeta = 0.5$.
- The frequency at which $|H|$ reaches its extremum depends on the value of ζ. For $\zeta = 0.707$ there really is no extremum (or we can say it occurs at $\omega = 0$). Then, as ζ decreases below 0.707, the extremum occurs at higher and higher frequencies, approaching $\omega = \omega_n$ for ζ approaching zero.
- The correction (value of $|H|$ at $\omega = \omega_n$) for $\zeta = 1$ is +6 dB, which is just twice the value of the similar correction in a first-order term.
- The angle curve changes more and more abruptly from $0°$ to $180°$ as ζ approaches zero.
- For a quadratic term in the denominator, we invert both these plots.

In general, we can determine the *frequency of the extremum* (minimum in Figure 11.9.6a) as follows. Rewriting equation (11.9.15) in terms of the normalized frequency

$$w = \frac{\omega}{\omega_n} \qquad (11.9.18)$$

we have $\qquad 20\log|H(jw)| = 20\log|1 - w^2 + j2\zeta w| \qquad$ dB $\qquad$ (11.9.19)

Since $|H(jw)|$ will have an extremum at the frequency where $|H(jw)|^2$ has one, we can set the derivative of $|H(jw)|^2$ equal to zero and solve for w

$$|H|^2 = (1 - w^2)^2 + (2\zeta w)^2 \qquad (11.9.20)$$

$$\frac{d|H|^2}{dw} = 2(1 - w^2)(-2w) + (4\zeta w)(2\zeta)$$

$$= (1 - w^2)(-4w) + 8\zeta^2 w = -4w + 4w^3 + 8\zeta^2 w \qquad (11.9.21)$$

which we set equal to 0. One solution of equation (11.9.21) is $w = 0$. This is perfectly valid. All the curves of $|H|$ approach zero slope at $w = 0$. The other solution is the one we seek:

$$-4 + 4w^2 + 8\zeta^2 = 0 \qquad w^2 = 1 - 2\zeta^2$$

$$w = \sqrt{1 - 2\zeta^2} \qquad (11.9.22a)$$

so that the actual radian frequency ω_x of the extremum is

$$\boxed{\omega_x = \omega_n\sqrt{1 - 2\zeta^2}} \qquad (11.9.22b)$$

We note that, for an extremum to exist, according to equation (11.9.22),

$$2\zeta^2 \le 1 \qquad \text{and} \qquad \zeta \le 0.707$$

Inserting (11.9.22a) into (11.9.20) enables us to solve for the *magnitude at the extremum*

$$|H|_{max} = \sqrt{1 - w_x^4} \qquad (11.9.23)$$

or substituting (11.9.22a) into (11.9.23) yields the magnitude of the extremum in terms of ζ

$$\boxed{|H|_{max} = 2\zeta\sqrt{1 - \zeta^2}} \qquad (11.9.24)$$

For a second-order *pole* the answer given by these equations is simply inverted and then converted into decibels. Equations (11.9.22) and (11.9.23) are valid only for a second-order critical frequency, where the asymptote breaks up *after being horizontal at low frequencies*. For example, a second-order pole that produces a break from a rising $+20$ dB/decade asymptote to a falling -20 dB/decade asymptote will have its $|H(j\omega)|_{max}$ at the break frequency.

EXAMPLE 11.9.1

Given the circuit of Figure 11.9.7, sketch $|H(j\omega)|$ and $\underline{/H(j\omega)}$ versus log ω; that is, find the Bode plot.

ANS.: $H(s)$ is obtained by voltage division

$$H(s) = \frac{LsR/(Ls + R)}{1/Cs + LsR/(Ls + R)} = \frac{s^2}{s^2 + (1/RC)s + 1/LC}$$

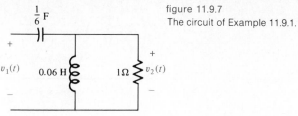

figure 11.9.7
The circuit of Example 11.9.1.

figure 11.9.8
(a) $|H(j\omega)|$ versus ω for the circuit of Example
11.9.1. (b) The angle $\underline{/H(j\omega)}$ versus ω.

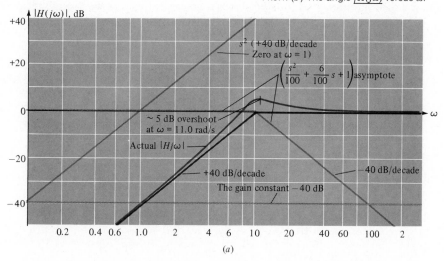

(a)

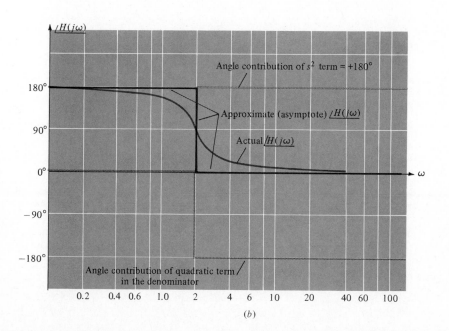

(b)

and with $R = 1$, $L = 0.06$ H, and $C = \frac{1}{6}$ F

$$H(s) = \frac{s^2}{s^2 + 6s + 100} = \frac{0.01s^2}{s^2/100 + (6/100)s + 1}$$

The 0.01 term's contribution to the magnitude plot is $20 \log 0.01 = -40$ dB at all frequencies. Its angle contribution is zero. The s^2 term becomes $-\omega^2$. Its magnitude contribution is a straight line with a slope of $+40$ dB/decade for all ω. Its angle contribution is $180°$ for all ω.

The quadratic term has $\omega_n = 10$ rad/s and $\zeta = 0.3$. Its magnitude contribution is zero up to $\omega = 10$ and then it breaks downward at -40 dB/decade. The complete plot is shown in Figure 11.9.8.

Since this quadratic term is in the denominator (and thus produces a *downward* break), we use the reciprocals of the values yielded by equations (11.9.22) and (11.9.24) to find that the maximum overshoot is $1.75 = 4.85$ dB at $10(1/0.91) = 11$ rad/s.

EXAMPLE 11.9.2

It is suggested that a notch filter (one that rejects only a small band of frequencies) could be designed by placing a pair of poles and a pair of zeros each with the same imaginary part near the $j\omega$ axis, as shown in Figure 11.9.9a.

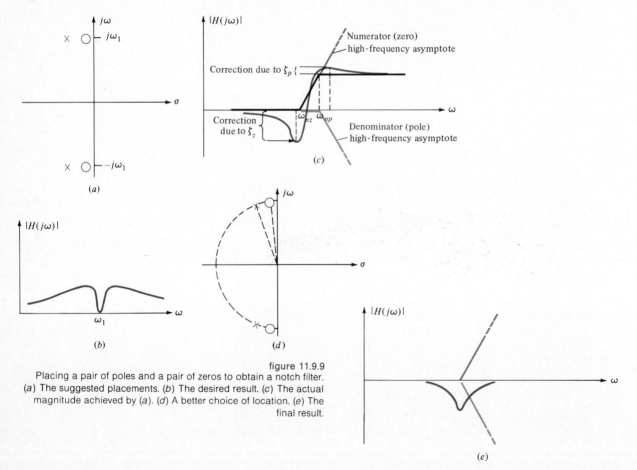

figure 11.9.9
Placing a pair of poles and a pair of zeros to obtain a notch filter.
(a) The suggested placements. (b) The desired result. (c) The actual magnitude achieved by (a). (d) A better choice of location. (e) The final result.

a. What would the $|H(j\omega)|$ plot look like?

b. Can you improve the placement of the pole pair so that the filter will pass high and low frequencies with equal gain?

ANS.:   From the rubber-sheet analogy the suggestion appears to be a good one. We should expect something like the plot shown in Figure 11.9.9b. What about the Bode plot? The transfer function would be of the form

$$H(s) = \frac{s^2 + 2\zeta_z \omega_{nz} s + \omega_{nz}^2}{s^2 + 2\zeta_p \omega_{np} s + \omega_{np}^2}$$

where the subscripts indicate whether the variables ζ and ω_n are associated with the zeros or with the poles. Since ω_n is the radial distance of the critical frequency from the origin, $\omega_{np} > \omega_{nz}$. Similarly we note that $\zeta_p > \zeta_z$. (The angle from the negative real axis up to the pole is less than that up to the zero.) The Bode magnitude plot is shown in Figure 11.9.9c. We use our knowledge that $\zeta_p > \zeta_z$ and indicate that the downward overshoot due to ζ_z is larger than the upward overshoot due to ζ_p. We see that this filter will pass frequencies higher than ω_{np} with higher gain (amplification) than it will frequencies lower than ω_{nz}, which can be corrected by forcing both the upward and the downward break to occur at the same frequency $\omega_{nz} = \omega_{np}$, that is, placing both the poles and zeros equidistant from the origin. Such placement is shown in Figure 11.9.9d, and the resulting Bode magnitude plot is shown in Figure 11.9.9e. With the notch placed at 60 Hz this type of filter is used to minimize the effect of power-line noise on sensitive measurements such as electrocardiograms. There are many other uses for this interesting circuit that is capable of rejecting a narrow band of unwanted frequencies.

The descriptive names of the basic categories of filters are most easily associated with the forms of their transfer functions through their associated Bode magnitude plots (Table 11.9.2).

Type	$H(s)$	Bode plot
Low-pass	$\dfrac{a}{s+a}$	Has single breakdown at $\omega = a$
High-pass	$\dfrac{s}{s+b}$	Rising at +6 dB/octave at low frequencies and breaks over at $\omega = b$
Band-pass	$\dfrac{ks}{(s+c)(s+d)}$	Rising at low frequencies, breaks over at $\omega = c$, and then breaks down at $\omega = d$
Band-reject	$\dfrac{k(s+e)(s+f)}{s}$	Has a slope of -6 dB/octave at low frequencies, breaks over at $\omega = e$, and then breaks up at $\omega = f$

table 11.9.2
Bode magnitude plots for filters.

summary **11.10**

In this chapter we have expanded the applicability of the notion of impedances, admittances, and transfer functions to make them quantitatively useful in a wider class of problems. We have also showed how the locations of the poles and zeros of the complex transfer function $H(s)$ determine not only a system's particular response to sinusoidal (and other types of complex exponential) inputs but its natural response as well. We have examined how to evaluate graphically the magnitude and angle of any $H(s)$ for any given value of s. Perhaps most important, we have used two techniques, the rubber-sheet analogy and the Bode plot, to estimate the approximate behavior of a system without painstakingly solving for its complete response and/or plotting out its $|H(j\omega)|$-versus-ω and $\underline{/H(j\omega)}$-versus-ω curves point by point. The best engineer is often the one who can take a quick look at a system and then estimate with a minimum of time and effort its properties and how it will behave. These are important techniques used in many different areas of electrical engineering.

problems

1. For a circuit whose input is $4e^{-3t}$ find the particular response $y(t)$ given that

$$H(p) = \frac{p+1}{p+7}$$

2. The series combination of $R = 10\ \Omega$, $L = 1$ F, and $C = 2$ H is driven by a voltage source $v_s(t) = 10e^{-2t} \cos{(3t + 20°)}u(t)$ V. Find an expression for the forced (particular) component of the voltage across the capacitor.

3. Given the transfer function

$$H(p) = \frac{10p}{p+3}$$

find the particular response $y(t)$ to the source $x(t) = 10 \cos{(3t + 50°)}$. Write your answer in the form $y(t) = A \sin{(bt + \theta)}$ with $\theta < 90°$.

4. A system has no finite zeros and a single finite pole at $s = -1$. Find the form of the particular response $y_p(t)$ for $t > 0$ if the input is $x(t) = e^{-t}u(t)$.

5. Find the particular (forced) component of the response for $t > 0$ to the input $4e^{-3t}u(t)$ from a system whose transfer function is

$$H(p) = \frac{2(p+2)}{p+1}$$

6. Given the transfer function

$$H(p) = \frac{10p(p+1)}{p^2 + p + 25}$$

(a) What are the locations of all the critical frequencies? (b) What is the value of the undamped natural frequency? (c) What is the value of the damping factor ζ? (d) Write the expression for the natural response (do not evaluate unknown coefficients). (e) How would the answers to parts (a) to (d) change if the lone p were missing from the numerator?

7. When all damping is removed from a certain network containing two energy-storage elements, the natural response oscillates at a frequency $\omega = 4$ rad/s. When the damping element(s) are replaced, the natural-response envelope decays to half its initial value in 0.231 s. Find the pole locations of this network.

8. In a certain circuit any complex exponential input is related to the corresponding output by

$$H(s) = \frac{2(s+1)}{s^2 + s + 1}$$

Give the locations of *all* the poles and zeros of this system.

9. Find the transfer-function operator $H(s)$ in rational polynomial form of a system that has poles at $s = -5 \pm j12$ and a zero at $s = 0$ and has $H(-5) = 1$. Find the steady-state response if the system's input is $= 10 \cos (13t + 37°)$.

10. A 2-F capacitor C_1 is in series with the parallel combination of $C_2 = 1$ F and $L = 1$ H. Find the complex exponential input impedance $Z(s)$ of the overall circuit. Sketch and numerically label the pole-zero constellation of $Z(s)$.

11. Given

$$H(s) = \frac{s+3}{s(s+2)}$$

plot the pole-zero constellation. Label the locations of all finite poles and zeros. Evaluate $H(-1 + j)$. Show the graphical construction (label distances and angles).

12. Suppose a system's complex exponential transfer function has $K = 1$, poles at $s = -5 \pm j1000$, and zeros at $s = -5 \pm j990$. It is excited by a sinusoidal source whose frequency is 1000 rad/s. What is the approximate value of the transfer function? (Assume that the effects of the lower-half-plane pole and zero cancel each other.)

13. Given a series RLC circuit ($R = 10$ Ω, $L = 1$ H, and $C = 0.1$ F). The input is the applied current $i(t) = 10 \cos \omega t$. The output is $v_T(t)$, the overall voltage across the entire circuit. Sketch $|H(j\omega)|$ and $\underline{/H(j\omega)}$ versus ω. Evaluate these magnitude and angle plots at $\omega = 0$, 1, 10, 15, and at the value of ω for which $v_T(t)$ is minimum.

14. Given a linear circuit with poles at $s = -3 \pm j4$, zeros at $s = 0$ and -3, and $H(\infty) = 0.1$. (a) Find $H(s)$. (b) Find the particular response to the input $50e^{-6t}$.

15. For an $H(s)$ having poles at $s = -1 \pm j2$, no finite zeros, and $K = 1$, find the response due to the source $x(t) = 30e^{-4t} \cos 2t$.

16. Given the transfer function

$$H(s) = \frac{(s-1)(s^2 + 4)}{s+1}$$

what is the steady-state sinusoidal response to an input signal $= 2 \sin (t + 30°)$? Put your answer in the form $A \cos (\omega t + \theta)$.

Given the network shown in Figure P11.17. Consider the inductor current $i_L(t)$ to be the output and $v(t)$ to be the input. (a) Find the transfer function $H(s)$ in rational polynomial form. (b) Find the resonant frequency. (c) Plot and label the pole-zero constellation. What are the values of $H(0)$ and $H(\infty)$? (d) Sketch $|H|$ versus ω and $\underline{/H}$ versus ω. (Precise determination of maximum values and their locations is not necessary.)

18. In Figure P11.18 find the transfer function relating the output $v_2(t)$ to the general input $v_1(t) = V_1 e^{st}$. Plot and numerically label the pole-zero constellation.

19. A certain system variable is described by the differential equation

$$2\frac{d^2x}{dt^2} + 8\frac{dx}{dt} + 32x = 4\frac{dy}{dt} + y$$

(a) Find the form of the complete response to $y(t) = 8e^{-2t}u(t)$. (Do not evaluate coefficients

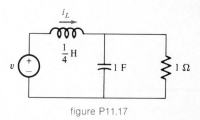

figure P11.17

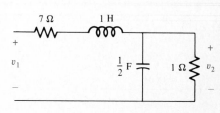

figure P11.18

in the natural-response component but *do* evaluate the particular component numerically.)
(b) Plot the pole-zero constellation. Sketch and label lengths and angles of all *s*-plane vectors needed to evaluate the particular response in part (a).

20. Given the transfer function

$$H(p) = \frac{4}{p^2 + 2p + 4}$$

(a) If the input is $x(t) = 10 \cos 20t$ and the output is $y(t)$, write the corresponding differential equation in d/dt notation. (b) Find ζ and ω_n and tell what type of damping will occur. (c) Find the form of the natural response. (d) Find the particular response and evaluate all its coefficients.

21. A series RC circuit ($R = 2\ \Omega$ and $C = \frac{1}{2}$ F) is driven by a voltage source $v_s(t)$. The output variable is the resulting voltage across the capacitor. (a) Find $H(s)$. Plot its pole-zero constellation. (b) Solve for the zero-state response to each of the following inputs using $H(s)$ as much as possible: (i) $v_s(t) = 10e^{-0.9t}u(t)$, (ii) $v_s(t) = 10e^{-t}u(t)$, (iii) $v_s(t) = 10e^{-1.1t}u(t)$. (c) Plot each of the responses in part (b) versus time for $0 < t < 2$ s. (d) In each case determine the time at which the response is maximum.

22. A parallel RLC circuit ($R = 1\ \Omega$, $L = 2$ H, $C = 1$ F) is driven by a current source $i_s(t) = 10e^{-t}u(t)$ A. (a) Determine the form of the natural response $v_n(t)$ of the voltage across the circuit. (b) Determine the particular response $v_p(t)$ for $t > 0$.

23. The unit-impulse response of a system having no finite zeros is given as $h(t) = 12e^{-t} \cos(t + 30°)u(t)$. Find the phase lag produced by this system if a pure sinusoidal input of frequency 1 rad/s is used.

24. A transfer function has poles at $s = -1 \pm j2$ and has no finite zeros (assume $K = 1$). (a) What is the form of the unit-impulse response $h(t)$? (b) If, in addition to the above, the system has a zero at $s = -0.5$, what will $h(t)$ be?

25. Find the unit-impulse response of the system whose transfer function is $H(s) = 3/(s + 3)$.

26. The driving-point impedance operator of a certain circuit is

$$Z(p) = \frac{p^2 + 1}{p^2 + p + 1}$$

(a) At what frequency will this system's natural response oscillate? Find the particular response to the following input functions: (b) $2 \cos t$, (c) $2 \cos 0.866t$, and (d) $u(t)$.

27. The voltage response of the source-free RLC circuit shown in Figure P11.27 for $t > 0$ is $v_C = 100e^{-300t} \cos 400t$. If the initial energy stored in the capacitor is 0.05 J, find R, L, and C and the initial inductor current $i_L(0+)$.

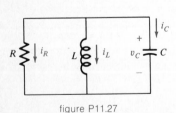

figure P11.27

28. A 3-Ω resistor is in series with a 0.1-H inductor and the combination is driven by a voltage source $v_s(t) = 10e^{-25t} \cos 12t\ u(t)$. Find the zero-state response $i(t)$.

29. For the circuit of Figure P11.29 (a) find the transfer function $H(s)$ that relates the two voltages v_1 and v_2. Plot its pole-zero constellation. (b) Find the complete response $v_2(t)$ to $v_1(t) = 10 \cos 5t\ u(t)$. (c) Find the input impedance, $Z_{in}(s)$. Where are its poles and zeros?

30. A 450-kg mass is supported by the parallel combination of a spring (compliance, $K = 27.43\ \mu\text{m/N}$) and a shock absorber (damper D). The other end (from the mass) of the spring and damper is driven by a velocity source $\mathcal{V}_s(t)$. (a) Find the transfer function $H(s)$ that relates the velocity of the mass to $\mathcal{V}_s$. (b) Find the value of D that will make

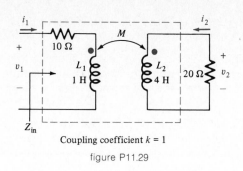

Coupling coefficient $k = 1$

figure P11.29

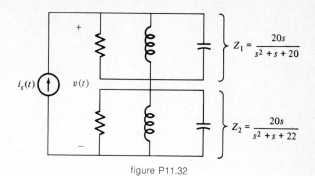

$$Z_1 = \frac{20s}{s^2 + s + 20}$$

$$Z_2 = \frac{20s}{s^2 + s + 22}$$

figure P11.32

$\zeta = 0.707$. What are the resulting pole and zero locations of $H(s)$? (c) Under the conditions achieved in part (b), find the form of the natural response. (d) If $\mathcal{V}_s(t) = \cos 6.36t$, find the resulting steady-state velocity of the mass.

31. A series circuit consisting of $R = 20\ \Omega$, $L = 1$ mH, and $C = 10\ \mu$F is driven by voltage source $v_s(t)$. (a) Find ζ for this circuit. What type of damping is this? (b) What additional resistance R_x must be placed in the circuit to make $\zeta = 0.5$? (Should your additional R_x go in series or parallel with the 20 Ω?) (c) With the circuit adjusted as in part (a), what will the time between zero crossings of the natural response be? (d) If all resistance could be removed from the circuit, leaving a pure LC circuit, what would be the frequency in hertz of the natural response? (e) If $v_s(t) = u(t)\sin 10^4 t$, what is the form of the particular response? (Do not evaluate constants, and again assume that all resistance is removed.)

32. In the circuit shown in Figure P11.32 the input is $i_s(t)$ and the output is the voltage $v(t)$. Plot the pole-zero constellation of the appropriate transfer function. Keeping in mind the rubber-sheet analogy, sketch the magnitude of the transfer function as a function of input frequency. Label the approximate locations of any peaks and/or dips in this plot.

33. Consider the input admittance of a series circuit with $R = 6\ \Omega$, $L = \frac{1}{2}$ H, and $C = \frac{1}{50}$ F. Plot the pole-zero constellation for $Y(s)$. What is the value of Q_0, the resonant quality factor?

34. Consider the input impedance of a parallel circuit with $R = 20$ kΩ, $L = 10$ mH, and $C = 0.25\ \mu$F. Plot the pole-zero constellation for $Z(s)$. What is the value of Q_0, the resonant quality factor?

35. An inductor ($L = 1$ mH) with series parasitic resistance R is to be used with a capacitor C in a series resonant circuit such that $\omega_r = 10^6$ rad/s. If the resulting Q_0 must be greater than 10, what is the restriction on the value of R?

36. Find the resonant frequency and the bandwidth, both in hertz, of a series circuit consisting of $R = 2\ \Omega$, $L = 1$ mH, and $C = 2\ \mu$F.

37. Given a series circuit with $R = \frac{1}{4}\ \Omega$, $L = \frac{1}{9}$ H, and $C = 1$ F, consider the input $Y(s)$, and find: (a) ζ and ω_n, (b) pole locations, (c) Q_0 and B.

38. Given a system for which $H(s) = s/(s^2 + s + 100)$. Find Q_0 and the bandwidth B.

39. Consider a series circuit with $R = 5\ \Omega$, $L = 0.2$ H, and $C = 2\ \mu$F. (a) Find the Q_0 and the bandwidth. (b) If the bandwidth is to be halved, keeping the same resonant frequency and the same impedance level, find the necessary new values of the elements.

40. A manufacturer guarantees the inherent series resistance of his 1-H inductors to be, at most, 10 Ω. If we series-resonate such an inductor (at 1000 rad/s), what will the resulting bandwidth be? What value of capacitance will accomplish this?

41. Given an unknown coil (series combination of pure L and pure R), we find that we get series resonance at 10 kHz with $C = 0.1$ μF. Find the new resonant frequency, the Q_0, and the resulting bandwidth if this coil is placed in series with a capacitor whose value is 0.001 μF?

42. The admittance of a series RLC circuit has two poles located at $s = -0.5 \pm j4$ and has a zero at the origin. Find (a) the undamped natural frequency, (b) the value of Q_0, (c) the bandwidth, and (d) the frequency of the maximum of the magnitude of $Y(j\omega)$.

43. An actual coil (L in series with parasitic resistance R_s) is placed in series with a pure capacitor whose value is variable. The combination is driven by a pure sinusoidal voltage source $v(t) = 15 \cos 200\pi t$, at which time $v_C = 40$ V (as read on a true rms voltmeter) and $C = 70$ pF. (a) What is the Q_0 of the coil? (b) What is the value of R_s?

44. When an inductive impedance $Z = R + jX$ is connected across a 100-V rms sinusoidal source, the delivered apparent power is 400 VA and the real power is 200 W. What is the series resonant quality factor Q_0 of the impedance at that frequency?

45. Sketch the Bode plot (both the log magnitude and the angle) of $H(s) = \frac{1}{2}(s + 2)/(s + 1)$. First use asymptotes and then sketch in the approximate actual curves. Label all important points with their numerical values in radians per second, decibels, and/or degrees.

46. Sketch the Bode plot (magnitude and phase) for the voltage transfer function of the circuit shown in Figure P11.46.

47. Sketch the Bode plot (magnitude and phase) for the transfer function

$$H(s) = \frac{100s(s + 100)}{(s + 10)(s + 1000)}$$

48. (a) Over what range of radian frequency is the magnitude-versus-frequency plot of the following transfer function greater than unity (log magnitude > 0)?

$$H(s) = \frac{16s^2}{(s + 1)(s^2 + 4s + 16)}$$

(b) Without making any numerical substitutions, estimate the maximum value of this magnitude and tell at what radian frequency it occurs.

49. Sketch the Bode plot (log magnitude and phase) for the V_2/V_1 transfer function of the circuit in Figure P11.49.

50. Repeat problem 49 for the circuit in Figure P11.50 but using $H(s) = V_2/I$.

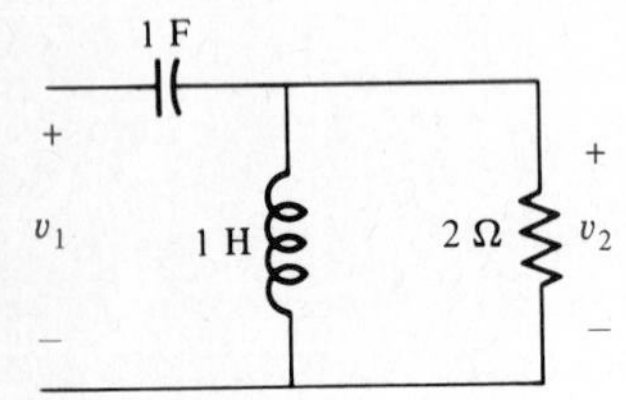

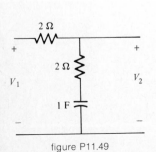

figure P11.46

figure P11.49

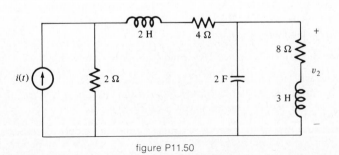

figure P11.50

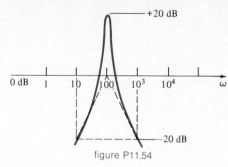

figure P11.54

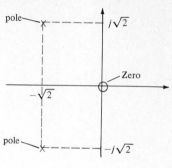

figure P11.55

51. Sketch the Bode plot (log magnitude and phase) for a system whose input is $i(t)$ and whose differential equation is

$$\frac{d^2v}{dt^2} + 110\frac{dv}{dt} + 1000v = 1000\frac{di}{dt}$$

52. A certain system has poles at $s = 10$ and $s = -10$; $H(0) = 1$. Sketch the Bode log-magnitude and phase plots for this system. Label all important values.

53. Given

$$H(s) = \frac{10^4 s}{(s + 1)(s^2 + 60s + 10^4)}$$

(a) Plot the magnitude of $H(j\omega)$ in decibels versus $\log \omega$. Label every important frequency, magnitude, and slope. (b) At what radian frequency does the overshoot occur? How many decibels is the overshoot? (c) Plot the angle of $H(j\omega)$ versus ω. Label the important points. (d) Plot the pole-zero constellation. Label all pole and zero locations.

54. Given the log-magnitude function plotted in Figure P11.54, find the corresponding transfer function $H(s)$. Sketch the accompanying phase angle plot and the pole-zero constellation.

55. When a certain transfer function $H(s)$ is arranged in the $s/a + 1$ form, the gain constant $K = 1$. The pole-zero plot of this $H(s)$ is shown in Figure P11.55. Sketch, as accurately as possible, the log-magnitude and phase functions that constitute the Bode plot.

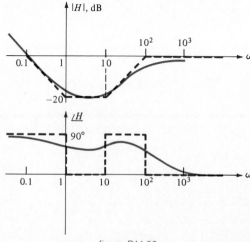

figure P11.56

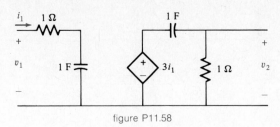

figure P11.58

56. Find the transfer function whose Bode plot is given in Figure P11.56.

57. A transfer function $H(s)$ has a Bode plot with asymptotes as follows:

Frequency	Log magnitude	Phase
$\omega < 1$	+20 dB/decade	$+90°$
$1 < \omega < 10$	Constant at 40 dB	$0°$
$\omega > 10$	-20 dB/decade	$-90°$

(a) Write an expression for $H(s)$. (b) Sketch and numerically label the pole-zero constellation of this system.

58. Find the voltage transfer function that relates v_1 and v_2 in the circuit of Figure P11.58. Sketch the Bode log magnitude and phase plots.

59. Repeat problem 58 for the circuit of Figure P11.59.

60. (a) Repeat problem 58 for the circuit shown in Figure P11.60. (b) Find the input frequency at which the output voltage $v_2(t)$ will be 180° out of phase with $v_1(t)$. (c) Sketch the pole-zero constellation.

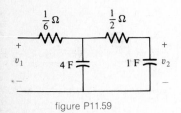

figure P11.59

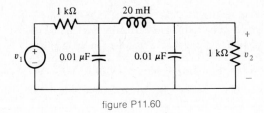

figure P11.60

chapter 12
TWO-PORT NETWORKS

introduction 12.1

Many electrical and mechanical systems have the form of Figure 12.1.1. This is called a *two-port* or *four-terminal network*. Examples of such systems are mechanical levers, gears, electric transformers, hi-fi amplifiers, voltage divider networks, and electric filter networks.

There are obviously four variables of interest: e_1, e_2, f_1, and f_2. Although we might already have in mind, when we look at any given two-port, which variable(s) we will make the input(s) and which will be output(s), it is not at all necessary or desirable to specify them in advance in order to describe the network via the following techniques. That is why we define both through variables as positive inward flowing. If, when the circuit is put into use, one of those currents turns out to be negative—fine!

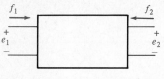

figure 12.1.1
A two-port (four-terminal) network.

Let us assume there are no independent sources in the network and that all the elements in the network are linear. For convenience we use the electrical units V_1, V_2, I_1, and I_2 in place of the more general across and through variables e_1, e_2, f_1, and f_2. It should be clear, however, that what follows is valid for *any* linear two-port (not only electric ones).

Because the systems that we will consider are linear, it follows that the relationships among the four terminal quantities must be linear. We can write equations for any two of these variables in terms of the other two. So we might write

$$V_1 = f_a(I_1, I_2)$$

$$V_2 = f_b(I_1, I_2)$$

where f_a and f_b are linear equations in the two variables I_1 and I_2. Or we could write

$$I_1 = f_c(V_1, V_2)$$

$$I_2 = f_d(V_1, V_2)$$

or

$$V_1 = f_e(I_1, V_2)$$

$$I_2 = f_f(I_1, V_2)$$

There are six ways (combinations) of choosing two variables at a time from a total of four variables, assuming that the order in which we write the two equations is not important. Therefore, we have six possible ways of describing a two-port network by means of writing equations for two of its terminal quantities in terms of the other two.

Since all these equations are linear, we can use matrices to write them. Also, in all the discussions in this chapter we will assume that all initial conditions are zero and all input functions are complex exponentials so that each element may be described by its complex exponential immitances, for example, $Z_R(s) = R$, $Z_L(s) = Ls$, $Z_C(s) = 1/Cs$, $Y_R(s) = 1/R$, $Y_L(s) = 1/Ls$, and $Y_C(s) = Cs$.

12.2   open-circuit impedance parameters $[z]$

One way that we can describe the interrelationships among the four terminal variables of any two-port is to write expressions for both of the voltages in terms of the currents:

$$V_1 = z_{11}I_1 + z_{12}I_2 \tag{12.2.1}$$

$$V_2 = z_{21}I_1 + z_{22}I_2 \tag{12.2.2}$$

or

$$[V] = [z][I] \tag{12.2.3}$$

where

$$[V] = \begin{bmatrix} V_1 \\ V_2 \end{bmatrix} \qquad [I] = \begin{bmatrix} I_1 \\ I_2 \end{bmatrix}$$

and
$$[z] = \begin{bmatrix} z_{11} & z_{12} \\ z_{21} & z_{22} \end{bmatrix}$$

Each of the z quantities has the units of impedance (is a function of s, where s is determined by the excitation function waveform). From (12.2.1) and (12.2.2):

$$z_{11} = \left. \frac{V_1}{I_1} \right|_{I_2=0} \qquad z_{12} = \left. \frac{V_1}{I_2} \right|_{I_1=0}$$

$$z_{21} = \left. \frac{V_2}{I_1} \right|_{I_2=0} \qquad z_{22} = \left. \frac{V_2}{I_2} \right|_{I_1=0}$$

(12.2.4)

Since each of these parameters is defined when some current or other is zero, they are called the *open-circuit impedance parameters* (for example, z_{21} equals the ratio of V_2 to I_1 when I_2 is zero—i.e., when port 2 is *open-circuited*).

EXAMPLE 12.2.1

Find the z matrix for the two-port in Figure 12.2.1a.

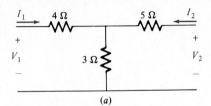

(a)

figure 12.2.1
(a) The two-port of Example 12.2.1. (b) Finding z_{12}. (c) Finding z_{21}.

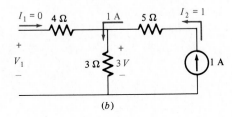

(b)

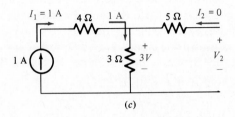

(c)

ANS.:
$$z_{11} = \left. \frac{V_1}{I_1} \right|_{I_2=0} = 4 + 3 = 7 \ \Omega \qquad \text{and} \qquad z_{22} = \left. \frac{V_2}{I_2} \right|_{I_1=0} = 5 + 3 = 8 \ \Omega$$

by inspection. But z_{12} and z_{21} are less obvious:

$$z_{12} = \left.\frac{V_1}{I_2}\right|_{I_1=0}$$

Let $I_2 = 1$ A and solve for V_1.

Since $I_1 = 0$, there is no voltage across the 4-Ω resistor and therefore $V_1 = 3$ V. So

$$z_{12} = \left.\frac{V_1}{I_2}\right|_{I_1=0} = \frac{3}{1} = 3 \ \Omega$$

This is a fairly standard and reliable way to find any of these parameters; i.e., insert the denominator quantity with unity amplitude. Solve for the numerator. This is the numerical value of the parameter. So to find z_{21} we insert $I_1 = 1$ A as in Figure 12.2.1c and solve for V_2.

$$z_{21} = \left.\frac{V_2}{I_1}\right|_{I_2=0} = 3 \ \Omega \qquad \text{so} \qquad [z] = \begin{bmatrix} 7 & 3 \\ 3 & 8 \end{bmatrix}$$

EXAMPLE 12.2.2

Find the z matrix for the two-port in Figure 12.2.2.

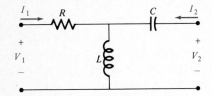

figure 12.2.2
The circuit of Example 12.2.2.

ANS.:   By applying the definitions of the z parameters, Equation 12.2.4, we find

$$z = \begin{bmatrix} R + Ls & Ls \\ Ls & \dfrac{1}{Cs} + Ls \end{bmatrix}$$

Notice that in both Examples 12.2.1 and 12.2.2 the value of z_{12} was equal to the value of z_{21}. Networks in which this is the case are called *reciprocal networks*. All two-ports made up of only linear, passive, *RLC* elements (or their analogs) are reciprocal. Very often, circuits with active elements (dependent sources) are nonreciprocal, i.e., circuits containing transistors, diodes, etc.

The reciprocal T network is a particularly easy one for which to obtain the z parameters. Consider the general T network in Figure 12.2.3a.

Because of the definition of z_{12} we see that

$$z_{12} = \left.\frac{V_1}{I_2}\right|_{I_1=0} = Z_B \tag{12.2.5}$$

or

$$Z_B = z_{12} \tag{12.2.6}$$

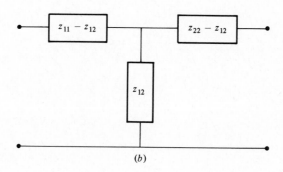

figure 12.2.3
Two equivalent ways of labeling a reciprocal
T-network.

And so we label it thus in Figure 12.2.3b.

Since z_{11} is the input impedance of the network as seen from the left port (while the right port is open-circuited).

$$z_{11} = Z_A + Z_B \qquad (12.2.7)$$

Using (12.2.6) in (12.2.7):

$$Z_A = z_{11} - z_{12} \qquad (12.2.8)$$

Similarly, we can write

$$z_{22} = Z_D + Z_B \qquad (12.2.9)$$

and substituting (12.2.6) into (12.2.9):

$$Z_D = z_{22} - z_{12} \qquad (12.2.10)$$

Figure 12.2.3b shows the results of equations (12.2.6), (12.2.8), and (12.2.10).

EXAMPLE 12.2.3

Synthesize a T network, the z matrix of which is defined by the parameters

$$z_{11}(s) = \frac{8s^2 + 10s + 1}{2s} \qquad z_{12}(s) = z_{21}(s) = 4s + 5 \qquad z_{22}(s) = \frac{12s^2 + 15s + 1}{3s}$$

ANS.:   Referring to Figure 12.2.3, we can immediately write

$$Z_B = z_{12}(s)$$

$$= 4s + 5$$

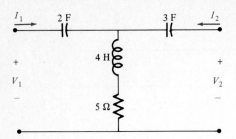

figure 12.2.4
The solution to Example 12.2.3.

Then using equation (12.2.8),

$$Z_A = z_{11} - z_{12}$$

$$= \frac{8s^2 + 10s + 1}{2s} - (4s + 5)$$

$$= \frac{8s^2 + 10s + 1 - 8s^2 - 10s}{2s}$$

$$= \frac{1}{2s}$$

Similarly, via equation 12.2.10,

$$Z_D = z_{22} - z_{12}$$

$$= \frac{12s^2 + 15s + 1}{3s} - (4s + 5)$$

$$= \frac{1}{3s}$$

The circuit is shown in Figure 12.2.4.

A two-port that has $z_{11} = z_{22}$ is called *symmetrical*. For example, the network in Figure 12.2.5 is symmetrical (as well as reciprocal). The network in Figure 12.2.4 is reciprocal but not symmetrical.

The z parameters of any passive linear two-port may be obtained by writing node equations for that network. Figure 12.2.6 shows a general two-port network. If current sources I_1 and I_2 are applied, the node equations have the

figure 12.2.5
A symmetrical two-port: $z_{11} = z_{22}$.

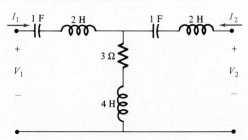

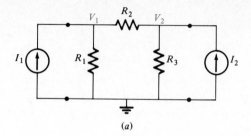

figure 12.2.6
(a) The two-port of Example 12.2.4. (b)
Finding z_{12} directly.

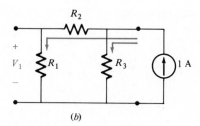

(b)

form:

$$b_{11}V_1 + b_{12}V_2 + b_{13}V_3 + \cdots + b_{1n}V_n = I_1$$

$$b_{21}V_1 + b_{22}V_2 + b_{23}V_3 + \cdots + b_{2n}V_n = I_2$$

$$b_{31}V_1 + b_{32}V_2 + b_{33}V_3 + \cdots + b_{3n}V_n = 0 \qquad (12.2.11)$$

$$\cdots\cdots\cdots\cdots\cdots\cdots\cdots\cdots\cdots\cdots\cdots\cdots\cdots$$

$$b_{n1}V_1 + b_{n2}V_2 + b_{n3}V_3 + \cdots + b_{nn}V_n = 0$$

where the b_{ij} are, in general, complex admittances $b_{ij}(s)$. Note that in every equation, other than the first two, the right-hand side is zero. (I_1 and I_2 deliver current to nodes 1 and 2, respectively, and the remainder of the network is passive.) Solving† for the inverse set of equations, we get:

$$\frac{\Delta_{11}}{\Delta}I_1 + \frac{\Delta_{21}}{\Delta}I_2 \left\lvert\, + \frac{\Delta_{31}}{\Delta}(0) + \frac{\Delta_{41}}{\Delta}(0) + \cdots + 0 \right\rvert = V_1$$

$$\frac{\Delta_{12}}{\Delta}I_1 + \frac{\Delta_{22}}{\Delta}I_2 \left\lvert\, + \frac{\Delta_{32}}{\Delta}(0) + \frac{\Delta_{42}}{\Delta}(0) + \cdots + 0 \right\rvert = V_2$$

$$\cdots\cdots\cdots\cdots\cdots \left\lvert\cdots\cdots\cdots\cdots\cdots\cdots\cdots\cdots\cdots\right\rvert = V_3 \qquad (12.2.12)$$

$$\frac{\Delta_{1n}}{\Delta}I_1 + \frac{\Delta_{2n}}{\Delta}I_2 \left\lvert\, + \frac{\Delta_{3n}}{\Delta}(0) + \frac{\Delta_{4n}}{\Delta}(0) + \cdots + 0 \right\rvert = V_n$$

Terms all $= 0$

From equations (12.2.4) and (12.2.12) we see that:

† Via matrix inversion.

$$z_{11} = \frac{\Delta_{11}}{\Delta} \qquad z_{12} = \frac{\Delta_{21}}{\Delta}$$

$$z_{21} = \frac{\Delta_{12}}{\Delta} \qquad z_{22} = \frac{\Delta_{22}}{\Delta}$$

(12.2.13)

where the determinants and cofactors in (12.2.13) are from the *coefficient matrix* in the *node equations* [equation (12.2.11)].

EXAMPLE 12.2.4

Write the node equations for the circuit in Figure 12.2.6a. Use the coefficient matrix of this pair of equations to find the z parameters of this network.

ANS.: The node equations are

$$V_1\left(\frac{1}{R_1} + \frac{1}{R_2}\right) - V_2\left(\frac{1}{R_2}\right) = I_1$$

$$-V_1\left(\frac{1}{R_2}\right) + V_2\left(\frac{1}{R_2} + \frac{1}{R_3}\right) = I_2$$

or

$$V_1\left(\frac{R_2 + R_1}{R_1 R_2}\right) + V_2\left(\frac{-1}{R_2}\right) = I_1$$

$$V_1\left(\frac{-1}{R_2}\right) + V_2\left(\frac{R_2 + R_3}{R_2 R_3}\right) = I_2$$

The determinant of the coefficient matrix of these equations is

$$\Delta = \frac{(R_1 + R_2)(R_2 + R_3)}{R_1 R_2^2 R_3} - \frac{1}{R_2^2}$$

$$= \frac{R_1 R_2 + R_2^2 + R_1 R_3 + R_2 R_3 - R_1 R_3}{R_1 R_2^2 R_3}$$

$$= \frac{R_1 + R_2 + R_3}{R_1 R_2 R_3}$$

Then from equation (12.2.13)

$$z_{11} = \frac{\Delta_{11}}{\Delta} = \frac{R_2 + R_3}{R_2 R_3} \frac{R_1 R_2 R_3}{R_1 + R_2 + R_3}$$

= parallel combination of R_1 and $R_2 + R_3$ (Correct by inspection)

and

$$z_{22} = \frac{R_2 + R_1}{R_1 R_2} \frac{R_1 R_2 R_3}{R_1 + R_2 + R_3} = R_3 \text{ in parallel with } R_1 + R_2 \quad \text{(Correct by inspection)}$$

and

$$z_{12} = \frac{1}{R_2} \frac{R_1 R_2 R_3}{R_1 + R_2 + R_3}$$

$$= \frac{R_1 R_3}{R_1 + R_2 + R_3}$$

This last value is not so obviously correct. Let us check this result by finding z_{12} from the definition—see

Figure 12.2.6b. Inserting $I_2 = 1$ A and solving for V_1, we find

$$z_{12} = \frac{V_1}{I_2}\bigg|_{I_1 = 0}$$

$$= R_1 \times (\text{current in } R_1)$$

By current division

$$z_{12} = R_1 \left(\frac{R_3}{R_1 + R_2 + R_3} \right) (1) \qquad \text{QED}$$

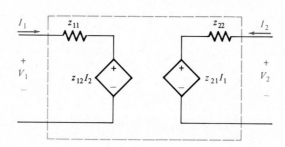

figure 12.2.7
The z-parameter equivalent circuit of a two-port network.

The z parameters of a two-port can tell us much about that network. For example, consider the ratio z_{21}/z_{11}. Since

$$z_{21} = \frac{V_2}{I_1}\bigg|_{I_2 = 0}$$

and

$$z_{11} = \frac{V_1}{I_1}\bigg|_{I_2 = 0}$$

We see that

$$\frac{z_{21}}{z_{11}} = \frac{V_2}{V_1}\bigg|_{I_2 = 0} = \text{open-circuit voltage gain} \qquad (12.2.14)$$

Similarly,

$$\frac{z_{12}}{z_{22}} = \frac{V_1/I_2}{V_2/I_2}\bigg|_{I_1 = 0} = \frac{V_1}{V_2}\bigg|_{I_1 = 0} = \text{reverse open-circuit voltage gain} \quad (12.2.15)$$

In addition, the defining equations, (12.2.1) and (12.2.2), suggest the equivalent circuit shown in Figure 12.2.7.

short-circuit admittance parameters $[y]$ **12.3**

It is possible to express the interrelationships among the four terminal quantities of any two-port by solving for the currents in terms of the voltages:

$$I_1 = y_{11}V_1 + y_{12}V_2 \qquad (12.3.1)$$

$$I_2 = y_{21}V_1 + y_{22}V_2 \qquad (12.3.2)$$

or
$$[I] = [y][V] \qquad (12.3.3)$$

where

$$[I] = \begin{bmatrix} I_1 \\ I_2 \end{bmatrix} \qquad [V] = \begin{bmatrix} V_1 \\ V_2 \end{bmatrix}$$

and
$$[y] = \begin{bmatrix} y_{11} & y_{12} \\ y_{21} & y_{22} \end{bmatrix}$$

So, from (12.3.1) and (12.3.2),

$$
y_{11} = \frac{I_1}{V_1}\bigg|_{V_2=0} \qquad y_{12} = \frac{I_1}{V_2}\bigg|_{V_1=0}
$$
$$
y_{21} = \frac{I_2}{V_1}\bigg|_{V_2=0} \qquad y_{22} = \frac{I_2}{V_2}\bigg|_{V_1=0}
$$
$$(12.3.4)$$

Since each of these parameters is defined when some voltage or other is set equal to zero, the y's are called the short-circuit admittance parameters.

From equation (12.2.3) in Section 12.2

$$[V] = [z][I]$$

and from equation (12.3.3)

$$[I] = [y][V]$$

Multiplying this latter equation by $[z]$ yields

$$[z][I] = [z][y][V]$$

which, on the left, is $[V]$, thus

$$[z][y]$$

is the identity matrix. Therefore,

$$[z] = [y]^{-1} \qquad \text{and} \qquad [y] = [z]^{-1}$$

Performing these inverse operations yields

$$
z_{11} = \frac{y_{22}}{\Delta_y} \qquad z_{12} = \frac{-y_{12}}{\Delta_y}
$$

$$
z_{21} = \frac{-y_{21}}{\Delta_y} \qquad z_{22} = \frac{y_{11}}{\Delta_y}
$$

and

$$
y_{11} = \frac{z_{22}}{\Delta_z} \qquad y_{12} = \frac{-z_{12}}{\Delta_z}
$$

$$
y_{21} = \frac{-z_{21}}{\Delta_z} \qquad y_{22} = \frac{z_{11}}{\Delta_z}
$$

EXAMPLE 12.3.1

Find the y matrix for the two-port in Figure 12.3.1, then take its inverse to find the z matrix.

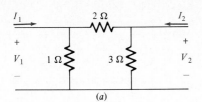

(a)

figure 12.3.1
(a) The two-port of Example 12.3.1. (b) Finding y_{11}. (c) Finding y_{22}. (d) Finding y_{12}.
(e) Finding y_{21}.

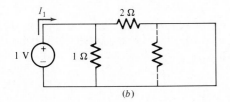

(b)

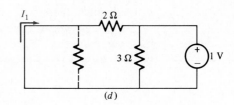

(d)

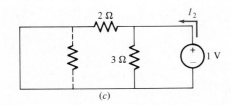

(c)

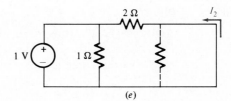

(e)

ANS.:

$$y_{11} = \frac{I_1}{V_1}\bigg|_{V_2 = 0}$$

Insert $V_1 = 1$ and solve for I_1. See Figure 12.3.1b.

$$I_1 = \frac{1}{2/3} = \frac{3}{2} = y_{11}$$

$$y_{22} = \frac{I_2}{V_2}\bigg|_{V_1 = 0}$$

Set $V_2 = 1$ and find I_2. See Figure 12.3.1c.

$$I_2 = \frac{1}{6/5} = \frac{5}{6} = y_{22}$$

$$y_{12} = \frac{I_1}{V_2}\bigg|_{V_1 = 0}$$

$V_2 = 1$, find I_1. See Figure 12.3.1d.

$$y_{12} = -\tfrac{1}{2}$$

$$y_{21} = \frac{I_2}{V_1}\bigg|_{V_2 = 0}$$

$V_1 = 1$, find I_2. See Figure 12.3.1e

$$I_2 = -\tfrac{1}{2} = y_{21}$$

Thus $\quad [y] = \begin{bmatrix} \tfrac{3}{2} & -\tfrac{1}{2} \\ -\tfrac{1}{2} & \tfrac{5}{6} \end{bmatrix} \quad$ and $\quad [z] = [y]^{-1} = \begin{bmatrix} \tfrac{5}{6} & \tfrac{1}{2} \\ \tfrac{1}{2} & \tfrac{3}{2} \end{bmatrix}$

The negative sign in the y matrix of the previous problem arises simply because we defined both currents inward. Remember that neither y_{12} nor y_{21} is a driving-point admittance—the input admittance of a two-terminal (one-port) device. Rather, they are ratios of a current in one place to a voltage in another place. As such, there is no reason they cannot be negative.

The pi network is a particularly convenient network when it comes to finding its y parameters. Consider the general pi network of Figure 12.3.2a. Because of the definition of y_{12} we see that

$$y_{12} = \frac{I_1}{V_2}\bigg|_{V_1 = 0} \tag{12.3.5}$$

or $\qquad\qquad y_{12} = -Y_B \tag{12.3.6}$

And so we label it thus in Figure 12.3.2b

Since y_{11} is the input admittance of the network as seen from the left port (while the right port is short-circuited),

figure 12.3.2
Two equivalent ways of labeling a general reciprocal pi network.

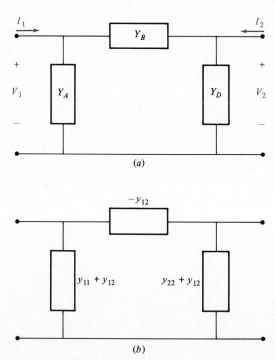

(a)

(b)

$$y_{11} = Y_A + Y_B \tag{12.3.7}$$

Using (12.3.6) in (12.3.7), we find

$$Y_A = y_{11} + y_{12} \tag{12.3.8}$$

Similarly we can write

$$y_{22} = Y_B + Y_D \tag{12.3.9}$$

and substituting (12.3.6) into (12.3.9), we find

$$Y_D = y_{22} + y_{12} \tag{12.3.10}$$

Figure 12.3.2*b* shows the results of equations (12.3.6), (12.3.8), and (12.3.10).

The *y* parameters of any passive, linear two-port may be obtained by writing mesh equations for that network. Figure 12.3.3 shows a typical two-port network. If voltage sources V_1 and V_2 are applied to any two-port, the mesh equations have the following form. (Allow I_1 and I_2 to define the first two mesh currents.)

$$
\begin{aligned}
a_{11}I_1 + a_{12}I_2 + a_{13}I_3 + \cdots + a_{1n}I_n &= V_1 \\
a_{21}I_1 + a_{22}I_2 + a_{23}I_3 + \cdots + a_{2n}I_n &= V_2 \\
a_{31}I_1 + a_{32}I_2 + a_{33}I_3 + \cdots + a_{3n}I_n &= 0 \\
\cdots\cdots\cdots\cdots\cdots\cdots\cdots\cdots\cdots & \\
a_{n1}I_1 + a_{n2}I_2 + a_{n3}I_3 + \cdots + a_{nn}I_n &= 0
\end{aligned}
\tag{12.3.11}
$$

where the a_{ij} are, in general, complex impedances $a_{ij}(s)$. Note that in every equation, other than the first two, the right-hand side is zero. (V_1 and V_2 supply voltage to the first two meshes and the remainder of the network is passive.) Solving for the inverse set of equations we get

$$
\begin{aligned}
\frac{\Delta_{11}}{\Delta} V_1 + \frac{\Delta_{21}}{\Delta} V_2 + \left[\frac{\Delta_{31}}{\Delta}(0) + \frac{\Delta_{41}}{\Delta}(0) + \cdots + 0 \right] &= I_1 \\
\frac{\Delta_{12}}{\Delta} V_1 + \frac{\Delta_{22}}{\Delta} V_2 + \left[\frac{\Delta_{32}}{\Delta}(0) + \frac{\Delta_{42}}{\Delta}(0) + \cdots + 0 \right] &= I_2 \\
\cdots\cdots\cdots\cdots\cdots\cdots\cdots\cdots\cdots\cdots\cdots\cdots &= I_3 \\
\frac{\Delta_{1n}}{\Delta} V_1 + \frac{\Delta_{2n}}{\Delta} V_2 + \left[\frac{\Delta_{3n}}{\Delta}(0) + \frac{\Delta_{4n}}{\Delta}(0) + \cdots + 0 \right] &= I_n
\end{aligned}
\tag{12.3.12}
$$

Terms all = 0

From equations (12.3.4) and (12.3.12) we see that

$$y_{11} = \frac{\Delta_{11}}{\Delta} \qquad y_{12} = \frac{\Delta_{21}}{\Delta}$$

$$y_{21} = \frac{\Delta_{12}}{\Delta} \qquad y_{22} = \frac{\Delta_{22}}{\Delta} \tag{12.3.13}$$

where the determinants and cofactors in (12.3.13) are from the *coefficient matrix* in the *mesh equations* [equations (12.3.11)].

EXAMPLE 12.3.2

Write the mesh equations for the two-port in Figure 12.3.3. Choose I_1 and I_2 as two of the mesh currents. Use the coefficient matrix of this set of equations to obtain the y parameters of the network.

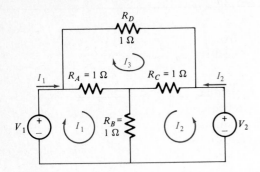

figure 12.3.3
The two-port of Example 12.3.2.

ANS.: The mesh equations are

$$2I_1 + I_2 - I_3 = V_1$$
$$I_1 + 2I_2 + I_3 = V_2$$
$$-I_1 + I_2 + 3I_3 = 0$$

The system determinant is

$$\Delta = \begin{vmatrix} 2 & 1 & -1 \\ 1 & 2 & 1 \\ -1 & 1 & 3 \end{vmatrix}$$

Adding the second row to the third, we have

$$\Delta = \begin{vmatrix} 2 & 1 & -1 \\ 1 & 2 & 1 \\ 0 & 3 & 4 \end{vmatrix}$$

Multiplying the second row by -2 and adding it to row 1 gives

$$\Delta = \begin{vmatrix} 0 & -3 & -3 \\ 1 & 2 & 1 \\ 0 & 3 & 4 \end{vmatrix} = (-1)\begin{vmatrix} -3 & -3 \\ 3 & 4 \end{vmatrix} = (-1)(-12 + 9) = 3$$

Then by equations (12.3.13)

$$y_{11} = \frac{\Delta_{11}}{\Delta} = \frac{5}{3} \qquad y_{12} = \frac{\Delta_{21}}{\Delta} = -\frac{4}{3}$$

$$y_{21} = \frac{\Delta_{12}}{\Delta} = -\frac{4}{3} \qquad y_{22} = \frac{\Delta_{22}}{\Delta} = \frac{5}{3}$$

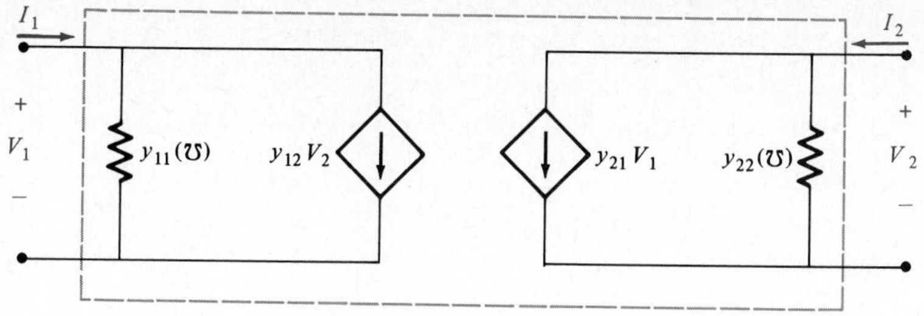

figure 12.3.4
The y-parameter equivalent circuit of a two-port network.

The y parameters of a two-port can tell us much about that network. For example, consider the ratio $-y_{21}/y_{11}$:

$$\frac{-y_{21}}{y_{11}} = \left.\frac{-I_2/V_1}{I_1/V_1}\right|_{V_2=0} = \left.\frac{-I_2}{I_1}\right|_{V_2=0}$$

$$= \text{short-circuit current gain} \qquad (12.3.14)$$

Also

$$\frac{-y_{12}}{y_{22}} = \left.\frac{-I_1/V_2}{I_2/V_2}\right|_{V_1=0} = \left.\frac{-I_1}{I_2}\right|_{V_1=0}$$

$$= \text{reverse short-circuit current gain} \qquad (12.3.15)$$

In addition, the defining equations, (12.3.1) and (12.3.2), suggest the general equivalent circuit shown in Figure 12.3.4.

hybrid parameters [h] and [g]   12.4

Solving for V_1 and I_2 in terms of I_1 and V_2 yields another matrix called [h], the *hybrid* parameters.

$$V_1 = h_{11}I_1 + h_{12}V_2 \qquad (12.4.1)$$

$$I_2 = h_{21}I_1 + h_{22}V_2 \qquad (12.4.2)$$

so

$$h_{11} = \left.\frac{V_1}{I_1}\right|_{V_2=0} \qquad h_{12} = \left.\frac{V_1}{V_2}\right|_{I_1=0}$$

$$\qquad (12.4.3)$$

$$h_{21} = \left.\frac{I_2}{I_1}\right|_{V_2=0} \qquad h_{22} = \left.\frac{I_2}{V_2}\right|_{I_1=0}$$

or in matrix form,

$$\begin{bmatrix} V_1 \\ I_2 \end{bmatrix} = [h]\begin{bmatrix} I_1 \\ V_2 \end{bmatrix} \qquad (12.4.4)$$

We notice from the definitions of these h parameters [equation (12.4.3)] that the h parameters may be obtained in terms of the z's and y's:

$$h_{12} = \left.\frac{V_1}{V_2}\right|_{I_1=0} = \left.\frac{V_1/I_2}{V_2/I_2}\right|_{I_1=0} = \frac{z_{12}}{z_{22}} \tag{12.4.5}$$

and, since $z_{12} = -y_{12}/\Delta_y$ and $z_{22} = y_{11}/\Delta_y$, equation (12.4.5) also tells us that

$$h_{12} = -\frac{y_{12}}{y_{11}} \tag{12.4.6}$$

Similarly,

$$h_{22} = \left.\frac{I_2}{V_2}\right|_{I_1=0} = \frac{1}{z_{22}} = \frac{\Delta_y}{y_{11}} \tag{12.4.7}$$

$$h_{11} = \left.\frac{V_1}{I_1}\right|_{V_2=0} = \frac{1}{y_{11}} = \frac{\Delta_z}{z_{22}} \tag{12.4.8}$$

and

$$h_{21} = \left.\frac{I_2}{I_1}\right|_{V_2=0} = \left.\frac{I_2/V_1}{I_1/V_1}\right|_{V_2=0} = \frac{y_{21}}{y_{11}} = -\frac{z_{21}}{z_{22}} \tag{12.4.9}$$

Also, from equations (12.4.5) through (12.4.9),

$$\Delta_h = h_{11}h_{22} - h_{12}h_{21} \tag{12.4.10}$$

$$= \frac{\Delta_z - z_{12}(-z_{21})}{z_{22}^2}$$

$$= \frac{z_{11}z_{22} - z_{12}z_{21} + z_{12}z_{21}}{z_{22}^2}$$

$$\Delta_h = \frac{z_{11}}{z_{22}} \tag{12.4.11}$$

Similarly, using the y matrix parameters from equations (12.4.5) to (12.4.9) in equation (12.4.10) gives

$$\Delta_h = \frac{(1)\Delta_y - (-y_{12})(y_{21})}{y_{11}^2} \tag{12.4.12}$$

$$\Delta_h = \frac{y_{22}}{y_{11}} \tag{12.4.13}$$

Notice from equations (12.4.6) and (12.4.9) that, if the two-port has the *reciprocity* property ($y_{12} = y_{21}$), then

$$h_{12} = -h_{21} \tag{12.4.14}$$

If the two-port is symmetrical ($y_{11} = y_{22}$), then, from equation (12.4.13),

$$\Delta_h = 1 \tag{12.4.15}$$

Note the definitions of the various h parameters:

- h_{11} is a short-circuit input impedance.
- h_{12} is the reverse voltage gain with port 1 open-circuited.
- h_{21} is the short-circuit current gain.
- h_{22} is the admittance seen looking back into port 2 with port 1 open-circuited.

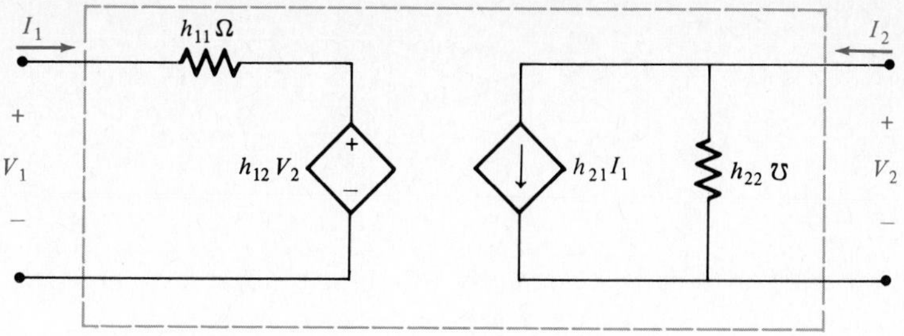

figure 12.4.1
The hybrid equivalent circuit for any two-port network.

Thus a possible equivalent circuit is shown in Figure 12.4.1

It is very convenient when working with transistor design problems to use the h parameters to describe the action of circuits. Electronics engineers use slightly different nomenclature for the h's:

$$h_{11} = h_i \qquad (input \text{ impedance})$$

$$h_{12} = h_r \qquad (r = reverse)$$

$$h_{21} = h_f \qquad (f = forward)$$

$$h_{22} = h_o \qquad (output \text{ admittance})$$

The inverse of equation (12.4.4) gives rise to the g matrix. That is,

$$\begin{bmatrix} I_1 \\ V_2 \end{bmatrix} = [h]^{-1} \begin{bmatrix} V_1 \\ I_2 \end{bmatrix} = [g] \begin{bmatrix} V_1 \\ I_2 \end{bmatrix} \tag{12.4.16}$$

or

$$I_1 = g_{11} V_1 + g_{12} I_2$$
$$V_2 = g_{21} V_1 + g_{22} I_2 \tag{12.4.17}$$

where

$$g_{11} = \frac{I_1}{V_1}\bigg|_{I_2=0} \qquad g_{12} = \frac{I_1}{I_2}\bigg|_{V_1=0}$$

$$g_{21} = \frac{V_2}{V_1}\bigg|_{I_2=0} \qquad g_{22} = \frac{V_2}{I_2}\bigg|_{V_1=0} \tag{12.4.18}$$

Since g is the inverse of the h matrix,

$$[h]^{-1} = [g] = \begin{bmatrix} \dfrac{h_{22}}{\Delta_h} & \dfrac{-h_{12}}{\Delta_h} \\ \dfrac{-h_{21}}{\Delta_h} & \dfrac{h_{11}}{\Delta_h} \end{bmatrix} \tag{12.4.19}$$

where $\Delta_h = h_{11} h_{22} - h_{12} h_{21}$. And, of course, from the left side of (12.4.19),

$$[h] = [g]^{-1} \tag{12.4.20}$$

so that

$$[h] = \begin{bmatrix} \dfrac{g_{22}}{\Delta_g} & \dfrac{-g_{12}}{\Delta_g} \\[2mm] \dfrac{-g_{21}}{\Delta_g} & \dfrac{g_{11}}{\Delta_g} \end{bmatrix} \qquad (12.4.21)$$

where

$$\Delta_g = g_{11}g_{22} - g_{21}g_{12}$$

which from equation (12.4.19) is

$$\Delta_g = \frac{h_{22}h_{11} - h_{12}h_{21}}{\Delta_h^2} = \frac{1}{\Delta_h} \qquad (12.4.22)$$

If the two-port is *reciprocal*, then, from the right side of equation (12.4.19), we note that

$$g_{12} = -g_{21} \qquad (12.4.23)$$

If the network is *symmetrical*, then, from equations (12.4.15) and (12.4.22),

$$\Delta_g = 1 \qquad (12.4.24)$$

12.5 the transmission parameters $\begin{bmatrix} a_{11} & a_{12} \\ a_{21} & a_{22} \end{bmatrix}$

One remaining way to write two of the terminal variables of a linear two-port in terms of the others is

$$V_1 = f_1(V_2, I_2) \qquad (12.5.1)$$

$$I_1 = f_2(V_2, I_2) \qquad (12.5.2)$$

For various reasons, it was decided to write these relationships as

$$V_1 = a_{11}V_2 - a_{12}I_2 \qquad (12.5.3)$$

$$I_1 = a_{21}V_2 - a_{22}I_2 \qquad (12.5.4)$$

$$\begin{bmatrix} V_1 \\ I_1 \end{bmatrix} = \begin{bmatrix} a_{11} & a_{12} \\ a_{21} & a_{22} \end{bmatrix} \begin{bmatrix} V_2 \\ -I_2 \end{bmatrix} \qquad (12.5.5)$$

The minus sign causes us minor annoyance but we will cope with it. We will call the coefficient matrix in (12.5.5) the *a* matrix. From (12.5.3) and (12.5.4) we have the definitions

$$\frac{1}{a_{11}} = \frac{V_2}{V_1}\bigg|_{I_2=0} = \text{open-circuit voltage gain}$$

$$-\frac{1}{a_{12}} = \frac{I_2}{V_1}\bigg|_{V_2=0} = \text{short-circuit transfer admittance} \qquad (12.5.6)$$

$$\frac{1}{a_{21}} = \frac{V_2}{I_1}\bigg|_{I_2=0} = \text{open-circuit transfer impedance}$$

$$-\frac{1}{a_{22}} = \frac{I_2}{I_1}\bigg|_{V_2=0} = \text{short-circuit current gain}$$

We write these definitions in the somewhat unusual form of equations (12.5.6) to emphasize the actual meanings of these parameters as indicated after each definition above. Also in this form, our general method for evaluating such parameters in any given two-port is still valid: in the two-port, set either $I_2 = 0$ or $V_2 = 0$ as indicated, then apply the denominator quantity as a unity-valued source and solve for the numerator quantity.

Thus in the circuit of Figure 12.5.1, for example, we find the a_{12} parameter by setting $V_2 = 0$ and $V_1 = 1$ V. Then

$$-I_2 = \frac{1}{R} \quad \text{A}$$

or

$$-\frac{1}{a_{12}} = -\frac{1}{R}$$

$$a_{12} = R \tag{12.5.7}$$

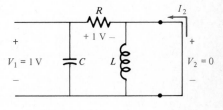

figure 12.5.1
Finding the parameter a_{12} in a simple two-port network.

Therefore, the evaluation of each of the a parameters from equations (12.5.6) is straightforward. It is the same method that we use for finding any of the z's, y's, h's, or g's from their corresponding definitions.

We can solve for the a parameters in terms of the z parameters of the same two-port as follows. The z parameters are used in equations relating the voltages to the currents:

$$V_1 = z_{11}I_1 + z_{12}I_2 \tag{12.5.8}$$

$$V_2 = z_{21}I_1 + z_{22}I_2 \tag{12.5.9}$$

Solve for I_1 in (12.5.9):

$$I_1 = \frac{1}{z_{21}}V_2 - \frac{z_{22}}{z_{21}}I_2 \tag{12.5.10}$$

and substitute it into (12.5.8), thus obtaining

$$V_1 = \frac{z_{11}}{z_{21}}V_2 - \left(\frac{z_{11}z_{22}}{z_{21}} - z_{12}\right)I_2 \tag{12.5.11}$$

Putting the term in parentheses in (12.5.11) over a common denominator and copying (12.5.10), we have the a-parameter equations we seek:

$$V_1 = \left(\frac{z_{11}}{z_{21}}\right)V_2 - \left(\frac{\Delta_z}{z_{21}}\right)I_2 \tag{12.5.12}$$

$$I_1 = \frac{1}{z_{21}}V_2 - \left(\frac{z_{22}}{z_{21}}\right)I_2 \tag{12.5.13}$$

Comparing (12.5.12) with (12.5.3) and (12.5.13) with (12.5.4), we see that

$$a_{11} = \frac{z_{11}}{z_{21}} \qquad a_{12} = \frac{\Delta_z}{z_{21}}$$

$$(12.5.14)$$

$$a_{21} = \frac{1}{z_{21}} \qquad a_{22} = \frac{z_{22}}{z_{21}}$$

The system determinant of the *a* matrix may be written from (12.5.14) as

$$\Delta_a = a_{11}a_{22} - a_{12}a_{21}$$

$$= \frac{z_{11}z_{22} - \Delta_z}{z_{21}^2}$$

$$\Delta_a = \frac{z_{12}}{z_{21}} \qquad\qquad (12.5.15)$$

For *reciprocal* networks $z_{12} = z_{21}$, so

$$\Delta_a = 1 \qquad\qquad (12.5.16)$$

For symmetrical networks $z_{11} = z_{22}$, so from (12.5.14)

$$a_{11} = a_{22} \qquad\qquad (12.5.17)$$

To get the *z* parameters in terms of the *a* parameters, we solve equation (12.5.4) for V_2:

$$V_2 = \frac{1}{a_{21}}(I_1 + a_{22}I_2)$$

$$= \frac{1}{a_{21}}I_1 + \frac{a_{22}}{a_{21}}I_2 \qquad\qquad (12.5.18)$$

and substitute it into (12.5.3):

$$V_1 = a_{11}\left(\frac{1}{a_{21}}I_1 + \frac{a_{22}}{a_{21}}I_2\right) - a_{12}I_2$$

$$= \frac{a_{11}}{a_{21}}I_1 + \left(\frac{a_{11}a_{22}}{a_{21}} - a_{12}\right)I_2$$

$$V_1 = \frac{a_{11}}{a_{21}}I_1 + \frac{\Delta_a}{a_{21}}I_2 \qquad\qquad (12.5.19)$$

Recopying (12.5.18) as (12.5.20)

$$V_2 = \frac{1}{a_{21}}I_1 + \frac{a_{22}}{a_{21}}I_2 \qquad\qquad (12.5.20)$$

we see that

$$z_{11} = \frac{a_{11}}{a_{21}} \qquad z_{12} = \frac{\Delta_a}{a_{21}}$$

$$(12.5.21)$$

$$z_{21} = \frac{1}{a_{21}} \qquad z_{22} = \frac{a_{22}}{a_{21}}$$

The last set of parameters we will discuss is obtained by solving equation (12.5.5)   **THE a' MATRIX**
for V_2 and I_2 in terms of V_1 and $-I_1$. First, a simple inversion yields

$$\begin{bmatrix} V_2 \\ -I_2 \end{bmatrix} = \begin{bmatrix} a_{11} & a_{12} \\ a_{21} & a_{22} \end{bmatrix}^{-1} \begin{bmatrix} V_1 \\ I_1 \end{bmatrix} \tag{12.5.22}$$

where the indicated inverse matrix is obtained in the usual way:

$$\begin{bmatrix} a_{11} & a_{12} \\ a_{21} & a_{22} \end{bmatrix}^{-1} = \begin{bmatrix} \dfrac{a_{22}}{\Delta_a} & \dfrac{-a_{12}}{\Delta_a} \\ \dfrac{-a_{21}}{\Delta_a} & \dfrac{a_{11}}{\Delta_a} \end{bmatrix} \tag{12.5.23}$$

So (12.5.22) becomes

$$V_2 = \frac{a_{22}}{\Delta_a} V_1 - \frac{a_{12}}{\Delta_a} I_1 \tag{12.5.24}$$

and

$$-I_2 = -\frac{a_{21}}{\Delta_a} V_1 + \frac{a_{11}}{\Delta_a} I_1 \tag{12.5.25}$$

But we want I_2, not $-I_2$. Multiplying (12.5.25) by -1 and writing this and (12.5.24) in matrix form,

$$\begin{bmatrix} V_2 \\ I_2 \end{bmatrix} = \begin{bmatrix} \dfrac{a_{22}}{\Delta_a} & \dfrac{a_{12}}{\Delta_a} \\ \dfrac{a_{21}}{\Delta_a} & \dfrac{a_{11}}{\Delta_a} \end{bmatrix} \begin{bmatrix} V_1 \\ -I_1 \end{bmatrix} \tag{12.5.26}$$

$$= \begin{bmatrix} a'_{11} & a'_{12} \\ a'_{21} & a'_{22} \end{bmatrix} \begin{bmatrix} V_1 \\ -I_1 \end{bmatrix}$$

or

$$V_2 = a'_{11} V_1 - a'_{12} I_1$$

$$I_2 = a'_{21} V_1 - a'_{22} I_1$$

So

$$a'_{11} = \frac{V_2}{V_1}\bigg|_{I_1=0} \qquad a'_{12} = \frac{-V_2}{I_1}\bigg|_{V_1=0}$$

$$a'_{21} = \frac{I_2}{V_1}\bigg|_{I_1=0} \qquad a'_{22} = \frac{-I_2}{I_1}\bigg|_{V_1=0} \tag{12.5.27}$$

From (12.5.26) we write the a' parameters in terms of the a parameters.

$$a'_{11} = \frac{a_{22}}{\Delta_a} \qquad a'_{12} = \frac{a_{12}}{\Delta_a}$$

$$a'_{21} = \frac{a_{21}}{\Delta_a} \qquad a'_{22} = \frac{a_{11}}{\Delta_a} \tag{12.5.27a}$$

Note carefully that $[a] \neq [a']^{-1}$.

The system determinant of the a' matrix is

Table 12.5.1

	z	y	h	g	a	a'	Definitions
z	$z_{11}\quad z_{12}$ $z_{21}\quad z_{22}$	$\frac{y_{22}}{\Delta_y}\quad -\frac{y_{12}}{\Delta_y}$ $-\frac{y_{21}}{\Delta_y}\quad \frac{y_{11}}{\Delta_y}$	$\frac{\Delta_h}{h_{22}}\quad \frac{h_{12}}{h_{22}}$ $-\frac{h_{21}}{h_{22}}\quad \frac{1}{h_{22}}$	$\frac{1}{g_{11}}\quad -\frac{g_{12}}{g_{11}}$ $\frac{g_{21}}{g_{11}}\quad \frac{\Delta_g}{g_{11}}$	$\frac{a_{11}}{a_{21}}\quad \frac{\Delta_a}{a_{21}}$ $\frac{1}{a_{21}}\quad \frac{a_{22}}{a_{21}}$	$\frac{a'_{22}}{a'_{21}}\quad \frac{1}{a'_{21}}$ $\frac{\Delta'_a}{a'_{21}}\quad \frac{a'_{11}}{a'_{21}}$	$\frac{V_1}{I_1}\Big|_{I_2=0}\quad \frac{V_1}{I_2}\Big|_{I_1=0}$ $\frac{V_2}{I_1}\Big|_{I_2=0}\quad \frac{V_2}{I_2}\Big|_{I_1=0}$
y	$\frac{z_{22}}{\Delta_z}\quad -\frac{z_{12}}{\Delta_z}$ $-\frac{z_{21}}{\Delta_z}\quad \frac{z_{11}}{\Delta_z}$	$y_{11}\quad y_{12}$ $y_{21}\quad y_{22}$	$\frac{1}{h_{11}}\quad -\frac{h_{12}}{h_{11}}$ $\frac{h_{21}}{h_{11}}\quad \frac{\Delta_h}{h_{11}}$	$\frac{\Delta_g}{g_{22}}\quad \frac{g_{12}}{g_{22}}$ $-\frac{g_{21}}{g_{22}}\quad \frac{1}{g_{22}}$	$\frac{a_{22}}{a_{12}}\quad -\frac{\Delta_a}{a_{12}}$ $-\frac{1}{a_{12}}\quad \frac{a_{11}}{a_{12}}$	$\frac{a'_{11}}{a'_{12}}\quad -\frac{1}{a_{12}}$ $-\frac{\Delta'_a}{a'_{12}}\quad \frac{a'_{22}}{a'_{12}}$	$\frac{I_1}{V_1}\Big|_{V_2=0}\quad \frac{I_1}{V_2}\Big|_{V_1=0}$ $\frac{I_2}{V_1}\Big|_{V_2=0}\quad \frac{I_2}{V_2}\Big|_{V_1=0}$
h	$\frac{\Delta_z}{z_{22}}\quad \frac{z_{12}}{z_{22}}$ $-\frac{z_{21}}{z_{22}}\quad \frac{1}{z_{22}}$	$\frac{1}{y_{11}}\quad -\frac{y_{12}}{y_{11}}$ $\frac{y_{21}}{y_{11}}\quad \frac{\Delta_y}{y_{11}}$	$h_{11}\quad h_{12}$ $h_{21}\quad h_{22}$	$\frac{g_{22}}{\Delta_g}\quad -\frac{g_{12}}{\Delta_g}$ $-\frac{g_{21}}{\Delta_g}\quad \frac{g_{11}}{\Delta_g}$	$\frac{a_{12}}{a_{22}}\quad \frac{\Delta_a}{a_{22}}$ $-\frac{1}{a_{22}}\quad \frac{a_{21}}{a_{22}}$	$\frac{a'_{12}}{a'_{11}}\quad \frac{1}{a'_{11}}$ $-\frac{\Delta'_a}{a'_{11}}\quad \frac{a'_{21}}{a'_{11}}$	$\frac{V_1}{I_1}\Big|_{V_2=0}\quad \frac{V_1}{V_2}\Big|_{I_1=0}$ $\frac{I_2}{I_1}\Big|_{V_2=0}\quad \frac{I_2}{V_2}\Big|_{I_1=0}$
g	$\frac{1}{z_{11}}\quad -\frac{z_{12}}{z_{11}}$ $\frac{z_{21}}{z_{11}}\quad \frac{\Delta_z}{z_{11}}$	$\frac{\Delta_y}{y_{22}}\quad \frac{y_{12}}{y_{22}}$ $-\frac{y_{21}}{y_{22}}\quad \frac{1}{y_{22}}$	$\frac{h_{22}}{\Delta_h}\quad -\frac{h_{12}}{\Delta_h}$ $-\frac{h_{21}}{\Delta_h}\quad \frac{h_{11}}{\Delta_h}$	$g_{11}\quad g_{12}$ $g_{21}\quad g_{22}$	$\frac{a_{21}}{a_{11}}\quad -\frac{\Delta_a}{a_{11}}$ $\frac{1}{a_{11}}\quad \frac{a_{12}}{a_{11}}$	$\frac{a'_{21}}{a'_{22}}\quad -\frac{1}{a'_{22}}$ $\frac{\Delta'_a}{a'_{22}}\quad \frac{a'_{12}}{a'_{22}}$	$\frac{I_1}{V_1}\Big|_{I_2=0}\quad \frac{I_1}{I_2}\Big|_{V_1=0}$ $\frac{V_2}{V_1}\Big|_{I_2=0}\quad \frac{V_2}{I_2}\Big|_{V_1=0}$
a	$\frac{z_{11}}{z_{21}}\quad \frac{\Delta_z}{z_{21}}$ $\frac{1}{z_{21}}\quad \frac{z_{22}}{z_{21}}$	$-\frac{y_{22}}{y_{21}}\quad -\frac{1}{y_{21}}$ $-\frac{\Delta_y}{y_{21}}\quad -\frac{y_{11}}{y_{21}}$	$-\frac{\Delta_h}{h_{21}}\quad -\frac{h_{11}}{h_{21}}$ $-\frac{h_{22}}{h_{21}}\quad -\frac{1}{h_{21}}$	$\frac{1}{g_{21}}\quad \frac{g_{22}}{g_{21}}$ $\frac{g_{11}}{g_{21}}\quad \frac{\Delta_g}{g_{21}}$	$a_{11}\quad a_{12}$ $a_{21}\quad a_{22}$	$\frac{a'_{22}}{\Delta'_a}\quad \frac{a'_{12}}{\Delta'_a}$ $\frac{a'_{21}}{\Delta'_a}\quad \frac{a'_{11}}{\Delta'_a}$	$\frac{V_1}{V_2}\Big|_{I_2=0}\quad -\frac{V_1}{I_2}\Big|_{V_2=0}$ $\frac{I_1}{V_2}\Big|_{I_2=0}\quad -\frac{I_1}{I_2}\Big|_{V_2=0}$
a'	$\frac{z_{22}}{z_{12}}\quad \frac{\Delta_z}{z_{12}}$ $\frac{1}{z_{12}}\quad \frac{z_{11}}{z_{12}}$	$-\frac{y_{11}}{y_{12}}\quad -\frac{1}{y_{12}}$ $-\frac{\Delta_y}{y_{12}}\quad -\frac{y_{22}}{y_{12}}$	$\frac{1}{h_{12}}\quad \frac{h_{11}}{h_{12}}$ $\frac{h_{22}}{h_{12}}\quad \frac{\Delta_h}{h_{12}}$	$-\frac{\Delta_g}{g_{12}}\quad -\frac{g_{22}}{g_{12}}$ $-\frac{g_{11}}{g_{12}}\quad -\frac{1}{g_{12}}$	$\frac{a_{22}}{\Delta_a}\quad \frac{a_{12}}{\Delta_a}$ $\frac{a_{21}}{\Delta_a}\quad \frac{a_{11}}{\Delta_a}$	$a'_{11}\quad a'_{12}$ $a'_{21}\quad a'_{22}$	$\frac{V_2}{V_1}\Big|_{I_1=0}\quad -\frac{V_2}{I_1}\Big|_{V_1=0}$ $\frac{I_2}{V_1}\Big|_{I_1=0}\quad -\frac{I_2}{I_1}\Big|_{V_1=0}$

Matrix	For reciprocal two-ports	For symmetrical two-ports
z	$z_{12} = z_{21}$	$z_{11} = z_{22}$
y	$y_{12} = y_{21}$	$y_{11} = y_{22}$
h	$h_{12} = -h_{21}$	$\Delta_h = 1$
g	$g_{12} = -g_{21}$	$\Delta_g = 1$
a	$\Delta_a = 1$	$a_{11} = a_{22}$
a'	$\Delta'_a = 1$	$a'_{11} = a'_{22}$

Some additional general relationships

$$\Delta_h = \frac{z_{11}}{z_{22}} = \frac{y_{22}}{y_{11}} = \frac{1}{\Delta_g}$$

$$\Delta_a = \frac{z_{12}}{z_{21}} = \frac{1}{\Delta'_a}$$

$$\Delta_a' = \frac{a_{22}\,a_{11} - a_{12}\,a_{21}}{\Delta_a^2} = \frac{1}{\Delta_a} \tag{12.5.28}$$

For a *reciprocal* network, (12.5.16) inserted into (12.5.28) tells us that

$$\Delta_a' = 1 \tag{12.5.29}$$

For a *symmetrical* network, (12.5.17) inserted into (12.5.27a) yields

$$a_{11}' = a_{22}' \tag{12.5.30}$$

In order to find the a parameters in terms of the parameters a_{11}', a_{12}', a_{21}', a_{22}' we can invert equation (12.5.26) and get back to equation (12.5.22) (after multiplying through one of the resulting equations by -1). The other remaining interrelationships are found via the definitions.

Table 12.5.1 shows the interrelationships among all the two-ports.

some special uses of two-port matrices 12.6

The method of interconnecting two 2-port networks shown in Figure 12.6.1 is called a parallel connection of two-ports because the corresponding ports of each subnetwork are connected in parallel. Now, with

$$\begin{bmatrix} I_1 \\ I_2 \end{bmatrix} = [y^{(1)}]\begin{bmatrix} V_1 \\ V_2 \end{bmatrix} \quad \text{and} \quad \begin{bmatrix} I_3 \\ I_4 \end{bmatrix} = [y^{(2)}]\begin{bmatrix} V_3 \\ V_4 \end{bmatrix} \tag{12.6.1}$$

and assuming

$$\begin{bmatrix} V_1 \\ V_2 \end{bmatrix} = \begin{bmatrix} V_3 \\ V_4 \end{bmatrix} = \begin{bmatrix} V_A \\ V_B \end{bmatrix} \tag{12.6.2}$$

we have

$$\begin{bmatrix} I_A \\ I_B \end{bmatrix} = \begin{bmatrix} I_1 + I_3 \\ I_2 + I_4 \end{bmatrix} = [y^{(1)} + y^{(2)}]\begin{bmatrix} V_A \\ V_B \end{bmatrix} \tag{12.6.3}$$

So the overall y matrix is the sum of the two component y matrices. This is often an extremely useful way of obtaining the y matrix of a complicated two-port.

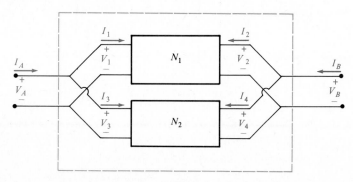

figure 12.6.1
Two 2-port networks connected in *parallel*.

EXAMPLE 12.6.1

  a. Find the open-circuit voltage gain of the twin-T circuit shown in Figure 12.6.2a.
  b. What kind of network is this?

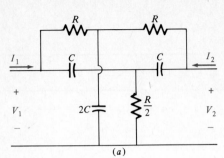

figure 12.6.2
(a) The twin-T circuit of Example 12.6.1. (b) One component two-port of the twin T.
(c) The other component two-port of the twin T.

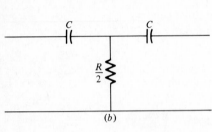

(b)

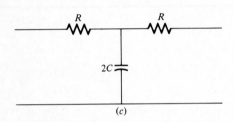

(c)

ANS.: (a) First find the y matrix of each component T network. Then add them together. Finally use

$$\frac{-y_{21}}{y_{22}} = \frac{V_2}{V_1}\bigg|_{I_2=0}$$

For the T circuit of Figure 12.6.2b we easily write the z matrix

$$[z]^{(1)} = \begin{bmatrix} \dfrac{1}{Cs} + \dfrac{R}{2} & \dfrac{R}{2} \\[3mm] \dfrac{R}{2} & \dfrac{1}{Cs} + \dfrac{R}{2} \end{bmatrix} = \begin{bmatrix} \dfrac{RCs+2}{2Cs} & \dfrac{R}{2} \\[3mm] \dfrac{R}{2} & \dfrac{RCs+2}{2Cs} \end{bmatrix}$$

Now invert to get the y matrix. The system determinant is

$$\Delta_z = \frac{(RCs+2)^2}{4C^2s^2} - \frac{R^2}{4} = \frac{RCs+1}{C^2s^2}$$

Thus

$$y_{11}^{(1)} = \frac{RCs+2}{2Cs}\frac{C^2s^2}{RCs+1} = \frac{Cs(RCs+2)}{2(RCs+1)} \tag{12.6.4}$$

Similarly,

$$y_{12}^{(1)} = -\frac{RC^2s^2}{2(RCs+1)} \tag{12.6.5}$$

Also

$$y_{22}^{(1)} = y_{11}^{(1)} \quad\text{and}\quad y_{21}^{(1)} = y_{12}^{(1)} \tag{12.6.6}$$

For the circuit of Figure 12.6.2c the z matrix is

$$[z]^{(2)} = \begin{bmatrix} R + \dfrac{1}{2Cs} & \dfrac{1}{2Cs} \\[3mm] \dfrac{1}{2Cs} & R + \dfrac{1}{2Cs} \end{bmatrix} = \begin{bmatrix} \dfrac{2RCs + 1}{2Cs} & \dfrac{1}{2Cs} \\[3mm] \dfrac{1}{2Cs} & \dfrac{2RCs + 1}{2Cs} \end{bmatrix}$$

Inverting, we find

$$\Delta_z = \frac{(2RCs + 1)^2}{4C^2 s^2} - \frac{1}{4C^2 s^2} = \frac{R(RCs + 1)}{Cs}$$

Therefore

$$y_{11} = y_{22} = \frac{2RCs + 1}{2Cs}\frac{Cs}{R(RCs + 1)} = \frac{2RCs + 1}{2R(RCs + 1)} \tag{12.6.7}$$

Similarly,

$$y_{12} = y_{21} = \frac{-1}{2R(RCs + 1)} \tag{12.6.8}$$

Adding corresponding matrix elements, we have

$$y_{21} = y_{21}^{(1)} + y_{21}^{(2)} = -\frac{R^2 C^2 s^2 + 1}{2R(RCs + 1)} \tag{12.6.9}$$

and

$$y_{22} = y_{22}^{(1)} + y_{22}^{(2)} = \frac{RCs(RCs + 2) + 2RCs + 1}{2R(RCs + 1)}$$

So that

$$H(s) = \frac{V_2}{V_1}\bigg|_{I_2 = 0} = \frac{-y_{21}}{y_{22}} = \frac{R^2 C^2 s^2 + 1}{R^2 C^2 s^2 + 4RCs + 1}$$

(b) Find the Bode plot. Zeros are

$$(RC)^2 s^2 + 1 = 0$$

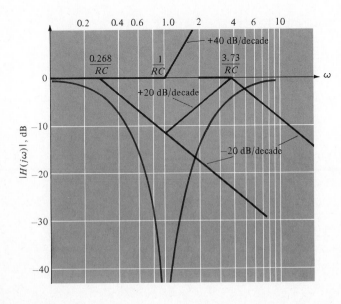

figure 12.6.3
The Bode plot [magnitude of $H(j\omega)$ versus ω] for the two-port of Example 12.6.1.

$$s^2 = -\frac{1}{(RC)^2}$$

$$s = \pm j\,\frac{1}{RC} \qquad \text{(with } \zeta = 0)$$

Poles are

$$s = \frac{-4RC \pm \sqrt{16R^2C^2 - 4R^2C^2}}{2R^2C^2}$$

$$= \frac{-2 \pm \sqrt{3}}{RC} = \frac{-3.73}{RC},\ \frac{-0.268}{RC}$$

So the magnitude plot has a double-ordered break up at $\omega = 1/RC$ and two single-ordered breaks down—one at $0.268/RC$ and one at $3.73/RC$. The sketch is shown in Figure 12.6.3. Clearly this is a circuit which infinitely attenuates a sinusoidal signal whose frequency is $\omega = 1/RC$. This circuit is therefore a notch filter.

The definition of the a parameters as the matrix that relates $[V_1, I_1]$ to $[V_2, -I_2]$ ideally suits it to describe two-port networks connected in *cascade*. In Figure 12.6.4 we have

$$\begin{bmatrix} V_1 \\ I_1 \end{bmatrix} = [a^{(1)}] \begin{bmatrix} V_2 \\ -I_2 \end{bmatrix} \tag{12.6.10}$$

and

$$\begin{bmatrix} V_3 \\ I_3 \end{bmatrix} = [a^{(2)}] \begin{bmatrix} V_4 \\ -I_4 \end{bmatrix} \tag{12.6.11}$$

etc. Also

$$\begin{bmatrix} V_2 \\ -I_2 \end{bmatrix} = \begin{bmatrix} V_3 \\ I_3 \end{bmatrix} \quad \begin{bmatrix} V_4 \\ -I_4 \end{bmatrix} = \begin{bmatrix} V_5 \\ I_5 \end{bmatrix} \tag{12.6.12}$$

etc. Therefore, substituting (12.6.12) into (12.6.11) and then (12.6.11) into (12.6.10), we have

$$\begin{bmatrix} V_1 \\ I_1 \end{bmatrix} = [a^{(1)}][a^{(2)}][a^{(3)}] \begin{bmatrix} V_6 \\ -I_6 \end{bmatrix} \tag{12.6.13}$$

So the overall a matrix of a two-port made up of component two-ports in cascade is the product of each of the component a matrices.

figure 12.6.4
A two-port that consists of component two-ports connected in *cascade*.

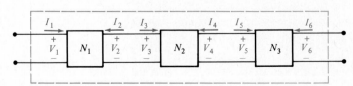

EXAMPLE 12.6.2

Find the transmission matrix of the circuit shown in Figure 12.6.5a. The transformer is an ideal one whose terminal quantities are related by $V_1 = nV_2$ and $I_1 = -(1/n)I_2$ in Figure 12.6.5b, where n is the turns ratio.

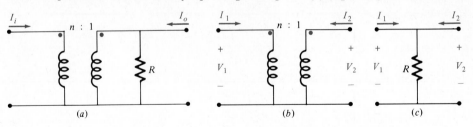

figure 12.6.5
(a) The two-port network of Example 12.6.2. (b) The ideal transformer component two-port. (c) The resistor shown as a component two-port.

ANS.:   We first write the a matrix of the transformer from its defining equations

$$[a^{(1)}] = \begin{bmatrix} n & 0 \\ 0 & \dfrac{1}{n} \end{bmatrix}$$

(This type of a matrix, diagonal with $a_{11} = 1/a_{22}$, is the transmission matrix of any *perfect coupler*—lever, gears, etc.) The a matrix of the resistor two-port of Figure 12.6.5c is obtained from the definitions

$$a_{11} = \left.\frac{V_1}{V_2}\right|_{I_2=0} = 1 \qquad a_{12} = \left.\frac{-V_1}{I_2}\right|_{V_2=0} = -\frac{1}{-\infty} = 0$$

$$a_{21} = \left.\frac{I_1}{V_2}\right|_{I_2=0} = \frac{1}{R} \qquad a_{22} = -\left.\frac{I_1}{I_2}\right|_{V_2=0} = 1$$

Therefore

$$[a^{(2)}] = \begin{bmatrix} 1 & 0 \\ \dfrac{1}{R} & 1 \end{bmatrix}$$

The overall transmission matrix is

$$[a] = [a^{(1)}][a^{(2)}] = \begin{bmatrix} n & 0 \\ 0 & \dfrac{1}{n} \end{bmatrix}\begin{bmatrix} 1 & 0 \\ \dfrac{1}{R} & 1 \end{bmatrix}$$

$$= \begin{bmatrix} n & 0 \\ \dfrac{1}{nR} & \dfrac{1}{n} \end{bmatrix}$$

The transmission matrices of simple series elements and simple shunt elements can be multiplied to obtain the overall a matrix of any ladder network. See Figure 12.6.6. By direct use of the a-parameter definitions we find the a matrices

Series impedance:
$$[a] = \begin{bmatrix} 1 & Z \\ 0 & 1 \end{bmatrix}$$

figure 12.6.6
(a) A simple two-port consisting of a single series impedance. (b) A simple two-port consisting of a single shunt admittance.

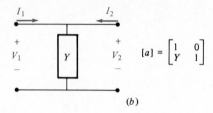

$$[a] = \begin{bmatrix} 1 & Z \\ 0 & 1 \end{bmatrix}$$

(a)

$$[a] = \begin{bmatrix} 1 & 0 \\ Y & 1 \end{bmatrix}$$

(b)

Shunt admittance:

$$[a] = \begin{bmatrix} 1 & 0 \\ Y & 1 \end{bmatrix}$$

EXAMPLE 12.6.3

a. Find the transmission matrix of the two-port shown in Figure 12.6.7.
b. What is the overall voltage transfer function V_2/V_1 when a 1-Ω resistor is connected to port 2?

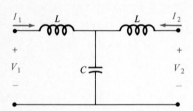

figure 12.6.7
The two-port network of Example 12.6.3.

ANS.: (a) Simply write the product of the three individual a matrices:

$$a = \begin{bmatrix} 1 & Ls \\ 0 & 1 \end{bmatrix} \begin{bmatrix} 1 & 0 \\ Cs & 1 \end{bmatrix} \begin{bmatrix} 1 & Ls \\ 0 & 1 \end{bmatrix}$$

Working from the right gives

$$\begin{bmatrix} 1 & 0 \\ Cs & 1 \end{bmatrix} \begin{bmatrix} 1 & Ls \\ 0 & 1 \end{bmatrix} = \begin{bmatrix} 1 & Ls \\ Cs & LCs^2 + 1 \end{bmatrix}$$

Then

$$[a] = \begin{bmatrix} 1 & Ls \\ 0 & 1 \end{bmatrix} \begin{bmatrix} 1 & Ls \\ Cs & LCs^2 + 1 \end{bmatrix}$$

$$= \begin{bmatrix} 1 + LCs^2 & Ls + L^2Cs^3 + Ls \\ Cs & LCs^2 + 1 \end{bmatrix}$$

or
$$= \begin{bmatrix} LCs^2 + 1 & Ls(LCs^2 + 2) \\ Cs & LCs^2 + 1 \end{bmatrix}$$

(b) Since the shunt 1-Ω resistor has the *a* matrix

$$[a] = \begin{bmatrix} 1 & 0 \\ 1 & 1 \end{bmatrix}$$

the overall *a* matrix is given by the product

$$[a] = \begin{bmatrix} LCs^2 + 1 & Ls(LCs^2 + 2) \\ Cs & LCs^2 + 1 \end{bmatrix} \begin{bmatrix} 1 & 0 \\ 1 & 1 \end{bmatrix}$$

thus the overall a_{11} parameter is determined:

$$a_{11} = (LCs^2 + 1)(1) + Ls(LCs^2 + 2)(1)$$
$$= L^2Cs^3 + LCs^2 + 2Ls + 1$$

and
$$\frac{1}{a_{11}} = \frac{V_2}{V_1}\bigg|_{I_2 = 0} = \frac{1}{L^2Cs^3 + LCs^2 + 2Ls + 1}$$

passive filters 12.7

In many branches of electrical engineering, we very often have need of recovering sinusoidal signals in one band (range) of frequencies while rejecting all other sinusoidal signals. For example, when we listen to channel 2† on our TV set we want to allow frequencies in the range 54 MHz to 60 MHz to be amplified by the receiver and all others to be rejected. A two-port network that treats some frequencies differently from the way it treats others is called an electric filter network. For example, low-pass filters permit sinusoidal signals with frequencies lower than ω_0, the cutoff frequency, to pass through to the output terminals essentially unchanged from the way they appear at the input port. All signals with frequencies greater than ω_0 are attenuated. There are also high-pass, bandpass, bandstop and many other types of filters. There are phase-shifter filters that affect the phase but not the magnitude; there are others that are designed to give a response only when a particular waveform appears at their inputs (*matched filters*). There are indeed many different types of filters. Many books are written on this one subject. In this section we will briefly examine some of the properties of a certain class of filter and present a method for designing these filters.

The method that we will use is the following:

1. We will synthesize (build) filters that have a specified overall $Z_{21}(s) = V_2/I_1$ or $Y_{21}(s) = I_2/V_1$. We will build low-pass filters—if another type of filter (e.g., bandpass) is really what is desired, we will later convert our low-pass filter into that other type. But in any event we will first synthesize the *low-pass prototype*.

† Actually 55.25 to 59.75 MHz.

figure 12.7.1
Different types of filters: (*a*) low-pass, (*b*) high-pass, (*c*) bandpass, and (*d*) band-reject (also called band-elimination or stopband).

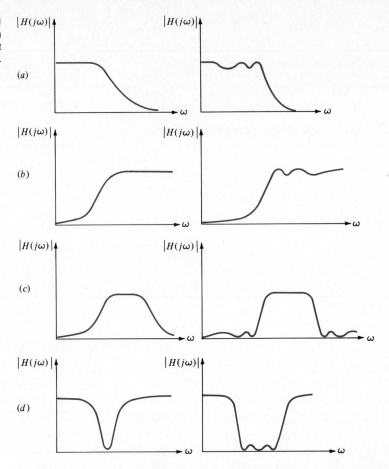

2. The low-pass prototypes that we will build will be *LC* ladder networks terminated in 1-Ω resistors, and since they are to be low-pass filters, the series elements will be inductors and the shunt elements will be capacitors. (Any shunt inductor would short out a dc signal and any series capacitor would block it.)

3. We will build an *LC* ladder that has the correct driving-point immittance (z_{22} or y_{22}) necessary to produce a specified overall $Y_{21}(s)$ or $Z_{21}(s)$ when that ladder is loaded by a 1-Ω resistor.

For a moment let us consider some of the properties of the *z* and *y* parameters of an *LC* two-port. The *driving-point immittances*, z_{11}, z_{22}, y_{11}, and y_{22}, approach either *ks* or *k/s* in the limit as *s* approaches zero (and also as *s* approaches infinity). In other words, any *LC* circuit's input impedance will look like either an inductance or a capacitance at very low frequencies and at very high frequencies (i.e., as *s* approaches zero and as *s* approaches infinity). Therefore, *both the highest and lowest powers of s in numerator and denominator must differ by 1.*

EXAMPLE 12.7.1

Given the circuit shown in Figure 12.7.2, find $z_{11}(s)$ as $s \to \infty$ and also as $s \to 0$.

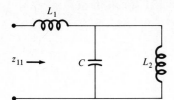

figure 12.7.2
The open-circuit input impedance of an LC network. See Example 12.7.1.

ANS.: Allow s to approach infinity along the $j\omega$ axis. For $\omega \to \infty$, $z_{11}(j\omega)$ is simply the inductive impedance $j\omega L_1$ (the capacitive reactance approaches zero and this shorts out L_2).

$$z_{11}(s) = L_1 s + \frac{\dfrac{1}{Cs} L_2 s}{\dfrac{1}{Cs} + L_2 s}$$

$$= \frac{L_1 L_2 C s^3 + (L_1 + L_2)s}{L_2 C s^2 + 1}$$

which, in the limit as $s \to \infty$, is equal to

$$\lim z_{11}(s)\bigg|_{s \to \infty} = \lim L_1 s \bigg|_{s \to \infty} = \infty$$

And at $s = 0$

$$\lim z_{11}(s)\bigg|_{s \to 0} = 0$$

i.e., at dc, the inductors form a short circuit to ground (the capacitor is an open circuit). The highest powers of s in numerator and denominator differ by 1. The lowest powers also differ by 1—as they must in an LC circuit.

Another property of LC two-ports is that *all z and y parameters are ratios of polynomials where every other term is missing.* Such polynomials contain only even powers of s or only odd powers of s. Such polynomials are called even (Ev) and odd (Od) respectively. So all z's and y's are ratios of either Ev/Od or Od/Ev polynomials. For instance, in Example 12.6.3 we found the a matrix of the LC two-port of Figure 12.6.7 to be

$$[a] = \begin{bmatrix} LCs^2 + 1 & Ls(LCs^2 + 2) \\ Cs & LCs^2 + 1 \end{bmatrix} \tag{12.7.1}$$

Using $z_{11} = a_{11}/a_{21}$ and $y_{11} = a_{22}/a_{12}$ we have

$$z_{11}(s) = \frac{LCs^2 + 1}{Cs} \tag{12.7.2}$$

and
$$y_{11}(s) = \frac{LCs^2 + 1}{L^2Cs^3 + 2Ls} \qquad (12.7.3)$$

These are both either Od/Ev or Ev/Od rational functions of s. [Only even-ordered terms appear in the numerators and only odd-ordered terms in the denominators of equations (12.7.2) and (12.7.3).] Also, the highest- and lowest-powered terms differ by *exactly* 1. Now, as to the off-diagonal parameters:

$$z_{12}(s) = \frac{\Delta_a}{a_{21}} = \frac{1}{Cs}$$

and
$$y_{12}(s) = \frac{-\Delta_a}{a_{12}} = \frac{1}{L^2Cs^3 + 2Ls}$$

These *transfer* immitances also are ratios of Od/Ev or Ev/Od. However, there is *no* requirement that their highest and lowest powers differ by unity [it happens to be true for the z_{12} here, but this is not generally the case—e.g., see $y_{12}(s)$].

A convenient method for obtaining the schematic of an *LC* ladder network if we know its $z_{11}(s)$ or $y_{11}(s)$ is called Cauer's *continued fraction expansion*. A ladder network such as the one in Figure 12.7.3 has a driving-point impedance $z_{11}(s)$ that can be written in the form:

$$z_{11}(s) = Z_1 + \cfrac{1}{Y_2 + \cfrac{1}{Z_3 + \cfrac{1}{Y_4}}} \qquad (12.7.4)$$
$$\phantom{z_{11}(s) = Z_1 + \cfrac{1}{Y_2 + \cfrac{1}{Z_3 + \cfrac{1}{Y_4}}}}\text{etc.}$$

Conversely, given a $z_{11}(s)$ function, we can put it into the form of equation (12.7.4) and recognize the value of each of the elements: Z_1, Y_2, Z_3, etc. Similarly, the method is also valid for synthesizing a ladder network with specified $z_{22}(s)$, or $y_{11}(s)$, or $y_{22}(s)$:

$$y_{22}(s) = Y_1 + \cfrac{1}{Z_2 + \cfrac{1}{Y_3 + \cfrac{1}{Z_4}}} \qquad (12.7.5)$$
$$\phantom{y_{22}(s) = Y_1 + \cfrac{1}{Z_2 + \cfrac{1}{Y_3 + \cfrac{1}{Z_4}}}}\text{etc.}$$

figure 12.7.3
A ladder network with specified driving-point impedance $z_{11}(s)$.

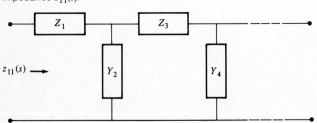

EXAMPLE 12.7.2

Synthesize a ladder network having the LC input admittance given by equation (12.7.2). Let $L = 2$ H and $C = 3$ F. Thus

$$z_{11}(s) = \frac{6s^2 + 1}{3s}$$

ANS.:   The continued fraction expansion of equation (12.7.4) is realized by doing one step of long division and then inverting the remainder and dividing again in a continuous process. This can be written $z_{11}(s)$:

$$
\begin{array}{r}
2s \longleftarrow Z_1 \\
3s\ \overline{)\ 6s^2 + 1} \\
\underline{6s^2} \qquad \begin{array}{r} 3s \longleftarrow Y_2 \\ 1\ \overline{)\ 3s} \end{array} \\
\underline{3s} \\
0 \longleftarrow Y \text{ (remainder)}
\end{array}
$$

$Y = 0$ is an open circuit in parallel with Y_2

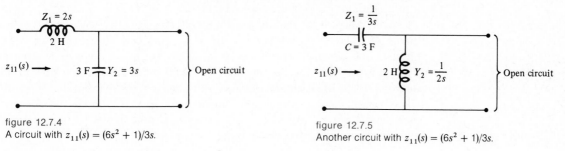

figure 12.7.4
A circuit with $z_{11}(s) = (6s^2 + 1)/3s$.

figure 12.7.5
Another circuit with $z_{11}(s) = (6s^2 + 1)/3s$.

The result is shown in Figure 12.7.4. Note that the third element $Z_3 = Ls$ in Figure 12.6.7 is not used in the computation of $z_{11}(s)$. The circuits of Figures 12.7.4 and 12.6.7 have the same input impedance.

In Example 12.7.2 note that $z_{11}(s)$ has zeros at $s = \pm j1/\sqrt{6}$ and poles at $s = 0$ and $s = \infty$.

By dividing the highest powers of s terms first, we are *removing the pole at infinity*. In other words, if we do one step of long division and stop, we have first built a series 2-H inductor (in series with the remainder function $1/3s$); i.e.,

$$z_{11}(s) = 2s + \frac{1}{3s}$$

The quantity $2s$ is indeed

$$z_{11}(\infty) = \lim \left. \frac{6s^2 + 1}{3s} \right|_{s \to \infty} = 2s$$

We could have removed the pole at $s = 0$ first by simply dividing the lowest-order terms first; i.e., we could rewrite $z_{11}(s)$ as

$$z_{11}(s) = \frac{1 + 6s^2}{3s}$$

Then

$$
\begin{array}{r}
\dfrac{1}{3s} \longleftarrow Z_1 \\[4pt]
3s \overline{\smash{)}\,1 + 6s^2} \\[4pt]
\underline{1} \qquad \dfrac{1}{2s} \longleftarrow Y_2 \\[4pt]
6s^2 \overline{\smash{)}\,3s} \\[4pt]
\underline{3s} \\[4pt]
0 \longleftarrow Y \text{ (remainder)}
\end{array}
$$

This gives rise to the circuit of Figure 12.7.5. If the $z_{11}(s)$ does not happen to have a pole at zero (infinity) then attempting to divide lowest (highest) powers of s first will leave a nonrealizable remaining function.

Clearly, the circuits of Figures 12.7.4 and 12.7.5 have the same $z_{11}(s)$, but they most certainly do *not* have the same $z_{12}(s)$ or $z_{22}(s)$. They are *not equivalent two-ports*—they simply look the same at their input terminals (port 1) when the output terminals (port 2) are open-circuited.

For our purpose of building *low-pass* filters we clearly wish to have the first element of $z_{11}(s)$ or $z_{22}(s)$ a series inductive impedance. This means we must first remove a pole of *impedance* at $s = \infty$ (i.e., *divide the highest power of s first* in z_{11} or z_{22}). Or we want the first element to be a shunt capacitor—a pole of admittance at $s = \infty$. In summary then, we will always remove poles at $s = \infty$ (divide *highest powers first*) in order to produce a low-pass *LC* ladder. If the driving-point immittance we are given does not *have* a pole at $s = \infty$, we can invert it and synthesize that inverse function.

EXAMPLE 12.7.3

Synthesize a low-pass *LC* ladder that has the driving-point impedance

$$z_{22}(s) = \frac{2s^2 + 1}{s^3 + 2s}$$

ANS.: This z_{22} has a zero at $s = \infty$, not a pole. If we attempt to build it blindly by dividing denominator into numerator, this is what happens:

$$
\begin{array}{r}
\dfrac{2}{s} \longleftarrow Z_1 \\[4pt]
s^3 + 2s \overline{\smash{)}\,2s^2 + 1} \\[4pt]
\underline{2s^2 + 4} \qquad -\dfrac{s^3}{3} \qquad (?) \\[4pt]
-3 \overline{\smash{)}\,s^3 + 2s}
\end{array}
$$

What kind of passive element has an admittance $Y(s) = -s^3/3$? None. But if we build $1/z_{22}$ (which is an admittance and has a pole at $s = \infty$) we get

$$
\begin{array}{r}
\dfrac{s}{2} \qquad\qquad \longleftarrow Y_1 \\
2s^2 + 1 \overline{\smash{\big)}\, s^3 + 2s} \\
\underline{s^3 + \tfrac{1}{2}s} \qquad \tfrac{4}{3}s \quad \longleftarrow Z_2 \\
\tfrac{3}{2}s \,\overline{\smash{\big)}\, 2s^2 + 1} \\
\underline{2s^2} \qquad \dfrac{3s}{2} \longleftarrow Y_3 \\
1 \,\overline{\smash{\big)}\, \tfrac{3}{2}s} \\
\underline{\tfrac{3}{2}s} \\
0 \longleftarrow Y \text{ (remainder)}
\end{array}
$$

$Y = 0$ is an open circuit in parallel with Y_3

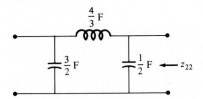

figure 12.7.6
A low-pass LC ladder whose $z_{22} = (2s^2 + 1)/(s^3 + 2s)$.

The resulting circuit is shown in Figure 12.7.6.

Suppose, instead of being given $z_{11}(s)$ or $y_{11}(s)$, we are asked to build a network to realize the overall *transfer impedance* $Z_{21} = V_2/I_1$ of the two-port shown in Figure 12.7.7a. We can solve for this Z_{21} in terms of the z parameters of the LC ladder. We do this by finding the Thevenin equivalent of everything to the left of the 1-Ω resistor and then reconnecting the resistor to that equivalent circuit.

In Figure 12.7.7b we see that the open-circuit (Thevenin) voltage is given by

$$
\left.\frac{V_2}{I_1}\right|_{I_2=0} = z_{21} \tag{12.7.6}
$$

$$
\left.V_2\right|_{I_2=0} = V_{\text{Th}} = z_{21}I_1 \tag{12.7.7}
$$

The output (Thevenin) impedance is, by definition

$$
z_{22} = \left.\frac{V_2}{I_2}\right|_{I_1=0} \tag{12.7.8}
$$

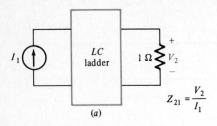

$$Z_{21} = \frac{V_2}{I_1}$$

(a)

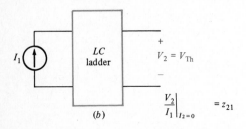

$$\left. \frac{V_2}{I_1} \right|_{I_2=0} = z_{21}$$

(b)

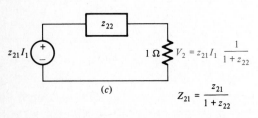

$$V_2 = z_{21}I_1 \frac{1}{1+z_{22}}$$

$$Z_{21} = \frac{z_{21}}{1+z_{22}}$$

(c)

figure 12.7.7
(a) An LC ladder terminated in 1 Ω. (b) Finding the Thevenin (open-circuit) voltage with the 1 Ω removed. (c) The Thevenin circuit in terms of the z parameters of the LC ladder.

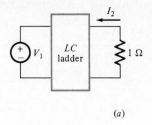

(a)

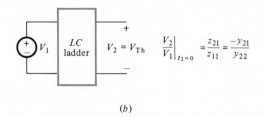

$$\left. \frac{V_2}{V_1} \right|_{I_2=0} = \frac{z_{21}}{z_{11}} = \frac{-y_{21}}{y_{22}}$$

(b)

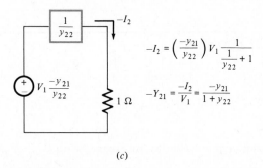

$$-I_2 = \left(\frac{-y_{21}}{y_{22}} \right) V_1 \frac{1}{\frac{1}{y_{22}}+1}$$

$$-Y_{21} = \frac{-I_2}{V_1} = \frac{-y_{21}}{1+y_{22}}$$

(c)

figure 12.7.8
(a) An LC ladder terminated in 1 Ω. (b) Finding the Thevenin (open-circuit) voltage with the 1 Ω removed. (c) The Thevenin circuit in terms of the y parameters of the LC ladder.

Thus in Figure 12.7.7c we have

$$V_2 = (z_{21}I_1) \frac{1}{1+z_{22}} \tag{12.7.9}$$

$$Z_{21}(s) = \frac{V_2}{I_1} = \frac{z_{21}}{1+z_{22}} \tag{12.7.10}$$

We can build the z_{22} via a continued fraction expansion. Similarly, if the overall transfer *admittance* Y_{21} is specified, we can get this in terms of the y parameters of the LC ladder. This similar derivation is shown in Figure 12.7.8. It also involves Theveninizing everything to the left of port 2 of the LC ladder. The result is shown in Figure 12.7.8c as

$$-Y_{21} = \frac{-I_2}{V_1} = \frac{-y_{21}}{1+y_{22}} \tag{12.7.11}$$

We can build the y_{22} via a continued fraction expansion. But first, given a Z_{21}

or Y_{21}, we must see how to get it into the form of either equation (12.7.10) or equation (12.7.11).

We have already said that the z_{21} and y_{21} parameters of a pure LC network are always ratios of even/odd or odd/even polynomials. This is because in an LC circuit, with sinusoidal input, the currents are $90°$ out of phase with corresponding voltages. So, the input current and output voltage must be $\pm 90°$ out of phase. This means the real part of $z_{21}(j\omega)$ and $y_{21}(j\omega)$ must be equal to zero. For this to be true, it follows in turn that $y_{21}(s)$ and $z_{21}(s)$ are in the form odd/even or even/odd rational functions.

Therefore, given a general overall transfer immittance such as $Z_{21}(s)$, we can put this into the form of equation (12.7.10) simply by dividing both numerator and denominator of $Z_{21}(s)$ by either the even part or the odd part of the denominator such that the resulting z_{21} comes out odd/even or even/odd. However, if the $Z_{21}(s)$ or $Y_{21}(s)$ are low-pass filters that have all their zeros at $s = \infty$, i.e., the numerator is a simple constant (an *even* power of s), then we shall *always divide by the odd part of the denominator*. The denominator polynomials of $-y_{21}$ and y_{22} are thus identical. Also z_{21} and z_{22} have identical denominators. Therefore, z_{21} and z_{22} have identical poles. Similarly, so do y_{21} and y_{22}. In using our synthesis technique (continued fraction expansion) which realizes the poles of z_{22} (or y_{22}), we automatically realize the poles of z_{21} and y_{21} at the same time, since they are the same.

EXAMPLE 12.7.4

Synthesize a filter network of the form shown in Figure 12.7.7a such that

$$Z_{21}(s) = \frac{1}{s^3 + 2s^2 + 2s + 1}$$

ANS.:   We put this into the form of equation (12.7.10)

$$Z_{21}(s) = \frac{z_{21}}{1 + z_{22}}$$

by dividing numerator and denominator by the odd-powered terms in the denominator (the unity numerator is even):

$$Z_{21} = \frac{\dfrac{1}{s^3 + 2s}}{1 + \dfrac{2s^2 + 1}{s^3 + 2s}}$$

This gives us a $z_{22}(s)$ and a $z_{21}(s)$ which are ratios of Ev/Od polynomials which must be true if the ladder is to be pure LC. Thus we must synthesize

$$z_{22} = \frac{2s^2 + 1}{s^3 + 2s}$$

Exactly this procedure was carried out in Example 12.7.3. The overall filter specified by the given $Z_{21}(s)$ is shown in Figure 12.7.9.

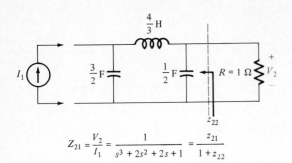

figure 12.7.9
The realization of the overall $Z_{21}(s)$ given in Example 12.7.4.

$$Z_{21} = \frac{V_2}{I_1} = \frac{1}{s^3 + 2s^2 + 2s + 1} = \frac{z_{21}}{1 + z_{22}}$$

We check to make sure that the LC ladder network has $z_{21} = 1/(s^3 + 2s)$.

$$z_{21} = \left.\frac{V_2}{I_1}\right|_{I_2 = 0}$$

So insert $I_1 = 1$ and solve for V_2. V_2 is the voltage across the $\frac{1}{2}$-F capacitor. The current through the capacitor is found via current division. Hence,

$$V_2 = \frac{\frac{1}{\frac{3}{2}s} I_1}{\frac{1}{\frac{3}{2}s} + \frac{4}{3}s + \frac{1}{\frac{1}{2}s}} \frac{1}{\frac{1}{2}s} = \frac{1}{s^3 + 2s} \qquad \text{checks}$$

BUTTERWORTH (MAXIMALLY FLAT) FILTERS

Consider how we might go about specifying $H(s)$ of a low-pass filter so that its $|H(j\omega)|$ versus ω plot would be as flat as possible in the low-frequency passband. One (of many) possible answers to this question is to set not only the first derivative (slope) of the magnitude $|H(j\omega)|$ at $\omega = 0$ equal to zero, but also to demand that the second derivative (upward acceleration of the plot with increasing frequency) be zero. And while we are at it, let us demand that as many such derivatives as possible be equal to zero at $\omega = 0$:

$$\frac{d^n |H(j\omega)|}{d\omega^n} = 0 \qquad \text{for all } n \tag{12.7.12}$$

(That should ensure *maximum flatness* of $|H(j\omega)|$ at $\omega = 0$!) A transfer function magnitude that at least has the property that the first $2n - 1$ (not quite all, but a lot) of its derivatives are equal to zero at $\omega = 0$ is

$$|H(j\omega)| = \frac{1}{\sqrt{1 + \omega^{2n}}} \tag{12.7.13}$$

Clearly, this is a low-pass filter with $|H(0)| = 1$, $|H(\infty)| = 0$, and $|H(j)| = 1/\sqrt{2}$. So this seems to be a very good low-pass filter characteristic. How can we synthesize a network that will have this characteristic? We must develop the complex transfer function, $H(s)$. We can do this by considering the square of the magnitude $|H(j\omega)|^2$.

From (12.7.13),

$$|H(j\omega)|^2 = \frac{1}{1 + \omega^{2n}} \tag{12.7.14}$$

Substituting $s = j\omega$ gives

$$|H(s)|^2 = \frac{1}{1 + \left(\dfrac{s}{j}\right)^{2n}} \tag{12.7.15}$$

Remember that for any $s = j\omega$ the product of $H(s)$ times $H(-s)$ equals the square of the magnitude of $H(s)$. (Think of drawing vectors from each critical frequency to the points $s = j\omega$ and $s = -j\omega$.)

$$|H(s)|^2 = H(s)H(-s) \qquad \text{for } s = j\omega \tag{12.7.16}$$

Therefore we may write (12.7.15) as

$$|H(s)|^2 = H(s)H(-s) = \frac{1}{1 + \left(\dfrac{s}{j}\right)^{2n}} \tag{12.7.17}$$

The poles of this function are located at

$$\left(\frac{s}{j}\right)^{2n} = -1 \tag{12.7.18}$$

$$s^{2n} = -(j)^{2n}$$

$$s^{2n} = 1 = 1\underline{/0^\circ} \text{ for } n \text{ odd} \tag{12.7.19}$$

$$s^{2n} = -1 = 1\underline{/180^\circ} \text{ for } n \text{ even} \tag{12.7.20}$$

For n odd, from (12.7.19), we have to find the 2nd, 6th, 10th, etc. roots of $1\underline{/0^\circ}$. The magnitudes of these roots will all be unity and the angle will be any multiple of 360° divided by $2n$. So tabulating gives

n	s
1	$1\underline{/0^\circ}$, $1\underline{/180^\circ}$
3	$1\underline{/60^\circ}$, $1\underline{/120^\circ}$, $1\underline{/180^\circ}$, $1\underline{/240^\circ}$, $1\underline{/300^\circ}$, $1\underline{/360^\circ}$
5	$1\underline{/36^\circ}$, $1\underline{/72^\circ}$, $1\underline{/108^\circ}$, $1\underline{/144^\circ}$, $1\underline{/180^\circ}$
	$1\underline{/216^\circ}$, $1\underline{/252^\circ}$, $1\underline{/288^\circ}$, $1\underline{/324^\circ}$, $1\underline{/360^\circ}$
etc.	

For even values of n we get from (12.7.20)

$$s^{2n} = 1\underline{/180^\circ}$$

So tabulating gives

n	s
2	$1\underline{/45^\circ}$, $1\underline{/135^\circ}$, $1\underline{/225^\circ}$, $1\underline{/315^\circ}$
4	$1\underline{/22.5^\circ}$, $1\underline{/67.5^\circ}$, $1\underline{/112.5^\circ}$, $1\underline{/157.5^\circ}$
	$1\underline{/202.5^\circ}$, $1\underline{/247.5^\circ}$, $1\underline{/292.5^\circ}$, $1\underline{/337.5^\circ}$
etc.	

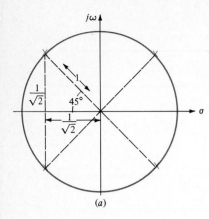

(a)

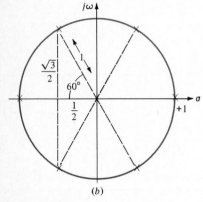

(b)

figure 12.7.10
Pole locations for Butterworth (maximally flat) filters: (a) $n = 2$, (b) $n = 3$.

From the tabulations it follows that the poles are uniformly distributed on the unit circle. In light of equation (12.7.17) we arbitrarily associate poles in the left half plane with $H(s)$ and those in the right half plane with $H(-s)$. [We want to build stable systems, so $H(s)$ cannot have right-half-plane poles.]

For example, suppose the $H(s)$ was specified as having to be a second-order $Y_{21}(s)$. For $n = 2$, $s = -1/\sqrt{2} \pm j1/\sqrt{2}$ and thus

$$Y_{21}(s) = \frac{1}{\left(s + \dfrac{1}{\sqrt{2}} + j\,\dfrac{1}{\sqrt{2}}\right)\left(s + \dfrac{1}{\sqrt{2}} - j\,\dfrac{1}{\sqrt{2}}\right)}$$

$$Y_{21}(s) = \frac{1}{s^2 + \sqrt{2}\,s + 1} \tag{12.7.21}$$

Dividing as usual by the odd part of the denominator, we have

$$Y_{21}(s) = \frac{\dfrac{1}{\sqrt{2}\,s}}{1 + \dfrac{s^2 + 1}{\sqrt{2}\,s}} = \frac{y_{21}}{1 + y_{22}}$$

Building y_{22} by removing its pole at $s = \infty$, we have

$$
\begin{array}{r}
\dfrac{1}{\sqrt{2}}s \longleftarrow Y_1 \\[4pt]
\sqrt{2}s\,\overline{)\,s^2 + 1} \\[2pt]
\underline{s^2} \qquad\quad \sqrt{2}s \longleftarrow Z_2 \\[2pt]
1\,\overline{)\sqrt{2}s} \\[2pt]
\underline{\sqrt{2}s} \\[2pt]
0
\end{array}
$$

The final realization is shown in Figure 12.7.11.

The student should consider the implications of the fact that all n of the poles of the nth-order Butterworth filter's transfer function lie on the unit circle; i.e., each such pole is at radial distance $\omega_n = 1$. Therefore insofar as the $|H(j\omega)|$ Bode plot of this nth-order transfer function is concerned, all n downward breaks occur at $\omega_n = 1$ rad/s and the high-frequency asymptote's slope is the nth multiple of -20 dB/decade.

figure 12.7.11
A second-order Butterworth low-pass filter whose overall $Y_{21} = I_2/V_1$ is specified in equation 12.7.21

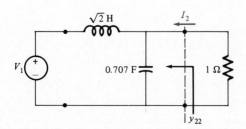

table 12.7.1
Quadratic factors and corresponding pole
locations of Butterworth polynomials. The
absolute value of the real part of each
pole-pair location is the damping factor ζ for
that quadratic.

Order n	The value of A in $s^2 + As + 1$	Pole locations
2	1.414214	$-0.707107 \pm j0.707107$
3	1.0	$-0.5 \quad \pm j0.866025$
4	0.76537	$-0.382684 \pm j0.923880$
	1.84776	$-0.923880 \pm j0.382684$
5	0.61803	$-0.309017 \pm j0.951057$
	1.61803	$-0.809017 \pm j0.587785$
6	0.51764	$-0.258819 \pm j0.965926$
	1.41421	$-0.707107 \pm j0.707107$
	1.93185	$-0.965926 \pm j0.258819$
7	0.44504	$-0.222521 \pm j0.974928$
	1.24698	$-0.623490 \pm j0.781832$
	1.80194	$-0.900969 \pm j0.433884$
8	0.39018	$-0.195090 \pm j0.980785$
	1.11114	$-0.555570 \pm j0.831470$
	1.66294	$-0.831470 \pm j0.555570$
	1.96157	$-0.980785 \pm j0.195090$
9	0.34730	$-0.173648 \pm j0.984808$
	1.0	$-0.5 \quad \pm j0.866025$
	1.53209	$-0.766045 \pm j0.642788$
	1.87939	$-0.939693 \pm j0.342020$

Table 12.7.1 shows the pole locations of Butterworth low-pass prototype filters. The order n of the filter indicates the number of poles, the number of energy-storage elements in the filter, and which multiple of -20 dB/decade the slope of the high-frequency asymptote has. Of course, every odd polynomial also contains the factor $(s + 1)$, i.e., has a pole at $s = -1$. When these factors are multiplied together, they result in Butterworth polynomials as listed in Table 12.7.2.

table 12.7.2
Butterworth polynomials in the form
$s^n + a_1 s^{n-1} + a_2 s^{n-2} + \cdots + a_2 s^2 + a_1 s + 1$.

n	a_1	a_2	a_3	a_4
2	1.41421			
3	2.0			
4	2.61313	3.41421		
5	3.23607	5.23607		
6	3.86370	7.46410	9.14162	
7	4.49396	10.09784	14.5918	
8	5.12583	13.13707	21.84615	25.68836
9	5.75877	16.58172	31.16344	41.9864

In conclusion, we design each Butterworth low-pass filter using an LC ladder network terminated in a 1-Ω resistor. We always design the filter to have its break frequency at $\omega = 1$ rad/s. (We will see in the next section how to change this to a different frequency.) If we want to drive the filter with a voltage source V_1 we specify its $-Y_{21} = -I_2/V_1 = -y_{21}/(1 + y_{22})$. The result of synthesizing y_{22} yields one of the two filters shown in Figure 12.7.12. If we intend to drive our filter with a current source I_1, then we specify that the Z_{21} have a Butterworth characteristic:

figure 12.7.12
The filter configurations for voltage-source
excitation. $-Y_{21}$ is specified; y_{22} or $1/y_{22}$ is
synthesized. (a) Even order. (b) Odd order.

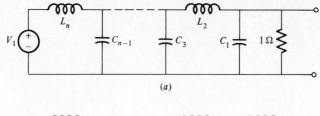

(a)

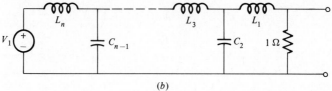

(b)

$$Z_{21} = \frac{z_{21}}{1 + z_{22}}$$

We synthesize z_{22} and the result is one of the two filters shown in Figure 12.7.13.

The element values in the circuits of Figures 12.7.12 and 12.7.13 are given in Table 12.7.3.

figure 12.7.13
The filter configurations for current-source
excitation. Z_{21} is specified; z_{22} or $1/z_{22}$ is
synthesized. (a) Even order. (b) Odd order.

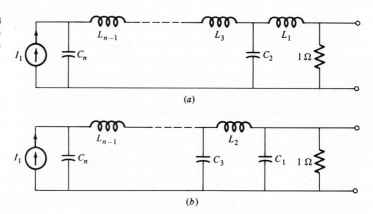

table 12.7.3
Element values for nth-order Butterworth
low-pass prototype filters.

	Elements in Figures 12.7.12a (even) and 12.7.13b (odd)								
n	C_1	L_2	C_3	L_4	C_5	L_6	C_7	L_8	C_9
2	0.7071	1.4142							
3	0.5	1.3333	1.5						
4	0.3827	1.0824	1.5772	1.5307					
5	0.3090	0.8944	1.3820	1.6944	1.5451				
6	0.2588	0.7579	1.2016	1.5529	1.7593	1.5529			
7	0.2225	0.6560	1.0550	1.3972	1.6588	1.7988	1.5577		
8	0.1951	0.5776	0.9371	1.2588	1.5283	1.7287	1.8246	1.5607	
9	0.1736	0.5155	0.8414	1.1408	1.4037	1.6202	1.7772	1.8424	1.56284
n	L_1	C_2	L_3	C_4	L_5	C_6	L_7	C_8	L_9
	Elements in Figures 12.7.12b (odd) and 12.7.13a (even)								

scaling and transformations 12.8

Although it is most convenient to use values of inductance and capacitance in the range of, say, 1 through 10 or 100 to do problems, it is unfortunately the case that in practice we use values like 0.0015 or 0.5×10^{-7}. This is because in the real world we want resonant frequencies and so forth to occur at frequencies in the kilohertz or megahertz range rather than at values like $\omega_0 = 1$ rad/s or 10 rad/s.

FREQUENCY SCALING

We can design a prototype circuit in the low-frequency range (where the numbers are easy to work with) and then *transform the values of the elements so that in the resulting network, each impedance (or admittance) will have the same ohmic value at the higher frequency that it has at the design frequency.* For example, let

$$\omega_{\text{old}} = \text{frequency at which we design}$$

$$\omega_{\text{new}} = \text{frequency actually desired}$$

Now define a frequency normalization factor ω_ϕ such that

$$\omega_\phi \, \omega_{\text{old}} = \omega_{\text{new}}$$

At the new frequency ω_{new} we want

$$j\omega_{\text{new}} L_{\text{new}} = j\omega_{\text{old}} L_{\text{old}}$$

$$\omega_\phi \, \omega_{\text{old}} L_{\text{new}} = \omega_{\text{old}} L_{\text{old}}$$

$$\boxed{L_{\text{new}} = \frac{L_{\text{old}}}{\omega_\phi}}$$

Thus L_{new} is an inductance that will have the same reactance at ω_{new} as L_{old} does at the design frequency ω_{old}:

Similarly we find a new capacitance value C_{new} that will have the same reactance at ω_{new} as C_{old} does at ω_{old}:

$$-j\,\frac{1}{\omega_{\text{new}} C_{\text{new}}} = -j\,\frac{1}{\omega_{\text{old}} C_{\text{old}}}$$

$$\omega_{\text{new}} C_{\text{new}} = \omega_{\text{old}} C_{\text{old}}$$

$$\omega_\phi \, \omega_{\text{old}} C_{\text{new}} = \omega_{\text{old}} C_{\text{old}}$$

$$\boxed{C_{\text{new}} = \frac{C_{\text{old}}}{\omega_\phi}}$$

Resistors have the same resistance at any frequency: $R_{\text{new}} = R_{\text{old}}$.

EXAMPLE 12.8.1

Scale the network of Example 12.7.4 so that the half-power frequency ω_n occurs at $f = 20,000$ Hz rather than $\omega = 1$ rad/s.

ANS.:

$$\omega_\phi = \frac{2\pi(20{,}000)}{1} = 4\pi \times 10^4$$

Thus

$$C_1 = \frac{\frac{1}{2}}{4\pi \times 10^4} = 3.98 \ \mu\text{F}$$

$$L_2 = \frac{\frac{4}{3}}{4\pi \times 10^4} = 10.6 \ \mu\text{H}$$

$$C_3 = \frac{\frac{3}{2}}{4\pi \times 10^4} = 11.9 \ \mu\text{F}$$

IMPEDANCE SCALING Suppose we raised the value of every element's impedance magnitude by some factor R_ϕ. Such a change in the *impedance level* would be independent of frequency; i.e., it would not shift the location of any break frequencies, etc.

To accomplish such a change in the level of any circuit's impedance we simply multiply the resistors and inductors by R_ϕ and divide the capacitors by R_ϕ. This process and frequency scaling can be accomplished simultaneously.

EXAMPLE 12.8.2

Suppose we have a third-order low-pass prototype filter with $C_1 = \frac{1}{2}$ F, $L_2 = \frac{4}{3}$ H, and $C_3 = \frac{3}{2}$ F. What should the value of each of these elements be if we want the break to occur at $f = 1600$ kHz and the resistor terminating the filter to be 10 kΩ.

ANS.: In general,

$$R_{\text{new}} = R_{\text{old}} R_\phi \qquad L_{\text{new}} = L_{\text{old}} \frac{R_\phi}{\omega_\phi} \qquad C_{\text{new}} = \frac{C_{\text{old}}}{R_\phi \omega_\phi}$$

In this problem $R_\phi = 10^4$ Ω and $\omega_\phi = 2\pi \times 1.6 \times 10^6$ rad/s.

$$R_L = (1)10^4 = 10{,}000 \ \Omega$$

$$C_1 = \frac{\frac{1}{2}}{10^4 \times 2\pi \times 1.6 \times 10^6} = 4.97 \ \text{pF}$$

$$L_2 = \frac{\frac{4}{3} \times 10^4}{2\pi \times 1.6 \times 10^6} = 1.33 \ \text{mH}$$

$$C_3 = \frac{\frac{3}{2}}{10^4 \times 2\pi \times 1.6 \times 10^6} = 14.9 \ \text{pF}$$

LOW-PASS TO HIGH-PASS
TRANSFORMATION The passband of our low-pass prototype extends, by definition, from $\omega = 0$ to the half-power point $\omega = 1$. Suppose we want to make a high-pass filter—one that passes sinusoidal waveforms whose frequencies are *greater than* some specified

(half-power) frequency $\omega = \omega_0$ equally as well as the prototype passes sinusoids with frequencies *less than* $\omega = 1$.

A moment's thought leads us to the conclusion that, for any low-pass prototype with transfer function $H(j\omega)$, we would like a new filter having the transfer function $H(j\omega_h)$, where

$$\omega_h = \frac{\omega_0}{\omega} \tag{12.8.1}$$

That is to say, we wish to produce a mapping of the passband from frequencies less than 1 into those greater than ω_0—an *inverse frequency transformation*. Such a mapping is obtained from the relationship

$$s_h = \frac{\omega_0}{s_p} \tag{12.8.2}$$

where the subscripts h and p indicate *high*-pass and *prototype*, respectively. This complex frequency transformation not only performs the mapping of Figure 12.8.1 but also turns inductors into capacitors and vice versa, as follows: The impedance of any element in the prototype should be unchanged under the frequency transformation equation (12.8.2). So for any inductor L_p in the prototype,

$$L_p s_p = L_p \frac{\omega_0}{s_h} = \frac{1}{C_h s_h}$$

where

$$C_h = \frac{1}{L_p \omega_0} \tag{12.8.3}$$

So every prototype inductor turns into a capacitor via equation (12.8.3). Similarly, via equation (12.8.2) prototype capacitors turn into inductors:

$$\frac{1}{C_p s_p} = \frac{s_h}{C_p \omega_0} = L_h s_h$$

where

$$L_h = \frac{1}{C_p \omega_0} \tag{12.8.4}$$

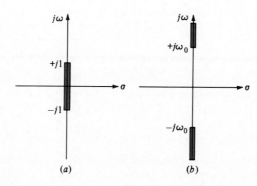

figure 12.8.1
The passbands of (a) the low-pass prototype and (b) a high-pass filter.

EXAMPLE 12.8.3

Given the third-order low-pass prototype of Figure 12.7.9, transform it into a high-pass filter with half-power frequency equal to 1 MHz.

ANS.:
$$\omega_0 = 2\pi \times 10^6$$

From equations (12.8.3) and (12.8.4)

$$L_{h_1} = \frac{1}{0.5 \times 2\pi \times 10^6} = 0.318 \ \mu\text{H}$$

$$C_{h_2} = \frac{1}{1.33 \times 2\pi \times 10^6} = 0.119 \ \mu\text{F}$$

and

$$L_{h_3} = \frac{1}{1.5 \times 2\pi \times 10^6} = 0.106 \ \mu\text{H}$$

The network is shown in Figure 12.8.2.

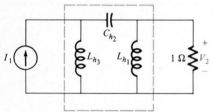

figure 12.8.2
The high-pass filter of Example 12.8.3.

Complex frequency mappings into bandpass and band-reject filters are similarly used to transform the prototype elements as shown in Figure 12.8.3. In the bandpass and band-reject filters, the upper and lower half-power frequencies are denoted by ω_2 and ω_1. Their difference is the bandwidth $B = \omega_2 - \omega_1$. The *center frequency* of the band is called ω_0:

$$\omega_0 = \sqrt{\omega_1 \omega_2} \qquad (12.8.5)$$

Note that ω_0 is the *geometric mean* of ω_1 and ω_2 and, as such, *does appear halfway* between ω_1 and ω_2 on the *logarithmic frequency axis* of the Bode plot.

For the bandpass transformation

$$s_p = \frac{\omega_0}{B}\left(\frac{s_b}{\omega_0} + \frac{\omega_0}{s_b}\right) \qquad (12.8.6)$$

Substituting s_p, the complex frequency of the prototype, into the expression for inductive impedance in the prototype yields

$$Z_L(s_p) = L_p s_p = \frac{L_p}{B} s_b + \frac{L_p \omega_0^2}{B s_b} \qquad (12.8.7)$$

so that, in the transformation to a bandpass filter, each prototype inductor

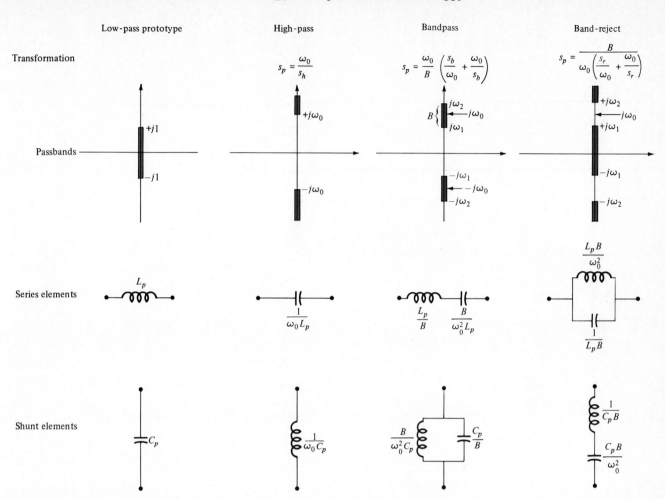

figure 12.8.3
A summary of filter transformations.

becomes the series connection of an inductor, L_p/B henries, and a capacitor, $B/(L_p\,\omega_0^2)$ farads.

Similarly, inserting (12.8.6) into the expression for capacitive impedance

$$Z_c(s_p) = \frac{1}{C_p s_p} = \frac{B}{C_p \omega_0\left(\dfrac{s_b}{\omega_0} + \dfrac{\omega_0}{s_b}\right)} = \frac{1}{\dfrac{C_p}{B}s_b + \dfrac{\omega_0^2 C_p}{Bs_b}} \quad (12.8.8)$$

The denominator of equation (12.8.8) is an admittance which is the sum (parallel combination) of a capacitor, C_p/B farads, and an inductor, $B/\omega_0^2 C_p$ henries.

For the band-reject transformation, substitute the frequency transformation

$$s_p = \frac{B}{\omega_0\left(\dfrac{s_r}{\omega_0} + \dfrac{\omega_0}{s_r}\right)} \quad (12.8.9)$$

into the expressions for prototype impedances

$$Z_L = L_p s_p \quad \text{and} \quad Z_c = \frac{1}{C_p s_p}$$

The results are

$$Z_L(s_p) = L_p s_p = \frac{1}{\dfrac{s_r}{L_p B} + \dfrac{\omega_0^2}{L_p B s_r}} \tag{12.8.10}$$

and

$$Z_c(s_p) = \frac{1}{C_p s_p} = \frac{\omega_0}{C_p B}\left(\frac{s_r}{\omega_0} + \frac{\omega_0}{s_r}\right) = \frac{s_r}{C_p B} + \frac{\omega_0^2}{C_p B s_r} \tag{12.8.11}$$

Thus the band-reject filter is obtained as follows: From equation (12.8.10) we note that each inductor in the low-pass prototype should be replaced by the parallel combination of a capacitor, $1/(L_p B)$ farads, and an inductor, $L_p B/\omega_0^2$ henries. Equation (12.8.11) tells us that each capacitor in the low-pass prototype should be replaced by the series combination of an inductor, $1/(C_p B)$ henries, and a capacitor, $C_p B/\omega_0^2$ farads.

EXAMPLE 12.8.4

Design a third-order Butterworth bandpass filter that will pass frequencies from $f_1 = 200$ Hz to $f_2 = 2$ kHz. The filter is to be driven with a voltage source and the load resistor should be 1000 Ω.

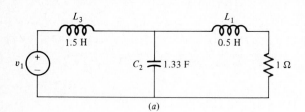

(a)

figure 12.8.4
The filter of Example 12.8.4: (a) the low-pass prototype, (b) the bandpass transformation, (c) the impedance-scaled filter.

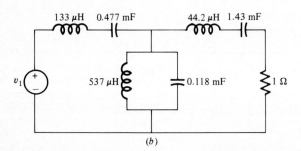

(b)

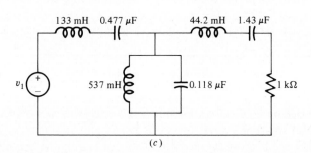

(c)

ANS.: From Table 12.7.3 we find the prototype $L_1 = 0.5$ H, $C_2 = 1.33$ F, and $L_3 = 1.5$ H. See Figure 12.8.4a. Then by equations (12.8.7) and (12.8.8) we make the transformations shown in Figure 12.8.3 where

$$\omega_1 = 2\pi 200 = 1257 \text{ rad/s}$$

$$\omega_2 = 2\pi 2000 = 12{,}566 \text{ rad/s}$$

$$\omega_0 = \sqrt{\omega_1 \omega_2} = 3974 \text{ rad/s}$$

$$B = \omega_2 - \omega_1 = 11{,}309 \text{ rad/s}$$

L_p	$L = L_p/B$	in series with	$C = B/(\omega_0^2 L_p)$
$L_1 = 0.5$	44.2 μH		1.43 mF
$L_3 = 1.5$	133 μH		0.477 mF
C_p	$L = B/(\omega_0^2 C_p)$	in parallel with	$C = C_p/B$
$C_2 = 1.33$	537 μH		0.118 mF

The result is shown in Figure 12.8.4*b*. Impedance level scaling simply requires that all inductances and resistances be multiplied by the scaling factor (1000) and all capacitances be divided by this same amount. The complete and final result is shown in Figure 12.8.4*c*.

active filters 12.9

Although all the filter designs we have discussed so far are quite valid and nicely operating networks, they have the disadvantage of incorporating inductors. Inductors are things to be avoided if possible. They are big and heavy, radiate a magnetic field into the volume around them, are typically nonlinear, waste energy (hysteresis), and—worst of all—are almost impossible to generate physically on an integrated circuit. Fortunately methods exist for realizing all the filters we have discussed (and many more) *without using inductors*. Instead of inductors we will incorporate operational amplifiers into our designs. The inclusion of an *active circuit* element like an op amp is why these are called *active filters*. Op amps in practice are quite complicated networks containing many transistors and resistors and a few capacitors, but they are easy (and quite cheap) to build on an integrated circuit chip. They are vastly preferable when compared to all the problems associated with actual inductors. The only time inductors are used in modern filter designs is when the power levels that are to be transmitted through the filter are too high for op amps to handle; for example, the output stages of radio and television transmitters.

Consider the usual ladder network but with the slightly unusual labeling of series elements by their admittances and the shunt elements by their impedances as shown in Figure 12.9.1. We can write

THE LEAPFROG CONNECTION†

$$I_1 = Y_1(V_1 - V_2) \tag{12.9.1}$$

$$V_2 = Z_2(I_1 - I_3) \tag{12.9.2}$$

† F. E. J. Girling and E. F. Good, "The Leapfrog or Active Ladder Synthesis" *Wireless World*, vol. 76, pp. 341–345, July 1970. Also two other papers in the same journal: September 1970 and October 1970.

figure 12.9.1
A ladder network.

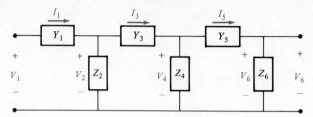

figure 12.9.2
The block diagrams of equations (12.9.1)
through (12.9.6).

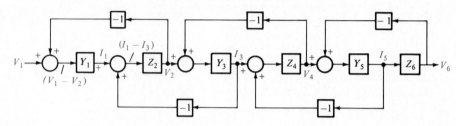

$$I_3 = Y_3(V_2 - V_4) \tag{12.9.3}$$

$$V_4 = Z_4(I_3 - I_5) \tag{12.9.4}$$

$$I_5 = Y_5(V_4 - V_6) \tag{12.9.5}$$

$$V_6 = Z_6 I_5 \tag{12.9.6}$$

A block diagram of equations (12.9.1) through (12.9.6) is shown in Figure 12.9.2

Alternatively, Figure 12.9.2 could be rewritten in either of the forms shown in Figure 12.9.3, thus eliminating the need for the inverters (-1 blocks) in each feedback path. Note that these alternative forms have every forward transfer function entered with an associated minus sign. We shall realize these with inverting operational amplifier circuits. The difference between the two forms is created

figure 12.9.3
Two alternative realizations of the leapfrog connection for either a fifth- or sixth-order *LC* ladder network that is to be terminated in a resistor R_L. (*a*) Requires nine op-amp stages. (*b*) Requires ten op-amp stages.

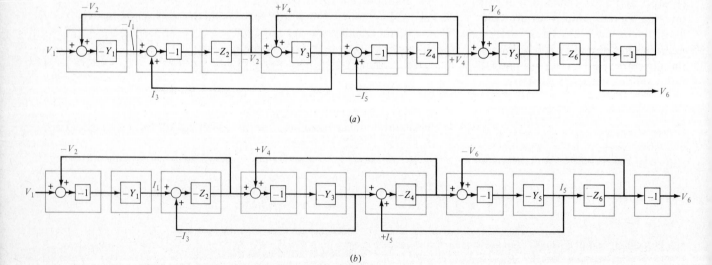

by the presence or absence of the inverter immediately after the first summing junction—all other differences stem from that.

The *LC* ladder filters we are interested in building all terminate in a resistor. If the last branch of the filter is in shunt, then the resistor can be combined with it into a single shunt immittance. If the last *LC* branch is connected in series, then in general (if the impedance level is to be scaled), the resistor is not combined with the last series branch.

EXAMPLE 12.9.1

Find the leapfrog realization of the third-order Butterworth low-pass filter.

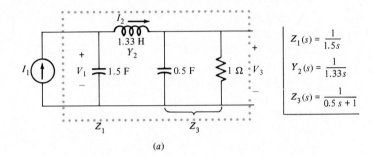

$$Z_1(s) = \frac{1}{1.5s}$$

$$Y_2(s) = \frac{1}{1.33s}$$

$$Z_3(s) = \frac{1}{0.5\,s + 1}$$

(a)

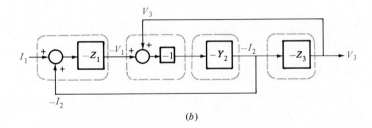

(b)

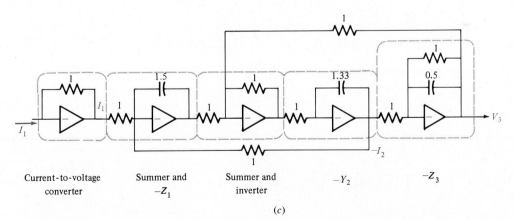

Current-to-voltage converter   Summer and $-Z_1$ Summer and inverter $-Y_2$ $-Z_3$

(c)

figure 12.9.4
(a) The Butterworth filter to be realized. (b) The block diagram. (c) The full leapfrog realization of this filter.

ANS.: The standard form of the filter appears in Figure 12.9.4a. The branch equations are

$$V_1 = Z_1(I_1 - I_2)$$
$$I_2 = Y_2(V_1 - V_3)$$
$$V_3 = Z_3(I_2)$$

We realize these via a leapfrog connection that utilizes a summing integrator as its first stage. See Figure 12.9.4b. The realization is in Figure 12.9.4c. Note that the leapfrog realization is preceded by a current-to-voltage converter. This is needed because we are using *voltage-driven* inverting op amps.

The last comment in Example 12.9.1 *must be clearly understood*. The outputs of all the various op-amp circuits we use to realize these filters are *voltages*. If we label an output as being, say, $-I_2$, we mean that at this node there exists a *voltage* which varies in the same way as (is an analog of) the current $-I_2$. An actual current I_1 flows into the current-to-voltage converter at the left of Figure 12.9.4c. At every other node that is labeled with an I, there exists a *voltage* which has the same time variation as does that corresponding actual current in the prototype circuit (shown in Figure 12.9.4a).

Methods exist for extending the leapfrog realization procedure to include bandpass filters. And, most important of all, we note that in Figure 12.9.4c (or in any leapfrog realization, for that matter) *there is not an inductor in sight!*

THE SALLEN AND KEY FILTER A simple and useful active filter was developed by R. P. Sallen and E. L. Key† which uses a pair of resistors, a pair of capacitors, and an op amp in order to synthesize a transfer function that has one finite pair of poles and has two zeros at infinity. The equivalent circuit is shown in Figure 12.9.5. We develop the voltage transfer function of this network via nodal analysis as follows. At node b,

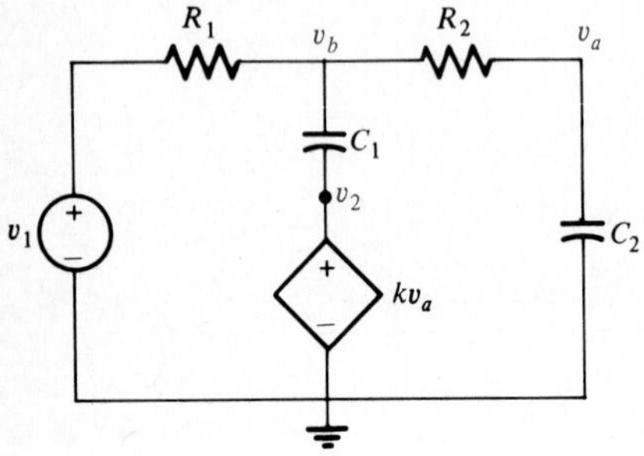

$$\frac{v_1 - v_b}{R_1} + \frac{v_a - v_b}{R_2} = C_1 s(v_b - v_2) \tag{12.9.7}$$

$$v_1\left(\frac{1}{R_1}\right) + v_2(C_1 s) - v_b\left(\frac{1}{R_1} + \frac{1}{R_2} + C_1 s\right) + \frac{v_a}{R_2} = 0 \tag{12.9.8}$$

Now substituting

$$v_a = \frac{v_2}{k} \tag{12.9.9}$$

into (12.9.8) yields

$$\frac{v_1}{R_1} + v_2\left(C_1 s + \frac{1}{kR_2}\right) - v_b\left(\frac{R_2 + R_1 + R_1 R_2 C_1 s}{R_1 R_2}\right) = 0 \tag{12.9.10}$$

At node a,

figure 12.9.5
The Sallen and Key circuit.

† "A Practical Method of Designing *RC* Active Filters," *IRE Transactions on Circuit Theory*, vol. CT-2, 1955, pp. 74–85.

$$\frac{v_a - v_b}{R_2} + v_a C_2 s = 0$$

Again substituting equation (12.9.9) here and then solving for v_b yields

$$v_b = \frac{v_2}{k}(1 + R_2 C_2 s) \qquad (12.9.11)$$

Inserting (12.9.11) into (12.9.10) results in

$$\frac{v_2}{v_1} = H(s) = \frac{\dfrac{k}{R_1 R_2 C_1 C_2}}{s^2 + \left(\dfrac{1-k}{R_2 C_2} + \dfrac{1}{R_1 C_1} + \dfrac{1}{R_2 C_1}\right)s + \dfrac{1}{R_1 R_2 C_1 C_2}} \qquad (12.9.12)$$

which is in the form

$$H(s) = \frac{k\omega_n^2}{s^2 + 2\zeta\omega_n s + \omega_n^2} \qquad (12.9.13)$$

which has the aforementioned pair of finite poles and two zeros at $s = \infty$. The schematic diagram of a practical realization of this circuit is shown in Figure 12.9.6.

Let us again seek a low-pass prototype filter with cutoff frequency $\omega_0 = 1$ rad/s. From equation (12.9.12) this dictates that

$$R_1 R_2 C_1 C_2 = 1 \qquad (12.9.14)$$

Clearly, one possibility (among an infinity of possibilities) is

$$R_1 = R_2 = C_1 = C_2 = 1 \qquad (12.9.15)$$

In this case equation (12.9.12) becomes

$$H(s) = \frac{k}{s^2 + (3-k)s + 1} \qquad (12.9.16)$$

so that

$$\zeta = \frac{3-k}{2} \qquad (12.9.17)$$

In general the pole locations are plotted as a function of k in Figure 12.9.7.

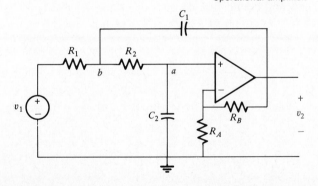

figure 12.9.6
The Sallen and Key active filter. Note the presence of the noninverting operational amplifier.

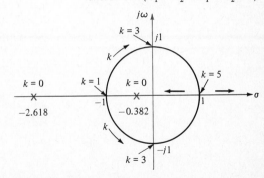

figure 12.9.7
Locus of the poles of equation (12.9.16) as a function of k ($R_1 = R_2 = C_1 = C_2 = 1$).

figure 12.9.8
Choosing $R_1 = R_2 = C_1 = C_2 = 1$ yields this
circuit. It has voltage transfer function
$H(s) = (3 - 2\zeta)/(s^2 + 2\zeta s + 1)$.

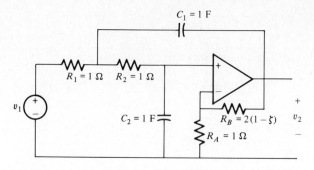

Solving equation (12.9.17) for k, we have

$$k = 3 - 2\zeta \tag{12.9.18}$$

So, knowing the value of the damping factor ζ that we wish to have, we can specify k. This in turn specifies the R_B/R_A ratio. For example, for a second-order Butterworth characteristic, $\zeta = 0.707$. Therefore

$$k = 3 - 2(0.707) = 1.59$$

and since

$$k = 1 + \frac{R_B}{R_A} \tag{12.9.19}$$

$$\frac{R_B}{R_A} = 0.59$$

Combining (12.9.18) and (12.9.19) gives us the general relationship

$$\frac{R_B}{R_A} = 2(1 - \zeta) \tag{12.9.20}$$

Remember that equations (12.9.16) through (12.9.20) depend on the element choice indicated in equation (12.9.15). See Figure 12.9.8.

This choice of element values leads to the overall circuit having a low-frequency voltage gain of

$$H(0) = 3 - 2\zeta \tag{12.9.21}$$

figure 12.9.9
The voltage divider used to restore unity
low-frequency gain to the Sallen and Key
circuit of Figure 12.9.8.

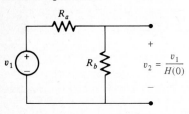

If this is unacceptable, a voltage divider can be used to reduce this value to unity gain. Consider the (Thevenin) circuit that consists of v_1 and R_1 in Figure 12.9.8. This could be replaced by the circuit of Figure 12.9.9 with no loss in generality. We simply require

$$\frac{R_b}{R_a + R_b} = \frac{1}{H(0)} \tag{12.9.22}$$

and the Thevenin equivalent output resistance must be unity as it was in the original circuit. Thus

$$\frac{R_a R_b}{R_a + R_b} = 1 \tag{12.9.23}$$

Multiplying equation (12.9.22) by R_a and setting the result equal to 1 [as indicated in equation (12.9.23)] yields

$$H(0) = R_a \qquad (12.9.24)$$

Inserting (12.9.24) into (12.9.23) yields

$$R_b = \frac{H(0)}{H(0) - 1} \qquad (12.9.25)$$

So

$$R_a = 3 - 2\zeta \quad \text{and} \quad R_b = 1 + \frac{1}{2(1 - \zeta)} \qquad (12.9.26)$$

See Figure 12.9.10 for the complete network.

Another choice of element values results if we demand a priori that $k = 1$, i.e.,

$$k = 1 + \frac{R_B}{R_A} = 1$$

This requires

$$R_B = 0 \qquad (12.9.27)$$

and R_A is unnecessary.

Also let us demand

$$R_1 = R_2 = 1 \ \Omega \qquad (12.9.28)$$

From equation (12.9.12) we see that, in order for ω_n to equal 1,

$$C_1 C_2 = 1 \qquad (12.9.29)$$

From the middle term in the denominator of (12.9.12) we see that (with $k = 1$)

$$C_1 = \frac{1}{\zeta} \qquad (12.9.30)$$

and from (12.9.29)

$$C_2 = \zeta \qquad (12.9.31)$$

This realization is shown in Figure 12.9.11.

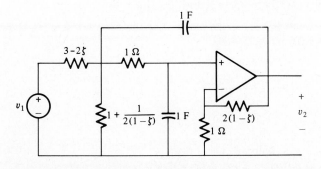

figure 12.9.10
A Sallen and Key circuit with unity low-frequency gain $\omega_0 = 1$ rad/s as a function of damping factor ζ.

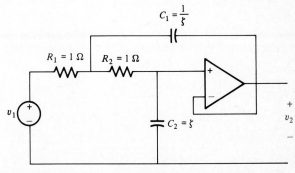

figure 12.9.11
Choosing $R_B = 0$ yields a voltage follower $k = 1$ and the further choice of $R_1 = R_2 = 1$ Ω leads to the requirement that $C_2 = 1/C_1 = \zeta$. $H(0) = 1$.

It inherently has

$$H(0) = 1 \qquad\qquad (12.9.32)$$

Check this by inserting (12.9.28), (12.9.30), and (12.9.31) into (12.9.12).

12.10   summary

In this chapter we have examined in detail a particular circuit configuration called the two-port network. We have confined our attention to linear networks, thus the equations that describe them are linear equations and we may use matrices to write them. We have discussed six different matrix descriptions of the interrelationships among the terminal variables. We have seen how to convert from any one equivalent two-port to any of the others. Certain of the matrix descriptions were shown to be particularly useful in building more complicated two-ports from two or more component two-ports connected, say, in parallel or cascade, etc.

The second half of the chapter was devoted to the design of practical filter networks. First we designed low-pass Butterworth (maximally flat) prototypes as LC ladders terminated in 1-Ω resistors. High-pass, bandpass, and band-reject filters were discussed and methods were demonstrated for obtaining each from the low-pass prototype. Frequency and impedance scaling were discussed. Finally, two examples of active (inductorless) filters were discussed in detail: the leapfrog-connected filter and the Sallen and Key filter.

problems   **1.** Find the open-circuit impedance z matrix of the two-port shown in Figure P12.1. Is this a reciprocal network?

2. Find the open-circuit impedance z matrix of an ideal transformer that has n_1 turns on the input winding and n_2 turns on the secondary. Both I_1 and I_2 enter dotted terminals.

3. For the two-port network in Figure P12.3: (a) Find the z matrix. (b) Find the inverse of the z matrix. Is your answer consistent with the definitions of the elements in the y matrix?

4. (a) Find the open-circuit impedance matrix z for a T circuit similar to that shown in Figure 12.2.3a where $Z_A = Z_D = 3s$ and $Z_B = 1/2s$. (b) Find V_2/V_1 when $I_2 = 0$ as a ratio of polynomials in s.

5. Find the open-circuit impedance parameters of the two-port shown in Figure P12.5.

6. (a) Use mesh equations to find the z parameters of the circuit shown in Figure P12.6. (b) Find the z parameters directly by applying their definitions to this circuit.

7. (a) Find the open-circuit impedance parameters for the two-port network shown in Figure P12.7. (b) Is this network reciprocal?

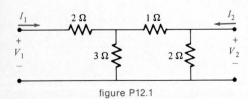

figure P12.1

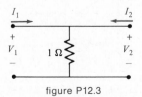

figure P12.3

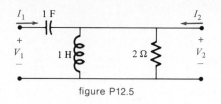

figure P12.5

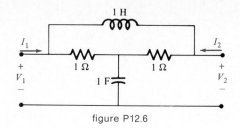

figure P12.6

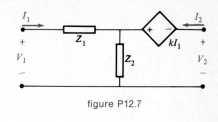

figure P12.7

8. The z matrix for a certain two-port network is

$$[z] = \begin{bmatrix} 6 & 3 \\ 3 & 6 \end{bmatrix}$$

Find the characteristic impedance Z_0, that is, that value of load impedance Z_L which will make Z_{in} appear to be Z_L.

9. Consider a T network made up of three 1-Ω resistors. Find the equivalent pi network via $[y] = [z]^{-1}$.

10. (*a*) Find the inverse matrix of

$$[y] = \begin{bmatrix} 2 & -1 \\ -1 & 2 \end{bmatrix}$$

(*b*) Synthesize a T network having the y parameters given in part (*a*).

11. Find an equivalent pi network for the two-port network in Figure P12.11.

12. Find the four short-circuit admittance parameters for the circuit shown in Figure P12.12.

13. If $y_{22} = 5$ ℧ in Figure P12.13, find the value of R.

14. Two experiments are performed on a symmetrical, reciprocal two-port. The partial results are tabulated below. Complete the table.

Experiment	V_1	I_1	V_2	I_2
1	2	0	5	1
2	0	2		

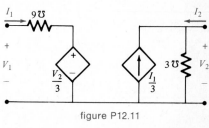

figure P12.11

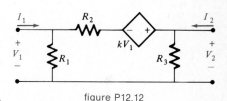

figure P12.12

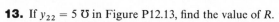

figure P12.13

15. Find the pi-to-T transformation for the general pi circuit shown in Figure 12.3.2*a*; i.e., find the values of the elements in the equivalent T network in terms of Y_A, Y_B, and Y_D.

16. Find the T-to-pi transformation for the general T circuit shown in Figure 12.2.3*a*; i.e., find the values of the elements in the equivalent pi network in terms of Z_A, Z_B, and Z_D.

17. Given the network shown in Figure 12.2.3*a*, let $Z_A = 2$ Ω, $Z_B = 6$ Ω, and $Z_D = 3$ Ω (all pure resistors). (*a*) Find the g matrix. (*b*) Find the values of the elements and source gains in the g matrix equivalent two-port network (which includes a CDCS and a VDVS).

18. Find the h parameters of the circuit shown in Figure P12.18.

19. Repeat problem 18 for the two-port shown in Figure P12.19.

20. Repeat problem 18 for the two-port shown in Figure P12.20.

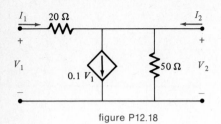

figure P12.18

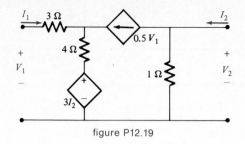

figure P12.19

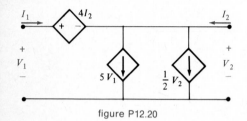

figure P12.20

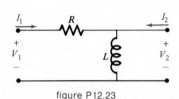

figure P12.23

21. Consider a transistor whose h parameters are $h_{11} = 50 \ \Omega$, $h_{12} = 3 \times 10^{-3}$, $h_{21} = -0.98$, and $h_{22} = 4 \times 10^{-6} \ \mho$. This transistor is to be used as a voltage amplifier between a 2-kΩ load resistor and a 10-mV source having an internal (Thevenin) resistance of 100 Ω. Determine the output voltage that will result across the load.

22. A simplified linear equivalent circuit for a transistor used at audio frequencies in its common-emitter mode is completely specified by the hybrid matrix

$$[h] = \begin{bmatrix} 10^3 & 0 \\ 10^2 & 10^{-5} \end{bmatrix}$$

Two identical such transistor stages are used in cascade (output of the first connected to the input of the second). If the combination is terminated in $R_L = 10 \ \text{k}\Omega$, find V_L/V_1, where V_1 is the input to the first stage and V_L is the output voltage across R_L.

23. (a) Find the transmission a parameters of the two-port in Figure P12.23. (b) Show that $\Delta_a = 1$.

24. Replace the inductor in Figure P12.23 with a 1-F capacitor and let $R = 1 \ \Omega$. Find the transmission a matrix of this two-port. Use this to find the overall open-circuit voltage gain of a cascade combination of two such stages.

25. Given a linear, reciprocal two-port. If $V_1 = 1$ V and $I_2 = 0$, then $V_2 = 3$ V. Question: If $V_1 = 0$ and $I_1 = 3$ A, what will the value of I_2 be? Is more information needed in order to answer this?

26. The transmission matrix of an ideal transformer is given in Example 12.6.2 as

$$[a] = \begin{bmatrix} n & 0 \\ 0 & \dfrac{1}{n} \end{bmatrix}$$

where n is the turns ratio. Find the matrices $[z]$, $[y]$, $[h]$, $[g]$, and $[a']$ for this perfect coupler.

27. Consider two mutually coupled inductors with self-inductances L_1, L_2 and mutual inductance M. Assume i_1 and i_2 both enter a dotted terminal. (a) Find the equivalent T network. (b) Find the equivalent pi network.

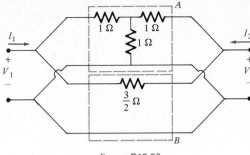

figure P12.30

28. Use the defining equations of two mutually coupled coils to obtain the h parameters of that system. Sketch the h parameter equivalent two-port (include the effect of L_1, L_2, and M).

29. (a) Find the transmission a matrix of a pair of coupled inductors. Include the effects of L_1, L_2, and M. (b) What must be true about L_1, L_2, and M in order for this set of inductors to be an ideal transformer? (See Example 12.6.2.)

30. Given the network in Figure P12.30. (a) Find the y matrix of network A. (b) Find the y matrix of network B. (c) Find the overall y matrix. (d) What is the value of g_{21} for the overall two-port?

31. Given a symmetrical lattice network as shown in Figure P12.31. (a) Find the y matrix for this two-port in terms of Y_a and Y_b. (b) Find Y_a and Y_b in terms of the y parameters. (c) Let Y_a be a 1-H inductor and Y_b a 1-F capacitor. Assume the network is driven by an ideal voltage source $v_1(t)$ and feeds into a short circuit. Find the transfer function $I_2(s)/V_1(s)$ under these conditions. (d) Plot the pole-zero constellation of this transfer function and also its Bode plot. (e) Suppose this lattice is loaded by an open circuit. Repeat part (d) for $V_2(s)/V_1(s)$.

32. Verify the results of Example 12.3.2 by adding the y matrices of two parallel-connected two-ports: one a resistive T and the other the single bridging resistor.

33. Consider the bridged-T network shown in Figure P12.6. Find the value of the y-matrix determinant Δ_y. What can you tell from this regarding z_{11} and y_{11}?

34. Redraw the LC ladder shown in Figure 12.7.9. Call the $\frac{3}{2}$-F capacitor C_1, the inductor L, and the $\frac{1}{2}$-F capacitor C_2. Connect a pair of output terminals across R. Consider the network to be a cascade of three separate two-ports: the first contains only C_1, the second contains only L, and the third contains C_2 and R. (a) Find the transmission matrix of each of these two-ports and use them to find the overall a matrix of the ladder. (b) Use your answer to part (a) to determine an expression for the overall $Z_{21} = V_2/I_1$ of the network. Verify the result of Example 12.7.4 by substituting numerical values into this expression. (c) If $R = 1\ \Omega$, find the values of the other elements such that this filter becomes a third-order Butterworth low-pass prototype. (d) Sketch the Bode plot of the input impedance of this Butterworth filter.

35. Design a network that has the following input driving-point impedance:

$$Z_{in} = \frac{s^3 + 2s}{s^4 + 3s^2 + 1}$$

Use the continued fraction expansion method and remove poles at (a) infinity, (b) the origin ($s = 0$).

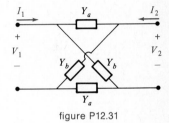

figure P12.31

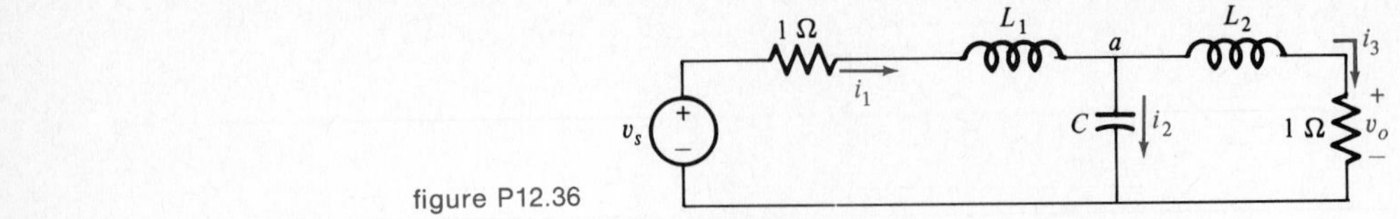

figure P12.36

36. The circuit shown in Figure P12.36 is known as a doubly terminated filter (because of the presence of the two resistors). (a) Find $H(s) = v_o(s)/v_s(s)$. (b) If $L_1 = L_2 = 1$ H and $C = 2$ F, then $H(s)$ will have a pole at $s = -1$. Find all the other poles and zeros of this $H(s)$ under these conditions.

37. Synthesize an LC low-pass prototype filter that is to be loaded by a 1-Ω resistor, whose overall transfer impedance is

$$Z_{21}(s) = \frac{V_2(s)}{I_1(s)} = \frac{0.7155}{(s + 0.6264)(s^2 + 0.6264s + 1.1423)}$$

Sketch the circuit and label each element.

38. Design an LC ladder network (terminated in a 1-Ω resistor and driven by a voltage source) whose overall transfer admittance is

$$Y_{21}(s) = \frac{1}{12s^2 + 3s + 1}$$

Sketch the circuit and label each element.

39. Design an LC ladder network (terminated in a 1-Ω resistor and driven by a current source) whose overall transfer admittance is

$$Z_{21}(s) = \frac{1}{24s^3 + 12s^2 + 5s + 1}$$

Sketch the circuit and label each element.

40. Using the quadratic terms given in Table 12.7.1 (and $s + 1$ for odd n), evaluate all eight Butterworth polynomials given there at $s = j\omega = j1$, the cutoff frequency. (a) What does this demonstrate about the value of all Butterworth $H(j\omega)$ at $\omega = 1$? (b) What does this indicate about the product of all values of A for any given n?

41. Suppose you have a low-pass prototype filter whose cutoff frequency is 1 rad/s, and whose termination resistor is 1 Ω. There is an inductor in this prototype whose value is 3 H. What would the new value of this inductor be if we wanted a similar low-pass filter except that its cutoff frequency is to be 2000 rad/s and the terminating resistor is to be 1 kΩ.

42. Given a parallel RLC circuit with $R = 1$ kΩ, $L = 1000$ H, and $C = 0.1$ F. Frequency-scale the circuit elements so that resonance occurs at $f_0 = 63$ MHz. What is the bandwidth of the resulting circuit?

43. We wish to convert a second-order low-pass prototype filter of the type shown in Figure 12.7.12a, where $C_1 = 0.7$ F and $L_2 = 0.94$ H, into a bandpass filter with $\omega_1 = 5000$ rad/s and $\omega_2 = 10,000$ rad/s. We have available a 100-mH inductor and a 76-mH inductor. What value of load resistance will be needed?

44. The "40-m" amateur radio band extends from 7.000 to 7.300 MHz. Design a third-order LC Butterworth bandpass filter for that range of frequencies. It is to be voltage-source-driven and is to be loaded by a 52-Ω resistor.

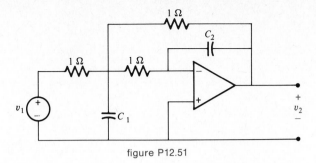

figure P12.51

45. Design a second-order bandstop filter that will reject signals in the standard broadcast band (550 to 1600 kHz). The filter is to be driven by a voltage source and is to work into a 300-Ω resistor.

46. Design a Butterworth LC filter that is to be driven by a current source, that will pass frequencies between 88 and 108 MHz, and whose attenuation will be greater than 10 dB for frequencies higher than 216 MHz. The filter is to be loaded by a 1-kΩ resistor.

47. Design the leapfrog equivalent of the prototype described in problem 43. (*a*) Sketch and label the block diagram. (*b*) Sketch the actual op-amp circuit and label all elements with their values.

48. Design a leapfrog realization of a fourth-order Butterworth low-pass filter whose cutoff frequency is $\omega = 6283$ rad/s (1000 Hz). The LC prototype should be voltage-source-driven and the final op-amp circuit should utilize 10-kΩ resistors.

49. Design a Sallen and Key filter that is equivalent to the low-pass filter in problem 43.

50. Design a low-pass prototype Sallen and Key filter (Figure 12.9.6) assuming $C_1 = 1$ F, $R_1 C_1 = R_2 C_2$, and $R_A = R_B$. Find the values of each element (either the absolute numerical value or in terms of ζ and/or Q_0).

51. The circuit in Figure P12.51 is a Sallen and Key type second-order low-pass prototype with $\omega = 1$ rad/s. Find the values of C_1 and C_2 in terms of ζ and/or Q_0.

52. Design a fourth-order Butterworth low-pass filter using two cascaded Sallen and Key circuits. Utilize voltage followers in both stages. The desired cutoff frequency is 2000 Hz; 1-kΩ resistors are to be used in the final realization.

53. A third-order (non-Butterworth) voltage-driven low-pass prototype filter has the transfer function

$$H(s) = \frac{1}{(s + 0.494)(s^2 + 0.494s + 0.994)}$$

Design a two-stage active filter (you may also incorporate an inverter) that will have this transfer function. One stage should be a Sallen and Key type of the form shown in Figure 12.9.11. Use $R_1 = R_2 = 10^6$. (*a*) Sketch the Bode plot. (*b*) Suppose we want to design a low-pass filter similar to this one (except that the new one should perform in the same way at $\omega = 1000$ rad/s as the prototype does at $\omega = 1$ rad/s). What is the new transfer function $H_{new}(s)$? Changing only capacitance values, what is the new design?

54. Use Figure 8.5.4*b* to design a second-order low-pass Butterworth filter prototype. Use only three inverting op amps in your design. (*Hint*: The differential amplifier in Figure P8.4 is useful.) Suppose other output terminals were derived from the outputs of each of the other two op amps. What type filter would each of those yield?

chapter 13
FOURIER SERIES

introduction 13.1

Up to this point in our study of circuits and systems we have used only one type of periodic input function: the sinusoid. We have seen many different types of sources: step functions, impulses, sinusoids, ramps, etc. But, none of these have been periodic except the sinusoid. In this chapter we will develop a method for describing nonsinusoidal periodic waveforms and using them as inputs (sources) in circuits and systems. This method is called the Fourier series.

Most nonsinusoidal periodic waveforms can be expressed as a sum of sinusoidal components each having a different frequency, magnitude, and phase shift. Except in special cases, an infinite number of such components are theoretically required. Practically, though, only a few terms usually give a reasonable approximation of the desired nonsinusoidal waveform.

Since we know how to find the particular response of any linear system to a single sinusoidal source, we can utilize the superposition principle by first finding the response of the system to each individual sinusoidal component and then summing these individual responses. Thus, the Fourier series gives us a way of breaking up a complicated periodic waveform into a number of simple sinusoids with which we already know how to work. Superposition tells us to sum all these component responses to get the particular response to the nonsinusoidal input.

Two important questions must be answered any time we are confronted by a nonsinusoidal periodic source driving a linear system:

1. Does this nonsinusoidal waveform have a Fourier series equivalent? In other words, can it be broken down into sinusoidal components?
2. If the answer to (1) is yes, then what are the magnitude, frequency, and phase of each of those components?

The answer to the first question, above, is that any waveform $f(t)$ with period T that has the following two properties has a legitimate Fourier series equivalent.

Property 1: $f(t)$ has a finite number of discontinuities, maxima, and minima in any period, T.

figure 13.1.1
Using the Fourier series and superposition to obtain the particular (forced) response of a linear system to a nonsinusoidal, periodic input.

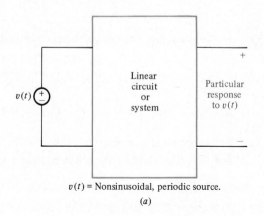

$v(t)$ = Nonsinusoidal, periodic source.

(a)

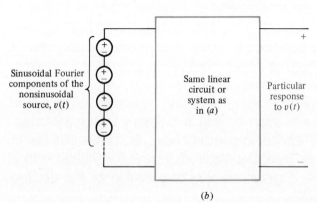

(b)

Property 2:

$$\int_{t_0}^{t_0+t} |f(t)|\, dt < \omega \qquad \text{for any } t_0$$

[If this is true we say that $f(t)$ is absolutely integrable.] These two qualifications are called the *Dirichlet conditions* and are sufficient (but not necessary) for a function to have a Fourier series description.

The answer to the second question (what *are* the values of magnitude, frequency, and phase of each component) we will develop in detail in Section 12.3. However, it should be understood that the component sinusoids of a Fourier series are each *harmonically* related to the *first harmonic* component. The first harmonic component is a sinusoid that has the same period T as the non-sinusoidal waveform that we are attempting to describe. Its frequency is therefore $\omega_0 = 2\pi/T$. The frequency of each higher harmonic is an *integer multiple* of ω_0.

EXAMPLE 13.1.1

The following is an example of a Fourier series.

$$v(t) = \underbrace{3}_{\text{dc component}} + \underbrace{6\cos(7t+14°)}_{\text{1st harmonic (or fundamental)}} + \underbrace{140\cos(14t+3°)}_{\text{2d harmonic (or 1st overtone)}} + \underbrace{2\cos(21t-18°)}_{\text{3d harmonic (or 2d overtone)}}$$

$$+ \underbrace{1.3\cos(28t+87°)}_{\text{4th harmonic (or 3d overtone)}} + \cdots$$

The waveform $v(t)$ has a first harmonic frequency of $\omega_0 = 7$ rad/s. Thus $v(t)$ will periodically repeat itself every T seconds, where

$$\omega_0 = 7$$
$$2\pi f_0 = 7$$
$$\frac{2\pi}{T} = 7$$
$$T = \frac{2\pi}{7}\ \text{s}$$

Note that the constant term tells us the time average value of the waveform (all the remaining terms are pure sinusoids having zero average values). Engineers, and most mathematicians and physicists, talk about the various terms as being the *first harmonic, second harmonic*, etc. Musicians use the words fundamental, first overtone, etc.

Even if the magnitude of one or more of the components (including the first harmonic) is zero, the resulting waveform still is periodic at the fundamental or first harmonic frequency.

EXAMPLE 13.1.2

What is the period of

$$v(t) = 6 \cos (9t) - 7 \cos (12t + 14°)$$

ANS.:   We must find the greatest common divisor of the two frequencies 9 rad/s and 12 rad/s, i.e.,

$$9 = \text{3d harmonic frequency of } \omega_0 = 3$$

and

$$12 = \text{4th harmonic frequency of } \omega_0 = 3$$

Therefore the period of $v(t)$ is

$$\omega = 2\pi f = \frac{2\pi}{T} = 3 \qquad \text{or} \qquad T = \frac{2\pi}{3} \text{ s}$$

Frequencies which are not harmonically related are termed *incommensurate*. The ratio of two incommensurate frequencies is an irrational number. Sums of components with incommensurate frequencies are not periodic. (They never repeat.)

EXAMPLE 13.1.3

$\omega_a = 4$ rad/s and $\omega_b = \pi$ rad/s are incommensurate frequencies.

13.2 three forms of the Fourier series

A Fourier series is a sum of harmonically related sinusoids, each such sinusoid having its own magnitude and phase shift. Thus the series may be written:

$$f(t) = \frac{c_0}{2} + \sum_{k=1}^{k=\infty} c_k \cos (k\omega_0 t + \phi_k)$$

$$\text{Period} = \frac{2\pi}{\omega_0}$$

(13.2.1)

The reason for writing the constant term in the form $c_0/2$ will become apparent later; suffice it now just to realize that this term is the average value of the function $f(t)$.

It is a simple matter to prove that harmonically related components sum to yield a result that is periodic with the same period as the fundamental. In other words, if m and n are integers and

$$f(t) = b_1 \sin \omega_0 t + b_n \sin (n\omega_0 t) + b_m \sin (m\omega_0 t)$$

where $\omega_0 = 2\pi/T$ or $T = 2\pi/\omega_0$, then does $f(t) = f(t + 2\pi/\omega_0)$?

To answer this, write out the expression for $f(t + 2\pi/\omega_0)$ by replacing t with $(t + 2\pi/\omega_0)$ in $f(t)$. Thus,

$$f\left(t + \frac{2\pi}{\omega_0}\right) = b_1 \sin (\omega_0 t + 2\pi) + b_n \sin (n\omega_0 t + 2\pi n) + b_m \sin (m\omega_0 t + 2\pi m)$$

Adding integer multiples of 2π to the arguments of trigonometric functions does nothing to the value of those functions. So $f(t) = f(t + 2\pi/\omega_0)$ (even if b_1 happens to = 0)!

EXAMPLE 13.2.1

Given that a balanced square wave of magnitude V_m has a Fourier series that consists of odd-numbered harmonics:

$$v(t) = \frac{4V_m}{\pi} \left(\cos \omega_0 t - \tfrac{1}{3} \cos 3\omega_0 t + \tfrac{1}{5} \cos 5\omega_0 t - \tfrac{1}{7} \cos 7\omega_0 t + \cdots\right)$$

a. Sketch the first and third harmonics (first two terms) and note how their sum roughly approximates the square wave.
b. Sketch the fifth harmonic and note how adding it in will improve the approximation achieved in **a.**

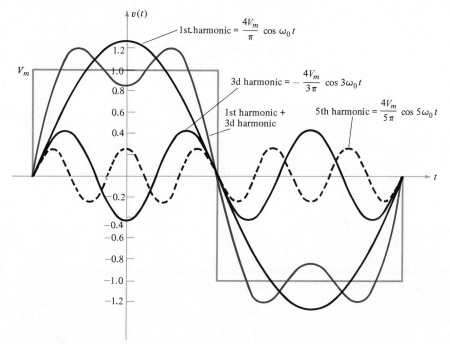

figure 13.2.1
The first three odd harmonics of a balanced square wave.

ANS.:   (a) See Figure 13.2.1. Note that the sum of the first and third harmonics is already beginning to look something like the square wave. (b) Where the sum of the first and third harmonics is less than V_m, the fifth harmonic (the third term in the series) will add to that sum and more closely approximate the square wave. Similarly, where the first two terms' sum is greater than the square-wave magnitude, the fifth harmonic will diminish this overshoot. Notice, however, that the first half-cycle of the fifth harmonic ends (at a zero crossing) that occurs later than when the first two terms' sum exceeds V_m. So, although the addition of the fifth harmonic will help to improve the accuracy of our approximation of the square wave, it will still not be perfect. Each additional harmonic term that is added in will reduce but not eliminate the error (difference) between our approximation and the actual square wave. A perfect approximation therefore would require a series containing an infinite number of component harmonic terms.

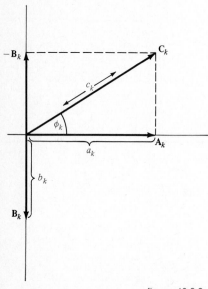

figure 13.2.2
Interrelationships among parameters of two
forms of the Fourier series.

An alternative form which is equivalent to equation (13.2.1) is

$$f(t) = \frac{a_0}{2} + \sum_{k=1}^{\infty} a_k \cos k\omega_0 t + \sum_{k=1}^{\infty} b_k \sin k\omega_0 t$$

(13.2.2)

$$b_0 = 0$$

This alternative form may be obtained by applying the following conversion term by term:

$$c_k \cos (k\omega_0 t + \phi_k) = a_k \cos k\omega_0 t + b_k \sin k\omega_0 t$$

which implies the phasor relationships shown in Fig. 13.2.2. Note that a_k and b_k are the respective signed magnitudes of phasors $\mathbf{A}_k$ and $\mathbf{B}_k$.

Thus†

$$c_k = \sqrt{a_k^2 + b_k^2}$$

$$\phi_k = \tan^{-1}\left(\frac{-b_k}{a_k}\right)$$

and

$$a_k = c_k \cos \phi_k$$

$$b_k = -c_k \sin \phi_k$$

EXAMPLE 13.2.2

Convert the following term of a Fourier series into the corresponding terms in the sin-cos form:

$$9.6 \cos (377t + 31°)$$

ANS.:

$$c_k = 9.6 \qquad \phi_k = 31°$$

Thus

$$a_k = c_k \cos \phi_k = 9.6 \cos 31° = 8.23$$

$$b_k = -c_k \sin \phi_k = -9.6 \sin 31° = -4.94$$

so the equivalent second-form terms are

† Great care must be taken in applying these formulae. Hand calculators and computers often give answers that are ambivalent when computing arctan functions; e.g., for $b_k = -12$, $a_k = -5$, the correct angle is 112.6°, not $-67.4°$ (which is what many calculators will give.)

$$8.23 \cos 377t - 4.94 \sin 377t$$

The equivalent phasor diagram is shown in Figure 13.2.2.

There is a third form of the Fourier series that results from applying Euler's relationships

$$\sin x = \frac{e^{jx} + e^{-jx}}{2j}$$

and

$$\cos x = \frac{e^{jx} + e^{-jx}}{2}$$

to either of the first two forms.

Consider the kth term in the sin-cos form of the series. It can be written in terms of complex exponentials as follows:

$$a_k \cos k\omega_0 t + b_k \sin k\omega_0 t = a_k \frac{e^{jk\omega_0 t} + e^{-jk\omega_0 t}}{2} + b_k \frac{e^{jk\omega_0 t} - e^{-jk\omega_0 t}}{2j}$$

$$= \underbrace{\left(\frac{a_k - jb_k}{2}\right)}_{\alpha_k} e^{jk\omega_0 t} + \underbrace{\left(\frac{a_k + jb_k}{2}\right)}_{\alpha_{-k}} e^{-jk\omega_0 t}$$

If we denote the complex coefficients in parentheses as α_{+k} and α_{-k} then the series can be written in a single summation. Note that

$$\alpha_{+k} = \alpha^*_{-k}$$

and

$$\alpha_0 = \frac{a_0}{2} = \frac{c_0}{2}$$

Thus

$$f(t) = \frac{a_0}{2} + \sum_{k=1}^{\infty} a_k \cos k\omega_0 t + \sum_{k=1}^{\infty} b_k \sin k\omega_0 t$$

or

$$f(t) = \sum_{k=-\infty}^{\infty} \alpha_k e^{jk\omega_0 t} \qquad (13.2.3)$$

where $\alpha_0 = a_0/2$.

We can also show that this exponential form of the series is directly equivalent to the form of the series

$$f(t) = \frac{c_0}{2} \sum_{k=1}^{\infty} c_k \cos (k\omega_0 t + \phi_k)$$

via the following steps:

$$f(t) = \frac{c_0}{2} + \sum_{1}^{\infty} \frac{c_k}{2} [e^{j(k\omega_0 t + \phi_k)} + e^{-j(k\omega_0 t + \phi_k)}]$$

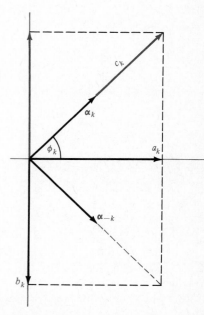

figure 13.2.3
Interrelationships among parameters of three forms of the Fourier series.

$$= \frac{c_0}{2} + \sum_1^\infty \underbrace{\frac{c_k}{2} e^{j\phi_k} e^{jk\omega_0 t}}_{\alpha_k} + \underbrace{\frac{c_k}{2} e^{-j\phi_k} e^{-jk\omega_0 t}}_{\alpha_{-k}}$$

so, as before,

$$f(t) = \sum_{k=-\infty}^\infty \alpha_k e^{jk\omega_0 t}$$

where $\alpha_0 = c_0/2$.

Generally, in the pages that follow, we will drop the phasor notation on the variable α_k and simply denote it as α_k. The student therefore must remember that α_k is a complex quantity without special notation (in exactly the same way that the complex frequency variable s is).

Table 13.2.1 is a summary of interrelationships among the three forms of the Fourier series. (Also see Figure 13.2.3.)

a_k	$c_k = \sqrt{a_k^2 + b_k^2}$	$\alpha_k = \dfrac{a_k - jb_k}{2}$
b_k	$\phi_k = \tan^{-1}\left(\dfrac{-b_k}{a_k}\right)$	
$a_k = c_k \cos \phi_k$	c_k	$\alpha_k = \dfrac{c_k}{2} e^{j\phi_k}$
$b_k = -c_k \sin \phi_k$	ϕ_k	$= \dfrac{c_k}{2} \underline{/\phi_k}$
$a_k = 2 \operatorname{Re} [\alpha_k]$ $-b_k = 2 \operatorname{Im} [\alpha_k]$	$c_k = 2|\alpha_k|$ $\phi_k = \underline{/\alpha_k}$	α_k $\alpha_{-k} = \alpha_{+k}^*$

EXAMPLE 13.2.3

Given the following partial Fourier series:

$$v(t) = 10 + 8 \cos \omega_0 t + 6 \sin \omega_0 t + 4 \cos 2\omega_0 t - 5 \sin 2\omega_0 t$$

find the other two forms of this series.

ANS.: Dc term: The constant term is the same in all forms of the series. It represents the average value of the function, thus:

$$\frac{a_0}{2} = \frac{c_0}{2} = \alpha_0 = 10$$

First harmonic: We are given that $a_1 = 8$, $b_1 = 6$. Thus

$$c_1 = \sqrt{a_1^2 + b_1^2} = \sqrt{(8)^2 + (6)^2} = 10$$

$$\phi_1 = \tan^{-1}\left(\frac{-b_1}{a_1}\right) = \tan^{-1}\left(\frac{-6}{8}\right) = -36.8°$$

and

$$\alpha_1 = \frac{a_1 - jb_1}{2} = 4 - j3 = 5\underline{/-36.8°} = 5\underline{/-0.644} \text{ rad}$$

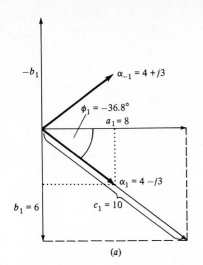

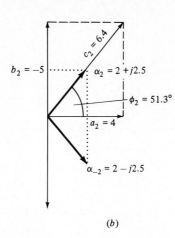

figure 13.2.4
Phasor diagrams for Example 13.2.3.

$$\alpha_{-1} = 4 + j3 = 5\underline{/+36.8°} = 5\underline{/+0.644} \text{ rad}$$

Therefore the other two equivalent forms are

$$v_1(t) = 10 \cos(\omega_0 t - 36.8°)$$

and

$$v_1(t) = 5e^{-j0.644}e^{j\omega_0 t} + 5e^{j0.644}e^{-j\omega_0 t}$$

See Figure 13.2.4a.

Second harmonic $a_2 = 4$, $b_2 = -5$. Using the same relationships as above we find that

$$c_2 = \sqrt{4^2 + (-5)^2} = 6.4$$

$$\phi_2 = \tan^{-1}\left(\tfrac{5}{4}\right) = 51.3°$$

$$\alpha_2 = \frac{4 + j5}{2} = 2 + j2.5 = 3.2\underline{/51.3°} = 3.2\underline{/0.896} \text{ rad}$$

$$\alpha_{-2} = 2 - j2.5 = 3.2\underline{/-51.3°} = 3.2\underline{/-0.896} \text{ rad}$$

So

$$v_2(t) = 6.4 \cos(2\omega_0 t + 51.3°)$$

or

$$v_2(t) = 3.2e^{j0.896}e^{j2\omega_0 t} + 3.2e^{-j0.896}e^{-j2\omega_0 t}$$

In general, then, we may write

$$v(t) = 10 + 8 \cos \omega_0 t + 6 \sin \omega_0 t + 4 \cos 2\omega_0 t - 5 \sin 2\omega_0 t$$

$$= 10 + 10 \cos(\omega_0 t - 36.8) + 6.4 \cos(2\omega_0 t + 51.3°)$$

$$= 3.2e^{-j0.896}e^{-j2\omega_0 t} + 5e^{j0.644}e^{-j\omega_0 t} + 10 + 5e^{-j0.644}e^{+j\omega_0 t} + 3.2e^{j0.896}e^{j2\omega_0 t}$$

In the exponential form of the Fourier series two terms are necessary to describe each harmonic. In Example 13.2.3 the first harmonic is given by

$$(5e^{+j0.644})e^{-j\omega_0 t} + (5e^{-j0.644})e^{j\omega_0 t}$$

figure 13.2.5
Counterrotating complex exponentials.

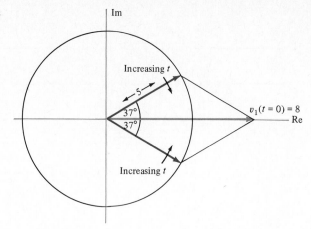

or

$$5\underline{/+0.644}\,e^{-j\omega_0 t} + 5\underline{/-0.644}\,e^{+j\omega_0 t}$$

$$5\underline{/+37°}\,e^{-j\omega_0 t} + 5\underline{/-37°}\,e^{+j\omega_0 t}$$

These two terms represent counterrotating complex exponentials. They are shown in Figure 13.2.5 at $t = 0$. The sum of these two complex exponentials will always be a real quantity. This quantity is the instantaneous value of the first harmonic $v_1(t)$.

13.3 evaluation of the coefficients

Suppose we are given some periodic function of time, $f(t)$, and we want to find a Fourier series representation for it. How do we determine the proper values of the coefficients in the series? In this section we will develop a general answer to this question.

First consider the sin-cos general form:

$$f(t) = \frac{a_0}{2} + a_1 \cos \omega_0 t + a_2 \cos 2\omega_0 t + a_3 \cos 3\omega_0 t + \cdots$$

$$+ b_1 \sin \omega_0 t + b_2 \sin 2\omega_0 t + b_3 \sin 3\omega_0 t + \cdots \qquad (13.3.1)$$

Suppose we were to integrate both sides of equation (13.3.1) over one (or any integer number) period. Since the time integral of each term on the right-hand side of the equation is simply the area under that expression, we note that the cosine and sine terms all integrate to zero because they each have as much positive area as they do negative area in an integer number of periods. The integral of the constant term on the right is not zero, however, i.e.,

$$\int_{t_0}^{t_0+T} f(t)\,dt = \frac{a_0}{2} \int_{t_0}^{t_0+T} dt + a_1 \int_{t_0}^{t_0+T} \cos \omega_0 t\,dt + a_2 \int_{t_0}^{t_0+T} \cos 2\omega_0 t\,dt + \cdots$$

$$+ b_1 \int_{t_0}^{t_0+T} \sin \omega_0 t \; dt + b_2 \int_{t_0}^{t_0+T} \sin 2\omega_0 t \; dt + \cdots$$

$$= \frac{a_0}{2} T$$

Therefore

$$a_0 = \frac{1}{T/2} \int_{t_0}^{t_0+T} f(t) \; dt \qquad (13.3.2)$$

We note that a_0 *equals twice the time average value* of $f(t)$.

A similar method can be used to evaluate all the other coefficients: a_k and b_k for $k > 0$. In order to do this we use a three-step method:

1. Multiply both sides of equations (13.3.1) by a suitable factor.
2. Integrate the result term by term over one period of the fundamental. (If we have picked the correct factor in step 1, all terms except the one containing the coefficient we want to evaluate will integrate to zero.)
3. Simplify.

An expression for the coefficient a_k of the kth harmonic can be found by these three steps. Multiply equation (13.3.1) through by $\cos k\omega_0 t$, then integrate from any arbitrary time, t_0, to $t_0 + T$, and finally simplify. Thus,

$$\int_{t_0}^{t_0+T} f(t) \cos k\omega_0 t \; dt = \frac{a_0}{2} \int_{t_0}^{t_0+T} \cos k\omega_0 t \; dt \qquad \text{(type 1)}$$

$$+ a_1 \int_{t_0}^{t_0+T} \cos \omega_0 t \cos k\omega_0 t \; dt \qquad \text{(type 2)}$$

$$+ \cdots + a_k \int_{t_0}^{t_0+T} \cos^2 k\omega_0 t \; dt + \cdots \qquad \text{(type 3)}$$

$$+ b_1 \int_{t_0}^{t_0+T} \sin \omega_0 t \cos k\omega_0 t \; dt + \cdots \qquad \text{(type 4)}$$

$$+ b_k \int_{t_0}^{t_0+T} \sin k\omega_0 t \cos k\omega_0 t \; dt + \cdots \qquad \text{(type 5)}$$

There are several types of integrals involved here:

- *Type 1.* Area under one period of a cosine wave = 0. (Or work it out.)

- *Type 2.*

$$\int_{t_0}^{t_0+T} \cos a \cos b \; dt$$

where $a = m\omega_0 t$ and $b = n\omega_0 t$. Use

$$\cos (a + b) = \cos a \cos b - \sin a \sin b$$

$$\cos (a - b) = \cos a \cos b + \sin a \sin b$$

$$\overline{\cos (a + b) + \cos (a - b) = 2 \cos a \cos b}$$

So the type 2 integral is

$$\int_{t_0}^{t_0+T} \tfrac{1}{2}[\cos\,(m+n)\omega_0\,t + \cos\,(m-n)\omega_0\,t]\,dt = 0$$

- *Type 3.*

$$a_k \int_{t_0}^{t_0+T} \cos^2 k\omega_0\,t\,dt$$

Use

$$\cos^2 k\omega_0\,t = \frac{\cos 2k\omega_0\,t + 1}{2}$$

So the type 3 integral is

$$\frac{a_k}{2}\int_{t_0}^{t_0+T} \cos 2k\omega_0\,t\,dt + \frac{a_k}{2}\int_{t_0}^{t_0+T} dt = \frac{a_k}{2}\,T$$

- *Type 4.*

$$b_n \int_{t_0}^{t_0+T} \sin n\omega_0\,t\,\cos k\omega_0\,t\,dt$$

Use

$$\sin\,(a+b) = \sin a \cos b + \cos a \sin b$$
$$\sin\,(a-b) = \sin a \cos b - \cos a \sin b$$

$$\overline{\sin\,(a+b) + \sin\,(a-b) = 2 \sin a \cos b}$$

So the type 4 integral is

$$\frac{b_n}{2}\int_{t_0}^{t_0+T} [\sin\,(n+k)\omega_0\,t + \sin\,(n-k)\omega_0\,t]\,dt = 0$$

- *Type 5.*   This is the same as type 4 except that $n = k$, so the second integral equals zero immediately. But again the integral is equal to zero.

The net result is that the only nonzero-valued integral is type 3. Thus

$$\int_{t_0}^{t_0+T} f(t) \cos k\omega_0\,t\,dt = \frac{a_k}{2}\,T$$

or

$$\boxed{a_k = \frac{1}{T/2}\int_{t_0}^{t_0+T} f(t) \cos k\omega_0\,t\,dt}$$

(13.3.3)

Note that this is valid for all harmonics $0 \le k$, including a_0.

By a similar procedure, multiplying through by $\sin k\omega_0\,t$ and then by

$$\int_{t_0}^{t_0+T} dt$$

we find

$$b_k = \frac{1}{T/2} \int_{t_0}^{t_0 + T} f(t) \sin k\omega_0 t \, dt \qquad (13.3.4)$$

This is valid for $k \geq 1$ since $b_0 \triangleq 0$, by definition.

EXAMPLE 13.3.1

Find the Fourier series for a full-wave-rectified sine wave of amplitude A.

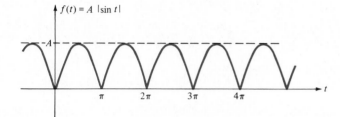

figure 13.3.1
The waveform of Example 13.3.1.

ANS.:
$$f(t) = A |\sin t|$$

$$T = \pi \text{ seconds} = \text{period} \qquad \text{so } \omega_0 = 2\pi f = \frac{2\pi}{T} = 2 \text{ rad/s}$$

$$a_0 = \frac{1}{\pi/2} \int_0^\pi A \sin t \, dt = \frac{2A}{\pi} \left(-\cos t \right) \Big|_0^\pi = \frac{2A}{\pi} (1 + 1) = \frac{4A}{\pi}$$

$$a_k = \frac{1}{\pi/2} \int_0^\pi A \sin \underbrace{t}_{b} \cos \underbrace{k2t}_{a} \, dt = \frac{2A}{\pi} \int_0^\pi \sin b \cos a \, dt$$

$$= \frac{2A}{\pi 2} \int_0^\pi \sin (a + b) - \sin (a - b) \, dt$$

$$= \frac{A}{\pi} \int_0^\pi \sin (2k + 1)t - \sin (2k - 1)t \, dt$$

$$= \frac{A}{\pi} \left[\frac{-\cos (2k + 1)t}{2k + 1} \Big|_0^\pi + \frac{\cos (2k - 1)t}{2k - 1} \Big|_0^\pi \right]$$

$$= \frac{A}{\pi} \left[\frac{-1}{2k + 1} (-1 - 1) + \frac{1}{2k - 1} (-1 - 1) \right]$$

$$= \frac{A}{\pi} \left(\frac{2}{2k + 1} - \frac{2}{2k - 1} \right) = \frac{A}{\pi} \left(\frac{4k - 2 - 4k - 2}{4k^2 - 1} \right)$$

$$= \frac{A}{\pi} \left(\frac{4}{1 - 4k^2} \right) \qquad \text{[Note this gives the correct value for } a_0 \text{ also}$$
$$\text{(so we did } not \text{ have to compute } a_0 \text{ separately).]}$$

$$b_k = \frac{1}{\pi/2} \int_0^\pi A \underbrace{\sin t}_{b} \underbrace{\sin k2t}_{a} \, dt$$

$$= \frac{A}{\pi} \int_0^\pi \cos(2k-1)t - \cos(2k+1)t \, dt$$

$$= \frac{A}{\pi} \left[\frac{\sin(2k-1)t}{2k-1} \bigg|_0^\pi - \frac{\sin(2k+1)t}{2k+1} \bigg|_0^\pi \right]$$

$$= \frac{A}{\pi} [0 - 0 - (0 - 0)] = 0 \qquad \text{for all } k$$

The coefficients α_k in the complex exponential form of the Fourier series may be evaluated via a technique similar to that used to find a_k and b_k:

1. Multiply through by a suitable factor.
2. Integrate the result term by term over the period of the fundamental.
3. Simplify.

The exponential form of the Fourier series is

$$f(t) = \sum_{k=-\infty}^{\infty} \alpha_k e^{jk\omega_0 t} \tag{13.3.5}$$

Multiply by $(e^{jn\omega_0 t})^* = e^{-jn\omega_0 t}$:

$$f(t)e^{-jn\omega_0 t} = \sum_{k=-\infty}^{\infty} \alpha_k e^{j(k-n)\omega_0 t} \tag{13.3.6}$$

Integrate over t_0, $t_0 + T$:

$$\int_{t_0}^{t_0+T} f(t)e^{-jn\omega_0 t} \, dt = \int_{t_0}^{t_0+T} \sum_{k=-\infty}^{\infty} \alpha_k e^{j(k-n)\omega_0 t} \, dt \tag{13.3.7}$$

or
$$\int_{t_0}^{t_0+T} f(t)e^{-jn\omega_0 t} \, dt = \sum_{k=-\infty}^{\infty} \int_{t_0}^{t_0+T} \alpha_k e^{j(k-n)\omega_0 t} \, dt \tag{13.3.8}$$

Let $k - n = m$:

$$\int_{t_0}^{t_0+T} f(t)e^{-jn\omega_0 t} \, dt = \sum_{k=-\infty}^{\infty} \frac{\alpha_k}{jm\omega_0} e^{jm\omega_0 t} \bigg|_{t_0}^{t_0+T} \tag{13.3.9}$$

$$= \sum_{k=-\infty}^{\infty} \frac{\alpha_k}{jm\omega_0} [e^{jm\omega_0(t_0+T)} - e^{jm\omega_0 t_0}]$$

$$= \sum_{k=-\infty}^{\infty} \frac{\alpha_k}{jm\omega_0} e^{jm\omega_0 t_0}(e^{jm\omega_0 T} - 1)$$

Recalling that $\omega_0 = 2\pi/T$ we have

$$\int_{t_0}^{t_0+T} f(t)e^{-jn\omega_0 t} \, dt = 0 \qquad \text{for } k \neq n \tag{13.3.10}$$

If $k = n$, then in equation (13.3.8)

$$\int_{t_0}^{t_0 + T} f(t)e^{-jk\omega_0 t}\, dt = \int_{t_0}^{t_0 + T} \alpha_k 1\, dt = \alpha_k T \qquad (13.3.11)$$

Thus

$$\boxed{\alpha_k = \frac{1}{T} \int_{t_0}^{t_0 + T} f(t)e^{-jk\omega_0 t}\, dt} \qquad (13.3.12)$$

EXAMPLE 13.3.2

Find the exponential form of the Fourier series for the waveform shown in Figure 13.3.2.

$$f(t) = \begin{cases} A & 0 < t < T_p \\ 0 & T_p < t < T \end{cases}$$

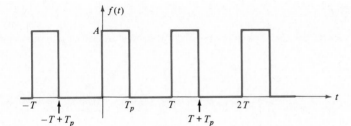

figure 13.3.2
The waveform of Example 13.3.2.

ANS.:

$$\alpha_k = \frac{1}{T} \int_0^{T_p} A e^{-jk\omega_0 t}\, dt$$

where $\omega_0 = 2\pi/T$.

$$\alpha_k = \frac{A}{T}\frac{1}{-jk\omega_0} e^{-jk\omega_0 t}\Big|_0^{T_p}$$

$$= -\frac{A}{jk2\pi}(e^{-jk(2\pi/T)T_p} - 1)$$

$$= \frac{A}{k\pi 2j} e^{-jk\pi(T_p/T)}(e^{jk\pi(T_p/T)} - e^{-jk\pi(T_p/T)})$$

$$= \frac{A}{k\pi} e^{-jk\pi(T_p/T)} \sin\left(k\pi\, \frac{T_p}{T}\right) \qquad \text{for all } k \neq 0$$

Therefore

$$f(t) = \sum_{k=-\infty}^{\infty} \alpha_k e^{jk(2\pi/T)t}$$

becomes

$$f(t) = \sum_{k=-\infty}^{\infty} \left[\underbrace{\frac{A}{\pi k} \sin\left(k\pi \frac{T_p}{T}\right) \bigg/ -\frac{k\pi T_p}{T}}_{\alpha_k} \right] e^{jk\omega_0 t}$$

Note that α_0 is of indeterminant form. Use L'Hospital's rule:

$$\alpha_0 = \lim_{k \to 0} \frac{A}{\pi k} \sin\left(k\pi \frac{T_p}{T}\right)$$

$$= \lim_{k \to 0} \frac{A}{\pi} \frac{(d/dk)[\sin k\pi(T_p/T)]}{(d/dk)[k]} = \frac{A}{\pi} \frac{\pi(T_p/T) \cos k\pi(T_p/T)}{1} \bigg|_{k=0}$$

$$= A \frac{T_p}{T}$$

The last expression is obviously the average value of the waveform.

13.4 symmetry

Certain types of symmetrical properties are often observed in periodic waveforms.

1. *Even* symmetry: $f(t) = f(-t)$.

EXAMPLE 13.4.1

$f(t) = \cos \omega_0 t$ has even symmetry.

---

2. *Odd* symmetry: $f(t) = -f(-t)$.

EXAMPLE 13.4.2

$f(t) = \sin \omega_0 t$ has odd symmetry.

---

3. *Half-wave* symmetry: $f(t + T/2) = -f(t)$. Waveforms with half-wave sym-

figure 13.4.1
$f(t) = f(-t)$.

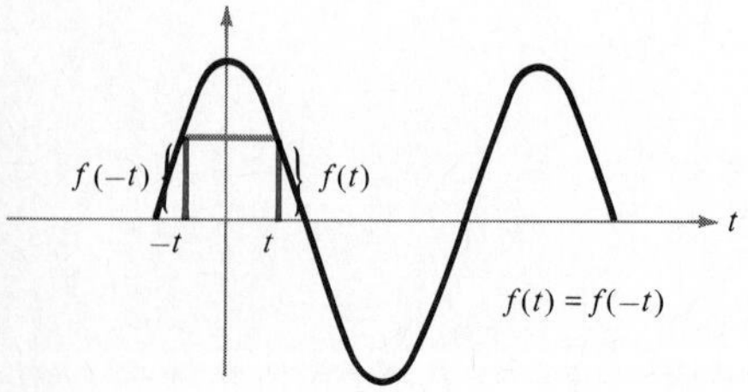

$f(t) = f(-t)$

figure 13.4.2
$f(t) = -f(-t)$.

$f(t) = -f(-t)$

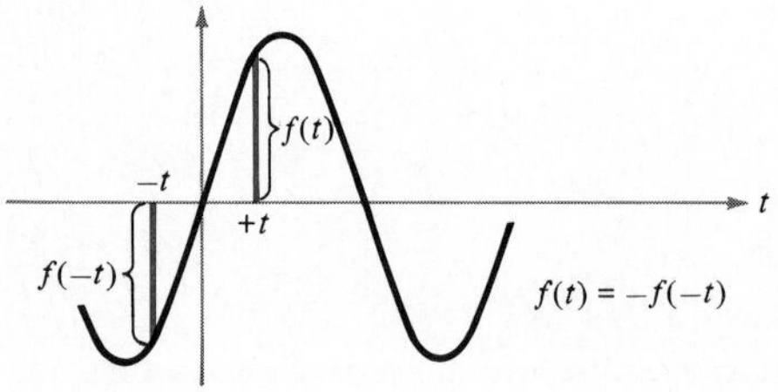

metry have identical upper and lower half-cycles (except that the last half is inverted).

EXAMPLE 13.4.3

a. The waveform in Figure 13.4.3a *has* half-wave symmetry.
b. The waveform in Figure 13.4.3b does *not* have half-wave symmetry.

figure 13.4.3
$(a)\, f_a(t) = -f_a(T/2 + t),\ (b)\, f_b(t) = -f_b(T/2 + t).$

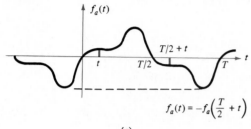

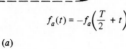

$$f_a(t) = -f_a\left(\frac{T}{2} + t\right)$$

(a)

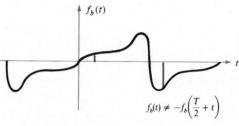

$$f_b(t) \neq -f_b\left(\frac{T}{2} + t\right)$$

(b)

---

4. *Quarter-wave* symmetry: A wave that has half-wave symmetry (as in item 3 above) may also be symmetrical about the midordinate of its positive and negative half-cycles (see Figure 13.4.4).

The Fourier series of a waveform that has one or more of the types of symmetries just mentioned will have certain resulting properties.

Consider the sin-cos form of the series:

$$f(t) = \frac{a_0}{2} + \sum_{k=1}^{\infty} a_k \cos k\omega_0 t + \sum_{k=1}^{\infty} b_k \sin k\omega_0 t$$

For odd functions $f(t)$, all $a_k = 0$. No even symmetry components (cosines) are allowed. For even functions $f(t)$, all $b_k = 0$. No odd symmetry components (sines) are allowed. For half-wave symmetrical $f(t)$, no even-numbered harmonics are allowed and $a_0 = 0$. For quarter-wave symmetrical $f(t)$, no even-numbered harmonics are allowed, $a_0 = 0$, and all nonzero harmonics pass through zero at the same instant of time.

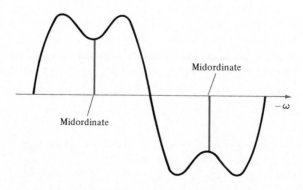

figure 13.4.4
A waveform possessing quarter-wave symmetry.

EXAMPLE 13.4.4

What types of symmetry does the waveform in Figure 13.4.5 possess? What can be stated about its Fourier series?

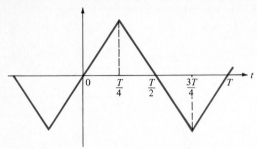

figure 13.4.5
The waveform of Example 13.4.4.

ANS.:   The function possesses odd symmetry: $f(t) = -f(-t)$, so all $a_k = 0$. The function possesses half-wave symmetry: $f(t + T/2) = -f(t)$, so all even harmonics are missing. The function possesses quarter-wave symmetry, so all harmonics go through zero simultaneously.

EXAMPLE 13.4.5

Sketch a balanced square wave with even symmetry and its first harmonic. Show how the presence of a nonzero sine (odd symmetric) component would destroy the even symmetry of the square wave.

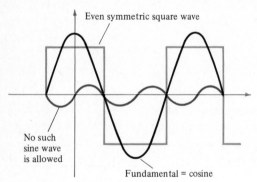

Even symmetric square wave

No such
sine wave
is allowed

Fundamental = cosine

figure 13.4.6
The answer to Example 13.4.5.

ANS.:   See Figure 13.4.6.

Any signal (periodic or not) may be broken down into the sum of its even and odd parts: $f(t) = f_e(t) + f_o(t)$. We can find a method for obtaining $f_e(t)$ and $f_o(t)$ as follows:
If

$$f(t) = f_e(t) + f_o(t) \qquad (13.4.1)$$

then

$$f(-t) = f_e(-t) + f_o(-t) \qquad (13.4.2)$$

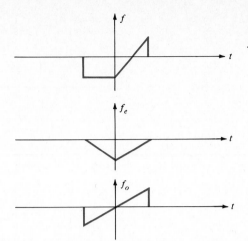

figure 13.4.7
The even and odd symmetry components,
$f_e(t)$ and $f_o(t)$ respectively, of a function $f(t)$.

or
$$f(-t) = f_e(t) - f_o(t) \qquad (13.4.3)$$

Adding equations (13.4.1) and (13.4.3) yields

$$f_e(t) = \tfrac{1}{2}[f(t) + f(-t)]$$

Subtracting (13.4.3) from (13.4.1) yields

$$f_o(t) = \tfrac{1}{2}[f(t) - f(-t)]$$

Breaking any arbitrary sinusoid down into the sum of a sine wave plus a cosine wave is a special case of this procedure.

EXAMPLE 13.4.6

Find the even and odd components of

$$f(t) = 2 \cos (3t + 30°)$$

ANS.: First find $f(-t)$:

$$f(-t) = 2 \cos (-3t + 30°)$$
$$= 2 \cos [-(3t - 30°)]$$
$$= 2 \cos (3t - 30°)$$

Now (via phasors), $f_e(t) = \tfrac{1}{2}[f(t) + f(-t)]$ and $f_o(t) = \tfrac{1}{2}[f(t) - f(-t)]$. So using

$$F_+ = 2\underline{/30°} \qquad \text{and} \qquad F_- = 2\underline{/-30°}$$

we have

$$F_e = \tfrac{1}{2}(F_+ + F_-)$$
$$= \tfrac{1}{2}(2\underline{/30°} + 2\underline{/-30})$$
$$= \sqrt{3}\underline{/0°}$$

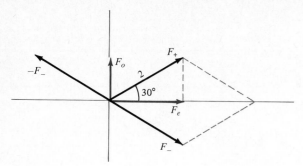

figure 13.4.8
The even and odd symmetry components of $f(t) = 2 \cos (3t + 30°)$.

and

$$F_o = \tfrac{1}{2}(F_+ - F_-)$$
$$= \tfrac{1}{2}(2\underline{/30°} - 2\underline{/-30°})$$
$$= 1\underline{/90°}$$

Therefore

$$f_e(t) = \sqrt{3} \ \cos 3t$$

and

$$f_o(t) = -\sin 3t$$

Any time we must calculate the Fourier coefficients of a periodic $f(t)$ we should first check what type (if any) symmetry the waveform possesses. This is because of the two following properties of symmetrical functions.

1. The integral of an even function over symmetrical limits is equal to twice the integral from zero to the upper limit:

$$\int_{-t_0}^{+t_0} f_e(t) \ dt = 2 \int_0^{t_0} f_e(t) \ dt$$

2. The integral of an odd function over symmetrical limits is equal to zero.

EXAMPLE 13.4.7

Take advantage of any symmetrical properties of $f(t)$ and find the sin-cos form of the Fourier series for

$$f(t) = \begin{cases} \dfrac{2A}{T} t & 0 \le t \le T/2 \\[2mm] -\dfrac{2A}{T} t & -\dfrac{T}{2} \le t \le 0 \end{cases} \qquad \text{and} \qquad f(t) = f(t + T)$$

ANS.:

$$f(t) = \frac{a_0}{2} + \sum_{k=1}^{\infty} (a_k \cos k\omega_0 t + b_k \sin k\omega_0 t)$$

$f(t) = f(-t)$ is an even function. Therefore, $b_k = 0$ and

$$a_k = \frac{2}{T} \int_{t_0}^{t_0 + T} f(t) \cos k\omega_0 t \ dt$$

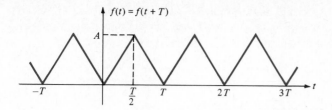

figure 13.4.9
The waveform in Example 13.4.7.

Both $f(t)$ and $\cos k\omega_0 t$ are even functions. The product of two even functions is also even.

$$\frac{2\pi}{T} = \omega_0$$

So

$$a_k = \frac{4}{T} \int_0^{T/2} \frac{2A}{T} t \cos k\omega_0 t \, dt$$

$$= \frac{8A}{T^2} \int_0^{\pi/\omega_0} t \cos k\omega_0 t \, dt$$

Integrating by parts gives

$$a_k = \frac{8A}{T^2} \left[\frac{1}{k^2 \omega_0^2} \cos k\omega_0 t + \underbrace{\frac{1}{k\omega_0} t \sin k\omega_0 t}_{} \right]_0^{\pi/\omega_0}$$

$$= \frac{-4A}{(k\pi)^2} \qquad \text{This term} = 0 \quad \text{for all } k = 1, 2, 3, \ldots$$

We find a_0 separately as follows:

$$a_0 = \frac{2}{T} \int_{t_0}^{t_0 + T} f(t) \, dt$$

Remember that

$$\int_{-l}^{+l} [\text{even } f(t)] = 2 \int_0^l [\text{even } f(t)] \, dt$$

Therefore

$$a_0 = \frac{4}{T} \int_0^{T/2} \frac{2At}{T} \, dt = \frac{8A}{T^2} \frac{t^2}{2} \bigg|_0^{T/2}$$

and so the average value of the waveform is

$$\frac{a_0}{2} = \frac{A}{2}$$

Thus $a_0 = A$ and

$$a_k = \begin{cases} 0 & \text{for } k \text{ even and } > 0 \\ \dfrac{-4A}{(k\pi)^2} & k \text{ odd} \end{cases}$$

$$b_k = 0$$

Since the waveform possesses half-wave symmetry, we could have predicted that its Fourier series would contain only odd harmonics. (But, unfortunately, that does not save us work because we have to compute a_k in general anyway. It does, however, help us check that our answer is a reasonable one.)

13.5 time displacement theorem

If a function $f(t)$ is delayed in time by t_1 seconds, then its Fourier series will have all t replaced by $(t - t_1)$, i.e.,

$$f(t) = \sum_{k=-\infty}^{\infty} \alpha_k e^{jk\omega_0 t}$$

becomes

$$f(t - t_1) = \sum_{k=-\infty}^{\infty} (\alpha_k e^{-jk\omega_0 t_1}) e^{jk\omega_0 t}$$

Thus

$$\underset{\text{function}}{(\alpha_k)_{\text{delayed}}} = \underset{\text{function}}{(\alpha_k)_{\text{original}}} \times \underbrace{(e^{-jk\omega_0 t_1})}$$

$\uparrow$
Represents a phase
shift of $-k\omega_0 t_1$ radians

So a time shift (lag) of t_1 seconds produces no change in the magnitude of the complex coefficients, but does produce a *phase shift* of $-k\omega_0 t_1$ radians in those coefficients (α_k and/or ϕ_k). Thus the phase shift introduced into each harmonic is proportional to the harmonic number of that term. In reverse this theorem says: if two functions have the same† harmonic amplitudes and if the phase angles differ so that the phase shift between corresponding harmonics is proportional to the number of that harmonic, then the two functions have the same waveform (except for time delay).

EXAMPLE 13.5.1

Do these two signals have same waveform?

$$v(t) = 100 \sin(\omega_0 t + 30°) - 50 \sin(3\omega_0 t - 60°) + 25 \sin(5\omega_0 t + 40°)$$

$$i(t) = 10 \sin(\omega_0 t - 60°) + 5 \sin(3\omega_0 t - 150°) + 2.5 \cos(5\omega_0 t - 140°)$$

ANS.: Magnitudes of corresponding harmonics all have a common ratio, 10, so $v(t)$ and $i(t)$ *may* have the same shape. In regard to phase, in the first harmonic, v leads i by 90°. In the third harmonic, v must lead i by $3 \times 90° = 270°$. (It does.) In the fifth harmonic, v must lead i by $5 \times 90° = 450°$. (It does.) See Figure 13.5.1. Conclusion: the waveforms have the same shape except for size and phase shift.

† If the magnitudes of each pair of corresponding Fourier coefficients have a common ratio, then the two waveforms may have the same shape but with different amplitudes.

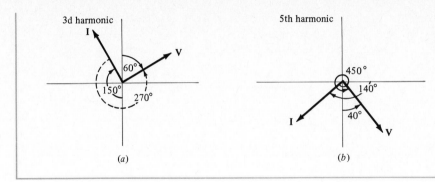

figure 13.5.1
Phasor diagrams for (a) the third and (b) the fifth harmonic terms of the waveforms in Example 13.5.1.

line spectra 13.6

In the complex exponential form of the Fourier series, in each exponential term that rotates on the complex plane in the *negative* direction, the minus sign can be thought of as being associated with the radian frequency, i.e.,

$$\alpha_{-k} e^{-jk\omega_0 t} = \alpha_{-k} e^{jk(-\omega_0)t}$$

The magnitude of the complex coefficient α_k of each exponential term can be plotted versus frequency. Any such plots will therefore have to be made on a horizontal axis that contains negative as well as positive values of ω. Similarly, the phase angle of each coefficient can be plotted versus frequency $(-\infty < \omega < +\infty)$. These plots are called, respectively, the *magnitude spectrum* and *phase spectrum* of the waveform $f(t)$. This nomenclature is consistent with the use of the word spectrum in describing the color (frequency) of a visible light source. Light, however, is usually (with the exception of laser light) spread out continuously over a range of frequencies. The spectrum of a periodic waveform, on the other hand, contains components only at *discrete frequencies*, so we call it a *discrete* or *line spectrum*. The magnitude and phase of all counterclockwise-rotating complex exponentials are plotted on the negative ω axes of the respective spectra.

EXAMPLE 13.6.1

Given the Fourier series

$$f(t) = \frac{4}{\pi} \left(\cos \omega_0 t + \tfrac{1}{3} \cos 3\omega_0 t + \tfrac{1}{5} \cos 5\omega_0 t + \cdots \right)$$

$$= \begin{cases} \dfrac{4}{\pi} \displaystyle\sum_{k=1}^{\infty} \dfrac{1}{k} \cos k\omega_0 t & k \text{ odd} \\ 0 & k \text{ even} \end{cases}$$

a. Find the exponential form of the series.
b. Plot the discrete spectrum of this waveform (the magnitude and the angle of α_k, each versus ω).

ANS.:   (a) Use

$$\alpha_k = \frac{c_k}{2}\, e^{j\phi_k}$$

In this example all $\phi_k = 0$ and

$$c_k = \begin{cases} \dfrac{4}{\pi k} & k \text{ odd} \\[2mm] 0 & k \text{ even} \end{cases}$$

$$f(t) = \sum_{k=-\infty}^{\infty} \frac{2}{\pi k}\, e^{-jk\omega_0 t}$$

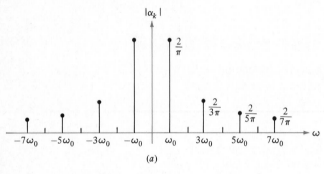

(a)

figure 13.6.1
(a) The amplitude spectrum and (b) the phase spectrum of the balanced square wave of Example 13.6.1.

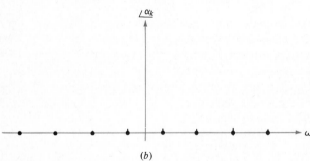

(b)

(b) The amplitude spectrum is shown in Figure 13.6.1a. The phase spectrum is zero because every $\phi_k = 0$.

It always should be remembered that each of these plots is defined only for $\pm\omega_0$, $\pm3\omega_0$, $\pm5\omega_0$, etc. The phase spectrum is equal to zero at these points.

EXAMPLE 13.6.2

Plot the magnitude and phase spectra for the periodic pulse waveform of Example 13.3.2. Choose a numerical value of $\frac{1}{6}$ for the *duty cycle*: $T_p/T = \frac{1}{6}$.

ANS.:   From Example 13.3.2 we see that

$$\alpha_k = \frac{A}{\pi k} \sin\left(k\pi\, \frac{T_p}{T} \right) \Big/ \!-k\pi\, \frac{T_p}{T}$$

$$= A\left(\frac{T_p}{T}\right)\left[\frac{\sin\left(k\pi\,\frac{T_p}{T}\right)}{k\pi\,\frac{T_p}{T}}\right]\bigg/\!-k\pi\,\frac{T_p}{T}$$

Tabulating values gives

k	$\dfrac{k\pi}{6}$	$\sin\dfrac{k\pi}{6}$	$\dfrac{\sin\dfrac{k\pi}{6}}{\dfrac{k\pi}{6}}$
0	0	0	1
1	0.524 (30°)	0.5	0.954
2	1.047 (60°)	0.866	0.827
3	1.571 (90°)	1	0.637
4	2.094 (120°)	0.866	0.414
5	2.618 (150°)	0.5	0.191
6	3.142 (180°)	0	0
7	3.665 (210°)	−0.5	−0.136
8	4.189 (240°)	−0.866	−0.207
9	4.712 (270°)	−1	−0.212
10	5.236 (300°)	−0.866	−0.165

Recall, as always, that the fundamental frequency ω_0 depends solely on the period T.

$$\omega_0 = \frac{2\pi}{T}$$

Note, particularly, that the frequency spacing between every pair of adjacent harmonics is ω_0. We plot the quantities

$$\frac{A}{6}\frac{\sin(k\pi/6)}{k\pi/6} \qquad \text{and} \qquad \frac{-k\pi}{6}$$

versus k and ω in Figure 13.6.2a. Or, in terms of the magnitude and angle of α_k, we obtain the magnitude spectrum and the phase spectrum as shown in Figure 13.6.2b.

There are several observations and conclusions which we may make about the line spectrum of a periodic sequence of rectangular pulses:

1. The width of the envelope of the magnitude spectrum is defined as the frequency interval between the nulls (frequencies at which it touches the horizontal axis). We see from the tabulation in Example 13.6.2 that the first null occurs when the argument of the sine function is $\pm180°$. In general an envelope null will occur whenever the sine function argument

$$k\pi\,\frac{T_p}{T} = n\pi$$

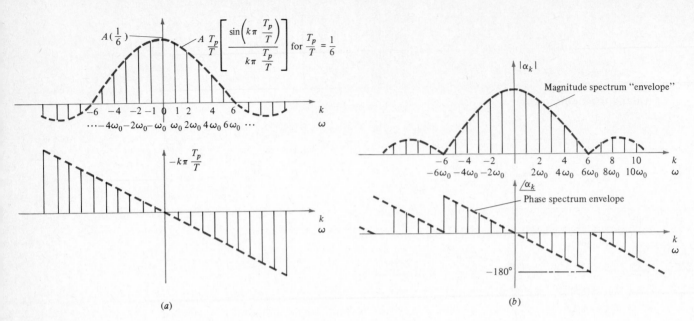

figure 13.6.2
(a) Plots of the tabulations made in Example 13.6.2. (b) The actual magnitude and phase spectra.

or
$$k_{\text{null}} = n \frac{T}{T_p} \quad \text{for any integer } -\infty < n < \infty$$

(In Example 13.6.2 envelope nulls are at $k = \pm 6, \pm 12, \pm 18, \ldots$.) In terms of frequency these nulls occur at

$$\omega_{\text{null}} = k_{\text{null}} \omega_0$$

$$= n \frac{T}{T_p} \frac{2\pi}{T}$$

$$= \frac{n 2\pi}{T_p} \quad \text{rad/s}$$

or
$$f_{\text{null}} = \frac{\omega_{\text{null}}}{2\pi}$$

$$= \frac{n}{T_p}$$

$$= \pm \frac{1}{T_p}, \pm \frac{2}{T_p}, \pm \frac{3}{T_p}, \ldots \quad \text{hertz}$$

Note that there is no null at $f = 0$. The nulls at $-1/T_p$ and $+1/T_p$ are twice as far apart as any other two adjacent nulls. Note carefully that the specific frequencies of the nulls in the magnitude envelope *do not depend in any way* on the period T of the waveform $f(t)$.

2. If T_p remains fixed but T increases, then since $\omega_0 = 2\pi/T$ is the spacing between adjacent harmonics, the number of spectral lines between $f = 0$ and the first null at $f = 1/T_p$ will increase. But the size and shape of the envelope will not change at all.

filtering 13.7

Suppose we are asked to find the output voltage (or current) from a linear system, given that the input is a periodic, nonsinusoidal signal. Using the Fourier series to break down the input signal into its sinusoidal components, we may then pass each component separately through the system. (The standard methods for handling sinusoidal inputs—phasors, impedances, etc.—can be used.) Superposition tells us that the total output signal is the summation of all the resulting sinusoidal output components.

This total output waveform is the *steady-state output* due to the nonsinusoidal input signal. It is the *particular response* of the system to this nonsinusoidal input. In other words, we assume that this input has been exciting the system for a long time so that any transient response (natural response due to initial conditions that might have been present when the input was first applied) has long since died away.

EXAMPLE 13.7.1

What is the output of a simple low-pass filter with transfer function

$$H(s) = \frac{1}{s + 1}$$

or

$$H(j\omega) = \frac{1}{1 + j\omega}$$

when the input signal is a balanced square wave of magnitude $\pm V_m$ having period $T = 4\pi$ seconds?
ANS.: The Fourier series for this input waveform (see Example 13.2.1) is

$$v_1(t) = \frac{4V_m}{\pi} \left(\cos \omega_0 t - \tfrac{1}{3} \cos 3\omega_0 t + \tfrac{1}{5} \cos 5\omega_0 t - \cdots \right)$$

i.e.,

$$c_k = \frac{-4V_m}{\pi k} (-1)^{(k+1)/2} \quad \text{and} \quad \phi_k = 0 \quad k \text{ odd}$$

and

$$\omega_0 = \frac{2\pi}{T} = 0.5 \text{ rad/s}$$

The output Fourier coefficients d_k and θ_k are obtained as follows: The value of each d_k is equal to the corresponding c_k multiplied by $|H(jk\omega_0)|$, the magnitude of the transfer function evaluated at the radian frequency $k\omega_0$. The phase angle θ_k of the kth harmonic of the output signal is the sum of ϕ_k plus $\underline{/H(jk\omega_0)}$, the angle of the transfer function evaluated at that same radian frequency, i.e.,

$$d_k = c_k |H(jk\omega_0)|$$

$$= \frac{-4V_m}{\pi k}(-1)^{(k+1)/2}\frac{1}{\sqrt{1+(k\omega_0)^2}}$$

$$= \frac{-4V_m(-1)^{(k+1)/2}}{\pi k\sqrt{1+(k/2)^2}} \quad k \text{ odd}$$

and

$$\theta_k = \phi_k + [-\tan^{-1}(k\omega_0)]$$

$$= -\tan^{-1}(k\omega_0)$$

$$= -\tan^{-1}\left(\frac{k}{2}\right) \quad k \text{ odd}$$

since $\phi_k = 0$ for all k.

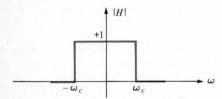

figure 13.7.1
An ideal low-pass filter's magnitude
response.

An ideal (nonrealizable) low-pass filter has a transfer function $H(j\omega)$ that would be as follows:

$$\underline{/H(j\omega)} = 0 \qquad |H(j\omega)| = \begin{cases} 1 & -\omega_c < \omega < +\omega_c \\ 0 & \text{elsewhere} \end{cases}$$

This says that, if a nonsinusoidal periodic signal is used as input for this filter, the output will consist of sinusoidal Fourier components of the input whose radian frequencies are lower than ω_c.

EXAMPLE 13.7.2

Given a periodic signal $f_{in}(t)$ as shown in Figure 13.7.2. $T_p = T/6$. Find $f_o(t)$. H is an ideal low-pass filter with cutoff frequency at

$$\omega_c = 2.5 \times 10^6 \text{ rad/s.} \quad T = 2\pi \times 10^{-6} \text{ s.}$$

figure 13.7.2
The system in Example 13.7.2 and its input waveform.

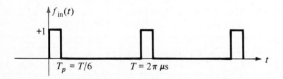

ANS.: From Examples 13.3.2 and 13.6.2 we know the values of α_k:

$k =$	1	2	3	4	$\cdots$
$|\alpha_k| =$	$\frac{1}{6}(0.954)$	$\frac{1}{6}(0.827)$	$\frac{1}{6}(0.637)$	$\frac{1}{6}(0.414)$	$\cdots$
$=$	0.159	0.138	0.106	0.069	$\cdots$
$\underline{/\alpha_k} =$	$-30°$	$-60°$	$-90°$	$-120°$	$\cdots$
$=$	$-\pi/6$	$-2\pi/6$	$-3\pi/6$	$-4\pi/6$	$\cdots$

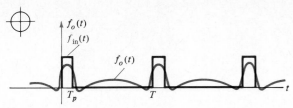

figure 13.7.3
The pulse train of Example 13.7.2 after passing through the ideal
filter of Figure 13.7.1.

where

$$\omega_0 = \frac{2\pi}{T} = \frac{2\pi}{2\pi \times 10^{-6}} = 10^6 \quad \text{and} \quad \alpha_0 = \frac{T_p}{T} = \frac{1}{6}$$

The filter lets the first and second harmonics and the dc value through, but no others. Therefore, the output will be

$$f_o(t) = 0.138e^{j\pi/3}e^{-j2 \times 10^6 t} + 0.159e^{j\pi/6}e^{-j10^6 t}$$

$$+ \tfrac{1}{6} + 0.159e^{-j\pi/6}e^{j10^6 t} + 0.138e^{-j\pi/3}e^{j2 \times 10^6 t}$$

or $\qquad f_o(t) = \tfrac{1}{6} + 2(0.159)\cos(10^6 t - 30°) + 2(0.138)\cos(2 \times 10^6 - 60°)$

EXAMPLE 13.7.3

Find the Fourier series description of the steady-state output voltage of the circuit in Figure 13.7.4a when the input current has the waveform shown in Figure 13.7.4b.

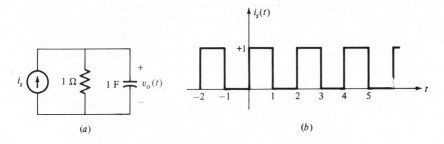

figure 13.7.4
(a) The circuit of Example 13.7.3 and
(b) its input signal.

ANS.: Reconstruct the $v_o(t)$ signal component by component.

$$V_o(j\omega) = I(j\omega)Z(j\omega)$$

where

$$Z(j\omega) = \frac{R\,\dfrac{1}{j\omega C}}{R + \dfrac{1}{j\omega C}} = \frac{1}{j\omega + 1} \quad \text{for } R = C = 1$$

The Fourier series for the input $i_s(t)$ is

$$i_s(t) = \sum_{-\infty}^{\infty} \alpha_k e^{jk\omega_0 t} \qquad \omega_0 = \frac{2\pi}{T} = \pi$$

where, from Example 13.6.2,

$$\alpha_k = \frac{1}{2} \frac{\sin \dfrac{k\pi}{2}}{\dfrac{k\pi}{2}} \left\lfloor -k\frac{\pi}{2} \right.$$

Therefore

$$v_o(t) = \sum_{k=-\infty}^{\infty} \beta_k e^{jk\omega_0 t}$$

where

$$\beta_k = \alpha_k \frac{1}{1+j\omega}$$

or

$$\beta_k = \frac{e^{-jk\pi/2}}{2(1+jk\pi)} \frac{\sin \dfrac{k\pi}{2}}{\dfrac{k\pi}{2}}$$

Based on the discussion in Section 13.5 it should be clear that for an arbitrarily shaped periodic waveform to pass undistorted through a linear system† (e.g., a hi-fi amplifier), any phase shifts introduced by the system must be proportional to harmonic number (to frequency). In other words it is necessary that the angle of the system's transfer function $\underline{/H(\omega)}$ be a linear function of frequency.

Suppose we have a system such as the one shown in Figure 13.7.5a wherein the input signal contains several harmonics, all of which are within the passband of the filter. In order for $v_{in}(t)$ and $v_o(t)$ to have the same shape (but allowing for a difference in their amplitudes and also a possible time shift between them), then $H(\omega)$ must be as shown in Figure 13.7.5b and c. The value of the slope of the straight-line phase plot (when plotted on a linear frequency axis) is unimportant; the steeper it is, the longer the time delay between $v_{in}(t)$ and $v_o(t)$. But it must be a straight line—otherwise the output will be distorted.

Incidentally, it would seem that this principle should be quite important in the field of high-fidelity sound reproduction. However, in practice it turns out that the human ear is less able to distinguish phase distortion over medium and high

† Called *distortionless transmission.*

figure 13.7.5
(a) A linear system, with magnitude response
(b), and phase response (c).

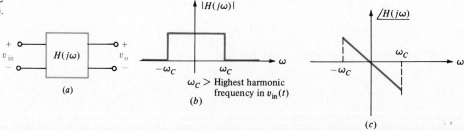

(a)

(b)

ω_C > Highest harmonic frequency in $v_{in}(t)$

(c)

audio frequencies than is the eye when such signals are displayed on an oscillo-scope. Therefore, the linear (straight line proportional to frequency) requirement can be relaxed somewhat for audio components—and usually is.

vectors in space related to functions 13.8
of time

Consider a three-dimensional orthogonal set of axes defined by the three unit vectors $\mathbf{u}_1$, $\mathbf{u}_2$, and $\mathbf{u}_3$:

$$\mathbf{u}_1 = \begin{bmatrix} 1 \\ 0 \\ 0 \end{bmatrix} \quad \mathbf{u}_2 = \begin{bmatrix} 0 \\ 1 \\ 0 \end{bmatrix} \quad \mathbf{u}_3 = \begin{bmatrix} 0 \\ 0 \\ 1 \end{bmatrix} \tag{13.8.1}$$

In some vector spaces (regardless of their dimensions) there is defined an oper-ation called the *inner product* (or *dot product*). For any two vectors such as $\mathbf{v}_1$ and $\mathbf{v}_2$ in geometrical three-dimensional space, the inner product is defined as the product of the magnitudes of the vectors and the cosine of the angle between them:

$$\langle \mathbf{v}_1, \mathbf{v}_2 \rangle = \mathbf{v}_1 \cdot \mathbf{v}_2 = |\mathbf{v}_1| \, |\mathbf{v}_2| \cos\sphericalangle \begin{smallmatrix} \mathbf{v}_1 \\ \mathbf{v}_2 \end{smallmatrix} \tag{13.8.2}$$

Orthogonal unit vectors such as $\mathbf{u}_1$, $\mathbf{u}_2$, and $\mathbf{u}_3$ in Figure 13.8.1 thereby have the property

$$\langle \mathbf{u}_i, \mathbf{u}_j \rangle = \mathbf{u}_i \cdot \mathbf{u}_j = \begin{cases} 0 & \text{if } i \neq j \\ 1 & \text{if } i = j \end{cases} \tag{13.8.3}$$

If the inner product of any two vectors turns out as it does in equation (13.8.3), we say the vectors are *orthonormal*. To be *orthogonal*, two vectors simply must have directions such that

$$\langle \mathbf{v}_i, \mathbf{v}_j \rangle = \mathbf{v}_i \cdot \mathbf{v}_j = \begin{cases} 0 & \text{if } i \neq j \\ a & \text{a constant } (\neq 0) \text{ if } i = j \end{cases}$$

and the constant need not be unity. If two vectors are orthogonal, and in addi-tion the value of the constant is equal to unity, then these vectors are called orthonormal. We see that our $\mathbf{u}_1$, $\mathbf{u}_2$, and $\mathbf{u}_3$ make up a set of orthonormal vectors.

Inner products are very useful for finding the projection (shadow) of one vector on another.

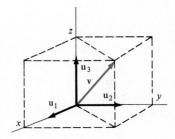

figure 13.8.1
$\mathbf{v} = x\mathbf{u}_1 + y\mathbf{u}_2 + z\mathbf{u}_3$

EXAMPLE 13.8.1

Find the projection of $\mathbf{v}$ on the $\mathbf{u}_1$ axis. (See Figure 13.8.1.)

ANS.:
$$\mathbf{u}_1 \cdot \mathbf{v} = \mathbf{u}_1 \cdot (x\mathbf{u}_1 + y\mathbf{u}_2 + z\mathbf{u}_3)$$
$$= x(\underbrace{\mathbf{u}_1 \cdot \mathbf{u}_1}_{1}) + y(\underbrace{\mathbf{u}_1 \cdot \mathbf{u}_2}_{0}) + z(\underbrace{\mathbf{u}_1 \cdot \mathbf{u}_3}_{0})$$
$$= x$$

Consider the properties of ordinary geometric vectors. Let **X**, **Y**, and **Z** be vectors and a and b scalars.

1. $(\mathbf{X} + \mathbf{Y}) + \mathbf{Z} = \mathbf{X} + (\mathbf{Y} + \mathbf{Z})$
2. There is a zero vector **0** such that $\mathbf{0} + \mathbf{X} = \mathbf{X}$
3. The number -1 multiplying **X** produces the negative of **X** such that $(-1)\mathbf{X} + \mathbf{X} = 0$
4. $\mathbf{X} + \mathbf{Y} = \mathbf{Y} + \mathbf{X}$
5. $a(b\mathbf{X}) = (ab)\mathbf{X}$
6. $(a + b)\mathbf{X} = a\mathbf{X} + b\mathbf{X}$
7. $a(\mathbf{X} + \mathbf{Y}) = a\mathbf{X} + a\mathbf{Y}$

Any set of objects **X**, **Y**, **Z**, ... that satisfies these seven properties is called a *vector space*. The objects themselves are called vectors. If, in addition to the above seven properties, a dot (inner) product is defined for the vectors, the set is then called an *inner product space*. Ordinary three-dimensional geometric vectors form an inner product space.

We can extend the concept of the inner product space, along with all of its ideas about orthogonality and projections of one vector on another, etc., to functions of time. We must redefine what we mean by the phrase inner product, because the definition used with geometrical vectors (product of the magnitudes and the cosine of the angle between them) has no meaning when applied to functions of time. There is no such thing as the angle between two functions.

The seven properties of a vector space apply to time functions (check it out). Therefore any set of time functions $f_i(t)$ may be considered to be the elements of a vector space. If we define an inner product operation in addition, then all $f_i(t)$ functions become members of an inner product space.

Let us define the inner product between two time functions to be

$$\langle f_1, f_1 \rangle = \int_{t_1}^{t_2} f_1(t) f_2^*(t)\, dt$$

where $f_2^*(t)$ denotes the complex conjugate of $f_2(t)$. Observe that if f_1 and f_2 are real-valued functions, we do not have to take the complex conjugate. The inner product thus defined is a *functional*. A functional is a function whose domain consists of functions and whose range is the real numbers (i.e., two functions are fed into the inner product functional and out comes a number $\langle f_1, f_2 \rangle$).

We now can talk about time functions that are orthogonal to one another. Of course, the same two functions that are orthogonal to each other over one interval of time may not be orthogonal over some other interval.

EXAMPLE 13.8.2

Are $f_1(t) = \sin t$ and $f_2(t) = \cos t$ orthogonal over the range $0 < t < \pi$?

ANS.:
$$\int_0^\pi f_1(t) f_2^*(t)\, dt = \int_0^\pi \sin t \cos t\, dt$$

$$= \frac{1}{2} \int_0^\pi \sin 2t\, dt = -\tfrac{1}{4} \cos 2t \Big|_0^\pi = -\tfrac{1}{4}(\cos 2\pi - \cos 0)$$

$$= 0$$

Therefore the answer is yes. How about over the range $0 < t < \pi/2$?

EXAMPLE 13.8.3

Consider the family of functions which are the components of the exponential form of the Fourier series:

$$f_k = e^{jk\omega_0 t}$$

where $\omega_0 = 2\pi/T$. Is this family of functions an orthogonal set of functions over the range $t_0 < t < t_0 + T$?

ANS.:
$$\langle f_k, f_m \rangle = \int_{t_0}^{t_0+T} e^{j(k-m)\omega_0 t} \, dt$$

Note that the minus sign in the exponent comes from taking the conjugate of $f_m(t)$.

For $k = m$,

$$\langle f_k, f_m \rangle = \int_{t_0}^{t_0+T} (1)dt = t_0 + T - t_0 = T$$

For $k \neq m$

$$\langle f_k, f_m \rangle = \int_{t_0}^{t_0+T} e^{jn\omega_0 t} \, dt = \frac{e^{jn\omega_0 t}}{jn\omega_0} \Big|_{t_0}^{t_0+T} \qquad \text{where } n = k - m$$

$$= \frac{1}{jn\omega_0} \left[e^{jn\omega_0(t_0 + T)} - e^{jn\omega_0 t_0} \right]$$

$$= \frac{e^{jn\omega_0 t_0}}{jn\omega_0} (e^{jn\omega_0 T} - 1)$$

$$= \frac{e^{jn\omega_0 t_0}}{jn\omega_0} \underbrace{(e^{jn2\pi} - 1)}_{= 0 \quad \text{for all integer values } n}$$

So these functions are an orthogonal set.

EXAMPLE 13.8.4

Are the family of functions that make up the sin-cos form of the Fourier series an orthogonal set of functions over the interval $t_0 < t < t_0 + T$?

ANS.: The set is $[1, \cos \omega_0 t, \cos 2\omega_0 t, \ldots, \cos n\omega_0 t, \ldots, \sin \omega_0 t, \sin 2\omega_0 t, \ldots, \sin n\omega_0 t, \ldots]$, where $\omega_0 = 2\pi/T$. The inner products among the functions are

$$\int_{t_0}^{t_0+T} (1) \cos (m\omega_0 t) \, dt = 0 \qquad \text{for } m \neq 0$$

but for $m = 0$ the result is T. Also

$$\int_{t_0}^{t_0+T} (1)\sin(m\omega_0 t)\,dt = 0 \qquad\qquad \text{for all } m$$

$$\int_{t_0}^{t_0+T} \cos(m\omega_0 t)\cos(n\omega_0 t)\,dt = \begin{cases} 0 & m \neq n \\ T/2 & m = n \neq 0 \end{cases}$$

$$\int_{t_0}^{t_0+T} \sin(m\omega_0 t)\sin(n\omega_0 t)\,dt = \begin{cases} 0 & m \neq n \\ T/2 & m = n \neq 0 \end{cases}$$

$$\int_{t_0}^{t_0+T} \sin(m\omega_0 t)\cos(n\omega_0 t)\,dt = 0 \qquad\qquad \text{for all } m \text{ and } n$$

Thus this set is an orthogonal set.

The property of orthogonality is of crucial importance in our method for finding the coefficients in the Fourier series. It is this property that makes all but the term of interest integrate to zero so that the coefficient of interest remains alone on the right-hand side of the equation. There are many sets of functions that have the orthogonality property over some interval or other.

EXAMPLE 13.8.5

The Legendre polynomials are a set of functions which are orthogonal over the interval -1 to $+1$:

$$\phi_0(t) = 1$$
$$\phi_1(t) = t$$
$$\phi_2(t) = \tfrac{3}{2}t^2 - \tfrac{1}{2}$$
$$\phi_3(t) = \tfrac{5}{2}t^3 - \tfrac{3}{2}t$$
$$\cdots\cdots\cdots\cdots$$

Find the first two terms of the series

$$f(t) = \sum_{n=0}^{\infty} k_n \phi_n(t) \qquad \text{for the function}$$

$$f(t) = \begin{cases} 0 & t < -1, t > 1 \\ 2\sin\pi t & -1 < t < 1 \end{cases}$$

ANS.: First find k_0 in $2\sin\pi t = k_0(1) + k_1 t + k_2(\tfrac{3}{2}t^2 - \tfrac{1}{2}) + \cdots$:

$$\int_{-1}^{+1} 2\sin\pi t\,dt = \int_{-1}^{+1} k_0\,dt + \underbrace{\cdots}$$

All equal 0 because of the orthogonal property:

$$2\left(-\frac{\cos\pi t}{\pi}\right)\Bigg|_{-1}^{+1} = 2k_0$$

or

$$-\frac{1}{\pi}[-1-(-1)] = 0 = k_0$$

Now find k_1:

$$\int_{-1}^{+1} 2t \sin \pi t \, dt = 0 + k_1 \int_{-1}^{+1} t^2 \, dt + \underset{\text{zero because of orthogonality}}{\ddots}$$

$$= \tfrac{2}{3} k_1$$

$$\int_{-1}^{+1} t \sin \pi t \, dt = \tfrac{1}{3} k_1$$

$$\frac{1}{\pi^2} \cancel{\sin \pi t}^{0} - \frac{1}{\pi} t \cos \pi t \Big|_{-1}^{+1} = \frac{k_1}{3}$$

$$-\frac{1}{\pi} [(1)(-1) - (-1)(-1)] = \frac{2}{\pi} = \frac{k_1}{3} \quad \text{or} \quad k_1 = \frac{6}{\pi}$$

Therefore

$$f(t) = 0 + \frac{6}{\pi} t + \cdots \qquad (-1 < t < +1)$$

A set of functions $f_i(t)$ forms a set of *orthonormal basis functions* if

$$\langle f_i(t), f_j(t) \rangle = 0 \qquad \text{for } i \neq j$$

and

$$\langle f_i(t), f_j(t) \rangle = 1 \qquad \text{for } i = j$$

Such a set of functions is analogous to the set of three-dimensional unit direction vectors $\mathbf{u}_1$, $\mathbf{u}_2$, and $\mathbf{u}_3$, that we discussed earlier. See equation (13.8.1) and Figure 13.8.1.

The *norm* of a function is the *square root of the inner product* of that function with itself. It is written as

$$\| f_k(t) \| = \langle f_k(t), f_k(t) \rangle^{1/2}$$

$$= \sqrt{\int_{t_1}^{t_2} f_k(t) f_k^*(t) \, dt}$$

We can *normalize* a set of orthogonal functions by dividing each member function by its norm.

EXAMPLE 13.8.6

Is the orthogonal set of basis functions $f(t) = e^{jk\omega_0 t}$ that make up the exponential form of the Fourier series also an *orthonormal* set? If not, normalize it.

ANS.:   We find the norm of each member of the set over the interval t_0 to $(t_0 + T)$ and check to see if it is unity.

$$\| f_k(t) \| = \int_{t_0}^{t_0 + T} e^{jk\omega_0 t} e^{-jk\omega_0 t} \, dt$$

$$= \int_{t_0}^{t_0+T} (1) \, dt$$

$$= T$$

So the functions $f_k(t) = e^{jk\omega_0 t}$ are seen to be *not* orthonormal over the interval t_0 to $(t_0 + T)$. We could make them orthonormal by dividing each member function by its norm. Thus the set of functions

$$g_k(t) = \frac{e^{jk\omega_0 t}}{T}$$

does constitute an orthonormal set of basis functions.

Recall that the projection of a three-dimensional vector $\mathbf{v}$ on one of its unit orthonormal direction basis vectors, say $\mathbf{u}_1$, is simply the inner product of $\mathbf{v}$ and $\mathbf{u}_1$. That is to say, for

$$\mathbf{v} = x\mathbf{u}_1 + y\mathbf{u}_2 + z\mathbf{u}_3$$

the inner product of $\mathbf{v}$ and $\mathbf{u}_1$ is x, i.e., $\langle \mathbf{v}, \mathbf{u}_1 \rangle = x$. (See Example 13.8.1.)

Another way of looking at this is to think of the scalar quantity x as the amount of $\mathbf{u}_1$ that is in $\mathbf{v}$. In other words $\mathbf{v}$ is made up of a certain amount x of $\mathbf{u}_1$; a certain amount y of $\mathbf{u}_2$; and a certain amount z of $\mathbf{u}_3$. The vector $\mathbf{v}$ is a linear sum of the unit vectors $\mathbf{u}_1$, $\mathbf{u}_2$, and $\mathbf{u}_3$. The scalar coefficients x, y, and z tell how much to weight each basis vector in the summation.

In exactly the same way, the Fourier coefficients α_k are the weights of each of the orthonormal basis functions

$$g_k(t) = \frac{e^{jk\omega_0 t}}{T}$$

in the Fourier series.

We can see that this is the case by examining the expression for α_k:

$$\alpha_k = \frac{1}{T} \int_{t_0}^{t_0+T} f(t) e^{-jk\omega_0 t} \, dt$$

$$= \int_{t_0}^{t_0+T} f(t) \left(\frac{e^{-jk\omega_0 t}}{T} \right) dt$$

$$= \langle f(t), g_k(t) \rangle$$

Thus we see that the kth coefficient α_k is the inner product of $f(t)$ with the kth orthonormal basis function $g_k(t)$, and thus is a measure of how much of $g_k(t)$ should be included in the linear Fourier series summation for $f(t)$.

13.9 truncated series

In Section 13.3 we developed a method for determining the value of each coefficient in the Fourier series. But is this the *best* method? In other words we have found one method (Fourier's method) to isolate and evaluate each coefficient.

There are, however, many other ways to do this that yield different numerical values for the coefficients. Which of all possible sets of numerical values is the best? And by what criterion do we judge what is meant by "best"?

Let us ask the question: Suppose, given an actual periodic $f(t)$, we attempt to write a series approximation of it by using only a finite number n of harmonic terms. Call this n-term approximation $f_n(t)$:

$$f_n(t) = \sum_{k=-n}^{+n} \alpha_k e^{jk\omega_0 t} \tag{13.9.1}$$

where the numerical value of each coefficient α_k is yet to be determined. *Only if* we choose

$$\alpha_k = \frac{1}{T} \int_{t_0}^{t_0+T} f(t) e^{-jk\omega_0 t} \, dt \tag{13.9.2}$$

to evaluate the coefficients in equation (13.9.1) may we call our approximation $f_n(t)$ a truncated *Fourier* series. But there are many other ways of choosing values for the α_k's. Only if we use the method of equation (13.9.2), proposed by Fourier, may we call $f_n(t)$ a Fourier series. The difference at any instant of time between any approximating $f_n(t)$ and the actual waveform $f(t)$ is the *error e(t)*.

$$e(t) = f(t) - f_n(t) \tag{13.9.3}$$

At some instants of time $e(t)$ might be a positive quantity; at other instants it might be negative—depending on whether our approximation $f_n(t)$ was too big or too small at that time.

One way of measuring how well any $f_n(t)$ approximates the actual $f(t)$ would be to compute the average value, over time, of $|e(t)|$ or $e^2(t)$ that results from using that particular $f_n(t)$. [If we simply computed the average value of $e(t)$, then positive errors at some instants would cancel out negative errors at other instants, giving a measure that would be misleadingly low.] Let us choose, as our figure of merit (of goodness of approximation), the *mean squared error* which we define as:

$$\overline{e^2(t)} = \frac{1}{T} \int_{t_0}^{t_0+T} e^2(t) \, dt \tag{13.9.4}$$

We only need to compute this time average over one complete cycle because every cycle of $e(t)$ is identical to every other. And squaring $e(t)$ eliminates the probability of falsely canceling out positive errors with negative errors. Thus

$$\overline{e^2(t)} = \frac{1}{T} \int_{t_0}^{t_0+T} [f(t) - f_n(t)]^2 \, dt$$

$$= \frac{1}{T} \int_{t_0}^{t_0+T} \left[f(t) - \sum_{k=-n}^{+n} \alpha_k e^{jk\omega_0 t} \right]^2 \, dt \tag{13.9.5}$$

Let us define† the "best" set of α_k coefficients as the set of α_k's that will *minimize* this mean squared error. In order to find the value of these coefficients,

---

† There are many other possible definitions: minimizing the *integral* of $e(t)$, for example.

perform the minimization: set the partial derivative of (13.9.5) with respect to each α_k equal to zero and then solve for each of those α_k's.

Consider the mth coefficient:

$$\frac{\partial \overline{e^2(t)}}{\partial \alpha_m} = \frac{\partial}{\partial \alpha_m} \left\{ \frac{1}{T} \int_{t_0}^{t_0+T} \left[f(t) - \sum_{k=-n}^{+n} \alpha_k e^{jk\omega_0 t} \right]^2 dt \right\} = 0 \qquad (13.9.6)$$

The partial derivative, with respect to the particular coefficient α_m, of all terms in equation (13.9.6) will be zero except for the term wherein $k = m$. Thus (13.9.6) becomes

$$\frac{1}{T} \int_{t_0}^{t_0+T} 2 \left[f(t) - \sum_{k=-n}^{+n} \alpha_k e^{jk\omega_0 t} \right] [-e^{jm\omega_0 t}] \, dt = 0$$

or
$$-\frac{2}{T} \int_{t_0}^{t_0+T} f(t) e^{jm\omega_0 t} \, dt + \frac{2}{T} \int_{t_0}^{t_0+T} \sum_{k=-n}^{n} \alpha_k e^{j(k+m)\omega_0 t} \, dt = 0 \qquad (13.9.7)$$

The integral of the summation in the second term may be computed as the sum of a series of individual integrals. Each such integral (except for the one where $k = -m$) is equal to zero because of the orthogonality property.

However, for $k = -m$, equation (13.9.7) may be written

$$\int_{t_0}^{t_0+T} f(t) e^{jm\omega_0 t} \, dt = \int_{t_0}^{t_0+T} \alpha_{-m}(1) \, dt$$

$$= \alpha_{-m}(t_0 + T - t_0) \qquad (13.9.8)$$

so that

$$\alpha_{-m} = \frac{1}{T} \int_{t_0}^{t_0+T} f(t) e^{jm\omega_0 t} \, dt$$

or
$$\alpha_k = \frac{1}{T} \int_{t_0}^{t_0+T} f(t) e^{-jk\omega_0 t} \, dt \qquad (13.9.9)$$

which is the definition of the Fourier coefficients.

Thereby we see that the Fourier coefficients (as opposed to any other possible set of coefficients) minimize the mean squared error between the actual function $f(t)$ and any *finite length* harmonic series approximation.

13.10   the Gibbs effect

It was pointed out in the last section that the Fourier coefficients yield the best (lowest value of mean squared error) finite harmonic series approximation for any practical periodic waveform that engineers might come across in their work. By induction, we realize that any additional Fourier terms added on to some finite-term Fourier approximation would therefore further reduce the mean squared error and thereby achieve a better fit. So we visualize an approximating waveform that has less and less ripple and error in it as we make it better and better by adding more and more terms—each time, of course, using the correct Fourier method for determining the coefficients.

Unfortunately, even if we use an infinite number of terms in our Fourier series

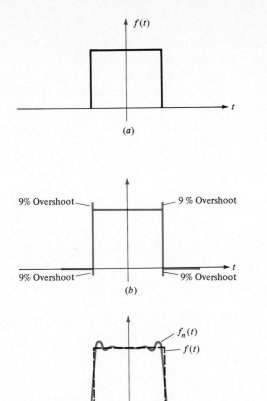

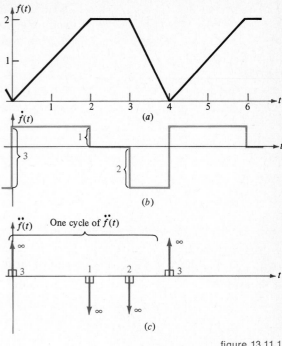

figure 13.11.1
(a) $f(t)$ with $T = 4$ s. Thus $\omega_0 = \pi/2$.
(b) df/dt and (c) d^2f/dt^2 for Example 13.11.1.

figure 13.10.1
(a) One cycle of a time function $f(t)$ that contains discontinuities. (b) Even a complete Fourier series yields an approximation that contains overshoots. (c) A truncated Fourier series will contain both ripples and overshoots.

approximation, we cannot achieve a perfect replica of the original $f(t)$ if it has any *discontinuities*. At any discontinuity even a complete (*infinite* number of harmonic terms) Fourier series will produce overshoots both at the top and bottom of each such discontinuity. These overshoots each amount to approximately 9 percent of the height of the discontinuity. This effect was noted by J. Willard Gibbs and is thus known as the *Gibbs effect*.

In addition, any *finite*-term Fourier approximation cuts through the average value (halfway between the top and bottom) of each discontinuity. See Figure 13.10.1.

finding Fourier coefficients by **13.11** differentiating

If a periodic function for which we would like to obtain the Fourier series has certain properties, such as containing only straight line segments (as in Figure 13.11.1), there exists a shortcut method for finding the coefficients. This method is predicated on our being able to differentiate $f(t)$ one or more times so that we end up with a train of impulses. It is an easy matter to find the Fourier series for such an impulse train because the expression for the coefficients involves an integral and it is easy to integrate functions that contain impulses. Finally, we convert the Fourier coefficients of the impulse train to the coefficients of the original function. We can do that final step in general as follows. If

$$f(t) = \sum_{k=-\infty}^{\infty} \alpha_k e^{jk\omega_0 t}$$

then

$$\frac{df}{dt} = \dot{f}(t) = \sum_{k=-\infty}^{\infty} jk\omega_0 \alpha_k e^{jk\omega_0 t} = \sum_{k=-\infty}^{\infty} \alpha'_k e^{jk\omega_0 t}$$

and

$$\frac{d^2 f}{dt^2} = \ddot{f}(t) = \sum_{k=-\infty}^{\infty} -k^2\omega_0^2 \alpha_k e^{jk\omega_0 t} = \sum_{k=-\infty}^{\infty} \alpha''_k e^{jk\omega_0 t}$$

and so forth, where

$$\alpha'_k = jk\omega_0 \alpha_k \qquad \text{or} \qquad \alpha_k = \left(\frac{1}{jk\omega_0}\right)\alpha'_k$$

$$\alpha''_k = -k^2\omega_0^2 \alpha_k \qquad \text{or} \qquad \alpha_k = \left(\frac{-1}{k^2\omega_0^2}\right)\alpha''_k$$

or, for the nth derivative,

$$\alpha_k^{[n]} = (jk\omega_0)^n \alpha_k \qquad \text{or} \qquad \alpha_k = \frac{\alpha_k^{[n]}}{(jk\omega_0)^n}$$

The average value, α_0, is calculated by the usual method.

EXAMPLE 13.11.1

Given the $f(t)$ shown in Figure 13.11.1a, find the Fourier series.

ANS.: Take enough time derivatives to get an impulse train. See Figure 13.11.1b and c. The second derivative is what we need. Then

$$\alpha''_k = \frac{1}{T} \int_0^T [3\delta(t) - \delta(t-2) - 2\delta(t-3)]e^{-jk\omega_0 t}\, dt$$

$$= \frac{1}{T}\left[\int_0^T 3\delta(t)e^{-jk\omega_0 t}\, dt - \int_0^T \delta(t-2)e^{-jk\omega_0 t}\, dt - 2\int_0^T \delta(t-3)e^{-jk\omega_0 t}\, dt\right]$$

$$= \frac{1}{T}[3e^{\emptyset} - e^{-jk\omega_0 2} - 2e^{-jk\omega_0 3}]$$

$$= \frac{1}{T}\left[3 - 1\underline{/-k\pi} - 2\underline{/-k\frac{3\pi}{2}}\right]$$

We obtain the required values of α_k by tabulation.

k	$-1\underline{/-k\pi}$	$\underline{/-k\dfrac{3\pi}{2}}$	$-2\underline{/-k\dfrac{3\pi}{2}}$	$\alpha_k'' T$	$\alpha_k = \left(\dfrac{-1}{k^2\omega_0^2}\right)\alpha_k''$
1	$+1$	$+j$	$-2j$	$4-2j$	$\dfrac{-1}{\pi^2}(4-2j)$
2	-1	-1	$+2$	4	$\dfrac{-1}{4\pi^2}(4)$
3	$+1$	$-j$	$+2j$	$4+2j$	$\dfrac{-1}{9\pi^2}(4+2j)$
4	-1	$+1$	-2	0	0
5	$+1$	$+j$	$-2j$	$4-2j$	$\dfrac{-1}{25\pi^2}(4-2j)$
6	-1	-1	$+2$	4	$\dfrac{-1}{36\pi^2}(4)$

power delivered to an impedance 13.12
by a Fourier series source

The average power delivered to any impedance $Z(s)$ by a periodic source (either voltage or current) is the sum of the average powers delivered by each of the individual harmonics.

The above statement is true and we will now prove it. But it should come as a bit of a surprise, because one should remember that, in general, the superposition theorem does not apply to power:

$$p(t) = \frac{v(t)^2}{R} = i(t)^2 R \qquad (13.12.1)$$

Power is a nonlinear function of $v(t)$ and of $i(t)$, and superposition applies only to linear quantities (e.g., voltages and currents in linear circuits). The fact that we can first compute the average power delivered by each harmonic component, then *add* those average *powers* together to get the total power delivered by the overall periodic waveform, is indeed a very special result. It occurs because of the mutual orthogonality (see Section 13.8) of the basis functions used in the Fourier series.

Let us assume we have an impedance $Z(s)$ driven by a nonsinusoidal periodic source. As a result, there will be a voltage $v(t)$ across and a current $i(t)$ through the impedance. [It is not important whether the source is a current source $i(t)$ which causes $v(t)$ to exist, or whether the source is a voltage source $v(t)$ which causes $i(t)$ to exist.] The question is: how much average power gets delivered to $Z(s)$?

We have

$$v(t) = \sum_{k=-\infty}^{\infty} \alpha_k e^{jk\omega_0 t} \qquad (13.12.2)$$

where

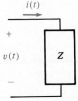

figure 13.12.1
An impedance Z excited by a nonsinusoidal periodic source.

$$\alpha_k = |\alpha_k| e^{j\phi_k}$$

and
$$i(t) = \sum_{n=-\infty}^{\infty} \beta_n e^{jn\omega_0 t} \tag{13.12.3}$$

where

$$\beta_n = |\beta_n| e^{j\psi_n}$$

Consider the kth harmonic of both the voltage and the current waveforms: the phase shift $(\phi_k - \psi_k)$ by which the sinusoidal harmonic current lags its corresponding harmonic voltage is equal to the phase angle of the impedance evaluated at that harmonic frequency, i.e.,

$$\phi_k - \psi_k = \underline{/Z(jk\omega_0)} \tag{13.12.4}$$

We calculate the total average power as the average power delivered over one cycle of the periodic waveform:

$$P_{av} = \frac{1}{T} \int_{t_0}^{t_0+T} v(t)i(t)\,dt \tag{13.12.5}$$

$$= \frac{1}{T} \int_{t_0}^{t_0+T} \left(\sum_{k=-\infty}^{\infty} \alpha_k e^{jk\omega_0 t} \right) \left(\sum_{n=-\infty}^{\infty} \beta_n e^{jn\omega_0 t} \right) dt \tag{13.12.6}$$

The integrand consists of the product of two sums. That is to say, each term in the first sum must multiply every term in the second sum. Thus the result is a sum of terms each of which itself contains a sum:

$$P_{av} = \frac{1}{T} \int_{t_0}^{t_0+T} \sum_{k=-\infty}^{\infty} \sum_{n=-\infty}^{\infty} \alpha_k \beta_n e^{jk\omega_0 t} e^{jn\omega_0 t}\,dt \tag{13.12.7}$$

$$= \frac{1}{T} \int_{t_0}^{t_0+T} \sum_{k=-\infty}^{\infty} \sum_{n=-\infty}^{\infty} \alpha_k \beta_n e^{j(k+n)\omega_0 t}\,dt \tag{13.12.8}$$

The integral of a sum may be calculated as a sum of integrals:

$$P_{av} = \frac{1}{T} \sum_{k=-\infty}^{\infty} \sum_{n=-\infty}^{\infty} \int_{t_0}^{t_0+T} \alpha_k \beta_n e^{j(k+n)\omega_0 t}\,dt \tag{13.12.9}$$

Unless $k = -n$, the value of the integral in equation (13.12.9) is zero because of orthogonality.

On the other hand, for $k = -n$, the integral becomes

$$\int_{t_0}^{t_0+T} \alpha_k \beta_n e^{j(k+n)\omega_0 t}\,dt = \alpha_k \beta_{-k} \int_{t_0}^{t_0+T} (1)\,dt = \alpha_k \beta_{-k} T \tag{13.12.10}$$

Therefore equation (13.12.9) becomes

$$P_{av} = \frac{1}{T} \sum_{k=-\infty}^{\infty} \alpha_k \beta_{-k} T$$

$$= \sum_{k=-\infty}^{\infty} \alpha_k \beta_{-k}$$

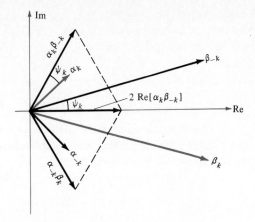

$$P_{av} = \alpha_0 \beta_0 + \alpha_1 \beta_{-1} + \alpha_2 \beta_{-2} + \alpha_3 \beta_{-3} + \cdots$$
$$+ \alpha_{-1}\beta_1 + \alpha_{-2}\beta_2 + \alpha_{-3}\beta_3 + \cdots \qquad (13.12.11)$$

Recalling that α_k and α_{-k} are complex conjugates, we can write the following equation (see also Figure 13.12.2):

$$P_{av} = \alpha_0 \beta_0 + 2\,\text{Re} \sum_{k=+1}^{\infty} \alpha_k \beta_{-k}$$

$$= \alpha_0 \beta_0 + 2\,\text{Re} \sum_{k=+1}^{\infty} |\alpha_k||\beta_k| e^{j(\phi_k - \psi_k)}$$

$$= \alpha_0 \beta_0 + 2 \sum_{k=1}^{\infty} |\alpha_k||\beta_k|\,\text{Re}\, e^{j(\phi_k - \psi_k)}$$

$$P_{av} = \alpha_0 \beta_0 + 2 \sum_{k=1}^{\infty} |\alpha_k||\beta_k|\cos(\phi_k - \psi_k) \qquad (13.12.12)$$

Now, using $c_k = 2|\alpha_k|$ and using superscripts to indicate which terms come from the voltage and which from the current waveforms, we have

$$P_{av} = \frac{c_0^v}{2}\frac{c_0^I}{2} + 2 \sum_{k=1}^{\infty} \frac{c_k^v}{2}\frac{c_k^I}{2}\cos(\phi_k - \psi_k) \qquad (13.12.13)$$

or finally,

$$P_{av} = \frac{c_0^v}{2}\frac{c_0^I}{2} + \sum_{k=1}^{\infty} \frac{c_k^v}{\sqrt{2}}\frac{c_k^I}{\sqrt{2}}\cos(\phi_k - \psi_k) \qquad (13.12.14)$$

dc values of voltage and current	rms values of voltage and current	power factor

So we see that the total average power delivered to any impedance by a Fourier series input is the *sum of the power delivered by each harmonic component.*

In light of this result (13.12.14), what, if anything, can we say about the rms value of any nonsinusoidal, periodic signal? We recall that the rms value of *any* waveform is the magnitude of the constant (dc) signal that delivers the same amount of average power to a resistor as does the periodic signal.

Assume we have a nonsinusoidal, periodic voltage described by its Fourier series:

$$v(t) = \frac{c_0}{2} + \sum_{k=1}^{\infty} c_k \cos\left(k\omega_0 t + \phi_k\right) \tag{13.12.15}$$

We have just seen that the average power delivered to any impedance (let us use R ohms for simplicity) is a complete sum, each term of which is the average power delivered by a harmonic:

$$P_{av} = \left[\left(\frac{c_0}{2}\right)^2 + \sum_{k=1}^{\infty} \left(\frac{c_k}{\sqrt{2}}\right)^2\right]\frac{1}{R} = \frac{V_{rms}^2}{R} \tag{13.12.16}$$

where $c_k/\sqrt{2}$ is the rms value of the kth harmonic and V_{rms} is the overall rms value of the voltage $v(t)$.

Thus, multiplying (13.12.16) by R gives

$$\left(\frac{c_0}{2}\right)^2 + \sum_{k=1}^{\infty} \left(\frac{c_k}{\sqrt{2}}\right)^2 = V_{rms}^2 = V_{ms} \tag{13.12.17}$$

or

$$\sqrt{\left(\frac{c_0}{2}\right)^2 + \sum_{k=1}^{\infty} \left(\frac{c_k}{\sqrt{2}}\right)^2} = V_{rms} \tag{13.12.18}$$

Because of equation (13.12.17), we can say that the *mean squared* value of $v(t)$ is the sum of the *mean squared* values of each of the harmonics.

Similarly, from (13.12.18) we can say that the effective or rms value of $v(t)$ is the square root of the sum of the harmonic mean squared values.

If our nonsinusoidal periodic signal is a current $i(t)$ rather than a voltage, then equation (13.12.16) becomes simply

$$P_{av} = \left[\left(\frac{c_0}{2}\right)^2 + \sum_{k=1}^{\infty} \left(\frac{c_k}{\sqrt{2}}\right)^2\right]R = I_{rms}^2 R \tag{13.12.19}$$

We then divide by R and proceed as above to a similar result:

$$\sqrt{\left(\frac{c_0}{2}\right)^2 + \sum_{k=1}^{\infty} \left(\frac{c_k}{\sqrt{2}}\right)^2} = I_{rms} \tag{13.12.20}$$

13.13   summary

In this chapter we have examined a method for analyzing networks that have *periodic nonsinusoidal functions* as inputs. We have seen that a Fourier series is a sum of harmonically related sinusoids and that there are three different forms in which we can write it. We saw how to calculate these forms directly from a given $f(t)$. Also, we saw that given any one form of the series we can find the other two. We discussed even and odd symmetry of waveforms and how to take advantage of this when calculating a Fourier series description. Time delay of a function $f(t)$

produces a phase lag in each Fourier component that is proportional to frequency. We discussed line spectra, filtering, and the requirements for distortionless transmission.

The relationship of vectors to time functions along with the concomitant ideas of orthogonality and orthonormality were discussed. What constitutes a set of basis functions was defined. The effects of truncating the series and having discontinuities in $f(t)$ were mentioned, as was a method for finding the series description of a certain class of function via differentiation rather than integration. Finally, we discussed the unusual fact that superposition applies to power calculations when the sources are harmonically related and drew the conclusion that the mean squared value of the $f(t)$ function is the sum of the mean squared values of the components.

problems

1. (*a*) What is meant by the phrase *incommensurate term* in a Fourier series? (*b*) What is the result of adding together two sinusoidal signals that are incommensurate? (*c*) What is the period (in seconds) of each of the following signals?

$$f_1(t) = 2 \cos (3t + 10°) + 4 \cos (8t + 20°) + 6 \cos (12.5t + 90°)$$

$$f_2(t) = 3 + 4 \sin 8t + 2 \sin 64t$$

2. Find all the coefficients α_k in the exponential form of the Fourier series for the periodic voltage defined by $v(t) = 160 \sin (377t)$ for $0 < t < T$.

3. Given $f(t) = 1 + 2e^{j45°}e^{jt} + 2e^{-j45°}e^{-jt} + e^{j30°}e^{j2t} + e^{-j30°}e^{-j2t}$, find the first five coefficients in the sin-cos form of the Fourier series: a_0, a_1, b_1, a_2, and b_2.

4. Find the $A \cos (k\omega_0 t + \theta)$ equivalent form of $f(t) = (3 + j4)e^{-j2t} + (5 - j12)e^{-jt} + 4 + (5 + j12)e^{jt} + (3 - j4)e^{j2t}$.

5. Given $f(t) = 10 + 3 \sin 2t + 5 \cos 2t$, what is the exponential form of this Fourier series?

6. Find the exponential form of the Fourier series for (*a*) $f(t) = \cos^2 t$; (*b*) $f(t) = 1 + 3 \cos 3t + 2 \cos 5t + 4 \sin 3t + \sin 5t$

7. Find the sin-cos form of the Fourier series for the periodic signal defined below:

$$f(t) = \begin{cases} 0 & \text{for } -5 < t < 0 \\ 3 & \text{for } 0 < t < 5 \end{cases}$$

$$f(t) = f(t + 10)$$

Demonstrate that the magnitudes of the nonzero terms in the series vary inversely with frequency.

8. Find the trigonometric Fourier series for the waveform described below. Can you tell, by inspection, which terms will be zero-valued? Can you evaluate a_0 by inspection?

$$v(t) = \begin{cases} V_m & \text{for } -1 < t < 1 \\ 0 & \text{for } 1 < t < 2 \end{cases}$$

$$v(t) = v(t + 3)$$

9. Find the first three terms (coefficients a_0, a_1, and b_1) of the sin-cos form of the Fourier series for the periodic current waveform shown in Figure P13.9.

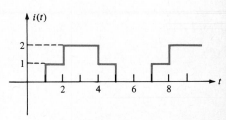

figure P13.9

10. Find a_0, a_1, and b_1 for the periodic signal defined by

$$f(t) = \begin{cases} +1 & \text{for} \quad -1 < t < 1 \\ 0 & \text{for} \quad 1 < t < 3 \text{ and } 5 < t < 7 \\ -1 & \text{for} \quad 3 < t < 5 \end{cases}$$

$$f(t) = f(t + 8)$$

11. Given $v(t) = \cos t$ for $-\pi/2 < t < +\pi/2$ and $f(t) = f(t + \pi)$. (*a*) Find an expression for a_k and b_k, the sin-cos form Fourier coefficients. (*b*) What is the amplitude of the fundamental?

12. Find the sin-cos form of the Fourier series for the periodic function

$$v(t) = \begin{cases} \cos t & \text{for} \quad -\pi < t < 0 \\ 1 & \text{for} \quad 0 < t < \pi \end{cases}$$

The period of $v(t)$ is 2π seconds.

13. Find expressions for coefficients a_k and b_k in the sin-cos form of the Fourier series for the periodic signal shown in Figure P13.13.

14. A signal that is periodic ($T = \pi$ seconds) is equal to $\cos t$ for $0 < t < \pi$. (*a*) Sketch this waveform. What is the value of ω_0? (*b*) Find general expressions in terms of k for a_k and b_k in the sin-cos form of the Fourier series for this signal.

15. Find the exponential form of the Fourier series for the periodic function defined by

$$f(t) = e^{-t} \qquad \text{for } 0 < t < 1 \qquad (T = 1)$$

16. Give all possible information about the Fourier series of each of the waveforms shown in Figure P13.16 *without* doing any integrals. (For example, what is the average value? $T = ?$ $\omega_0 = ?$ which terms, if any, will be missing? Which terms will go through zero simultaneously?)

17. Classify each of the functions defined below as being even, odd, or neither. For any

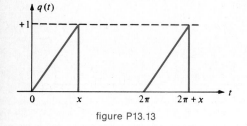

figure P13.13

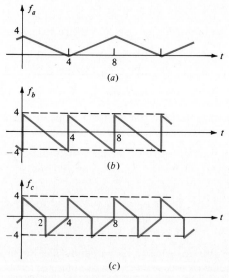

figure P13.16

that are neither even nor odd, sketch their even and odd parts. Tell as much about the Fourier coefficients of each $f(t)$ as possible (without actually finding them).

$$f(t) = \begin{cases} 2 & \text{for } 0 < t < 3 \\ -2 & \text{for } -3 < t < 0 \end{cases} \quad (T = 6)$$

$$f(t) = \begin{cases} \cos t & \text{for } 0 < t < \pi \\ 0 & \text{for } \pi < t < 2\pi \end{cases} \quad (T = 2\pi)$$

$$f(t) = t(10 - t) \quad \text{for } 0 < t < 10 \quad (T = 10)$$

18. Evaluate the dc component and the magnitude and phase (or a_k and b_k) of the first 10 harmonics of the half-wave-rectified sine wave of Example 13.3.1. Plot the approximation of the actual waveform that such a 10-harmonic truncated series provides. (*Suggestion:* write a computer program to perform the drudgery!)

19. Repeat problem 18 for the square wave of Example 13.2.1. For this 10-harmonic approximation, what is the percentage overshoot at the discontinuities? How different is this from the value predicted by Gibbs?

20. In order for the current and voltage waveforms given below to have the same general shape, what must be the values of A and θ? (Write A as a *positive* constant.)

$$v(t) = 10 \cos (4t + 20°) + 5 \sin (8t + 10°)$$

$$i(t) = -5 \sin (4t - 25°) + A \cos (8t + \theta)$$

21. Do the two voltages given below have the same shape? Why or why not?

$$v_1(t) = 3.4 + 4 \cos (8t + 14°) - 27 \sin (12t - 96°)$$

$$v_2(t) = 10.2 + 12 \sin (8t + 80°) + 9 \sin (12t + 36°)$$

22. If a zero-average-value periodic input voltage $v_{in}(t) = \sum_{n=-\infty}^{\infty} \alpha_n e^{jn\omega_0 t}$ is applied to an RC low-pass filter (v_{in} across series combination of $R = 1 \ \Omega$ and $C = 1$ F; v_o taken across C), the output $v_o(t) = \sum_{n=-\infty}^{\infty} \beta_n e^{jn\omega_0 t}$ will result. If the period of the fundamental of the input is 2π seconds and

$$\alpha_n = \left| \frac{1}{n} \right| \underline{/\frac{-n\pi}{3}} \quad \text{for } n \neq 0$$

find expressions for $|\beta_n|$ and $\underline{/\beta_n}$ as functions of n. Evaluate β_n for $n = -3, -2, -1, 0, 1, 2,$ and 3. Sketch the resulting line spectra.

23. (a) Find the coefficients α_{-3} through α_{+3} in the exponential form of the Fourier series, given that

$$f(t) = 1 - \tfrac{2}{3} \cos 2t + \tfrac{2}{15} \cos 3t + \tfrac{2}{35} \cos 6t + \cdots$$

Plot the line spectra (phase and magnitude). (b) What is the period of this waveform?

24. Find the corresponding first three terms of the Fourier series for $v_2(t)$ in the circuit shown in Figure P13.24, given that

$$v_1(t) = 5 + 3\sqrt{2} \cos t + \cos (2t + 15°) + \cdots$$

25. In the circuit of Figure P13.24, change C to $\tfrac{1}{6}$ F and R to $1 \ \Omega$. Then, given that

$$v_1(t) = \sum_{-\infty}^{\infty} \alpha_k e^{jk\omega_0 t}$$

where $\omega_0 = 3$, $\alpha_0 = 2$, $\alpha_1 = 4e^{j60°}$, $\alpha_2 = 2e^{j45°}$, and $\alpha_3 = (1)e^{j36°}$, find the dc, first, second, and third harmonic terms of the series for the output voltage $v_2(t)$.

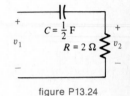

figure P13.24

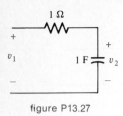

figure P13.27

26. Given a linear system with transfer function $H(s) = 1/(s + 1)$. (a) Find $v_o(t)$ when

$$v_{in}(t) = 10 + 3 \sin (t + 45°) + 7 \cos (\sqrt{3}t + 30°)$$

(b) What is the period of this output voltage?

27. The input voltage of the *RC* circuit shown in Figure P13.27 is described by the Fourier series

$$v_1(t) = 0.32 + 0.5 \cos (\pi t - 90°) + 0.21 \cos (2\pi t - 180°) + \cdots$$

Find the corresponding terms of the series for the output voltage $v_2(t)$.

28. If $f_1(t) = 10 + 3 \cos (2t + 30°) + 5 \cos (4t + 75°)$ is used as input to a system whose transfer function is

$$H(p) = \frac{2}{p + 2}$$

find $f_2(t)$, the output of the system.

29. In a certain class of transistor amplifier the collector (output) current contains some second-harmonic distortion. That is, when the input is a sinusoid of radian frequency ω_0, the output current is

$$i_c(t) = I_0 + I_1 \sin (\omega_0 t) - I_2 \cos (2\omega_0 t)$$

where $I_0 = I_Q + I_2$, I_Q being the constant value of $i_c(t)$ that exists at the *quiescent point*, i.e., when no input signal is applied to the amplifier. All harmonics beyond the second in the output current are assumed to be negligible. (a) Sketch $i_c(t)$ versus t for $I_0 = 0.2$ A, $I_1 = 0.1$ A, and $I_2 = 0.05$ A. (b) Does the average value of $i_c(t)$ satisfy the relationship, $0.5(I_{max} + I_{min})$? (c) Show that

$$\frac{[0.5(I_{max} + I_{min}) - I_Q]100}{I_{max} - I_{min}} = \frac{I_2}{I_1}$$

is a good measure of percent second-harmonic distortion. (d) What is the percent harmonic distortion for the values given in part (a)?

30. A certain operational amplifier circuit has the following relationship between its input voltage $x(t)$ and its output voltage $y(t)$: $y = x^2$. Find the Fourier series for $y(t)$ if $x(t) = 2 \cos 60t + 4 \cos 1000t$.

31. It is known that a certain amplifier produces a (distorted) output voltage $v_2 = k_1 v_1 + k_2 v_1^2 + k_3$, where $v_1(t)$ is the input voltage and the constants k_1, k_2, and k_3 are properties of the amplifier that we wish to determine. In order to do so, we apply a sinusoidal $v_1(t)$ whose magnitude is V_1. We measure both the true rms and the average values of output voltage $v_2(t)$ when $V_1 = 1$ V, $V_{2rms} = 3$ V, and $V_{2dc} = 1$ V and when $V_1 = 2$ V and $V_{2dc} = 1.5$ V. Find k_1, k_2, and k_3.

32. A voltage test signal $v_1(t) = \cos (\omega_0 t) + \cos (1.1\omega_0 t)$ is introduced into a nonlinear amplifier wherein the output voltage is given by $v_2 = v_1^2$. (a) Sketch the magnitude line spectrum of v_1. (b) Sketch the magnitude line spectrum of v_2.

33. (a) What is meant when it is said that two functions $f_1(t)$ and $f_2(t)$ are orthogonal? (b) Given $f_1(t) = t$ and $f_2(t) = 2t + 1$ for $-1 < t < 1$, are they orthogonal, orthonormal, or nonorthogonal?

34. Given two functions $f_1(t) = t$ and $f_2(t) = 2t^2 + 1$ defined over the interval $-1 < t < 1$. (a) Find the norm of each function; (b) Are these two functions orthogonal, orthonormal, or nonorthogonal?

35. The Legendre polynomials are a set of basis functions which are orthogonal (but not orthonormal) over the interval -1 to $+1$:

$$g_1(t) = 1$$

$$g_2(t) = t$$

$$g_3(t) = \tfrac{3}{2}t^2 - \tfrac{1}{2}$$

$$g_4(t) = \tfrac{5}{2}t^3 - \tfrac{3}{2}t$$

Find the first two terms of the series

$$f(t) = \sum_{n=0}^{\infty} k_n g_n(t) \quad \text{for} \quad f(t) = \begin{cases} 2\sin \pi t & \text{for } -1 < t < +1 \\ 0 & \text{elsewhere} \end{cases}$$

36. As an example of a relatively poor approximation, the constant π is used to approximate a ramp of unity slope, $r(t)$. What is the mean squared error of this approximation over the interval 0 to 2π?

37. Do problem 7 by the method of taking derivatives.

38. Do problem 8 by the method of taking derivatives.

39. Do problem 9 by the method of taking derivatives.

40. Do problem 10 by the method of taking derivatives.

41. Do problem 17a by the method of taking derivatives.

42. Show that the function $f(t)$ in Figure P13.42 can be written in the form

$$f(t) = \left(\frac{t}{2\pi} - \frac{1}{2} + \frac{1}{\pi} \sum_{k=1}^{\infty} \frac{\sin kt}{k} \right) u(t)$$

43. (a) What is the effective (rms) value of each of the functions in problem 1(c)? (b) The *energy content* (per cycle) of a sinusoid is defined as its mean squared value times its period. What is the energy content of each of the periodic functions in problem 1(c) in the range 0.5 through 2 Hz?

44. (a) How much average power is delivered to the resistor in the circuit shown in Figure P13.44? $v(t)$ is a square wave, as shown. (b) How much average power does the fundamental Fourier component of $v(t)$ deliver to the resistor?

45. At 60 Hz, a certain impedance Z_1 consists of 4 Ω resistance, 6 Ω capacitive reactance, and 3 Ω inductive reactance in series. An identical impedance Z_2 is in parallel with Z_1. A

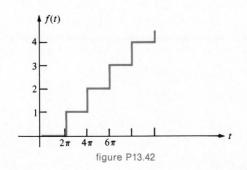

figure P13.42

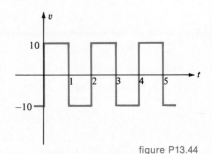

figure P13.44

third 60-Hz impedance Z_3 (consisting of 1.5 Ω resistance and 2 Ω inductive reactance in series) is connected in series with the parallel combination of Z_1 and Z_2. If a voltage,

$$v(t) = 100 \cos 377t - 50 \cos [3(377t + 30°)]$$

is impressed across the entire circuit, calculate: (a) the rms value of the total current entering the circuit, (b) the rms value of the current in each branch, (c) the expression for $i_1(t)$, the current in Z_1, (d) the total dissipated power, and (e) the power factor of the entire circuit.

46. A periodic voltage waveform has (only) harmonics of the following effective values: fundamental 100 V; third harmonic 50 V; fifth harmonic 25 V. (a) If this periodic voltage is placed across a 2-Ω resistor, how many watts of average power will be dissipated? (b) A dc component is added to the periodic voltage in part (a) and the measured average dissipated power increases 10 percent. What is the value of this new dc component?

chapter 14
THE FOURIER TRANSFORM

introduction 14.1

We have seen that the Fourier series affords us a technique for describing nonsinusoidal periodic input signals as summations of sinusoids. Each such component sinusoid can then be separately propagated through any linear system and its resulting particular (steady-state) sinusoidal response determined by means of phasors and $H(j\omega)$. We learned also that the summation of all such component responses is the particular response of the system to the original nonsinusoidal input (Figure 14.1.1).

Now let us investigate how we can modify that technique to enable us to work with *nonperiodic* waveforms as well (Figure 14.1.2). In other words, suppose we have a *single*, arbitrarily shaped *pulse* rather than a repetitive, periodic wavetrain. The phasor (for pure sinusoidal inputs)

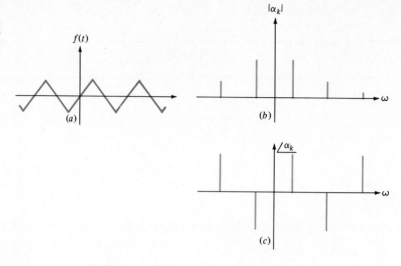

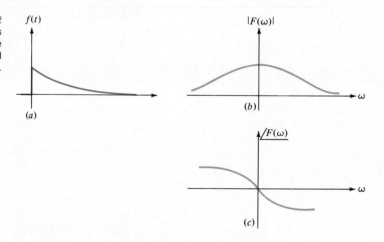

and Fourier series (for non-sinusoidal periodic inputs) are techniques unable to describe such pulses. To enable us to do so we shall generalize the Fourier *series* into the *Fourier transform* and in so doing see that it is capable of handling not only all the periodic input signals the series can handle but many† nonperiodic pulse shapes as well.

The Fourier transform is also the analytical tool we shall use to discover how such time functions as sinusoids, impulses, etc., can be described in the *frequency domain*,‡ which is nothing more esoteric than what we see when we look at the

---

† The same Dirichlet conditions apply to these pulses as applied to the periodic signals dealt with by the Fourier series (except that the word *period* is replaced with *finite interval*).

‡ The Fourier transform also forms an extremely good basis for understanding the topics covered in Chapter 15.

tuning dial of a radio receiver. One radio station's signal waveform is centered at one frequency and other stations' signals take up residence at other frequencies. We shall see that almost every practical time function has a frequency-domain description. Some signals are concentrated in a small part of the frequency domain; others spread out and take up a large range of frequencies.

Since one important thing the Fourier *series* enables us to do is to find the *line spectra* (magnitude and phase angle of the component sinusoids plotted versus frequency), we have seen that a *periodic* signal occupies *discrete frequencies*. In this chapter we shall find that the magnitude spectrum corresponding to any single *pulse* time function is spread out continuously over a range in the frequency domain. It will not be a line spectrum (like that for a periodic waveform) but instead will occupy a *continuum* of frequencies. The same is true of the corresponding phase spectrum.

the transition from Fourier series to 14.2
Fourier transform

To clarify how the change from discrete to continuous spectra takes place as we go from a periodic to a nonperiodic waveform, you should first review Examples 13.3.2 and 13.6.2. In them—and indeed in any discrete Fourier-series spectrum—the spacing between the lines is the fundamental frequency ω_0, where $\omega_0 = 2\pi/T$, and so a large value of T means closely spaced lines and vice versa. Now think of keeping the waveform of *one cycle* of a periodic $f(t)$ unchanged but carefully and intentionally delaying the onset of the next cycle until a much later time. In other words, keep one cycle (say the one that starts at $t = 0$) fixed but allow the period T to grow large without bound, as in Figure 14.2.1.

Since the spacing between adjacent harmonics is $\omega_0 = 2\pi/T$, if T approaches infinity, ω_0 will approach zero. This means that the spectral lines move closer and closer together, eventually becoming a *continuum*. The overall shapes of the magnitude and phase spectra are determined by the shape and duration of the single pulse (cycle) that remains in the now nonperiodic $f(t)$. To examine what happens mathematically we use the exponential form of the Fourier series

$$f(t) = \sum_{k=-\infty}^{\infty} \alpha_k e^{jk\omega_0 t} \tag{14.2.1}$$

where
$$\alpha_k = \frac{1}{T} \int_{t_0}^{t_0+T} f(t)e^{-jk\omega_0 t}\, dt \tag{14.2.2}$$

Choose $t_0 = -T/2$; then

$$\alpha_k = \frac{1}{T} \int_{-T/2}^{+T/2} f(t)e^{-jk\omega_0 t}\, dt \tag{14.2.3}$$

In the limit, as

$$T \to \infty \tag{14.2.4}$$

we see that

figure 14.2.1
Allowing the period T to get large without
bound in order to obtain a nonperiodic
waveform.

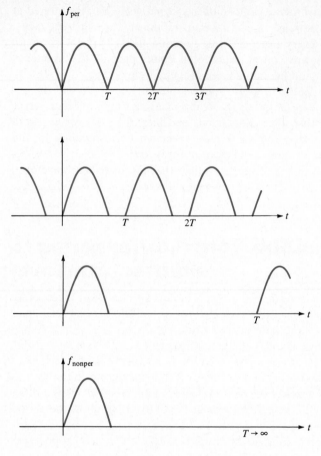

$$\omega_0 = \frac{2\pi}{T} \to d\omega \tag{14.2.5}$$

an infinitesimally small quantity, or

$$\frac{1}{T} \to \frac{d\omega}{2\pi} \tag{14.2.6}$$

Also since

$$k\omega_0 \to \omega \tag{14.2.7}$$

is a continuous frequency variable, the summation becomes an integral.
   Thus using (14.2.6) and (14.2.7) in (14.2.3) gives

$$\alpha_k = \frac{d\omega}{2\pi} \int_{t=-\infty}^{\infty} f(t)e^{-j\omega t}\, dt \tag{14.2.8}$$

Substitute (14.2.8) into (14.2.1), which, in the limit, then becomes

$$f(t) = \int_{\omega=-\infty}^{\infty} \left[\int_{t=-\infty}^{\infty} f(t)e^{-j\omega t}\, dt \right] e^{j\omega t}\, \frac{d\omega}{2\pi} \tag{14.2.9}$$

When the inner integral in brackets is evaluated, it is seen to be a function only of ω, not t. Therefore call it $F(\omega)$

$$F(\omega) = \int_{t=-\infty}^{\infty} f(t)e^{-j\omega t}\,dt \qquad (14.2.10)$$

and equation (14.2.9) can be written as

$$f(t) = \frac{1}{2\pi} \int_{\omega=-\infty}^{\infty} F(\omega)e^{j\omega t}\,d\omega \qquad (14.2.11)$$

Equations (14.2.10) and (14.2.11) denote the *Fourier-transform pair* that most electrical engineers use.†

When using equation (14.2.10), note that any given $f(t)$ has one and only one corresponding $F(\omega)$. Similarly, from (14.2.11), any given $F(\omega)$ has one and only one corresponding $f(t)$. Thus, for example, a voltage source $v(t)$ can be described in the time domain by writing out an expression for $v(t)$ but it can also be described just as completely and uniquely by its frequency-domain description $V(\omega)$.

nomenclature, symbols, and linearity 14.3

We use

$$\mathcal{F}[f(t)] = F(\omega) = \int_{-\infty}^{\infty} f(t)e^{-j\omega t}\,dt$$

to denote the *Fourier transform* of a time signal $f(t)$ and

$$\mathcal{F}^{-1}[F(\omega)] = f(t) = \frac{1}{2\pi} \int_{-\infty}^{\infty} F(\omega)e^{j\omega t}\,d\omega$$

to denote the *inverse process* of finding the time function $f(t)$ from its frequency-domain transform $F(\omega)$. The small letter denotes a time function whose transform is denoted by the same letter capitalized.

The Fourier transform is a *linear operation* in that the transform of a linear sum is that linear sum of transforms; i.e.,

$$\mathcal{F}[af_1(t) + bf_2(t)] = \int_{-\infty}^{\infty} [af_1(t) + bf_2(t)]e^{j\omega t}\,dt$$

† Some communications engineers prefer to write the frequency variable in hertz rather than radians per second. Equations (14.2.10) and (14.2.11) then are simply written (with obvious changes in variables)

$$S(f) = \int_{t=-\infty}^{\infty} s(t)e^{-j2\pi ft}\,dt \qquad \text{and} \qquad s(t) = \int_{f=-\infty}^{\infty} S(f)e^{jt2\pi f}\,df \qquad (14.2.12)$$

where S connotes the word *signal* and $f = \omega/2\pi$ (in hertz).

$$= a \int_{-\infty}^{\infty} f_1(t)e^{j\omega t}\, dt + b \int_{-\infty}^{\infty} f_2(t)e^{j\omega t}\, dt$$

$$= aF_1(\omega) + bF_2(\omega) \tag{14.3.1}$$

Thus homogeneity and additivity both apply.

14.4 some uses of the Fourier transform

Equation (14.2.10) yields a value for $F(\omega)$ which is, in general, complex. The *magnitude* of $F(\omega)$ plotted versus frequency is called the *magnitude spectrum*, and the angle of $F(\omega)$ plotted versus frequency is called the *phase spectrum*.

Using equations (14.2.10) and (14.2.11), we are now, for instance, able to tell how much space any given $f(t)$ waveform will occupy in the frequency domain; conversely, if we are allowed to use only a finite portion of the frequency domain, e.g., if our signal is *band-limited*, we are able to tell what effect this restriction will have on the shape of the $f(t)$ function in the time domain.

Also, suppose we want to use a nonperiodic pulse $f_1(t)$ as the input signal to a system with transfer function $H(s)$, where $H(s)$ might represent an amplifier, a transmission line between a CRT terminal and a computer, or even the effect of the atmosphere on a transmitted radar pulse. We can use the same method as with the Fourier series to find the system's response:

1. Find the Fourier transform of $f_1(t)$

$$F_1(\omega) = \int_{-\infty}^{\infty} f_1(t)e^{-j\omega t}\, dt$$

2. Multiply $F_1(\omega)$ by $H(j\omega)$ to find the output transform

$$F_2(\omega) = F_1(\omega)H(j\omega) \tag{14.4.1}$$

3. Take the inverse transform of $F_2(\omega)$ to get

$$f_2(t) = \frac{1}{2\pi} \int_{-\infty}^{\infty} F_2(\omega)e^{j\omega t}\, d\omega$$

We shall prove that these steps are the equivalent of convolving the input time function $f_1(t)$ with the system's impulse response $h(t)$ in order to find the zero-state output $f_2(t)$. Usually (not always) this Fourier-transform approach is easier because generally we need not actually perform the integration in step 3. Since there is a one-to-one correspondence between each possible function $f(t)$ and its $F(\omega)$, we often find that we already know the $f_2(t)$ that corresponds to $F_2(\omega)$ because we have used it before. Thus it pays to tabulate every $f(t)$ function together with its transform $F(\omega)$ as we go along doing problems.

One more comment should be made about the three steps listed above for finding the zero-state output $f_2(t)$ due to input $f_1(t)$ in a system $H(j\omega)$. This process is completely analogous to finding the product of two numbers by finding the logarithms of each, adding the logs, and then taking the inverse logarithm. If we already have a table of logarithms (and no calculator), addition is easier than multiplication (Figure 14.4.1).

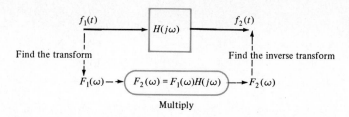

figure 14.4.1
Finding the particular response by means of
Fourier transforms.

Let us now prove that equation (14.4.1) is a valid method for finding the zero-state response of a linear system. We have already shown (Chapter 7) that the convolution of any input $f_1(t)$ with a system's unit-impulse response† $h(t)$ yields the system's zero-state response $f_2(t)$ to that input

$$f_2(t) = f_1(t) * h(t) = \int_{-\infty}^{\infty} f_1(\tau)h(t - \tau)\, d\tau$$

That this time-domain procedure yields the same answer as equation (14.4.1) in the frequency domain we prove by rewriting (14.4.1) as

$$F_2(\omega) = F_1(\omega)H(\omega) = \int_{-\infty}^{\infty} f_1(\tau)e^{-j\omega\tau}\, d\tau \int_{-\infty}^{\infty} h(\lambda)e^{-j\omega\lambda}\, d\lambda$$

$$= \int_{-\infty}^{\infty}\int_{-\infty}^{\infty} f_1(\tau)h(\lambda)e^{-j\omega(\tau+\lambda)}\, d\tau\, d\lambda \qquad (14.4.2)$$

Letting $t = \tau + \lambda$ gives

$$F_2(\omega) = \int_{-\infty}^{\infty}\int_{-\infty}^{\infty} f_1(\tau)h(t - \tau)\, d\tau\; e^{-j\omega t}\, dt$$

We recognize the outer integral to be the Fourier transform of the inner integral, which is

$$f_2(t) = \int_{-\infty}^{\infty} f_1(\tau)h(t - \tau)\, d\tau = f_1(t) * h(t) \qquad \text{QED}$$

Thus we have shown that the *Fourier-transform* technique of Figure 14.4.1 and equation (14.4.2) *yields the zero-state response $f_2(t)$*, just as convolving $f_1(t)$ with $h(t)$ does. Thus *convolution in the time domain is equivalent to multiplication in the frequency domain.*

the relationship between the impulse 14.5
response and the transfer function

The unit impulse $f(t) = \delta(t)$ has a transform that is particularly easy to calculate.

---

† The unit-impulse response $h(t)$ is defined as the zero-state (initial conditions equal to zero) response of a linear system to a unit-impulse-function $\delta(t)$ input.

EXAMPLE 14.5.1

Find $\mathcal{F}[\delta(t)]$.

ANS.:

$$\mathcal{F}[\delta(t)] = \int_{-\infty}^{\infty} \delta(t)e^{-j\omega t}\, dt$$

Because the impulse is zero-valued over so much of its range, the limits collapse to $0-$ and $0+$. Therefore

$$\mathcal{F}[\delta(t)] = \int_{0-}^{0+} \delta(t)e^{-j\omega t}\, dt = \int_{0-}^{0+} \delta(t)(1)\, dt = 1 \qquad (14.5.1)$$

The magnitude spectrum of a unit impulse is shown in Example 14.5.1 to be a constant (unity) for all values of frequency $-\infty < \omega < \infty$. The phase spectrum is everywhere equal to zero. An impulse, then, contains all frequencies; looking at it the other way around, we can say that in order to produce an impulse we must be able to produce frequencies that range all the way up to $\omega = \infty$.

From equations (14.4.2) and (14.5.1) we see that *the Fourier transform of the unit-impulse response of a linear system is its transfer function*; i.e., for a system $H(\omega)$ with zero initial conditions and input $f_1(t) = \delta(t)$ we know that the output, by definition, is $h(t)$. Equation (14.4.2) states that

$$F_2(\omega) = F_1(\omega)H(\omega) = \mathcal{F}[\delta(t)]H(\omega) = (1)H(\omega) = H(\omega)$$

Therefore

$$f_2(t) = \mathcal{F}^{-1}[F_2(\omega)] = \mathcal{F}^{-1}[H(\omega)]$$

which we know to be the unit-impulse response $h(t)$; that is why we have used the same symbol all along, $h(t)$ for the unit-impulse response and $H(\omega)$ for the transfer function.

This important and practically useful result means that if we can somehow experimentally determine the impulse response of an actual linear system [or, alternatively, its unit-step response and then take the time derivative, which thus gives us the $h(t)$], its Fourier transform $\mathcal{F}[h(t)] = H(\omega)$ is the transfer function. Thus by equation (14.4.2) we know how the system will respond to any input for which we can find the Fourier transform.

One other implication of the fact that the impulse response $h(t)$ and the transfer function $H(\omega)$ of any linear system are Fourier transforms of each other is that not all $H(\omega)$ functions can be *realized*. For the ideal low-pass filter of Example 13.7.2 and Figure 13.7.1

$$H(\omega) = \begin{cases} 1 & -\omega_c < \omega < \omega_c \\ 0 & \text{elsewhere} \end{cases}$$

where ω_c is the cutoff frequency. Therefore the impulse response must be given by

$$h(t) = \frac{1}{2\pi} \int_{-\infty}^{\infty} H(\omega)e^{j\omega t}\, d\omega = \frac{1}{2\pi} \int_{-\omega_c}^{\omega_c} (1)e^{j\omega t}\, d\omega$$

$$= \left. \frac{e^{j\omega t}}{2\pi j t} \right|_{-\omega_c}^{+\omega_c} = \frac{1}{\pi t} \frac{e^{j\omega_c t} - e^{-j\omega_c t}}{2j}$$

Multiplying and dividing by ω_c gives

$$h(t) = \frac{\omega_c}{\pi} \frac{\sin \omega_c t}{\omega_c t}$$

This† $h(t)$ is non-zero-valued for negative values of t. That means this system has to start responding before $t = 0$ *in anticipation* that an impulse will be delivered to it at $t = 0$. No such system can be built in practice. This ideal filter is, therefore, *unrealizable*.

some transforms and their properties 14.6

Some specific examples will enable us to investigate other properties of the Fourier transform.

EXAMPLE 14.6.1

a. Given that $\mathcal{F}[f_1(t)] = F_1(\omega)$, what is the effect on $f(t)$ of introducing a phase shift of $-\omega t_0$ rad in the frequency domain? That is, find

$$f_2(t) = \mathcal{F}^{-1}[F_1(\omega)e^{-j\omega t_0}]$$

b. Find $\mathcal{F}[\delta(t - 3)]$.

ANS.:   (a) $f_2(t) = \dfrac{1}{2\pi} \displaystyle\int_{-\infty}^{\infty} F_1(\omega)e^{-j\omega t_0}e^{j\omega t}\, d\omega = \dfrac{1}{2\pi} \displaystyle\int_{-\infty}^{\infty} F_1(\omega)e^{j(t-t_0)\omega}\, d\omega = f_1(t - t_0)$

Thus introducing a *frequency-dependent phase shift* $-\omega t_0$ rad into the phase spectrum corresponds to a *time delay* of t_0 s and vice versa.‡
    (b) Since, from Example 14.5.1,

$$\mathcal{F}[\delta(t)] = F_1(\omega) = 1$$

from (a),

$$\mathcal{F}[\delta(t - 3)] = F_2(\omega) = 1e^{-j\omega 3} = 1\underline{/-3\omega}$$

or $|F_2(\omega) = 1$ and $\underline{/F_2(\omega)} = -3\omega$

EXAMPLE 14.6.2

Find the Fourier transform, $F(\omega)$ corresponding to $f(t) = Ae^{-at}u(t)$.

ANS.: $F(\omega) = \mathcal{F}[Ae^{-at}u(t)] = \displaystyle\int_{-\infty}^{\infty} [Ae^{-at}u(t)]e^{-j\omega t}\, dt = A \displaystyle\int_{0}^{\infty} e^{-(a+j\omega)t}\, dt$

---

† The function $(\sin x)/x$, which often appears in the study of frequency-domain functions and techniques, is called *sinc x* by some authors and the *sampling function* by others.

‡ This is relevant to, and completely consistent with, the discussion in Section 13.7 of distortionless transmission.

$$= \frac{-A}{a + j\omega} e^{-(a+j\omega)t} \Big|_0^\infty = \frac{A}{a + j\omega}$$

so that the magnitude and phase spectra are

$$|F(\omega)| = \frac{A}{\sqrt{a^2 + \omega^2}} \quad \text{and} \quad \underline{/F(\omega)} = -\tan^{-1} \frac{\omega}{a}$$

see Figure 14.1.2.

EXAMPLE 14.6.3

a. Find the $f(t)$ that corresponds to $F(\omega) = \delta(\omega - \omega_0)$, a unit impulse in the frequency domain located at the specific frequency $\omega = \omega_0$.

b. Find $f(t)$ if $F(\omega) = \delta(\omega + \omega_0)$.

ANS.:   (a)
$$f(t) = \frac{1}{2\pi} \int_{\omega = -\infty}^{\infty} \delta(\omega - \omega_0) e^{j\omega t} \, d\omega = \frac{1}{2\pi} \int_{\omega_0 -}^{\omega_0 +} \delta(\omega - \omega_0) e^{j\omega t} \, d\omega$$

$$= \frac{1}{2\pi} e^{j\omega_0 t} \int_{\omega_0 -}^{\omega_0 +} \delta(\omega - \omega_0) \, d\omega = \frac{1}{2\pi} e^{j\omega_0 t}$$

(b) Following the steps of part (a) but with the impulse located at $\omega = -\omega_0$, we find

$$f(t) = \frac{1}{2\pi} e^{-j\omega_0 t}$$

14.7 Fourier transforms of sinusoidal waveforms

We can use the linearity property together with the results of Example 14.6.3 to find the Fourier transform of a pure cosine or sine time function.

EXAMPLE 14.7.1

a. Find $\mathcal{F}[A \cos \omega_0 t]$.
b. Find $\mathcal{F}[A \sin \omega_0 t]$.

ANS.:   (a)
$$A \cos \omega_0 t = A \frac{e^{j\omega_0 t} + e^{-j\omega_0 t}}{2} = \frac{A}{2} e^{j\omega_0 t} + \frac{A}{2} e^{-j\omega_0 t}$$

Therefore

$$\mathcal{F}[A \cos \omega_0 t] = \mathcal{F}\left[\frac{A}{2} e^{j\omega_0 t} + \frac{A}{2} e^{-j\omega_0 t} \right]$$

$$= \mathcal{F}\left[\frac{A2\pi}{2} \left(\frac{1}{2\pi} e^{j\omega_0 t} \right) + \frac{A2\pi}{2} \left(\frac{1}{2\pi} e^{-j\omega_0 t} \right) \right]$$

$$= A\pi[\delta(\omega - \omega_0) + \delta(\omega + \omega_0)]$$

Thus the frequency-domain description of $A \cos \omega_0 t$ is a pair of impulses, one at $\omega = \omega_0$ and one at $\omega = -\omega_0$, each with area $A\pi$. The phase spectrum is zero-valued for both $\omega = \pm\omega_0$.

(b) Similarly we can show that

$$\mathcal{F}[A \sin \omega_0 t] = \mathcal{F}\left[\frac{1}{2j}\left(e^{j\omega_0 t} - e^{-j\omega_0 t}\right)\right] = jA\pi[\delta(\omega + \omega_0) - \delta(\omega - \omega_0)]$$

The phase spectrum is $90°$ at $\omega = -\omega_0$, $-90°$ at $\omega = +\omega_0$, and zero elsewhere.

In Example 14.7.1 we found the Fourier transform not of a single pulse but of a periodic waveform. This result raises a subtle question. Equation (14.4.2) clearly gives the transform of the zero-state response to an arbitrary input $f_1(t)$; but if we use a sinusoidal input, say $f_1(t) = A \cos \omega_0 t$ that exists and is defined for $-\infty < t < +\infty$, what does equation (14.4.2) yield? What is the meaning of $F_2(\omega) = F_1(\omega)H(\omega)$ in this context?

It turns out that if we know the Fourier transform $F(\omega)$ of any function $f(t)$, it is easy to obtain the Fourier transform of the time derivative of $f(t)$, assuming that it exists. That is, given that $\mathcal{F}[f(t)] = F(\omega)$ and

$$f(t) = \frac{1}{2\pi} \int_{-\infty}^{\infty} F(\omega)e^{j\omega t} \, d\omega \tag{14.7.1}$$

taking the time derivative of both sides of (14.7.1) gives

$$\frac{d}{dt} f(t) = \frac{1}{2\pi} \int_{\omega=-\infty}^{\infty} j\omega F(\omega)e^{j\omega t} \, d\omega$$

The nth derivative is

$$\frac{d^n}{dt^n} f(t) = \frac{1}{2\pi} \int_{\omega=-\infty}^{\infty} [(j\omega)^n F(\omega)]e^{j\omega t} \, d\omega$$

Thus

$$\mathcal{F}\left[\frac{d^n}{dt^n} f(t)\right] = (j\omega)^n F(\omega) \tag{14.7.2a}$$

Suppose we have a linear time-invariant system described by its typical nth-order linear differential equation. Call the input $x(t)$ and the output $y(t)$

$$a_n \frac{d^n y}{dt^n} + a_{n-1} \frac{d^{n-1} y}{dt^{n-1}} + \cdots + a_2 \frac{d^2 y}{dt^2} + a_1 \frac{dy}{dt} + a_0 y$$

$$= b_m \frac{d^m x}{dt^m} + b_{m-1} \frac{d^{m-1} x}{dt^{m-1}} + \cdots + b_2 \frac{d^2 x}{dt^2} + b_1 \frac{dx}{dt} + b_0 x \tag{14.7.2b}$$

Using the linearity property and equation (14.7.2a), take the Fourier transform of (14.7.2b) term by term

$$a_n(j\omega)^n Y(\omega) + a_{n-1}(j\omega)^{n-1} Y(\omega) + \cdots + a_2(j\omega)^2 Y(\omega) + a_1 j\omega Y(\omega) + a_0 Y(\omega)$$

$$= b_m(j\omega)^m X(\omega) + b_{m-1}(j\omega)^{m-1} X(\omega) + \cdots + b_2(j\omega)^2 X(\omega) + b_1 j\omega X(\omega) + b_0 X(\omega)$$

or

$$Y(\omega) = \frac{b_m(j\omega)^m + b_{m-1}(j\omega)^{m-1} + \cdots + b_2(j\omega)^2 + b_1 j\omega + b_0}{a_n(j\omega)^n + a_{n-1}(j\omega)^{n-1} + \cdots + a_2(j\omega)^2 + a_1 j\omega + a_0} X(\omega) \quad (14.7.3)$$

Clearly, the ratio in equation (14.7.3) is $H(j\omega)$, sinusoidal transfer function, i.e., the ratio of the complex output to the complex input. In other words, if the system's input is $x(t) = Xe^{j\omega_0 t}$, we know the particular response will be of the form $y_p(t) = Ye^{j\omega_0 t}$. Substituting these two quantities into (14.7.2b) and collecting terms will also yield equation (14.7.3). We thus have shown in (14.7.3) that the Fourier transform of the *particular response to any sinusoidal input* is obtained by multiplying the Fourier transform of that input by the transfer function, or (once again)

$$F_2(\omega) = F_1(\omega)H(\omega) \quad (14.7.4)$$

Any periodic function that has a Fourier-series description can be treated as the sum of a set of sinusoids. Thus we see that equation (14.4.2) or (14.7.4) yields the system's *particular response to any periodic input* and the *zero-state response to a nonperiodic input*.

14.8 Fourier transforms of general periodic waveforms

Since the Fourier *series* of a periodic waveform $f_{per}(t)$ can be written as a sum of complex exponentials

$$f_{per}(t) = \sum_{k=-\infty}^{\infty} \alpha_k e^{jk\omega_0 t} \quad (14.8.1)$$

where

$$\alpha_k = \frac{1}{T} \int_{t_0}^{t_0+T} f_{per}(t)e^{-jk\omega_0 t}\, dt$$

we see that the Fourier *transform* of any periodic function describable by a Fourier series can be written in terms of its series coefficients

$$F_{per}(\omega) = \mathcal{F}[f_{per}(t)] = \mathcal{F}\left[\sum_{k=-\infty}^{\infty} \alpha_k e^{jk\omega_0 t}\right] = \sum_{k=-\infty}^{\infty} \alpha_k \mathcal{F}[e^{jk\omega_0 t}]$$

$$= \sum_{k=-\infty}^{\infty} \alpha_k 2\pi\delta(\omega - k\omega_0) \quad (14.8.2)$$

This equation indicates that a *periodic* function's magnitude spectrum $|F(\omega)|$ is a train of *impulses* in the frequency domain, each located at a multiple of the fundamental frequency ω_0. The area of each impulse is equal to 2π *times the magnitude of the corresponding Fourier-series coefficient* α_k. The phase spectrum is given by the angle of each α_k at $k\omega_0$. These impulse spectra correspond to the line spectra of the Fourier series. In discussing the magnitude and phase spectra of the Fourier transform of a periodic function the lines become impulse functions of the continuous frequency variable ω.

EXAMPLE 14.8.1

a. Find the Fourier transform (magnitude and phase spectra) of a periodic square-wave voltage signal $v(t)$ that has even symmetry and magnitude V as shown in Figure 14.8.1a.

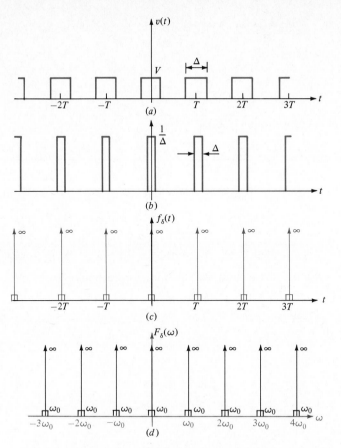

figure 14.8.1
(a) The square wave of Example 14.8.1 (b) Waveform of (a) with $V = 1/\Delta$ and Δ growing smaller. (c) The limit as Δ approaches zero. (d) The Fourier transform of (c).

b. Find the Fourier transform of a time-domain wavetrain of unit impulses $f_\delta(t) = \sum_{n=-\infty}^{\infty} \delta(t - nT)$. Use the result of part (a).

ANS.: (a) First find the Fourier-series coefficients

$$\alpha_k = \frac{1}{T} \int_{t_0}^{t_0 + T} f(t)e^{-jk\omega_0 t}\, dt = \frac{1}{T} \int_{-\Delta/2}^{\Delta/2} Ve^{-jk\omega_0 t}\, dt = \frac{V}{-jk\omega_0 T}(e^{-jk\omega_0\Delta/2} - e^{+jk\omega_0\Delta/2}) = \frac{V\Delta}{T} \frac{\sin k\omega_0 \Delta/2}{k\omega_0 \Delta/2}$$

Then, since from equation (14.8.2) the transform of any periodic waveform is

$$F_{\text{per}}(\omega) = \sum_{k=-\infty}^{\infty} \alpha_k 2\pi\delta(\omega - k\omega_0)$$

we have

$$F_v(\omega) = \sum_{k=-\infty}^{\infty} 2\pi \frac{V\Delta}{T} \frac{\sin k\omega_0 \Delta/2}{k\omega_0 \Delta/2} \delta(\omega - k\omega_0) \tag{14.8.3}$$

Thus, the magnitude spectrum of $v(t)$ consists of a train of impulses (in the frequency domain) with areas proportional to $(\sin x)/x$, where $x = k\omega_0 \Delta/2$. The phase spectrum is defined only at frequencies $k\omega_0$ and in this example is zero-valued at each such frequency.

(b) If we let V take on the value $1/\Delta$ and then allow $\Delta \to 0$, the time-domain waveform of Figure 14.8.1a becomes that of Figure 14.8.1b and, in the limit, the train of unit impulses shown in Figure 14.8.1c, $f_\delta(t) = \sum_{n=-\infty}^{\infty} \delta(t - nT)$. Thus the Fourier transform of the impulse train $f_\delta(t)$ is

$$F_\delta(\omega) = \mathcal{F}[(f_\delta(t)] \qquad \text{where } f_\delta(t + T) = f_\delta(t)$$

$$= \mathcal{F}\left[\lim_{\Delta \to 0} v(t)\right] \quad \text{with } V = \frac{1}{\Delta}$$

$$= \lim_{\Delta \to 0} F_v(\omega)$$

Since $2\pi/T = \omega_0$ and $\lim_{x \to 0}[(\sin x)/x] = 1$, we have from equation (14.8.3)

$$F_\delta(\omega) = \sum_{k=-\infty}^{\infty} \omega_0\, \delta(\omega - k\omega_0) \qquad \text{for all } k \qquad (14.8.4)$$

Thus the time-domain unit-impulse train has a frequency-domain description which is also a train of impulses (in ω); see Figure 14.8.1d.

EXAMPLE 14.8.2

Given the linear system of Figure 14.8.2, use Fourier transforms to find the time-domain response $v(t)$ to the input functions:

a. $i_1(t) = 10 \sin(9t + 30°)$
b. $i_2(t) = 4e^{-t}u(t)$

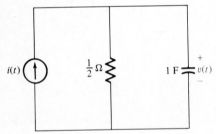

figure 14.8.2
The system of Example 14.8.2.

ANS.:   The transfer function is

$$\frac{V(\omega)}{I(\omega)} = H(\omega) = Z(\omega) \qquad \text{and} \qquad H(\omega) = \frac{1}{2 + j\omega}$$

(a) Since $i_1(t) = 10 \sin(9t + 30°) = 10 \cos\left(9t - \frac{\pi}{3}\right) = 10 \cos 9\left(t - \frac{\pi}{27}\right)$ we have

$$\mathcal{F}[i_1(t)] = I_1(\omega) = 10\pi[\delta(\omega - 9) + \delta(\omega + 9)]e^{-j\omega\pi/27}$$

and

$$V_1(\omega) = I_1(\omega)H(\omega) = 10\pi[\delta(\omega - 9) + \delta(\omega + 9)]e^{-j\omega\pi/27}\,\frac{1}{2 + j\omega}$$

$$\mathcal{F}^{-1}[V_1(\omega)] = v_1(t) = \frac{1}{2\pi}\int_{-\infty}^{\infty} V_1(\omega)e^{j\omega t}\, d\omega$$

This integral is particularly easy because of the impulses in the integrand. In other words, each integral is simply equal to the integrand evaluated at the value of ω at which the impulse occurs, or

$$v_1(t) = \frac{10\pi}{2\pi}\left[\int_{-\infty}^{\infty} \delta(\omega - 9)\frac{e^{-j\omega\pi/27}}{2 + j\omega}e^{j\omega t}\, d\omega + \int_{-\infty}^{\infty} \delta(\omega + 9)\frac{e^{-j\omega\pi/27}}{2 + j\omega}e^{j\omega t}\, d\omega\right]$$

$$= 5\left(\frac{e^{-j9\pi/27}}{2 + j9} e^{j9t} + \frac{e^{-j9\pi/27}}{2 - j9} e^{-j9t}\right)$$

$$= 5[H(j9)e^{j9t} + H(-j9)e^{-j9t}]e^{-j\pi/3}$$

$$= 10|H(j9)| \cos\left[9t - \frac{\pi}{3} + \underline{/H(j9)}\right]$$

Since $\pi/3 = -60°$ and $\underline{/H(j9)} = -75.5°$

$$v_1(t) = 1.08 \cos(9t - 137.5°)$$

This answer, which is obviously the steady-state sinusoidal (particular) response, can be obtained more easily with phasors

$$I_1 = 10\underline{/-60°} \qquad\qquad Z(j9) = \frac{1}{2 + j9} = \frac{1}{9.2\underline{/77.5°}} = 0.108\underline{/-77.5°}$$

so that $V_1 = I_1 Z = 1.08\underline{/-137.5°}$ or $v_1(t) = 1.08 \cos(9t - 137.5°)$.

The point to be remembered is that the Fourier-transform technique yields the *particular response* to any Fourier-transformable *periodic input* function.

(b) For $i(t) = 4e^{-t}u(t)$ we found in Example 14.6.2 that $I(\omega) = 4/(1 + j\omega)$. Thus

$$V(\omega) = I(\omega)H(\omega) = \frac{4}{1 + j\omega}\frac{1}{2 + j\omega} = \frac{4}{(1 + j\omega)(2 + j\omega)}$$

We manipulate this into more tractable form by the steps

$$V(\omega) = \frac{4}{(1 + j\omega)(2 + j\omega)} = \frac{k_1}{1 + j\omega} + \frac{k_2}{2 + j\omega}$$

where we can find k_1 by multiplying through by $1 + j\omega$ and then evaluating at $j\omega = -1$; that is,

$$\frac{4(1 + j\omega)}{(1 + j\omega)(2 + j\omega)} = k_1 + \frac{k_2(1 + j\omega)}{2 + j\omega}$$

$$\frac{4}{2 + j\omega}\bigg|_{j\omega = -1} = k_1 + 0$$

$$4 = k_1$$

Similarly

$$\frac{4(2 + j\omega)}{(1 + j\omega)(2 + j\omega)}\bigg|_{j\omega = -2} = \frac{k_1(2 + j\omega)}{1 + j\omega}\bigg|_{j\omega = -2} + k_2 \qquad -4 = 0 + k_2$$

Thus

$$V(\omega) = \frac{4}{1 + j\omega} - \frac{4}{2 + j\omega}$$

and so

$$v(t) = (4e^{-t} - 4e^{-2t})u(t)$$

which is the *zero-state response* of this circuit to the input $i(t) = 4e^{-t}u(t)$.

The method used to evaluate $V(\omega)$ in the previous example as the sum of two separate terms involving the constants k_1 and k_2 is called a *partial-fraction expan-*

sion. We develop this method fully in the next chapter, where we shall use it much more (see also Section 8.5).

14.9 another form of the Fourier transform

We can substitute for the exponential term in the Fourier transform as follows:

$$F(\omega) = \int_{-\infty}^{\infty} f(t)e^{-j\omega t}\,dt = \int_{-\infty}^{\infty} f(t)(\cos \omega t - j \sin \omega t)\,dt$$

$$F(\omega) = A(\omega) - jB(\omega) \tag{14.9.1a}$$

where

$$A(\omega) = \int_{-\infty}^{\infty} f(t)\cos \omega t\,dt \tag{14.9.1b}$$

and

$$B(\omega) = \int_{-\infty}^{\infty} f(t)\sin \omega t\,dt \tag{14.9.1c}$$

This form of the transform lends itself well to numerical (computer) computations of $F(\omega)$. One value of ω is chosen. Then for that value of ω both integrals are separately computed by summing areas under the integrands [over values of t for which $f(t)$ is appreciably greater than zero]. Then a new value of ω is chosen and the process is repeated: $A(\omega)$ and $B(\omega)$ are called, respectively, the *cosine and sine transforms*. Note also that if we replace ω with $-\omega$ in (14.9.1),

$$F(-\omega) = \int_{-\infty}^{\infty} f(t)e^{+j\omega t}\,dt = \int_{-\infty}^{\infty} f(t)(\cos \omega t + j \sin \omega t)\,dt = A(\omega) + jB(\omega)$$
$$\tag{14.9.2}$$

where $A(\omega)$ and $B(\omega)$ are defined as in equations (14.9.1). We note that for *any* $f(t)$ that has a Fourier transform

$$F(-\omega) = A(\omega) + jB(\omega) = F^*(\omega) \tag{14.9.3}$$

and therefore

$$F(\omega)F(-\omega) = F(\omega)F(\omega)^* = |F(\omega)|^2 \tag{14.9.4}$$

Note from the definitions of $A(\omega)$ and $B(\omega)$ in equations (14.9.1b) and (14.9.1c) that *if* $f(t)$ *is a real* function of time, then $A(\omega)$ *and* $B(\omega)$ *are both real* functions of ω. Moreover, $A(\omega)$ *has even symmetry* with respect to ω and $B(\omega)$ *is odd*. Consider

$$|F(\omega)|^2 = F(\omega)F(\omega)^* = [A(\omega) - jB(\omega)][A(\omega) + jB(\omega)] = A^2(\omega) + B^2(\omega)$$
$$\tag{14.9.5}$$

This will be a *real and even-symmetric* function of ω. (The square of an even function is even. The square of an odd function is also even.) All practical voltage and current waveforms actually used in real circuits are *real functions of time*, and their *amplitude spectra* are therefore *even functions of* ω. If, in addition to being real, $f(t)$ *has even symmetry*, then $B(\omega) = 0$. In other words $\mathcal{F}[f_e(t)]$ is real if $f_e(t)$ is real and even because in computing $B(\omega)$ the product of an even and odd function is odd. And the integral, over symmetrical limits, of an odd function is zero. If $f(t)$ is real and has odd symmetry, then $A = 0$ and $F(\omega)$ is pure imaginary.

energy density spectrum 14.10

If $f(t)$ is either the voltage across, or the current in, a 1-Ω resistor, the total energy in the pulse $f(t)$ is given by

$$\int_{-\infty}^{\infty} f^2(t)\, dt = \int_{-\infty}^{\infty} f(t)\, \frac{1}{2\pi} \int_{\omega=-\infty}^{\infty} F(\omega) e^{j\omega t}\, d\omega\, dt$$

When the order of integration is changed,

$$\frac{1}{2\pi} \int_{\omega=-\infty}^{\infty} F(\omega) \left[\int_{t=-\infty}^{\infty} f(t) e^{j\omega t}\, dt \right] d\omega = \frac{1}{2\pi} \int_{-\infty}^{\infty} F(\omega) F(-\omega)\, d\omega$$

and

$$\frac{1}{2\pi} \int_{-\infty}^{\infty} |F(\omega)|^2\, d\omega = \int_{-\infty}^{\infty} f^2(t)\, dt \qquad (14.10.1)$$

known as *Parseval's theorem*.

Thus we interpret the plot of $(1/2\pi)[\,|F(\omega)|^2]$ versus frequency as the *energy density spectrum*.

The integral of the square of the magnitude of $F(\omega)$ between any two frequencies is the energy contained between those two frequencies times 2π. The right-hand side of (14.10.1) implies a summing up of power which is distributed (perhaps unevenly) in time. By analogy, the left-hand side of (14.10.1) represents a summing up of energy which is distributed (perhaps unevenly) over the frequency domain.

additional properties of the Fourier 14.11
transform

Examples will help us investigate some of the other properties of the Fourier transform.

EXAMPLE 14.11.1

Show that if $\mathscr{F}[f(t)] = F(\omega)$, then $\mathscr{F}[f(-t)] = F(-\omega)$.
ANS.: Use the defining equation for the Fourier transform

$$\mathscr{F}[f(-t)] = \int_{-\infty}^{\infty} f(-t) e^{-j\omega t}\, dt$$

Let $-t = x$ (thus $dt = -dx$), and then

$$\mathscr{F}[f(-t)] = -\int_{x=\infty}^{x=-\infty} f(x) e^{j\omega x}\, dx = \int_{x=-\infty}^{x=+\infty} f(x) e^{-j(-\omega)x}\, dx$$

Recognizing that x is simply the variable of integration (which does not survive the integration process), we can write

$$\mathcal{F}[f(-t)] = \int_{t=-\infty}^{\infty} f(t)e^{-j(-\omega)t}\, dt = F(-\omega) \qquad (14.11.1)$$

With the result of Example 14.11.1 it is a simple matter to show in general that the transform of an even-symmetric $f(t)$ is also evenly symmetric; i.e., if $\mathcal{F}[f(t)] = F(\omega)$ and $f(t) = f(-t)$, then

$$\mathcal{F}[f(-t)] = \mathcal{F}[f(t)] = F(\omega) = F(-\omega) \qquad \text{(QED)}$$

EXAMPLE 14.11.2

If $\mathcal{F}[f(t)] = F(\omega)$, show that

$$\mathcal{F}[f(t)e^{j\omega_0 t}] = F(\omega - \omega_0) \qquad \omega_0 = \text{const}$$

ANS.:   Again use the defining equation

$$\mathcal{F}[f(t)e^{j\omega_0 t}] = \int_{-\infty}^{\infty} f(t)e^{j\omega_0 t}e^{-j\omega t}\, dt = \int_{-\infty}^{\infty} f(t)e^{-j(\omega-\omega_0)t}\, dt = F(\omega - \omega_0) \qquad (14.11.2)$$

This is called the *frequency-shift property* of the Fourier transform.

EXAMPLE 14.11.3

Using the frequency-shift property, find $\mathcal{F}[f(t)\cos \omega_0 t]$ given that $\mathcal{F}[f(t)] = F(\omega)$.

ANS.: $$\mathcal{F}[f(t)\cos \omega_0 t] = \mathcal{F}[f(t)\tfrac{1}{2}e^{j\omega_0 t} + f(t)\tfrac{1}{2}e^{-j\omega_0 t}] = \tfrac{1}{2}F(\omega - \omega_0) + \tfrac{1}{2}F(\omega + \omega_0) \qquad (14.11.3)$$

We can use the result of Example 14.11.3 to determine the spectrum of the amplitude-modulated (AM) signals found on ordinary commercial radio bands and the video portion of commercial television. An AM signal $f(t)$ is given by

$$f(t) = A[1 + m(t)]\cos \omega_0 t \qquad (14.11.4)$$

where $m(t)$ = modulating signal (voice, music, etc.)
A = strength of transmitted signal
ω_0 = the carrier frequency

Clearly, if $m(t) = 0$, no information is being broadcast although $A\cos \omega_0 t$ is being transmitted. This enables you to find the station by tuning your receiver to get a maximum on the signal strength (tuning) indicator and/or by minimizing the noise audible through the speakers (this is called *quieting*).

If some nonzero modulating signal $m(t)$ is present

$$\mathcal{F}[m(t)] = M(\omega) \qquad (14.11.5)$$

then

$$\mathcal{F}[A\{1 + m(t)\}\cos \omega_0 t] = \mathcal{F}[A\cos \omega t] + \mathcal{F}[Am(t)\cos \omega_0 t] \qquad (14.11.6)$$

(a)

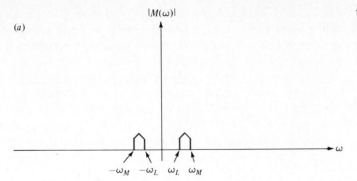

$|M(\omega)|$

$-\omega_M$ $-\omega_L$ ω_L ω_M

figure 14.11.1
(a) The magnitude spectrum of the
modulating signal. (b) The magnitude
spectrum of the resulting AM signal.

(b)

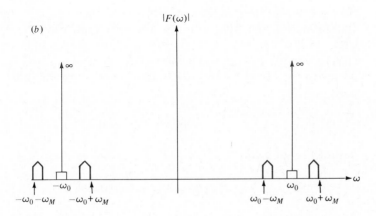

$|F(\omega)|$

$-\omega_0-\omega_M$ $-\omega_0$ $-\omega_0+\omega_M$ $\omega_0-\omega_M$ ω_0 $\omega_0+\omega_M$

$$F(\omega) = A\pi[\delta(\omega - \omega_0) + \delta(\omega + \omega_0)] + \frac{A}{2}[M(\omega - \omega_0) + M(\omega + \omega_0)]$$

$$(14.11.7)$$

Thus the AM technique consists of shifting the modulating spectrum to a
higher frequency range (near the carrier frequency) because those frequencies can
be broadcast efficiently by the radio station's transmitter antenna. Low frequencies (within the range of the human voice and hearing) cannot be efficiently
transmitted by electromagnetic radiation; even if they could, only one radio or
TV station would be possible in any one locality without them interfering with
one another.

The signal transmitted by any AM broadcast station has a magnitude spectrum similar to that in Figure 14.11.1b. The components of that spectrum due to
the modulating signal (with maximum frequency ω_M and minimum frequency ω_L)
lie in four ranges called *sidebands:*

$$(-\omega_0 - \omega_M) < \omega < (-\omega_0 - \omega_L)$$

$$(-\omega_0 + \omega_L) < \omega < (-\omega_0 + \omega_M)$$

$$(\omega_0 - \omega_M) < \omega < (\omega_0 - \omega_L)$$

$$(\omega_0 + \omega_L) < \omega < (\omega_0 + \omega_M)$$

The spectral shape of the sidebands is continually changing as the announcer talks and/or music plays. Since the carrier is simply a pair of impulses with fixed area $A\pi$, it contains no program information. A modulation scheme for communicating information developed to eliminate the two carrier impulses in Figure 14.11.1b is called *double sideband suppressed carrier* (DSBSC). Another scheme eliminates one of the two positive-frequency sidebands (and therefore the corresponding one of the two negative-frequency sidebands) and is termed *single sideband suppressed carrier* (SSBSC). Suffice it to say that these and many other techniques for transmitting information are of great interest to communications engineers, for whom the Fourier transform is one of the most useful analytical tools.

SYMMETRY PROPERTY   If we know that the transform of $f(t)$ is $F(\omega)$, then the time function $F(t)$ has the transform $2\pi f(-\omega)$. This means that every time we calculate the transform of an $f(t)$ we get a free answer to another problem at the same time, helping us to enlarge our table of $f(t)$'s and their $F(\omega)$'s at double speed. This *symmetry property* can be proved by examining the inverse transform (multiplied through by 2π)

$$2\pi f(t) = \int_{-\infty}^{\infty} F(\omega)e^{j\omega t}\, d\omega \tag{14.11.8}$$

Letting $t \to -t$ gives

$$2\pi f(-t) = \int_{-\infty}^{\infty} F(\omega)e^{-j\omega t}\, d\omega \tag{14.11.9}$$

Then let $\omega = \alpha$ and $t = \beta$, so that

$$2\pi f(-\beta) = \int_{\alpha = -\infty}^{\infty} F(\alpha)e^{-j\alpha\beta}\, d\alpha \tag{14.11.10}$$

Since α is the dummy variable of integration, we can call it anything we like, say t; then

$$2\pi f(-\beta) = \int_{t = -\infty}^{\infty} F(t)e^{-jt\beta}\, dt \tag{14.11.11}$$

Now let $\beta = \omega$

$$2\pi f(-\omega) = \int_{t = -\infty}^{\infty} F(t)e^{-j\omega t}\, dt = \mathcal{F}[F(t)] \tag{14.11.12}$$

EXAMPLE 14.11.4

From Example 14.5.1 we know that $\mathcal{F}[\delta(t)] = 1 = F(\omega)$; therefore, since $\mathcal{F}[F(t)] = 2\pi f(-\omega)$, it follows that

$$\mathcal{F}[1] = 2\pi\delta(-\omega) = 2\pi\delta(\omega) \tag{14.11.13}$$

which says that the frequency-domain spectrum of a 1-V dc signal is an impulse at zero frequency. The energy in the signal is infinite because it is of infinite duration in the time domain.

Another property of the Fourier transform tells us what happens in the frequency **TIME-SCALE COMPRESSION** domain if we compress a function's time scale. The property is written as follows. If $\mathcal{F}[f(t)] = F(\omega)$ then

$$\mathcal{F}[f(at)] = \frac{1}{|a|}\, F\!\left(\frac{\omega}{a}\right) \qquad (14.11.14)$$

In words this says that if a function $f(t)$ is altered so that is has a compressed time scale ($a > 1$ so that things happen in a *shorter time* than in the original signal), then $F(\omega)$ will *expand in the frequency domain* to higher frequencies. $F(\omega)$ will also shrink in magnitude.

EXAMPLE 14.11.5

Given that $\mathcal{F}[Ae^{-4t}u(t)] = A/(4 + j\omega)$, find the frequency-domain description of $Ae^{-24t}u(t)$.

ANS.: Let $f_1(t) = Ae^{-4t}u(t)$; then

$$f_2(t) = Ae^{-24t}u(t) = Ae^{-4(6t)}u(6t) = f_1(at) \qquad \text{where } a = 6$$

Therefore

$$F_2(\omega) = \tfrac{1}{6}F_1\!\left(\frac{\omega}{6}\right) = \frac{1}{6}\frac{A}{4 + j\omega/6} = \frac{A}{24 + j\omega}$$

which is the correct transform for $Ae^{-24t}u(t)$.

However, *if the transform* $F_1(\omega)$ *of the original function* $f_1(t)$ *contains any impulses, equation (14.11.14) will lead to an erroneous magnitude for* $F_2(\omega)$ *and its use should be avoided.*

EXAMPLE 14.11.6

Given that $\mathcal{F}[\cos 4t] = \pi\delta(\omega - 4) + \pi\delta(\omega + 4)$, try to find the frequency-domain description of $\cos 8t$ using the time-scale compression property, equation (14.11.14).

ANS.: Let $f_1(t) = \cos 4t$. Then

$$f_2(t) = \cos 8t = \cos 4(2t) = \cos 4(at) \qquad \text{where } a = 2$$

and

$$f_2(t) = f_1(at)$$

Therefore, according to equation (14.11.14),

$$F_2(\omega) = \mathcal{F}[f_1(at)] = \frac{1}{|a|}\, F_1\!\left(\frac{\omega}{a}\right) = \frac{\pi}{2}\left[\delta\!\left(\frac{\omega}{2} - 4\right) + \delta\!\left(\frac{\omega}{2} + 4\right)\right]$$

Recalling that an impulse occurs where its argument is zero-valued would give

$$F_2(\omega) = \frac{\pi}{2}\left[\delta(\omega - 8) + \delta(\omega + 8)\right]$$

but this is wrong! The Fourier transform of $\cos 8t$ is

$$\mathcal{F}[\cos 8t] = \pi\delta(\omega - 8) + \pi\delta(\omega + 8)$$

We can prove the time-scale-change property as follows. Since

$$\mathcal{F}[f(t)] = \int_{-\infty}^{\infty} f(t)e^{-j\omega t}\, dt \qquad (14.11.15)$$

we have

$$\mathcal{F}[f(at)] = \int_{-\infty}^{\infty} f(at)e^{-j\omega t}\, dt \qquad (14.11.16)$$

Letting $at = x$ on the right-hand side (and thus $a\, dt = dx$) leads to

$$\mathcal{F}[f(at)] = \frac{1}{a}\int_{x=-\infty}^{\infty} f(x)e^{-j\omega x/a}\, dx \qquad (14.11.17)$$

where in writing the limits as we have, we have assumed $a > 0$. Since x is the variable of integration which does not survive the integration process, we are free to call it anything we like; call it t:

$$\mathcal{F}[f(at)] = \frac{1}{a}\int_{t=-\infty}^{\infty} f(t)e^{-j(\omega/a)t}\, dt \qquad a > 0 \qquad (14.11.18)$$

If on the other hand, the constant a is negative $(a < 0)$, equation (14.11.17) becomes

$$\mathcal{F}[f(at)] = \frac{1}{a}\int_{x=\infty}^{-\infty} f(x)e^{-j(\omega/a)x}\, dx = \left(-\frac{1}{a}\right)\int_{x=-\infty}^{\infty} f(x)e^{-j(\omega/a)x}\, dx \qquad (14.11.19)$$

$$\mathcal{F}[f(at)] = \left(-\frac{1}{a}\right)\int_{-\infty}^{\infty} f(t)e^{-j(\omega/a)t}\, dt \qquad a < 0 \qquad (14.11.20)$$

Combining (14.11.18) and (14.11.20) into a single equation, we get

$$[f(at)] = \frac{1}{|a|}\int_{-\infty}^{\infty} f(t)e^{-j(\omega/a)t}\, dt = \frac{1}{|a|}F_1\left(\frac{\omega}{a}\right) \qquad \text{QED} \qquad (14.11.21)$$

14.12 convolution in the frequency domain

In equation (14.4.2) we showed that multiplication in the frequency domain is equivalent to convolution in the time domain; i.e., if

$$\mathcal{F}[f_1(t)] = F_1(\omega) \qquad \mathcal{F}[h(t)] = H(\omega) \qquad \mathcal{F}[f_2(t)] = F_2(\omega)$$

and

$$f_1(t) * h(t) = f_2(t)$$

then

$$F_1(\omega)H(\omega) = F_2(\omega)$$

The converse is also true: *multiplication in the time domain is equivalent to convolution in the frequency domain.* We show this as follows. Suppose we have three time functions $f_1(t)$, $f_2(t)$, and $f_3(t)$ such that

$$\mathcal{F}[f_1(t)] = F_1(\omega) \qquad \mathcal{F}[f_2(t)] = F_2(\omega)$$
$$\mathcal{F}[f_3(t)] = F_3(\omega) \qquad f_3(t) = f_1(t)f_2(t) \qquad (14.12.1)$$

We can solve for $f_3(t)$ in terms of $F_1(\omega)$ and $F_2(\omega)$. In other words, write $f_1(t)$ and

$f_2(t)$ in terms of inverse transforms (changing variables to avoid confusion between the two integrals)

$$f_3(t) = \frac{1}{(2\pi)^2} \int_{\Omega = -\infty}^{\infty} \int_{\psi = -\infty}^{\infty} F_1(\Omega)F_2(\psi)e^{j\Omega t}e^{j\psi t}\, d\psi\, d\Omega$$

$$= \frac{1}{(2\pi)^2} \int_{\Omega = -\infty}^{\infty} F_1(\Omega) \int_{\psi = -\infty}^{\infty} F_2(\psi)e^{j(\Omega + \psi)t}\, d\psi\, d\Omega$$

Let $\omega = \Omega + \psi$. Thus in the inner integral $d\psi = d\omega$, and so

$$f_3(t) = \frac{1}{(2\pi)^2} \int_{\Omega = -\infty}^{\infty} F_1(\Omega) \int_{\omega = -\infty}^{\infty} F_2(\omega - \Omega)e^{j\omega t}\, d\omega\, d\Omega$$

Interchanging the order of integration, we have

$$f_3(t) = \frac{1}{2\pi} \int_{\omega = -\infty}^{\infty} \left[\frac{1}{2\pi} \int_{\Omega = -\infty}^{\infty} F_1(\Omega)F_2(\omega - \Omega)\, d\Omega \right] e^{j\omega t}\, d\omega$$

Since this expression for $f_3(t)$ is nothing more than the inverse transform of $F_3(\omega)$, the bracketed inner integral must be $F_3(\omega)$

$$F_3(\omega) = \frac{1}{2\pi} \int_{\Omega = -\infty}^{\infty} F_1(\Omega)F_2(\omega - \Omega)\, d\Omega \qquad (14.12.2)$$

which is the convolution of $F_1(\omega)$ with $F_2(\omega)$ multiplied by $1/2\pi$ QED.

A good electrical engineer should be adept at graphical convolution. This qualitative technique affords an idea of the form of the spectrum of any waveform which itself is the time-domain product of two other waveforms. For instance, the discussion of AM above concerned in part the time-domain product of a modulating signal $f_m(t)$ and a carrier $f_c(t) = A \cos \omega_0 t$. This product $f_m(t)f_c(t)$ gave rise to the sidebands in the total AM signal. The spectrum $F(\omega) = \mathcal{F}[f_m(t)f_c(t)]$ can be qualitatively described by a simple graphical convolution (Figure 14.12.1) between $F_m(\omega)$ and $F_c(\omega)$. Recall that graphical convolution involves reversing the time axis of one function, shifting it, multiplying it by the other function, and integrating the result. If you have forgotten the process, you should review that earlier discussion. Another typical example of the usefulness of graphical frequency-domain convolution is in finding the spectrum of a *gated* cosine wave (Figure 14.12.2). An extremely important example of the usefulness of frequency-domain convolution is that of determining the spectrum of a sampled, band-limited waveform. *Sampling*, in electrical engineering, is the process of recording the instantaneous amplitude of a signal. Sampling done at a fixed rate, e.g., every T s, is called *periodic* or *fixed-rate sampling*. When the interval between samples is a random quantity, it is called *random sampling*. In either event, the mathematical equivalent of a single sampling of a function $f(t)$ is to multiply that $f(t)$ by a unit impulse. This product is simply another impulse occurring at the same time as the first but whose area is the amplitude of $f(t)$ at that instant. The numerical value can be recovered by integrating that impulse; i.e., we find the value of $f(t)$ at $t = t_0$ as follows:

$$f_s(t) = f(t)\delta(t - t_0)$$

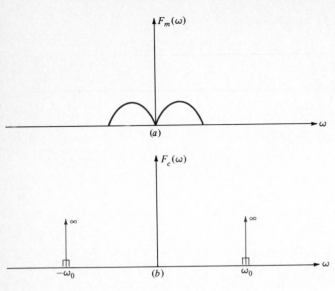

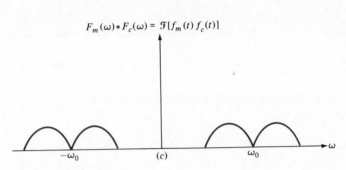

figure 14.12.1
(a) The (real) spectrum of a (symmetric) modulating signal $f_m(t)$. (b) The (real) spectrum of the carrier $f_c(t) = \cos \omega t$. (c) The spectrum of the product $f_m(t)f_c(t)$ which is obtained by convolving $F_m(\omega)$ and $F_c(\omega)$.

figure 14.12.2
(a) A pure cosine wave and (b) its spectrum. (c) A gate function and (d) its spectrum. (e) The symmetrically gated cosine obtained by multiplying the waveforms in (a) and (c). (f) The spectrum of the symmetrically gated cosine obtained by convolving the spectra in (b) and (d).

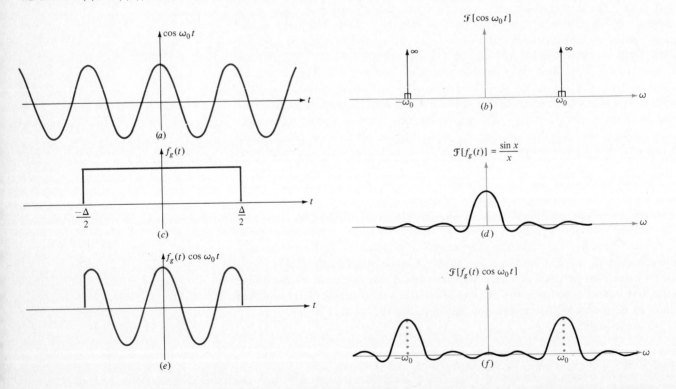

Integration of $f_s(t)$ yields

$$\int_{-\infty}^{\infty} f(t)\delta(t - t_0)\ dt = f(t_0)$$

We ask (1) what does the spectrum of a periodically sampled waveform look like and (2) what restrictions on the sampling rate are obvious from this spectrum?

Assume that the signal to be sampled $f(t)$ is band-limited; that is, $F(\omega)$ is zero-valued for $\omega > \omega_b$ (see Figure 14.12.3a). The sampling is accomplished by *multiplying* $f(t)$ by the unit-impulse train (in the time domain)

$$f_\delta(t) = \sum_{n=-\infty}^{\infty} \delta(t - nT)$$

whose spectrum [see part (b) of Example 14.8.1 and Figures 14.8.1c and d and 14.12.3b] is

$$F_\delta(\omega) = \sum_{k=-\infty}^{\infty} \omega_0\,\delta(\omega - k\omega_0)\qquad\text{where } \omega_0 = \frac{2\pi}{T}$$

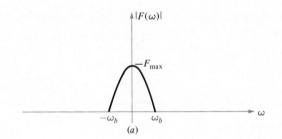

(a)

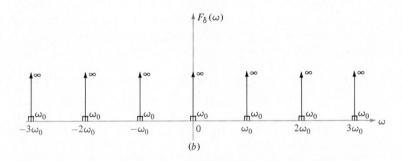

(b)

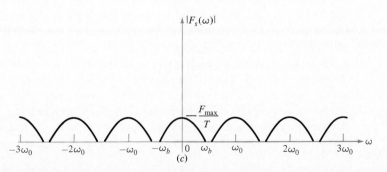

(c)

figure 14.12.3
(a) The (real) spectrum of a (symmetrical) band-limited signal. (b) The spectrum of a unit-impulse train. (c) The spectrum of the product of the waveforms in (a) and (b).

and T is the sampling interval. Thus by equation (14.12.2) the spectrum of the sampled waveform

$$f_s(t) = \sum_{n=-\infty}^{\infty} \delta(t - nT)f(t)$$

is given by

$$F_\delta(\omega) = \frac{1}{2\pi}\int_{\Omega=-\infty}^{\infty} F_\delta(\Omega)F(\omega-\Omega)\, d\Omega = \frac{1}{2\pi}\int_{-\infty}^{\infty}\sum_{k=-\infty}^{\infty} \omega_0\, \delta(\Omega-k\omega_0)F(\omega-\Omega)\, d\Omega$$

$$= \frac{\omega_0}{2\pi}\sum_{k=-\infty}^{\infty}\int_{\Omega=-\infty}^{\infty} \delta(\Omega-k\omega_0)F(\omega-\Omega)\, d\Omega = \frac{1}{T}\sum_{k=-\infty}^{\infty} F(\omega-k\omega_0)$$

which states that the spectrum is an infinite series of copies of the original $F(\omega)$ except with magnitude divided by T. Each such modified copy of $F(\omega)$ is situated at a multiple of ω_0 in the frequency domain (see Figure 14.12.3c).

The answer to the question whether there are any restrictions on the properties of $f(t)$ or $F(\omega)$ made obvious by an examination of the final $F_s(\omega)$ spectrum is now clear. From Figure 14.12.3c we note that if $\omega_b > \omega_0/2$, there will be interference between adjacent portions of the spectrum. Thus we conclude that in any fixed-rate sampling process the sampling rate must be at least twice the highest frequency at which energy is contained in the signal to be sampled. This rule, called *Nyquist's sampling theorem*, is one of the basic laws in designing sampled data control and communications systems.

14.13 summary

In this chapter we have developed a way of describing both nonperiodic and periodic signals in the frequency domain. We started by showing how to extend the Fourier-series representation of a periodic waveform into the Fourier-transform description of a nonperiodic signal. Several properties of the transform were discussed, including its linearity and the fact that its use together with $H(j\omega)$ enables us to solve for the particular response of a system to periodic inputs and the zero-state response to nonperiodic inputs. We showed that the transfer function $H(j\omega)$ is the transform of the unit-impulse response $h(t)$, and we used this idea to discuss why certain $H(j\omega)$ transfer functions cannot be built: $h(t)$ functions cannot predict the future.

We developed the partial list of Fourier transforms and their properties shown in Table 14.13.1. We discussed the sine-cosine form of the transform and concluded in passing that if $f(t)$ is real, then $|F(\omega)|$ is an even function of ω. Also, if in addition to being real, $f(t)$ has even symmetry in the time domain, then $F(\omega)$ is real. The energy density spectrum of a nonperiodic function was defined and discussed, as was AM. Finally we showed that multiplication of two functions in the time domain is equivalent to convolving their transforms in the frequency domain and used this fact to demonstrate Nyquist's sampling theorem.

Although we have showed that the Fourier transform is useful in solving for the zero-state response, it is much more widely used for analyzing the effects of

$f(t)$	$F(\omega)$
$af_1(t) + bf_2(t)$	$aF_1(\omega) + bF_2(\omega)$
$\delta(t)$	1
$f(t - t_0)$	$F(\omega)e^{-j\omega t_0}$
$Ae^{-at}u(t)$	$\dfrac{A}{a + j\omega}$
$\dfrac{1}{2\pi} e^{j\omega_0 t}$	$\delta(\omega - \omega_0)$
$\dfrac{1}{2\pi} e^{-j\omega_0 t}$	$\delta(\omega + \omega_0)$
$A \cos \omega_0 t$	$A\pi[\delta(\omega + \omega_0) + \delta(\omega - \omega_0)]$
$A \sin \omega_0 t$	$jA\pi[\delta(\omega + \omega_0) - \delta(\omega - \omega_0)]$
$f_{\text{per}}(t)$	$\displaystyle\sum_{k=-\infty}^{\infty} \alpha_k\, 2\pi\delta(\omega - k\omega_0)$
$f_{\text{per}}(t)\begin{cases} V & -\Delta/2 < t < +\Delta/2 \\ 0 & \text{elsewhere} \end{cases} f(t) = f(t + T)$	$\displaystyle\sum_{k=-\infty}^{\infty} 2\pi\, \frac{V\Delta}{T}\, \frac{\sin k\omega_0 \Delta/2}{k\omega_0 \Delta/2}\, \delta(\omega - k\omega_0)$
$\displaystyle\sum_{n=-\infty}^{\infty} \delta(t - nT)$	$\displaystyle\sum_{k=-\infty}^{\infty} \omega_0 \delta(\omega - k\omega_0) \quad \text{where } \omega_0 = \frac{2\pi}{T}$
$f(-t)$	$F(-\omega)$
$f(t)e^{j\omega_0 t}$	$F(\omega - \omega_0)$
If $\qquad f_1(t)$	$F_1(\omega)$
then $\qquad F_1(t)$	$2\pi f(-\omega)$
1	$2\pi\delta(\omega)$
$f(at)$	$\dfrac{1}{\|a\|} F\!\left(\dfrac{\omega}{a}\right) \quad$ no $\delta(\omega)$ in $F(\omega)$

table 14.13.1
Partial list of Fourier transforms and their properties

signal-processing schemes in the frequency domain. Communications engineers make frequent use of the Fourier transform, which is one of their primary tools in designing and analyzing communication networks and information channels. It turns out however, that its use in system analysis is more limited than that of another (really a modification of the Fourier) transform to be discussed in the next chapter.

problems

1. Use the defining integral to determine the Fourier transform of $f(t) = \delta(t + 1) + \delta(t - 1)$.

2. What is the Fourier transform of $f(t) = A \cos \omega_0\, tu(t)$?

3. Use the defining integral to determine the Fourier transform of a unit impulse that occurs at $t = t_0 > 0$.

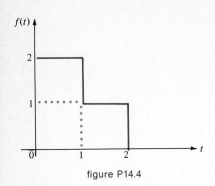

figure P14.4

4. Find the Fourier transform of the time function shown in Figure P14.4.

5. In Example 14.6.2 the Fourier transform of $Ae^{-at}u(t)$ was shown to be $A/(a + j\omega)$. Let $A = a$ and plot the magnitude and phase spectra. Describe the effect of allowing the quantity a to become large without bound. Is the result consistent with any other result that we have seen?

6. Find the Fourier transform of $f(t) = Ae^{at}u(-t)$.

7. Use the result of problem 6 together with the fact that the Fourier transform of $Ae^{-at}u(t) = A/(a + j\omega)$ to find the transform of $f(t) = Ae^{|a|t}$.

8. Use the result of problem 6 together with the fact that the Fourier transform of $Ae^{-at}u(t) = A/(a + j\omega)$ to find the Fourier transform of $f(t) = Ae^{-at}u(t) - Ae^{at}u(-t)$.

9. The *signum function* sgn t is defined as

$$\text{sgn } t = \begin{cases} 1 & t > 0 \\ -1 & t < 0 \end{cases}$$

(a) Use the result of problem 8 to find the Fourier transform of sgn t. (b) Use the result of part (a) to determine the Fourier transform of $u(t)$.

10. Given that $f_1(t) = e^t u(-t) + e^{-t}u(t)$, find the Fourier transform $F_2(\omega)$ of $f_2(t) = f_1(t) \cos 100t$.

11. Given that the Fourier transform of $f_1(t) = e^{j\omega_0 t}$ is $2\pi \, \delta(\omega - \omega_0)$, find the transform of $f_2(t) = e^{-j\omega_0 t}$.

12. A certain system's unit-impulse response is $h(t) = 2e^{-\sqrt{3}t}u(t)$. What will the response of this system be to inputs of (a) $\delta(t - 3)$, (b) $\cos t$, (c) $e^{-t}u(t)$?

13. A series combination of a 1-Ω resistor and a 2-H inductor is driven by a voltage source $v(t) = e^{-t}u(t)$. Find an expression for the voltage across the resistor $v_R(t)$ for $t > 0$. Assume that $v_R(0-) = 0$. Use Fourier transforms.

14. Use the Fourier-transform pair developed in Section 14.5 together with the symmetry property that if $f(t) \leftrightarrow F(\omega)$, then $F(t) \leftrightarrow 2\pi f(-\omega)$ to obtain the Fourier transform of $f(t) = u(t + t_0) - u(t - t_0)$.

15. (a) Find the Fourier transform of $\delta(t - t_0)$. (b) Find the Fourier transform of $d^n f(t)/dt^n$. (c) Use the defining integral to find the Fourier transform of $u(t) - u(t - T)$. (d) Take the derivative of the function in part (c) and find its transform. Then use the results of parts (a) and (b) to check your answer.

16. (a) Plot sinc t for $-4\pi < t < 4\pi$. (b) Find the exact value of sinc 0 via a limiting process.

17. The Fourier transform of $f(t)$ can be written $2 \int_0^\infty f(t) \cos \omega t \, dt$ if and only if $f(t)$ has what properties?

18. In Figure P14.18 if $f_1(t)$ transforms to $F_1(\omega)$, find $F_2(\omega)$ in terms of $F_1(\omega)$.

19. An exponential pulse whose Fourier transform is $F_1(\omega) = 10/(3 + j\omega)$ is passed through an ideal low-pass filter whose cutoff frequency is π rad/s. (a) What is the expression for $f_1(t)$, the input signal? (b) What percentage of the energy in the original pulse survives the filtering process and is contained in the output signal?

20. (a) Find the even and odd parts $f_{ev}(t)$ and $f_{od}(t)$ of $f(t) = u(t)u(1 - t)$. (b) Find $F_{ev}(\omega)$ and $F_{od}(\omega)$. (c) Find an expression for the energy density spectrum. (d) Plot the energy

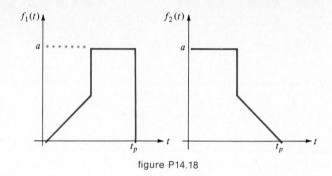

figure P14.18

density spectrum versus ω over the range $0 < \omega < 4\pi$. Can you estimate (compute) the percentage of the total energy in the pulse that is contained in the central node of the spectrum?

21. (*a*) Show that if a signal is of finite duration in the time domain, it is of infinite extent in the frequency domain. (*b*) Show that if a signal is band-limited (of finite extent) in the frequency domain, it must last for an infinite duration in the time domain.

22. Show how the spectrum of $\cos^2 \omega_0 t$ is obtained from the spectrum of $\cos \omega_0 t$ by convolution. Sketch and label a sufficient number of plots to demonstrate this procedure.

23. Suppose that an engineer has used a computer to obtain the Fourier transform of a finite (T-s duration) tape recording of the output signal from a seismograph. He or she notes several equally spaced peaks in the resulting energy density spectrum and concludes that what has been found is the resonant frequency of the earth! The engineer's boss remarks that the peaks are "bogus artifacts." (*a*) Explain what the engineer has probably seen and what the boss means. (*b*) What would the spacing of the peaks be if they are bogus? (*c*) Where would such peaks be most obvious (largest amplitude)?

chapter 15
LAPLACE TRANSFORMS

introduction 15.1

Many of the input functions we would like to use do not have Fourier transforms. For example, the function $v(t) = tu(t)$ does not have a legitimate corresponding $F(\omega)$ function in the frequency domain. If we are confronted, say, with an *RLC* circuit driven by a ramp-function source, is there any method of solution other than the classical time-domain method of Chapter 7, which involves adding the natural and particular responses? Happily, the answer to this is yes. Another transform which can handle all the time-domain input functions we are likely to encounter (and more) is called the *Laplace transform*, after its developer, Pierre Simon Laplace (1749−1827), a famous French mathematician. In this chapter we shall see how the Laplace transform is related to the Fourier transform.

The properties of the Laplace transform and its use in analyzing linear circuits and systems enable us to get closed form solutions to sets of state equations and zero-input, zero-state, and complete responses of linear systems by algebra; no integrations or assumed solutions are necessary. Initial conditions are automatically taken into account. In short, as long as the system we have to work with is linear, the ultimate analytical tool available is the Laplace transform. Moreover, it affords insight into other topics and methods. For example, the unit-impulse response, unit-step response, and the convolution process will be better understood after we have investigated the use of the Laplace transform in these areas.

15.2 from Fourier to Laplace

Let us find the Fourier transform of $f(t)u(t)$ multiplied by a damping term $e^{-\sigma t}$

$$g(t) = e^{-\sigma t}f(t)u(t) \tag{15.2.1}$$

Presumably we can find some large enough but finite value of the constant σ such that $g(t)$ is absolutely integrable for a wide class of functions $f(t)$

$$\int_{t_0}^{t_0 + t} |g(t)| \, dt < \infty \qquad \text{for any } t_0 \tag{15.2.2}$$

[This and the requirement that $g(t)$ have a finite number of discontinuities and extrema in any period of time T constitute the Dirichlet conditions, sufficient for transformability.] We write the Fourier transform of equation (15.2.1)

$$G(\omega) = \int_{t=-\infty}^{\infty} e^{-\sigma t}f(t)u(t)e^{-j\omega t} \, dt = \int_{0}^{\infty} e^{-\sigma t}f(t)e^{-j\omega t} \, dt = \int_{0}^{\infty} f(t)e^{-(\sigma + j\omega)t} \, dt$$

$$\tag{15.2.3}$$

Letting $s = \sigma + j\omega$ gives

$$G(\omega)\bigg|_{s=\sigma+j\omega} = \int_{0}^{\infty} f(t)e^{-st} \, dt$$

Call this function $F(s)$:

$$\boxed{F(s) = \int_{t=0-}^{\infty} f(t)e^{-st} \, dt} \tag{15.2.4}$$

where the lower limit is written as $t = 0-$ because we wish to allow for the inclusion of any possible impulse function in $f(t)$ at $t = 0$. We call $F(s)$ the Laplace transform of $f(t)$, $\mathcal{L}\{f(t)\} = F(s)$. The inverse Fourier transform of equation (15.2.3) is

$$e^{-\sigma t}f(t)u(t) = \frac{1}{2\pi} \int_{\omega=-\infty}^{\infty} G(\omega)e^{j\omega t} \, d\omega \tag{15.2.5}$$

Multiplying equation (15.2.5) through by $e^{\sigma t}$ gives

$$f(t)u(t) = \frac{1}{2\pi} \int_{\omega=-\infty}^{\infty} G(\omega)e^{(\sigma+j\omega)t} \, d\omega = \frac{1}{2\pi j} \int_{j\omega=-j\infty}^{+j\infty} G(\omega)e^{(\sigma+j\omega)t} \, dj\omega$$

Again, with $s = \sigma + j\omega$ and therefore $ds = dj\omega$,

$$f(t) = \frac{1}{2\pi j} \int_{s=\sigma-j\infty}^{\sigma+j\infty} F(s)e^{st} \, ds \qquad (15.2.6)$$

where $f(t)$ is assumed to be zero-valued for $t < 0$.

Equation (15.2.6) constitutes the inverse Laplace transform of $F(s)$, $\mathcal{L}^{-1}\{F(s)\} = f(t)$. A sufficient requirement for $f(t)$ to be Laplace-transformable is that

$$\lim_{T\to\infty} \int_0^T |f(t)| e^{-\sigma t} \, dt < \infty \qquad (15.2.7)$$

The Laplace transform is a linear transform; i.e., it has both the additivity and homogeneity properties

$$\mathcal{L}\{af_1(t) + bf_2(t)\} = \int_{0-}^{\infty} [af_1(t) + bf_2(t)]e^{-st} \, dt$$

$$= a \int_{0-}^{\infty} f_1(t)e^{-st} \, dt + b \int_{0-}^{\infty} f_2(t)e^{-st} \, dt$$

$$= aF_1(s) + bF_2(s) \qquad \text{QED} \qquad (15.2.8)$$

At this point we realize that thanks to the $e^{-\sigma t}$ damping term there may be many more time functions that are Laplace-transformable than are Fourier-transformable. However, although we have an intuitive grasp on the notion that the Fourier transform simply shows how any signal acts in the frequency domain, the result $F(s)$ of the Laplace transform has little meaning for us yet. We shall quickly see that the Laplace transform maps the time domain into the *s plane* in much the same way that the Fourier transform maps the time domain into the *jω axis* (what we have called the frequency domain). We shall use the Laplace transform of any input signal (assuming it is zero-valued for *t* less than zero) together with the transfer function $H(s)$ to get the output of any linear system. Moreover we shall find that whereas the Fourier transform gives us the zero-state response, the Laplace transform gives us the *complete response*—all initial conditions automatically taken into account.

This seems a tall order for any transform technique, and it is; but the Laplace transform is the ultimate tool for analyzing linear systems. First let us make a list of the Laplace transforms of some popular time functions (later we may add to it).

In all the following examples and discussions we assume that all the $f(t)$ functions we are working with are *zero-valued for negative t*. For this reason, the actual correct name for the expression in equation (15.2.4) is the *one-sided Laplace transform*. A two-sided Laplace transform exists, but its use requires a background in complex-variable theory beyond the scope of this text. The one-sided transform will suffice for the tasks we have set out to accomplish.

15.3 Laplace transforms of some typical time functions

Directly applying the definition of the Laplace transform gives many of the $F(s)$ functions that correspond to the typical time functions we have been using.

EXAMPLE 15.3.1

Find $\mathcal{L}\{e^{-at}u(t)\}$.

ANS.:
$$F(s) = \int_{0-}^{\infty} e^{-at}e^{-st}\, dt = \int_{0-}^{\infty} e^{-(s+a)t}\, dt = -\frac{-1}{s+a}e^{-(s+a)t}\Big|_{0-}^{\infty} = \frac{1}{s+a} \qquad (15.3.1)$$

The result of Example 15.3.1 enables us to find several other transform functions.

EXAMPLE 15.3.2

Find $F(s)$ for $f(t) = Ae^{at}u(t)$.
ANS.:   From the linearity property

$$\mathcal{L}\{Ae^{at}\} = A\mathcal{L}\{e^{at}\}$$

and from the result of Example 15.3.1

$$F(s) = \mathcal{L}\{Ae^{at}\} = \frac{A}{s-a} \qquad (15.3.2)$$

EXAMPLE 15.3.3

Find $F(s)$ for $f(t) = u(t)$.
ANS.:   Use the result of Example 15.3.2 but let $a = 0$. Thus

$$\mathcal{L}\{u(t)\} = \frac{1}{s} \qquad (15.3.3)$$

EXAMPLE 15.3.4

Find $\mathcal{L}\{u(t) \sin \omega_0 t\}$.

ANS.:
$$f(t) = \sin \omega_0 t = \frac{e^{j\omega_0 t} - e^{-j\omega_0 t}}{2j}$$

Thus
$$F(s) = \frac{1}{2j}\left(\frac{1}{s - j\omega_0} - \frac{1}{s + j\omega_0}\right) = \frac{\omega_0}{s^2 + \omega_0^2} \qquad (15.3.4)$$

EXAMPLE 15.3.5

Find $\mathcal{L}\{u(t) \cos \omega_0 t\}$.

ANS.:
$$f(t) = \cos \omega_0 t = \frac{e^{j\omega_0 t} + e^{-j\omega_0 t}}{2}$$

Thus
$$F(s) = \frac{1}{2}\left(\frac{1}{s - j\omega_0} + \frac{1}{s + j\omega_0}\right) = \frac{s}{s^2 + \omega_0^2} \qquad (15.3.5)$$

EXAMPLE 15.3.6

Find:
a. $\mathcal{L}\{u(t) \sinh bt\}$
b. $\mathcal{L}\{u(t) \cosh bt\}$

ANS.:   (a)
$$f_1(t) = \sinh bt = \frac{e^{bt} - e^{-bt}}{2}$$

$$F_1(s) = \frac{1}{2}\left(\frac{1}{s - b} - \frac{1}{s + b}\right) = \frac{b}{s^2 - b^2} \qquad (15.3.6)$$

(b)
$$f_2(t) = \cosh bt = \frac{e^{bt} + e^{-bt}}{2}$$

$$F_2(s) = \frac{1}{2}\left(\frac{1}{s - b} + \frac{1}{s + b}\right) = \frac{s}{s^2 - b^2} \qquad (15.3.7)$$

EXAMPLE 15.3.7

Find:
a. $\mathcal{L}\{e^{-\alpha t} \sin \omega_0 t \, u(t)\}$
b. $\mathcal{L}\{e^{-\alpha t} \cos \omega_0 t \, u(t)\}$

ANS.:   (a)
$$f_1(t) = e^{-\alpha t} \sin \omega_0 t = e^{-\alpha t} \frac{e^{j\omega_0 t} - e^{-j\omega_0 t}}{2j} = \frac{1}{2j}\left(e^{-(\alpha - j\omega_0)t} - e^{-(\alpha + j\omega_0)t}\right)$$

so that
$$F_1(s) = \frac{1}{2j}\left[\frac{1}{s + (\alpha - j\omega_0)} - \frac{1}{s + (\alpha + j\omega_0)}\right] = \frac{\omega_0}{(s + \alpha)^2 + \omega_0^2} \qquad (15.3.8)$$

(b) $F_2(t) = e^{-\alpha t} \cos \omega_0 t$

$$= e^{-\alpha t} \frac{e^{j\omega_0 t} + e^{-j\omega_0 t}}{2} = \tfrac{1}{2}(e^{-(\alpha - j\omega_0)t} + e^{-(\alpha + j\omega_0)t})$$

so that
$$F_2(s) = \frac{1}{2}\left[\frac{1}{s + (\alpha - j\omega_0)} + \frac{1}{s + (\alpha + j\omega_0)}\right]$$

$$= \frac{s + \alpha}{(s + \alpha)^2 + \omega_0^2} \qquad (15.3.9)$$

EXAMPLE 15.3.8

Find:
a. $\mathcal{L}\{A \sin (\omega_0 t + \theta)u(t)\}$
b. $\mathcal{L}\{A \cos (\omega_0 t + \theta)u(t)\}$
ANS.: (a) Using the trigonometric identity

$$\sin (a + b) = \sin a \cos b + \cos a \sin b$$

we can write

$$\sin (\omega_0 t + \theta) = \sin \omega_0 t \cos \theta + \cos \omega_0 t \sin \theta$$

Therefore, from Examples 15.3.4 and 15.3.5,

$$F_a(s) = \frac{A\omega_0 \cos \theta}{s^2 + \omega_0^2} + \frac{sA \sin \theta}{s^2 + \omega_0^2} \tag{15.3.10}$$

(b) Using

$$\cos (a + b) = \cos a \cos b - \sin a \sin b$$
$$\cos (\omega_0 t + \theta) = \cos \omega_0 t \cos \theta - \sin \omega_0 t \sin \theta$$

and thus

$$F_b(s) = \frac{sA \cos \theta}{s^2 + \omega_0^2} - \frac{\omega_0 A \sin \theta}{s^2 + \omega_0^2} \tag{15.3.11}$$

EXAMPLE 15.3.9

Find

$$\mathcal{L}^{-1}\left\{\frac{2.6(s + 5.8)}{s^2 + 100}\right\}$$

ANS.: $\qquad F(s) = \frac{2.6s + 15.08}{s^2 + 100} = \frac{A\omega_0 \cos \theta + sA \sin \theta}{s^2 + \omega_0^2}$

Hence $\qquad \omega_0^2 = 100 \quad$ and $\quad \omega_0 = 10$

$$A\omega_0 \cos \theta = 15.08$$
$$A \cos \theta = 1.508 \tag{15.3.12}$$

and $\qquad A \sin \theta = 2.6 \tag{15.3.13}$

Dividing (15.3.13) by (15.3.12) gives

$$\frac{A \sin \theta}{A \cos \theta} = \frac{2.6}{1.508} = \tan \theta = 1.724$$

whereby $\theta = 59.9°$, and from either (15.3.12) or (15.3.13) $A = 3$. Thus

$$f(t) = 3 \sin (10t + 59.9°) \tag{15.3.14}$$

EXAMPLE 15.3.10

Find $\mathcal{L}\{t^n u(t)\}$.

ANS.:   From the definition, equation (15.2.4),

$$F(s) = \int_{0-}^{\infty} t^n e^{-st}\, dt$$

Integrating by parts gives

$$\int u\, dv = uv - \int v\, du$$

Let $u = t^n$ and $dv = e^{-st}\, dt$; then

$$du = nt^{n-1}\, dt \qquad \text{and} \qquad v = -(1/s)e^{-st}$$

Thus

$$F(s) = -\frac{t^n}{s} e^{-st}\Big|_{0-}^{\infty} - \int_{0-}^{\infty} -\frac{1}{s} e^{-st} nt^{n-1}\, dt = 0 + \frac{n}{s}\,\mathcal{L}\{t^{n-1}\}$$

Continuing gives

$$F(s) = \frac{n}{s}\cdot\frac{n-1}{s}\cdot\frac{n-2}{s}\cdots\frac{3}{s}\cdot\frac{2}{s}\cdot\frac{1}{s}\cdot\mathcal{L}\{t^0\}$$

and

$$\mathcal{L}\{t^0 u(t)\} = \mathcal{L}\{u(t)\} = \frac{1}{s}$$

so that

$$\mathcal{L}\{t^0 u(t)\} = \frac{1}{s} \qquad \mathcal{L}\{t^1 u(t)\} = \frac{1}{s^2}$$

$$\mathcal{L}\{t^2 u(t)\} = \frac{2}{s^3} \qquad \mathcal{L}\{t^3 u(t)\} = \frac{6}{s^4}$$

and in general

$$\mathcal{L}\{t^n u(t)\} = \frac{n!}{s^{n+1}} \tag{15.3.15}$$

EXAMPLE 15.3.11

Find $\mathcal{L}\{\delta(t)\}$.

ANS.:   From the definition,

$$F(s) = \int_{0-}^{\infty} \delta(t)e^{-st}\, dt = 1 \tag{15.3.16}$$

table 15.3.1
Some time functions and their Laplace
transforms

Item	$f(t)$†	$F(s)$
1	$\delta(t)$	1
2	$ku(t)$	$\dfrac{k}{s}$
3	$tu(t)$	$\dfrac{1}{s^2}$
4	$t^n u(t)$	$\dfrac{n!}{s^{n+1}}$
5	$e^{-\alpha t} u(t)$	$\dfrac{1}{s+\alpha}$
6	$\sin \omega_0 t \; u(t)$	$\dfrac{\omega_0}{s^2 + \omega_0^2}$
7	$\cos \omega_0 t \; u(t)$	$\dfrac{s}{s^2 + \omega_0^2}$
8	$\sin (\omega_0 t + \theta) \, u(t)$	$\dfrac{s \sin \theta + \omega_0 \cos \theta}{s^2 + \omega_0^2}$
9	$\cos (\omega_0 t + \theta) \, u(t)$	$\dfrac{s \cos \theta - \omega_0 \sin \theta}{s^2 + \omega_0^2}$
10	$\sinh bt \; u(t)$	$\dfrac{b}{s^2 - b^2}$
11	$\cosh bt \; u(t)$	$\dfrac{s}{s^2 - b^2}$
12	$e^{-\alpha t} \sin (\omega_0 t + \theta) \, u(t)$	$\dfrac{(s+\alpha) \sin \theta + \omega_0 \cos \theta}{(s+\alpha)^2 + \omega_0^2}$
13	$e^{-\alpha t} \cos (\omega_0 t + \theta) \, u(t)$	$\dfrac{(s+\alpha) \cos \theta - \omega_0 \sin \theta}{(s+\alpha)^2 + \omega_0^2}$

† All these $f(t)$ functions are zero for all negative values of t.

We have now compiled a short list of some of the time functions often used in circuit and system analysis (Table 15.3.1).

15.4 some properties of the Laplace transform

Knowledge of several general properties of the Laplace transform (other than the homogeneity and additivity already discussed) will help lengthen our list of transform pairs.

COMPLEX-FREQUENCY-SHIFT PROPERTY   Multiplying the time function by a decaying exponential is the equivalent of introducing a shift in the s variable; i.e., if

$$\mathcal{L}\{f(t)\} = F(s) \qquad \text{then} \qquad \mathcal{L}\{e^{-at}f(t)\} = F(s + a)$$

This can be proved simply by using the definition of the transform

$$\mathcal{L}\{f(t)\} = F(s) = \int_{0-}^{\infty} f(t)e^{-st}\, dt$$

from which

$$\mathcal{L}\{e^{-at}f(t)\} = \int_{0-}^{\infty} e^{-at}f(t)e^{-st}\, dt = \int_{0-}^{\infty} f(t)e^{-(s+a)t}\, dt = F(s+a) \qquad \text{QED}$$

$$(15.4.1)$$

EXAMPLE 15.4.1

Find the Laplace transform of $f(t) = Ae^{-at}\cos(\omega_0 t + \theta)$ using item 9 in Table 15.3.1 and equation (15.4.1).
ANS.:   From Table 15.3.1 we note that

$$\mathcal{L}\{A \cos(\omega_0 t + \theta)\} = \frac{A(s \cos\theta - \omega_0 \sin\theta)}{s^2 + \omega_0^2}$$

Therefore

$$\mathcal{L}\{Ae^{-at}\cos(\omega_0 t + \theta)\} = \frac{A[(s+a)\cos\theta - \omega_0 \sin\theta]}{(s+a)^2 + \omega_0^2} \qquad (15.4.2)$$

Introducing a time shift (or *transport delay*) into an $f(t)$ function is equivalent in the complex domain to multiplying $F(s)$ by $e^{-t_0 s}$, that is, if   **TIME-SHIFT PROPERTY**

$$\mathcal{L}\{f(t)\} = F(s)$$

then $\qquad \mathcal{L}\{f(t-t_0)u(t-t_0)\} = e^{-t_0 s}F(s) \qquad (15.4.3)$

We prove this using the defining equation

$$\int_{0-}^{\infty} f(t-t_0)u(t-t_0)e^{-st}\, dt = \int_{t_0}^{\infty} f(t-t_0)e^{-st}\, dt$$

Let $\tau = t - t_0$; thus $t = \tau + t_0$ and $dt = d\tau$. When $t = t_0$, this means that $\tau = 0$, giving

$$\int_{0}^{\infty} f(t-t_0)u(t-t_0)e^{-st}\, dt = \int_{0}^{\infty} f(\tau)e^{-s(\tau+t_0)}\, d\tau$$

$$= e^{-t_0 s}\int_{0}^{\infty} f(\tau)e^{-s\tau}\, d\tau = e^{-t_0 s}F(s) \qquad \text{QED}$$

Note that in this last equation s is the variable and t_0 is a constant.

EXAMPLE 15.4.2

Find the Laplace transform of the single pulse $f(t)$ that consists of the first half cycle of a sine wave.

ANS.: This function can be constructed by adding a sin $\omega_0 t\, u(t)$ function and a delayed version

$$\sin \omega_0\left(t - \frac{T}{2}\right)u\left(t - \frac{T}{2}\right)$$

or

$$f(t) = \sin \omega_0 t\, u(t) + \sin \omega_0\left(t - \frac{T}{2}\right)u\left(t - \frac{T}{2}\right)$$

where $\omega_0 = 2\pi/T$. Therefore

$$F(s) = \frac{\omega_0}{s^2 + \omega_0^2} + \frac{\omega_0 e^{-Ts/2}}{s^2 + \omega_0^2} = \frac{\omega_0}{s^2 + \omega_0^2}(1 + e^{-Ts/2}) \tag{15.4.4}$$

PERIODIC FUNCTIONS The Laplace transform of a periodic function (one that is zero for $t < 0$) is given by

$$\mathcal{L}\{f_{per}(t)\} = \frac{1}{1 - e^{-Ts}}\mathcal{L}\{\text{first cycle}\} \tag{15.4.5}$$

This can be proved as follows. Let $f_{per}(t)$ have a first cycle called $f_1(t)$. Then

$$f_{per}(t) = f_1(t) + f_1(t - T)u(t - T) + f_1(t - 2T)u(t - 2T) + \cdots$$

By the time-delay theorem

$$F(s) = F_1(s) + F_1(s)e^{-Ts} + F_1(s)e^{-2Ts} + \cdots = \frac{1}{1 - e^{-Ts}}F_1(s) \qquad \text{QED}$$

That this last step is correct can be seen from the long-division process

$$
\begin{array}{r}
1 + e^{-Ts} + e^{-2Ts} + \cdots \\
1 - e^{-Ts})\overline{\,1} \\
\underline{1 - e^{-Ts}} \\
e^{-Ts} \\
\underline{e^{-Ts} - e^{-2Ts}} \\
+ e^{-2Ts} \\
\vdots
\end{array}
$$

EXAMPLE 15.4.3

What is the Laplace transform of a full-wave-rectified sine wave that begins at $t = 0$?
ANS.: From Example 15.4.2 we found the transform of the first half cycle of a sine wave (first full cycle of the wave we want to build in this example)

$$F_1(s) = \frac{\omega_0}{s^2 + \omega_0^2}(1 + e^{-Ts/2})$$

Therefore the complete full-wave-rectified wave has the transform

$$F(s) = \frac{1}{1 - e^{-Ts/2}}F_1(s) = \frac{\omega_0}{s^2 + \omega_0^2}\frac{1 + e^{-Ts/2}}{1 - e^{-Ts/2}} \tag{15.4.6}$$

where T is the period of the original unrectified sine wave.

Differentiating $F(s)$ with respect to s is equivalent to multiplying $-f(t)$ by t. If   **INTEGRALS AND DERIVATIVES**
$F(s) = \mathcal{L}\{f(t)\}$, then

$$\frac{dF(s)}{ds} = -\mathcal{L}\{tf(t)\} \qquad (15.4.7)$$

We can show this simply by taking the derivative of the defining equation. If

$$F(s) = \int_{0-}^{\infty} f(t)e^{-st}\,dt$$

then

$$\frac{dF}{ds} = -\int_{0-}^{\infty} tf(t)e^{-st}\,dt = -\mathcal{L}\{tf(t)\}$$

EXAMPLE 15.4.4

Find $\mathcal{L}\{te^{-at}\}$.
ANS.:   We already know that $\mathcal{L}\{e^{-at}\} = 1/(s+a)$. Differentiating gives

$$\frac{d}{ds}\frac{1}{s+a} = \frac{d}{ds}(s+a)^{-1} = -(s+a)^{-2}(1)$$

Thus

$$\mathcal{L}\{te^{-at}\} = \frac{1}{(s+a)^2} \qquad (15.4.8)$$

We can also show that

$$\int_{s=s}^{\infty} F(s)\,ds = \mathcal{L}\left\{\frac{f(t)}{t}\right\} \qquad \text{where } F(s) = \mathcal{L}\{f(t)\} \qquad (15.4.9)$$

From the defining equation

$$F(s) = \int_{0-}^{\infty} f(t)e^{-st}\,dt$$

Integrating gives

$$\int_{s=s}^{\infty} F(s)\,ds = \int_{s=s}^{\infty}\int_{t=0-}^{\infty} f(t)e^{-st}\,dt\,ds$$

Reversing the order of integration, we have

$$\int_{s=s}^{\infty} F(s)\,ds = \int_{t=0-}^{\infty}\int_{s=s}^{\infty} f(t)e^{-ts}\,ds\,dt$$

Performing the inner integration leads to

$$\int_{t=0-}^{\infty} \frac{f(t)}{-t} e^{-ts}\Big|_{s=s}^{\infty} dt = \int_{t=0-}^{\infty} \frac{f(t)}{t} e^{-st} dt$$

$$\mathcal{L}\left\{\frac{f(t)}{t}\right\} \qquad \text{QED}$$

In circuit and system analysis we often have to deal with integrals of the form

$$I = \int_{-\infty}^{t} f(t) \, dt \tag{15.4.10}$$

For example, the voltage on a capacitor C at time t is directly proportional to the charge that has accumulated on the capacitor. To find that charge we integrate the current (charge flow) into the capacitor from $t = -\infty$ up until the instant t

$$v_C(t) = \frac{1}{C} \int_{-\infty}^{t} i_C(t) \, dt \tag{15.4.11}$$

Similarly
$$i_L(t) = \frac{1}{L} \int_{-\infty}^{t} v_L(t) \, dt \tag{15.4.12}$$

Let us consider taking the Laplace transform of either (15.4.11) or (15.4.12). On the right we have the Laplace transform of an integral; i.e., we wish to evaluate an expression of the form

$$\mathcal{L}\left\{\int_{-\infty}^{t} f(t) \, dt\right\} \tag{15.4.13}$$

We can break this into two problems (thanks to the fact that the Laplace is a linear transform)

$$\mathcal{L}\left\{\int_{-\infty}^{t} f(t) \, dt\right\} = \mathcal{L}\left\{\int_{-\infty}^{0-} f(t) \, dt + \int_{0-}^{t} f(t) \, dt\right\}$$

$$= \mathcal{L}\left\{\int_{-\infty}^{0-} f(t) \, dt\right\} + \mathcal{L}\left\{\int_{0-}^{t} f(t) \, dt\right\} = F_1(s) + F_2(s) \tag{15.4.14}$$

Because the limits of the first integral on the right are constants, the integral will be a constant. Let

$$\int_{-\infty}^{0-} f(t) \, dt = k$$

Therefore
$$F_1(s) = \frac{k}{s} \tag{15.4.15}$$

We can evaluate $F_2(s)$ using integration by parts

$$F_2(s) = \mathcal{L}\left\{\int_{0-}^{t} f(t) \, dt\right\} = \int_{0-}^{\infty} \int_{0-}^{t} f(t) \, dt \, e^{-st} \, dt \tag{15.4.16}$$

which is of the form

$$\int_{0-}^{\infty} v \ du = uv - \int_{0-}^{\infty} u \ dv$$

where

$$v = \int_{0-}^{t} f(t) \ dt \qquad u = -\frac{1}{s} e^{-st}$$

$$dv = f(t) \ dt \qquad du = e^{-st} \ dt$$

Hence equation (15.4.16) becomes

$$F_2(s) = -\frac{1}{s} e^{-st} \int_{0-}^{t} f(t) \ dt \Big|_{0-}^{\infty} + \frac{1}{s} \int_{0-}^{\infty} e^{-st} f(t) \ dt = 0 + \frac{F(s)}{s} \qquad (15.4.17)$$

Thus using (15.4.15) and (15.4.17), equation (15.4.14) becomes

$$\mathcal{L}\left\{ \int_{-\infty}^{t} f(t) \ dt \right\} = \frac{F(s)}{s} + \frac{k}{s} \qquad (15.4.18)$$

where k is the *initial value* of the integral

$$k = \int_{-\infty}^{0-} f(t) \ dt = I(0-)$$

EXAMPLE 15.4.5

Find the Laplace transform of $v_C(t)$, the voltage across a capacitor C.
ANS.:   Rewriting equation (15.4.11) gives

$$v_C(t) = \frac{1}{C} \int_{-\infty}^{t} i(t) \ dt = \frac{1}{C} \int_{0-}^{t} i(t) \ dt + \frac{1}{C} \int_{-\infty}^{0-} i(t) \ dt$$

The integral in the second term is the charge on the capacitor at the instant $t = 0-$. The entire second term is therefore $v_C(0-)$, the voltage across the capacitor at the instant $t = 0-$. Taking the Laplace transform term by term, we get

$$V_C(s) = \frac{1}{C} \frac{I(s)}{s} + \frac{v_C(0-)}{s}$$

Now that we know how to take the Laplace transform of an integral, how about the derivative $df(t)/dt$? Let us call the derivative $\dot{f}(t)$; that is,

$$\dot{f}(t) = \frac{df(t)}{dt}$$

The Laplace transform of $\dot{f}(t)$ is, by definition,

$$\mathcal{L}\{\dot{f}(t)\} = \int_{0-}^{\infty} \dot{f}(t) e^{-st} \ dt$$

We use integration by parts

$$\int u\,dv = uv - \int v\,du \qquad \text{where} \qquad \begin{array}{ll} u = e^{-st} & v = f(t) \\ du = -se^{-st}\,dt & dv = \dot{f}(t)\,dt \end{array}$$

Thus
$$\mathcal{L}\{\dot{f}(t)\} = e^{-st}f(t)\Big|_{t=0-}^{\infty} + s\int_{0-}^{\infty} f(t)e^{-st}\,dt$$

Rearranging the order of the terms on the right gives

$$\mathcal{L}\{\dot{f}(t)\} = sF(s) - f(0-) \tag{15.4.19}$$

The second derivative, $\ddot{f}(t)$, is the derivative of $f(t)$; thus

$$\mathcal{L}\{\ddot{f}(t)\} = s[sF(s) - f(0-)] - \dot{f}(0-) = s^2F(s) - sf(0-) - \dot{f}(0-) \tag{15.4.20}$$

So that, in general,

$$\mathcal{L}\left\{\frac{d^n}{dt^n}f(t)\right\} = s^nF(s) - s^{n-1}f(0-) - s^{n-2}\frac{df}{dt}\Big|_{0-} - \cdots - \frac{d^{n-1}f}{dt^{n-1}}\Big|_{0-} \tag{15.4.21}$$

EXAMPLE 15.4.6

Solve the following homogeneous differential equation for the natural response $x_n(t)$ if the initial conditions are $x(0-) = 0$ and $\dot{x}(0-) = 3$:

$$\frac{d^2x}{dt^2} + 3\frac{dx}{dt} + 6x = 0$$

ANS.: Using Laplace transforms, equations (15.4.19) and (15.4.20), gives

$$s^2X(s) - sx(0-) - \dot{x}(0-) + 3sX(s) - 3x(0-) + 6X(s) = 0$$

Solving for $X(s)$ yields

$$X(s) = \frac{3}{s^2 + 3s + 6}$$

From item 6 in Table 15.3.1

$$\sin \omega_0 t = \frac{\omega_0}{s^2 + \omega_0^2}$$

Thus
$$Ae^{-\alpha t}\sin \omega_0 t = \frac{A\omega_0}{(s+\alpha)^2 + \omega_0^2} = \frac{A\omega_0}{s^2 + 2\alpha s + \alpha^2 + \omega_0^2}$$

whence
$$2\alpha = 3 \qquad \text{and} \qquad \alpha = 1.5$$

and
$$\alpha^2 + \omega_0^2 = 6 \qquad \omega_0 = \sqrt{6 - (1.5)^2} = 1.94$$

Thus
$$A\omega_0 = 3 \qquad \text{and} \qquad A = \frac{3}{1.94} = 1.55$$

so that $x(t) = 1.55e^{-1.5t}\sin 1.94t\, u(t)$.

This is an extremely important example. It demonstrates the fact that by using Laplace transforms differential equations can be solved for their *complete*

Item	$f(t)$	$F(s)$	Name of property
1	$af_1(t) + bf_2(t)$	$aF_1(s) + bF_2(s)$	Linearity
2	$e^{-at}f(t)$	$F(s + a)$	Complex-frequency shift
3	$f(t - t_0)u(t - t_0)$	$e^{-t_0 s}F(s)$	Time delay
4	$f_{per}(t)$	$\dfrac{\mathcal{L}\{\text{first cycle}\}}{1 - e^{-Ts}}$	Periodic function
5	$tf(t)$	$\dfrac{-d}{ds}F(s)$	
6	$\dfrac{f(t)}{t}$	$\displaystyle\int_s^\infty F(s)\,ds$	
7	$\displaystyle\int_0^t f(t)\,dt$	$\dfrac{1}{s}F(s)$	
8	$\dfrac{d}{dt}f(t)$	$sF(s) - f(0-)$	
9	$\dfrac{d^n}{dt^n}f(t)$	$s^nF(s) - s^{n-1}f(0-) - s^{n-2}\dfrac{df}{dt}\Big|_{0-} - \cdots - \dfrac{d^{n-1}f}{dt^{n-1}}\Big|_{0-}$	

table 15.4.1
Summary of the properties of the Laplace transform for $f(t) = F(s)$

response. All initial conditions are automatically taken into account in the process.

The properties of the Laplace transform are summarized in Table 15.4.1.

system analysis with the Laplace transform 15.5

We saw in Chapter 7 that *convolving* an input function $f_{in}(t)$ with a system's unit-impulse response† $h(t)$ yields the zero-state response of that system to that input. We shall now show that *convolving* any two functions, say $f_1(t)$ and $f_2(t)$, *in the time domain* is equivalent to *multiplying* their Laplace transforms *in the complex-frequency domain*. Suppose, in the time domain, a function $f_3(t)$ is defined as being the convolution of $f_1(t)$ and $f_2(t)$

$$f_3(t) = \int_{\tau = -\infty}^{\infty} f_1(\tau)f_2(t - \tau)\,d\tau \tag{15.5.1}$$

Take the Laplace transform of equation (15.5.1)

$$F_3(s) = \mathcal{L}\{f_3(t)\} = \int_{t=0-}^{\infty}\int_{\tau=-\infty}^{\infty} f_1(\tau)f_2(t - \tau)\,d\tau\;e^{-st}\,dt$$

Interchanging the order of integration leads to

† Recall that the unit-impulse response is the zero-state response of a (linear) system to a unit-impulse input.

$$F_3(s) = \int_{\tau=-\infty}^{\infty} \int_{t=0-}^{\infty} f_1(\tau) f_2(t-\tau) e^{-st}\, dt\, d\tau$$

$$= \int_{\tau=-\infty}^{\infty} f_1(\tau) \int_{t=0-}^{\infty} f_2(t-\tau) e^{-st}\, dt\, d\tau$$

Perform the inner integration. Let $x = t - \tau$. Thus $dx = dt$ and $x = -\tau$ when $t = 0-$ and

$$F_3(s) = \int_{\tau=-\infty}^{\infty} f_1(\tau) \int_{x=-\tau}^{\infty} f_2(x) e^{-s(x+\tau)}\, dx\, d\tau$$

$$= \int_{\tau=-\infty}^{\infty} f_1(\tau) e^{-s\tau} \int_{x=-\tau}^{\infty} f_2(x) e^{-sx}\, dx\, d\tau$$

If $f_1(t) = 0$ and $f_2(t) = 0$ for $t < 0$, this becomes

$$F_3(s) = F_1(s) F_2(s) \qquad \text{QED} \tag{15.5.2}$$

We know that if, in equation (15.5.1), $f_1(t)$ is an input function and $f_2(t)$ is the system impulse response $h(t)$, then $f_3(t)$ is the zero-state response. Equation (15.5.2) says that the Laplace transform of the *zero-state* response $F_3(s)$ is obtained by multiplying the Laplace transform of the input $F_1(s)$ by the Laplace transform of the impulse response $F_2(s) = H(s)$, where $H(s)$ is the transfer function of the system. Of course, we have been using $H(s)$ for quite a while, having been first introduced to it when we wanted to pass complex exponential inputs through a system. We saw that the ratio of the complex coefficients of the input and (particular-response) output complex exponentials is a complex number which we called $H(s)$. Now we see that $H(s)$ is the Laplace transform of the impulse response. Of course, $H(s)$ still has all those properties which we derived earlier. We can find the natural response from its denominator; if we let $s = j\omega$, then $H(j\omega)$ is the ratio of the output complex (phasor) quantity to the input complex (phasor) quantity, etc. But now, thanks to equations (15.5.1) and (15.5.2), we conclude that we can use $H(s)$ to find the zero-state response of the system to *any input function* that has a Laplace transform.

In other words, if we have the differential equation of any linear system

$$a_n \frac{d^n y}{dt^n} + a_{n-1} \frac{d^{n-1} y}{dt^{n-1}} + \cdots + a_1 \frac{dy}{dt} + a_0 y$$

$$= b_m \frac{d^m x}{dt^m} + b_{m-1} \frac{d^{m-1} x}{dt^{m-1}} + \cdots + b_1 \frac{dx}{dt} + b_0 x \tag{15.5.3}$$

where $x(t)$ is the input and $y(t)$ is the output, then if all initial conditions are zero, the Laplace transform of equation (15.5.3) is

$$(a_n s^n + a_{n-1} s^{n-1} + \cdots + a_1 s + a_0) Y(s)$$

$$= (b_m s^m + b_{m-1} s^{m-1} + \cdots + b_1 s + b_0) X(s) \tag{15.5.4}$$

We define the ratio $Y(s)/X(s)$ to be the transfer function

$$\frac{Y(s)}{X(s)} = \frac{b_m s^m + b_{m-1} s^{m-1} + \cdots + b_1 s + b_0}{a_n s^n + a_{n-1} s^{n-1} + \cdots + a_1 s + a_0} \equiv H(s) \tag{15.5.5}$$

and therefore write

$$Y(s) = X(s)H(s) \qquad (15.5.6)$$

Thus we say that the Laplace transform of the zero-state response [remember that in order to get equation (15.5.4) we set all initial conditions equal to zero] is the Laplace transform of the input times the transfer function $H(s)$. The derivation preceding equation (15.5.2) tells us that equation (15.5.6) is the complex-frequency-domain equivalent of

$$y(t) = x(t) * h(t) \qquad (15.5.7)$$

where, as before, the asterisk signifies the convolution process.

   If, on the other hand, the initial conditions are not zero, the Laplace transform of equation (15.5.3) is

$$(a_n s^n + a_{n-1} s^{n-1} + \cdots + a_0)Y(s) - (a_n s^{n-1} + a_{n-1} s^{n-2} + \cdots + a_1)y(0-)$$

$$- (a_n s^{n-2} + a_{n-1} s^{n-3} + \cdots + a_2)\left.\frac{dy}{dt}\right|_{0-} - \cdots - a_n \left.\frac{d^{n-1}}{dt^{n-1}}\right|_{0-} = F(s) \quad (15.5.8)$$

where $F(s)$ is the Laplace transform (taken term by term) of the right-hand side of equation (15.5.3). Solving equation (15.5.8) for $Y(s)$ yields

$$Y(s) = \frac{F(s) + \text{initial-condition terms}}{a_n s^n + a_{n-1} s^{n-1} + \cdots + a_1 s + a_0} \qquad (15.5.9)$$

$$Y(s) = \frac{F(s)}{a_n s^n + a_{n-1} s^{n-1} + \cdots + a_1 s + a_0} + \frac{\text{initial-condition terms}}{a_n s^n + a_{n-1} s^{n-1} + \cdots + a_1 s + a_0}$$

$$(15.5.10)$$

The first term in equation (15.5.10) is the Laplace transform of the *zero-state response*, and the second is the transform of the *zero-input response*. Hence equation (15.5.10) yields the transform of the complete response $\mathcal{L}\{y(t)\}$ of the system to the input $x(t)$.

EXAMPLE 15.5.1

Find the zero-state output from a system whose impulse response is $h(t) = e^{-t}u(t)$ when the input is $f_{in}(t) = tu(t)$.

ANS.:   Since $tu(t)$ is not a complex exponential input function, up to now we could not have made use of transfer-function techniques; however, we can now write

$$F_o(s) = F_{in}(s)H(s)$$

From items 3 and 5 in Table 15.3.1 we have

$$F_{in}(s) = \frac{1}{s^2} \qquad \text{and} \qquad H(s) = \frac{1}{s+1}$$

Thus

$$F_o(s) = \frac{1}{s^2(s+1)} = \frac{k_1}{s^2} + \frac{k_2}{s} + \frac{k_3}{s+1} \qquad (15.5.11)$$

The constant k_1 in the first term on the right can be found by multiplying through by its denominator s^2

$$\frac{s^2}{s^2(s+1)} = k_1 + \frac{k_2 s^2}{s} + \frac{k_3 s^2}{s+1}$$

$$\frac{1}{s+1} = k_1 + k_2 s + \frac{k_3 s^2}{s+1}$$

Then allowing $s = 0$ eliminates all other terms on the right except k_1

$$1 = k_1 \tag{15.5.12}$$

Similarly to find k_3 we multiply equation (15.5.11) by the denominator $s + 1$, which puts this factor into the numerator of all other terms on the right

$$\frac{s+1}{s^2(s+1)} = \frac{k_1(s+1)}{s^2} + \frac{k_2(s+1)}{s} + k_3$$

Now let $s = -1$, and all terms on the right except k_3 become zero-valued

$$\frac{1}{(-1)^2} = 0 + 0 + k_3$$

$$1 = k_3 \tag{15.5.13}$$

To find k_2 we can insert (15.5.12) and (15.5.13) into (15.5.11)

$$\frac{1}{s^2(s+1)} = \frac{1}{s^2} + \frac{k_2}{s} + \frac{1}{s+1}$$

$$\frac{1 - (s+1) - s^2}{s^2(s+1)} = \frac{k_2}{s}$$

$$-1 = k_2 \tag{15.5.14}$$

Thus

$$F_o(s) = \frac{1}{s^2} - \frac{1}{s} + \frac{1}{s+1} \qquad f_o(t) = (t - 1 + e^{-t})u(t)$$

Expanding equation (15.5.11) into a sum of three terms, called a *partial-fraction expansion*† (PFE), is a technique which we shall use a great deal. It is important that we realize at this point that if we start with the time-domain system integrodifferential equation (including the forcing function) then take its Laplace transform term by term, we get the *complete response* of the system. On the other hand, if we multiply the Laplace transform of the input signal by the transfer function and then take the inverse transform of that product, we get the *zero-state* response of the system to that input. The next section will show how to modify this transfer-function method to get the *complete response* from it as well.

† This method will be fully developed in Section 15.7. We have already made preliminary use of this expansion in Section 8.5 and Example 14.8.2.

initial-condition generators and the 15.6 Laplace-transform solution of circuits and systems

Our discussion of energy-storage elements (see page 124) showed that an initially charged capacitor is equivalent to the series combination of an equal-sized but uncharged capacitor and a constant voltage source whose magnitude is the initial condition $v_C(0-)$. In Example 15.4.5 we showed that the Laplace transform of any capacitor voltage $v_C(t)$ is

$$V_C(s) = \frac{1}{Cs} I_C(s) + \frac{v_C(0-)}{s} = Z_C(s)I_C(s) + \frac{v_C(0-)}{s} \qquad (15.6.1)$$

which is consistent with that earlier description. The first term on the right-hand side of equation (15.6.1) involves the transfer function (impedance) $Z_C(s)$. As a transfer function, it is a valid description of the system (the C-F capacitor) only if the initial condition is zero. The second term in equation (15.6.1) is a constant voltage (source) of magnitude $v_C(0-)$. It is called an *initial-condition generator*. Similarly, we can write for any inductor the Laplace transform of

$$i_L(t) = \frac{1}{L} \int_{0-}^{t} v(t)\, dt + i_L(0-)$$

$$I_L(s) = \frac{1}{Ls} V_L(s) + \frac{i_L(0-)}{s} = Y(s)V_L(s) + \frac{i_L(0-)}{s} \qquad (15.6.2)$$

Equations (15.6.1) and (15.6.2) describe the situations shown in Figure 15.6.1*a* and *b* respectively. Some students have difficulty in accepting the fact that the constant-valued sources shown in Figure 15.6.1 *remain active for all $t > 0$.* The circuits on the right-hand side of Figure 15.6.1*a* and *b* are terminal equivalent circuits—in exactly the same way that Thevenin and Norton circuits are. The

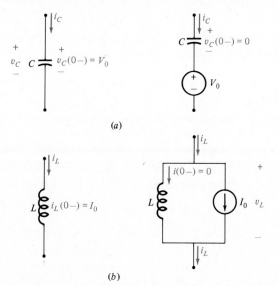

(a)

(b)

figure 15.6.1
(*a*) An initially charged capacitor and its equivalent, for which $V_C(s) = (1/Cs)I_C(s) + V_0/s$. (*b*) An initial current exists in the inductor on the left. In the equivalent, no initial current exists in the inductor and $I_L(s) = (1/Ls)V_L(s) + I_0/s$.

Thevenin output resistor may not exist in reality as a single fixed resistor; it is the *equivalent* of perhaps many different resistors inside the actual circuit. In the same way, the sources V_0 and I_0 in Figure 15.6.1a and b are convenient mathematical equivalents of what actually happens. For example if the actual capacitor voltage $v_C(t)$ approaches zero in some circuit as time approaches infinity, then in the equivalent circuit the capacitor will eventually develop a voltage $-V_0$ that will cancel the source. The total of the two will approach zero, as it should.

EXAMPLE 15.6.1

Solve for the current $i(t)$ in the circuit shown in Figure 15.6.2a. Use initial-condition generators, impedances, and the Laplace transform.

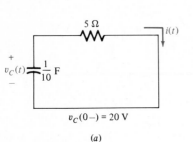
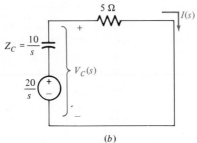

figure 15.6.2
(a) The circuit of Example 15.6.1 in which the capacitor has an initial charge of 20 V.
(b) The equivalent circuit. The capacitor has zero initial conditions and thus is describable by its impedance $Z_C(s)$.

(a)

(b)

ANS.:   We convert the initially charged capacitor in Figure 15.6.2a into an uncharged one in series with the constant voltage source, as shown in Figure 15.6.2b. Thence,

$$I(s) = \frac{20/s}{10/s + 5} = \frac{4}{s + 2}$$

so that

$$i(t) = 4e^{-2t}u(t)$$

When we write mesh equations, we sum voltages and so will generally have terms like equation (15.6.1), repeated here as (15.6.3)

$$V_C(s) = \frac{1}{Cs} I_C(s) + \frac{v_C(0-)}{s} \tag{15.6.3}$$

and equation (15.6.2) solved for $V_L(s)$

$$V_L(s) = L[sI_L(s) - i_L(0-)] \tag{15.6.4}$$

Similarly when writing node equations we are summing currents; thus we shall be working with terms like equation (15.6.3) solved for $I_C(s)$

$$I_C(s) = C[sV_C(s) - v_C(0-)] \tag{15.6.5}$$

and equation (15.6.2) repeated here as (15.6.6)

$$I_L(s) = \frac{1}{Ls} V_L(s) + \frac{i_L(0-)}{s} \tag{15.6.6}$$

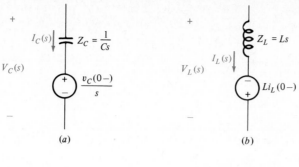

figure 15.6.3
Summary of initial-condition generators. In general (a) and (b) are most convenient when writing mesh equations; (c) and (d) are most useful when writing node equations. The sources in (b) and (c) are s-domain descriptions of *time-domain Dirac (impulse) functions*, but the s-domain sources in (a) and (d) correspond to constant (dc) time-domain sources.

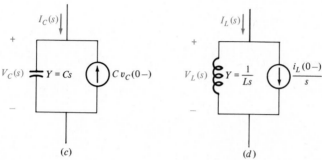

Equations (15.6.3) to (15.6.6) define the four initial-condition generator forms shown in Figure 15.6.3

Care must be taken to realize that these circuits are already in the complex-frequency domain; i.e., in equations (15.6.4) and (15.6.5) the constants $Li_L(0-)$ and $Cv_C(0-)$ represent *time-domain Dirac functions*. These forms of initial-condition-generator circuit thus contain impulse sources and are in every way terminally equivalent *for time greater than zero* to the elements having nonzero initial conditions and to the other initial-condition-generator circuits that incorporate constant (step-function) sources.

EXAMPLE 15.6.2

For the circuit shown in Figure 15.6.4a, which contains the nonzero initial conditions $v_C(0-)$ and $i_L(0-)$, use immittances and initial-condition generators to
a. Write the mesh equations
b. Write the node equations
ANS.:   (a) See Figure 15.6.4b

$$\left(R_1 + \frac{1}{Cs}\right)I_1(s) - \frac{1}{Cs}\,I_2(s) = +V_s(s) - \frac{v_C(0-)}{s} \tag{15.6.7}$$

$$-\frac{1}{Cs}\,I_1(s) + \left(Ls + R_2 + \frac{1}{Cs}\right)I_2(s) = +Li_L(0-) + \frac{v_C(0-)}{s} \tag{15.6.8}$$

(b) See Figure 15.6.4c

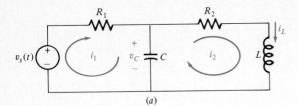

figure 15.6.4
(a) The circuit of Example 15.6.2. (b) The same circuit incorporating initial-condition generators appropriate for writing mesh equations. (c) The same circuit incorporating initial-condition generators appropriate for writing node equations.

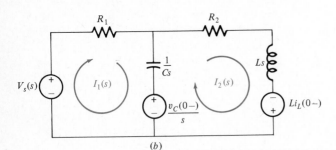

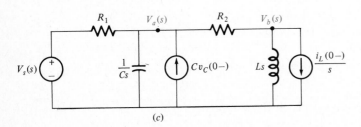

$$\left(\frac{1}{R_1} + \frac{1}{R_2} + Cs\right)V_a(s) - \frac{1}{R_2}V_b(s) - \frac{1}{R_1}V_s(s) = Cv_C(0-) \qquad (15.6.9)$$

$$-\frac{1}{R_2}V_a(s) + \left(\frac{1}{R_2} + \frac{1}{Ls}\right)V_b(s) = \frac{-i_L(0-)}{s} \qquad (15.6.10)$$

A final step would be to move the term involving the source $V_s(s)$ in equation (15.6.9) to the right-hand side.

15.7 finding the inverse transform by partial-fraction expansion

At the end of every analysis problem solved with Laplace transforms the task remains of getting the final answer back into the time domain. Theoretically, we could always use the inverse-transform formula, equation (15.2.6), but this is almost never done. Usually we can expand the $F(s)$ function into a sum of terms each of which is the recognizable Laplace transform of a simple $f(t)$ we have seen before. This sum of terms is the partial-fraction expansion (PFE), and we have already discussed how to obtain it (see Example 14.8.2 and Section 8.5). Here we discuss several additional points concerning the PFE used in conjunction with Laplace transforms.

We simply factor the denominator of the $F(s)$ function. Each term in the expansion contains one of these factors, and the coefficients are found as described earlier. So, for example,

$$F(s) = \frac{s^2 + 15s + 18}{s^3 + 5s^2 + 6s} = \frac{3}{s} + \frac{4}{s+2} - \frac{6}{s+3} \qquad (15.7.1a)$$

and then, by inspection,

$$f(t) = (3 + 4e^{-2t} - 6e^{-3t})u(t) \qquad (15.7.1b)$$

Another example is

$$F(s) = \frac{s-2}{s(s+1)^3} = -\frac{2}{s} + \frac{3}{(s+1)^3} + \frac{2}{(s+1)^2} + \frac{2}{s+1} \qquad (15.7.1c)$$

where, again by inspection,

$$f(t) = [-2 + (\tfrac{3}{2}t^2 + 2t + 2)e^{-t}]u(t) \qquad (15.7.1d)$$

Usually it is desirable to keep any underdamped quadratic term (one that gives rise to a pair of complex-conjugate poles) intact in the PFE. Such a term has the form

$$\frac{k_1 s + k_2}{s^2 + 2\zeta\omega_n s + \omega_n^2}$$

EXAMPLE 15.7.1

Find the inverse Laplace transform of

$$F(s) = \frac{s^2 + 3}{(s+2)(s^2 + 2s + 5)}$$

ANS.: The PFE has the form

$$F(s) = \frac{s^2 + 3}{(s+2)(s^2 + 2s + 5)} = \frac{k_1}{s+2} + \frac{k_2 s + k_3}{s^2 + 2s + 5}$$

and k_1 is found as usual by

$$\left.\frac{s^2 + 3}{s^2 + 2s + 5}\right|_{s=-2} = k_1 + 0$$

$$k_1 = \tfrac{7}{5}$$

Thus we have

$$\frac{s^2 + 3}{(s+2)(s^2 + 2s + 5)} = \frac{7/5}{s+2} + \frac{k_2 s + k_3}{s^2 + 2s + 5} \qquad (15.7.2)$$

Let $s = 0$ in equation (15.7.2). This yields

$$\frac{3}{2(5)} = \frac{7/5}{2} + \frac{k_3}{5}$$

$$k_3 = -2 \qquad (15.7.3)$$

We can find k_2 simply by multiplying through equation (15.7.2) by s and then evaluating the result at $s = \infty$

$$\left.\frac{s(s^2 + 3)}{(s + 2)(s^2 + 2s + 5)}\right|_{s=\infty} = \left.\frac{\frac{7}{5}s}{s + 2}\right|_{s=\infty} + \left.\frac{k_2 s^2 + k_3 s}{s^2 + 2s + 5}\right|_{s=\infty}$$

$$1 = \tfrac{7}{5} + k_2$$

$$k_2 = -\tfrac{2}{5} \tag{15.7.4}$$

Therefore the PFE is

$$F(s) = \frac{7/5}{s + 2} - \frac{\frac{2}{5}s + 2}{s^2 + 2s + 5} \tag{15.7.5}$$

Completing the square in the denominator of the last term, we have

$$s^2 + 2s + 5 = (s + \alpha)^2 + \omega^2 = s^2 + 2\alpha s + \alpha^2 + \omega^2$$

$$2 = 2\alpha \quad \text{and} \quad \alpha = 1$$

and $\alpha^2 + \omega^2 = 5$

$$\omega^2 = 5 - 1 = 4 \quad \text{and} \quad \omega = 2$$

so that

$$s^2 + 2s + 5 = (s + 1)^2 + (2)^2$$

So now we can write the second term in equation (15.7.5) as

$$\frac{\frac{2}{5}s + 2}{s^2 + 2s + 5} = \frac{\frac{2}{5}s + 2/5 - 2/5 + 2}{(s + 1)^2 + 2^2}$$

$$= \frac{\frac{2}{5}(s + 1)}{(s + 1)^2 + 2^2} + \frac{8/5}{(s + 1)^2 + 2^2}$$

$$= \frac{\frac{2}{5}(s + 1)}{(s + 1)^2 + 2^2} + \frac{\frac{4}{5}(2)}{(s + 1)^2 + 2^2} \tag{15.7.6}$$

So

$$F(s) = \frac{7/5}{s + 2} - \frac{\frac{2}{5}(s + 1)}{(s + 1)^2 + 2^2} - \frac{\frac{4}{5}(2)}{(s + 1)^2 + 2^2} \tag{15.7.7}$$

and therefore

$$f(t) = (\tfrac{7}{5}e^{-2t} - \tfrac{2}{5}e^{-t}\cos 2t - \tfrac{4}{5}e^{-t}\sin 2t)u(t)$$

$$= [\tfrac{7}{5}e^{-2t} - e^{-t}(\tfrac{2}{5}\cos 2t + \tfrac{4}{5}\sin 2t)]u(t)$$

$$= [\tfrac{7}{5}e^{-2t} - 0.894e^{-t}\cos(2t - 63.4°)]u(t) \tag{15.7.8}$$

EXAMPLE 15.7.2

Find the zero-state, zero-input, and complete responses of $v(t)$ in the circuit shown in Figure 15.7.1a. Given that $v(0+) = 4$ V, $i_L(0+) = 1$ A, $i_s(t) = t^2 u(t)$, $R = 4\ \Omega$, $L = 3$ H, and $C = \tfrac{1}{24}$ F.

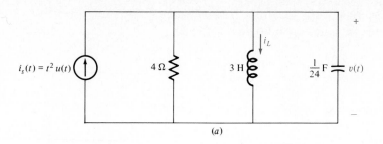

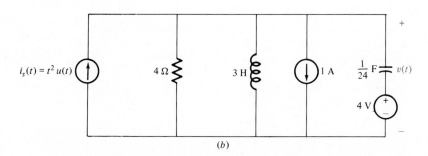

figure 15.7.1
(a) The circuit of part (a) of Example
15.7.2. (b) The circuit including the
initial-condition generators.

a. Derive the time-domain integrodifferential equation and take its Laplace transform.
b. Use initial-condition generators and impedances.

ANS.:   (a) Write a node equation for v

$$\frac{v}{R} + \frac{1}{L}\int_{-\infty}^{t} v\,dt + C\frac{dv}{dt} = i_s(t)$$

$$\frac{v}{4} + \frac{1}{3}\int_{-\infty}^{t} v\,dt + \frac{1}{24}\frac{dv}{dt} = t^2 \quad \text{for } t > 0 \tag{15.7.9}$$

Take the Laplace transform

$$\frac{V(s)}{4} + \frac{1}{3}\frac{V(s)}{s} + \frac{1}{3}\frac{\int_{-\infty}^{0-} v\,dt}{s} + \frac{1}{24}\left[sV(s) - v(0-)\right] = \frac{2}{s^3} \tag{15.7.10}$$

Inserting the initial conditions and rearranging and recalling that

$$\frac{1}{3}\int_{-\infty}^{0-} v\,dt = i_L(0-)$$

we have

$$\left(\frac{1}{4} + \frac{1}{3s} + \frac{s}{24}\right)V(s) = \frac{2}{s^3} - \frac{1}{s} + \frac{1}{6}$$

and

$$V(s) = \frac{24s(2/s^3 - 1/s + 1/6)}{s^2 + 6s + 8} \tag{15.7.11}$$

where $2/s^3$ is the forcing-function term and $-1/s$ and $\frac{1}{6}$ are the initial-condition terms. Then

$$V(s) = \frac{48/s^2}{(s+2)(s+4)} + \frac{4s - 24}{(s+2)(s+4)} = V_{zs}(s) + V_{zi}(s) \tag{15.7.12}$$

where $V_{zs}(s)$ and $V_{zi}(s)$ are the Laplace transforms of the zero-state and zero-input responses, respectively. $V_{zs}(s)$ contains the input $2/s^3$ and $V_{zi}(s)$ contains the initial-condition terms. Hence

$$V_{zs}(s) = \frac{k_1}{s^2} + \frac{k_2}{s} + \frac{k_3}{s+2} + \frac{k_4}{s+4}$$

$$\frac{48}{(s+2)(s+4)}\bigg|_{s=0} = k_1 + k_2 s + \frac{k_3 s^2}{s+2} + \frac{k_4 s^2}{s+4} \qquad k_1 = 6$$

Taking d/ds of this equation and again evaluating at $s = 0$

$$\frac{(s^2 + 6s + 8)(0) - 48(2s + 6)}{(s^2 + 6s + 8)^2}\bigg|_{s=0} = k_2 \qquad k_2 = \tfrac{9}{2}$$

$$\frac{48}{s^2(s+4)}\bigg|_{s=-2} = k_3 \qquad k_3 = 6$$

and

$$\frac{48}{s^2(s+2)}\bigg|_{s=-4} = k_4 \qquad k_4 = -\tfrac{3}{2}$$

Thus

$$V_{zs}(t) = (6t - \tfrac{9}{2} + 6e^{-2t} - \tfrac{3}{2}e^{-4t})u(t)$$

For $V_{zi}(s)$ we have

$$\frac{4s - 24}{(s+2)(s+4)} = \frac{k_1}{s+2} + \frac{k_2}{s+4}$$

where

$$\frac{4s - 24}{s+4}\bigg|_{s=-2} = k_1 \qquad k_1 = -16$$

and

$$\frac{4s - 24}{s+2}\bigg|_{s=-4} = k_2 \qquad k_2 = 20$$

so that

$$V_{zi}(t) = (-16e^{-2t} + 20e^{-4t})u(t)$$

Thus the complete response is

$$v(t) = v_{zs}(t) + v_{zi}(t) = (6t - \tfrac{9}{2} - 10e^{2t} + \tfrac{37}{2}e^{-4t})u(t)$$

See Example 7.8.1.

(b) Initial-condition generators are shown in Figure 15.7.1b. At the upper node KCL yields

$$\left(\frac{1}{4} + \frac{1}{3s}\right)V(s) + \frac{1}{24}s\left[V(s) - \frac{4}{s}\right] = \frac{2}{s^3} - \frac{1}{s}$$

which is identical to equation (15.7.11). The remainder of the solution is identical to that in part (a).

TWO SPECIAL CASES   Consider the meaning of the expression $F_o(s) = F_{in}(s)H(s)$ in the light of the fact that we can make a PFE for the resulting $F_o(s)$. For example, if

$$F_{in}(s) = \frac{N_1(s)}{D_1(s)} = \frac{N_1(s)}{(s+a)(s+b)\cdots}$$

and
$$H(s) = \frac{N_2(s)}{D_2(s)} = \frac{N_2(s)}{(s + c)(s + d) \cdots}$$

then
$$F_o(s) = \frac{N_1(s)N_2(s)}{(s + a)(s + b) \cdots (s + c)(s + d) \cdots} \qquad (15.7.13)$$

$$= \frac{k_1}{s + a} + \frac{k_2}{s + b} + \cdots + \frac{k_3}{s + c} + \frac{k_4}{s + d} + \cdots \qquad (15.7.14)$$

Clearly, when the inverse transform is taken, in equation (15.7.14) the terms that come from $H(s)$ give rise to the *natural response* and the terms that come from $F_{in}(s)$ give rise to the *particular response*.

In Chapter 7 we stated without proof that:

1. The particular response of a system possessing an eigenvalue at $s = 0$ is the time integral of what would normally be correct.

We realize now that an eigenvalue (pole) at $s = 0$ in equation (15.7.13) is due to the presence of a factor $1/s$. In this case (15.7.13) could be written

$$F_o(s) = \frac{N_2(s)}{(s + c)(s + d)} \frac{1}{s} \frac{N_1(s)}{(s + a)(s + b)} \qquad (15.7.15)$$

Multiplying any $F_{in}(s)$ by $1/s$ is equivalent to integrating the corresponding time function.

2. If a source has a term that is equal to one appearing in the natural response, the usual particular response should be multiplied by $k_1 t + k_2$.

We see now that if *both* $F_{in}(s)$ and $H(s)$ have, say, the factor $s + g$ in their denominators then

a. Both $f_{in}(t)$ and $f_n(t)$ will contain the term Ae^{-gt}.
b. $F_o(s)$ will have the factor $(s + g)^2$ in its denominator.

Thanks to b, the PFE for $F_o(s)$ will have the terms

$$\frac{k_1}{(s + g)^2} + \frac{k_2}{s + g} \qquad (15.7.16)$$

Because of property 5 in Table 15.4.1, this will result in the time-domain output function $f_o(t)$ containing the terms

$$k_1 t e^{-gt} + k_2 e^{-gt} \qquad (15.7.17)$$

3. It was pointed out that (as a result of 2 above) exciting a function at one of its eigenvalues (poles) would *not* result in the output growing infinitely large if the pole was within the finite s plane. This *would* result in the system's "blowing up" if the pole was *on* the $j\omega$ axis.

The former situation is described by equations (15.7.16) and (15.7.17). If the pole is *on* the $j\omega$-axis, we have

$$H(s) = \frac{N_2(s)}{s^2 + \omega_0^2 \cdots} \qquad (15.7.18)$$

and if

$$F_{in}(s) = \frac{N_1(s)}{s^2 + \omega_0^2} \cdots \qquad (15.7.19)$$

then

$$F_o(s) = \frac{N_1(s)N_2(s)}{(s^2 + \omega_0^2)^2} \cdots = \frac{f_1(s)}{(s^2 + \omega_0^2)^2} + \frac{f_2(s)}{s^2 + \omega_0^2} + \cdots \qquad (15.7.20)$$

where

$$\frac{f_1(s)}{(s^2 + \omega_0^2)^2} = -\frac{d}{ds}\left[f_3(s)(s^2 + \omega_0^2)^{-1} \right] \qquad (15.7.21)$$

so that its inverse transform contains terms of the form

$$t \cos (\omega_0 t + \theta) \qquad (15.7.22)$$

which increases without bound with time.

EXAMPLE 15.7.3

Find the zero-state response of the system described by

$$\frac{d^2y}{dt^2} + \frac{dy}{dt} = 2t \qquad \text{for } t > 0$$

ANS.:   The Laplace transform of the equation is

$$\left[s^2 Y(s) - sy(0-) - \frac{dy}{dt}\bigg|_{0-} \right] + [sY(s) - y(0-)] = \frac{2}{s^2}$$

Solving for $Y(s)$ and leaving the forcing function in the numerator gives

$$Y(s) = \frac{2/s^2}{s(s + 1)}$$

Note the presence of a zero eigenvalue (pole at the origin) in the system transfer function. Continuing, we have

$$Y(s) = \frac{2}{s^3(s + 1)} = \frac{k_1}{s^3} + \frac{k_2}{s^2} + \frac{k_3}{s} + \frac{k_4}{s + 1}$$

$$\frac{2}{s + 1} = k_1 + k_2 s + k_3 s^2 + \frac{k_4 s^3}{s + 1} \qquad (15.7.23)$$

and

$$k_1 = 2$$

Taking the derivative d/ds of (15.7.23) and evaluating at $s = 0$ leads to

$$\frac{(s + 1)(0) - (2)(1)}{(s + 1)^2}\bigg|_{s=0} = k_2 + 2k_3 s + \frac{d}{ds}\frac{k_4 s^3}{s + 1} \qquad (15.7.24)$$

and

$$-2 = k_2$$

Taking the derivative d/ds of equation (15.7.24) leads to

$$\frac{(s + 1)^2(0) - (-2)2(s + 1)(1)}{(s + 1)^4}\bigg|_{s=0} = 2k_3 \qquad k_3 = 2$$

and finally
$$k_4 = \frac{2}{s^3}\bigg|_{s=-1} = -2$$

Therefore
$$y(t) = (t^2 - 2t + 2 - 2e^{-t})u(t)$$

Note the presence of the term t^2 that is the integral of the forcing function (see Example 7.7.1).

EXAMPLE 15.7.4

Find the zero-state response of the system described by

$$\frac{d^2 y}{dt^2} - y = e^{-t} \qquad \text{for } t > 0$$

ANS.:   The Laplace transform (with zero initial conditions) is

$$s^2 Y(s) - Y(s) = \frac{1}{s+1}$$

$$Y(s) = \frac{1}{(s+1)(s^2-1)} = \frac{1}{(s+1)^2(s-1)} = \frac{k_1}{(s+1)^2} + \frac{k_2}{s+1} + \frac{k_3}{s-1}$$

Evaluating as usual, we have

$$\frac{1}{s-1} = k_1 + k_2(s+1) + \frac{k_3(s+1)^2}{s-1} \qquad\qquad (15.7.25)$$

With $s = -1$

$$k_1 = -\tfrac{1}{2}$$

Taking the derivative d/ds of equation (15.7.25) and evaluating at $s = -1$ gives

$$\frac{(s-1)(0) - (1)(1)}{(s-1)^2}\bigg|_{s=-1} = k_2 + 0$$

and
$$k_2 = -\tfrac{1}{4}$$

Then k_3 is evaluated

$$\frac{1}{(s+1)^2}\bigg|_{s=+1} = k_3 = \tfrac{1}{4}$$

Therefore
$$Y(s) = \frac{-1/2}{(s+1)^2} + \frac{-1/4}{s+1} + \frac{1/4}{s-1}$$

and
$$y(t) = (-\tfrac{1}{2}te^{-t} - \tfrac{1}{4}e^{-t} + \tfrac{1}{4}e^{+t})u(t) = \tfrac{1}{4}[e^t - e^{-t}(2t+1)]u(t)$$

(see Example 7.7.2).

The forcing function, which is also a mode of the natural response, appears in the output multiplied by but the resulting function is bounded. The presence of the e^t mode, which *does* increase unboundedly wit time, is due to the inherent instability of the system and has nothing to do with the input function.

GRAPHICAL EVALUATION OF THE PFE CONSTANTS†

Suppose we have a function $F(s)$ for which we would like to obtain a PFE. Assume that all the poles are simple

$$F(s) = \frac{K(s - z_1)(s - z_2) \cdots}{(s - p_1)(s - p_2) \cdots (s - p_n)} \qquad (15.7.26)$$

where the z_i are system zeros and the p_i are system poles. The expansion is

$$F(s) = \frac{k_1}{s - p_1} + \frac{k_2}{s - p_2} + \cdots + \frac{k_n}{s - p_n} \qquad (15.7.27)$$

and the ith constant is found from

$$k_i = (s - p_i)F(s)\bigg|_{s = p_i} = \frac{K(p_i - z_1)(p_i - z_2) \cdots}{(p_i - p_1) \cdots (p_i - p_n)} \qquad (15.7.28)$$

where the ith term in the denominator is missing because it cancelled out with the $s - p_i$ multiplier in the numerator. The term $p_i - z_1$ is a complex quantity whose

figure 15.7.2
Evaluation of the constants in the PFE.

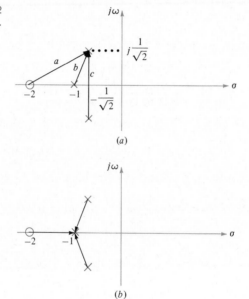

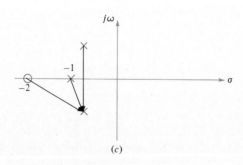

† These constants are sometimes called *residues*.

magnitude and angle are given by the vector that runs from the zero z_1 to the pole p_i. A similar vector quantity can be associated with each of the other factors in both numerator and denominator of equation (15.7.28).

Thus, evaluation of k_i is much the same process as evaluating the magnitude and angle of $H(s)$ graphically for any fixed value of s. Vectors are drawn to the pole for which we wish to evaluate k from all the other *finite* poles and zeros.

EXAMPLE 15.7.5

Graphically evaluate the constants in the PFE for

$$H(s) = \frac{s + 2}{(s + 1)(s^2 + 1.41s + 1)}$$

$$= \frac{k_1}{s + 1} + \frac{k_2}{s + 1/\sqrt{2} + j1/\sqrt{2}} + \frac{k_3}{s + 1/\sqrt{2} - j1/\sqrt{2}}$$

ANS.:   The construction of the vectors is shown in Figure 15.7.2. From Figure 15.7.2a

$$k_3 = \frac{1.47\underline{/28.7°}}{0.765\underline{/67.5°}\,1.41\underline{/90°}} = 1.36\underline{/-129°} = -0.86 - j1.06$$

From Figure 15.7.2b

$$k_1 = \frac{1\underline{/0}}{0.765\underline{/-112.5°}\,0.765\underline{/+112.5°}} = 1.71$$

Compare Figure 15.7.2a and c:

$$k_2 = k_3^*$$

An obvious but important conclusion drawn from this graphical technique is that the PFE constants of any pair of conjugate poles are complex conjugates.

If the order of the numerator polynomial equals the order of the denominator polynomial, $\mathcal{L}^{-1}\{F(s)\}$ contains an impulse (Dirac) function. Such an $F(s)$ is called an *improper fraction*. The PFE of an improper fraction is obtained by *first performing one step of long division* and finding the remainder of the PFE as usual.

EXAMPLE 15.7.6

Find the inverse transform of

$$F(s) = \frac{3s^2 + 18s + 28}{s^2 + 5s + 6}$$

ANS.:   Perform one step of long division (as in the first step of a continued-fraction expansion)

$$
\begin{array}{r}
3 \\
s^2 + 5s + 6\overline{)3s^2 + 18s + 28} \\
\underline{3s^2 + 15s + 18} \\
3s + 10
\end{array}
$$

so that

$$F(s) = 3 + \frac{3s + 10}{s^2 + 5s + 6}$$

Then, as usual,

$$\frac{3s + 10}{s^2 + 5s + 6} = \frac{3s + 10}{(s + 2)(s + 3)} = \frac{k_1}{s + 2} + \frac{k_2}{s + 3}$$

where

$$k_1 = \frac{3s + 10}{s + 3}\bigg|_{s = -2} = 4 \quad \text{and} \quad k_2 = \frac{3s + 10}{s + 2}\bigg|_{s = -3} = -1$$

Therefore

$$f(t) = 3\delta(t) + (4e^{-2t} - e^{-3t})u(t)$$

15.8 unit-impulse and unit-step responses determine $H(s)$

In practice, an engineer may often be presented with a system whose $H(s)$ is not known and be asked to determine how this system would react to some arbitrary input. The problem, of course, is to quickly determine the transfer function of the system $H(s)$. One way of doing this is to find the system's unit-impulse response $h(t)$ and take its Laplace transform. Another way is to find the zero-state response of the system $w(t)$ to a unit-step function. This may be the best alternative if an approximation of an impulse input might damage the actual system or force it into nonlinear behavior. Since

$$F_o(s) = F_{in}(s)H(s) \tag{15.8.1}$$

where $H(s)$ is the transfer function, when $f_{in}(t) = \delta(t)$, we have $F_{in}(s) = 1$ and so

$$F_o(s) = H(s)$$

But when the input is a unit-step function

$$f_{in}(t) = u(t) \qquad F_{in}(s) = \frac{1}{s}$$

$$F_o(s) = \frac{1}{s} H(s) = W(s)$$

we can find

$$H(s) = sW(s) \tag{15.8.2}$$

All this simply says that we can determine the transfer function $H(s)$ by first finding the system's unit-step response $w(t)$. Then take the Laplace transform of $w(t)$ to obtain $W(s)$. Finally, multiply by s; the result is $H(s)$.

EXAMPLE 15.8.1

Find the Bode plot of a system if its unit-step response is

$$w(t) = (1 - e^{-3t})u(t)$$

ANS.:

$$W(s) = \frac{1}{s} - \frac{1}{s+3}$$

From (15.8.2)

$$H(s) = sW(s) = 1 - \frac{s}{s+3} = \frac{3}{s+3}$$

which in Bode form is

$$H(s) = \frac{1}{s/3 + 1}$$

Therefore the system has a simple low-pass characteristic with a breakdown at $\omega = 3$ rad/s.

One simple test: either finding $h(t)$ or $w(t)$ is sufficient to determine the transfer function $H(s)$ of a linear system. And $H(s)$ is sufficient information to enable us to predict how that system will react to *any* input function that has a Laplace transform.

solution of mesh and node equations with the Laplace transform   15.9

For any general nth-order linear multimesh and/or multinode network, the mesh and node equations are sets of n coupled integrodifferential equations. These time-domain system equations can be Laplace-transformed term by term. The dependent transformed variables (mesh currents and node voltages) can then be obtained by inverting the system coefficient matrix. In other words, if we use ICT to stand for initial-condition terms, after taking the Laplace transform of a set of system equations (either node or mesh) the result is of the general form

$$\left(\frac{a_{11}}{s} + b_{11} + c_{11}s\right)X_1(s) + \left(\frac{a_{12}}{s} + b_{12} + c_{12}s\right)X_2(s) + \cdots$$

$$+ \left(\frac{a_{1n}}{s} + b_{1n} + c_{1n}s\right)X_n(s) = F_1(s) + \text{ICT}_1$$

$$\left(\frac{a_{21}}{s} + b_{21} + c_{21}s\right)X_1(s) + \left(\frac{a_{22}}{s} + b_{22} + c_{22}s\right)X_2(s) + \cdots$$

$$+ \left(\frac{a_{2n}}{s} + b_{2n} + c_{2n}s\right)X_n(s) = F_2(s) + \text{ICT}_2$$

. . . .

$$\left(\frac{a_{n1}}{s} + b_{n1} + c_{n1}s\right)X_1(s) + \left(\frac{a_{n2}}{s} + b_{n2} + c_{n2}s\right)X_2(s) + \cdots$$

$$+ \left(\frac{a_{nn}}{s} + b_{nn} + c_{nn}s\right)X_n(s) = F_n(s) + \text{ICT}_n$$

The $F_i(s)$ terms are due to the input forcing functions. In matrix form these transformed system equations are

$$[A(s)] \begin{bmatrix} X_1(s) \\ X_2(s) \\ \vdots \\ X_n(s) \end{bmatrix} = \begin{bmatrix} F_1(s) \\ F_2(s) \\ \vdots \\ F_n(s) \end{bmatrix} + \begin{bmatrix} \text{ICT}_1 \\ \text{ICT}_2 \\ \vdots \\ \text{ICT}_n \end{bmatrix}$$

Premultiplying by $[A]^{-1}$ yields

$$\begin{bmatrix} X_1(s) \\ X_2(s) \\ \vdots \\ X_n(s) \end{bmatrix} = [A]^{-1} \begin{bmatrix} F_1(s) \\ F_2(s) \\ \vdots \\ F_n(s) \end{bmatrix} + [A]^{-1} \begin{bmatrix} \text{ICT}_1 \\ \text{ICT}_2 \\ \vdots \\ \text{ICT}_n \end{bmatrix}$$

zero-state response   zero-input response

In performing the inverse, the system determinant Δ is computed. This is a polynomial in s which *appears in the denominator of every dependent variable* (mesh current or node voltage). It is this denominator which determines the poles (eigenvalues) of the zero-input response transform. The polynomial Δ is therefore the *characteristic polynomial*.

EXAMPLE 15.9.1

a. Solve the circuit in Figure 15.9.1a for its transformed mesh currents $I_1(s)$ and $I_2(s)$.
b. Let $i_{L_1}(0-) = I_A = 2$ A and $i_{L_2}(0-) = I_B = 1$ A and solve for the zero-input responses $i_1(t)_{zi}$ and $i_2(t)_{zi}$.

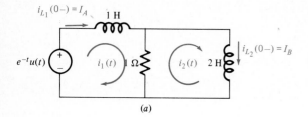

figure 15.9.1
(a) The circuit of Example 15.9.1. (b) The same circuit with initial-condition generators.

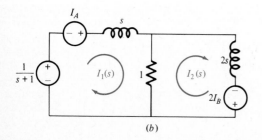

ANS.:   (a) Using initial-condition generators as in Figure 15.9.1b, we obtain mesh equations

$$(s + 1)I_1(s) - (1)I_2(s) = \frac{1}{s + 1} + I_A$$

and

$$-(1)I_1(s) + (2s + 1)I_2(s) = 2I_B$$

In matrix form these are

$$\begin{bmatrix} s + 1 & -1 \\ -1 & 2s + 1 \end{bmatrix} \begin{bmatrix} I_1(s) \\ I_2(s) \end{bmatrix} = \begin{bmatrix} \frac{1}{s + 1} \\ 0 \end{bmatrix} + \begin{bmatrix} I_A \\ 2I_B \end{bmatrix}$$

Solving for the $I(s)$ vector gives

$$\begin{bmatrix} I_1(s) \\ I_2(s) \end{bmatrix} = \begin{bmatrix} s + 1 & -1 \\ -1 & 2s + 1 \end{bmatrix}^{-1} \begin{bmatrix} \frac{1}{s + 1} \\ 0 \end{bmatrix} + \begin{bmatrix} s + 1 & -1 \\ -1 & 2s + 1 \end{bmatrix}^{-1} \begin{bmatrix} I_A \\ 2I_B \end{bmatrix}$$

$$= \begin{bmatrix} \frac{2s + 1}{\Delta} & \frac{1}{\Delta} \\ \frac{1}{\Delta} & \frac{s + 1}{\Delta} \end{bmatrix} \begin{bmatrix} \frac{1}{s + 1} \\ 0 \end{bmatrix} + \begin{bmatrix} \frac{2s + 1}{\Delta} & \frac{1}{\Delta} \\ \frac{1}{\Delta} & \frac{s + 1}{\Delta} \end{bmatrix} \begin{bmatrix} I_A \\ 2I_B \end{bmatrix}$$

where

$$\Delta = (s + 1)(2s + 1) - (1) = 2s^2 + 3s = 2s(s + \tfrac{3}{2})$$

Thus

$$I_1(s) = \frac{2(s + 1/2)}{2s(s + 3/2)(s + 1)} + \frac{2(s + 1/2)I_A + 2I_B}{2s(s + 3/2)}$$

$$I_2(s) = \frac{1}{2s(s + 3/2)(s + 1)} + \frac{I_A + 2I_B(s + 1)}{2s(s + 3/2)}$$

In each of the above expressions the first term is the transform of the zero-state response and the second is the transform of the zero-input response. Note that the modes of the zero-input response for both I_1 and I_2 are the same; i.e., both $i_{1n}(t)$ and $i_{2n}(t)$ have a constant term and a negative exponential term with time constant $\tau = \tfrac{2}{3}$ s.

(b) Solving specifically for the zero-input responses we get from PFE

$$I_1(s)_{zi} = \frac{2(s + 1/2)(2) + 2(1)}{2s(s + 3/2)} = \frac{2(s + 1)}{s(s + 3/2)} = \frac{4/3}{s} + \frac{2/3}{s + 3/2}$$

and

$$I_2(s)_{zi} = \frac{2 + 2(1)(s + 1)}{2s(s + 3/2)} = \frac{s + 2}{s(s + 3/2)} = \frac{4/3}{s} - \frac{1/3}{s + 3/2}$$

Thus $i_1(t)_{zi} = (\tfrac{4}{3} + \tfrac{2}{3}e^{-3/2t})u(t)$ and $i_2(t)_{zi} = (\tfrac{4}{3} - \tfrac{1}{3}e^{-3/2t})u(t)$

Incidentally, certain sets of initial conditions† will make one or more modes be absent in the zero-state response. For example, try $I_A + I_B = 1$ A in $I_2(s)_{zi}$. With the possibility of this one minor exception, we note that every dependent variable has the same modes (kinds of terms) in its zero-input (natural) response. The

---

† Any initial condition that lies on one of the system eigenvectors (see Chapter 9).

zero-state response also has all these modes, besides others due to the input functions.

EXAMPLE 15.9.2

Show that writing a node equation for the circuit of Example 15.9.1 gives a node voltage that has eigenvalues (modes) in its zero-input response that are identical to the mesh currents in Example 15.9.1.
ANS.: See Figure 15.9.1b. Let the voltage across the 1-Ω resistor be $V_a(s)$. Write a KCL summation at that upper right node

$$\frac{V_a - [1/(s+1) + I_A]}{s} + \frac{V_a}{1} + \frac{V_a - (-2I_B)}{2s} = 0$$

Solving for $V_a(s)$ is akin to inverting the coefficient matrix

$$V_a(s) = \frac{s}{s(s+1)(s+3/2)} + \frac{I_A - I_B}{s+3/2}$$

The zero-input response is due only to initial conditions and is given by the second term in the last expression. Thus, with $I_A = 2$ A and $I_B = 1$ A, we get

$$(V_a)(t)_{zi} = e^{-3/2t}u(t)$$

15.10 initial- and final-value theorems

When we solve a circuit or systems-analysis problem using Laplace-transform methods, it would be reassuring to have some kind of check on our work about halfway through the problem. For example, before we perform a PFE it would be helpful to be able to see whether our response function's initial and final values correspond with reality. It is usually fairly simple, given a system or circuit together with its forcing function, to predict the initial value $f(0+)$ and final value $f(\infty)$ of the response $f(t)$. If we cannot get a correct check at this point, there is no sense in proceeding; we should go back and check our analysis for errors. Let us see how to find $f(0+)$ from $F(s)$.

INITIAL-VALUE THEOREM

If the time functions $f(t)$ and df/dt are both Laplace-transformable, the initial value $f(0+)$ of $f(t)$ is given by

$$f(0+) = \lim_{t \to 0+} f(t) = \lim_{s \to \infty} sF(s)$$

Proof
Consider the Laplace transform of df/dt

$$\int_{0-}^{\infty} \frac{df}{dt} e^{-st} dt = sF(s) - f(0+)$$

Now allow s to approach infinity

$$\lim_{s\to\infty} \int_{0-}^{\infty} \frac{df}{dt}\, e^{-st}\, dt = \lim_{s\to\infty}[sF(s) - f(0+)]$$

$$\int_{0-}^{\infty} \frac{df}{dt} \lim_{s\to\infty}(e^{-st})\, dt = \lim_{s\to\infty}[sF(s)] - f(0+)$$

The limit on the left-hand side is zero. Thus

$$f(0+) = \lim_{s\to\infty} sF(s) \qquad \text{QED}$$

EXAMPLE 15.10.1

Find the value of $f(0+)$ from

$$F(s) = \frac{4}{s+3} - \frac{1}{s+2} = \frac{3s+5}{s^2+5s+6}$$

ANS.:
$$f(0+) = \lim_{s\to\infty} sF(s) = \lim_{s\to\infty} \frac{3s^2+5s}{s^2+5s+6} = 3$$

We check this answer by noting that

$$\mathcal{L}^{-1}\{F(s)\} = (4e^{-3t} - e^{-2t})u(t)$$

and thus has an initial value $f(0+) = 3$.

If $f(t)$ and df/dt are Laplace-transformable, then

FINAL-VALUE THEOREM

$$f(\infty) = \lim_{s\to 0} sF(s)$$

Proof
Again consider the transform of df/dt

$$\int_{0-}^{\infty} \frac{df}{dt}\, e^{-st}\, dt = sF(s) - f(0-) \qquad (15.10.1)$$

Take the limit of this expression as s approaches zero

$$\lim_{s\to 0} \int_{0-}^{\infty} \frac{df}{dt}\, e^{-st}\, dt = \lim_{s\to 0} sF(s) - f(0-) \qquad (15.10.2)$$

The left side of equation (15.10.2) becomes

$$\lim_{s\to 0} \int_{0-}^{\infty} \frac{df}{dt}\, e^{-st}\, dt = \int_{0-}^{\infty} \frac{df}{dt} \lim_{s\to 0}(e^{-st})\, dt = \int_{0-}^{\infty} \frac{df}{dt}\, dt$$

$$= \lim_{t\to\infty} \int_{0-}^{t} \frac{df}{dt}\, dt = f(\infty) - f(0-) \qquad (15.10.3)$$

Setting equation (15.10.3) equal to the right-hand side of equation (15.10.2) yields

$$f(\infty) = \lim_{s\to 0} sF(s) \qquad \text{QED}$$

EXAMPLE 15.10.2

Find the final value $f(\infty)$ of $f(t)$ where

$$F(s) = \frac{10}{s} + \frac{3}{s+2} = \frac{13s+20}{s^2+2s}$$

ANS.:

$$f(\infty) = \lim_{s\to 0} sF(s) = \lim_{s\to 0} \frac{13s^2+20s}{s^2+2s} = 10$$

We see that this is correct because

$$\mathcal{L}^{-1}\{F(s)\} = (10 + 3e^{-2t})u(t)$$

whose final value indeed is 10.

15.11 solution of linear state equations by Laplace transforms

In Chapter 9 we studied a method for describing systems known as state equations. In that method an nth-order linear system is described by a coupled set of n first-order ordinary differential equations. Linear equations can be written in matrix form

$$[\dot{x}] = [A][x] + [B][u] \tag{15.11.1}$$

$$[y] = [C][x] + [D][u] \tag{15.11.2}$$

where the dot denotes the first derivative with respect to time. The state (variable) vector can be found from equation (15.11.1) by direct application of the Laplace transform

$$[s][X(s)] - [x(0)] = [A]X(s) + [B]U(s)$$

$$(s[I] - [A])[X(s)] = [x(0)] + [B]U(s)$$

$$\boxed{X(s) = [sI - A]^{-1}[x(0)] + [sI - A]^{-1}[B]U(s)} \tag{15.11.3}$$

where the *first term* on the right represents the transform of the *zero-input* responses and the *second term* is the transform of the *zero-state* responses of all the individual state variables. The matrix $[sI - A]^{-1}$ is the Laplace transform of the state transition matrix $\Phi(t)$.

EXAMPLE 15.11.1

The state equations for the circuit shown in Figure 15.11.1 are

$$\dot{x}_1 = -4x_1 + x_2 \quad \text{and} \quad \dot{x}_2 = -2x_1 - 2x_2 + 2v_s$$

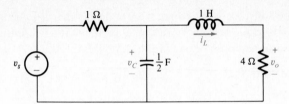

figure 15.11.1
The circuit of Example 15.11.1. The voltage v_o is the output.

where $$x_1 = i_L \qquad x_2 = v_c \qquad y = v_0 = 4x_1$$

Find the complete state and output responses when $x_1(0-) = 1$, $x_2(0-) = 2$, and $v_s(t) = \delta(t)$.

ANS.:   The appropriate state equations are

$$[\dot{x}] = [A][x] + [B][u] \qquad \text{and} \qquad [y] = [C][x] + [D][u]$$

where $$[A] = \begin{bmatrix} -4 & 1 \\ -2 & -2 \end{bmatrix} \qquad [B] = \begin{bmatrix} 0 \\ 2 \end{bmatrix} \qquad [C] = [4 \quad 0] \qquad [D] = [\varnothing]$$

so that $$s[I] - [A] = \begin{bmatrix} s+4 & -1 \\ 2 & s+2 \end{bmatrix}$$

and the system determinant is

$$\Delta = (s+4)(s+2) - (2)(-1) = s^2 + 6s + 10 = (s+3)^2 + 1$$

Therefore the inverse matrix is

$$(s[I] - [A])^{-1} = \begin{bmatrix} \dfrac{s+2}{\Delta} & \dfrac{1}{\Delta} \\[2ex] \dfrac{-2}{\Delta} & \dfrac{s+4}{\Delta} \end{bmatrix}$$

Equation (15.11.3) therefore yields

$$\begin{bmatrix} X_1(s) \\ X_2(s) \end{bmatrix} = \begin{bmatrix} \dfrac{s+2}{\Delta} & \dfrac{1}{\Delta} \\[2ex] \dfrac{-2}{\Delta} & \dfrac{s+4}{\Delta} \end{bmatrix} \begin{bmatrix} 1 \\ 2 \end{bmatrix} + \begin{bmatrix} \dfrac{s+2}{\Delta} & \dfrac{1}{\Delta} \\[2ex] \dfrac{-2}{\Delta} & \dfrac{s+4}{\Delta} \end{bmatrix} \begin{bmatrix} 0 \\ 2 \end{bmatrix} \qquad (1)$$

$$= \begin{bmatrix} \dfrac{s+3}{(s+3)^2+1} + \dfrac{3}{(s+3)^2+1} \\[3ex] \dfrac{4(s+3)}{(s+3)^2+1} + \dfrac{2}{(s+3)^2+1} \end{bmatrix} \qquad\qquad (15.11.4)$$

Then taking the inverse transforms, we get

$$x_1(t) = e^{-3t}(\cos t + 3 \sin t)u(t) = 3.16e^{-3t} \cos (t - 71.6°)u(t)$$

and

$$x_2(t) = e^{-3t}(4 \cos t + 2 \sin t)u(t) = 4.47e^{-3t} \cos (t - 26.6°)u(t)$$

Then $y(t) = 4x_1(t) = 12.64e^{-3t} \cos (t - 71.6°)u(t)$.

In Example 15.11.1 if the source v_s is set equal to zero, notice that the elements of the $(s[I] - [A])^{-1}$ matrix each, together with one of the initial conditions, determine a part of the natural response of one of the state variables

$$X_1(s)_{zi} = \Phi_{11}(s)x_1(0) + \Phi_{12}(s)x_2(0)$$

$$X_2(s)_{zi} = \Phi_{21}(s)x_1(0) + \Phi_{22}(s)x_2(0)$$

or

$$x_1(t)_{zi} = x_1(0)\phi_{11}(t) + x_2(0)\phi_{12}(t)$$

$$x_2(t)_{zi} = x_1(0)\phi_{21}(t) + x_2(0)\phi_{22}(t)$$

Hence one way of determining the $[\Phi(s)] = (s[I] - [A])^{-1}$ matrix is by finding the natural response of *each* state variable due to *each* initial condition. The Laplace transform of each such time response is one element in $[\Phi(s)]$.

In the same way that we took the Laplace transform of equation (15.11.1) to get equation (15.11.3) we can take the transform of equation (15.11.2) and get

$$\boxed{Y(s) = [C]X(s) + [D]U(s)} \qquad (15.11.5)$$

Inserting (15.11.3) into (15.11.5) yields a transform matrix equation for the output vector $Y(s)$. For the case where $x(0) = 0$, the (zero-state) output is

$$Y(s) = \{[C](s[I] - [A])^{-1}[B] + [D]\}U(s) \qquad (15.11.6)$$

For a *single-input–single-output* system $Y(s)$ and $U(s)$ are single (scalar) quantities. In that case we can divide equation (15.11.6) by $U(s)$ and have, as a result, an expression for the system transfer function $H(s) = Y(s)/U(s)$

$$\boxed{H(s) = [C](s[I] - [A])^{-1}[B] + [D]} \qquad (15.11.7)$$

In this way we can find, in Example 15.11.1,

$$H(s) = [4 \quad 0] \begin{bmatrix} \dfrac{s + 2}{(s + 3)^2 + 1} & \dfrac{1}{(s + 3)^2 + 1} \\ \dfrac{-2}{(s + 3)^2 + 1} & \dfrac{s + 4}{(s + 3)^2 + 1} \end{bmatrix} \begin{bmatrix} 0 \\ 2 \end{bmatrix} + [\varnothing]$$

$$= \frac{8}{s^2 + 6s + 10}$$

The procedures described in Chapters 8 and 9 enabled us, given a circuit (or its transfer function), to obtain its state-variable description. Equation (15.11.7) enables us to accomplish the inverse task for a single-input–single-output system. Now, given the A, B, C and D matrices, we can obtain the transfer function.

15.12   summary

In this final chapter we have studied a method which is not only a powerful analytical tool in its own right but also provides insight and gives unity to all that has gone before. You may ask: "Why have we waited until now?" Circuits are solved so much easier with the Laplace transform, and indeed some texts present this material much earlier. We have not done so for two major reasons:

1. Presented early, the Laplace transform would drop from the sky before you had any background in transform theory. You would use it without knowing (or possibly caring) where it came from. Engineers should understand the derivations of their analytical tools. A *technician* just uses things.
2. If you had learned about Laplace transforms early in this study of circuits and systems, you probably would have developed the bad habit of using it for everything. You do not use a sledgehammer to drive a thumbtack, and you should *not* use the Laplace transform to analyze simple *RL* and *RC* circuits.

In any event, in this chapter we have developed the Laplace transform from the Fourier transform, investigated some of its properties, and started a table of $f(t)$ functions and their corresponding $F(s)$ transforms. We have seen how to use these transforms to find the complete solutions of linear differential equations with constant coefficients and also how to get those same complete responses starting with the schematic diagram and then using immittances and initial-condition generators. All that we have discussed here is merely an introduction to some of the analytical and design tools engineers use in working with circuits and systems. You should now continue to expand your basic mathematical and engineering analysis skills and to study the applications of these fundamental techniques in all the various areas of electrical engineering.

problems

1. Find the Laplace transform of (a) $10e^{-3t} \cos(4t - 60°)$, (b) $(t - 3)^2 u(t - 3)$.

2. Find the Laplace transform of (a) $t \sin \omega_0 t$, (b) $t \cos \omega_0 t$.

3. Find the Laplace transform of the pulses shown in Figure P15.3.

4. Find the Laplace transform of the waveform shown in Figure P15.4.

5. Find the $f(t)$ whose transform is $F(s) = (s + 1)/(s + 2)^2$.

6. $I(s) = 2s/(s + 3)^2$ Find (a) its PFE and (b) its inverse transform $i(t)$.

7. If $F(s) = (s - 2)/[s(s + 1)^3]$, find $f(t)$.

8. Find $f(t)$ given that

$$F(s) = \frac{s^2 + 15s + 18}{s^3 + 5s^2 + 6s}$$

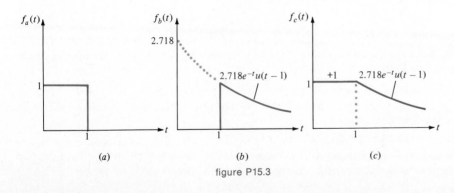

figure P15.3

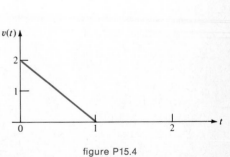

figure P15.4

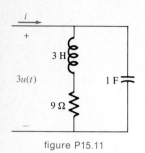

figure P15.11

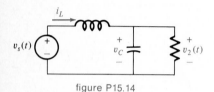

figure P15.14

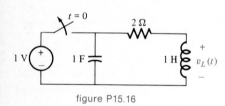

figure P15.16

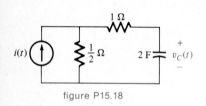

figure P15.18

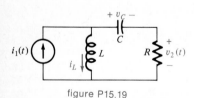

figure P15.19

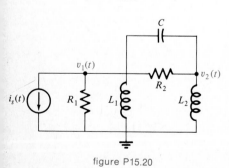

figure P15.20

9. Find the inverse Laplace transform of $F(s) = (s + 2)/(s + 1)$.

10. Find $f(t)$ for the following two $F(s)$ functions. Sketch $f(t)$ versus t in both cases. (a) $3/(s^2 + 5s + 6)$; (b) $s/(s + 4)$.

11. In the circuit shown in Figure P15.11 the initial conditions are zero. Find $i(t)$.

12. A system with unknown $H(s)$ is tested by inserting a unit-step-function input $v_{in}(t) = u(t)$. The resulting zero-state response is $v(t) = Ae^{-at}u(t)$. (a) What is the transfer function of the system? (b) A balanced (zero dc component) square wave with amplitude $\pm B$ and period T is used as input to this system. Find the resulting output $V(s)$. (c) Find an expression for the zero-state output $v(t)$ that results from the periodic input described in part (b).

13. Find the zero-state response $y(t)$ of a system whose input is $10e^{-3t}u(t)$ and whose transfer function is

$$H(s) = \frac{0.1(s + 1)}{(s + 4)(s + 2)}$$

14. In the circuit shown in Figure P15.14 solve for $V_2(s)$ in terms of $V_s(s)$, V_o, and I_o where $v_C(0+) = V_o$ and $i_L(0+) = I_o$; $R = L = C = 1$. Separate your answer into the zero-state and zero-input response transforms.

15. The natural response of a low-pass filter is $(A_1 t + A_2)e^{-2t}$ for $t > 0$. All the system zeros are known to be at infinity. It is also known that a constant input will produce that same constant as output; i.e., the system's dc gain is unity. If the input to this network is $3e^{-t}u(t)$, what is (a) the particular (forced) response? (b) The zero-state response?

16. In the circuit shown in Figure P15.16 find $v_L(t)$ for $t > 0$. The switch has remained closed for a long time before $t = 0$. At $t = 0$ the switch opens and stays open.

17. Find an expression for the current $i(t)$ that will flow out of a unit-impulse voltage source which is connected across the series combination of a 1-Ω resistor and a $\frac{1}{2}$-F capacitor. The capacitor voltage is initially zero.

18. Using Laplace transforms, solve for $v_C(t)$ in Figure P15.18, given that $i(t) = u(t)$ and $v_C(0-) = 2$ V.

19. In Figure P15.19 $R = L = C = 1$, $v_C(0-) = 1$ V, $i_L(0-) = 1$ A, and $i_1(t) = \delta(t)$. Using Laplace transforms, find $v_2(t)$ for $t > 0$.

20. Write a set of node equations for the circuit in Figure P15.20 in $V_1(s)$ and $V_2(s)$. Include, on the right, all possible initial-condition terms as well as the term for $I_s(s)$.

21. Determine the unit-step response as a function of ζ of a system whose transfer function is

$$H(s) = \frac{1}{s^2 + 2\zeta s + 1}$$

22. Solve the circuit in Figure P15.22 for current $i(t)$. Use Laplace transforms, impedances, and KCL.

23. A voltage source $v(t) = 10e^{-t}u(t)$ drives a series combination of $R = 5\ \Omega$ and $C = 0.1$ F. The capacitor has an initial voltage of 2 V that tends to oppose the source current. Find $I(s)$ and $i(t)$.

24. A shipping box contains a fragile mass M separated from the walls of the box by packing material characterized by damping D. The box is initially at rest. Find the force $f(t)$ on M if the box experiences a step velocity input $Au(t)$.

25. The zero-state response of an unknown system to a unit-step function is found to be

$$w(t) = \tfrac{1}{2}(1 + e^{-3t})u(t)$$

Find the transfer function $H(s)$ and sketch the Bode plot.

26. A network whose unit-impulse response is $e^{-t}u(t)$ is driven by an input $f_1(t) = u(t) - u(t-1)$. Find the resulting output $f_2(t)$. Use Laplace transforms.

27. Given a passive linear driving-point impedance $Z(s)$, where $v(t)$ and $i(t)$ are the time-domain descriptions of the terminal quantities. For $v(t) = 10u(t)$, it is noted that $i(t) = 10(1 - e^{-2t})u(t)$. (a) Find $Z(s)$. (b) Synthesize $Z(s)$ and sketch the circuit diagram.

28. For a network with a unit-step input $v_1(t) = u(t)$ the zero-state output response $v_2(t) = (1 - e^{-t})u(t)$. Find the unit-impulse response of this system and its transfer function.

29. Using Laplace transforms, find an expression for the voltage $v_1(t)$ in Figure P15.29.

30. A voltage source, $v(t) = {}^t u(t)$, drives the series combination of $R = 1\ \Omega$ and $L = 1$ H. Find the zero-state response $i(t)$. Use Laplace transforms.

31. A unit-impulse voltage source drives the series combination of $R = 4\ \Omega$ and $L = 3$ H. (a) Find $i(t)$ for all t. Assume $i(0-) = 0$. (b) Find $v_L(t)$.

32. A battery V_0, an open switch that closes at $t = 0$, and an impedance $Z(s) = 2(s+2)/(s+4)$ are in series. (a) Determine the form of the current in the circuit after $t = 0$. (b) Evaluate the unknown battery voltage by making use of the information that $i(0+) = 6$ A. (c) Assume that the switch after being closed for a long time opens at (redefined) $t = 0$. Determine $v(t)$, the voltage across the impedance.

33. Determine the voltage gain $V_o(s)/V_1(s)$ for the transistor amplifier circuit shown in Figure P15.33. (a) Locate all critical frequencies on the s plane and sketch the system's Bode plot. What is its high-frequency gain? (b) What is the unit-impulse response of this amplifier? (c) What is its response to $v_1(t) = 0.01e^{-444t}u(t)$? (d) If we define the effective gain of the amplifier to be $(V_o)_{\max}/(V_{\text{in}})_{\max}$, what is its effective gain in decibels for the test input pulse in part (c)?

34. An admittance $Y(s)$ has a zero at $s = -6$ and a pole at $s = -2$. Furthermore, $Y(\infty) = 0.2\ \mho$. (a) Find $Y(s)$. (b) A 12-V battery and a switch are placed in series with $Y(s)$. The switch is closed at $t = 0$. Find $i(t)$ using Laplace transforms. (c) Suppose the switch has been closed for a long time and then is opened at (redefined) $t = 0$. Find the value of the extremum of the voltage that will appear across $Y(s)$.

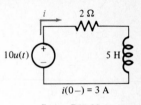

figure P15.22

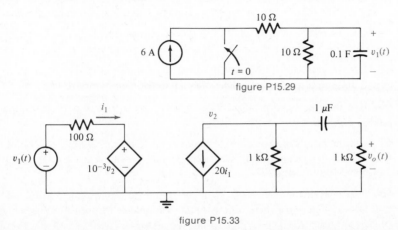

figure P15.29

figure P15.33

35. Find the initial and final values of $f(t)$ if (a) $F(s) = 4/(s + 2)$. (b) $F(s) = (3s^2 + 1)/[s(s^2 + 4s + 2)]$.

36. Repeat problem 35 for

$$F(s) = \frac{6(3s + 1)}{s(s + 1)}$$

37. What value will $v(t)$ approach as t approaches (a) ∞ (b) $0+$?

$$V(s) = \frac{8s^2 + 3}{s(2s^2 + 6)}$$

38. If $F(s)$ is given as

$$F(s) = \frac{s - 2}{(s + 1)^2}$$

(a) What is the initial value (at $t = 0+$) of $f(t)$? (b) What value does $f(t)$ approach after a very long time? (c) Find an expression for $f(t)$.

39. Consider the differential equation

$$\frac{d^2x}{dt^2} + 2\frac{dx}{dt} + x = 3 \sin tu(t)$$

(a) Write the Laplace-transformed equation. (b) If the derivative $\dot{x}(0-) = 1$ and $x(0-) = -2$, solve for $X(s)$. (c) Expand your answer to part (b) in a partial-fraction expansion. Do not evaluate the coefficients. (d) Write out the form that $x(t)$ will have. Use the unevaluated coefficients from part (c). (d) What will be the initial value of $x(t)$?

40. Given that $\ddot{x} + 3\dot{x} + 4x = 1$, where $x(0-) = -1$ and $\dot{x}(0-) = 2$; (a) solve for $X(s)$. (b) Find $f(0+)$ and $f(\infty)$.

41. Given that $p^2x + px + 2x = 0$, $x(0-) = 1$, and $\dot{x}(0-) = 2$, find $x(t)$ for $t > 0$.

42. Solve the following pair of simultaneous differential equations for $x_1(t)$ by simply taking their Laplace transforms. (Do not use matrices, state-variable methods, etc.) The initial conditions are $x_1(0-) = 1$ and $x_2(0-) = 2$.

$$\frac{dx_1}{dt} = -3x_1 + 2x_2 + \delta(t) \qquad \frac{dx_2}{dt} = -x_1 - x_2$$

43. Use the results of problems 9.1 and 9.21 to solve for the voltage $v_o(t)$ that results from a step-function input $v_s(t) = u(t)$ in that problem. (All initial conditions are zero.)

44. Use the results of problem 9.13 to find the transforms of the state variables $X_1(s)$ and $X_2(s)$ if $v_s(t)$ is a unit impulse and all initial conditions are zero.

45. Use the results of problem 9.20 to solve for state variables $x_1(t)$ and $x_2(t)$ that result from a unit-impulse input $v(t) = \delta(t)$ when all initial conditions are zero.

46. Use the results of problem 9.23 to solve for $x_1(t)$ and $x_2(t)$ given that $v(t) = u(t)$, $x_1(0-) = 1$, and $x_2(0-) = 0$.

47. Use the results of problem 9.24 to determine the transfer function $H(s)$ of that system. You were to determine two different sets of A, B, C, and D matrices in that problem. Do they both yield the same $H(s)$?

48. Use the results of problem 9.33 to determine the transfer function $H(s)$ of that system. You were to determine three different state variable descriptions of that system. Do they all yield the same $H(s)$?

appendix A.
MATRICES AND DETERMINANTS

matrices A.1

A matrix is an array of quantities (numbers or functions). It is enclosed in brackets. A single symbol standing for the matrix can be written $[a]$.

$$[a] = \begin{bmatrix} a_{11} & a_{12} & a_{13} & \ldots & a_{1n} \\ a_{21} & a_{22} & a_{23} & \ldots & a_{2n} \\ \hdashline a_{n1} & a_{n2} & a_{n3} & \ldots & a_{nn} \end{bmatrix} \qquad \text{(A.1.1)}$$

Equation (A.1.1) represents a square matrix: there are an equal number of rows and columns in a square matrix. Not all matrices are square. A matrix with only a single row or a single column is called a *vector*. A matrix is merely a shorthand notation; it has no numerical value.

A.2 determinants

A determinant is a *function* of an array of numbers. The array is enclosed by a pair of straight vertical lines. All determinants are square.

$$\det [a] = \begin{vmatrix} a_{11} & a_{12} & a_{13} & \cdots & a_{1n} \\ a_{21} & a_{22} & a_{23} & \cdots & a_{2n} \\ \multicolumn{5}{c}{\dotfill} \\ a_{n1} & a_{n2} & \cdots & \cdots & a_{nn} \end{vmatrix} \tag{A.2.1}$$

A determinant made from a matrix by simply changing the bracket into straight lines is called the *system determinant* and often given the symbol Δ.

Determinants have the following properties:

1. The value of the function is unchanged if the elements of any row (column) are replaced by the sums of the elements of that row (column) and the corresponding ones of another row (column), i.e., if $a_{11}, a_{12}, \ldots, a_{1n}$ are replaced by $a_{11} + a_{31}, a_{12} + a_{32}, \ldots, a_{1n} + a_{3n}$.

2. The value of the function is unchanged if the elements of any row (column) times an arbitrary factor are added to (or subtracted from) the corresponding elements of another row (column).

3. The value of the function is multiplied by the constant k if all the elements of any row or column are multiplied by k.

4. The value of the function is unity if all the elements on the major diagonal $a_{11}, a_{22}, a_{33}, \ldots, a_{nn}$ are unity and all the others are zero.

5. The algebraic sign of the function is reversed when any two rows (columns) are interchanged.

6. The value of the function is zero if all elements in a row (column) are zero *or* if the corresponding elements of any two rows (columns) are identical *or* have a common ratio.

* Minor determinant   A minor determinant M_{jk} is a determinant formed by striking out row j and column k in a larger determinant.

* Cofactor   A cofactor is a signed minor; for example, $\Delta_{jk} = -1^{j+k} M_{jk}$.

* Evaluation   A determinant can be evaluated by *Laplace's development*

$$\Delta = a_{j1}\Delta_{j1} + a_{j2}\Delta_{j2} + a_{j3}\Delta_{j3} + \cdots + a_{jn}\Delta_{jn} \qquad j = 1, 2, 3, \ldots, n$$

i.e., sum the elements in any row (column) after multiplying each times its cofactor.

Matrices are useful in writing a set of linear simultaneous algebraic equations in a brief and concise form. For example,

$$[a][x] = [y] \tag{A.2.2}$$

implies

$$\begin{bmatrix} a_{11} & a_{12} & a_{13} & \cdots & a_{1n} \\ a_{21} & a_{22} & a_{23} & \cdots & a_{2n} \\ \cdots\cdots\cdots\cdots\cdots\cdots\cdots \\ a_{n1} & a_{n2} & a_{n3} & \cdots & a_{nn} \end{bmatrix} \begin{bmatrix} x_1 \\ x_2 \\ \cdots \\ x_n \end{bmatrix} = \begin{bmatrix} y_1 \\ y_2 \\ \cdots \\ y_n \end{bmatrix} \qquad (A.2.3)$$

which in turn implies

$$a_{11}x_1 + a_{12}x_2 + \cdots + a_{1n}x_n = y_1$$
$$a_{21}x_1 + a_{22}x_2 + \cdots + a_{2n}x_n = y_2$$
$$\cdots\cdots\cdots\cdots\cdots\cdots\cdots\cdots\cdots\cdots \qquad (A.2.4)$$
$$a_{n1}x_1 + a_{n2}x_2 + \cdots + a_{nn}x_n = y_n$$

To multiply two matrices, proceed across the *i*th row in $[a]$ and down the *j*th column in $[x]$, adding the corresponding products, and place the result in the *i*th row and *j*th column of the answer matrix $[y]$. Thus,

$$[a][x] \neq [x][a] \qquad (A.2.5)$$

so the order of multiplication is important. For two matrices to be multipliable (*conformable*) the first (premultiplying) matrix must have the same number of columns as the second (postmultiplying) matrix has rows.

Equations such as (A.2.2) to (A.2.4) can be solved as follows: multiply the rows by the cofactors $\Delta_{11}, \Delta_{21}, \Delta_{31}, \ldots$, respectively

$$\Delta_{11}(a_{11}x_1 + a_{12}x_2 + \cdots + a_{1n}x_n) = y_1\Delta_{11}$$
$$\Delta_{21}(a_{21}x_1 + a_{22}x_2 + \cdots + a_{2n}x_n) = y_2\Delta_{21}$$
$$\cdots\cdots\cdots\cdots\cdots\cdots\cdots\cdots\cdots\cdots\cdots\cdots \qquad (A.2.6)$$
$$\Delta_{n1}(a_{n1}x_1 + a_{n2}x_2 + \cdots + a_{nn}x_n) = y_n\Delta_{n1}$$

Add all these equations

$$\begin{aligned} &(a_{11}\Delta_{11} & + a_{21}\Delta_{21} & + \cdots + a_{n1}\Delta_{n1})x_1 \\ + &(a_{12}\Delta_{11} & + a_{22}\Delta_{21} & + \cdots + a_{n2}\Delta_{n1})x_2 \\ + &\cdots\cdots\cdots\cdots\cdots\cdots\cdots\cdots\cdots \\ + &(a_{1n}\Delta_{11} & + a_{2n}\Delta_{21} & + \cdots + a_{nn}\Delta_{n1})x_n \\ = & \quad \Delta_{11}y_1 + & \Delta_{21}y_2 + & \cdots + \Delta_{n1}y_n \end{aligned} \qquad (A.2.7)$$

Here the coefficient of x_1 is Laplace's development of determinant Δ down the first column. The coefficient of x_2 is Laplace's development of Δ but with the first column replaced by the second column

$$\begin{vmatrix} a_{12} & a_{12} & a_{13} & \cdots & a_{1n} \\ a_{22} & a_{22} & a_{23} & \cdots & a_{2n} \\ \cdots\cdots\cdots\cdots\cdots\cdots\cdots \\ a_{n2} & a_{n2} & a_{n3} & \cdots & a_{nn} \end{vmatrix} \qquad (A.2.8)$$

From determinant property 6, the x_2 coefficient is zero. Similarly, the coefficient of x_3 is determinant Δ with its first column replaced by its third column and is therefore equal to zero. Also the x_4, x_5, coefficients are zero. Thus (A.2.7) can be written

$$\Delta x_1 = \Delta_{11} y_1 + \Delta_{21} y_2 + \cdots + \Delta_{n1} y_n$$

$$x_1 = \frac{\Delta_{11} y_1 + \Delta_{21} y_2 + \cdots + \Delta_{n1} y_n}{\Delta} \tag{A.2.9}$$

Similarly, the value of x_2 can be determined by multiplying the equations in (A.2.2) through by $\Delta_{12}, \Delta_{22}, \Delta_{32}, \ldots, \Delta_{n2}$ respectively, and then adding, etc., to obtain

$$x_2 = \frac{\Delta_{12} y_1 + \Delta_{22} y_2 + \cdots + \Delta_{n2} y_n}{\Delta} \tag{A.2.10}$$

or, in general,

$$x_k = \frac{\Delta_{1k} y_1 + \Delta_{2k} y_2 + \cdots + \Delta_{nk} y_n}{\Delta} \tag{A.2.11}$$

More concisely,

$$x_k = \sum_{j=1}^{n} \frac{\Delta_{jk} y_j}{\Delta} \tag{A.2.12}$$

The numerator in equation (A.2.11) is Laplace's development of the Δ determinant but with the y's substituted for the kth column. This is called *Cramer's rule*.

A.3   inverse of a matrix

Suppose we solve the original set of equations for x rather than y

$$\begin{aligned}
c_{11} y_1 + c_{12} y_2 + \cdots + c_{1n} y_n &= x_1 \\
c_{21} y_1 + c_{22} y_2 + \cdots + c_{2n} y_n &= x_2 \\
&\cdots\cdots\cdots\cdots\cdots\cdots\cdots\cdots \\
c_{n1} y_1 + c_{n2} y_2 + \cdots + c_{nn} y_n &= x_n
\end{aligned} \tag{A.3.1}$$

Comparing (A.2.12) and (A.3.1), we have

$$c_{rs} = \frac{\Delta_{sr}}{\Delta}$$

Thus to find $[c]$

1. Replace each element in $[a]$ by the quotient of its cofactor divided by the determinant of $[a]$.
2. Transpose the result (interchange rows for columns).

DRILL PROBLEM 1

Develop the following determinant:

$$\begin{vmatrix} 4 & 9 & 6 \\ 0 & 3 & 1 \\ 2 & 0 & 3 \end{vmatrix}$$

a. Down the first column.
b. Across the second row.
ANS.: det = 18.

Perform the following indicated operations on

$$[a] = \begin{bmatrix} 1 & 0 \\ 3 & 2 \end{bmatrix} \qquad [b] = \begin{bmatrix} 3 & 2 \\ 1 & 9 \end{bmatrix} \qquad [c] = [4 \quad 2]$$

If the operation is undefined for this case, write "undef."
a. $[a][c]$
b. $[c]^T$ = The transpose of $[c]$
c. $[c]^{-1}$
d. $[a][b]$
e. $[b]^{-1}$
ANS.:   (c) undef. (e) If $[d] = [b]^{-1}$, $d_{12} = -\frac{2}{25}$.

appendix B
SIMULATION PROGRAM FOR INTEGRATED CIRCUIT ELECTRONICS (SPICE)

introduction B.1

SPICE is a large and powerful general-purpose circuit/simulation program for dc, transient, ac, and Fourier analysis. Circuits may contain resistors, capacitors, inductors, mutual inductors, diodes, independent and dependent voltage and current sources, as well as elements such as semiconductor devices that are beyond the scope of this course.

types of analyses B.2

SPICE can perform any of the following types of circuit analyses:

> SPICE automatically solves for dc voltages at all nodes and also direct current in all voltage sources.

.DC   Solves for the dc output-input characteristic; i.e., it prints and/or plots the dc value of any specified node voltage as the dc magnitude of any specified input source is slowly increased.

.AC   Yields the magnitude (either numerical value or in decibels) and/or phase of any node voltage(s) versus sinusoidal input frequency f in hertz (incremented either linearly or logarithmically by decade or octave).

.TRAN   Yields the instantaneous value of any specified node voltage(s) versus time during a user-specified interval. In conjunction with this mode of analysis *SPICE can compute the Fourier series* of any node-voltage waveform.

SPICE *invariably* uses the standard relative polarities of voltage and current. Thus, *even for current sources*, the positive terminal is the node at which current flows *into* the source (not out, as might be assumed). Thus, for example, the "current in voltage source V1" defines the current *entering* the positive terminal of that source.

B.3   to analyze a circuit with SPICE

Circuits are typically analyzed by SPICE on a large computer system as follows:

1. Sign on to the computer.

2. Select BATCH mode.

3. Name and create (via XEDIT or some other EDITOR or via TEXT mode) a *primary local file* (1fn) in your workspace. We shall call this your SPICE file; it is actually a data file, so do *not* use line numbers.

4. Your SPICE file must begin with a title. This first line of the file will not be processed by SPICE but is simply printed out at the beginning of the output generated by each different type of requested analysis.

5. Next come the *element statements*, each of which tells SPICE about one element in your circuit and states the nodes to which it is connected and its numerical value.

6. Then come the *control statements*, which tell SPICE what type(s) of analysis you want it to perform on the circuit. Each of these control statements must begin with a leading period.

7. The last statement in the file must be

.END

8. Finally, attach SPICE to your local file. The command for doing so depends on the operating system of your computer. Typically it might be

FINDLIB, SPICE

Wait for an answer, then type

SPICE, 1fn

where 1fn is the name of your local SPICE file.

writing your first SPICE file B.4

To get SPICE up and going, try this simple example.

Consider the resistive circuit with two dc sources shown in Figure 1.11.10. First, in order to use SPICE, each node of your circuit must be numbered (the reference node, ground, is always called node 0). Doing this and then numbering the remaining nodes 1, 2, and 3 from left to right, we write

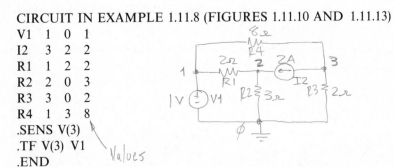

```
CIRCUIT IN EXAMPLE 1.11.8 (FIGURES 1.11.10 AND 1.11.13)
V1   1   0   1
I2   3   2   2
R1   1   2   2
R2   2   0   3
R3   3   0   2
R4   1   3   8
.SENS V(3)
.TF V(3) V1
.END
```

Values

to which SPICE replies by relisting this file and then giving the results shown in Figures B.4.1 and B.4.2.

```
*************** 85/04/21. ********************** SPICE 2D.2 (26SEP76)
CIRCUIT IN EXAMPLE 1.11.8 (FIGURE 1.11.10 AND 1.11.13)
                   SMALL SIGNAL BIAS SOLUTION
****************************************************************

 NODE    VOLTAGE      NODE    VOLTAGE      NODE    VOLTAGE

(  1)     1.0000     (  2)     3.0000     (  3)    -3.0000



      VOLTAGE SOURCE CURRENTS
      NAME         CURRENT

      V1           5.000E-01

      TOTAL POWER DISSIPATION   1.15E+01   WATTS
```

figure B.4.1

```
**** SMALL-SIGNAL CHARACTERISTICS

    V(3)/V1                        =  2.000E-01
    INPUT RESISTANCE AT V1         =  3.333E+00
    OUTPUT RESISTANCE AT V(3)      =  1.600E+00
```

figure B.4.2

```
*************** 85/04/21. ********************* SPICE 2D.2 (26SEP76)
CIRCUIT IN EXAMPLE 1.11.8 (FIGURE 1.11.10 AND 1.11.13)
                    DC SENSITIVITY ANALYSIS
****************************************************************************

DC SENSITIVITIES OF OUTPUT V(3)
        ELEMENT        ELEMENT        ELEMENT        NORMALIZED
         NAME           VALUE        SENSITIVITY    SENSITIVITY
                                    (VOLTS/UNIT)  (VOLTS/PERCENT)
         R1          2.000E+00         0.             0.
         R2          3.000E+00         0.             0.
         R3          2.000E+00      -1.200E+00     -2.400E-02
         R4          8.000E+00      -1.000E-01     -8.000E-03
         V1          1.000E+00       2.000E-01      2.000E-03
         I2          2.000E+00      -1.600E+00     -3.200E-02

    JOB CONCLUDED
    TOTAL JOB TIME          .063
    TOTAL JOB COST     $    .019
```

The .SENS statement produces the table listing the sensitivity of the voltage of node 3 with respect to node 0 as a function of every circuit element as well as the voltage source V1.

The .TF statement produces the transfer (voltage-gain) functions V(3)/V1 as well as the input resistance and output resistance. Neither of these control statements need be included in your first trial run. Now for the details.

B.5 element statements

Each element in the circuit is specified by an element statement that contains the element name, the circuit nodes to which the element is connected, and the values of the parameters that determine the electrical characteristics of the element. The first letter of the element name specifies the element type. The format for the SPICE element types is given in what follows. For example, a resistor name must begin with the letter R and can contain from one to eight characters; hence R, R1, RSE, ROUT, and R3AC2ZY are valid resistor names.

The circuit cannot contain a loop of voltage sources and/or inductors or a node to which only current sources or only capacitors are connected. Each node in the circuit must have *a dc path to ground*.

SPICE recognizes the following numerical suffixes on your element values:

$$G = 1.0E9 \quad MEG = 1.0E6 \quad K = 1.0E3 \quad M = 1.0E{-}3$$

$$U = 1.0E{-}6 \quad N = 1.0E{-}9 \quad P = 1.0E{-}12$$

Letters immediately following a number that are not scale factors are ignored, and letters immediately following a scale factor are ignored. Hence, 10, 10V, 10VOLTS, and 10HZ all represent the same number, and M, MA, MSEC, and MMHOS all represent the same scale factor. Note that 1000, 1000.0, 1000HZ, 1E3, 1.0E3, 1KHZ, and 1K all represent the same number. Do *not* leave a space between 1000 and HZ; the program will try to use HZ as another numerical input and not be able to interpret its value.

| General form: | RXXXXXXX N1 N2 VALUE | **RESISTORS** |
| Example: | RC1 12 17 1K | |

N1 and N2 are the two element nodes. VALUE is the resistance (in ohms) and may be positive or negative but not zero.

In the statements that follow some entries are optional. These are indicated by brackets, e.g., [P1 P2 ...].

General form:	CXXXXXXX N+ N− P0 [P1 P2 ...] [IC = INCOND]	**CAPACITORS AND INDUCTORS**
	LYYYYYYY N+ N− P0 [P1 P2 ...] [IC = INCOND]	
Examples:	CBYP 13 0 1UF	
	COSC 17 23 10U 5U IC = 3V	
	LLINK 42 69 1UH	
	LSHUNT 23 51 10U 32U IC = 15.7A	

N+ and N− are the positive and negative element nodes, respectively. P0, P1, P2, ... are the coefficients of a polynomial describing the element value. For capacitors, the capacitance (in farads) is expressed as a function of the voltage across the element. For inductors, the inductance (in henrys) is expressed as a function of the current through the element. To illustrate, the second example above describes a capacitor with a value defined by

$$C = 10E{-}6 + 5E{-}6*V$$

where V is the voltage across the capacitor.

For the capacitor, the (optional) initial condition is the initial (time-zero) value of capacitor voltage (in volts). For the inductor, the (optional) initial condition is the initial (time-zero) value of inductor current (in amperes) that flows from N+ through the inductor to N−. Note that the initial conditions (if any) apply *only* if the UIC option is specified in the .TRAN statement.

| General form: | KXXXXXXX LYYYYYYY LZZZZZZZ VALUE | **COUPLED (MUTUAL) INDUCTORS** |
| Example: | KXFRMR L1 L2 0.87 | |

LYYYYYYY and LZZZZZZZ are the names of the two coupled inductors, and VALUE is the coefficient of coupling, $K = M/\sqrt{L_1 L_2}$, which must satisfy the relation $0 < K \leq 1$. The coupled inductors can be entered in either order and must be constant (only P0 can be specified in the inductor statement). In terms of the dot convention, this coupling puts a dot by the first node of each of the coupled inductors.

INDEPENDENT SOURCES

General form:

VXXXXXXX N+ N [[DC] DC/TRAN value][AC[ACMAG [ACPHASE]]]
IYYYYYYY N+ N [[DC] DC/TRAN value][AC[ACMAG [ACPHASE]]]

Examples: VCC 10 0 DC 6
                VIN 13 2 0.001 AC 1 SIN(0 1 1MEG)
                VMEAS 12 9

N+ and N− are the positive and negative nodes, respectively. Note that voltage sources need not be grounded. Current is assumed to flow from the positive node through the source to the negative node.

DC/TRAN is the dc and transient analysis value of the source. If the source value is zero both for dc and transient analyses, this value may be omitted. If the source value is time-invariant (e.g., a power supply), the value may optionally be preceded by the letters DC.

ACMAG is the ac magnitude and ACPHASE is the ac phase. The source is set to this value in the ac analysis. If ACMAG is omitted following the keyword AC, a value of unity is assumed. If ACPHASE is omitted, a value of zero is assumed. If the source is not an ac small-signal input, the keyword AC and the ac values are omitted.

Any independent source can be assigned a time-dependent value for transient analysis. If a source is assigned a time-dependent value, the time-zero value is used for dc analysis. There are four independent source functions: pulse, exponential, sinusoidal, and piecewise linear. If parameters others than source values are omitted or set to zero, the default values shown will be assumed. (TSTEP is the printing increment and TSTOP is the final time; see the .TRAN statement for explanation.)

Pulse: PULSE(V1 V2 TD TR TF PW PER)

Example: VIN 3 0 PULSE(−1 1 2NS 2NS 2NS 50NS 100NS)

Parameters, default values, and units

V1	Initial value	....	Volts or amperes
V2	Pulsed value	....	Volts or amperes
TD	Delay time	0.0	Seconds
TR	Rise time	TSTEP	Seconds
TF	Fall time	TSTEP	Seconds
PW	Pulse width	TSTOP	Seconds
PER	Period	TSTOP	Seconds

A single pulse so specified is described by Figure B.5.1.

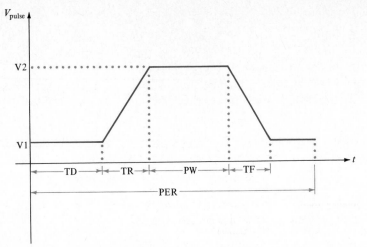

| Sinusoidal: | SIN(VO VA FREQ TD ZW) |
| Example: | VIN 3 0 SIN(0 1 100MEG 1NS 1E10) |

					Parameters, default values, and units
VO	Offset	...	...	Volts or amperes	
VA	Amplitude	...	...	Volts or amperes	
FREQ	Frequency		1/TSTOP	Hertz	
TD	Delay		0.0	Seconds	
ZW	Damping		0.0	Seconds^{-1}	

The shape of the waveform is described by the following table:

Time	Value
0–TD	VO
TD–TSTOP	VO + VA*EXP(−(T − TD)*ZW)*SINE(TWOPI*FREQ*(T − TD))

| Exponential: | EXP(V1 V2 TD1 TAU1 TD2 TAU2) |
| Example: | VIN 3 0 EXP(−4 −1 2NS 30NS 50NS 40NS) |

				Parameters, default values, and units
V1	Initial value	...	Volts or amperes	
V2	Pulsed value	...	Volts or amperes	
TD1	Rise delay time	0.0	Seconds	
TAU1	Rise-time constant	TSTEP	Seconds	
TD2	Fall delay time	TSTEP	Seconds	
TAU2	Fall-time constant	TSTEP	Seconds	

The shape of the waveform is described by the following table:

Time	Value
0–TD1	V1
TD1–TD2	V1 + (V2 − V1)(1 − EXP(−(T − TD1)/(TAU1))
TD2–TSTOP	V1 + (V2 − V1)(1 − EXP(−(T − TD1)/TAU1)) + (V1 − V2)(1 − EXP(−(T − TD2)/TAU2))

Piecewise linear: PWL(T1 V1 [T2 V2 T3 V3 T4 V4 ...])

Example: VCLOCK 7 5 PWL(0 −7 10NS −7 11NS
−3 17NS −3 18NS −7 50NS −7)

Each pair of values (TI, VI) specifies that the value of the source is VI (in volts or amperes) at time = TI. The value of the source at intermediate values of time is determined by using linear interpolation on the input values.

LINEAR DEPENDENT SOURCES   SPICE allows circuits to contain linear dependent sources characterized by any of the four equations

$$I = G*V \qquad V = E*V \qquad I = F*I \qquad V = H*I$$

where G, E, F, and H are constants representing transconductance, voltage gain, current gain, and transresistance, respectively.

Linear Voltage-Controlled Current Sources

General form: GXXXXXXX N+ N− NC+ NC− VALUE

Example: G1 2 0 5 0 0.1MMHO

N+ and N− are the positive and negative nodes, respectively. Current flow is from the positive node through the source to the negative node. NC+ and NC− are the positive and negative controlling nodes, respectively. VALUE is the transconductance (in mhos).

Linear Voltage-Controlled Voltage Sources

General form: EXXXXXXX N+ N− NC+ NC− VALUE

Example: E1 2 3 14 1 2.0

N+ is the positive node, and N− is the negative node. NC+ and NC− are the positive and negative controlling nodes, respectively. VALUE is the voltage gain.

Linear Current-Controlled Current Sources

General form: FXXXXXXX N+ N− VNAM VALUE

Example: F1 13 4 VSENS 5

N+ and N− are the positive and negative nodes, respectively. Current flow is from the positive node, through the source, to the negative node. VNAM is the name of a voltage source through which the controlling current flows. The direction of positive controlling current flow is from the positive node through the source to the negative node of VNAM. VALUE is the current gain.

Linear Current-Controlled Voltage Sources

General form: HXXXXXXX N+ N− VNAM VALUE

Example: HX 5 17 VZ 0.5K

N+ and N− are the positive and negative nodes, respectively. VNAM is the name of a voltage source through which the controlling current flows. The direction of positive controlling current flow is from the positive node through the source to the negative node of VNAM. VALUE is the transresistance (in ohms).

control statements B.6

Example: POWER AMPLIFIER CIRCUIT

This statement must be the first line in the file. Its contents are printed verbatim as the heading for each section of output.

Example: .END

This statement must always be the last line in the file. Note that the leading period is an essential part of the statement.

General form: *ANY COMMENTS

Example: *RF = 1K   GAIN SHOULD BE 100

This line is printed out in the input listing but is otherwise ignored.

General form: .DC SORNAM VSTART VSTOP VINCR

Example: .DC VIN 0.25 5.0 0.25

This statement defines the dc transfer-curve source and sweep limits. SORNAM is the name of an independent voltage or current source. VSTART, VSTOP, and VINCR are the starting, final, and incrementing values respectively. The above example will cause the value of the voltage source VIN to be swept from 0.25 to 5.0 V in increments of 0.25 V.

General form: .TF OUTVAR INSOR

Example: .TF V(5,3) VIN

This statement defines the small-signal output and input for the dc small-signal analysis. OUTVAR is the small-signal output variable and INSOR is the small-signal input source. If this statement is included, SPICE will compute the dc small-signal value of the transfer function (output or input), input resistance, and output resistance. For the above example, SPICE would compute the ratio of V(5,3) to VIN, the small-signal input resistance at VIN, and the small-signal output resistance measured across nodes 5 and 3.

General form: .SENS OV1 [OV2 ...]

Example: .SENS V(9) V(4,3) V(17)

If a .SENS statement is included in the input file, SPICE will determine the dc small-signal sensitivities of each specified output variable with respect to every circuit parameter. Note that for large circuits, large amounts of output can be generated.

.AC STATEMENT   General form: .AC DEC ND FSTART FSTOP
.AC OCT NO FSTART FSTOP
.AC LIN NP FSTART FSTOP

Examples: .AC DEC 10 1 10KHZ
.AC OCT 5 1 64HZ
.AC LIN 100 10 600

DEC stands for decade variation, and ND is the number of points per decade. OCT stands for octave variation, and NO is the number of points per octave. LIN stands for linear variation, and NP is the total number of points. FSTART is the starting frequency, and FSTOP is the final frequency. If this statement is included in the file, SPICE will perform an ac analysis of the circuit over the specified frequency range. For this analysis to be meaningful, at least one independent source must have been specified with an ac value.

.TRAN STATEMENT   General form: .TRAN TSTEP TSTOP [TSTART[TMAX]] [UIC]

Examples: .TRAN 1NS 100NS
.TRAN 1NS 100NS 500NS
.TRAN 10NS 1US UIC

TSTEP is the printing increment, TSTOP is the final time, and TSTART is the initial time. If TSTART is omitted, it is assumed to be zero. The transient analysis always begins at time zero. In the interval (ZERO, TSTART), the circuit is analyzed (to reach a steady state), but no outputs are stored. In the interval (TSTART, TSTOP), the circuit is analyzed and outputs are stored. TMAX is the maximum step size that SPICE will use (default value is TSTOP/50.0).

UIC (use initial conditions) is an optional keyword which indicates that the user does not want SPICE to solve for the quiescent operating point before beginning the transient analysis. If this keyword is specified, SPICE uses the values specified using $IC = \cdots$ on the various elements as the initial transient condition and proceeds with the analysis.

.FOUR STATEMENT   General form: .FOUR FREQ OV1 [OV2 OV3 ...]

Example: .FOUR 100KHZ V(5)

This controls whether SPICE performs a Fourier analysis as a part of the transient analysis. FREQ is the fundamental frequency, and OV1, ... are the output variables for which the analysis is desired. The Fourier analysis is performed over the interval (TSTOP-PERIOD, TSTOP), where TSTOP is the final time specified for the transient analysis and PERIOD is one period of the fundamental frequency. The dc component and the first nine components are determined. For maximum accuracy, TMAX (see the .TRAN statement) should be set to period/100.0 (or less for very high-Q circuits).

General form:	.PRINT PRTYPE OV1 [OV2 ... OV8]	**.PRINT STATEMENT**
Examples:	.PRINT TRAN V(4) I(VIN)	
	.PRINT AC VM(4,2) VR(7) VP(8,3)	
	.PRINT DC V(2) I(VSRC) V(23,17)	

This statement defines the contents of a tabular listing of one to eight output variables. PRTYPE is the type of the analysis (DC, AC, or TRAN) for which the specified outputs are desired. The form for voltage or current output variables is as follows:

V(N1[,N2])   Specifies the voltage difference between nodes N1 and N2. If N2 (and the preceding comma) is omitted, ground (0) is assumed. For the ac analysis, five additional outputs can be accessed by replacing the letter V by

> VR for Real part
> VI for Imaginary part
> VM for Magnitude
> VP for Phase
> VDB for 20 * log(Magnitude)

I(VXXXXXXX)   Specifies the current flowing in the independent voltage source named VXXXXXXX. Positive current flows from the positive node, through the source, to the negative node. For the ac analysis, corresponding replacements for the letter I can be made in the same way as described for voltage outputs.

General form:	.PLOT PLTYPE OV1 [(PLO1,PHI1)]	**.PLOT STATEMENT**
	[OV2 [(PLO2,PHI2)] ... OV8]	
Examples:	.PLOT DC V(4) V(5) V(1)	
	.PLOT TRAN V(17,5) (2,5) I(VIN) V(17) (1,9)	
	.PLOT AC VM(5) VM(31,24) VDB(5) VP(5)	

This statement defines the contents of one plot of from one to eight output variables. PLTYPE is the type of analysis (DC, AC, TRAN) for which the specified outputs are desired. The syntax for the OV1 is identical to that for the .PRINT statement described above.

The (optional) plot limits PLO/PHI may be specified after any of the output variables. All output variables to the left of a pair of plot limits (PLO, PHI) will be plotted using the same lower and upper plot bounds. If plot limits are not specified, SPICE will automatically determine the minimum and maximum values of all output variables being plotted and scale the plot to fit. More than one scale will be used if the output variable values warrant; i.e., mixing output variables with values which are orders-of-magnitude different still gives readable plots.

The overlap of two or more traces on any plot is indicated by the letter X.

B.7 ac small-signal analysis

Almost all electronic elements—resistors, transistors, etc.—are nonlinear but can be considered essentially linear if the applied voltages and currents are not too large. Therefore, if we assume linearity in such circuits, we realize that the signals cannot be very large in amplitude and we call such analyses *small-signal analyses*.

The ac small-signal portion of SPICE computes the ac output variables as a function of input frequency. The program first computes the dc operating point of the circuit and determines linearized small-signal models for all the nonlinear devices in the circuit. The resultant linear circuit is then analyzed over a user-specified range of frequencies. The desired output of an ac small-signal analysis is usually a transfer function (voltage gain, transimpedance, etc.). If the circuit has only one ac input, it is convenient to set that input to unity and zero phase, so that output variables have the same value as the transfer function of the output variable with respect to the input.

EXAMPLE B.7.1

In Fig. 10.5.14 find the voltage (magnitude and phase) across the 2-Ω resistor.
ANS.:   The SPICE file in Figure B.7.1 yields the desired quantities. (Do not forget that the title must be the first line.)

---

figure B.7.1

```
CIRCUIT IN EXAMPLE 10.5.11 (FIG. 10.5.14)
IS 0 1 AC 10 30
R1 1 0 1
R2 1 2 1
L1 2 0 2
C1 1 3 2
R3 2 3 1
R4 3 0 2
.AC LIN 1 0.3183 0.3183
.PRINT AC VM(3) VP(3)
.END
```

```
*************** 85/04/21. **********************
CIRCUIT IN EXAMPLE 10.5.11 (FIG. 10.5.14)
                    AC ANALYSIS
****************************************************

      FREQ        VM(3)       VP(3)
  3.183E-01     6.468E+00   4.404E+01

              JOB CONCLUDED
              TOTAL JOB TIME          .073
              TOTAL JOB COST    $     .023
```

SPICE can also plot magnitude (linearly and in decibels) and phase both versus frequency.

EXAMPLE B.7.2

Find the Bode plot (magnitude in decibels and phase both versus frequency on the logarithmic scale) for the output voltage of the circuit in Figure 11.9.7.

ANS.:   This is done as shown in Figures B.7.2 to B.7.4, the file as relisted by the computer. (Remember that the first line of the file, as you enter it, is the title.)

```
***** 85/04/21. ********  SPICE 2D.2 (26SEP76) ******** 14.39.49.*****     figure B.7.2
CIRCUIT IN FIGURE 11.9.7 (BODE PLOT)
      INPUT LISTING            TEMPERATURE =   27.000 DEG C
*********************************************************************


V1 1 0 AC
C1 1 2 0.1667
L1 2 0 .06
R1 2 0 1
.AC DEC 15 .1 100
.PLOT AC VDB(2) VP(2)
.PRINT AC VDB(2) VP(2)
.END



*************** 85/04/21. ********************** SPICE 2D.2 (26SEP76)
CIRCUIT IN FIGURE 11.9.7 (BODE PLOT)
                SMALL SIGNAL BIAS SOLUTION
*********************************************************************

  NODE    VOLTAGE     NODE    VOLTAGE

 ( 1)    0.0000    ( 2)    0.0000



     VOLTAGE SOURCE CURRENTS
     NAME        CURRENT

     V1        0.

     TOTAL POWER DISSIPATION   0.         WATTS
```

figure B.7.3

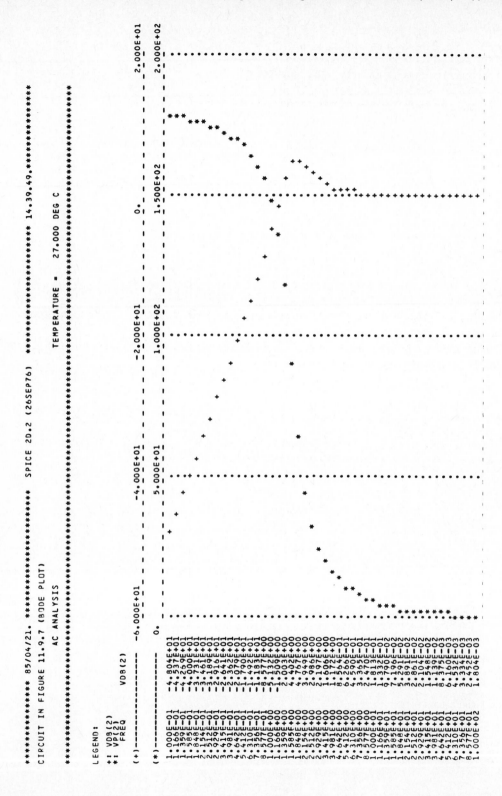

```
************** 85/04/21. *************
CIRCUIT IN FIGURE 11.9.7 (BODE PLOT)
                AC ANALYSIS
*************************************
```

FREQ	VDB(2)	VP(2)
1.000E-01	-4.804E+01	1.778E+02
1.166E-01	-4.537E+01	1.775E+02
1.359E-01	-4.269E+01	1.770E+02
1.585E-01	-4.000E+01	1.765E+02
1.848E-01	-3.731E+01	1.760E+02
2.154E-01	-3.461E+01	1.753E+02
2.512E-01	-3.189E+01	1.745E+02
2.929E-01	-2.916E+01	1.735E+02
3.415E-01	-2.641E+01	1.723E+02
3.981E-01	-2.362E+01	1.709E+02
4.642E-01	-2.079E+01	1.692E+02
5.412E-01	-1.790E+01	1.670E+02
6.310E-01	-1.492E+01	1.642E+02
7.356E-01	-1.183E+01	1.606E+02
8.577E-01	-8.577E+00	1.555E+02
1.000E+00	-5.132E+00	1.481E+02
1.166E+00	-1.509E+00	1.365E+02
1.359E+00	2.003E+00	1.178E+02
1.585E+00	4.402E+00	9.078E+01
1.848E+00	4.767E+00	6.344E+01
2.154E+00	3.949E+00	4.428E+01
2.512E+00	2.986E+00	3.241E+01
2.929E+00	2.197E+00	2.482E+01
3.415E+00	1.606E+00	1.966E+01
3.981E+00	1.172E+00	1.593E+01
4.642E+00	8.564E-01	1.312E+01
5.412E+00	6.266E-01	1.093E+01
6.310E+00	4.590E-01	9.179E+00
7.356E+00	3.365E-01	7.753E+00
8.577E+00	2.470E-01	6.576E+00
1.000E+01	1.813E-01	5.595E+00
1.166E+01	1.332E-01	4.770E+00
1.359E+01	9.790E-02	4.073E+00
1.585E+01	7.196E-02	3.482E+00
1.848E+01	5.291E-02	2.980E+00
2.154E+01	3.891E-02	2.551E+00
2.512E+01	2.861E-02	2.185E+00
2.929E+01	2.104E-02	1.873E+00
3.415E+01	1.548E-02	1.605E+00
3.981E+01	1.139E-02	1.376E+00
4.642E+01	8.375E-03	1.180E+00
5.412E+01	6.160E-03	1.012E+00
6.310E+01	4.532E-03	8.675E-01
7.356E+01	3.334E-03	7.439E-01
8.577E+01	2.452E-03	6.380E-01
1.000E+02	1.804E-03	5.471E-01

B.8 transient analysis

The transient-analysis portion of SPICE computes the transient output variables as a function of time over a user-specified time interval. The initial conditions are automatically determined by a dc analysis. All sources which are not time-dependent, e.g., power supplies, are set to their dc value.

EXAMPLE B.8.1

The circuit in Examples 7.9.1 and 7.9.2 is solved as shown in Figures B.8.1 and B.8.2.

figure B.8.1

```
***** 85/04/21. ********   SPICE 2D.2 (26SEP76)  ******** 14.40.57.*****
CIRCUIT IN EXAMPLES 7.9.1 AND 7.9.2 (FIG. 7.9.1)
        INPUT LISTING                    TEMPERATURE =   27.000 DEG C
*****************************************************************************


VS 1 0 PULSE 0 1 0 0 0
L1 1 2 1
R1 2 3 12
C1 3 0 .01
.TRAN .04 1
.PLOT TRAN V(3)
.END




**************** 85/04/21. ********************** SPICE 2D.2 (26SEP76)
CIRCUIT IN EXAMPLES 7.9.1 AND 7.9.2 (FIG. 7.9.1)
                    INITIAL TRANSIENT SOLUTION
*****************************************************************************

    NODE  VOLTAGE     NODE  VOLTAGE     NODE  VOLTAGE

    ( 1)   0.0000     ( 2)   0.0000     ( 3)   0.0000



       VOLTAGE SOURCE CURRENTS
       NAME        CURRENT

       VS        0.

       TOTAL POWER DISSIPATION   0.       WATTS
```

EXAMPLE B.8.2

Consider the circuit in Figure 13.7.2. Find the Fourier series of the periodic output-voltage waveform.

ANS.:

```
CIRCUIT OF FIGURE 13.7.2
VS 1 0 PULSE(1 0 .16  0  0  .83 1)
R1 1 0 1
.TRAN .005      1
.FOUR 1       V(1)
.END
```

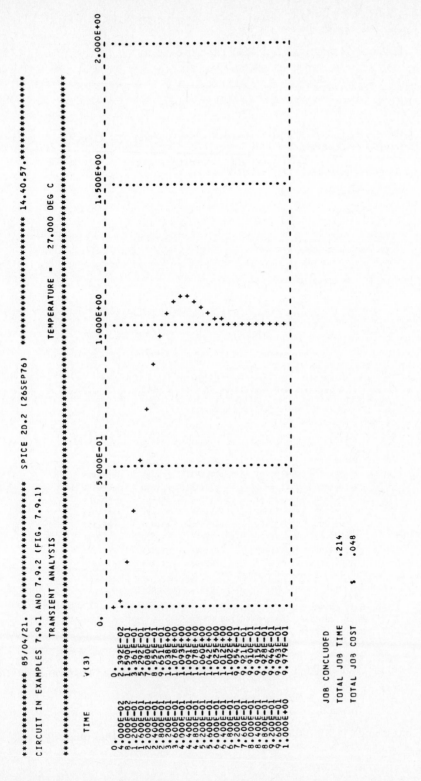

figure B.8.2

figure B.8.3
```
***** 85/05/04. ******** SPICE 2D.2 (26SEP76) ******** 11.21.02.*****
CIRCUIT OF EXAMPLE 13.7.2
       INPUT LISTING                      TEMPERATURE =   27.000 DEG C
**********************************************************************

VS 1 0 PULSE(1 0 .166666 0 0 .833333 1)
R1 1 0 1
.TRAN .005 1
.FOUR 1 V(1)
.END



*************** 85/05/04. ********************** SPICE 2D.2 (26SEP76)
CIRCUIT OF EXAMPLE 13.7.2
                      INITIAL TRANSIENT SOLUTION
**********************************************************************

   NODE   VOLTAGE

 ( 1)    1.0000



       VOLTAGE SOURCE CURRENTS
       NAME       CURRENT

       VS      -1.000E+00

       TOTAL POWER DISSIPATION   1.00E+00   WATTS
```

figure B.8.4
```
***** 85/05/04. ******** SPICE 2D.2 (26SEP76) ******** 11.21.02.*****
CIRCUIT OF EXAMPLE 13.7.2
       FOURIER ANALYSIS                   TEMPERATURE =   27.000 DEG C
**********************************************************************

FOURIER COMPONENTS OF TRANSIENT RESPONSE V(1)
DC COMPONENT =   1.749E-01
HARMONIC  FREQUENCY    FOURIER    NORMALIZED    PHASE    NORMALIZED
   NO       (HZ)      COMPONENT   COMPONENT    (DEG)    PHASE (DEG)

    1     1.000E+00   3.325E-01   1.000000    60.271      0.000
    2     2.000E+00   2.835E-01    .852593    30.557    -29.714
    3     3.000E+00   2.115E-01    .635913     .880     -59.391
    4     4.000E+00   1.287E-01    .387040   -28.697    -88.969
    5     5.000E+00   4.874E-02    .146564   -57.813   -118.085
    6     6.000E+00   1.652E-02    .049681    38.923     28.652
    7     7.000E+00   5.888E-02    .177055    61.302      1.031
    8     8.000E+00   7.542E-02    .226801    32.029    -28.242
    9     9.000E+00   6.850E-02    .205986     2.798    -57.473


   TOTAL HARMONIC DISTORTION =    119.593752  PERCENT

        JOB CONCLUDED
        TOTAL JOB TIME          .259
        TOTAL JOB COST      $   .055
```

The output that results is shown in Figures B.8.3 and B.8.4. SPICE yields Fourier harmonics in the somewhat unusual form

$$f_k(t) = c_k \sin(k\omega_0 t + \theta)$$

where $c_k = 2|\alpha_k|$ as usual but the phase angle is 90° different from the standard form of that quantity.

index

index

IMMITTANCE LOCI OF SIMPLE *RL* AND *RC* CIRCUITS

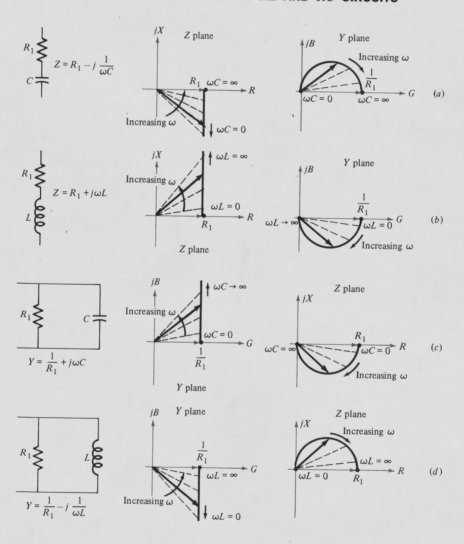